U0181658

Principles of Quantum Mechanics

SECOND EDITION

Principles of Quantum Mechanics

SECOND EDITION

R. Shankar
Yale University
New Haven, Connecticut

Library of Congress Cataloging–in–Publication Data

Shankar, Ramamurti.
Principles of quantum mechanics / R. Shankar. – 2nd ed.
p. cm.
Includes bibliographical references and index.

1. Quantum theory. I. Title.
QC174. 12.S52 1994
530. 1'2–dc20 94–26837
CIP

ISBN 978-1-4757-0578-2 ISBN 978-1-4757-0576-8 (eBook)
DOI 10.1007/978-1-4757-0576-8

19 (corrected printing, 2014)

springer.com

To

My Parents

and to

Uma, Umesh, Ajeet, Meera, and Maya

To

My Parents

and to

Lorna, [illegible] and Maya

Preface to the Second Edition

Over the decade and a half since I wrote the first edition, nothing has altered my belief in the soundness of the overall approach taken here. This is based on the response of teachers, students, and my own occasional rereading of the book. I was generally quite happy with the book, although there were portions where I felt I could have done better and portions which bothered me by their absence. I welcome this opportunity to rectify all that.

Apart from small improvements scattered over the text, there are three major changes. First, I have rewritten a big chunk of the mathematical introduction in Chapter 1. Next, I have added a discussion of time-reversal invariance. I don't know how it got left out the first time—I wish I could go back and change it. The most important change concerns the inclusion of Chaper 21, "Path Integrals: Part II." The first edition already revealed my partiality for this subject by having a chapter devoted to it, which was quite unusual in those days. In this one, I have cast off all restraint and gone all out to discuss many kinds of path integrals and their uses. Whereas in Chapter 8 the path integral recipe was simply given, here I start by deriving it. I derive the configuration space integral (the usual Feynman integral), phase space integral, and (oscillator) coherent state integral. I discuss two applications: the derivation and application of the Berry phase and a study of the lowest Landau level with an eye on the quantum Hall effect. The relevance of these topics is unquestionable. This is followed by a section of imaginary time path integrals—its description of tunneling, instantons, and symmetry breaking, and its relation to classical and quantum statistical mechanics. An introduction is given to the transfer matrix. Then I discuss spin coherent state path integrals and path integrals for fermions. These were thought to be topics too advanced for a book like this, but I believe this is no longer true. These concepts are extensively used and it seemed a good idea to provide the students who had the wisdom to buy this book with a head start.

How are instructors to deal with this extra chapter given the time constraints? I suggest omitting some material from the earlier chapters. (No one I know, myself included, covers the whole book while teaching any fixed group of students.) A realistic option is for the instructor to teach part of Chapter 21 and assign the rest as reading material, as topics for take-home exams, term papers, etc. To ignore it,

I think, would be to lose a wonderful opportunity to expose the student to ideas that are central to many current research topics and to deny them the attendant excitement. Since the aim of this chapter is to guide students toward more frontline topics, it is more concise than the rest of the book. Students are also expected to consult the references given at the end of the chapter.

Over the years, I have received some very useful feedback and I thank all those students and teachers who took the time to do so. I thank Howard Haber for a discussion of the Born approximation; Harsh Mathur and Ady Stern for discussions of the Berry phase; Alan Chodos, Steve Girvin, Ilya Gruzberg, Martin Gutzwiller, Ganpathy Murthy, Charlie Sommerfeld, and Senthil Todari for many useful comments on Chapter 21. I am most grateful to Captain Richard F. Malm, U.S.C.G. (Retired), Professor Dr. D. Schlüter of the University of Kiel, and Professor V. Yakovenko of the University of Maryland for detecting numerous errors in the first printing and taking the trouble to bring them to my attention. I thank Amelia McNamara of Plenum for urging me to write this edition and Plenum for its years of friendly and warm cooperation. I thank Ron Johnson, Editor at Springer for his tireless efforts on behalf of this book, and Chris Bostock, Daniel Keren and Jimmy Snyder for their generous help in correcting errors in the 14th printing. Finally, I thank my wife Uma for shielding me as usual from real life so I could work on this edition, and my battery of kids (revised and expanded since the previous edition) for continually charging me up.

R. Shankar

New Haven, Connecticut

Preface to the First Edition

Publish and perish—*Giordano Bruno*

Given the number of books that already exist on the subject of quantum mechanics, one would think that the public needs one more as much as it does, say, the latest version of the Table of Integers. But this does not deter me (as it didn't my predecessors) from trying to circulate my own version of how it ought to be taught. The approach to be presented here (to be described in a moment) was first tried on a group of Harvard undergraduates in the summer of '76, once again in the summer of '77, and more recently at Yale on undergraduates ('77–'78) and graduates ('78–'79) taking a year-long course on the subject. In all cases the results were very satisfactory in the sense that the students seemed to have learned the subject well and to have enjoyed the presentation. It is, in fact, their enthusiastic response and encouragement that convinced me of the soundness of my approach and impelled me to write this book.

The basic idea is to develop the subject from its postulates, after addressing some indispensable preliminaries. Now, most people would agree that the best way to teach any subject that has reached the point of development where it can be reduced to a few postulates is to start with the latter, for it is this approach that gives students the fullest understanding of the foundations of the theory and how it is to be used. But they would also argue that whereas this is all right in the case of special relativity or mechanics, a typical student about to learn quantum mechanics seldom has any familiarity with the mathematical language in which the postulates are stated. I agree with these people that this problem is real, but I differ in my belief that it should and can be overcome. This book is an attempt at doing just this.

It begins with a rather lengthy chapter in which the relevant mathematics of vector spaces developed from simple ideas on vectors and matrices the student is assumed to know. The level of rigor is what I think is needed to make a practicing quantum mechanic out of the student. This chapter, which typically takes six to eight lecture hours, is filled with examples from physics to keep students from getting too fidgety while they wait for the "real physics." Since the math introduced has to be taught sooner or later, I prefer sooner to later, for this way the students, when they get to it, can give quantum theory their fullest attention without having to battle with the mathematical theorems at the same time. Also, by segregating the mathematical theorems from the physical postulates, any possible confusion as to which is which is nipped in the bud.

This chapter is followed by one on classical mechanics, where the Lagrangian and Hamiltonian formalisms are developed in some depth. It is for the instructor to decide how much of this to cover; the more students know of these matters, the better they will understand the connection between classical and quantum mechanics. Chapter 3 is devoted to a brief study of idealized experiments that betray the inadequacy of classical mechanics and give a glimpse of quantum mechanics.

Having trained and motivated the students I now give them the postulates of quantum mechanics of a single particle in one dimension. I use the word "postulate" here to mean "that which cannot be deduced from pure mathematical or logical reasoning, and given which one can formulate and solve quantum mechanical problems and interpret the results." This is not the sense in which the true axiomatist would use the word. For instance, where the true axiomatist would just postulate that the dynamical variables are given by Hilbert space operators, I would add the operator identifications, i.e., specify the operators that represent coordinate and momentum (from which others can be built). Likewise, I would not stop with the statement that there is a Hamiltonian operator that governs the time evolution through the equation $i\hbar\partial|\psi\rangle/\partial t=H|\psi\rangle$; I would say the H is obtained from the classical Hamiltonian by substituting for x and p the corresponding operators. While the more general axioms have the virtue of surviving as we progress to systems of more degrees of freedom, with or without classical counterparts, students given just these will not know how to calculate anything such as the spectrum of the oscillator. Now one can, of course, try to "derive" these operator assignments, but to do so one would have to appeal to ideas of a postulatory nature themselves. (The same goes for "deriving" the Schrödinger equation.) As we go along, these postulates are generalized to more degrees of freedom and it is for pedagogical reasons that these generalizations are postponed. Perhaps when students are finished with this book, they can free themselves from the specific operator assignments and think of quantum mechanics as a general mathematical formalism obeying certain postulates (in the strict sense of the term).

The postulates in Chapter 4 are followed by a lengthy discussion of the same, with many examples from fictitious Hilbert spaces of three dimensions. Nonetheless, students will find it hard. It is only as they go along and see these postulates used over and over again in the rest of the book, in the setting up of problems and the interpretation of the results, that they will catch on to how the game is played. It is hoped they will be able to do it on their own when they graduate. I think that any attempt to soften this initial blow will be counterproductive in the long run.

Chapter 5 deals with standard problems in one dimension. It is worth mentioning that the scattering off a step potential is treated using a wave packet approach. If the subject seems too hard at this stage, the instructor may decide to return to it after Chapter 7 (oscillator), when students have gained more experience. But I think that sooner or later students must get acquainted with this treatment of scattering.

The classical limit is the subject of the next chapter. The harmonic oscillator is discussed in detail in the next. It is the first realistic problem and the instructor may be eager to get to it as soon as possible. If the instructor wants, he or she can discuss the classical limit after discussing the oscillator.

We next discuss the path integral formulation due to Feynman. Given the intuitive understanding it provides, and its elegance (not to mention its ability to give the full propagator in just a few minutes in a class of problems), its omission from so many books is hard to understand. While it is admittedly hard to actually evaluate a path integral (one example is provided here), the notion of expressing the propagator as a sum over amplitudes from various paths is rather simple. The importance of this point of view is becoming clearer day by day to workers in statistical mechanics and field theory. I think every effort should be made to include at least the first three (and possibly five) sections of this chapter in the course.

The content of the remaining chapters is standard, in the first approximation. The style is of course peculiar to this author, as are the specific topics. For instance, an entire chapter (11) is devoted to symmetries and their consequences. The chapter on the hydrogen atom also contains a section on how to make numerical estimates starting with a few mnemonics. Chapter 15, on addition of angular momenta, also contains a section on how to understand the "accidental" degeneracies in the spectra of hydrogen and the isotropic oscillator. The quantization of the radiation field is discussed in Chapter 18, on time-dependent perturbation theory. Finally the treatment of the Dirac equation in the last chapter (20) is intended to show that several things such as electron spin, its magnetic moment, the spin–orbit interaction, etc. which were introduced in an ad hoc fashion in earlier chapters, emerge as a coherent whole from the Dirac equation, and also to give students a glimpse of what lies ahead. This chapter also explains how Feynman resolves the problem of negative-energy solutions (in a way that applies to bosons and fermions).

For Whom Is this Book Intended?

In writing it, I addressed students who are trying to learn the subject by themselves; that is to say, I made it as self-contained as possible, included a lot of exercises and answers to most of them, and discussed several tricky points that trouble students when they learn the subject. But I am aware that in practice it is most likely to be used as a class text. There is enough material here for a full year graduate course. It is, however, quite easy so adapt it to a year-long undergraduate course. Several sections that may be omitted without loss of continuity are indicated. The sequence of topics may also be changed, as stated earlier in this preface. I thought it best to let the instructor skim through the book and chart the course for his or her class, given their level of preparation and objectives. Of course the book will not be particularly useful if the instructor is not sympathetic to the broad philosophy espoused here, namely, that first comes the mathematical training and then the development of the subject from the postulates. To instructors who feel that this approach is all right in principle but will not work in practice, I reiterate that it has been found to work in practice, not just by me but also by teachers elsewhere.

The book may be used by nonphysicists as well. (I have found that it goes well with chemistry majors in my classes.) *Although I wrote it for students with no familiarity with the subject, any previous exposure can only be advantageous.*

Finally, I invite instructors and students alike to communicate to me any suggestions for improvement, whether they be pedagogical or in reference to errors or misprints.

Acknowledgments

As I look back to see who all made this book possible, my thoughts first turn to my brother R. Rajaraman and friend Rajaram Nityananda, who, around the same time, introduced me to physics in general and quantum mechanics in particular. Next come my students, particularly Doug Stone, but for whose encouragement and enthusiastic response I would not have undertaken this project. I am grateful to Professor Julius Kovacs of Michigan State, whose kind words of encouragement assured me that the book would be as well received by my peers as it was by my students. More recently, I have profited from numerous conversations with my colleagues at Yale, in particular Alan Chodos and Peter Mohr. My special thanks go to Charles Sommerfield, who managed to make time to read the manuscript and made many useful comments and recommendations. The detailed proofreading was done by Tom Moore. I thank you, the reader, in advance, for drawing to my notice any errors that may have slipped past us.

The bulk of the manuscript production costs were borne by the J. W. Gibbs fellowship from Yale, which also supported me during the time the book was being written. Ms. Laurie Liptak did a fantastic job of typing the first 18 chapters and Ms. Linda Ford did the same with Chapters 19 and 20. The figures are by Mr. J. Brosious. Mr. R. Badrinath kindly helped with the index.‡

On the domestic front, encouragement came from my parents, my in-laws, and most important of all from my wife, Uma, who cheerfully donated me to science for a year or so and stood by me throughout. Little Umesh did his bit by tearing up all my books on the subject, both as a show of support and to create a need for this one.

R. Shankar

New Haven, Connecticut

‡ It is a pleasure to acknowledge the help of Mr. Richard Hatch, who drew my attention to a number of errors in the first printing.

Prelude

Our description of the physical world is dynamic in nature and undergoes frequent change. At any given time, we summarize our knowledge of natural phenomena by means of certain laws. These laws adequately describe the phenomenon studied up to that time, to an accuracy then attainable. As time passes, we enlarge the domain of observation and improve the accuracy of measurement. As we do so, we constantly check to see if the laws continue to be valid. Those laws that do remain valid gain in stature, and those that do not must be abandoned in favor of new ones that do.

In this changing picture, the laws of classical mechanics formulated by Galileo, Newton, and later by Euler, Lagrange, Hamilton, Jacobi, and others, remained unaltered for almost three centuries. The expanding domain of classical physics met its first obstacles around the beginning of this century. The obstruction came on two fronts: at large velocities and small (atomic) scales. The problem of large velocities was successfully solved by Einstein, who gave us his relativistic mechanics, while the founders of quantum mechanics—Bohr, Heisenberg, Schrödinger, Dirac, Born, and others—solved the problem of small-scale physics. The union of relativity and quantum mechanics, needed for the description of phenomena involving simultaneously large velocities and small scales, turns out to be very difficult. Although much progress has been made in this subject, called quantum field theory, there remain many open questions to this date. We shall concentrate here on just the small-scale problem, that is to say, on non-relativistic quantum mechanics.

The passage from classical to quantum mechanics has several features that are common to all such transitions in which an old theory gives way to a new one:

(1) There is a domain D_n of phenomena described by the new theory and a subdomain D_o wherein the old theory is reliable (to a given accuracy).
(2) Within the subdomain D_o either theory may be used to make quantitative predictions. It might often be more expedient to employ the old theory.
(3) In addition to numerical accuracy, the new theory often brings about radical conceptual changes. Being of a qualitative nature, these will have a bearing on all of D_n.

For example, in the case of relativity, D_o and D_n represent (macroscopic) phenomena involving small and arbitrary velocities, respectively, the latter, of course, being bounded by the velocity of light. In addition to giving better numerical predictions for high-velocity phenomena, relativity theory also outlaws several cherished notions of the Newtonian scheme, such as absolute time, absolute length, unlimited velocities for particles, etc.

In a similar manner, quantum mechanics brings with it not only improved numerical predictions for the microscopic world, but also conceptual changes that rock the very foundations of classical thought.

This book introduces you to this subject, starting from its postulates. Between you and the postulates there stand three chapters wherein you will find a summary of the mathematical ideas appearing in the statement of the postulates, a review of classical mechanics, and a brief description of the empirical basis for the quantum theory. In the rest of the book, the postulates are invoked to formulate and solve a variety of quantum mechanical problems. It is hoped that, by the time you get to the end of the book, you will be able to do the same yourself.

Note to the Student

Do as many exercises as you can, especially the ones marked * or whose results carry equation numbers. The answer to each exercise is given either with the exercise or at the end of the book.

The first chapter is very important. Do not rush through it. Even if you know the math, read it to get acquainted with the notation.

I am not saying it is an easy subject. But I hope this book makes it seem reasonable.

Good luck.

Contents

1

Mathematical Introduction

The aim of this book is to provide you with an introduction to quantum mechanics, starting from its axioms. It is the aim of this chapter to equip you with the necessary mathematical machinery. All the math you will need is developed here, starting from some basic ideas on vectors and matrices that you are assumed to know. Numerous examples and exercises related to classical mechanics are given, both to provide some relief from the math and to demonstrate the wide applicability of the ideas developed here. The effort you put into this chapter will be well worth your while: not only will it prepare you for this course, but it will also unify many ideas you may have learned piecemeal. To really learn this chapter, you must, as with any other chapter, work out the problems.

1.1. Linear Vector Spaces: Basics

In this section you will be introduced to *linear vector spaces*. You are surely familiar with the arrows from elementary physics encoding the magnitude and direction of velocity, force, displacement, torque, etc. You know how to add them and multiply them by scalars and the rules obeyed by these operations. For example, you know that scalar multiplication is distributive: the multiple of a sum of two vectors is the sum of the multiples. What we want to do is abstract from this simple case a set of basic features or axioms, and say that any set of objects obeying the same forms a linear vector space. The cleverness lies in deciding which of the properties to keep in the generalization. If you keep too many, there will be no other examples; if you keep too few, there will be no interesting results to develop from the axioms.

The following is the list of properties the mathematicians have wisely chosen as requisite for a vector space. As you read them, please compare them to the world of arrows and make sure that these are indeed properties possessed by these familiar vectors. But note also that conspicuously missing are the requirements that every vector have a magnitude and direction, which was the first and most salient feature drilled into our heads when we first heard about them. So you might think that in dropping this requirement, the baby has been thrown out with the bath water. However, you will have ample time to appreciate the wisdom behind this choice as you go along and see a great unification and synthesis of diverse ideas under the heading of vector spaces. You will see examples of vector spaces that involve entities that you cannot intuitively perceive as having either a magnitude or a direction. While you should be duly impressed with all this, remember that it does not hurt at all to think of these generalizations in terms of arrows and to use the intuition to prove theorems or at the very least anticipate them.

Definition 1. A linear vector space $\mathbb{V}$ *is a collection of objects* $|1\rangle$, $|2\rangle, \ldots, |V\rangle, \ldots, |W\rangle, \ldots$, called vectors, for which there exists

1. A definite rule for forming the vector sum, denoted $|V\rangle+|W\rangle$
2. A definite rule for multiplication by scalars $a, b, \ldots$, denoted $a|V\rangle$ with the following features:

- The result of these operations is another element of the space, a feature called *closure*: $|V\rangle+|W\rangle \in \mathbb{V}$.
- Scalar multiplication is *distributive in the vectors*: $a(|V\rangle+|W\rangle)=a|V\rangle+a|W\rangle$.
- Scalar multiplication is *distributive in the scalars*: $(a+b)|V\rangle=a|V\rangle+b|V\rangle$.
- Scalar multiplication is *associative*: $a(b|V\rangle)=ab|V\rangle$.
- Addition is *commutative*: $|V\rangle+|W\rangle=|W\rangle+|V\rangle$.
- Addition is *associative*: $|V\rangle+(|W\rangle+|Z\rangle)=(|V\rangle+|W\rangle)+|Z\rangle$.
- There exists a *null vector* $|0\rangle$ obeying $|V\rangle+|0\rangle=|V\rangle$.
- For every vector $|V\rangle$ there exists an *inverse under addition*, $|-V\rangle$, such that $|V\rangle+|-V\rangle=|0\rangle$.

There is a good way to remember all of these; *do what comes naturally.*

Definition 2. The numbers $a, b, \ldots$ are called the *field* over which the vector space is defined.

If the field consists of all real numbers, we have a *real vector space*, if they are complex, we have a *complex vector space*. The vectors themselves are neither real nor complex; the adjective applies only to the scalars.

Let us note that the above axioms imply

- $|0\rangle$ is unique, i.e., if $|0'\rangle$ has all the properties of $|0\rangle$, then $|0\rangle=|0'\rangle$.
- $0|V\rangle=|0\rangle$.
- $|-V\rangle=-|V\rangle$.
- $|-V\rangle$ is the unique additive inverse of $|V\rangle$.

The proofs are left as to the following exercise. You don't have to know the proofs, but you do have to know the statements.

Exercise 1.1.1. Verify these claims. For the first consider $|0\rangle+|0'\rangle$ and use the advertised properties of the two null vectors in turn. For the second start with $|0\rangle=(0+1)|V\rangle+|-V\rangle$. For the third, begin with $|V\rangle+(-|V\rangle)=0|V\rangle=|0\rangle$. For the last, let $|W\rangle$ also satisfy $|V\rangle+|W\rangle=|0\rangle$. Since $|0\rangle$ is unique, this means $|V\rangle+|W\rangle=|V\rangle+|-V\rangle$. Take it from here.

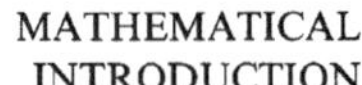

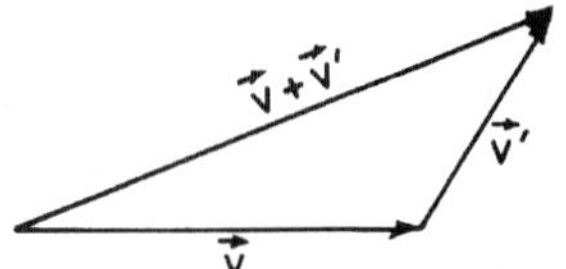

Figure 1.1. The rule for vector addition. Note that it obeys axioms (i)–(iii).

Exercise 1.1.2. Consider the set of all entities of the form (a, b, c) where the entries are real numbers. Addition and scalar multiplication are defined as follows:

$$(a, b, c)+(d, e, f)=(a+d, b+e, c+f)$$

$$\alpha(a, b, c)=(\alpha a, \alpha b, \alpha c).$$

Write down the null vector and inverse of (a, b, c). Show that vectors of the form $(a, b, 1)$ do not form a vector space.

Observe that we are using a new symbol $|V\rangle$ to denote a generic vector. This object is called *ket V* and this nomenclature is due to Dirac whose notation will be discussed at some length later. We do not purposely use the symbol $\vec{V}$ *to denote the vectors as the first step in weaning you away from the limited concept of the vector as an arrow. You are however not discouraged from associating with* $|V\rangle$ the arrowlike object till you have seen enough vectors that are not arrows and are ready to drop the crutch.

You were asked to verify that the set of arrows qualified as a vector space as you read the axioms. Here are some of the key ideas you should have gone over. The vector space consists of arrows, typical ones being $\vec{V}$ *and* $\vec{V}'$. The rule for addition is familiar: take the tail of the second arrow, put it on the tip of the first, and so on as in Fig. 1.1.

Scalar multiplication by a corresponds to stretching the vector by a factor a. This is a real vector space since stretching by a complex number makes no sense. (If a is negative, we interpret it as changing the direction of the arrow as well as rescaling it by $|a|$.) Since these operations acting on arrows give more arrows, we have closure. Addition and scalar multiplication clearly have all the desired associative and distributive features. The null vector is the arrow of zero length, while the inverse of a vector is the vector reversed in direction.

So the set of all arrows qualifies as a vector space. But we cannot tamper with it. For example, the set of all arrows with positive z-components do not form a vector space: there is no inverse.

Note that so far, no reference has been made to magnitude or direction. The point is that while the arrows have these qualities, members of a vector space need not. This statement is pointless unless I can give you examples, so here are two.

Consider the set of all 2×2 matrices. We know how to add them and multiply them by scalars (multiply all four matrix elements by that scalar). The corresponding rules obey closure, associativity, and distributive requirements. The null matrix has all zeros in it and the inverse under *addition* of a matrix is the matrix with all elements negated. You must agree that here we have a genuine vector space consisting of things which don't have an obvious length or direction associated with them. When we want to highlight the fact that the matrix M is an element of a vector space, we may want to refer to it as, say, ket number 4 or: $|4\rangle$.

As a second example, consider all functions $f(x)$ defined in an interval $0 \leq x \leq L$. We define scalar multiplication by a simply as $af(x)$ and addition as pointwise addition: the sum of two functions f and g has the value $f(x)+g(x)$ at the point x. The null function is zero everywhere and the additive inverse of f is $-f$.

Exercise 1.1.3. Do functions that vanish at the end points $x=0$ and $x=L$ form a vector space? How about *periodic functions* obeying $f(0)=f(L)$? How about functions that obey $f(0)=4$? If the functions do not qualify, list the things that go wrong.

The next concept is that of *linear independence* of a set of vectors $|1\rangle, |2\rangle \ldots |n\rangle$. First consider a linear relation of the form

$$\sum_{i=1}^{n} a_i |i\rangle = |0\rangle \tag{1.1.1}$$

We may assume without loss of generality that the left-hand side does not contain any multiple of $|0\rangle$, for if it did, it could be shifted to the right, and combined with the $|0\rangle$ there to give $|0\rangle$ once more. (We are using the fact that any multiple of $|0\rangle$ equals $|0\rangle$.)

Definition 3. The set of vectors is said to be *linearly independent* if the only such linear relation as Eq. (1.1.1) is the trivial one with all $a_i=0$. If the set of vectors is not linearly independent, we say they are *linearly dependent*.

Equation (1.1.1) tells us that it is not possible to write any member of the linearly independent set in terms of the others. On the other hand, if the set of vectors is linearly dependent, such a relation will exist, and it must contain at least two nonzero coefficients. Let us say $a_3 \neq 0$. Then we could write

$$|3\rangle = \sum_{i=1, \neq 3}^{n} \frac{-a_i}{a_3} |i\rangle \tag{1.1.2}$$

thereby expressing $|3\rangle$ in terms of the others.

As a concrete example, consider two nonparallel vectors $|1\rangle$ and $|2\rangle$ in a plane. These form a linearly independent set. There is no way to write one as a multiple of the other, or equivalently, no way to combine them to get the null vector. On the other hand, if the vectors are parallel, we can clearly write one as a multiple of the other or equivalently play them against each other to get 0.

Notice I said 0 and not $|0\rangle$. This is, strictly speaking, incorrect since a set of vectors can only add up to a vector and not a number. It is, however, common to represent the null vector by 0.

Suppose we bring in a third vector $|3\rangle$ also in the plane. If it is parallel to either of the first two, we already have a linearly dependent set. So let us suppose it is not. But even now the three of them are *linearly dependent*. This is because we can write one of them, say $|3\rangle$, as a linear combination of the other two. To find the combination, draw a line from the tail of $|3\rangle$ in the direction of $|1\rangle$. Next draw a line antiparallel to $|2\rangle$ from the tip of $|3\rangle$. These lines will intersect since $|1\rangle$ and $|2\rangle$ are not parallel by assumption. The intersection point P will determine how much of $|1\rangle$ and $|2\rangle$ we want: we go from the tail of $|3\rangle$ to P using the appropriate multiple of $|1\rangle$ and go from P to the tip of $|3\rangle$ using the appropriate multiple of $|2\rangle$.

Exercise 1.1.4. Consider three elements from the vector space of real 2×2 matrices:

$$|1\rangle=\begin{bmatrix}0 & 1\\ 0 & 0\end{bmatrix}\qquad |2\rangle=\begin{bmatrix}1 & 1\\ 0 & 1\end{bmatrix}\qquad |3\rangle=\begin{bmatrix}-2 & -1\\ 0 & -2\end{bmatrix}$$

Are they linearly independent? Support your answer with details. (Notice we are calling these matrices vectors and using kets to represent them to emphasize their role as elements of a vector space.)

Exercise 1.1.5. Show that the following row vectors are linearly dependent: (1, 1, 0), (1, 0, 1), and (3, 2, 1). Show the opposite for (1, 1, 0), (1, 0, 1), and (0, 1, 1).

Definition 4. A vector space has *dimension n* if it can accommodate a maximum of n linearly independent vectors. It will be denoted by $\mathbb{V}^n(R)$ if the field is real and by $\mathbb{V}^n(C)$ if the field is complex.

In view of the earlier discussions, the plane is two-dimensional and the set of all arrows not limited to the plane define a three-dimensional vector space. How about 2×2 matrices? They form a four-dimensional vector space. Here is a proof. The following vectors are linearly independent:

$$|1\rangle=\begin{bmatrix}1 & 0\\ 0 & 0\end{bmatrix}\qquad |2\rangle=\begin{bmatrix}0 & 1\\ 0 & 0\end{bmatrix}\qquad |3\rangle=\begin{bmatrix}0 & 0\\ 1 & 0\end{bmatrix}\qquad |4\rangle=\begin{bmatrix}0 & 0\\ 0 & 1\end{bmatrix}$$

since it is impossible to form linear combinations of any three of them to give the fourth any three of them will have a zero in the one place where the fourth does not. So the space is at least four-dimensional. Could it be bigger? No, since any arbitrary 2×2 matrix can be written in terms of them:

$$\begin{bmatrix}a & b\\ c & d\end{bmatrix}=a|1\rangle+b|2\rangle+c|3\rangle+d|4\rangle$$

If the scalars a, b, c, d are real, we have a *real four-dimensional space*, if they are complex we have a *complex four-dimensional space*.

Theorem 1. Any vector $|V\rangle$ in an n-dimensional space can be written as a linear combination of n linearly independent vectors $|1\rangle \ldots |n\rangle$.

The proof is as follows: if there were a vector $|V\rangle$ for which this were not possible, it would join the given set of vectors and form a set of $n+1$ linearly independent vectors, which is not possible in an n-dimensional space by definition.

Definition 5. A set of n linearly independent vectors in an n-dimensional space is called a *basis*.

Thus we can write, on the strength of the above

$$|V\rangle=\sum_{i=1}^{n} v_i|i\rangle \tag{1.1.3}$$

where the vectors $|i\rangle$ form a basis.

Definition 6. The coefficients of expansion v_i of a vector in terms of a linearly independent basis ($|i\rangle$) are called the *components of the vector in that basis.*

Theorem 2. The expansion in Eq. (1.1.3) is unique.

Suppose the expansion is not unique. We must then have a second expansion:

$$|V\rangle=\sum_{i=1}^{n} v_i'|i\rangle \tag{1.1.4}$$

Subtracting Eq. (1.1.4) from Eq. (1.1.3) (i.e., multiplying the second by the scalar -1 and adding the two equations) we get

$$|0\rangle=\sum_{i}(v_i-v_i')|i\rangle \tag{1.1.5}$$

which implies that

$$v_i=v_i' \tag{1.1.6}$$

since the basis vectors are linearly independent and only a trivial linear reiation between them can exist. Note that given a basis the components are unique, but if we change the basis, the components will change. We refer to $|V\rangle$ as the vector in the abstract, having an existence of its own and satisfying various relations involving other vectors. When we choose a basis the vectors assume concrete forms in terms of their components and the relation between vectors is satisfied by the components. Imagine for example three arrows in the plane, $\vec{A}$, $\vec{B}$, $\vec{C}$ satisfying $\vec{A}+\vec{B}=\vec{C}$ according to the laws for adding arrows. So far no basis has been chosen and we do not need a basis to make the statement that the vectors from a closed triangle. Now we choose a basis and write each vector in terms of the components. The components will satisfy $C_i=A_i+B_i$, $i=1, 2$. If we choose a different basis, the components will change in numerical value, but the relation between them expressing the equality of $\vec{C}$ to the sum of the other two will still hold between the new set of components.

In the case of nonarrow vectors, adding them in terms of components proceeds as in the elementary case thanks to the axioms. If

$$|V\rangle=\sum_i v_i|i\rangle \quad \text{and} \tag{1.1.7}$$

$$|W\rangle=\sum_i w_i|i\rangle \quad \text{then} \tag{1.1.8}$$

$$|V\rangle+|W\rangle=\sum_i (v_i+w_i)|i\rangle \tag{1.1.9}$$

where we have used the axioms to carry out the regrouping of terms. Here is the conclusion:

To add two vectors, add their components.

There is no reference to taking the tail of one and putting it on the tip of the other, etc., since in general the vectors have no head or tail. Of course, if we are dealing with arrows, we can add them either using the tail and tip routine or by simply adding their components in a basis.

In the same way, we have:

$$a|V\rangle=a\sum_i v_i|i\rangle=\sum_i av_i|i\rangle \tag{1.1.10}$$

In other words,

To multiply a vector by a scalar, multiply all its components by the scalar.

1.2. Inner Product Spaces

The matrix and function examples must have convinced you that we can have a vector space with no preassigned definition of length or direction for the elements. However, we can make up quantities that have the same properties that the lengths and angles do in the case of arrows. The first step is to define a sensible analog of the dot product, for in the case of arrows, from the dot product

$$\vec{A}\cdot\vec{B}=|A||B|\cos\theta \tag{1.2.1}$$

we can read off the length of say $\vec{A}$ as $\sqrt{|A|\cdot|A|}$ and the cosine of the angle between two vectors as $\vec{A}\cdot\vec{B}/|A||B|$. Now you might rightfully object: how can you use the dot product to define the length and angles, if the dot product itself requires knowledge of the lengths and angles? The answer is this. Recall that the dot product has a second

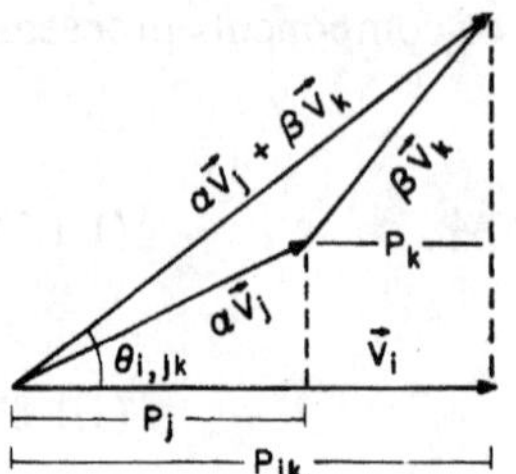

Figure 1.2. Geometrical proof that the dot product obeys axiom (3) for an inner product. The axiom requires that the projections obey $P_k + P_j = P_{jk}$.

equivalent expression in terms of the components:

$$\vec{A}\cdot\vec{B} = A_xB_x + A_yB_y + A_zB_z \tag{1.2.2}$$

Our goal is to define a similar formula for the general case where we *do* have the notion of components in a basis. To this end we recall the main features of the above dot product:

1. $\vec{A}\cdot\vec{B}=\vec{B}\cdot\vec{A}$ (symmetry)
2. $\vec{A}\cdot\vec{A}\geq 0 \quad 0$ *iff* $\vec{A}=0$ (positive semidefiniteness)
3. $\vec{A}\cdot(b\vec{B}+c\vec{C})=b\vec{A}\cdot\vec{B}+c\vec{A}\cdot\vec{C}$ (linearity)

The linearity of the dot product is illustrated in Fig. 1.2.

We want to invent a generalization called the *inner product* or *scalar product* between any two vectors $|V\rangle$ and $|W\rangle$. We denote it by the symbol $\langle V|W\rangle$. It is once again a number (generally complex) dependent on the two vectors. We demand that it obey the following axioms:

- $\langle V|W\rangle = \langle W|V\rangle^*$ (skew-symmetry)
- $\langle V|V\rangle \geq 0 \quad 0$ *iff* $|V\rangle = |0\rangle$ (positive semidefiniteness)
- $\langle V|(a|W\rangle + b|Z\rangle) \equiv \langle V|aW+bZ\rangle = a\langle V|W\rangle + b\langle V|Z\rangle$ (linearity in ket)

Definition 7. A vector space with an inner product is called an *inner product space.*

Notice that we have not yet given an explicit rule for actually evaluating the scalar product, we are merely demanding that any rule we come up with must have these properties. With a view to finding such a rule, let us familiarize ourselves with the axioms. The first differs from the corresponding one for the dot product and makes the inner product sensitive to the order of the two factors, with the two choices leading to complex conjugates. In a real vector space this axioms states the symmetry of the dot product under exchange of the two vectors. For the present, let us note that this axiom ensures that $\langle V|V\rangle$ is real.

The second axiom says that $\langle V|V\rangle$ is not just real but also positive semidefinite, vanishing only if the vector itself does. If we are going to define the length of the vector as the square root of its inner product with itself (as in the dot product) this quantity had better be real and positive for all nonzero vectors.

The last axiom expresses the linearity of the inner product when a linear superposition $a|W\rangle+b|Z\rangle\equiv|aW+bZ\rangle$ appears as the second vector in the scalar product. We have discussed its validity for the arrows case (Fig. 1.2).

What if the first factor in the product is a linear superposition, i.e., what is $\langle aW+bZ|V\rangle$? This is determined by the first axiom:

$$\begin{aligned}\langle aW+bZ|V\rangle &= \langle V|aW+bZ\rangle^* \\ &= (a\langle V|W\rangle+b\langle V|Z\rangle)^* \\ &= a^*\langle V|W\rangle^*+b^*\langle V|Z\rangle^* \\ &= a^*\langle W|V\rangle+b^*\langle Z|V\rangle \qquad (1.2.3)\end{aligned}$$

which expresses the *antilinearity* of the inner product with respect to the first factor in the inner product. In other words, the inner product of a linear superposition with another vector is the corresponding superposition of inner products if the superposition occurs in the second factor, while it is the superposition with all coefficients conjugated if the superposition occurs in the first factor. This asymmetry, unfamiliar in real vector spaces, is here to stay and you will get used to it as you go along.

Let us continue with inner products. Even though we are trying to shed the restricted notion of a vector as an arrow and seeking a corresponding generalization of the dot product, we still use some of the same terminology.

Definition 8. We say that two vectors are *orthogonal* or perpendicular if their inner product vanishes.

Definition 9. We will refer to $\sqrt{\langle V|V\rangle}\equiv|V|$ as the *norm* or length of the vector. A *normalized vector* has unit norm.

Definition 10. A set of basis vectors all of unit norm, which are pairwise orthogonal will be called an *orthonormal basis*.

We will also frequently refer to the inner or scalar product as the dot product.

We are now ready to obtain a concrete formula for the inner product in terms of the components. Given $|V\rangle$ and $|W\rangle$

$$|V\rangle=\sum_i v_i|i\rangle$$

$$|W\rangle=\sum_j w_j|j\rangle$$

we follow the axioms obeyed by the inner product to obtain:

$$\langle V|W\rangle=\sum_i\sum_j v_i^* w_j\langle i|j\rangle \qquad (1.2.4)$$

To go any further we have to know $\langle i|j\rangle$, the inner product between basis vectors. That depends on the details of the basis vectors and all we know for sure is that they are linearly independent. This situation exists for arrows as well. Consider a two-dimensional problem where the basis vectors are two linearly independent but nonperpendicular vectors. If we write all vectors in terms of this basis, the dot product of any two of them will likewise be a double sum with four terms (determined by the four possible dot products between the basis vectors) as well as the vector components. However, if we use an orthonormal basis such as $\vec{i}, \vec{j}$, only diagonal terms like $\langle i|i\rangle$ will survive and we will get the familiar result $\vec{A}\cdot\vec{B}=A_xB_x+A_yB_y$ depending only on the components.

For the more general nonarrow case, we invoke Theorem 3.

Theorem 3 (Gram-Schmidt). Given a linearly independent basis we can form linear combinations of the basis vectors to obtain an orthonormal basis.

Postponing the proof for a moment, *let us assume that the procedure has been implemented and that the current basis is orthonormal*:

$$\langle i|j\rangle=\begin{cases}1 & \text{for } i=j\\ 0 & \text{for } i\neq j\end{cases}\equiv\delta_{ij}$$

where δ_{ij} is called the *Kronecker delta symbol.* Feeding this into Eq. (1.2.4) we find the double sum collapses to a single one due to the Kronecker delta, to give

$$\langle V|W\rangle=\sum_i v_i^* w_i \tag{1.2.5}$$

This is the form of the inner product we will use from now on.

You can now appreciate the first axiom; but for the complex conjugation of the components of the first vector, $\langle V|V\rangle$ would not even be real, not to mention positive. But now it is given by

$$\langle V|V\rangle=\sum_i |v_i|^2\geq 0 \tag{1.2.6}$$

and vanishes only for the null vector. This makes it sensible to refer to $\langle V|V\rangle$ as the length or norm squared of a vector.

Consider Eq. (1.2.5). Since the vector $|V\rangle$ is uniquely specified by its components in a given basis, we may, in this basis, write it as a column vector:

$$|V\rangle\rightarrow\begin{bmatrix}v_1\\ v_2\\ \vdots\\ v_n\end{bmatrix}\quad \text{in this basis} \tag{1.2.7}$$

Likewise

$$|W\rangle \to \begin{bmatrix} w_1 \\ w_2 \\ \vdots \\ w_n \end{bmatrix} \quad \text{in this basis} \tag{1.2.8}$$

The inner product $\langle V|W\rangle$ is given by the matrix product of the transpose conjugate of the column vector representing $|V\rangle$ with the column vector representing $|W\rangle$:

$$\langle V|W\rangle = [v_1^*, v_2^*, \ldots, v_n^*] \begin{bmatrix} w_1 \\ w_2 \\ \vdots \\ w_n \end{bmatrix} \tag{1.2.9}$$

1.3. Dual Spaces and the Dirac Notation

There is a technical point here. The inner product is a number we are trying to generate from two kets $|V\rangle$ and $|W\rangle$, which are both represented by column vectors in some basis. Now there is no way to make a number out of two columns by direct matrix multiplication, but there is a way to make a number by matrix multiplication of a row times a column. Our trick for producing a number out of two columns has been to associate a unique row vector with one column (its transpose conjugate) and form its matrix product with the column representing the other. This has the feature that the answer depends on which of the two vectors we are going to convert to the row, the two choices ($\langle V|W\rangle$ and $\langle W|V\rangle$) leading to answers related by complex conjugation.

But one can also take the following alternate view. Column vectors are concrete manifestations of an abstract vector $|V\rangle$ or ket in a basis. We can also work backward and go from the column vectors to the abstract kets. But then it is similarly possible to work backward and associate with each *row vector* an abstract object $\langle W|$, called *bra-W*. Now we can name the bras as we want but let us do the following. Associated with every ket $|V\rangle$ is a column vector. Let us take its *adjoint*, or transpose conjugate, and form a row vector. The abstract bra associated with this will bear the same label, i.e., it will be called $\langle V|$. Thus there are two vector spaces, the space of kets and a dual space of bras, with a ket for every bra and vice versa (the components being related by the adjoint operation). Inner products are really defined only between bras and kets and hence from elements of two distinct but related vector spaces. There is a basis of vectors $|i\rangle$ for expanding kets and a similar basis $\langle i|$ for expanding bras. The basis ket $|i\rangle$ is represented in the basis we are using by a column vector with all zeros except for a 1 in the ith row, while the basis bra $\langle i|$ is a row vector with all zeros except for a 1 in the ith column.

All this may be summarized as follows:

$$|V\rangle \leftrightarrow \begin{bmatrix} v_1 \\ v_2 \\ \vdots \\ \\ v_n \end{bmatrix} \leftrightarrow [v_1^*, v_2^*, \ldots v_n^*] \leftrightarrow \langle V| \tag{1.3.1}$$

where $\leftrightarrow$ means "within a basis."

There is, however, nothing wrong with the first viewpoint of associating a scalar product with a pair of columns or kets (making no reference to another dual space) and living with the asymmetry between the first and second vector in the inner product (which one to transpose conjugate?). If you found the above discussion heavy going, you can temporarily ignore it. The only thing you must remember is that in the case of a general nonarrow vector space:

- Vectors can still be assigned components in some orthonormal basis, just as with arrows, but these may be complex.
- The inner product of any two vectors is given in terms of these components by Eq. (1.2.5). This product obeys all the axioms.

1.3.1. Expansion of Vectors in an Orthonormal Basis

Suppose we wish to expand a vector $|V\rangle$ in an orthonormal basis. To find the components that go into the expansion we proceed as follows. We take the dot product of both sides of the assumed expansion with $|j\rangle$: (or $\langle j|$ if you are a purist)

$$|V\rangle = \sum_i v_i |i\rangle \tag{1.3.2}$$

$$\langle j|V\rangle = \sum_i v_i \underbrace{\langle j|i\rangle}_{\delta_{ij}} \tag{1.3.3}$$

$$= v_j \tag{1.3.4}$$

i.e., to find the jth component of a vector we take the dot product with the jth unit vector, exactly as with arrows. Using this result we may write

$$|V\rangle = \sum_i |i\rangle\langle i|V\rangle \tag{1.3.5}$$

Let us make sure the basis vectors look as they should. If we set $|V\rangle = |j\rangle$ in Eq. (1.3.5), we find the correct answer: the ith component of the jth basis vector is δ_{ij}. Thus for example the column representing basis vector number 4 will have a 1 in the 4th row and zero everywhere else. The abstract relation

$$|V\rangle = \sum_i v_i |i\rangle \tag{1.3.6}$$

becomes in this basis

$$\begin{bmatrix} v_1 \\ v_2 \\ \vdots \\ v_n \end{bmatrix} = v_1 \begin{bmatrix} 1 \\ 0 \\ \vdots \\ 0 \end{bmatrix} + v_2 \begin{bmatrix} 0 \\ 1 \\ 0 \\ \vdots \\ 0 \end{bmatrix} + \cdots v_n \begin{bmatrix} 0 \\ 0 \\ \vdots \\ 1 \end{bmatrix} \tag{1.3.7}$$

1.3.2. Adjoint Operation

We have seen that we may pass from the column representing a ket to the row representing the corresponding bra by the adjoint operation, i.e., transpose conjugation. Let us now ask: if $\langle V|$ is the bra corresponding to the ket $|V\rangle$ what bra corresponds to $a|V\rangle$ where a is some scalar? By going to any basis it is readily found that

$$a|V\rangle \to \begin{bmatrix} av_1 \\ av_2 \\ \vdots \\ av_n \end{bmatrix} \to [a^*v_1^*, a^*v_2^*, \ldots, a^*v_n^*] \to \langle V|a^* \tag{1.3.8}$$

It is customary to write $a|V\rangle$ as $|aV\rangle$ and the corresponding bra as $\langle aV|$. What we have found is that

$$\langle aV| = \langle V|a^* \tag{1.3.9}$$

Since the relation between bras and kets is linear we can say that if we have an equation among kets such as

$$a|V\rangle = b|W\rangle + c|Z\rangle + \cdots \tag{1.3.10}$$

this implies another one among the corresponding bras:

$$\langle V|a^* = \langle W|b^* + \langle Z|c^* + \cdots \tag{1.3.11}$$

The two equations above are said to be *adjoints of each other*. Just as any equation involving complex numbers implies another obtained by taking the complex conjugates of both sides, an equation between (bras) kets implies another one between (kets) bras. If you think in a basis, you will see that this follows simply from the fact that if two columns are equal, so are their transpose conjugates.

Here is the rule for taking the adjoint:

To take the adjoint of a linear equation relating kets (bras), replace every ket (bra) by its bra (ket) and complex conjugate all coefficients.

We can extend this rule as follows. Suppose we have an expansion for a vector:

$$|V\rangle = \sum_{i=1} v_i |i\rangle \tag{1.3.12}$$

in terms of basis vectors. The adjoint is

$$\langle V| = \sum_{i=1} \langle i| v_i^*$$

Recalling that $v_i = \langle i|V\rangle$ and $v_i^* = \langle V|i\rangle$, it follows that the adjoint of

$$|V\rangle = \sum_{i=1} |i\rangle\langle i|V\rangle \tag{1.3.13}$$

is

$$\langle V| = \sum_{i=1} \langle V|i\rangle\langle i| \tag{1.3.14}$$

from which comes the rule:

To take the adjoint of an equation involving bras and kets and coefficients, reverse the order of all factors, exchanging bras and kets and complex conjugating all coefficients.

Gram–Schmidt Theorem

Let us now take up the Gram–Schmidt procedure for converting a linearly independent basis into an orthonormal one. The basic idea can be seen by a simple example. Imagine the two-dimensional space of arrows in a plane. Let us take two nonparallel vectors, which qualify as a basis. To get an orthonormal basis out of these, we do the following:

- Rescale the first by its own length, so it becomes a unit vector. This will be the first basis vector.
- Subtract from the second vector its projection along the first, leaving behind only the part perpendicular to the first. (Such a part will remain since by assumption the vectors are nonparallel.)
- Rescale the left over piece by its own length. We now have the second basis vector: it is orthogonal to the first and of unit length.

This simple example tells the whole story behind this procedure, which will now be discussed in general terms in the Dirac notation.

Let $|I\rangle, |II\rangle, \ldots$ be a linearly independent basis. The first vector of the orthonormal basis will be

$$|1\rangle=\frac{|I\rangle}{|I|} \quad \text{where} \quad |I|=\sqrt{\langle I|I\rangle}$$

Clearly

$$\langle 1|1\rangle=\frac{\langle I|I\rangle}{|I|^2}=1$$

As for the second vector in the basis, consider

$$|2'\rangle=|II\rangle-|1\rangle\langle 1|II\rangle$$

which is $|II\rangle$ minus the part pointing along the first unit vector. (Think of the arrow example as you read on.) Not surprisingly it is orthogonal to the latter:

$$\langle 1|2'\rangle=\langle 1|II\rangle-\langle 1|1\rangle\langle 1|II\rangle=0$$

We now divide $|2'\rangle$ by its norm to get $|2\rangle$ which will be orthogonal to the first and normalized to unity. Finally, consider

$$|3'\rangle=|III\rangle-|1\rangle\langle 1|III\rangle-|2\rangle\langle 2|III\rangle$$

which is orthogonal to both $|1\rangle$ and $|2\rangle$. Dividing by its norm we get $|3\rangle$, the third member of the orthogonal basis. There is nothing new with the generation of the rest of the basis.

Where did we use the linear independence of the original basis? What if we had started with a linearly dependent basis? Then at some point a vector like $|2'\rangle$ or $|3'\rangle$ would have vanished, putting a stop to the whole procedure. On the other hand, linear independence will assure us that such a thing will never happen since it amounts to having a nontrivial linear combination of linearly independent vectors that adds up the null vector. (Go back to the equations for $|2'\rangle$ or $|3'\rangle$ and satisfy yourself that these are linear combinations of the old basis vectors.)

Exercise 1.3.1. Form an orthonormal basis in two dimensions starting with $\vec{A}=3\vec{i}+4\vec{j}$ and $\vec{B}=2\vec{i}-6\vec{j}$. Can you generate another orthonormal basis starting with these two vectors? If so, produce another.

Exercise 1.3.2. Show how to go from the basis

$$|I\rangle=\begin{bmatrix}3\\0\\0\end{bmatrix}\quad |II\rangle=\begin{bmatrix}0\\1\\2\end{bmatrix}\quad |III\rangle=\begin{bmatrix}0\\2\\5\end{bmatrix}$$

to the orthonormal basis

$$|1\rangle=\begin{bmatrix}1\\0\\0\end{bmatrix}\quad |2\rangle=\begin{bmatrix}0\\1/\sqrt{5}\\2/\sqrt{5}\end{bmatrix}\quad |3\rangle=\begin{bmatrix}0\\-2/\sqrt{5}\\1/\sqrt{5}\end{bmatrix}$$

When we first learn about dimensionality, we associate it with the number of perpendicular directions. In this chapter we defined it in terms of the maximum number of linearly independent vectors. The following theorem connects the two definitions.

Theorem 4. The dimensionality of a space equals $n_\perp$, the maximum number of mutually orthogonal vectors in it.

To show this, first note that any mutually orthogonal set is also linearly independent. Suppose we had a linear combination of orthogonal vectors adding up to zero. By taking the dot product of both sides with any one member and using the orthogonality we can show that the coefficient multiplying that vector had to vanish. This can clearly be done for all the coefficients, showing the linear combination is trivial.

Now $n_\perp$ can only be equal to, greater than or lesser than n, the dimensionality of the space. The Gram–Schmidt procedure eliminates the last case by explicit construction, while the linear independence of the perpendicular vectors rules out the penultimate option.

Schwarz and Triangle Inequalities

Two powerful theorems apply to any inner product space obeying our axioms:

Theorem 5. The Schwarz Inequality

$$|\langle V|W\rangle|\leq|V||W| \tag{1.3.15}$$

Theorem 6. The Triangle Inequality

$$|V+W|\leq|V|+|W| \tag{1.3.16}$$

The proof of the first will be provided so you can get used to working with bras and kets. The second will be left as an exercise.

Before proving anything, note that the results are obviously true for arrows: the *Schwarz inequality* says that the dot product of two vectors cannot exceed the product of their lengths and the *triangle inequality* says that the length of a sum cannot exceed the sum of the lengths. This is an example which illustrates the merits of thinking of abstract vectors as arrows and guessing what properties they might share with arrows. The proof will of course have to rely on just the axioms.

To prove the Schwarz inequality, consider axiom $\langle Z|Z\rangle \geq 0$ *applied to*

$$|Z\rangle = |V\rangle - \frac{\langle W|V\rangle}{|W|^2}|W\rangle \tag{1.3.17}$$

We get

$$\begin{aligned}\langle Z|Z\rangle &= \langle V - \frac{\langle W|V\rangle}{|W|^2} W | V - \frac{\langle W|V\rangle}{|W|^2} W\rangle \\ &= \langle V|V\rangle - \frac{\langle W|V\rangle\langle V|W\rangle}{|W|^2} - \frac{\langle W|V\rangle^*\langle W|V\rangle}{|W|^2} \\ &\quad + \frac{\langle W|V\rangle^*\langle W|V\rangle\langle W|W\rangle}{|W|^4} \\ &\geq 0\end{aligned} \tag{1.3.18}$$

where we have used the antilinearity of the inner product with respect to the bra. Using

$$\langle W|V\rangle^* = \langle V|W\rangle$$

we find

$$\langle V|V\rangle \geq \frac{\langle W|V\rangle\langle V|W\rangle}{|W|^2} \tag{1.3.19}$$

Cross-multiplying by $|W|^2$ and taking square roots, the result follows.

Exercise 1.3.3. When will this equality be satisfied? Does this agree with your experience with arrows?

Exercise 1.3.4. Prove the triangle inequality starting with $|V+W|^2$. You must use $\text{Re}\langle V|W\rangle \leq |\langle V|W\rangle|$ and the Schwarz inequality. Show that the final inequality becomes an equality only if $|V\rangle = a|W\rangle$ where a is a real positive scalar.

1.4. Subspaces

Definition 11. Given a vector space $\mathbb{V}$, a subset of its elements that form a vector space among themselves‡ is called a *subspace*. We will denote a particular subspace i of dimensionality n_i by $\mathbb{V}_i^{n_i}$.

‡ Vector addition and scalar multiplication are defined the same way in the subspace as in $\mathbb{V}$.

Example 1.4.1. In the space $\mathbb{V}^3(R)$, the following are some examples of subspaces: (a) all vectors along the x axis, the space $\mathbb{V}_x^1$; (b) all vectors along the y axis, the space $\mathbb{V}_y^1$; (c) all vectors in the $x-y$ plane, the space $\mathbb{V}_{xy}^2$. Notice that all subspaces contain the null vector and that each vector is accompanied by its inverse to fulfill axioms for a vector space. Thus the set of all vectors along the positive x axis alone do not form a vector space. □

Definition 12. Given two subspaces $\mathbb{V}_i^{n_i}$ and $\mathbb{V}_j^{m_j}$, we define their sum $\mathbb{V}_i^{n_i} \oplus \mathbb{V}_j^{m_j} = \mathbb{V}_k^{m_k}$ as the set containing (1) all elements of $\mathbb{V}_i^{n_i}$, (2) all elements of $\mathbb{V}_j^{m_j}$, (3) all possible linear combinations of the above. But for the elements (3), closure would be lost.

Example 1.4.2. If, for example, $\mathbb{V}_x^1 \oplus \mathbb{V}_y^1$ contained only vectors along the x and y axes, we could, by adding two elements, one from each direction, generate one along neither. On the other hand, if we also included all linear combinations, we would get the correct answer, $\mathbb{V}_x^1 \oplus \mathbb{V}_y^1 = \mathbb{V}_{xy}^2$. □

*Exercise 1.4.1.** In a space $\mathbb{V}^n$, prove that the set of all vectors $\{|V_\perp^1\rangle, |V_\perp^2,\rangle, \ldots\}$, orthogonal to any $|V\rangle \neq |0\rangle$, form a subspace $\mathbb{V}^{n-1}$.

Exercise 1.4.2. Suppose $\mathbb{V}_1^{n_1}$ and $\mathbb{V}_2^{n_2}$ are two subspaces such that any element of $\mathbb{V}_1$ is orthogonal to any element of $\mathbb{V}_2$. Show that the dimensionality of $\mathbb{V}_1 \oplus \mathbb{V}_2$ is $n_1 + n_2$. (Hint: Theorem 4.)

1.5. Linear Operators

An operator Ω is an instruction for transforming any given vector $|V\rangle$ into another, $|V'\rangle$. The action of the operator is represented as follows:

$$\Omega|V\rangle = |V'\rangle \tag{1.5.1}$$

One says that the operator Ω has transformed the ket $|V\rangle$ into the ket $|V'\rangle$. We will restrict our attention throughout to operators Ω that do not take us out of the vector space, i.e., if $|V\rangle$ is an element of a space $\mathbb{V}$, so is $|V'\rangle = \Omega|V\rangle$.

Operators can also act on bras:

$$\langle V'|\Omega = \langle V''| \tag{1.5.2}$$

We will only be concerned with *linear operators*, i.e., ones that obey the following rules:

$$\Omega\alpha|V_i\rangle = \alpha\Omega|V_i\rangle \tag{1.5.3a}$$

$$\Omega\{\alpha|V_i\rangle + \beta|V_j\rangle\} = \alpha\Omega|V_i\rangle + \beta\Omega|V_j\rangle \tag{1.5.3b}$$

$$\langle V_i|\alpha\Omega = \langle V_i|\Omega\alpha \tag{1.5.4a}$$

$$(\langle V_i|\alpha + \langle V_j|\beta)\Omega = \alpha\langle V_i|\Omega + \beta\langle V_j|\Omega \tag{1.5.4b}$$

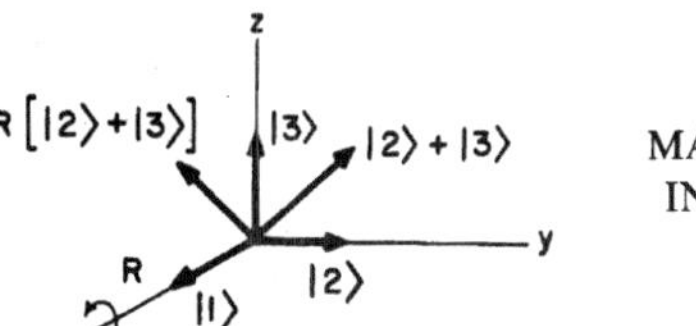

Figure 1.3. Action of the operator $R(\frac{1}{2}\pi\mathbf{i})$. Note that $R[|2\rangle+|3\rangle]=R|2\rangle+R|3\rangle$ as expected of a linear operator. (We will often refer to $R(\frac{1}{2}\pi\mathbf{i})$ as R if no confusion is likely.)

Example 1.5.1. The simplest operator is the identity operator, I, which carries the instruction:

$$I \rightarrow \text{Leave the vector alone!}$$

Thus,

$$I|V\rangle=|V\rangle \quad \text{for all kets } |V\rangle \tag{1.5.5}$$

and

$$\langle V|I=\langle V| \quad \text{for all bras } \langle V| \tag{1.5.6}$$

We next pass on to a more interesting operator on $\mathbb{V}^3(R)$:

$$R(\tfrac{1}{2}\pi\mathbf{i}) \rightarrow \text{Rotate vector by } \tfrac{1}{2}\pi \text{ about the unit vector } \mathbf{i}$$

[More generally, $R(\boldsymbol{\theta})$ stands for a rotation by an angle $\theta=|\boldsymbol{\theta}|$ about the axis parallel to the unit vector $\hat{\theta}=\boldsymbol{\theta}/\theta$.] Let us consider the action of this operator on the three unit vectors **i**, **j**, and **k**, which in our notation will be denoted by $|1\rangle$, $|2\rangle$, and $|3\rangle$ (see Fig. 1.3). From the figure it is clear that

$$R(\tfrac{1}{2}\pi\mathbf{i})|1\rangle=|1\rangle \tag{1.5.7a}$$

$$R(\tfrac{1}{2}\pi\mathbf{i})|2\rangle=|3\rangle \tag{1.5.7b}$$

$$R(\tfrac{1}{2}\pi\mathbf{i})|3\rangle=-|2\rangle \tag{1.5.7c}$$

Clearly $R(\frac{1}{2}\pi\mathbf{i})$ is linear. For instance, it is clear from the same figure that $R[|2\rangle+|3\rangle]=R|2\rangle+R|3\rangle$. □

The nice feature of linear operators is that once their action on the basis vectors is known, their action on any vector in the space is determined. If

$$\Omega|i\rangle=|i'\rangle$$

for a basis $|1\rangle, |2\rangle, \ldots, |n\rangle$ in $\mathbb{V}^n$, then for any $|V\rangle=\sum v_i|i\rangle$

$$\Omega|V\rangle=\sum_i \Omega v_i|i\rangle=\sum_i v_i\Omega|i\rangle=\sum_i v_i|i'\rangle \tag{1.5.8}$$

This is the case in the example $\Omega=R(\frac{1}{2}\pi\mathbf{i})$. If

$$|V\rangle=v_1|1\rangle+v_2|2\rangle+v_3|3\rangle$$

is any vector, then

$$R|V\rangle=v_1R|1\rangle+v_2R|2\rangle+v_3R|3\rangle=v_1|1\rangle+v_2|3\rangle-v_3|2\rangle$$

The *product of two operators* stands for the instruction that the instructions corresponding to the two operators be carried out in sequence

$$\Lambda\Omega|V\rangle=\Lambda(\Omega|V\rangle)=\Lambda|\Omega V\rangle \tag{1.5.9}$$

where $|\Omega V\rangle$ is the ket obtained by the action of Ω on $|V\rangle$. The order of the operators in a product is very important: in general,

$$\Omega\Lambda-\Lambda\Omega\equiv[\Omega,\Lambda]$$

called the *commutator* of Ω and Λ isn't zero. For example $R(\frac{1}{2}\pi\mathbf{i})$ and $R(\frac{1}{2}\pi\mathbf{j})$ do not commute, i.e., their commutator is nonzero.

Two useful identities involving commutators are

$$[\Omega,\Lambda\theta]=\Lambda[\Omega,\theta]+[\Omega,\Lambda]\theta \tag{1.5.10}$$

$$[\Lambda\Omega,\theta]=\Lambda[\Omega,\theta]+[\Lambda,\theta]\Omega \tag{1.5.11}$$

Notice that apart from the emphasis on ordering, these rules resemble the chain rule in calculus for the derivative of a product.

The *inverse* of Ω, denoted by Ω^{-1}, satisfies‡

$$\Omega\Omega^{-1}=\Omega^{-1}\Omega=I \tag{1.5.12}$$

Not every operator has an inverse. The condition for the existence of the inverse is given in Appendix A.1. The operator $R(\frac{1}{2}\pi\mathbf{i})$ has an inverse: it is $R(-\frac{1}{2}\pi\mathbf{i})$. The inverse of a product of operators is the product of the inverses in reverse:

$$(\Omega\Lambda)^{-1}=\Lambda^{-1}\Omega^{-1} \tag{1.5.13}$$

for only then do we have

$$(\Omega\Lambda)(\Omega\Lambda)^{-1}=(\Omega\Lambda)(\Lambda^{-1}\Omega^{-1})=\Omega\Lambda\Lambda^{-1}\Omega^{-1}=\Omega\Omega^{-1}=I$$

1.6. Matrix Elements of Linear Operators

We are now accustomed to the idea of an abstract vector being represented in a basis by an n-tuple of numbers, called its components, in terms of which all vector

‡ In $\mathbb{V}^n(C)$ with n finite, $\Omega^{-1}\Omega=I\Leftrightarrow\Omega\Omega^{-1}=I$. Prove this using the ideas introduced toward the end of Theorem A.1.1., Appendix A.1.

operations can be carried out. We shall now see that in the same manner a linear operator can be represented in a basis by a set of n^2 numbers, written as an $n \times n$ matrix, and called its *matrix elements* in that basis. Although the matrix elements, just like the vector components, are basis dependent, they facilitate the computation of all basis-independent quantities, by rendering the abstract operator more tangible.

Our starting point is the observation made earlier, that the action of a linear operator is fully specified by its action on the basis vectors. If the basis vectors suffer a change

$$\Omega|i\rangle = |i'\rangle$$

(where $|i'\rangle$ is known), then any vector in this space undergoes a change that is readily calculable:

$$\Omega|V\rangle = \Omega\sum_i v_i|i\rangle = \sum_i v_i\Omega|i\rangle = \sum_i v_i|i'\rangle$$

When we say $|i'\rangle$ is known, we mean that its components in the original basis

$$\langle j|i'\rangle = \langle j|\Omega|i\rangle \equiv \Omega_{ji} \tag{1.6.1}$$

are known. The n^2 numbers, Ω_{ij}, are the *matrix elements* of Ω in this basis. If

$$\Omega|V\rangle = |V'\rangle$$

then the components of the transformed ket $|V'\rangle$ are expressable in terms of the Ω_{ij} and the components of $|V\rangle$:

$$\begin{aligned} v_i' = \langle i|V'\rangle = \langle i|\Omega|V\rangle &= \langle i|\Omega\left(\sum_j v_j|j\rangle\right) \\ &= \sum_j v_j\langle i|\Omega|j\rangle \\ &= \sum_j \Omega_{ij}v_j \end{aligned} \tag{1.6.2}$$

Equation (1.6.2) can be cast in matrix form:

$$\begin{bmatrix} v_1' \\ v_2' \\ \vdots \\ v_n' \end{bmatrix} = \begin{bmatrix} \langle 1|\Omega|1\rangle & \langle 1|\Omega|2\rangle & \cdots & \langle 1|\Omega|n\rangle \\ \langle 2|\Omega|1\rangle & & & \\ \vdots & & & \vdots \\ \langle n|\Omega|1\rangle & & \cdots & \langle n|\Omega|n\rangle \end{bmatrix} \begin{bmatrix} v_1 \\ v_2 \\ \vdots \\ v_n \end{bmatrix} \tag{1.6.3}$$

A mnemonic: the elements of the first column are simply the components of the first transformed basis vector $|1'\rangle = \Omega|1\rangle$ in the given basis. *Likewise, the elements of the* j*th column represent the image of the* j*th basis vector after* Ω *acts on it.*

Convince yourself that the same matrix Ω_{ij} acting to the *left* on the row vector corresponding to any $\langle v'|$ gives the row vector corresponding to $\langle v''| = \langle v'|\Omega$.

Example 1.6.1. Combining our mnemonic with the fact that the operator $R(\frac{1}{2}\pi\mathbf{i})$ has the following effect on the basis vectors:

$$R(\tfrac{1}{2}\pi\mathbf{i})|1\rangle = |1\rangle$$

$$R(\tfrac{1}{2}\pi\mathbf{i})|2\rangle = |3\rangle$$

$$R(\tfrac{1}{2}\pi\mathbf{i})|3\rangle = -|2\rangle$$

we can write down the matrix that represents it in the $|1\rangle, |2\rangle, |3\rangle$ basis:

$$R(\tfrac{1}{2}\pi\mathbf{i}) \leftrightarrow \begin{bmatrix} 1 & 0 & 0 \\ 0 & 0 & -1 \\ 0 & 1 & 0 \end{bmatrix} \tag{1.6.4}$$

For instance, the -1 in the third column tells us that R rotates $|3\rangle$ into $-|2\rangle$. One may also ignore the mnemonic altogether and simply use the definition $R_{ij} = \langle i|R|j\rangle$ to compute the matrix. □

Exercise 1.6.1. An operator Ω is given by the matrix

$$\begin{bmatrix} 0 & 0 & 1 \\ 1 & 0 & 0 \\ 0 & 1 & 0 \end{bmatrix}$$

What is its action?

Let us now consider certain specific operators and see how they appear in matrix form.

(1) The Identity Operator I.

$$I_{ij} = \langle i|I|j\rangle = \langle i|j\rangle = \delta_{ij} \tag{1.6.5}$$

Thus I is represented by a diagonal matrix with 1's along the diagonal. You should verify that our mnemonic gives the same result.

(2) The Projection Operators. Let us first get acquainted with *projection operators*. Consider the expansion of an arbitrary ket $|V\rangle$ in a basis:

$$|V\rangle = \sum_{i=1}^{n} |i\rangle\langle i|V\rangle$$

In terms of the objects $|i\rangle\langle i|$, which are linear operators, and which, by definition, act on $|V\rangle$ to give $|i\rangle\langle i|V\rangle$, we may write the above as

$$|V\rangle=\left(\sum_{i=1}^{n}|i\rangle\langle i|\right)|V\rangle \tag{1.6.6}$$

Since Eq. (1.6.6) is true for all $|V\rangle$, the object in the brackets must be identified with the identity (operator)

$$I=\sum_{i=1}^{n}|i\rangle\langle i|=\sum_{i=1}^{n}\mathbb{P}_i \tag{1.6.7}$$

The object $\mathbb{P}_i=|i\rangle\langle i|$ is called the *projection operator* for the ket $|i\rangle$. Equation (1.6.7), which is called the *completeness relation*, expresses the identity as a sum over projection operators and will be invaluable to us. (If you think that any time spent on the identity, which seems to do nothing, is a waste of time, just wait and see.)

Consider

$$\mathbb{P}_i|V\rangle=|i\rangle\langle i|V\rangle=|i\rangle v_i \tag{1.6.8}$$

Clearly $\mathbb{P}_i$ is linear. Notice that whatever $|V\rangle$ is, $\mathbb{P}_i|V\rangle$ is a multiple of $|i\rangle$ with a coefficient (v_i) which is the component of $|V\rangle$ along $|i\rangle$. Since $\mathbb{P}_i$ projects out the component of any ket $|V\rangle$ along the direction $|i\rangle$, it is called a *projection operator*. The completeness relation, Eq. (1.6.7), says that the sum of the projections of a vector along all the n directions equals the vector itself. Projection operators can also act on bras in the same way:

$$\langle V|\mathbb{P}_i=\langle V|i\rangle\langle i|=v_i^*\langle i| \tag{1.6.9}$$

Projection operators corresponding to the basis vectors obey

$$\mathbb{P}_i\mathbb{P}_j=|i\rangle\langle i|j\rangle\langle j|=\delta_{ij}\mathbb{P}_j \tag{1.6.10}$$

This equation tells us that (1) once $\mathbb{P}_i$ projects out the part of $|V\rangle$ along $|i\rangle$, further applications of $\mathbb{P}_i$ make no difference; and (2) the subsequent application of $\mathbb{P}_j (j\neq i)$ will result in zero, since a vector entirely along $|i\rangle$ cannot have a projection along a perpendicular direction $|j\rangle$.

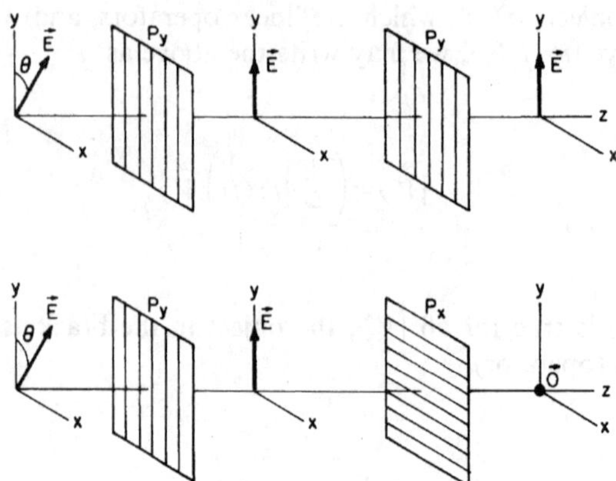

Figure 1.4. P_x and P_y are polarizers placed in the way of a beam traveling along the z axis. The action of the polarizers on the electric field **E** obeys the law of combination of projection operators: $P_i P_j = \delta_{ij} P_j$.

The following example from optics may throw some light on the discussion. Consider a beam of light traveling along the z axis and polarized in the $x-y$ plane at an angle θ with respect to the y axis (see Fig. 1.4). If a polarizer P_y, that only admits light polarized along the y axis, is placed in the way, the projection $E\cos\theta$ along the y axis is transmitted. An additional polarizer P_y placed in the way has no further effect on the beam. We may equate the action of the polarizer to that of a projection operator $\mathbb{P}_y$ that acts on the electric field vector **E**. If P_y is followed by a polarizer P_x the beam is completely blocked. Thus the polarizers obey the equation $P_i P_j = \delta_{ij} P_j$ expected of projection operators.

Let us next turn to the matrix elements of $\mathbb{P}_i$. There are two approaches. The first one, somewhat indirect, gives us a feeling for what kind of an object $|i\rangle\langle i|$ is. We know

$$|i\rangle \leftrightarrow \begin{bmatrix} 0 \\ 0 \\ 0 \\ \vdots \\ 1 \\ \vdots \\ 0 \end{bmatrix}$$

and

$$\langle i| \leftrightarrow [0, 0, \ldots, 1, 0, 0, \ldots, 0]$$

so that

$$|i\rangle\langle i| \leftrightarrow \begin{bmatrix} 0 \\ 0 \\ \vdots \\ 1 \\ 0 \\ \vdots \\ 0 \end{bmatrix} [0, 0, \ldots, 1, 0, \ldots, 0] = \begin{bmatrix} 0 & & & \cdots & & & 0 \\ & \ddots & & & & & \\ & & 0 & & & & \\ \vdots & & & 1 & & & \\ & & & & 0 & & \\ & & & & & \ddots & \\ 0 & & & & & & 0 \end{bmatrix} \tag{1.6.11}$$

by the rules of matrix multiplication. Whereas $\langle V|V'\rangle = (1\times n \text{ matrix}) \times (n\times 1 \text{ matrix}) = (1\times 1 \text{ matrix})$ is a scalar, $|V\rangle\langle V'| = (n\times 1 \text{ matrix}) \times (1\times n \text{ matrix}) = (n\times n \text{ matrix})$ is an operator. The inner product $\langle V|V'\rangle$ represents a bra and ket which have found each other, while $|V\rangle\langle V'|$, sometimes called the *outer product*, has the two factors looking the other way for a bra or a ket to dot with.

The more direct approach to the matrix elements gives

$$(\mathbb{P}_i)_{kl} = \langle k|i\rangle\langle i|l\rangle = \delta_{ki}\delta_{il} = \delta_{kl}\delta_{li} \tag{1.6.12}$$

which is of course identical to Eq. (1.6.11). The same result also follows from mnemonic. Each projection operator has only one nonvanishing matrix element, a 1 at the ith element on the diagonal. The completeness relation, Eq. (1.6.7), says that when all the $\mathbb{P}_i$ are added, the diagonal fills out to give the identity. If we form the sum over just some of the projection operators, we get the operator which projects a given vector into the subspace spanned by just the corresponding basis vectors.

Matrices Corresponding to Products of Operators

Consider next the matrices representing a product of operators. These are related to the matrices representing the individual operators by the application of Eq. (1.6.7):

$$\begin{aligned}(\Omega\Lambda)_{ij} &= \langle i|\Omega\Lambda|j\rangle = \langle i|\Omega I\Lambda|j\rangle \\ &= \sum_k \langle i|\Omega|k\rangle\langle k|\Lambda|j\rangle = \sum_k \Omega_{ik}\Lambda_{kj}\end{aligned} \tag{1.6.13}$$

Thus the matrix representing the product of operators is the product of the matrices representing the factors.

The Adjoint of an Operator

Recall that given a ket $\alpha|V\rangle \equiv |\alpha V\rangle$ the corresponding bra is

$$\langle \alpha V| = \langle V|\alpha^* \quad (\text{and } not\ \langle V|\alpha)$$

In the same way, given a ket

$$\Omega|V\rangle = |\Omega V\rangle$$

the corresponding bra is

$$\langle \Omega V| = \langle V|\Omega^\dagger \tag{1.6.14}$$

which *defines* the operator $\Omega^\dagger$. One may state this equation in words: if Ω turns a ket $|V\rangle$ to $|V'\rangle$, then $\Omega^\dagger$ turns the bra $\langle V|$ into $\langle V'|$. Just as a and a^*, $|V\rangle$ and $\langle V|$ are related but distinct objects, so are Ω and $\Omega^\dagger$. The relation between Ω, and $\Omega^\dagger$, called the *adjoint* of Ω or "omega dagger," is best seen in a basis:

$$\begin{aligned}(\Omega^\dagger)_{ij} &= \langle i|\Omega^\dagger|j\rangle = \langle \Omega i|j\rangle \\ &= \langle j|\Omega i\rangle^* = \langle j|\Omega|i\rangle^*\end{aligned}$$

so

$$\Omega^\dagger_{ij} = \Omega^*_{ji} \tag{1.6.15}$$

In other words, the matrix representing $\Omega^\dagger$ is the transpose conjugate of the matrix representing Ω. (Recall that the row vector representing $\langle V|$ is the transpose conjugate of the column vector representing $|V\rangle$. *In a given basis, the adjoint operation is the same as taking the transpose conjugate.*)

The adjoint of a product is the product of the adjoints in reverse:

$$(\Omega\Lambda)^\dagger = \Lambda^\dagger\Omega^\dagger \tag{1.6.16}$$

To prove this we consider $\langle \Omega\Lambda V|$. First we treat $\Omega\Lambda$ as one operator and get

$$\langle \Omega\Lambda V| = \langle (\Omega\Lambda)V| = \langle V|(\Omega\Lambda)^\dagger$$

Next we treat (ΛV) as just another vector, and write

$$\langle \Omega\Lambda V| = \langle \Omega(\Lambda V)| = \langle \Lambda V|\Omega^\dagger$$

We next pull out Λ, pushing $\Omega^\dagger$ further out:

$$\langle \Lambda V|\Omega^\dagger = \langle V|\Lambda^\dagger\Omega^\dagger$$

Comparing this result with the one obtained a few lines above, we get the desired result.

Consider now an equation consisting of kets, scalars, and operators, such as

$$\alpha_1|V_1\rangle = \alpha_2|V_2\rangle + \alpha_3|V_3\rangle\langle V_4|V_5\rangle + \alpha_4\Omega\Lambda|V_6\rangle \tag{1.6.17a}$$

What is its adjoint? Our old rule tells us that it is

$$\langle V_1|\alpha_1^* = \langle V_2|\alpha_2^* + \langle V_5|V_4\rangle\langle V_3|\alpha_3^* + \langle \Omega\Lambda V_6|\alpha_4^*$$

In the last term we can replace $\langle \Omega\Lambda V_6|$ by

$$\langle V_6|(\Omega\Lambda)^\dagger = \langle V_6|\Lambda^\dagger\Omega^\dagger$$

so that finally we have the adjoint of Eq. (1.6.17a):

$$\langle V_1|\alpha_1^* = \langle V_2|\alpha_2^* + \langle V_5|V_4\rangle\langle V_3|\alpha_3^* + \langle V_6|\Lambda^\dagger\Omega^\dagger\alpha_4^* \tag{1.6.17b}$$

The final rule for taking the adjoint of the most general equation we will ever encounter is this:

When a product of operators, bras, kets, and explicit numerical coefficients is encountered, reverse the order of all factors and make the substitutions $\Omega \leftrightarrow \Omega^\dagger$, $|\rangle \leftrightarrow \langle|$, $a \leftrightarrow a^*$.

(Of course, there is no real need to reverse the location of the scalars a except in the interest of uniformity.)

Hermitian, Anti-Hermitian, and Unitary Operators

We now turn our attention to certain special classes of operators that will play a major role in quantum mechanics.

Definition 13. An operator Ω is *Hermitian* if $\Omega^\dagger = \Omega$.

Definition 14. An operator Ω is *anti-Hermitian* if $\Omega^\dagger = -\Omega$.

The adjoint is to an operator what the complex conjugate is to numbers. Hermitian and anti-Hermitian operators are like pure real and pure imaginary numbers. Just as every number may be decomposed into a sum of pure real and pure imaginary parts,

$$\alpha = \frac{\alpha + \alpha^*}{2} + \frac{\alpha - \alpha^*}{2}$$

we can decompose every operator into its Hermitian and anti-Hermitian parts:

$$\Omega = \frac{\Omega + \Omega^\dagger}{2} + \frac{\Omega - \Omega^\dagger}{2} \tag{1.6.18}$$

*Exercise 1.6.2.** Given Ω and Λ are Hermitian what can you say about (1) $\Omega\Lambda$; (2) $\Omega\Lambda + \Lambda\Omega$; (3) $[\Omega, \Lambda]$; and (4) $i[\Omega, \Lambda]$?

Definition 15. An operator U is *unitary* if

$$UU^\dagger = I \tag{1.6.19}$$

This equation tells us that U and $U^\dagger$ are inverses of each other. Consequently, from Eq. (1.5.12),

$$U^\dagger U = I \tag{1.6.20}$$

Following the analogy between operators and numbers, unitary operators are like complex numbers of unit modulus, $u = e^{i\theta}$. Just as $u^* u = 1$, so is $U^\dagger U = I$.

*Exercise 1.6.3.** Show that a product of unitary operators is unitary.

Theorem 7. Unitary operators preserve the inner product between the vectors they act on.

Proof. Let

$$|V_1'\rangle = U|V_1\rangle$$

and

$$|V_2'\rangle = U|V_2\rangle$$

Then

$$\begin{aligned}\langle V_2'|V_1'\rangle &= \langle UV_2|UV_1\rangle \\ &= \langle V_2|U^\dagger U|V_1\rangle = \langle V_2|V_1\rangle\end{aligned} \tag{1.6.21}$$

(Q.E.D.)

Unitary operators are the generalizations of rotation operators from $\mathbb{V}^3(R)$ to $\mathbb{V}^n(C)$, for just like rotation operators in three dimensions, they preserve the lengths of vectors and their dot products. In fact, on a real vector space, the unitarity condition becomes $U^{-1} = U^T$ (T means transpose), which defines an *orthogonal* or rotation matrix. [$R(\frac{1}{2}\pi\mathbf{i})$ is an example.]

Theorem 8. If one treats the columns of an $n \times n$ unitary matrix as components of n vectors, these vectors are orthonormal. In the same way, the rows may be interpreted as components of n orthonormal vectors.

Proof 1. According to our mnemonic, the jth column of the matrix representing U is the image of the jth basis vector after U acts on it. Since U preserves inner products, the rotated set of vectors is also orthonormal. Consider next the rows. We now use the fact that $U^\dagger$ is also a rotation. (How else can it neutralize U to give $U^\dagger U = I$?) Since the rows of U are the columns of $U^\dagger$ (but for an overall complex conjugation which does not affect the question of orthonormality), the result we already have for the columns of a unitary matrix tells us the rows of U are orthonormal.

Proof 2. Since $U^\dagger U=I$,

$$\begin{aligned}\delta_{ij}&=\langle i|I|j\rangle=\langle i|U^\dagger U|j\rangle\\&=\sum_k\langle i|U^\dagger|k\rangle\langle k|U|j\rangle\\&=\sum_k U^\dagger_{ik}U_{kj}=\sum_k U^*_{ki}U_{kj}\end{aligned}\tag{1.6.22}$$

which proves the theorem for the columns. A similar result for the rows follows if we start with the equation $UU^\dagger=I$. Q.E.D.

Note that $U^\dagger U=I$ and $UU^\dagger=I$ are not independent conditions.

*Exercise 1.6.4.** It is assumed that you know (1) what a *determinant* is, (2) that det $\Omega^T=$ det Ω (T denotes transpose), (3) that the determinant of a product of matrices is the product of the determinants. [If you do not, verify these properties for a two-dimensional case

$$\Omega=\begin{pmatrix}\alpha & \beta\\ \gamma & \delta\end{pmatrix}$$

with det $\Omega=(\alpha\delta-\beta\gamma)$.] Prove that the determinant of a unitary matrix is a complex number of unit modulus.

*Exercise 1.6.5.** Verify that $R(\frac{1}{2}\pi\mathbf{i})$ is unitary (orthogonal) by examining its matrix.

Exercise 1.6.6. Verify that the following matrices are unitary:

$$\frac{1}{2^{1/2}}\begin{bmatrix}1 & i\\ i & 1\end{bmatrix},\qquad \frac{1}{2}\begin{bmatrix}1+i & 1-i\\ 1-i & 1+i\end{bmatrix}$$

Verify that the determinant is of the form $e^{i\theta}$ in each case. Are any of the above matrices Hermitian?

1.7. Active and Passive Transformations

Suppose we subject all the vectors $|V\rangle$ in a space to a unitary transformation

$$|V\rangle\to U|V\rangle\tag{1.7.1}$$

Under this transformation, the matrix elements of any operator Ω are modified as follows:

$$\langle V'|\Omega|V\rangle\to\langle UV'|\Omega|UV\rangle=\langle V'|U^\dagger\Omega U|V\rangle\tag{1.7.2}$$

It is clear that the same change would be effected if we left the vectors alone and subjected all operators to the change

$$\Omega \to U^\dagger \Omega U \tag{1.7.3}$$

The first case is called an *active transformation* and the second a *passive transformation*. The present nomenclature is in reference to the vectors: they are affected in an active transformation and left alone in the passive case. The situation is exactly the opposite from the point of view of the operators.

Later we will see that the physics in quantum theory lies in the matrix elements of operators, and that active and passive transformations provide us with two equivalent ways of describing the same physical transformation.

*Exercise 1.7.1.** The *trace* of a matrix is defined to be the sum of its diagonal matrix elements

$$\mathrm{Tr}\,\Omega = \sum_i \Omega_{ii}$$

Show that

(1) $\mathrm{Tr}(\Omega\Lambda) = \mathrm{Tr}(\Lambda\Omega)$
(2) $\mathrm{Tr}(\Omega\Lambda\theta) = \mathrm{Tr}(\Lambda\theta\Omega) = \mathrm{Tr}(\theta\Omega\Lambda)$ (The permutations are *cyclic*).
(3) The trace of an operator is unaffected by a unitary change of basis $|i\rangle \to U|i\rangle$. [Equivalently, show $\mathrm{Tr}\,\Omega = \mathrm{Tr}(U^\dagger \Omega U)$.]

Exercise 1.7.2. Show that the determinant of a matrix is unaffected by a unitary change of basis. [Equivalently show $\det \Omega = \det(U^\dagger \Omega U)$.]

1.8. The Eigenvalue Problem

Consider some linear operator Ω acting on an arbitrary *nonzero* ket $|V\rangle$:

$$\Omega|V\rangle = |V'\rangle \tag{1.8.1}$$

Unless the operator happens to be a trivial one, such as the identity or its multiple, the ket will suffer a nontrivial change, i.e., $|V'\rangle$ will not be simply related to $|V\rangle$. So much for an arbitrary ket. Each operator, however, has certain kets of its own, called its *eigenkets*, on which its action is simply that of rescaling:

$$\Omega|V\rangle = \omega|V\rangle \tag{1.8.2}$$

Equation (1.8.2) is an eigenvalue equation: $|V\rangle$ is an *eigenket* of Ω with *eigenvalue* ω. In this chapter we will see how, given an operator Ω, one can systematically determine all its eigenvalues and eigenvectors. How such an equation enters physics will be illustrated by a few examples from mechanics at the end of this section, and once we get to quantum mechanics proper, it will be eigen, eigen, eigen all the way.

Example 1.8.1. To illustrate how easy the eigenvalue problem really is, we will begin with a case that will be completely solved: the case $\Omega = I$. Since

$$I|V\rangle = |V\rangle$$

for all $|V\rangle$, we conclude that

(1) the only eigenvalue of I is 1;
(2) all vectors are its eigenvectors with this eigenvalue. □

Example 1.8.2. After this unqualified success, we are encouraged to take on a slightly more difficult case: $\Omega = \mathbb{P}_V$, the projection operator associated with a *normalized* ket $|V\rangle$. Clearly

(1) any ket $\alpha|V\rangle = |\alpha V\rangle$, parallel to $|V\rangle$ is an eigenket with eigenvalue 1:

$$\mathbb{P}_V|\alpha V\rangle = |V\rangle\langle V|\alpha V\rangle = \alpha|V\rangle|V|^2 = 1 \cdot |\alpha V\rangle$$

(2) any ket $|V_\perp\rangle$, perpendicular to $|V\rangle$, is an eigenket with eigenvalue 0:

$$\mathbb{P}_V|V_\perp\rangle = |V\rangle\langle V|V_\perp\rangle = 0 = 0|V_\perp\rangle$$

(3) kets that are neither, i.e., kets of the form $\alpha|V\rangle + \beta|V_\perp\rangle$, are simply not eigenkets:

$$\mathbb{P}_V(\alpha|V\rangle + \beta|V_\perp\rangle) = |\alpha V\rangle \neq \gamma(\alpha|V\rangle + \beta|V_\perp\rangle)$$

Since every ket in the space falls into one of the above classes, we have found all the eigenvalues and eigenvectors. □

Example 1.8.3. Consider now the operator $R(\frac{1}{2}\pi\mathbf{i})$. We already know that it has one eigenket, the basis vector $|1\rangle$ along the x axis:

$$R(\tfrac{1}{2}\pi\mathbf{i})|1\rangle = |1\rangle$$

Are there others? Of course, any vector $\alpha|1\rangle$ along the x axis is also unaffected by the x rotation. This is a general feature of the eigenvalue equation and reflects the linearity of the operator:

if

$$\Omega|V\rangle = \omega|V\rangle$$

then

$$\Omega\alpha|V\rangle = \alpha\Omega|V\rangle = \alpha\omega|V\rangle = \omega\alpha|V\rangle$$

for any multiple α. Since the eigenvalue equation fixes the eigenvector only up to an overall scale factor, we will not treat the multiples of an eigenvector as distinct eigenvectors. With this understanding in mind, let us ask if $R(\frac{1}{2}\pi\mathbf{i})$ has any eigenvectors besides $|1\rangle$. Our intuition says no, for any vector not along the x axis necessarily gets rotated by $R(\frac{1}{2}\pi\mathbf{i})$ and cannot possibly transform into a multiple of itself. Since every vector is either parallel to $|1\rangle$ or isn't, we have fully solved the eigenvalue problem.

The trouble with this conclusion is that it is wrong! $R(\frac{1}{2}\pi\mathbf{i})$ has two other eigenvectors besides $|1\rangle$. But our intuition is not to be blamed, for these vectors are in $\mathbb{V}^3(C)$ and not $\mathbb{V}^3(R)$. It is clear from this example that we need a reliable and systematic method for solving the eigenvalue problem in $\mathbb{V}^n(C)$. We now turn our attention to this very question. □

The Characteristic Equation and the Solution to the Eigenvalue Problem

We begin by rewriting Eq. (1.8.2) as

$$(\Omega-\omega I)|V\rangle=|0\rangle \tag{1.8.3}$$

Operating both sides with $(\Omega-\omega I)^{-1}$, assuming it exists, we get

$$|V\rangle=(\Omega-\omega I)^{-1}|0\rangle \tag{1.8.4}$$

Now, any finite operator (an operator with finite matrix elements) acting on the null vector can only give us a null vector. It therefore seems that in asking for a nonzero eigenvector $|V\rangle$, we are trying to get something for nothing out of Eq. (1.8.4). This is impossible. It follows that our assumption that the operator $(\Omega-\omega I)^{-1}$ exists (as a finite operator) is false. So we ask when this situation will obtain. Basic matrix theory tells us (see Appendix A.1) that the inverse of any matrix M is given by

$$M^{-1}=\frac{\text{cofactor } M^T}{\det M} \tag{1.8.5}$$

Now the cofactor of M is finite if M is. Thus what we need is the vanishing of the determinant. The condition for nonzero eigenvectors is therefore

$$\det(\Omega-\omega I)=0 \tag{1.8.6}$$

This equation will determine the eigenvalues ω. To find them, we project Eq. (1.8.3) onto a basis. Dotting both sides with a basis bra $\langle i|$, we get

$$\langle i|\Omega-\omega I|V\rangle=0$$

and upon introducing the representation of the identity [Eq. (1.6.7)], to the left of $|V\rangle$, we get the following image of Eq. (1.8.3):

$$\sum_j (\Omega_{ij} - \omega\delta_{ij})v_j = 0 \tag{1.8.7}$$

Setting the determinant to zero will give us an expression of the form

$$\sum_{m=0}^{n} c_m \omega^m = 0 \tag{1.8.8}$$

Equation (1.8.8) is called the *characteristic equation* and

$$P^n(\omega) = \sum_{m=0}^{n} c_m \omega^m \tag{1.8.9}$$

is called the *characteristic polynomial*. Although the polynomial is being determined in a particular basis, the eigenvalues, which are its roots, are basis independent, for they are defined by the abstract Eq. (1.8.3), which makes no reference to any basis.

Now, a fundamental result in analysis is that every nth-order polynomial has n roots, not necessarily distinct and not necessarily real. Thus every operator in $\mathbb{V}^n(C)$ has n eigenvalues. Once the eigenvalues are known, the eigenvectors may be found, at least for Hermitian and unitary operators, using a procedure illustrated by the following example. [Operators on $\mathbb{V}^n(C)$ that are not of the above variety may not have n eigenvectors—see Exercise 1.8.4. Theorems 10 and 12 establish that Hermitian and unitary operators on $\mathbb{V}^n(C)$ will have n eigenvectors.]

Example 1.8.4. Let us use the general techniques developed above to find all the eigenvectors and eigenvalues of $R(\frac{1}{2}\pi\mathbf{i})$. Recall that the matrix representing it is

$$R(\tfrac{1}{2}\pi\mathbf{i}) \leftrightarrow \begin{bmatrix} 1 & 0 & 0 \\ 0 & 0 & -1 \\ 0 & 1 & 0 \end{bmatrix}$$

Therefore the characteristic equation is

$$\det(R - \omega I) = \begin{vmatrix} 1-\omega & 0 & 0 \\ 0 & -\omega & -1 \\ 0 & 1 & -\omega \end{vmatrix} = 0$$

i.e.,

$$(1-\omega)(\omega^2 + 1) = 0 \tag{1.8.10}$$

with roots $\omega=1, \pm i$. We know that $\omega=1$ corresponds to $|1\rangle$. Let us see this come out of the formalism. Feeding $\omega=1$ into Eq. (1.8.7) we find that the components x_1, x_2, and x_3 of the corresponding eigenvector must obey the equations

$$\begin{bmatrix} 1-1 & 0 & 0 \\ 0 & 0-1 & -1 \\ 0 & 1 & 0-1 \end{bmatrix}\begin{bmatrix} x_1 \\ x_2 \\ x_3 \end{bmatrix}=\begin{bmatrix} 0 \\ 0 \\ 0 \end{bmatrix} \rightarrow \begin{matrix} 0=0 \\ \left.\begin{matrix} -x_2-x_3=0 \\ x_2-x_3=0 \end{matrix}\right\} \rightarrow x_2=x_3=0 \end{matrix}$$

Thus any vector of the form

$$x_1|1\rangle \leftrightarrow \begin{bmatrix} x_1 \\ 0 \\ 0 \end{bmatrix}$$

is acceptable, as expected. It is conventional to use the freedom in scale to normalize the eigenvectors. Thus in this case a choice is

$$|\omega=1\rangle=|1\rangle=\begin{bmatrix} 1 \\ 0 \\ 0 \end{bmatrix}$$

I say *a* choice, and not *the* choice, since the vector may be multiplied by a number of modulus unity without changing the norm. There is no universally accepted convention for eliminating this freedom, except perhaps to choose the vector with real components when possible.

Note that of the three simultaneous equations above, the first is not a real equation. In general, there will be only $(n-1)$ LI equations. This is the reason the norm of the vector is not fixed and, as shown in Appendix A.1, the reason the determinant vanishes.

Consider next the equations corresponding to $\omega=i$. The components of the eigenvector obey the equations

$$(1-i)x_1=0 \qquad \text{(i.e., } x_1=0\text{)}$$

$$-ix_2-x_3=0 \qquad \text{(i.e., } x_2=ix_3\text{)}$$

$$x_2-ix_3=0 \qquad \text{(i.e., } x_2=ix_3\text{)}$$

Notice once again that we have only $n-1$ useful equations. A properly normalized solution to the above is

$$|\omega=i\rangle \leftrightarrow \frac{1}{2^{1/2}}\begin{bmatrix} 0 \\ i \\ 1 \end{bmatrix}$$

A similar procedure yields the third eigenvector:

$$|\omega=-i\rangle \leftrightarrow \frac{1}{2^{1/2}}\begin{bmatrix} 0 \\ -i \\ 1 \end{bmatrix}$$ □

In the above example we have introduced a popular convention: labeling the eigenvectors by the eigenvalue. For instance, the ket corresponding to $\omega=\omega_i$ is labeled $|\omega=\omega_i\rangle$ or simply $|\omega_i\rangle$. This notation presumes that to each ω_i there is just one vector labeled by it. Though this is not always the case, only a slight change in this notation will be needed to cover the general case.

The phenomenon of a single eigenvalue representing more than one eigenvector is called *degeneracy* and corresponds to repeated roots for the characteristic polynomial. In the face of degeneracy, we need to modify not just the labeling, but also the procedure used in the example above for finding the eigenvectors. Imagine that instead of $R(\frac{1}{2}\pi\mathbf{i})$ we were dealing with another operator Ω on $\mathbb{V}^3(R)$ with roots ω_1 and $\omega_2=\omega_3$. It appears as if we can get two eigenvectors, by the method described above, one for each distinct ω. How do we get a third? Or is there no third? These questions will be answered in all generality shortly when we examine the question of degeneracy in detail. We now turn our attention to two central theorems on Hermitian operators. These play a vital role in quantum mechanics.

Theorem 9. The eigenvalues of a Hermitian operator are real.

Proof. Let

$$\Omega|\omega\rangle=\omega|\omega\rangle$$

Dot both sides with $\langle\omega|$:

$$\langle\omega|\Omega|\omega\rangle=\omega\langle\omega|\omega\rangle \tag{1.8.11}$$

Take the adjoint to get

$$\langle\omega|\Omega^\dagger|\omega\rangle=\omega^*\langle\omega|\omega\rangle$$

Since $\Omega=\Omega^\dagger$, this becomes

$$\langle\omega|\Omega|\omega\rangle=\omega^*\langle\omega|\omega\rangle$$

Subtracting from Eq. (1.8.11)

$$0=(\omega-\omega^*)\langle\omega|\omega\rangle$$

$$\omega=\omega^* \quad \text{Q.E.D.}$$

Theorem 10. To every Hermitian operator Ω, there exists (at least) a basis consisting of its orthonormal eigenvectors. It is diagonal in this eigenbasis and has its eigenvalues as its diagonal entries.

Proof. Let us start with the characteristic equation. It must have at least one root, call it ω_1. Corresponding to ω_1 there must exist at least one nonzero eigenvector $|\omega_1\rangle$. [If not, Theorem (A.1.1) would imply that $(\Omega-\omega_1 I)$ is invertible.] Consider the subspace $\mathbb{V}_{\perp 1}^{n-1}$ of all vectors orthogonal to $|\omega_1\rangle$. Let us choose as our basis the vector $|\omega_1\rangle$ (normalized to unity) and any $n-1$ orthonormal vectors $\{V_{\perp 1}^1, V_{\perp 1}^2, \ldots, V_{\perp 1}^{n-1}\}$ in $\mathbb{V}_{\perp 1}^{n-1}$. In this basis Ω has the following form:

$$\Omega \leftrightarrow \begin{bmatrix} \omega_1 & 0 & 0 & 0 & 0 & \cdots & 0 \\ 0 & & & & & & \\ 0 & & & \boxed{} & & & \\ \vdots & & & & & & \\ 0 & & & & & & \end{bmatrix} \tag{1.8.12}$$

The first column is just the image of $|\omega_1\rangle$ after Ω has acted on it. Given the first column, the first row follows from the Hermiticity of Ω.

The characteristic equation now takes the form

$$(\omega_1-\omega)\cdot(\text{determinant of boxed submatrix})=0$$

$$(\omega_1-\omega)\sum_{0}^{n-1} c_m\omega^m=(\omega_1-\omega)P^{n-1}(\omega)=0$$

Now the polynomial P^{n-1} must also generate one root, ω_2, and a normalized eigenvector $|\omega_2\rangle$. Define the subspace $\mathbb{V}_{\perp 1,2}^{n-2}$ of vectors in $\mathbb{V}_{\perp 1}^{n-1}$ orthogonal to $|\omega_2\rangle$ (and automatically to $|\omega_1\rangle$) and repeat the same procedure as before. Finally, the matrix Ω becomes, in the basis $|\omega_1\rangle, |\omega_2\rangle, \ldots, |\omega_n\rangle$,

$$\Omega \leftrightarrow \begin{bmatrix} \omega_1 & 0 & 0 & \cdots & 0 \\ 0 & \omega_2 & 0 & & 0 \\ 0 & 0 & \omega_3 & & 0 \\ \vdots & & & \ddots & \\ 0 & 0 & 0 & & \omega_n \end{bmatrix}$$

Since every $|\omega_i\rangle$ was chosen from a space that was orthogonal to the previous ones, $|\omega_1\rangle, |\omega_2\rangle, \ldots, |\omega_{i-1}\rangle$; the basis of eigenvectors is orthonormal. (Notice that nowhere did we have to assume that the eigenvalues were all distinct.) Q.E.D.

[The analogy between real numbers and Hermitian operators is further strengthened by the fact that in a certain basis (of eigenvectors) the Hermitian operator can be represented by a matrix with all real elements.]

In stating Theorem 10, it was indicated that there might exist more than one basis of eigenvectors that diagonalized Ω. This happens if there is any degeneracy. Suppose $\omega_1=\omega_2=\omega$. Then we have two orthonormal vectors obeying

$$\Omega|\omega_1\rangle=\omega|\omega_1\rangle$$

$$\Omega|\omega_2\rangle=\omega|\omega_2\rangle$$

It follows that

$$\Omega[\alpha|\omega_1\rangle+\beta|\omega_2\rangle]=\alpha\omega|\omega_1\rangle+\beta\omega|\omega_2\rangle=\omega[\alpha|\omega_1\rangle+\beta|\omega_2\rangle]$$

for any α and β. Since the vectors $|\omega_1\rangle$ and $|\omega_2\rangle$ are orthogonal (and hence LI), we find that there is a whole two-dimensional subspace spanned by $|\omega_1\rangle$ and $|\omega_2\rangle$, the elements of which are eigenvectors of Ω with eigenvalue ω. One refers to this space as an *eigenspace* of Ω with eigenvalue ω. Besides the vectors $|\omega_1\rangle$ and $|\omega_2\rangle$, there exists an infinity of orthonormal pairs $|\omega_1'\rangle$, $|\omega_2'\rangle$, obtained by a rigid rotation of $|\omega_1\rangle$, $|\omega_2\rangle$, from which we may select any pair in forming the eigenbasis of Ω. In general, if an eigenvalue occurs m_i times, that is, if the characteristic equation has m_i of its roots equal to some ω_i, there will be an eigenspace $\mathbb{V}_{\omega_i}^{m_i}$ from which we may choose any m_i orthonormal vectors to form the basis referred to in Theorem 10.

In the absence of degeneracy, we can prove Theorem 9 and 10 very easily. Let us begin with two eigenvectors:

$$\Omega|\omega_i\rangle=\omega_i|\omega_i\rangle \tag{1.8.13a}$$

$$\Omega|\omega_j\rangle=\omega_j|\omega_j\rangle \tag{1.8.13b}$$

Dotting the first with $\langle\omega_j|$ and the second with $\langle\omega_i|$, we get

$$\langle\omega_j|\Omega|\omega_i\rangle=\omega_i\langle\omega_j|\omega_i\rangle \tag{1.8.14a}$$

$$\langle\omega_i|\Omega|\omega_j\rangle=\omega_j\langle\omega_i|\omega_j\rangle \tag{1.8.14b}$$

Taking the adjoint of the last equation and using the Hermitian nature of Ω, we get

$$\langle\omega_j|\Omega|\omega_i\rangle=\omega_j^*\langle\omega_j|\omega_i\rangle$$

Subtracting this equation from Eq. (1.8.14a), we get

$$0=(\omega_i-\omega_j^*)\langle\omega_j|\omega_i\rangle \tag{1.8.15}$$

If $i=j$, we get, since $\langle\omega_i|\omega_i\rangle\neq0$,

$$\omega_i=\omega_i^* \tag{1.8.16}$$

If $i \neq j$, we get

$$\langle \omega_i | \omega_j \rangle = 0 \tag{1.8.17}$$

since $\omega_i - \omega_j^* = \omega_i - \omega_j \neq 0$ by assumption. That the proof of orthogonality breaks down for $\omega_i = \omega_j$ is not surprising, for two vectors labeled by a degenerated eigenvalue could be any two members of the degenerate space which need not necessarily be orthogonal. The modification of this proof in this case of degeneracy calls for arguments that are essentially the ones used in proving Theorem 10. The advantage in the way Theorem 10 was proved first is that it suffers no modification in the degenerate case.

Degeneracy

We now address the question of degeneracy as promised earlier. Now, our general analysis of Theorem 10 showed us that in the face of degeneracy, we have not one, but an infinity of orthonormal eigenbases. Let us see through an example how this variety manifests itself when we look for eigenvectors and how it is to be handled.

Example 1.8.5. Consider an operator Ω with matrix elements

$$\Omega \leftrightarrow \begin{bmatrix} 1 & 0 & 1 \\ 0 & 2 & 0 \\ 1 & 0 & 1 \end{bmatrix}$$

in some basis. The characteristic equation is

$$(\omega - 2)^2 \omega = 0$$

i.e.,

$$\omega = 0, 2, 2$$

The vector corresponding to $\omega = 0$ is found by the usual means to be

$$|\omega = 0\rangle \leftrightarrow \frac{1}{2^{1/2}} \begin{bmatrix} 1 \\ 0 \\ -1 \end{bmatrix}$$

The case $\omega = 2$ leads to the following equations for the components of the eigenvector:

$$-x_1 + x_3 = 0$$

$$0 = 0$$

$$x_1 - x_3 = 0$$

Now we have just one equation, instead of the two $(n-1)$ we have grown accustomed to! This is a reflection of the degeneracy. For every extra appearance (besides the first) a root makes, it takes away one equation. Thus degeneracy permits us extra degrees of freedom besides the usual one (of normalization). The conditions

$$x_1 = x_3$$

$$x_2 \text{ arbitrary}$$

define an ensemble of vectors that are perpendicular to the first, $|\omega=0\rangle$, i.e., lie in a plane perpendicular to $|\omega=0\rangle$. This is in agreement with our expectation that a twofold degeneracy should lead to a two-dimensional eigenspace. The freedom in x_2 (or more precisely, the ratio x_2/x_3) corresponds to the freedom of orientation in this plane. Let us arbitrarily choose $x_2=1$, to get a normalized eigenvector corresponding to $\omega=2$:

$$|\omega=2\rangle \leftrightarrow \frac{1}{3^{1/2}}\begin{bmatrix}1\\1\\1\end{bmatrix}$$

The third vector is now chosen to lie in this plane and to be orthogonal to the second (being in this plane automatically makes it perpendicular to the first $|\omega=0\rangle$):

$$|\omega=2, \text{second one}\rangle \leftrightarrow \frac{1}{6^{1/2}}\begin{bmatrix}1\\-2\\1\end{bmatrix}$$

Clearly each distinct choice of the ratio, x_2/x_3, gives us a distinct doublet of orthonormal eigenvectors with eigenvalue 2. □

Notice that in the face of degeneracy, $|\omega_i\rangle$ no longer refers to a single ket but to a generic element of the eigenspace $\mathbb{V}_{\omega_i}^{m_i}$. To refer to a particular element, we must use the symbol $|\omega_i, \alpha\rangle$, where α labels the ket within the eigenspace. A natural choice of the label α will be discussed shortly.

We now consider the analogs of Theorems 9 and 10 for unitary operators.

Theorem 11. The eigenvalues of a unitary operator are complex numbers of unit modulus.

Theorem 12. The eigenvectors of a unitary operator are mutually orthogonal. (We assume there is no degeneracy.)

Proof of Both Theorems (assuming no degeneracy). Let

$$U|u_i\rangle = u_i|u_i\rangle \tag{1.8.18a}$$

and

$$U|u_j\rangle = u_j|u_j\rangle \tag{1.8.18b}$$

If we take the adjoint of the second equation and dot each side with the corresponding side of the first equation, we get

$$\langle u_j|U^\dagger U|u_i\rangle = u_i u_j^* \langle u_j|u_i\rangle$$

so that

$$(1 - u_i u_j^*)\langle u_j|u_i\rangle = 0 \tag{1.8.19}$$

If $i=j$, we get, since $\langle u_i|u_i\rangle \neq 0$,

$$u_i u_i^* = 1 \tag{1.8.20a}$$

while if $i \neq j$,

$$\langle u_i|u_j\rangle = 0 \tag{1.8.20b}$$

since $|u_i\rangle \neq |u_j\rangle \Rightarrow u_i \neq u_j \Rightarrow u_i u_i^* \neq u_i u_j^* \Rightarrow u_i u_j^* \neq 1$. (Q.E.D.)

If U is degenerate, we can carry out an analysis parallel to that for the Hermitian operator Ω, with just one difference. Whereas in Eq. (1.8.12), the zeros of the first row followed from the zeros of the first column and $\Omega^\dagger = \Omega$, here they follow from the requirement that the sum of the modulus squared of the elements in each row adds up to 1. Since $|u_1| = 1$, all the other elements in the first row must vanish.

Diagonalization of Hermitian Matrices

Consider a Hermitian operator Ω on $\mathbb{V}^n(C)$ represented as a matrix in some orthonormal basis $|1\rangle, \ldots, |i\rangle, \ldots, |n\rangle$. If we trade this basis for the eigenbasis $|\omega_1\rangle, \ldots, |\omega_i\rangle, \ldots, |\omega_n\rangle$, the matrix representing Ω will become diagonal. Now the operator U inducing the change of basis

$$|\omega_i\rangle = U|i\rangle \tag{1.8.21}$$

is clearly unitary, for it "rotates" one orthonormal basis into another. (If you wish you may apply our mnemonic to U and verify its unitary nature: its columns contain the components of the eigenvectors $|\omega_i\rangle$ that are orthonormal.) This result is often summarized by the statement:

Every Hermitian matrix on $\mathbb{V}^n(C)$ may be diagonalized by a unitary change of basis.

We may restate this result in terms of passive transformations as follows:

If Ω is a Hermitian matrix, there exists a unitary matrix U (built out of the eigenvectors of Ω) such that $U^\dagger \Omega U$ is diagonal.

Thus the problem of finding a basis that diagonalizes Ω is equivalent to solving its eigenvalue problem.

Exercise 1.8.1. (1) Find the eigenvalues and normalized eigenvectors of the matrix

$$\Omega=\begin{bmatrix} 1 & 3 & 1 \\ 0 & 2 & 0 \\ 0 & 1 & 4 \end{bmatrix}$$

(2) Is the matrix Hermitian? Are the eigenvectors orthogonal?

*Exercise 1.8.2.** Consider the matrix

$$\Omega=\begin{bmatrix} 0 & 0 & 1 \\ 0 & 0 & 0 \\ 1 & 0 & 0 \end{bmatrix}$$

(1) Is it Hermitian?
(2) Find its eigenvalues and eigenvectors.
(3) Verify that $U^\dagger \Omega U$ is diagonal, U being the matrix of eigenvectors of Ω.

*Exercise 1.8.3.** Consider the Hermitian matrix

$$\Omega=\frac{1}{2}\begin{bmatrix} 2 & 0 & 0 \\ 0 & 3 & -1 \\ 0 & -1 & 3 \end{bmatrix}$$

(1) Show that $\omega_1=\omega_2=1$; $\omega_3=2$.
(2) Show that $|\omega=2\rangle$ is any vector of the form

$$\frac{1}{(2a^2)^{1/2}}\begin{bmatrix} 0 \\ a \\ -a \end{bmatrix}$$

(3) Show that the $\omega=1$ eigenspace contains all vectors of the form

$$\frac{1}{(b^2+2c^2)^{1/2}}\begin{bmatrix} b \\ c \\ c \end{bmatrix}$$

either by feeding $\omega=1$ into the equations or by requiring that the $\omega=1$ eigenspace be orthogonal to $|\omega=2\rangle$.

Exercise 1.8.4. An arbitrary $n\times n$ matrix need not have n eigenvectors. Consider as an example

$$\Omega=\begin{bmatrix} 4 & 1 \\ -1 & 2 \end{bmatrix}$$

(1) Show that $\omega_1=\omega_2=3$.
(2) By feeding in this value show we get only one eigenvector of the form

$$\frac{1}{(2a^2)^{1/2}}\begin{bmatrix} +a \\ -a \end{bmatrix}$$

We cannot find another one that is LI.

*Exercise 1.8.5.** Consider the matrix

$$\Omega=\begin{bmatrix} \cos\theta & \sin\theta \\ -\sin\theta & \cos\theta \end{bmatrix}$$

(1) Show that it is unitary.
(2) Show that its eigenvalues are $e^{i\theta}$ and $e^{-i\theta}$.
(3) Find the corresponding eigenvectors; show that they are orthogonal.
(4) Verify that $U^\dagger\Omega U=$(diagonal matrix), where U is the matrix of eigenvectors of Ω.

*Exercise 1.8.6.** (1) We have seen that the determinant of a matrix is unchanged under a unitary change of basis. Argue now that

$$\det\Omega=\text{product of eigenvalues of }\Omega=\prod_{i=1}^{n}\omega_i$$

for a Hermitian or unitary Ω.

(2) Using the invariance of the trace under the same transformation, show that

$$\operatorname{Tr}\Omega=\sum_{i=1}^{n}\omega_i$$

Exercise 1.8.7. By using the results on the trace and determinant from the last problem, show that the eigenvalues of the matrix

$$\Omega=\begin{bmatrix} 1 & 2 \\ 2 & 1 \end{bmatrix}$$

are 3 and -1. Verify this by explicit computation. Note that the Hermitian nature of the matrix is an essential ingredient.

*Exercise 1.8.8.** Consider Hermitian matrices M^1, M^2, M^3, M^4 that obey

$$M^i M^j + M^j M^i = 2\delta^{ij} I, \quad i,j = 1,\ldots,4$$

(1) Show that the eigenvalues of M^i are ± 1. (Hint: go to the eigenbasis of M^i, and use the equation for $i=j$.)

(2) By considering the relation

$$M^i M^j = -M^j M^i \quad \text{for } i \neq j$$

show that M^i are traceless. [Hint: $\text{Tr}(ACB) = \text{Tr}(CBA)$.]

(3) Show that they cannot be odd-dimensional matrices.

Exercise 1.8.9. A collection of masses m_α, located at $\mathbf{r}_\alpha$ and rotating with angular velocity $\boldsymbol{\omega}$ around a common axis has an angular momentum

$$\mathbf{l} = \sum_\alpha m_\alpha (\mathbf{r}_\alpha \times \mathbf{v}_\alpha)$$

where $\mathbf{v}_\alpha = \boldsymbol{\omega} \times \mathbf{r}_\alpha$ is the velocity of m_α. By using the identity

$$\mathbf{A} \times (\mathbf{B} \times \mathbf{C}) = \mathbf{B}(\mathbf{A} \cdot \mathbf{C}) - \mathbf{C}(\mathbf{A} \cdot \mathbf{B})$$

show that each Cartesian component l_i of $\mathbf{l}$ is given by

$$l_i = \sum_j M_{ij} \omega_j$$

where

$$M_{ij} = \sum_\alpha m_\alpha \left[r_\alpha^2 \delta_{ij} - (\mathbf{r}_\alpha)_i (\mathbf{r}_\alpha)_j \right]$$

or in Dirac notation

$$|l\rangle = M|\omega\rangle$$

(1) Will the angular momentum and angular velocity always be parallel?

(2) Show that the moment of inertia matrix M_{ij} is Hermitian.

(3) Argue now that there exist three directions for $\boldsymbol{\omega}$ such that $\mathbf{l}$ and $\boldsymbol{\omega}$ will be parallel. How are these directions to be found?

(4) Consider the moment of inertia matrix of a sphere. Due to the complete symmetry of the sphere, it is clear that every direction is its eigendirection for rotation. What does this say about the three eigenvalues of the matrix M?

Simultaneous Diagonalization of Two Hermitian Operators

Let us consider next the question of simultaneously diagonalizing two Hermitian operators.

Theorem 13. If Ω and Λ are two commuting Hermitian operators, there exists (at least) a basis of common eigenvectors that diagonalizes them both.

Proof. Consider first the case where at least one of the operators is nondegenerate, i.e., to a given eigenvalue, there is just one eigenvector, up to a scale. Let us assume Ω is nondegenerate. Consider any one of its eigenvectors:

$$\Omega|\omega_i\rangle = \omega_i|\omega_i\rangle$$

$$\Lambda\Omega|\omega_i\rangle = \omega_i\Lambda|\omega_i\rangle$$

Since $[\Lambda, \Omega]=0$,

$$\Omega\Lambda|\omega_i\rangle = \omega_i\Lambda|\omega_i\rangle \tag{1.8.22}$$

i.e., $\Lambda|\omega_i\rangle$ is an eigenvector of Ω with eigenvalue ω_i. Since this vector is unique up to a scale,

$$\Lambda|\omega_i\rangle = \lambda_i|\omega_i\rangle \tag{1.8.23}$$

Thus $|\omega_i\rangle$ is also an eigenvector of Λ with eigenvalue λ_i. Since every eigenvector of Ω is an eigenvector of Λ, it is evident that the basis $|\omega_i\rangle$ will diagonalize both operators. Since Ω is nondegenerate, there is only one basis with this property.

What if both operators are degenerate? By ordering the basis vectors such that the elements of each eigenspace are adjacent, we can get one of them, say Ω, into the form (Theorem 10)

$$\Omega \leftrightarrow \begin{bmatrix} \omega_1 & & & & & & & \\ & \omega_1 & & & & & & \\ & & \ddots & & & & & \\ & & & \omega_1 & & & & \\ & & & & \omega_2 & & & \\ & & & & & \ddots & & \\ & & & & & & \omega_m & \\ & & & & & & & \omega_m \end{bmatrix}$$

Now this basis is not unique: in every eigenspace $\mathbb{V}_{\omega_i}^{m_i} \equiv \mathbb{V}_i^{m_i}$ corresponding to the eigenvalue ω_i, there exists an infinity of bases. Let us arbitrarily pick in $\mathbb{V}_{\omega_i}^{m_i}$ a set $|\omega_i, \alpha\rangle$ where the additional label α runs from 1 to m_i.

How does Λ appear in the basis? Although we made no special efforts to get Λ into a simple form, it already has a simple form by virtue of the fact that it commutes with Ω. Let us start by mimicking the proof in the nondegenerate case:

$$\Omega\Lambda|\omega_i, \alpha\rangle = \Lambda\Omega|\omega_i, \alpha\rangle = \omega_i\Lambda|\omega_i, \alpha\rangle$$

However, due to the degeneracy of Ω, we can only conclude that

$$\Lambda|\omega_i, \alpha\rangle \text{ lies in } \mathbb{V}_i^{m_i}$$

Now, since vectors from different eigenspaces are orthogonal [Eq. (1.8.15)],

$$\langle\omega_j, \beta|\Lambda|\omega_i, \alpha\rangle = 0$$

if $|\omega_i, \alpha\rangle$ and $|\omega_j, \beta\rangle$ are basis vectors such that $\omega_i \neq \omega_j$. Consequently, in this basis,

$$\Lambda \leftrightarrow \begin{bmatrix} \boxed{\Lambda_1} & & & 0 \\ & \boxed{\Lambda_2} & & \\ & & \ddots & \\ 0 & & & \boxed{\Lambda_k} \end{bmatrix}$$

which is called a *block diagonal matrix* for obvious reasons. The block diagonal form of Λ reflects the fact that when Λ acts on some element $|\omega_i, \alpha\rangle$ of the eigenspace $\mathbb{V}_i^{m_i}$, it turns it into another element of $\mathbb{V}_i^{m_i}$. Within each subspace i, Λ is given by a matrix Λ_i, which appears as a block in the equation above. Consider a matrix Λ_i in $\mathbb{V}_i^{m_i}$. It is Hermitian since Λ is. It can obviously be diagonalized by trading the basis $|\omega_i, 1\rangle, |\omega_i, 2\rangle, \ldots, |\omega_i, m_i\rangle$ in $\mathbb{V}_i^{m_i}$ that we started with, for the eigenbasis of Λ_i. Let us make such a change of basis in each eigenspace, thereby rendering Λ diagonal. Meanwhile what of Ω? It remains diagonal of course, since it is indifferent to the choice of orthonormal basis in each degenerate eigenspace. If the eigenvalues of Λ_i are $\lambda_i^{(1)}\ \lambda_i^{(2)}, \ldots, \lambda_i^{(m_i)}$ then we end up with

$$\Lambda \leftrightarrow \begin{bmatrix} \lambda_1^{(1)} & & & & & & \\ & \lambda_1^{(2)} & & & & & \\ & & \ddots & & & & \\ & & & \lambda_1^{(m_1)} & & & \\ & & & & \lambda_2^{(1)} & & \\ & & & & & \ddots & \\ & & & & & & \lambda_k^{(m_k)} \end{bmatrix},$$

$$\Omega \leftrightarrow \begin{bmatrix} \omega_1 & & & & & & \\ & \omega_1 & & & & & \\ & & \ddots & & & & \\ & & & \omega_1 & & & \\ & & & & \omega_2 & & \\ & & & & & \ddots & \\ & & & & & & \omega_m \end{bmatrix}$$

Q.E.D.

If Λ is not degenerate *within any given subspace*, $\lambda_i^{(k)} \neq \lambda_i^{(l)}$, for any k, l, and i, the basis we end up with is unique: the freedom Ω gave us in each eigenspace is fully eliminated by Λ. The elements of this basis may be named uniquely by the pair of indices ω and λ as $|\omega, \lambda\rangle$, with λ playing the role of the extra label α. If Λ is degenerate within an eigenspace of Ω, if say $\lambda_1^{(1)} = \lambda_1^{(2)}$, there is a two-dimensional eigenspace from which we can choose any two orthonormal vectors for the common basis. It is then necessary to bring in a third operator Γ, that commutes with both Ω and Λ, and which will be nondegenerate in this subspace. In general, one can always find, for finite n, a set of operators $\{\Omega, \Lambda, \Gamma, \dots\}$ that commute with each other and that nail down a unique, common, eigenbasis, the elements of which may be labeled unambiguously as $|\omega, \lambda, \gamma, \dots\rangle$. In our study of quantum mechanics it will be assumed that such a *complete set of commuting operators* exists if n is infinite.

*Exercise 1.8.10.** By considering the commutator, show that the following Hermitian matrices may be simultaneously diagonalized. Find the eigenvectors common to both and verify that under a unitary transformation to this basis, both matrices are diagonalized.

$$\Omega = \begin{bmatrix} 1 & 0 & 1 \\ 0 & 0 & 0 \\ 1 & 0 & 1 \end{bmatrix}, \quad \Lambda = \begin{bmatrix} 2 & 1 & 1 \\ 1 & 0 & -1 \\ 1 & -1 & 2 \end{bmatrix}$$

Since Ω is degenerate and Λ is not, you must be prudent in deciding which matrix dictates the choice of basis.

Example 1.8.6. We will now discuss, in some detail, the complete solution to a problem in mechanics. It is important that you understand this example thoroughly, for it not only illustrates the use of the mathematical techniques developed in this chapter but also contains the main features of the central problem in quantum mechanics.

The mechanical system in question is depicted in Fig. 1.5. The two masses m are coupled to each other and the walls by springs of force constant k. If x_1 and x_2 measure the displacements of the masses from their equilibrium points, these coordinates obey the following equations, derived through an elementary application of Newton's laws:

$$\ddot{x}_1 = -\frac{2k}{m} x_1 + \frac{k}{m} x_2 \tag{1.8.24a}$$

$$\ddot{x}_2 = \frac{k}{m} x_1 - \frac{2k}{m} x_2 \tag{1.8.24b}$$

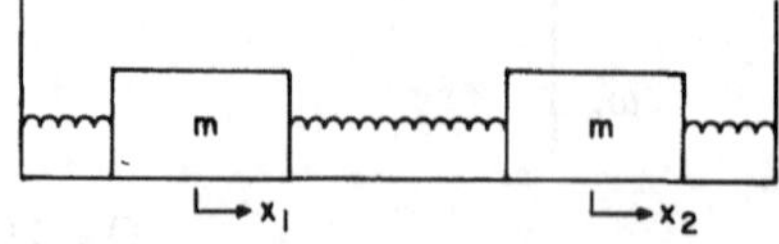

Figure 1.5. The coupled mass problem. All masses are m, all spring constants are k, and the displacements of the masses from equilibrium are x_1 and x_2.

The problem is to find $x_1(t)$ and $x_2(t)$ given the initial-value data, which in this case consist of the initial positions and velocities. If we restrict ourselves to the case of zero initial velocities, our problem is to find $x_1(t)$ and $x_2(t)$, given $x_1(0)$ and $x_2(0)$.

In what follows, we will formulate the problem in the language of linear vector spaces and solve it using the machinery developed in this chapter. As a first step, we rewrite Eq. (1.8.24) in matrix form:

$$\begin{bmatrix} \ddot{x}_1 \\ \ddot{x}_2 \end{bmatrix} = \begin{bmatrix} \Omega_{11} & \Omega_{12} \\ \Omega_{21} & \Omega_{22} \end{bmatrix} \begin{bmatrix} x_1 \\ x_2 \end{bmatrix} \tag{1.8.25a}$$

where the elements of the *Hermitian* matrix Ω_{ij} are

$$\Omega_{11} = \Omega_{22} = -2k/m, \qquad \Omega_{12} = \Omega_{21} = k/m \tag{1.8.25b}$$

We now view x_1 and x_2 as components of an abstract vector $|x\rangle$, and Ω_{ij} as the matrix elements of a Hermitian operator Ω. Since the vector $|x\rangle$ has two real components, it is an element of $\mathbb{V}^2(R)$, and Ω is a Hermitian operator on $\mathbb{V}^2(R)$. The abstract form of Eq. (1.8.25a) is

$$|\ddot{x}(t)\rangle = \Omega|x(t)\rangle \tag{1.8.26}$$

Equation (1.8.25a) is obtained by projecting Eq. (1.8.26) on the basis vectors $|1\rangle$, $|2\rangle$, which have the following physical significance:

$$|1\rangle \leftrightarrow \begin{bmatrix} 1 \\ 0 \end{bmatrix} \leftrightarrow \begin{bmatrix} \text{first mass displaced by unity} \\ \text{second mass undisplaced} \end{bmatrix} \tag{1.8.27a}$$

$$|2\rangle \leftrightarrow \begin{bmatrix} 0 \\ 1 \end{bmatrix} \leftrightarrow \begin{bmatrix} \text{first mass undisplaced} \\ \text{second mass displaced by unity} \end{bmatrix} \tag{1.8.27b}$$

An arbitrary state, in which the masses are displaced by x_1 and x_2, is given in this basis by

$$\begin{bmatrix} x_1 \\ x_2 \end{bmatrix} = \begin{bmatrix} 1 \\ 0 \end{bmatrix} x_1 + \begin{bmatrix} 0 \\ 1 \end{bmatrix} x_2 \tag{1.8.28}$$

The abstract counterpart of the above equation is

$$|x\rangle = |1\rangle x_1 + |2\rangle x_2 \tag{1.8.29}$$

It is in this $|1\rangle$, $|2\rangle$ basis that Ω is represented by the matrix appearing in Eq. (1.8.25), with elements $-2k/m$, k/m, etc.

The basis $|1\rangle$, $|2\rangle$ is very desirable physically, for the components of $|x\rangle$ in this basis (x_1 and x_2) have the simple interpretation as displacements of the masses. However, from the standpoint of finding a mathematical solution to the initial-value problem, it is not so desirable, for the components x_1 and x_2 obey the *coupled* differential equations (1.8.24a) and (1.8.24b). The coupling is mediated by the off-diagonal matrix elements $\Omega_{12}=\Omega_{21}=k/m$.

Having identified the problem with the $|1\rangle$, $|2\rangle$ basis, we can now see how to get around it: we must switch to a basis in which Ω is diagonal. The components of $|x\rangle$ in this basis will then obey another uncoupled differential equations which may be readily solved. Having found the solution, we can return to the physically preferable $|1\rangle$, $|2\rangle$ basis. This, then, is our broad strategy and we now turn to the details.

From our study of Hermitian operators we know that the basis that diagonalizes Ω is the basis of its normalized eigenvectors. Let $|\mathrm{I}\rangle$ and $|\mathrm{II}\rangle$ be its eigenvectors defined by

$$\Omega|\mathrm{I}\rangle=-\omega_{\mathrm{I}}^2|\mathrm{I}\rangle \tag{1.8.30a}$$

$$\Omega|\mathrm{II}\rangle=-\omega_{\mathrm{II}}^2|\mathrm{II}\rangle \tag{1.8.30b}$$

We are departing here from our usual notation: the eigenvalue of Ω is written as $-\omega^2$ rather than as ω in anticipation of the fact that Ω has eigenvalues of the form $-\omega^2$, with ω real. We are also using the symbols $|\mathrm{I}\rangle$ and $|\mathrm{II}\rangle$ to denote what should be called $|-\omega_{\mathrm{I}}^2\rangle$ and $|-\omega_{\mathrm{II}}^2\rangle$ in our convention.

It is a simple exercise (which you should perform) to solve the eigenvalue problem of Ω in the $|1\rangle$, $|2\rangle$ basis (in which the matrix elements of Ω are known) and to obtain

$$\omega_{\mathrm{I}}=\left(\frac{k}{m}\right)^{1/2}, \qquad |\mathrm{I}\rangle\leftrightarrow\frac{1}{2^{1/2}}\begin{bmatrix}1\\1\end{bmatrix} \tag{1.8.31a}$$

$$\omega_{\mathrm{II}}=\left(\frac{3k}{m}\right)^{1/2}, \qquad |\mathrm{II}\rangle\leftrightarrow\frac{1}{2^{1/2}}\begin{bmatrix}1\\-1\end{bmatrix} \tag{1.8.31b}$$

If we now expand the vector $|x(t)\rangle$ in this new basis as

$$|x(t)\rangle=|\mathrm{I}\rangle x_{\mathrm{I}}(t)+|\mathrm{II}\rangle x_{\mathrm{II}}(t) \tag{1.8.32}$$

[in analogy with Eq. (1.8.29)], the components x_{I} and x_{II} will evolve as follows:

$$\begin{bmatrix}\ddot{x}_{\mathrm{I}}\\ \ddot{x}_{\mathrm{II}}\end{bmatrix}=\begin{bmatrix}-\omega_{\mathrm{I}}^2 & 0\\ 0 & -\omega_{\mathrm{II}}^2\end{bmatrix}\begin{bmatrix}x_{\mathrm{I}}\\ x_{\mathrm{II}}\end{bmatrix}$$
$$=\begin{bmatrix}-\omega_{\mathrm{I}}^2 x_{\mathrm{I}}\\ -\omega_{\mathrm{II}}^2 x_{\mathrm{II}}\end{bmatrix} \tag{1.8.33}$$

We obtain this equation by rewriting Eq. (1.8.26) in the $|\mathrm{I}\rangle$, $|\mathrm{II}\rangle$ basis in which Ω has its eigenvalues as the diagonal entries, and in which $|x\rangle$ has components x_{I} and

x_{II}. Alternately we can apply the operator

$$\frac{d^2}{dt^2}-\Omega$$

to both sides of the expansion of Eq. (1.8.32), and get

$$|0\rangle=|\text{I}\rangle(\ddot{x}_{\text{I}}+\omega_{\text{I}}^2 x_{\text{I}})+|\text{II}\rangle(\ddot{x}_{\text{II}}+\omega_{\text{II}}^2 x_{\text{II}}) \tag{1.8.34}$$

Since $|\text{I}\rangle$ and $|\text{II}\rangle$ are orthogonal, each coefficient is zero.

The solution to the *decoupled* equations

$$\ddot{x}_i+\omega_i^2 x_i=0, \qquad i=\text{I, II} \tag{1.8.35}$$

subject to the condition of vanishing initial velocities, is

$$x_i(t)=x_i(0)\cos\omega_i t, \qquad i=\text{I, II} \tag{1.8.36}$$

As anticipated, the components of $|x\rangle$ in the $|\text{I}\rangle$, $|\text{II}\rangle$ basis obey decoupled equations that can be readily solved. Feeding Eq. (1.8.36) into Eq. (1.8.32) we get

$$|x(t)\rangle=|\text{I}\rangle x_{\text{I}}(0)\cos\omega_{\text{I}}t+|\text{II}\rangle x_{\text{II}}(0)\cos\omega_{\text{II}}t \tag{1.8.37a}$$

$$=|\text{I}\rangle\langle\text{I}|x(0)\rangle\cos\omega_{\text{I}}t+|\text{II}\rangle\langle\text{II}|x(0)\rangle\cos\omega_{\text{II}}t \tag{1.8.37b}$$

Equation (1.8.37) provides the explicit solution to the initial-value problem. It corresponds to the following algorithm for finding $|x(t)\rangle$ given $|x(0)\rangle$.

Step (1). Solve the eigenvalue problem of Ω.

Step (2). Find the coefficients $x_{\text{I}}(0)=\langle\text{I}|x(0)\rangle$ and $x_{\text{II}}(0)=\langle\text{II}|x(0)\rangle$ in the expansion

$$|x(0)\rangle=|\text{I}\rangle x_{\text{I}}(0)+|\text{II}\rangle x_{\text{II}}(0)$$

Step (3). Append to each coefficient $x_i(0)$ ($i=\text{I, II}$) a time dependence $\cos\omega_i t$ to get the coefficients in the expansion of $|x(t)\rangle$.

Let me now illustrate this algorithm by solving the following (general) initial-value problem: Find the future state of the system given that at $t=0$ the masses are displaced by $x_1(0)$ and $x_2(0)$.

Step (1). We can ignore this step since the eigenvalue problem has been solved [Eq. (1.8.31)].

Step (2).

$$x_{\rm I}(0)=\langle {\rm I}|x(0)\rangle=\frac{1}{2^{1/2}}(1,\,1)\begin{bmatrix}x_1(0)\\x_2(0)\end{bmatrix}=\frac{x_1(0)+x_2(0)}{2^{1/2}}$$

$$x_{\rm II}(0)=\langle {\rm II}|x(0)\rangle=\frac{1}{2^{1/2}}(1,\,-1)\begin{bmatrix}x_1(0)\\x_2(0)\end{bmatrix}=\frac{x_1(0)-x_2(0)}{2^{1/2}}$$

Step (3).

$$|x(t)\rangle=|{\rm I}\rangle\frac{x_1(0)+x_2(0)}{2^{1/2}}\cos\omega_{\rm I}t+|{\rm II}\rangle\frac{x_1(0)-x_2(0)}{2^{1/2}}\cos\omega_{\rm II}t$$

The explicit solution above can be made even more explicit by projecting $|x(t)\rangle$ onto the $|1\rangle$, $|2\rangle$ basis to find $x_1(t)$ and $x_2(t)$, the displacements of the masses. We get (feeding in the explicit formulas for $\omega_{\rm I}$ and $\omega_{\rm II}$)

$$\begin{aligned}x_1(t)&=\langle 1|x(t)\rangle\\&=\langle 1|{\rm I}\rangle\frac{x_1(0)+x_2(0)}{2^{1/2}}\cos\left[\left(\frac{k}{m}\right)^{1/2}t\right]+\langle 1|{\rm II}\rangle\frac{x_1(0)-x_2(0)}{2^{1/2}}\cos\left[\left(\frac{3k}{m}\right)^{1/2}t\right]\\&=\frac{1}{2}[x_1(0)+x_2(0)]\cos\left[\left(\frac{k}{m}\right)^{1/2}t\right]+\frac{1}{2}[x_1(0)-x_2(0)]\cos\left[\left(\frac{3k}{m}\right)^{1/2}t\right]\end{aligned}\tag{1.8.38a}$$

using the fact that

$$\langle 1|{\rm I}\rangle=\langle 1|{\rm II}\rangle=1/2^{1/2}$$

It can likewise be shown that

$$x_2(t)=\frac{1}{2}[x_1(0)+x_2(0)]\cos\left[\left(\frac{k}{m}\right)^{1/2}t\right]-\frac{1}{2}[x_1(0)-x_2(0)]\cos\left[\left(\frac{3k}{m}\right)^{1/2}t\right]\tag{1.8.38b}$$

We can rewrite Eq. (1.8.38) in matrix form as

$$\begin{bmatrix}x_1(t)\\x_2(t)\end{bmatrix}=\begin{bmatrix}\dfrac{\cos[(k/m)^{1/2}t]+\cos[(3k/m)^{1/2}t]}{2} & \dfrac{\cos[(k/m)^{1/2}t]-\cos[(3k/m)^{1/2}t]}{2}\\ \dfrac{\cos[(k/m)^{1/2}t]-\cos[(3k/m)^{1/2}t]}{2} & \dfrac{\cos[(k/m)^{1/2}t]+\cos[(3k/m)^{1/2}t]}{2}\end{bmatrix}\times\begin{bmatrix}x_1(0)\\x_2(0)\end{bmatrix}\tag{1.8.39}$$

This completes our determination of the future state of the system given the initial state.

The Propagator

There are two remarkable features in Eq. (1.8.39):

(1) The final-state vector is obtained from the initial-state vector upon multiplication by a matrix.
(2) This matrix is independent of the initial state. We call this matrix the *propagator*. Finding the propagator is tantamount to finding the complete solution to the problem, for given any other initial state with displacements $\tilde{x}_1(0)$ and $\tilde{x}_2(0)$, we get $\tilde{x}_1(t)$ and $\tilde{x}_2(t)$ by applying the same matrix to the initial-state vector.

We may view Eq. (1.8.39) as the image in the $|1\rangle$, $|2\rangle$ basis of the abstract relation

$$|x(t)\rangle = U(t)|x(0)\rangle \tag{1.8.40}$$

By comparing this equation with Eq. (1.8.37b), we find the abstract representation of U:

$$U(t) = |\mathrm{I}\rangle\langle\mathrm{I}| \cos\omega_{\mathrm{I}} t + |\mathrm{II}\rangle\langle\mathrm{II}| \cos\omega_{\mathrm{II}} t \tag{1.8.41a}$$

$$= \sum_{i=\mathrm{I}}^{\mathrm{II}} |i\rangle\langle i| \cos\omega_i t \tag{1.8.41b}$$

You may easily convince yourself that if we take the matrix elements of this operator in the $|1\rangle$, $|2\rangle$ basis, we regain the matrix appearing in Eq. (1.8.39). For example

$$\begin{aligned}
U_{11} &= \langle 1|U|1\rangle \\
&= \langle 1| \left\{ |\mathrm{I}\rangle\langle\mathrm{I}| \cos\left[\left(\frac{k}{m}\right)^{1/2} t\right] + |\mathrm{II}\rangle\langle\mathrm{II}| \cos\left[\left(\frac{3k}{m}\right)^{1/2} t\right]\right\} |1\rangle \\
&= \langle 1|\mathrm{I}\rangle\langle\mathrm{I}|1\rangle \cos\left[\left(\frac{k}{m}\right)^{1/2} t\right] + \langle 1|\mathrm{II}\rangle\langle\mathrm{II}|1\rangle \cos\left[\left(\frac{3k}{m}\right)^{1/2} t\right] \\
&= \frac{1}{2}\left\{\cos\left[\left(\frac{k}{m}\right)^{1/2} t\right] + \cos\left[\left(\frac{3k}{m}\right)^{1/2} t\right]\right\}
\end{aligned}$$

Notice that $U(t)$ [Eq. (1.8.41)] is determined completely by the eigenvectors and eigenvalues of Ω. We may then restate our earlier algorithm as follows. To solve the equation

$$|\ddot{x}\rangle = \Omega|x\rangle$$

(1) Solve the eigenvalue problem of Ω.
(2) Construct the propagator U in terms of the eigenvalues and eigenvectors.
(3) $|x(t)\rangle = U(t)|x(0)\rangle$.

The Normal Modes

There are two initial states $|x(0)\rangle$ for which the time evolution is particularly simple. Not surprisingly, these are the eigenkets $|\mathrm{I}\rangle$ and $|\mathrm{II}\rangle$. Suppose we have $|x(0)\rangle = |\mathrm{I}\rangle$. Then the state at time t is

$$\begin{aligned}|\mathrm{I}(t)\rangle &\equiv U(t)|\mathrm{I}\rangle \\ &= (|\mathrm{I}\rangle\langle \mathrm{I}| \cos \omega_{\mathrm{I}} t + |\mathrm{II}\rangle\langle \mathrm{II}| \cos \omega_{\mathrm{II}} t)|\mathrm{I}\rangle \\ &= |\mathrm{I}\rangle \cos \omega_{\mathrm{I}} t \end{aligned} \tag{1.8.42}$$

Thus the system starting off in $|\mathrm{I}\rangle$ is only modified by an overall factor $\cos \omega_{\mathrm{I}} t$. A similar remark holds with $\mathrm{I} \to \mathrm{II}$. These two modes of vibration, in which all (two) components of a vector oscillate in step are called *normal modes*.

The physics of the normal modes is clear in the $|1\rangle, |2\rangle$ basis. In this basis

$$|\mathrm{I}\rangle \leftrightarrow \frac{1}{2^{1/2}} \begin{bmatrix} 1 \\ 1 \end{bmatrix}$$

and corresponds to a state in which both masses are displaced by equal amounts. The middle spring is then a mere spectator and each mass oscillates with a frequency $\omega_{\mathrm{I}} = (k/m)^{1/2}$ in response to the end spring nearest to it. Consequently

$$|\mathrm{I}(t)\rangle \leftrightarrow \frac{1}{2^{1/2}} \begin{bmatrix} \cos[(k/m)^{1/2} t] \\ \cos[(k/m)^{1/2} t] \end{bmatrix}$$

On the other hand, if we start with

$$|\mathrm{II}\rangle \leftrightarrow \frac{1}{2^{1/2}} \begin{bmatrix} 1 \\ -1 \end{bmatrix}$$

the masses are displaced by equal and opposite amounts. In this case the middle spring is distorted by *twice* the displacement of each mass. If the masses are adjusted by Δ and $-\Delta$, respectively, each mass feels a restoring force of $3k\Delta$ ($2k\Delta$ from the middle spring and $k\Delta$ from the end spring nearest to it). Since the effective force constant is $k_{\mathrm{eff}} = 3k\Delta/\Delta = 3k$, the vibrational frequency is $(3k/m)^{1/2}$ and

$$|\mathrm{II}(t)\rangle \leftrightarrow \frac{1}{2^{1/2}} \begin{bmatrix} \cos[(3k/m)^{1/2} t] \\ -\cos[(3k/m)^{1/2} t] \end{bmatrix}$$

If the system starts off in a linear combination of $|\mathrm{I}\rangle$ and $|\mathrm{II}\rangle$ it evolves into the corresponding linear combination of the normal modes $|\mathrm{I}(t)\rangle$ and $|\mathrm{II}(t)\rangle$. This is the content of the propagator equation

$$
\begin{aligned}
|x(t)\rangle &= U(t)|x(0)\rangle \\
&= |\mathrm{I}\rangle\langle \mathrm{I}|x(0)\rangle \cos \omega_{\mathrm{I}} t + |\mathrm{II}\rangle\langle \mathrm{II}|\, x(0)\rangle \cos \omega_{\mathrm{II}} t \\
&= |I(t)\rangle\langle \mathrm{I}|x(0)\rangle + |\mathrm{II}(t)\rangle\langle \mathrm{II}|x(0)\rangle
\end{aligned}
$$

Another way to see the simple evolution of the initial states $|\mathrm{I}\rangle$ and $|\mathrm{II}\rangle$ is to determine the matrix representing U in the $|\mathrm{I}\rangle$, $|\mathrm{II}\rangle$ basis:

$$
U \underset{\substack{\mathrm{I,II} \\ \text{basis}}}{\longleftrightarrow} \begin{bmatrix} \cos \omega_{\mathrm{I}} t & 0 \\ 0 & \cos \omega_{\mathrm{II}} t \end{bmatrix} \tag{1.8.43}
$$

You should verify this result by taking the appropriate matrix elements of $U(t)$ in Eq. (1.8.41b). Since each column above is the image of the corresponding basis vectors ($|\mathrm{I}\rangle$ or $|\mathrm{II}\rangle$) after the action of $U(t)$, (which is to say, after time evolution), we see that the initial states $|\mathrm{I}\rangle$ and $|\mathrm{II}\rangle$ evolve simply in time.

The central problem in quantum mechanics is very similar to the simple example that we have just discussed. The state of the system is described in quantum theory by a ket $|\psi\rangle$ which obeys the Schrödinger equation

$$
i\hbar|\dot{\psi}\rangle = H|\psi\rangle
$$

where $\hbar$ is a constant related to Planck's constant h by $\hbar = h/2\pi$, and H is a *Hermitian* operator called the Hamiltonian. The problem is to find $|\psi(t)\rangle$ given $|\psi(0)\rangle$. [Since the equation is first order in t, no assumptions need be made about $|\dot{\psi}(0)\rangle$, which is determined by the Schrödinger equation to be $(-i/\hbar)H|\psi(0)\rangle$.]

In most cases, H is a time-independent operator and the algorithm one follows in solving this initial-value problem is completely analogous to the one we have just seen:

Step (1). Solve the eigenvalue problem of H.

Step (2). Find the propagator $U(t)$ in terms of the eigenvectors and eigenvalues of H.

Step (3). $|\psi(t)\rangle = U(t)|\psi(0)\rangle$.

You must of course wait till Chapter 4 to find out the physical interpretation of $|\psi\rangle$, the actual form of the operator H, and the precise relation between $U(t)$ and the eigenvalues and eigenvectors of H. □

Exercise 1.8.11. Consider the coupled mass problem discussed above.

(1) Given that the initial state is $|1\rangle$, in which the first mass is displaced by unity and the second is left alone, calculate $|1(t)\rangle$ by following the algorithm.

(2) Compare your result with that following from Eq. (1.8.39).

Exercise 1.8.12. Consider once again the problem discussed in the previous example. (1) Assuming that

$$|\ddot{x}\rangle = \Omega|x\rangle$$

has a solution

$$|x(t)\rangle = U(t)|x(0)\rangle$$

find the differential equation satisfied by $U(t)$. Use the fact that $|x(0)\rangle$ is arbitrary.

(2) Assuming (as is the case) that Ω and U can be simultaneously diagonalized, solve for the elements of the matrix U in this common basis and regain Eq. (1.8.43). Assume $|\dot{x}(0)\rangle = 0$.

1.9. Functions of Operators and Related Concepts

We have encountered two types of objects that act on vectors: scalars, which commute with each other and with all operators; and operators, which do not generally commute with each other. It is customary to refer to the former as c numbers and the latter as q numbers. Now, we are accustomed to functions of c numbers such as $\sin(x)$, $\log(x)$, etc. We wish to examine the question whether functions of q numbers can be given a sensible meaning. We will restrict ourselves to those functions that can be written as a power series. Consider a series

$$f(x) = \sum_{n=0}^{\infty} a_n x^n \tag{1.9.1}$$

where x is a c number. We define the same function of an operator or q number to be

$$f(\Omega) = \sum_{n=0}^{\infty} a_n \Omega^n \tag{1.9.2}$$

This definition makes sense only if the sum converges to a definite limit. To see what this means, consider a common example:

$$e^{\Omega} = \sum_{n=0}^{\infty} \frac{\Omega^n}{n!} \tag{1.9.3}$$

Let us restrict ourselves to Hermitian Ω. By going to the eigenbasis of Ω we can readily perform the sum of Eq. (1.9.3). Since

$$\Omega = \begin{bmatrix} \omega_1 & & & \\ & \omega_2 & & \\ & & \ddots & \\ & & & \omega_n \end{bmatrix} \tag{1.9.4}$$

and

$$\Omega^m = \begin{bmatrix} \omega_1^m & & & \\ & \omega_2^m & & \\ & & \ddots & \\ & & & \omega_n^m \end{bmatrix} \tag{1.9.5}$$

$$e^{\Omega} = \begin{bmatrix} \sum_{m=0}^{\infty} \frac{\omega_1^m}{m!} & & \\ & \ddots & \\ & & \sum_{m=0}^{\infty} \frac{\omega_n^m}{m!} \end{bmatrix} \tag{1.9.6}$$

Since each sum converges to the familiar limit e^{ω_i}, the operator e^{Ω} is indeed well defined by the power series in this basis (and therefore in any other).

*Exercise 1.9.1.** We know that the series

$$f(x) = \sum_{n=0}^{\infty} x^n$$

may be equated to the function $f(x) = (1-x)^{-1}$ if $|x| < 1$. By going to the eigenbasis, examine when the q number power series

$$f(\Omega) = \sum_{n=0}^{\infty} \Omega^n$$

of a Hermitian operator Ω may be identified with $(1-\Omega)^{-1}$.

*Exercise 1.9.2.** If H is a Hermitian operator, show that $U = e^{iH}$ is unitary. (Notice the analogy with c numbers: if θ is real, $u = e^{i\theta}$ is a number of unit modulus.)

Exercise 1.9.3. For the case above, show that det $U = e^{i\mathrm{Tr}H}$.

Derivatives of Operators with Respect to Parameters

Consider next an operator $\theta(\lambda)$ that depends on a parameter λ. Its derivative with respect to λ is defined to be

$$\frac{d\theta(\lambda)}{d\lambda} = \lim_{\Delta\lambda \to 0} \left[\frac{\theta(\lambda + \Delta\lambda) - \theta(\lambda)}{\Delta\lambda} \right]$$

If $\theta(\lambda)$ is written as a matrix in some basis, then the matrix representing $d\theta(\lambda)/d\lambda$ is obtained by differentiating the matrix elements of $\theta(\lambda)$. A special case of $\theta(\lambda)$ we are interested in is

$$\theta(\lambda)=e^{\lambda\Omega}$$

where Ω is Hermitian. We can show, by going to the eigenbasis of Ω, that

$$\frac{d\theta(\lambda)}{d\lambda}=\Omega e^{\lambda\Omega}=e^{\lambda\Omega}\Omega=\theta(\lambda)\Omega \tag{1.9.7}$$

The same result may be obtained, even if Ω is not Hermitian, by working with the power series, provided it exists:

$$\frac{d}{d\lambda}\sum_{n=0}^{\infty}\frac{\lambda^n\Omega^n}{n!}=\sum_{n=1}^{\infty}\frac{n\lambda^{n-1}\Omega^n}{n!}=\Omega\sum_{n=1}^{\infty}\frac{\lambda^{n-1}\Omega^{n-1}}{(n-1)!}=\Omega\sum_{m=0}^{\infty}\frac{\lambda^m\Omega^m}{m!}=\Omega e^{\lambda\Omega}$$

Conversely, we can say that if we are confronted with the differential Eq. (1.9.7), its solution is given by

$$\theta(\lambda)=c\exp\left(\int_0^\lambda \Omega\, d\lambda'\right)=c\exp(\Omega\lambda)$$

(It is assumed here that the exponential exists.) In the above, c is a constant (operator) of integration. The solution $\theta=e^{\Omega\lambda}$ corresponds to the choice $c=I$.

In all the above operations, we see that Ω behaves as if it were just a c number. Now, the real difference between c numbers and q numbers is that the latter do not generally commute. However, if only one q number (or powers of it) enter the picture, everything commutes and we can treat them as c numbers. If one remembers this mnemonic, one can save a lot of time.

If, on the other hand, more than one q number is involved, the order of the factors is all important. For example, it is true that

$$e^{\alpha\Omega}e^{\beta\Omega}=e^{(\alpha+\beta)\Omega}$$

as may be verified by a power-series expansion, while it is not true that

$$e^{\alpha\Omega}e^{\beta\theta}=e^{\alpha\Omega+\beta\theta}$$

or that

$$e^{\alpha\Omega}e^{\beta\theta}e^{-\alpha\Omega}=e^{\beta\theta}$$

unless $[\Omega,\theta]=0$. Likewise, in differentiating a product, the chain rule is

$$\frac{d}{d\lambda}e^{\lambda\Omega}e^{\lambda\theta}=\Omega e^{\lambda\Omega}e^{\lambda\theta}+e^{\lambda\Omega}e^{\lambda\theta}\theta \tag{1.9.8}$$

We are free to move Ω through $e^{\lambda\Omega}$ and write the first term as

$$e^{\lambda\Omega}\Omega e^{\lambda\theta}$$

but not as

$$e^{\lambda\Omega}e^{\lambda\theta}\Omega$$

unless $[\Omega, \theta]=0$.

1.10. Generalization to Infinite Dimensions

In all of the preceding discussions, the dimensionality (n) of the space was unspecified but assumed to be some finite number. We now consider the generalization of the preceding concepts to infinite dimensions.

Let us begin by getting acquainted with an infinite-dimensional vector. Consider a function defined in some interval, say, $a\le x\le b$. A concrete example is provided by the displacement $f(x, t)$ of a string clamped at $x=0$ and $x=L$ (Fig. 1.6).

Suppose we want to communicate to a person on the moon the string's displacement $f(x)$, at some time t. One simple way is to divide the interval $0-L$ into 20 equal parts, measure the displacement $f(x_i)$ at the 19 points $x=L/20, 2L/20, \ldots, 19L/20$, and transmit the 19 values on the wireless. Given these $f(x_i)$, our friend on the moon will be able to reconstruct the approximate picture of the string shown in Fig. 1.7.

If we wish to be more accurate, we can specify the values of $f(x)$ at a larger number of points. Let us denote by $f_n(x)$ the discrete approximation to $f(x)$ that coincides with it at n points and vanishes in between. Let us now interpret the ordered n-tuple $\{f_n(x_1), f_n(x_2), \ldots, f_n(x_n)\}$ as components of a ket $|f_n\rangle$ in a vector space $\mathbb{V}^n(R)$:

$$|f_n\rangle \leftrightarrow \begin{bmatrix} f_n(x_1) \\ f_n(x_2) \\ \vdots \\ f_n(x_n) \end{bmatrix} \tag{1.10.1}$$

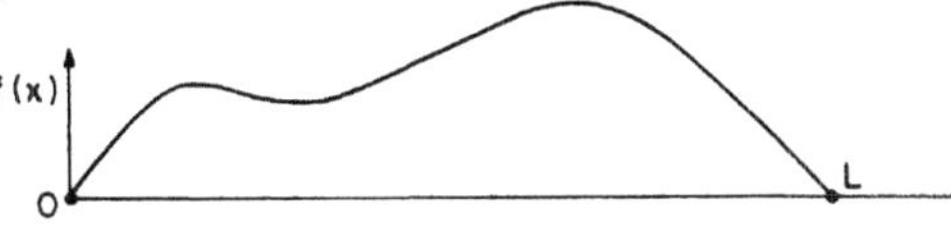

Figure 1.6. The string is clamped at $x=0$ and $x=L$. It is free to oscillate in the plane of the paper.

Figure 1.7. The string as reconstructed by the person on the moon.

The basis vectors in this space are

$$|x_i\rangle \leftrightarrow \begin{bmatrix} 0 \\ 0 \\ \vdots \\ 1 \\ 0 \\ \vdots \\ 0 \end{bmatrix} \leftarrow i\text{th place} \tag{1.10.2}$$

corresponding to the discrete function which is unity at $x=x_i$ and zero elsewhere. The basis vectors satisfy

$$\langle x_i|x_j\rangle = \delta_{ij} \text{ (orthogonality)} \tag{1.10.3}$$

$$\sum_{i=1}^{n} |x_i\rangle\langle x_i| = I \text{ (completeness)} \tag{1.10.4}$$

Try to imagine a space containing n mutually perpendicular axes, one for each point x_i. Along each axis is a unit vector $|x_i\rangle$. The function $f_n(x)$ is represented by a vector whose projection along the ith direction is $f_n(x_i)$:

$$|f_n\rangle = \sum_{i=1}^{n} f_n(x_i)|x_i\rangle \tag{1.10.5}$$

To every possible discrete approximation $g_n(x)$, $h_n(x)$, etc., there is a corresponding ket $|g_n\rangle$, $|h_n\rangle$, etc., and vice versa. You should convince yourself that if we define vector addition as the addition of the components, and scalar multiplication as the multiplication of each component by the scalar, then the set of all kets representing discrete functions that vanish at $x=0$, L and that are specified at n points in between, forms a vector space.

We next define the inner product in this space:

$$\langle f_n|g_n\rangle = \sum_{i=1}^{n} f_n(x_i)g_n(x_i) \tag{1.10.6}$$

Two functions $f_n(x)$ and $g_n(x)$ will be said to be orthogonal if $\langle f_n|g_n\rangle = 0$.

Let us now forget the man on the moon and consider the maximal specification of the string's displacement, by giving its value at every point in the interval $0-L$. In this case $f_\infty(x)\equiv f(x)$ is specified by an ordered infinity of numbers: an $f(x)$ for each point x. Each function is now represented by a ket $|f_\infty\rangle$ in an infinite-dimensional vector space and vice versa. Vector addition and scalar multiplication are defined just as before. Consider, however, the inner product. For finite n it was defined as

$$\langle f_n|g_n\rangle=\sum_{i=1}^{n} f_n(x_i)g_n(x_i)$$

in particular

$$\langle f_n|f_n\rangle=\sum_{i=1}^{n} [f_n(x_i)]^2$$

If we now let n go to infinity, so does the sum, for practically any function. What we need is the redefinition of the inner product for finite n in such a way that as n tends to infinity, a smooth limit obtains. The natural choice is of course

$$\langle f_n|g_n\rangle=\sum_{i=1}^{n} f_n(x_i)g_n(x_i)\Delta, \qquad \Delta=L/(n+1) \tag{1.10.6'}$$

If we now let n go to infinity, we get, by the usual definition of the integral,

$$\langle f|g\rangle=\int_0^L f(x)g(x)\,dx \tag{1.10.7}$$

$$\langle f|f\rangle=\int_0^L f^2(x)\,dx \tag{1.10.8}$$

If we wish to go beyond the instance of the string and consider complex functions of x as well, in some interval $a\le x\le b$, the only modification we need is in the inner product:

$$\langle f|g\rangle=\int_a^b f^*(x)g(x)\,dx \tag{1.10.9}$$

What are the basis vectors in this space and how are they normalized? We know that each point x gets a basis vector $|x\rangle$. The orthogonality of two different axes requires that

$$\langle x|x'\rangle=0, \qquad x\neq x' \tag{1.10.10}$$

What if $x=x'$? Should we require, as in the finite-dimensional case, $\langle x|x\rangle=1$? The answer is no, and the best way to see it is to deduce the correct normalization. We start with the natural generalization of the completeness relation Eq. (1.10.4) to the case where the kets are labeled by a continuous index x':

$$\int_a^b |x'\rangle\langle x'|\,dx'=\mathrm{I} \tag{1.10.11}$$

where, as always, the identity is required to leave each ket unchanged. Dotting both sides of Eq. (1.10.11) with some arbitrary ket $|f\rangle$ from the right and the basis bra $\langle x|$ from the left,

$$\int_a^b \langle x|x'\rangle\langle x'|f\rangle\, dx' = \langle x|I|f\rangle = \langle x|f\rangle \tag{1.10.12}$$

Now, $\langle x|f\rangle$, the projection of $|f\rangle$ along the basis ket $|x\rangle$, is just $f(x)$. Likewise $\langle x'|f\rangle = f(x')$. Let the inner product $\langle x|x'\rangle$ be some unknown function $\delta(x, x')$. Since $\delta(x, x')$ vanishes if $x \neq x'$ we can restrict the integral to an infinitesimal region near $x' = x$ in Eq. (1.10.12):

$$\int_{x-\varepsilon}^{x+\varepsilon} \delta(x, x') f(x')\, dx' = f(x) \tag{1.10.13}$$

In this infinitesimal region, $f(x')$ (for any reasonably smooth f) can be approximated by its value at $x' = x$, and pulled out of the integral:

$$f(x) \int_{x-\varepsilon}^{x+\varepsilon} \delta(x, x')\, dx' = f(x) \tag{1.10.14}$$

so that

$$\int_{x-\varepsilon}^{x+\varepsilon} \delta(x, x')\, dx' = 1 \tag{1.10.15}$$

Clearly $\delta(x, x')$ cannot be finite at $x' = x$, for then its integral over an infinitesimal region would also be infinitesimal. In fact $\delta(x, x')$ should be infinite in such a way that its integral is unity. Since $\delta(x, x')$ depends only on the difference $x - x'$, let us write it as $\delta(x - x')$. The "function," $\delta(x - x')$, with the properties

$$\begin{gathered} \delta(x-x') = 0, \qquad x \neq x' \\ \int_a^b \delta(x-x')\, dx' = 1, \qquad a < x < b \end{gathered} \tag{1.10.16}$$

is called the *Dirac delta function* and fixes the normalization of the basis vectors:

$$\langle x|x'\rangle = \delta(x-x') \tag{1.10.17}$$

It will be needed any time the basis kets are labeled by a continuous index such as x. Note that it is defined only in the context of an integration: the integral of the delta function $\delta(x - x')$ with any smooth function $f(x')$ is $f(x)$. One sometimes calls

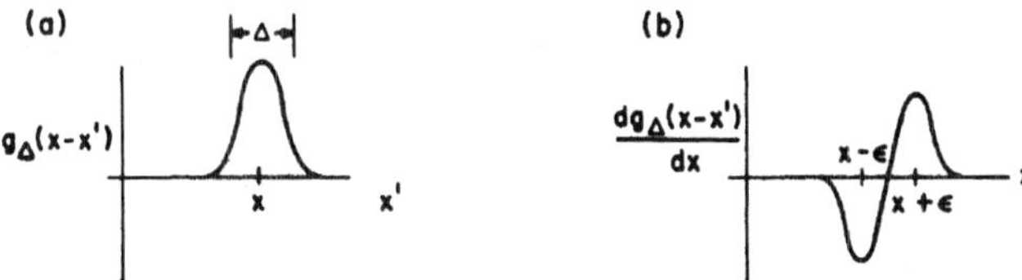

Figure 1.8. (a) The Gaussian g_Δ approaches the delta function as $\Delta \to 0$. (b) Its derivative $(dg/dx)(x-x')$ approaches $\delta'(x-x')$ as $\Delta \to 0$.

the delta function the sampling function, since it samples the value of the function $f(x')$ at one point‡

$$\int \delta(x-x')\, f(x')\, dx' = f(x) \tag{1.10.18}$$

The delta function does not look like any function we have seen before, its values being either infinite or zero. It is therefore useful to view it as the limit of a more conventional function. Consider a Gaussian

$$g_\Delta(x-x') = \frac{1}{(\pi\Delta^2)^{1/2}} \exp\left[-\frac{(x-x')^2}{\Delta^2}\right] \tag{1.10.19}$$

as shown in Fig. 1.8a. The Gaussian is centered at $x'=x$, has width Δ, maximum height $(\pi\Delta^2)^{-1/2}$, and unit area, independent of Δ. As Δ approaches zero, g_Δ becomes a better and better approximation to the delta function.§

It is obvious from the Gaussian model that the delta function is even. This may be verified as follows:

$$\delta(x-x') = \langle x|x'\rangle = \langle x'|x\rangle^* = \delta(x'-x)^* = \delta(x'-x)$$

since the delta function is real.

Consider next an object that is even more peculiar than the delta function: its derivative with respect to the *first* argument x:

$$\delta'(x-x') = \frac{d}{dx}\delta(x-x') = -\frac{d}{dx'}\delta(x-x') \tag{1.10.20}$$

What is the action of this function under the integral? The clue comes from the Gaussian model. Consider $dg_\Delta(x-x')/dx = -dg_\Delta(x-x')/dx'$ as a function of x'. As g_Δ shrinks, each bump at $\pm\varepsilon$ will become, up to a scale factor, the δ function. The

‡ We will often omit the limits of integration if they are unimportant.

§ A fine point that will not concern you till Chapter 8: This formula for the delta function is valid even if Δ^2 is pure imaginary, say, equal to $i\beta^2$. First we see from Eq. (A.2.5) that g has unit area. Consider next the integral of g times $f(x')$ over a region in x' that includes x. For the most part, we get zero because f is smooth and g is wildly oscillating as $\beta \to 0$. However, at $x=x'$, the derivative of the phase of g vanishes and the oscillations are suspended. Pulling $f(x'=x)$ out of the integral, we get the desired result.

first one will sample $-f(x-\varepsilon)$ and the second one $+f(x+\varepsilon)$, again up to a scale, so that

$$\int \delta'(x-x')f(x')\,dx' \propto f(x+\varepsilon)-f(x-\varepsilon)=2\varepsilon \left.\frac{df}{dx'}\right|_{x'=x}$$

The constant of proportionality happens to be $1/2\varepsilon$ so that

$$\int \delta'(x-x')f(x')\,dx' = \left.\frac{df}{dx'}\right|_{x'=x} = \frac{df(x)}{dx} \tag{1.10.21}$$

This result may be verified as follows:

$$\int \delta'(x-x')f(x')\,dx' = \int \frac{d\delta(x-x')}{dx} f(x')\,dx' = \frac{d}{dx}\int \delta(x-x')f(x')\,dx'$$
$$= \frac{df(x)}{dx}$$

Note that $\delta'(x-x')$ is an odd function. This should be clear from Fig. 1.8b or Eq. (1.10.20). An equivalent way to describe the action of the δ' function is by the equation

$$\delta'(x-x') = \delta(x-x')\frac{d}{dx'} \tag{1.10.22}$$

where it is understood that both sides appear in an integral over x' and that the differential operator acts on any function that accompanies the δ' function in the integrand. In this notation we can describe the action of higher derivatives of the delta function:

$$\frac{d^n\delta(x-x')}{dx^n} = \delta(x-x')\frac{d^n}{dx'^n} \tag{1.10.23}$$

We will now develop an alternate representation of the delta function. We know from basic Fourier analysis that, given a function $f(x)$, we may define its transform

$$f(k) = \frac{1}{(2\pi)^{1/2}}\int_{-\infty}^{\infty} e^{-ikx} f(x)\,dx \tag{1.10.24}$$

and its inverse

$$f(x')=\frac{1}{(2\pi)^{1/2}}\int_{-\infty}^{\infty} e^{ikx'} f(k)\, dk \tag{1.10.25}$$

Feeding Eq. (1.10.24) into Eq. (1.10.25), we get

$$f(x')=\int_{-\infty}^{\infty}\left(\frac{1}{2\pi}\int_{-\infty}^{\infty} dk\, e^{ik(x'-x)}\right) f(x)\, dx$$

Comparing this result with Eq. (1.10.18), we see that

$$\frac{1}{2\pi}\int_{-\infty}^{\infty} dk\, e^{ik(x'-x)}=\delta(x'-x) \tag{1.10.26}$$

*Exercise 1.10.1.** Show that $\delta(ax)=\delta(x)/|a|$. [Consider $\int \delta(ax)\, d(ax)$. Remember that $\delta(x)=\delta(-x)$.]

*Exercise 1.10.2.** Show that

$$\delta(f(x))=\sum_i \frac{\delta(x-x_i)}{|df/dx|_{x_i}}$$

where x_i are the zeros of $f(x)$. Hint: Where does $\delta(f(x))$ blow up? Expand $f(x)$ near such points in a Taylor series, keeping the first nonzero term.

*Exercise 1.10.3.** Consider the *theta function* $\theta(x-x')$ which vanishes if $x-x'$ is negative and equals 1 if $x-x'$ is positive. Show that $\delta(x-x')=d/dx\, \theta(x-x')$.

Operators in Infinite Dimensions

Having acquainted ourselves with the elements of this function space, namely, the kets $|f\rangle$ and the basis vectors $|x\rangle$, let us turn to the (linear) operators that act on them. Consider the equation

$$\Omega|f\rangle=|\tilde{f}\rangle$$

Since the kets are in correspondence with the functions, Ω takes the function $f(x)$ into another, $\tilde{f}(x)$. Now, one operator that does such a thing is the familiar differential operator, which, acting on $f(x)$, gives $\tilde{f}(x)=df(x)/dx$. In the function space we can describe the action of this operator as

$$D|f\rangle=|df/dx\rangle$$

where $|df/dx\rangle$ is the ket corresponding to the function df/dx. What are the matrix elements of D in the $|x\rangle$ basis? To find out, we dot both sides of the above equation with $\langle x|$,

$$\langle x|D|f\rangle=\left\langle x\left|\frac{df}{dx}\right.\right\rangle=\frac{df(x)}{dx}$$

and insert the resolution of identity at the right place

$$\int \langle x|D|x'\rangle\langle x'|f\rangle\,dx'=\frac{df}{dx} \tag{1.10.27}$$

Comparing this to Eq. (1.10.21), we deduce that

$$\langle x|D|x'\rangle=D_{xx'}=\delta'(x-x')=\delta(x-x')\frac{d}{dx'} \tag{1.10.28}$$

It is worth remembering that $D_{xx'}=\delta'(x-x')$ *is to be integrated over the second index* (x') *and pulls out the derivative of f at the first index* (x). Some people prefer to integrate $\delta'(x-x')$ over the first index, in which case it pulls out $-df/dx'$. Our convention is more natural if one views $D_{xx'}$ as a matrix acting to the *right* on the components $f_{x'}\equiv f(x')$ of a vector $|f\rangle$. Thus the familiar differential operator is an infinite-dimensional matrix with the elements given above. Normally one doesn't think of D as a matrix for the following reason. Usually when a matrix acts on a vector, there is a sum over a common index. In fact, Eq. (1.10.27) contains such a sum over the index x'. If, however, we feed into this equation the value of $D_{xx'}$, the delta function renders the integration trivial:

$$\int \delta(x-x')\frac{d}{dx'}f(x')\,dx'=\left.\frac{df}{dx'}\right|_{x'=x}=\frac{df}{dx}$$

Thus the action of D is simply to apply d/dx to $f(x)$ with no sum over a common index in sight. Although we too will drop the integral over the common index ultimately, we will continue to use it for a while to remind us that D, like all linear operators, is a matrix.

Let us now ask if D is Hermitian and examine its eigenvalue problem. If D were Hermitian, we would have

$$D_{xx'}=D^*_{x'x}$$

But this is not the case:

$$D_{xx'}=\delta'(x-x')$$

while

$$D^*_{x'x}=\delta'(x'-x)^*=\delta'(x'-x)=-\delta'(x-x')$$

But we can easily convert D to a Hermitian matrix by multiplying it with a pure imaginary number. Consider

$$K=-iD$$

which satisfies

$$K^*_{x'x}=[-i\delta'(x'-x)]^*=+i\delta'(x'-x)=-i\delta'(x-x')=K_{xx'}$$

It turns out that despite the above, the operator K is not guaranteed to be Hermitian, as the following analysis will indicate. Let $|f\rangle$ and $|g\rangle$ be two kets in the function space, whose images in the X basis are two functions $f(x)$ and $g(x)$ in the interval $a-b$. If K is Hermitian, it must also satisfy

$$\langle g|K|f\rangle=\langle g|Kf\rangle=\langle Kf|g\rangle^*=\langle f|K^\dagger|g\rangle^*=\langle f|K|g\rangle^*$$

So we ask

$$\int_a^b\int_a^b \langle g|x\rangle\langle x|K|x'\rangle\langle x'|f\rangle\,dx\,dx'$$

$$\stackrel{?}{=}\left(\int_a^b\int_a^b \langle f|x\rangle\langle x|K|x'\rangle\langle x'|g\rangle\,dx\,dx'\right)^*$$

$$\int_a^b g^*(x)\left[-\frac{i\,df(x)}{dx}\right]dx \stackrel{?}{=} \left\{\int_a^b f^*(x)\left[-\frac{i\,dg(x)}{dx}\right]dx\right\}^* = i\int_a^b \frac{dg^*}{dx}f(x)\,dx$$

Integrating the left-hand side by parts gives

$$-ig^*(x)f(x)\Big|_a^b + i\int_a^b \frac{dg^*(x)}{dx}f(x)\,dx$$

So K is Hermitian only if the *surface term* vanishes:

$$-ig^*(x)f(x)\Big|_a^b=0 \tag{1.10.29}$$

In contrast to the finite-dimensional case, $K_{xx'}=K^*_{x'x}$ is not a sufficient condition for K to be Hermitian. One also needs to look at the behavior of the functions at the end points a and b. Thus K is Hermitian if the space consists of functions that obey Eq. (1.10.29). One set of functions that obey this condition are the possible configurations $f(x)$ of the string clamped at $x=0$, L, since $f(x)$ vanishes at the end points. But condition (1.10.29) can also be fulfilled in another way. Consider functions in our own three-dimensional space, parametrized by r, θ, and ϕ (ϕ is the angle measured around the z axis). Let us require that these functions be single valued. In particular, if we start at a certain point and go once around the z axis, returning to the original point, the function must take on its original value, i.e.,

$$f(\phi) = f(\phi + 2\pi)$$

In the space of such periodic functions, $K = -i d/d\phi$ is a Hermitian operator. The surface term vanishes because the contribution from one extremity cancels that from the other:

$$-ig^*(\phi)f(\phi)\Big|_0^{2\pi} = -i[g^*(2\pi)f(2\pi) - g^*(0)f(0)] = 0$$

In the study of quantum mechanics, we will be interested in functions defined over the full interval $-\infty \le x \le +\infty$. They fall into two classes, those that vanish as $|x| \to \infty$, and those that do not, the latter behaving as e^{ikx}, k being a real parameter that labels these functions. It is clear that $K = -i d/dx$ is Hermitian when sandwiched between two functions of the first class or a function from each, since in either case the surface term vanishes. When sandwiched between two functions of the second class, the Hermiticity hinges on whether

$$e^{ikx}e^{-ik'x}\Big|_{-\infty}^{\infty} \stackrel{?}{=} 0$$

If $k = k'$, the contribution from one end cancels that from the other. If $k \neq k'$, the answer is unclear since $e^{i(k-k')x}$ oscillates, rather than approaching a limit as $|x| \to \infty$. Now, there exists a way of defining a limit for such functions that cannot make up their minds: the limit as $|x| \to \infty$ is defined to be the average over a large interval. According to this prescription, we have, say as $x \to \infty$,

$$\lim_{x\to\infty} e^{ikx}e^{-ik'x} = \lim_{\substack{L\to\infty \\ \Delta\to\infty}} \frac{1}{\Delta}\int_L^{L+\Delta} e^{i(k-k')x}\,dx = 0 \qquad \text{if } k \neq k'$$

and so K is Hermitian in this space.

We now turn to the eigenvalue problem of K. The task seems very formidable indeed, for we have now to find the roots of an infinite-order characteristic polynomial and get the corresponding eigenvectors. It turns out to be quite simple and you might have done it a few times in the past without giving yourself due credit. Let us begin with

$$K|k\rangle = k|k\rangle \tag{1.10.30}$$

Following the standard procedure,

$$\langle x|K|k\rangle = k\langle x|k\rangle$$

$$\int \langle x|K|x'\rangle\langle x'|k\rangle\, dx' = k\psi_k(x) \qquad (1.10.31)$$

$$-i\frac{d}{dx}\psi_k(x) = k\psi_k(x)$$

where by definition $\psi_k(x) = \langle x|k\rangle$. This equation could have been written directly had we made the immediate substitution $K = -i\, d/dx$ in the X basis. From now on we shall resort to this shortcut unless there are good reasons for not doing so.

The solution to the above equation is simply

$$\psi_k(x) = A\, e^{ikx} \qquad (1.10.32)$$

where A, the overall scale, is a free parameter, unspecified by the eigenvalue problem. So the eigenvalue problem of K is fully solved: any real number k is an eigenvalue, and the corresponding eigenfunction is given by $A\, e^{ikx}$. As usual, the freedom in scale will be used to normalize the solution. We choose A to be $(1/2\pi)^{-1/2}$ so that

$$|k\rangle \leftrightarrow \frac{1}{(2\pi)^{1/2}}\, e^{ikx}$$

and

$$\langle k|k'\rangle = \int_{-\infty}^{\infty} \langle k|x\rangle\langle x|k'\rangle\, dx = \frac{1}{2\pi}\int_{-\infty}^{\infty} e^{-i(k-k')x}\, dx = \delta(k-k') \qquad (1.10.33)$$

(Since $\langle k|k\rangle$ is infinite, no choice of A can normalize $|k\rangle$ to unity. The delta function normalization is the natural one when the eigenvalue spectrum is continuous.)

The attentive reader may have a question at this point.

"Why was it assumed that the eigenvalue k was real? It is clear that the function $A\, e^{ikx}$ with $k = k_1 + ik_2$ also satisfies Eq. (1.10.31)."

The answer is, yes, there are eigenfunctions of K with complex eigenvalues. If, however, our space includes such functions, K must be classified a non-Hermitian operator. (The surface term no longer vanishes since e^{ikx} blows up exponentially as x tends to either $+\infty$ or $-\infty$, depending on the sign of the imaginary part k_2.) In restricting ourselves to real k we have restricted ourselves to what we will call the *physical Hilbert space*, which is of interest in quantum mechanics. This space is defined as the space of functions that can be either normalized to unity or to the Dirac delta function and plays a central role in quantum mechanics. (We use the qualifier "physical" to distinguish it from the Hilbert space as defined by mathematicians, which contains only *proper vectors*, i.e., vectors normalizable to unity. The role of the *improper vectors* in quantum theory will be clear later.)

We will assume that the theorem proved for finite dimensions, namely, that the eigenfunctions of a Hermitian operator form a complete basis, holds in the Hilbert‡ space. (The trouble with infinite-dimensional spaces is that even if you have an infinite number of orthonormal eigenvectors, you can never be sure you have them all, since adding or subtracting a few still leaves you with an infinite number of them.)

Since K is a Hermitian operator, functions that were expanded in the X basis with components $f(x) = \langle x|f\rangle$ must also have an expansion in the K basis. To find the components, we start with a ket $|f\rangle$, and do the following:

$$f(k) = \langle k|f\rangle = \int_{-\infty}^{\infty} \langle k|x\rangle\langle x|f\rangle dx = \frac{1}{(2\pi)^{1/2}} \int_{-\infty}^{\infty} e^{-ikx} f(x)\, dx \tag{1.10.34}$$

The passage back to the X basis is done as follows:

$$f(x) = \langle x|f\rangle = \int_{-\infty}^{\infty} \langle x|k\rangle\langle k|f\rangle dk = \frac{1}{(2\pi)^{1/2}} \int_{-\infty}^{\infty} e^{ikx} f(k)\, dk \tag{1.10.35}$$

Thus the familiar Fourier transform is just the passage from one complete basis $|x\rangle$ to another, $|k\rangle$. Either basis may be used to expand functions that belong to the Hilbert space.

The matrix elements of K are trivial in the K basis:

$$\langle k|K|k'\rangle = k'\langle k|k'\rangle = k'\delta(k-k') \tag{1.10.36}$$

Now, we know where the K basis came from: it was generated by the Hermitian operator K. Which operator is responsible for the orthonormal X basis? Let us call it the operator X. The kets $|x\rangle$ are its eigenvectors with eigenvalue x:

$$X|x\rangle = x|x\rangle \tag{1.10.37}$$

Its matrix elements in the X basis are

$$\langle x'|X|x\rangle = x\delta(x'-x) \tag{1.10.38}$$

To find its action on functions, let us begin with

$$X|f\rangle = |\tilde{f}\rangle$$

and follow the routine:

$$\langle x|X|f\rangle = \int \langle x|X|x'\rangle\langle x'|f\rangle dx' = xf(x) = \langle x|\tilde{f}\rangle = \tilde{f}(x)$$

$$\therefore \quad \tilde{f}(x) = xf(x)$$

‡ Hereafter we will omit the qualifier "physical."

Thus the effect of X is to multiply $f(x)$ by x. As in the case of the K operator, one generally suppresses the integral over the common index since it is rendered trivial by the delta function. We can summarize the action of X in Hilbert space as

$$X|f(x)\rangle = |xf(x)\rangle \tag{1.10.39}$$

where as usual $|xf(x)\rangle$ is the ket corresponding to the function $xf(x)$.

There is a nice reciprocity between the X and K operators which manifests itself if we compute the matrix elements of X in the K basis:

$$\begin{aligned}\langle k|X|k'\rangle &= \frac{1}{2\pi}\int_{-\infty}^{\infty} e^{-ikx}\, x\, e^{ik'x}\, dx \\ &= +i\frac{d}{dk}\left(\frac{1}{2\pi}\int_{-\infty}^{\infty} e^{i(k'-k)x}\, dx\right) = i\delta'(k-k')\ddagger\end{aligned}$$

Thus if $|g(k)\rangle$ is a ket whose image in the k basis is $g(k)$, then

$$X|g(k)\rangle = \left|\frac{i\, dg(k)}{dk}\right\rangle \tag{1.10.40}$$

In summary then, in the X basis, X acts as x and K as $-i\, d/dx$ [on the functions $f(x)$], while in the K basis, K acts like k and X like $i\, d/dk$ [on $f(k)$]. Operators with such an interrelationship are said to be *conjugate* to each other.

The conjugate operators X and K do not commute. Their commutator may be calculated as follows. Let us operate X and K in both possible orders on some ket $|f\rangle$ and follow the action in the X basis:

$$X|f\rangle \to xf(x)$$

$$K|f\rangle \to -i\frac{df(x)}{dx}$$

So

$$XK|f\rangle \to -ix\frac{df(x)}{dx}$$

$$KX|f\rangle \to -i\frac{d}{dx}xf(x)$$

Therefore

$$[X, K]|f\rangle \to -ix\frac{df}{dx} + ix\frac{df}{dx} + if = if \to iI|f\rangle$$

‡ In the last step we have used the fact that $\delta(k'-k) = \delta(k-k')$.

Since $|f\rangle$ is an arbitrary ket, we now have the desired result:

$$[X, K]=iI \tag{1.10.41}$$

This brings us to the end of our discussion on Hilbert space, except for a final example. Although there are many other operators one can study in this space, we restricted ourselves to X and K since almost all the operators we will need for quantum mechanics are functions of X and $P=\hbar K$, where $\hbar$ is a constant to be defined later.

Example 1.10.1: A Normal Mode Problem in Hilbert Space. Consider a string of length L clamped at its two ends $x=0$ and L. The displacement $\psi(x, t)$ obeys the differential equation

$$\frac{\partial^2\psi}{\partial t^2}=\frac{\partial^2\psi}{\partial x^2} \tag{1.10.42}$$

Given that at $t=0$ the displacement is $\psi(x, 0)$ and the velocity $\dot{\psi}(x, 0)=0$, we wish to determine the time evolution of the string.

But for the change in dimensionality, the problem is identical to that of the two coupled masses encountered at the end of Section 1.8 [see Eq. (1.8.26)]. It is recommended that you go over that example once to refresh your memory before proceeding further.

We first identify $\psi(x, t)$ as components of a vector $|\psi(t)\rangle$ in a Hilbert space, the elements of which are in correspondence with possible displacements ψ, i.e., functions that are continuous in the interval $0\le x\le L$ and vanish at the end points. You may verify that these functions do form a vector space.

The analog of the operator Ω in Eq. (1.8.26) is the operator $\partial^2/\partial x^2$. We recognize this to be minus the square of the operator $K\leftrightarrow -i\partial/\partial x$. Since K acts on a space in which $\psi(0)=\psi(L)=0$, it is Hermitian, and so is K^2. Equation (1.10.42) has the abstract counterpart

$$|\ddot{\psi}(t)\rangle=-K^2|\psi(t)\rangle \tag{1.10.43}$$

We solve the initial-value problem by following the algorithm developed in Example 1.8.6:

Step (1). Solve the eigenvalue problem of $-K^2$.

Step (2). Construct the propagator $U(t)$ in terms of the eigenvectors and eigenvalues.

Step (3).

$$|\psi(t)\rangle=U(t)|\psi(0)\rangle \tag{1.10.44}$$

The equation to solve is

$$K^2|\psi\rangle = k^2|\psi\rangle \tag{1.10.45}$$

In the X basis, this becomes

$$-\frac{d^2}{dx^2}\psi_k(x) = k^2\psi_k(x) \tag{1.10.46}$$

the general solution to which is

$$\psi_k(x) = A\cos kx + B\sin kx \tag{1.10.47}$$

where A and B are arbitrary. However, not all these solutions lie in the Hilbert space we are considering. We want only those that vanish at $x=0$ and $x=L$. At $x=0$ we find

$$\psi_k(0) = 0 = A \tag{1.10.48a}$$

while at $x=L$ we find

$$0 = B\sin kL \tag{1.10.48b}$$

If we do not want a trivial solution $(A=B=0)$ we must demand

$$\sin kL = 0, \quad kL = m\pi, \qquad m = 1, 2, 3, \ldots \tag{1.10.49}$$

We do not consider negative m since it doesn't lead to any further LI solutions $[\sin(-x) = -\sin x]$. The allowed eigenvectors thus form a discrete set labeled by an integer m:

$$\psi_m(x) = \left(\frac{2}{L}\right)^{1/2}\sin\left(\frac{m\pi x}{L}\right) \tag{1.10.50}$$

where we have chosen $B=(2/L)^{1/2}$ so that

$$\int_0^L \psi_m(x)\psi_{m'}(x)\,dx = \delta_{mm'} \tag{1.10.51}$$

Let us associate with each solution labeled by the integer m an abstract ket $|m\rangle$:

$$|m\rangle \xrightarrow[X\text{ basis}]{} (2/L)^{1/2}\sin\left(\frac{m\pi x}{L}\right) \tag{1.10.52}$$

If we project $|\psi(t)\rangle$ on the $|m\rangle$ basis, in which K is diagonal with eigenvalues $(m\pi/L)^2$, the components $\langle m|\psi(t)\rangle$ will obey the decoupled equations

$$\frac{d^2}{dt^2}\langle m|\psi(t)\rangle = -\left(\frac{m^2\pi^2}{L^2}\right)\langle m|\psi(t)\rangle, \qquad m=1, 2, \ldots \tag{1.10.53}$$

in analogy with Eq. (1.8.33). These equations may be readily solved (subject to the condition of vanishing initial velocities) as

$$\langle m|\psi(t)\rangle = \langle m|\psi(0)\rangle \cos\left(\frac{m\pi t}{L}\right) \tag{1.10.54}$$

Consequently

$$\begin{aligned} |\psi(t)\rangle &= \sum_{m=1}^{\infty} |m\rangle\langle m|\psi(t)\rangle \\ &= \sum_{m=1}^{\infty} |m\rangle\langle m|\psi(0)\rangle \cos\omega_m t, \qquad \omega_m = \frac{m\pi}{L} \end{aligned} \tag{1.10.55}$$

or

$$U(t) = \sum_{m=1}^{\infty} |m\rangle\langle m| \cos\omega_m t, \qquad \omega_m = \frac{m\pi}{L} \tag{1.10.56}$$

The propagator equation

$$|\psi(t)\rangle = U(t)|\psi(0)\rangle$$

becomes in the $|x\rangle$ basis

$$\begin{aligned} \langle x|\psi(t)\rangle &= \psi(x, t) \\ &= \langle x|U(t)|\psi(0)\rangle \\ &= \int_0^L \langle x|U(t)|x'\rangle\langle x'|\psi(0)\rangle\, dx' \end{aligned} \tag{1.10.57}$$

It follows from Eq. (1.10.56) that

$$\begin{aligned} \langle x|U(t)|x'\rangle &= \sum_m \langle x|m\rangle\langle m|x'\rangle \cos\omega_m t \\ &= \sum_m \left(\frac{2}{L}\right)\sin\left(\frac{m\pi x}{L}\right)\sin\left(\frac{m\pi x'}{L}\right)\cos\omega_m t \end{aligned} \tag{1.10.58}$$

Thus, given any $\psi(x', 0)$, we can get $\psi(x, t)$ by performing the integral in Eq. (1.10.57), using $\langle x|U(t)|x'\rangle$ from Eq. (1.10.58). If the propagator language seems too abstract, we can begin with Eq. (1.10.55). Dotting both sides with $\langle x|$, we get

$$\begin{aligned}\psi(x, t) &= \sum_{m=1}^{\infty} \langle x|m\rangle\langle m|\psi(0)\rangle \cos \omega_m t \\ &= \sum_{m=1}^{\infty} \left(\frac{2}{L}\right)^{1/2} \sin\left(\frac{m\pi x}{L}\right) \cos \omega_m t \langle m|\psi(0)\rangle \end{aligned} \tag{1.10.59}$$

Given $|\psi(0)\rangle$, one must then compute

$$\langle m|\psi(0)\rangle = \left(\frac{2}{L}\right)^{1/2} \int_0^L \sin\left(\frac{m\pi x}{L}\right) \psi(x, 0)\, dx$$

Usually we will find that the coefficients $\langle m|\psi(0)\rangle$ fall rapidly with m so that a few leading terms may suffice to get a good approximation. □

Exercise 1.10.4. A string is displaced as follows at $t=0$:

$$\begin{aligned}\psi(x, 0) &= \frac{2xh}{L}, & 0 \le x \le \frac{L}{2} \\ &= \frac{2h}{L}(L-x), & \frac{L}{2} \le x \le L\end{aligned}$$

Show that

$$\psi(x, t) = \sum_{m=1}^{\infty} \sin\left(\frac{m\pi x}{L}\right) \cos \omega_m t \cdot \left(\frac{8h}{\pi^2 m^2}\right) \sin\left(\frac{\pi m}{2}\right)$$

2

Review of Classical Mechanics

In this chapter we will develop the Lagrangian and Hamiltonian formulations of mechanics starting from Newton's laws. These subsequent reformulations of mechanics bring with them a great deal of elegance and computational ease. But our principal interest in them stems from the fact that they are the ideal springboards from which to make the leap to quantum mechanics. The passage from the Lagrangian formulation to quantum mechanics was carried out by Feynman in his path integral formalism. A more common route to quantum mechanics, which we will follow for the most part, has as its starting point the Hamiltonian formulation, and it was discovered mainly by Schrödinger, Heisenberg, Dirac, and Born.

It should be emphasized, and it will soon become apparent, that all three formulations of mechanics are essentially the same theory, in that their domains of validity and predictions are identical. Nonetheless, in a given context, one or the other may be more inviting for conceptual, computational, or simply aesthetic reasons.

2.1. The Principle of Least Action and Lagrangian Mechanics

Let us take as our prototype of the Newtonian scheme a point particle of mass m moving along the x axis under a potential $V(x)$. According to Newton's Second Law,

$$m\frac{d^2x}{dt^2}=-\frac{dV}{dx} \tag{2.1.1}$$

If we are given the initial state variables, the position $x(t_i)$ and velocity $\dot{x}(t_i)$, we can calculate the classical trajectory $x_{cl}(t)$ as follows. Using the initial velocity and acceleration [obtained from Eq. (2.1.1)] we compute the position and velocity at a time $t_i+\Delta t$. For example,

$$x_{\text{cl}}(t_i+\Delta t)=x(t_i)+\dot{x}(t_i)\Delta t$$

Having updated the state variables to the time $t_i+\Delta t$, we can repeat the process again to inch forward to $t_i+2\Delta t$ and so on.

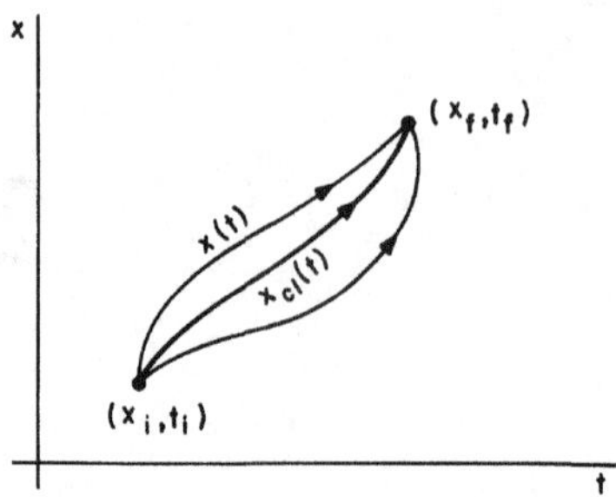

Figure 2.1. The Lagrangian formalism asks what distinguishes the actual path $x_{\text{cl}}(t)$ taken by the particle from all possible paths connecting the end points (x_i, t_i) and (x_f, t_f).

The equation of motion being second order in time, two pieces of data, $x(t_i)$ and $\dot{x}(t_i)$, are needed to specify a unique $x_{\text{cl}}(t)$. An equivalent way to do the same and one that we will have occasion to employ, is to specify two space-time points (x_i, t_i) and (x_f, t_f) on the trajectory.

The above scheme readily generalizes to more than one particle and more than one dimension. If we use n Cartesian coordinates $(x_1, x_2, \ldots, x_n)$ to specify the positions of the particles, the spatial configuration of the system may be visualized as a point in an n-dimensional *configuration space*. (The term "configuration space" is used even if the n coordinates are not Cartesian.) The motion of the representative point is given by

$$m_j \frac{d^2 x_j}{dt^2} = -\frac{\partial V}{\partial x_j} \tag{2.1.2}$$

where m_j stands for the mass of the particle whose coordinate is x_j. These equations can be integrated step by step, just as before, to determine the trajectory.

In the Lagrangian formalism, the problem of a single particle in a potential $V(x)$ is posed in a different way: given that the particle is at x_i and x_f at times t_i and t_f, respectively, what is it that distinguishes the actual trajectory $x_{\text{cl}}(t)$ from all other trajectories or paths that connect these points? (See Fig. 2.1.)

The Lagrangian approach is thus global, in that it tries to determine at one stroke the entire trajectory $x_{\text{cl}}(t)$, in contrast to the local approach of the Newtonian scheme, which concerns itself with what the particle is going to do in the next infinitesimal time interval.

The answer to the question posed above comes in three parts:

(1) Define a function $\mathscr{L}$, called the *Lagrangian*, given by $\mathscr{L} = T - V$, T and V being the kinetic and potential energies of the particle. Thus $\mathscr{L} = \mathscr{L}(x, \dot{x}, t)$. The explicit t dependence may arise if the particle is in an external time-dependent field. We will, however, assume the absence of this t dependence.

(2) For each path $x(t)$ connecting (x_i, t_i) and (x_f, t_f), calculate the *action* $S[x(t)]$ defined by

$$S[x(t)] = \int_{t_i}^{t_f} \mathscr{L}(x, \dot{x})\, dt \tag{2.1.3}$$

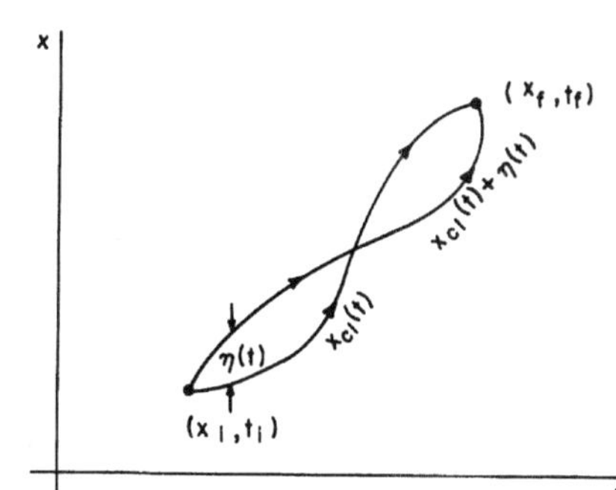

Figure 2.2. If $x_{\mathrm{cl}}(t)$ minimizes S, then $\delta S^{(1)}=0$ if we go to any nearby path $x_{\mathrm{cl}}(t)+\eta(t)$.

We use square brackets to enclose the argument of S to remind us that the function S depends on an entire path or function $x(t)$, and not just the value of x at some time t. One calls S a *functional* to signify that it is a function of a function.

(3) The classical path is one on which S is a minimum. (Actually we will only require that it be an extremum. It is, however, customary to refer to this condition as the *principle of least action*.)

We will now verify that this principle reproduces Newton's Second Law.

The first step is to realize that a functional $S[x(t)]$ is just a function of n variables as $n\to\infty$. In other words, the function $x(t)$ simply specifies an infinite number of values $x(t_i), \ldots, x(t), \ldots, x(t_f)$, one for each instant in time t in the interval $t_i \le t \le t_f$, and S is a function of these variables. To find its minimum we simply generalize the procedure for the finite n case. Let us recall that if $f=f(x_1, \ldots, x_n)=f(\mathbf{x})$; the minimum $\mathbf{x}^0$ is characterized by the fact that if we move away from it by a small amount $\boldsymbol{\eta}$ in any direction, the first-order change $\delta f^{(1)}$ in f vanishes. That is, if we make a Taylor expansion:

$$f(\mathbf{x}^0+\boldsymbol{\eta})=f(\mathbf{x}^0)+\sum_{i=1}^{n}\left.\frac{\partial f}{\partial x_i}\right|_{\mathbf{x}^0}\eta_i+\text{higher-order terms in }\eta \tag{2.1.4}$$

then

$$\delta f^{(1)} \equiv \sum_{i=1}^{n}\left.\frac{\partial f}{\partial x_i}\right|_{\mathbf{x}^0}\eta_i=0 \tag{2.1.5}$$

From this condition we can deduce an equivalent and perhaps more familiar expression of the minimum condition: every first-order partial derivative vanishes at $\mathbf{x}^0$. To prove this, for say, $\partial f/\partial x_i$, we simply choose $\boldsymbol{\eta}$ to be along the ith direction. Thus

$$\left.\frac{\partial f}{\partial x_i}\right|_{\mathbf{x}^0}=0, \qquad i=1, \ldots, n \tag{2.1.6}$$

Let us now mimic this procedure for the action S. Let $x_{\mathrm{cl}}(t)$ be the path of least action and $x_{\mathrm{cl}}(t)+\eta(t)$ a "nearby" path (see Fig. 2.2). The requirement that all paths coincide at t_i and t_f means

$$\eta(t_i)=\eta(t_f)=0 \tag{2.1.7}$$

Now

$$S[x_{cl}(t)+\eta(t)]=\int_{t_i}^{t_f}\mathscr{L}(x_{cl}(t)+\eta(t);\dot{x}_{cl}(t)+\dot{\eta}(t))\,dt$$

$$=\int_{t_i}^{t_f}\left[\mathscr{L}(x_{cl}(t),\dot{x}_{cl}(t))+\left.\frac{\partial\mathscr{L}}{\partial x(t)}\right|_{x_{cl}}\cdot\eta(t)\right.$$

$$\left.+\left.\frac{\partial\mathscr{L}}{\partial\dot{x}(t)}\right|_{x_{cl}}\cdot\dot{\eta}(t)+\cdots\right]dt$$

$$=S[x_{cl}(t)]+\delta S^{(1)}+\text{higher-order terms}$$

We set $\delta S^{(1)}=0$ in analogy with the finite variable case:

$$0=\delta S^{(1)}=\int_{t_i}^{t_f}\left[\left.\frac{\partial\mathscr{L}}{\partial x(t)}\right|_{x_{cl}}\cdot\eta(t)+\left.\frac{\partial\mathscr{L}}{\partial\dot{x}(t)}\right|_{x_{cl}}\cdot\dot{\eta}(t)\right]dt$$

If we integrate the second term by parts, it turns into

$$\left.\frac{\partial\mathscr{L}}{\partial\dot{x}(t)}\right|_{x_{cl}}\cdot\eta(t)\Bigg|_{t_i}^{t_f}-\int_{t_i}^{t_f}\left[\frac{d}{dt}\frac{\partial\mathscr{L}}{\partial\dot{x}(t)}\right]_{x_{cl}}\cdot\eta(t)\,dt$$

The first of these terms vanishes due to Eq. (2.1.7). So that

$$0=\delta S^{(1)}=\int_{t_i}^{t_f}\left[\frac{\partial\mathscr{L}}{\partial x(t)}-\frac{d}{dt}\frac{\partial\mathscr{L}}{\partial\dot{x}(t)}\right]_{x_{cl}}\cdot\eta(t)\,dt\tag{2.1.8}$$

Note that the condition $\delta S^{(1)}=0$ implies that S is extremized and not necessarily minimized. We shall, however, continue the tradition of referring to this extremum as the minimum. This equation is the analog of Eq. (2.1.5): the discrete variable η_i is replaced by $\eta(t)$; the sum over i is replaced by an integral over t, and $\partial f/\partial x_i$ is replaced by

$$\frac{\partial\mathscr{L}}{\partial x(t)}-\frac{d}{dt}\frac{\partial\mathscr{L}}{\partial\dot{x}(t)}$$

There are two terms here playing the role of $\partial f/\partial x_i$ since $\mathscr{L}$ (or equivalently S) has both explicit and implicit (through the $\dot{x}$ terms) dependence on $x(t)$. Since $\eta(t)$ is arbitrary, we may extract the analog of Eq. (2.1.6):

$$\left\{\frac{\partial\mathscr{L}}{\partial x(t)}-\frac{d}{dt}\left[\frac{\partial\mathscr{L}}{\partial\dot{x}(t)}\right]\right\}_{x_{cl}(t)}=0\quad\text{for } t_i\le t\le t_f\tag{2.1.9}$$

To deduce this result for some specific time t_0, we simply choose an $\eta(t)$ that vanishes everywhere except in an infinitesimal region around t_0.

Equation (2.1.9) is the celebrated *Euler–Lagrange equation*. If we feed into it $\mathscr{L}=T-V$, $T=\frac{1}{2}m\dot{x}^2$, $V=V(x)$, we get

$$\frac{\partial \mathscr{L}}{\partial \dot{x}}=\frac{\partial T}{\partial \dot{x}}=m\dot{x}$$

and

$$\frac{\partial \mathscr{L}}{\partial x}=-\frac{\partial V}{\partial x}$$

so that the Euler-Lagrange equation becomes just

$$\frac{d}{dt}(m\dot{x})=-\frac{\partial V}{\partial x}$$

which is just Newton's Second Law, Eq. (2.1.1).

If we consider a system described by n Cartesian coordinates, the same procedure yields

$$\frac{d}{dt}\left(\frac{\partial \mathscr{L}}{\partial \dot{x}_i}\right)=\frac{\partial \mathscr{L}}{\partial x_i} \quad (i=1,\ldots,n) \tag{2.1.10}$$

Now

$$T=\frac{1}{2}\sum_{i=1}^{n} m_i(\dot{x}_i)^2$$

and

$$V=V(x_1,\ldots,x_n)$$

so that Eq. (2.1.10) becomes

$$\frac{d}{dt}(m_i\dot{x}_i)=-\frac{\partial V}{\partial x_i}$$

which is identical to Eq. (2.1.2). Thus the minimum (action) principle indeed reproduces Newtonian mechanics if we choose $\mathscr{L}=T-V$.

Notice that we have assumed that V is velocity-independent in the above proof. An important force, that of a magnetic field $\mathbf{B}$ on a moving charge is excluded by this restriction, since $\mathbf{F}_B=q\mathbf{v}\times\mathbf{B}$, q being the charge of the particle and $\mathbf{v}=\dot{\mathbf{r}}$ its velocity. We will show shortly that this force too may be accommodated in the Lagrangian formalism, in the sense that we can find an $\mathscr{L}$ that yields the correct force law when Eq. (2.1.10) is employed. But this $\mathscr{L}$ no longer has the form $T-V$. One therefore frees oneself from the notion that $\mathscr{L}=T-V$; and views $\mathscr{L}$ as some function $\mathscr{L}(x_i, \dot{x}_i)$ which yields the correct Newtonian dynamics when fed into the Euler–Lagrange equations. To the reader who wonders why one bothers to even deal with a Lagrangian when all it does is yield Newtonian force laws in the end, I present a few of its main attractions besides its closeness to quantum mechanics. These will then be illustrated by means of an example.

(1) In the Lagrangian scheme one has merely to construct a single *scalar* $\mathscr{L}$ and all the equations of motion follow by simple differentiation. This must be contrasted with the Newtonian scheme, which deals with vectors and is thus more complicated.

(2) The Euler–Lagrange equations (2.1.10) have the same *form* if we use, instead of the n Cartesian coordinates $x_1, \ldots, x_n$, *any* general set of n independent coordinates $q_1, q_2, \ldots, q_n$. To remind us of this fact we will rewrite Eq. (2.1.10) as

$$\frac{d}{dt}\left(\frac{\partial \mathscr{L}}{\partial \dot{q}_i}\right)=\frac{\partial \mathscr{L}}{\partial q_i} \tag{2.1.11}$$

One can either verify this by brute force, making a change of variables in Eq. (2.1.10) and seeing that an identical equation with x_i replaced by q_i follows, or one can simply go through our derivation of the minimum action condition and see that nowhere were the coordinates assumed to be Cartesian. Of course, at the next stage, in showing that the Euler–Lagrange equations were equivalent to Newton's, Cartesian coordinates *were* used, for in these coordinates the kinetic energy T and the Newtonian equations have simple forms. But once the principle of least action is seen to generate the correct dynamics, we can forget all about Newton's laws and use Eq. (2.1.11) as the equations of motion. What is being emphasized is that these equations, which express the condition for least action, are form invariant under an arbitrary change of coordinates. This form invariance must be contrasted with the Newtonian equation (2.1.2), which presumes that the x_i are Cartesian. If one trades the x_i for another non-Cartesian set of q_i, Eq. (2.1.2) will have a different form (see Example 2.1.1 at the end of this section).

Equation (2.1.11) can be made to resemble Newton's Second Law if one defines a quantity

$$p_i=\frac{\partial \mathscr{L}}{\partial \dot{q}_i} \tag{2.1.12}$$

called the *canonical momentum conjugate to* q_i and the quantity

$$F_i=\frac{\partial \mathscr{L}}{\partial q_i} \tag{2.1.13}$$

called the *generalized force conjugate to* q_i. Although the rate of change of the canonical momentum equals the generalized force, one must remember that neither is p_i always a linear momentum (mass times velocity or "mv" momentum), nor is F_i always a force (with dimensions of mass times acceleration). For example, if q_i is an angle θ, p_i will be an angular momentum and F_i a torque.

(3) Conservation laws are easily obtained in this formalism. Suppose the Lagrangian depends on a certain velocity $\dot{q}_i$ but not on the corresponding coordinate q_i. The latter is then called a *cyclic coordinate*. It follows that the corresponding p_i is conserved:

$$\frac{d}{dt}\left(\frac{\partial \mathscr{L}}{\partial \dot{q}_i}\right) = \frac{dp_i}{dt} = \frac{\partial \mathscr{L}}{\partial q_i} = 0 \qquad (2.1.14)$$

Although Newton's Second Law, Eq. (2.1.2), also tells us that if a Cartesian coordinate x_i is cyclic, the corresponding momentum $m_i\dot{x}_i$ is conserved, Eq. (2.1.14) is more general. Consider, for example, a potential $V(x, y)$ in two dimensions that depends only upon $\rho = (x^2+y^2)^{1/2}$, and not on the polar angle ϕ, so that $V(\rho, \phi) = V(\rho)$. It follows that ϕ is a cyclic coordinate, as T depends only on $\dot{\phi}$ (see Example 2.1.1 below). Consequently $\partial\mathscr{L}/\partial\dot{\phi} = p_\phi$ is conserved. In contrast, no obvious conservation law arises from the Cartesian Eqs. (2.1.2) since neither x nor y is cyclic. If one rewrites Newton's laws in polar coordinates to exploit $\partial V/\partial\dot{\phi} = 0$, the corresponding equations get complicated due to centrifugal and Coriolis terms. It is the Lagrangian formalism that allows us to choose coordinates that best reflect the symmetry of the potential, without altering the simple form of the equations.

Example 2.1.1. We now illustrate the above points through an example. Consider a particle moving in a plane. The Lagrangian, in Cartesian coordinates, is

$$\begin{aligned}\mathscr{L} &= \tfrac{1}{2}m(\dot{x}^2 + \dot{y}^2) - V(x, y) \\ &= \tfrac{1}{2}m\mathbf{v}\cdot\mathbf{v} - V(x, y)\end{aligned} \qquad (2.1.15)$$

where $\mathbf{v}$ is the velocity of the particle, with $\mathbf{v} = \dot{\mathbf{r}}, \mathbf{r}$ being its position vector. The corresponding equations of motion are

$$m\ddot{x} = -\frac{\partial V}{\partial x} \qquad (2.1.16)$$

$$m\ddot{y} = -\frac{\partial V}{\partial y} \qquad (2.1.17)$$

which are identical to Newton's laws. If one wants to get the same Newton's laws in terms of polar coordinates ρ and ϕ, some careful vector analysis is needed to unearth the centrifugal and Coriolis terms:

$$m\ddot{\rho} = -\frac{\partial V}{\partial \rho} + m\rho(\dot{\phi})^2 \qquad (2.1.18)$$

$$m\ddot{\phi} = -\frac{1}{\rho^2}\frac{\partial V}{\partial \phi} - \frac{2m\dot{\rho}\dot{\phi}}{\rho} \qquad (2.1.19)$$

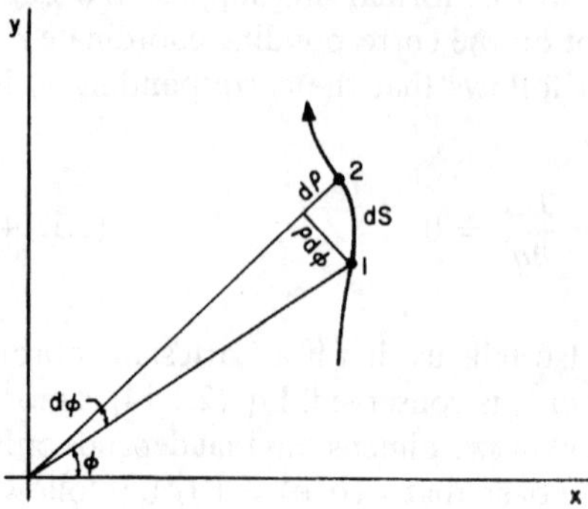

Figure 2.3. Points (1) and (2) are positions of the particle at times differing by Δt.

Notice the difference in form between Eqs. (2.1.16) and (2.1.17) on the one hand and Eqs. (2.1.18) and (2.1.19) on the other.

In the Lagrangian scheme one has only to recompute $\mathscr{L}$ in polar coordinates. From Fig. 2.3 it is clear that the distance traveled by the particle in time Δt is

$$dS = [(d\rho)^2 + (\rho\, d\phi)^2]^{1/2}$$

so that the magnitude of velocity is

$$v = \frac{dS}{dt} = [(\dot{\rho})^2 + \rho^2(\dot{\phi})^2]^{1/2}$$

and

$$\mathscr{L} = \tfrac{1}{2}m(\dot{\rho}^2 + \rho^2\dot{\phi}^2) - V(\rho, \phi) \tag{2.1.20}$$

(Notice that in these coordinates T involves not just the velocities $\dot{\rho}$ and $\dot{\phi}$ but also the coordinate ρ. This does not happen in Cartesian coordinates.) The equations of motion generated by this $\mathscr{L}$ are

$$\frac{d}{dt}(m\dot{\rho}) = -\frac{\partial V}{\partial \rho} + m\rho\dot{\phi}^2 \tag{2.1.21}$$

$$\frac{d}{dt}(m\rho^2\dot{\phi}) = -\frac{\partial V}{\partial \phi} \tag{2.1.22}$$

which are the same as Eqs. (2.1.18) and (2.1.19). In Eq. (2.1.22) the canonical momentum $p_\phi = m\rho^2\dot{\phi}$ is the angular momentum and the generalized force $-\partial V/\partial\phi$ is the torque, both along the z axis. Notice how easily the centrifugal and Coriolis forces came out.

Finally, if $V(\rho, \phi) = V(\rho)$, the conservation of p_ϕ is obvious in Eq. (2.1.22). The conservation of p_ϕ follows from Eq. (2.1.19) only after some manipulations and is practically invisible in Eqs. (2.1.16) and (2.1.17). Both the conserved quantity and its conservation law arise naturally in the Lagrangian scheme. □

*Exercise 2.1.1.** Consider the following system, called a *harmonic oscillator*. The block has a mass m and lies on a frictionless surface. The spring has a force constant k.

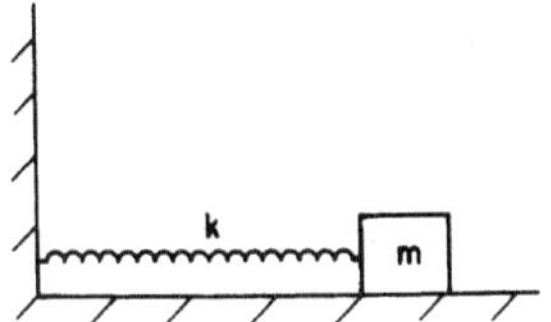

Write the Lagrangian and get the equations of motion.

*Exercise 2.1.2.** Do the same for the coupled-mass problem discussed at the end of Section 1.8. Compare the equations of motion with Eqs. (1.8.24) and (1.8.25).

*Exercise 2.1.3.** A particle of mass m moves in three dimensions under a potential $V(r, \theta, \phi) = V(r)$. Write its $\mathscr{L}$ and find the equations of motion.

2.2. The Electromagnetic Lagrangian‡

Recall that the force on a charge q due to an electric field $\mathbf{E}$ and magnetic field $\mathbf{B}$ is given by

$$\mathbf{F} = q\left(\mathbf{E} + \frac{\mathbf{v}}{c} \times \mathbf{B}\right) \tag{2.2.1}$$

where $\mathbf{v} = \dot{\mathbf{r}}$ is the velocity of the particle. Since the force is velocity-dependent, we must analyze the problem afresh, not relying on the preceding discussion, which was restricted to velocity-independent forces.

Now it turns out that if we use

$$\mathscr{L}_{e\cdot m} = \tfrac{1}{2} m\mathbf{v}\cdot\mathbf{v} - q\phi + \frac{q}{c}\mathbf{v}\cdot\mathbf{A} \tag{2.2.2}$$

we get the correct electromagnetic force laws. In Eq. (2.2.2) c is the velocity of light, while ϕ and $\mathbf{A}$ are the scalar and vector potentials related to $\mathbf{E}$ and $\mathbf{B}$ via

$$\mathbf{E} = -\nabla\phi - \frac{1}{c}\frac{\partial\mathbf{A}}{\partial t} \tag{2.2.3}$$

and

$$\mathbf{B} = \nabla \times \mathbf{A} \tag{2.2.4}$$

‡ See Section 18.4 for a review of classical electromagnetism.

The Euler–Lagrange equations corresponding to $\mathscr{L}_{e\text{-}m}$ are

$$\frac{d}{dt}\left(m\dot{x}_i+\frac{q}{c}A_i\right)=-q\frac{\partial\phi}{\partial x_i}+\frac{q}{c}\frac{\partial(\mathbf{v}\cdot\mathbf{A})}{\partial x_i}, \qquad i=1, 2, 3 \tag{2.2.5}$$

Combining the three equations above into a single vector equation we get

$$\frac{d}{dt}\left(m\mathbf{v}+\frac{q\mathbf{A}}{c}\right)=-q\nabla\phi+\frac{q}{c}\nabla(\mathbf{v}\cdot\mathbf{A}) \tag{2.2.6}$$

The canonical momentum is

$$\mathbf{p}=m\mathbf{v}+\frac{q\mathbf{A}}{c} \tag{2.2.7}$$

Rewriting Eq. (2.2.6), we get

$$\frac{d}{dt}(m\mathbf{v})=-q\nabla\phi+\frac{q}{c}\left[-\frac{d\mathbf{A}}{dt}+\nabla(\mathbf{v}\cdot\mathbf{A})\right] \tag{2.2.8}$$

Now, the total derivative $d\mathbf{A}/dt$ has two parts: an explicit time dependence $\partial A/\partial t$, plus an implicit one $(\mathbf{v}\cdot\nabla)\mathbf{A}$ which represents the fact that a spatial variation in $\mathbf{A}$ will appear as a temporal variation to the moving particle.Now Eq. (2.2.8) becomes

$$\frac{d}{dt}(m\mathbf{v})=-q\nabla\phi-\frac{q}{c}\frac{\partial\mathbf{A}}{\partial t}+\frac{q}{c}[\nabla(\mathbf{v}\cdot\mathbf{A})-(\mathbf{v}\cdot\nabla)\mathbf{A}] \tag{2.2.9}$$

which is identical to Eq. (2.2.1) by virtue of the identity

$$\mathbf{v}\times(\nabla\times\mathbf{A})=\nabla(\mathbf{v}\cdot\mathbf{A})-(\mathbf{v}\cdot\nabla)\mathbf{A}$$

Notice that $\mathscr{L}_{e\text{-}m}$ is not of the form $T-V$, for the quantity $U=q\phi-(q/c)\mathbf{v}\cdot\mathbf{A}$ (sometimes called the *generalized potential*) cannot be interpreted as the potential energy of the charged particle. First of all, the force due to a time-dependent electromagnetic field is not generally conservative and does not admit a path-independent work function to play the role of a potential. Even in the special cases when the force is conservative, only $q\phi$ can be interpreted as the electrical potential energy. The $[-q(\mathbf{v}\cdot\mathbf{A})/c]$ term is not a magnetic potential energy, since the magnetic force $\mathbf{F}_B=q(\mathbf{v}\times\mathbf{B})/c$ never does any work, being always perpendicular to the velocity. To accommodate forces such as the electro-magnetic, we must, therefore, redefine $\mathscr{L}$ to be that function $\mathscr{L}(q, \dot{q}, t)$ which, when fed into the Euler–Lagrange equations, reproduces the correct dynamics. The rule $\mathscr{L}=T-V$ becomes just a useful mnemonic for the case of conservative forces.

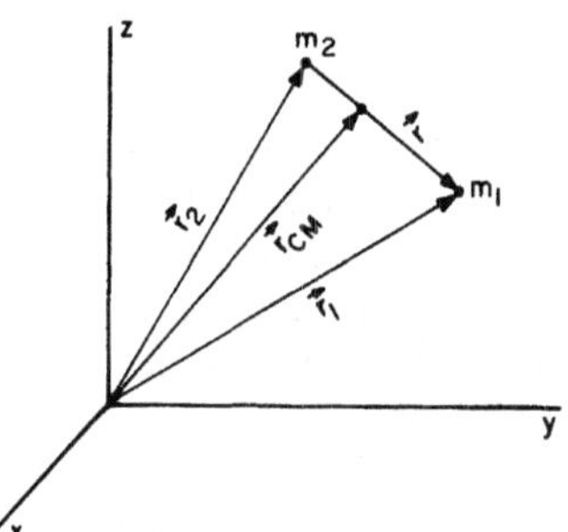

Figure 2.4. The relation between $\mathbf{r}_1$, $\mathbf{r}_2$ and $\mathbf{r}_{CM}$, $\mathbf{r}$.

2.3. The Two-Body Problem

We discuss here a class of problems that plays a central role in classical physics: that of two masses m_1 and m_2 exerting equal and opposite forces on each other. Since the particles are responding to each other and nothing external, it follows that the potential between them depends only on the *relative coordinate* $\mathbf{r}=\mathbf{r}_1-\mathbf{r}_2$ and not the individual positions $\mathbf{r}_1$ and $\mathbf{r}_2$. But $V(\mathbf{r}_1, \mathbf{r}_2)=V(\mathbf{r}_1-\mathbf{r}_2)$ means in turn that there are three cyclic coordinates, for V depends on only three variables rather than the possible six. (In Cartesian coordinates, since T is a function only of velocities, a coordinate missing in V is also cyclic.) The corresponding conserved momenta will of course be the three components of the total momentum, which are conserved in the absence of external forces. To bring out these features, it is better to trade $\mathbf{r}_1$ and $\mathbf{r}_2$ in favor of

$$\mathbf{r}=\mathbf{r}_1-\mathbf{r}_2 \tag{2.3.1}$$

and

$$\mathbf{r}_{CM}=\frac{m_1\mathbf{r}_1+m_2\mathbf{r}_2}{m_1+m_2} \tag{2.3.2}$$

where $\mathbf{r}_{CM}$ is called the *center-of-mass (CM) coordinate*. One can invert Eqs. (2.3.1) and (2.3.2) to get (see Fig. 2.4)

$$\mathbf{r}_1=\mathbf{r}_{CM}+\frac{m_2\mathbf{r}}{m_1+m_2} \tag{2.3.3}$$

$$\mathbf{r}_2=\mathbf{r}_{CM}-\frac{m_1\mathbf{r}}{m_1+m_2} \tag{2.3.4}$$

If one rewrites the Lagrangian

$$\mathscr{L}=\tfrac{1}{2}m_1|\dot{\mathbf{r}}_1|^2+\tfrac{1}{2}m_2|\dot{\mathbf{r}}_2|^2-V(\mathbf{r}_1-\mathbf{r}_2) \tag{2.3.5}$$

in terms of $\mathbf{r}_{CM}$ and $\mathbf{r}$, one gets

$$\mathscr{L}=\frac{1}{2}(m_1+m_2)|\dot{\mathbf{r}}_{CM}|^2+\frac{1}{2}\frac{m_1m_2}{m_1+m_2}|\dot{\mathbf{r}}|^2-V(\mathbf{r}) \tag{2.3.6}$$

The main features of Eq. (2.3.6) are the following.

(1) The problem of two mutually interacting particles has been transformed to that of two fictitious particles that do not interact with each other. In other words, the equations of motion for $\mathbf{r}$ do not involve $\mathbf{r}_{\text{CM}}$ and vice versa, because $\mathscr{L}(\mathbf{r}, \dot{\mathbf{r}}; \mathbf{r}_{\text{CM}}, \dot{\mathbf{r}}_{\text{CM}}) = \mathscr{L}(\mathbf{r}, \dot{\mathbf{r}}) + \mathscr{L}(\mathbf{r}_{\text{CM}}, \dot{\mathbf{r}}_{\text{CM}})$.

(2) The first fictitious particle is the CM, of mass $M = m_1 + m_2$. Since $\mathbf{r}_{\text{CM}}$ is a cyclic variable, the momentum $\mathbf{p}_{\text{CM}} = M\dot{\mathbf{r}}_{\text{CM}}$ (which is just the total momentum) is conserved as expected. Since the motion of the CM is uninteresting one usually ignores it. One clear way to do this is to go to the CM frame in which $\dot{\mathbf{r}}_{\text{CM}} = 0$, so that the CM is completely eliminated in the Lagrangian.

(3) The second fictitious particle has mass $\mu = m_1 m_2/(m_1 + m_2)$ (called the *reduced mass*), momentum $\mathbf{p} = \mu\dot{\mathbf{r}}$ and moves under a potential $V(\mathbf{r})$. One has just to solve this one-body problem. If one chooses, one may easily return to the coordinates $\mathbf{r}_1$ and $\mathbf{r}_2$ at the end, using Eqs. (2.3.1) and (2.3.2).

*Exercise 2.3.1.** Derive Eq. (2.3.6) from (2.3.5) by changing variables.

2.4. How Smart Is a Particle?

The Lagrangian formalism seems to ascribe to a particle a tremendous amount of foresight: a particle at (x_i, t_i) destined for (x_f, t_f) manages to calculate ahead of time the action for every possible path linking these points, and takes the one with the least action. But this, of course, is an illusion. The particle need not know its entire trajectory ahead of time, it needs only to obey the Euler–Lagrange equations at each instant in time to minimize the action. This in turn means just following Newton's law, which is to say, the particle has to sample the potential in its immediate vicinity and accelerate in the direction of greatest change.

Our esteem for the particle will sink further when we learn quantum mechanics. We will discover that far from following any kind of strategy, the particle, in a sense, goes from (x_i, t_i) to (x_f, t_f) along all possible paths, giving equal weight to each! How it is that despite this, classical particles do seem to follow $x_{\text{cl}}(t)$ is an interesting question that will be answered when we come to the path integral formalism of quantum mechanics.

2.5. The Hamiltonian Formalism

In the Lagrangian formalism, the independent variables are the coordinates q_i and velocities $\dot{q}_i$. The momenta are derived quantities defined by

$$p_i = \frac{\partial \mathscr{L}}{\partial \dot{q}_i} \tag{2.5.1}$$

In the Hamiltonian formalism one exchanges the roles of $\dot{q}$ and p: one replaces the Lagrangian $\mathscr{L}(q, \dot{q})$‡ by a Hamiltonian $\mathscr{H}(q, p)$ which generates the equations of motion, and $\dot{q}$ becomes a derived quantity,

$$\dot{q}_i = \frac{\partial \mathscr{H}}{\partial p_i} \tag{2.5.2}$$

thereby completing the role reversal of the $\dot{q}$'s and the p's.

There exists a standard procedure for effecting such a change, called a *Legendre transformation*, which is illustrated by the following simple example. Suppose we have a function $f(x)$ with

$$u(x) = \frac{df}{dx} \tag{2.5.3}$$

Let it be possible to invert $u(x)$ to get $x(u)$. [For example if $u(x) = x^3$, $x(u) = u^{1/3}$, etc.] If we define a function

$$g(u) = x(u)u - f(x(u)) \tag{2.5.4}$$

then

$$\frac{dg}{du} = \frac{dx}{du} \cdot u + x(u) - \frac{df}{dx} \cdot \frac{dx}{du} = x(u) \tag{2.5.5}$$

That is to say, in going from f to g (or vice versa) we exchange the roles of x and u. One calls Eq. (2.5.4) a *Legendre transformation* and f and g *Legendre transforms* of each other.

More generally, if $f = f(x_1, x_2, \ldots, x_n)$, one can eliminate a subset $\{x_i, i = 1 \text{ to } j\}$ in favor of the partial derivatives $u_i = \partial f/\partial x_i$ by the transformation

$$g(u_1, \ldots, u_j, x_{j+1}, \ldots, x_n) = \sum_{i=1}^{j} u_i x_i - f(x_1, \ldots, x_n) \tag{2.5.6}$$

It is understood in the right-hand side of Eq. (2.5.6) that all the x_i's to be eliminated have been rewritten as functions of the allowed variables in g. It can be easily verified that

$$\frac{\partial g}{\partial u_i} = x_i \tag{2.5.7}$$

where in taking the above partial derivative, one keeps all the other variables in g constant.

‡ We will often refer to $q_1, \ldots, q_n$ as q and $p_1, \ldots, p_n$ as p.

Table 2.1. Comparison of the Lagrangian and Hamiltonian Formalisms

Lagrangian formalism	Hamiltonian formalism
(1) The state of a system with n degrees of freedom is described by n coordinates $(q_1,\ldots,q_n)$ and n velocities $(\dot{q}_1,\ldots,\dot{q}_n)$, or in a more compact notation by $(q,\dot{q})$.	(1) The state of a system with n degrees of freedom is described by n coordinates and n momenta $(q_1,\ldots,q_n;p_1,\ldots,p_n)$ or, more succinctly, by (q,p).
(2) The state of the system may be represented by a point moving with a definite velocity in an *n-dimensional configuration space*.	(2) The state of the system may be represented by a point in a *2n-dimensional phase space*, with coordinates $(q_1,\ldots,q_n;p_1,\ldots,p_n)$.
(3) The n coordinates evolve according to n second-order equations.	(3) The $2n$ coordinates and momenta obey $2n$ first-order equations.
(4) For a given $\mathscr{L}$, several trajectories may pass through a given point in configuration space depending on $\dot{q}$.	(4) For a given $\mathscr{H}$ only one trajectory passes through a given point in phase space.

Applying these methods to the problem in question, we define

$$\mathscr{H}(q,p)=\sum_{i=1}^{n} p_i\dot{q}_i-\mathscr{L}(q,\dot{q}) \tag{2.5.8}$$

where the $\dot{q}$'s are to be written as functions of q's and p's. This inversion is generally easy since $\mathscr{L}$ is a polynomial of rank 2 in $\dot{q}$, and $p_i=\partial\mathscr{L}/\partial\dot{q}_i$ is a polynomial of rank 1 in the $\dot{q}$'s, e.g., Eq. (2.2.7). Consider now

$$\frac{\partial\mathscr{H}}{\partial p_i}=\frac{\partial}{\partial p_i}\left(\sum_j p_j\dot{q}_j-\mathscr{L}\right) \tag{2.5.9}$$

$$=\dot{q}_i+\sum_j p_j\frac{\partial\dot{q}_j}{\partial p_i}-\sum_j\frac{\partial\mathscr{L}}{\partial\dot{q}_j}\frac{\partial\dot{q}_j}{\partial p_i}$$

$$=\dot{q}_i\quad\left(\text{since } p_j=\frac{\partial\mathscr{L}}{\partial\dot{q}_j}\right) \tag{2.5.10}$$

[There are no $(\partial\mathscr{L}/\partial q_j)(\partial q_j/\partial p_i)$ terms since q is held constant in $\partial\mathscr{H}/\partial p_i$; that is, q and p are independent variables.] Similarly,

$$\frac{\partial\mathscr{H}}{\partial q_i}=\sum_j p_j\frac{\partial\dot{q}_j}{\partial q_i}-\frac{\partial\mathscr{L}}{\partial q_i}-\sum_j\frac{\partial\mathscr{L}}{\partial\dot{q}_j}\frac{\partial\dot{q}_j}{\partial q_i}=-\frac{\partial\mathscr{L}}{\partial q_i} \tag{2.5.11}$$

We now feed in the dynamics by replacing $(\partial\mathscr{L}/\partial q_i)$ by $\dot{p}_i$, and obtain *Hamilton's canonical equations*:

$$\frac{\partial\mathscr{H}}{\partial p_i}=\dot{q}_i,\qquad -\frac{\partial\mathscr{H}}{\partial q_i}=\dot{p}_i \tag{2.5.12}$$

Note that we have altogether $2n$ first-order equations (in time) for a system with n degrees of freedom. Given the initial-value data, $(q_i(0),p_i(0))$, $i=1,\ldots,n$, we can integrate the equations to get $(q_i(t),p_i(t))$.

Table 2.1 provides a comparison of the Lagrangian and Hamiltonian formalisms.

Now, just as $\mathscr{L}$ may be interpreted as $T-V$ if the force is conservative, so there exists a simple interpretation for $\mathscr{H}$ in this case. Consider the sum $\sum_i p_i \dot{q}_i$. Let us use Cartesian coordinates, in terms of which

$$T=\sum_{i=1}^{n} \tfrac{1}{2} m_i \dot{x}_i^2$$

$$p_i=\frac{\partial \mathscr{L}}{\partial \dot{x}_i}=\frac{\partial T}{\partial \dot{x}_i}=m_i \dot{x}_i$$

and

$$\sum_{i=1}^{n} p_i \dot{x}_i=\sum_{i=1}^{n} m_i \dot{x}_i^2=2T \tag{2.5.13}$$

so that

$$\mathscr{H}=\sum_i p_i \dot{x}_i-\mathscr{L}=T+V \tag{2.5.14}$$

the total energy. Notice that although we used Cartesian coordinates along the way, the resulting equation (2.5.14) is a relation among scalars and thus coordinate independent.

Exercise 2.5.1. Show that if $T=\sum_i \sum_j T_{ij}(q) \dot{q}_i \dot{q}_j$, where $\dot{q}$'s are generalized velocities, $\sum_i p_i \dot{q}_i=2T$.

The Hamiltonian method is illustrated by the simple example of a harmonic oscillator, for which

$$\mathscr{L}=\tfrac{1}{2} m \dot{x}^2-\tfrac{1}{2} k x^2$$

The canonical momentum is

$$p=\frac{\partial \mathscr{L}}{\partial \dot{x}}=m \dot{x}$$

It is easy to invert this relation to obtain $\dot{x}$ as a function of p:

$$\dot{x}=p/m$$

and obtain

$$\mathscr{H}(x,p)=T+V=\tfrac{1}{2}m[\dot{x}(p)]^2+\tfrac{1}{2}kx^2$$

$$=\frac{p^2}{2m}+\frac{1}{2}kx^2 \tag{2.5.15}$$

The equations of motion are

$$\frac{\partial\mathscr{H}}{\partial p}=\dot{q}\rightarrow\frac{p}{m}=\dot{x} \tag{2.5.16}$$

$$-\frac{\partial\mathscr{H}}{\partial q}=\dot{p}\rightarrow -kx=\dot{p} \tag{2.5.17}$$

These equations can be integrated in time, given the initial q and p. If, however, we want the familiar second-order equation, we differentiate Eq. (2.5.16) with respect to time, and feed it into Eq. (2.5.17) to get

$$m\ddot{x}+kx=0$$

Exercise 2.5.2. Using the conservation of energy, show that the trajectories in phase space for the oscillator are ellipses of the form $(x/a)^2+(p/b)^2=1$, where $a^2=2E/k$ and $b^2=2mE$.

Exercise 2.5.3. Solve Exercise 2.1.2 using the Hamiltonian formalism.

*Exercise 2.5.4.** Show that $\mathscr{H}$ corresponding to $\mathscr{L}$ in Eq. (2.3.6) is $\mathscr{H}=|\mathbf{p}_{\rm CM}|^2/2M+|\mathbf{p}|^2/2\mu+V(\mathbf{r})$, where M is the total mass, μ is the reduced mass, $\mathbf{p}_{\rm CM}$ and $\mathbf{p}$ are the momenta conjugate to $\mathbf{r}_{\rm CM}$ and $\mathbf{r}$, respectively.

2.6. The Electromagnetic Force in the Hamiltonian Scheme

The passage from $\mathscr{L}_{e\cdot m}$ to its Legendre transform $\mathscr{H}_{e\cdot m}$ is not sensitive in any way to the velocity-dependent nature of the force. If $\mathscr{L}_{e\cdot m}$ generated the correct force laws, so will $\mathscr{H}_{e\cdot m}$, the dynamical content of the schemes being identical. In contrast, the velocity independence of the force was assumed in showing that the numerical value of $\mathscr{H}$ is $T+V$, the total energy. Let us therefore repeat the analysis for the electromagnetic case. As

$$\mathscr{L}_{e\cdot m}=\tfrac{1}{2}m\mathbf{v}\cdot\mathbf{v}-q\phi+\frac{q}{c}\mathbf{v}\cdot\mathbf{A}$$

and‡

‡ Note that in this discussion, q is the charge and not the coordinate. The (Cartesian) coordinate $\mathbf{r}$ is hidden in the functions $\mathbf{A}(\mathbf{r},t)$ and $\phi(\mathbf{r},t)$.

$$\mathbf{p} = m\mathbf{v} + \frac{q\mathbf{A}}{c}$$

we have

$$\begin{aligned}
\mathscr{H}_{e\cdot m} &= \mathbf{p}\cdot\mathbf{v} - \mathscr{L}_{e\cdot m} \\
&= m\mathbf{v}\cdot\mathbf{v} + q\frac{\mathbf{v}\cdot\mathbf{A}}{c} - \frac{1}{2}m\mathbf{v}\cdot\mathbf{v} + q\phi - \frac{q\mathbf{v}\cdot\mathbf{A}}{c} \\
&= \tfrac{1}{2}m\mathbf{v}\cdot\mathbf{v} + q\phi = T + q\phi
\end{aligned} \tag{2.6.1}$$

Now, there is something very disturbing about Eq. (2.6.1): the vector potential $\mathbf{A}$ seems to have dropped out along the way. How is $\mathscr{H}_{e\cdot m}$ to generate the correct dynamics without knowing what $\mathbf{A}$ is? The answer is, of course, the $\mathscr{H}$ is more than just $T+q\phi$; it is $T+q\phi$ written in terms of the correct variables, in particular, in terms of $\mathbf{p}$ and not $\mathbf{v}$. Making the change of variables, we get

$$\mathscr{H}_{e\cdot m} = \frac{|(\mathbf{p} - q\mathbf{A}/c)|^2}{2m} + q\phi \tag{2.6.2}$$

with the vector potential very much in the picture.

2.7. Cyclic Coordinates, Poisson Brackets, and Canonical Transformations

Cyclic coordinates are defined here just as in the Lagrangian case and have the same significance: if a coordinate q_i is missing in $\mathscr{H}$, then

$$\dot{p}_i = -\frac{\partial\mathscr{H}}{\partial q_i} = 0 \tag{2.7.1}$$

Now, there will be other quantities, such as the energy, that may be conserved in addition to the canonical momenta.§ There exists a nice method of characterizing these in the Hamiltonian formalism. Let $\omega(p, q)$ be some function of the state variables, *with no explicit dependence on t*. Its time variation is given by

$$\begin{aligned}
\frac{d\omega}{dt} &= \sum_i \left(\frac{\partial\omega}{\partial q_i}\dot{q}_i + \frac{\partial\omega}{\partial p_i}\dot{p}_i\right) \\
&= \sum_i \left(\frac{\partial\omega}{\partial q_i}\frac{\partial\mathscr{H}}{\partial p_i} - \frac{\partial\omega}{\partial p_i}\frac{\partial\mathscr{H}}{\partial q_i}\right) \\
&\equiv \{\omega, \mathscr{H}\}
\end{aligned} \tag{2.7.2}$$

§ Another example is the conservation of $l_z = xp_y - yp_x$ when $V(x, y) = V(x^2+y^2)$. There are no cyclic coordinates here. Of course, if we work in polar coordinates, $V(\rho, \phi) = V(\rho)$, and $p_\phi = m\rho^2\dot{\phi} = l_z$ is conserved because it is the momentum conjugate to the cyclic coordinate ϕ.

where we have defined the Poisson bracket (PB) between two variables $\omega(p, q)$ and $\lambda(p, q)$ to be

$$\{\omega, \lambda\} \equiv \sum_i \left(\frac{\partial \omega}{\partial q_i} \frac{\partial \lambda}{\partial p_i} - \frac{\partial \omega}{\partial p_i} \frac{\partial \lambda}{\partial q_i} \right) \tag{2.7.3}$$

It follows from Eq. (2.7.2) that any variable whose PB with $\mathscr{H}$ vanishes is constant in time, i.e., conserved. In particular $\mathscr{H}$ itself is a constant of motion (identified as the total energy) if it has no explicit t dependence.

*Exercise 2.7.1.** Show that

$$\{\omega, \lambda\} = -\{\lambda, \omega\}$$

$$\{\omega, \lambda + \sigma\} = \{\omega, \lambda\} + \{\omega, \sigma\}$$

$$\{\omega, \lambda\sigma\} = \{\omega, \lambda\}\sigma + \lambda\{\omega, \sigma\}$$

Note the similarity between the above and Eqs. (1.5.10) and (1.5.11) for commutators.

Of fundamental importance are the PB between the q's and the p's. Observe that

$$\{q_i, q_j\} = \{p_i, p_j\} = 0 \tag{2.7.4a}$$

$$\{q_i, p_j\} = \delta_{ij} \tag{2.7.4b}$$

since $(q_i, \ldots, p_n)$ are independent variables ($\partial q_i/\partial q_j = \delta_{ij}$, $\partial q_i/\partial p_k = 0$, etc.). Hamilton's equations may be written in terms of PB as

$$\dot{q}_i = \{q_i, \mathscr{H}\} \tag{2.7.5a}$$

$$\dot{p}_i = \{p_i, \mathscr{H}\} \tag{2.7.5b}$$

by setting $\omega = q_i$ or p_i in Eq. (2.7.2).

*Exercise 2.7.2.** (i) Verify Eqs. (2.7.4) and (2.7.5). (ii) Consider a problem in two dimensions given by $\mathscr{H} = p_x^2 + p_y^2 + ax^2 + by^2$. Argue that if $a = b$, $\{l_z, \mathscr{H}\}$ must vanish. Verify by explicit computation.

Canonical Transformations

We have seen that the Euler–Lagrange equations are form invariant under an arbitrary‡ change of coordinates in configuration space

$$q_i \to \bar{q}_i(q_1, \ldots, q_n), \qquad i = 1, \ldots, n \tag{2.7.6a}$$

‡ We assume the transformation is invertible, so we may write q in terms of $\bar{q}$: $q = q(\bar{q})$. The transformation may also depend on time explicitly [$\bar{q} = q(q, t)$], but we do not consider such cases.

or more succinctly

$$q \to \bar{q}(q) \tag{2.7.6b}$$

The response of the velocities to this transformation follows from Eq. (2.7.6a):

$$\dot{\bar{q}}_i = \dot{\bar{q}}_i = \frac{d\bar{q}_i}{dt} = \sum_j \left(\frac{\partial \bar{q}_i}{\partial q_j}\right) \dot{q}_j \tag{2.7.7}$$

The response of the canonical momenta may be found by rewriting $\mathscr{L}$ in terms of $(\bar{q}, \dot{\bar{q}})$ and taking the derivative with respect to $\dot{\bar{q}}$:

$$\bar{p}_i = \frac{\partial \mathscr{L}(\bar{q}, \dot{\bar{q}})}{\partial \dot{\bar{q}}_i} \tag{2.7.8}$$

The result is (Exercise 2.7.8):

$$\bar{p}_i = \sum_j \left(\frac{\partial q_j}{\partial \bar{q}_i}\right) p_j \tag{2.7.9}$$

Notice that although $\mathscr{L}$ enters Eq. (2.7.8), it drops out in Eq. (2.7.9), which connects $\bar{p}$ to the old variables. This is as it should be, for we expect that the response of the momenta to a coordinate transformation (say, a rotation) is a purely kinematical question.

A word of explanation about $\mathscr{L}(\bar{q}, \dot{\bar{q}})$. By $\mathscr{L}(\bar{q}, \dot{\bar{q}})$ we mean the Lagrangian (say $T-V$, for definiteness) written in terms of $\bar{q}$ and $\dot{\bar{q}}$. Thus the numerical value of the Lagrangian is unchanged under $(q, \dot{q}) \to (\bar{q}, \dot{\bar{q}})$; for $(q, \dot{q})$ and $(\bar{q}, \dot{\bar{q}})$ refer to the *same physical state*. The functional form of the Lagrangian, however, *does* change and so we should really be using two different symbols $\mathscr{L}(q, \dot{q})$ and $\bar{\mathscr{L}}(\bar{q}, \dot{\bar{q}})$. Nonetheless we follow the convention of denoting a given dynamical variable, such as the Lagrangian, by a fixed symbol in all coordinate systems.

The invariance of the Euler–Lagrange equations under $(q, \dot{q}) \to (\bar{q}, \dot{\bar{q}})$ implies the invariance of Hamilton's equation under $(q, p) \to (\bar{q}, \bar{p})$, i.e., $(\bar{q}, \bar{p})$ obey

$$\dot{\bar{q}}_i = \partial \mathscr{H}/\partial \bar{p}_i, \qquad \dot{\bar{p}}_i = -(\partial \mathscr{H}/\partial \bar{q}_i) \tag{2.7.10}$$

where $\mathscr{H} = \mathscr{H}(\bar{q}, \bar{p})$ is the Hamiltonian written in terms of $\bar{q}$ and $\bar{p}$. The proof is simple: we start with $\mathscr{L}(\bar{q}, \dot{\bar{q}})$, perform a Legendre transform, and use the fact that $\bar{q}$ obeys Euler–Lagrange equations.

The transformation

$$q_i \to \bar{q}_i(q_1, \ldots, q_n), \qquad \bar{p}_i = \sum_j \left(\frac{\partial q_j}{\partial \bar{q}_i}\right) p_j \tag{2.7.11}$$

is called a *point transformation*. If we view the Hamiltonian formalism as something derived from the Lagrangian scheme, which is formulated in n-dimensional configuration space, this is the most general (time-independent) transformation which preserves the form of Hamilton's equations (that we can think of). On the other hand, if we view the Hamiltonian formalism in its own right, the backdrop is the $2n$-dimensional phase space. In this space, the point transformation is unnecessarily restrictive. One can contemplate a more general transformation of phase space coordinates:

$$\begin{aligned} q &\to \bar{q}(q,p) \\ p &\to \bar{p}(q,p) \end{aligned} \tag{2.7.12}$$

Although all sets of $2n$ independent coordinates $(\bar{q},\bar{p})$ are formally adequate for describing the state of the system, not all of them will preserve the canonical form of Hamilton's equations. (This is like saying that although Newton's laws may be written in terms of any complete set of coordinates, the simple form $m\ddot{q}_i=-\partial V/\partial q_i$ is valid only if the q_i are Cartesian). If, however, $(\bar{q},\bar{p})$ obey the canonical equations (2.7.10), we say that they are *canonical coordinates* and that Eq. (2.7.12) defines a *canonical transformation*. Any set of coordinates $(q_1,\ldots,q_n)$, and the corresponding momenta generated in the Lagrangian formalism $(p_i=\partial\mathscr{L}/\partial\dot{q}_i)$, are canonical coordinates. Given one set, (q,p), we can get another, $(\bar{q},\bar{p})$, by the point transformation, which is a special case of the canonical transformation. This does not, however, exhaust the possibilities. Let us now ask the following question. *Given a new set of coordinates* $(\bar{q}(q,p),\bar{p}(q,p))$, *how can we tell if they are canonical* [*assuming* (q,p) *are*]? Now it is true for any $\omega(q,p)$ that

$$\dot{\omega}=\{\omega,\mathscr{H}\}=\sum_i\left(\frac{\partial\omega}{\partial q_i}\frac{\partial\mathscr{H}}{\partial p_i}-\frac{\partial\omega}{\partial p_i}\frac{\partial\mathscr{H}}{\partial q_i}\right) \tag{2.7.13}$$

Applying this to $\bar{q}_j(q,p)$ we find

$$\dot{\bar{q}}_j=\sum_i\left(\frac{\partial\bar{q}_j}{\partial q_i}\frac{\partial\mathscr{H}}{\partial p_i}-\frac{\partial\bar{q}_j}{\partial p_i}\frac{\partial\mathscr{H}}{\partial q_i}\right) \tag{2.7.14}$$

If we view $\mathscr{H}$ as a function of $(\bar{q},\bar{p})$ and use the chain rule, we get

$$\frac{\partial\mathscr{H}(q,p)}{\partial p_i}=\frac{\partial\mathscr{H}(\bar{q},\bar{p})}{\partial p_i}=\sum_k\left(\frac{\partial\mathscr{H}}{\partial\bar{q}_k}\frac{\partial\bar{q}_k}{\partial p_i}+\frac{\partial\mathscr{H}}{\partial\bar{p}_k}\frac{\partial\bar{p}_k}{\partial p_i}\right) \tag{2.7.15a}$$

and

$$\frac{\partial\mathscr{H}(q,p)}{\partial q_i}=\frac{\partial\mathscr{H}(\bar{q},\bar{p})}{\partial q_i}=\sum_k\left(\frac{\partial\mathscr{H}}{\partial\bar{q}_k}\frac{\partial\bar{q}_k}{\partial q_i}+\frac{\partial\mathscr{H}}{\partial\bar{p}_k}\frac{\partial\bar{p}_k}{\partial q_i}\right) \tag{2.7.15b}$$

Feeding all this into Eq. (2.7.14) we find, upon regrouping terms,

$$\dot{\bar{q}}_j = \sum_k \left(\frac{\partial \mathscr{H}}{\partial \bar{q}_k} \{\bar{q}_j, \bar{q}_k\} + \frac{\partial \mathscr{H}}{\partial \bar{p}_k} \{\bar{q}_j, \bar{p}_k\} \right) \tag{2.7.16}$$

It can similarly be established that

$$\dot{\bar{p}}_j = \sum_k \left(\frac{\partial \mathscr{H}}{\partial \bar{q}_k} \{\bar{p}_j, \bar{q}_k\} + \frac{\partial \mathscr{H}}{\partial \bar{p}_k} \{\bar{p}_j, \bar{p}_k\} \right) \tag{2.7.17}$$

If Eqs. (2.7.16) and (2.7.17) are to reduce to the canonical equations (2.7.10) for any $\mathscr{H}(q, p)$, we must have

$$\begin{gathered} \{\bar{q}_j, \bar{q}_k\} = 0 = \{\bar{p}_j, \bar{p}_k\} \\ \{\bar{q}_j, \bar{p}_k\} = \delta_{jk} \end{gathered} \tag{2.7.18}$$

These then are the conditions to be satisfied by the new variables if they are to be canonical. Notice that these constraints make no reference to the specific functional form of $\mathscr{H}$: the equations defining canonical variables are purely kinematical and true for any $\mathscr{H}(q, p)$.

Exercise 2.7.3. Fill in the missing steps leading to Eq. (2.7.18) starting from Eq. (2.7.14).

Exercise 2.7.4. Verify that the change to a rotated frame

$$\bar{x} = x \cos\theta - y \sin\theta$$

$$\bar{y} = x \sin\theta + y \cos\theta$$

$$\bar{p}_x = p_x \cos\theta - p_y \sin\theta$$

$$\bar{p}_y = p_x \sin\theta + p_y \cos\theta$$

is a canonical transformation.

Exercise 2.7.5. Show that the polar variables $\rho = (x^2+y^2)^{1/2}$, $\phi = \tan^{-1}(y/x)$,

$$p_\rho = \hat{e}_\rho \cdot \mathbf{p} = \frac{xp_x + yp_y}{(x^2+y^2)^{1/2}}, \qquad p_\phi = xp_y - yp_x (= l_z)$$

are canonical. ($\hat{e}_\rho$ is the unit vector in the radial direction.)

*Exercise 2.7.6.** Verify that the change from the variables $\mathbf{r}_1, \mathbf{r}_2, \mathbf{p}_1, \mathbf{p}_2$ to $\mathbf{r}_{CM}, \mathbf{p}_{CM}, \mathbf{r}$, and $\mathbf{p}$ is a canonical transformation. (See Exercise 2.5.4).

Exercise 2.7.7. Verify that

$$\bar{q} = \ln(q^{-1} \sin p)$$

$$\bar{p} = q \cot p$$

is a canonical transformation.

Exercise 2.7.8. We would like to derive here Eq. (2.7.9), which gives the transformation of the momenta under a coordinate transformation in configuration space:

$$q_i \to \bar{q}_i(q_1, \ldots, q_n)$$

(1) Argue that if we invert the above equation to get $q = q(\bar{q})$, we can derive the following counterpart of Eq. (2.7.7):

$$\dot{q}_i = \sum_j \frac{\partial q_i}{\partial \bar{q}_j} \dot{\bar{q}}_j$$

(2) Show from the above that

$$\left(\frac{\partial \dot{q}_i}{\partial \dot{\bar{q}}_j}\right)_{\bar{q}} = \frac{\partial q_i}{\partial \bar{q}_j}$$

(3) Now calculate

$$\bar{p}_i = \left[\frac{\partial \mathscr{L}(\bar{q}, \dot{\bar{q}})}{\partial \dot{\bar{q}}_i}\right]_{\bar{q}} = \left[\frac{\partial \mathscr{L}(q, \dot{q})}{\partial \dot{\bar{q}}_i}\right]_{\bar{q}}$$

Use the chain rule and the fact that $q = q(\bar{q})$ and not $q(\bar{q}, \dot{\bar{q}})$ to derive Eq. (2.7.9).

(4) Verify, by calculating the PB in Eq. (2.7.18), that the point transformation is canonical.

If (q, p) and $(\bar{q}, \bar{p})$ are both canonical, we must give them both the same status, for Hamilton's equations have the same appearance when expressed in terms of either set. Now, we have defined the PB of two variables ω and σ in terms of (q, p) as

$$\{\omega, \sigma\} = \sum_i \left(\frac{\partial \omega}{\partial q_i}\frac{\partial \sigma}{\partial p_i} - \frac{\partial \omega}{\partial p_i}\frac{\partial \sigma}{\partial q_i}\right) \equiv \{\omega, \sigma\}_{q,p}$$

Should we not also define a PB, $\{\omega, \sigma\}_{\bar{q},\bar{p}}$ for every canonical pair $(\bar{q}, \bar{p})$? Fortunately it turns out that *the PB are invariant under canonical transformations*:

$$\{\omega, \sigma\}_{q,p} = \{\omega, \sigma\}_{\bar{q},\bar{p}} \tag{2.7.19}$$

(It is understood that ω and σ are written as functions of $\bar{q}$ and $\bar{p}$ on the right-hand side.)

Exercise 2.7.9. Verify Eq. (2.7.19) by direct computation. Use the chain rule to go from q, p derivatives to $\bar{q}, \bar{p}$ derivatives. Collect terms that represent PB of the latter.

Besides the proof by direct computation (as per Exercise 2.7.9 above) there is an alternate way to establish Eq. (2.7.19).

Consider first $\sigma=\mathscr{H}$. We know that since (q, p) obey canonical equations,

$$\dot{\omega}=\{\omega, \mathscr{H}\}_{q,p}$$

But then $(\bar{q}, \bar{p})$ also obey canonical equations, so

$$\dot{\omega}=\{\omega, \mathscr{H}\}_{\bar{q},\bar{p}}$$

Now ω is some physical quantity such as the kinetic energy or the component of angular momentum in some fixed direction, so its rate of change is independent of the phase space coordinates used, i.e., $\dot{\omega}$ is $\dot{\omega}$, whether $\omega=\omega(q, p)$ or $\omega(\bar{q}, \bar{p})$. So

$$\{\omega, \mathscr{H}\}_{q,p}=\{\omega, \mathscr{H}\}_{\bar{q},\bar{p}} \tag{2.7.20}$$

Having proved the result for what seems to be the special case $\sigma=\mathscr{H}$, we now pull the following trick. Note that nowhere in the derivation did we have to assume that $\mathscr{H}$ was any particular function of q and p. In fact, Hamiltonian dynamics, as a consistent mathematical scheme, places no restriction on $\mathscr{H}$. It is the physical requirement that the time evolution generated by $\mathscr{H}$ coincide with what is *actually* observed, that restricts $\mathscr{H}$ to be $T+V$. Thus $\mathscr{H}$ could have been any function at all in the preceding argument and in the result Eq. (2.7.20) (which is just a relation among partial derivatives.) If we understand that $\mathscr{H}$ is not $T+V$ *in this argument* but an arbitrary function, call it σ, we get the desired result.

Active Transformations

So far, we have viewed the transformation

$$\bar{q}=\bar{q}(q, p)$$

$$\bar{p}=\bar{p}(q, p)$$

as passive: both (q, p) and $(\bar{q}, \bar{p})$ refer to the same point in phase space described in two different coordinate systems. Under the transformation $(q, p)\to(\bar{q}, \bar{p})$, the numerical values of all dynamical variables are unchanged (for we are talking about the same physical state), but their functional form is changed. For instance, under a change from Cartesian to spherical coordinates, $\omega(x, y, z)=x^2+y^2+z^2\to\omega(r, \theta, \phi)=r^2$. As mentioned earlier, we use the same symbol for a given variable even if its functional dependence on the coordinates changes when we change coordinates.

Consider now a restricted class of transformations, called *regular transformations*, which preserve the range of the variables: (q, p) and $(\bar{q}, \bar{p})$ have the same range. A change from one Cartesian coordinate to a translated or rotated one is regular (each variable goes from $-\infty$ to $+\infty$ before and after), whereas a change to spherical coordinates (where some coordinates are nonnegative, some are bounded by 2π, etc.) is not.

A regular transformation $(q, p) \to (\bar{q}, \bar{p})$ permits an alternate interpretation: instead of viewing $(\bar{q}, \bar{p})$ as the same phase space point in a new coordinate system, we may view it as a new point in the same coordinate system. This corresponds to an active transformation which changes the state of the system. Under this change, the numerical value of any dynamical variable $\omega(q, p)$ will generally change: $\omega(q, p) \neq \omega(\bar{q}, \bar{p})$, though its functional dependence will not: $\omega(\bar{q}, \bar{p})$ is the same function $\omega(q, p)$ evaluated at the new point $(q = \bar{q}, p = \bar{p})$.

We say that ω is *invariant* under the regular transformation $(q, p) \to (\bar{q}, \bar{p})$ if

$$\omega(q, p) = \omega(\bar{q}, \bar{p}) \tag{2.7.21}$$

(This equation has content only if we are talking about the active transformations, for it is true for any ω under a passive transformation.)

Whether we view the transformation $(q, p) \to (\bar{q}, \bar{p})$ as active or passive, it is called canonical if $(\bar{q}, \bar{p})$ obey Eq. (2.7.18). As we shall see, only regular *canonical* transformations are physically interesting.

2.8. Symmetries and Their Consequences

Let us begin our discussion by examining what the word "symmetry" means in daily usage. We say that a sphere is a very symmetric object because it looks the same when seen from many directions. Or, equivalently, a sphere looks the same before and after it is subjected to a rotation around *any* axis passing through its center. A cylinder has symmetry too, but not as much: the rotation must be performed around its axis. Generally then, the symmetry of an object implies its invariance under some transformations, which in our examples are rotations.

A symmetry can be discrete or continuous, as illustrated by the example of a hexagon and a circle. While the rotation angles that leave a hexagon unchanged form a discrete set, namely, multiples of 60°, the corresponding set for a circle is a continuum. We may characterize the continuous symmetry of the circle in another way. Consider the *identity transformation*, which does nothing, i.e., rotates by 0° in our example. This leaves both the circle and the hexagon invariant. Consider next an *infinitesimal transformation*, which is infinitesimally "close" to the identity; in our example this is a rotation by an infinitesimal angle ε. The infinitesimal rotation leaves the circle invariant but not the hexagon. The circle is thus characterized by its invariance under infinitesimal rotations. Given this property, its invariance under finite rotations follows, for any finite rotation may be viewed as a sequence of infinitesimal rotations (each of which leaves it invariant).

It is also possible to think of functions of some variables as being symmetric in the sense that if one changes the values of the variables in a certain way, the value of the function is invariant. Consider for example

$$f(x, y) = x^2 + y^2$$

If we make the following change

$$\begin{aligned} x &\to \bar{x} = x\cos\theta - y\sin\theta \\ y &\to \bar{y} = x\sin\theta + y\cos\theta \end{aligned} \tag{2.8.1}$$

in the arguments, we find that f is invariant. We say that f is symmetric under the above transformation. In the terminology introduced earlier, the transformation in question is continuous: its infinitesimal version is

$$\begin{aligned} x &\to \bar{x} = x\cos\varepsilon - y\sin\varepsilon = x - y\varepsilon \\ y &\to \bar{y} = x\sin\varepsilon + y\cos\varepsilon = x\varepsilon + y \qquad (\textit{to order } \varepsilon) \end{aligned} \tag{2.8.2}$$

Consider now the function $\mathscr{H}(q, p)$. There are two important dynamical consequences that follow from its invariance under *regular canonical* transformations.

I. If $\mathscr{H}$ is invariant under the following *infinitesimal* transformation (which you may verify is canonical, Exercise 2.8.2),

$$\begin{aligned} q_i &\to \bar{q}_i = q_i + \varepsilon\frac{\partial g}{\partial p_i} \equiv q_i + \delta q_i \\ p_i &\to \bar{p}_i = p_i - \varepsilon\frac{\partial g}{\partial q_i} \equiv p_i + \delta p_i \end{aligned} \tag{2.8.3}$$

where $g(q, p)$ is any dynamical variable, *then g is conserved*, i.e., a constant of motion. One calls g the *generator of the transformation.*

II. If $\mathscr{H}$ is invariant under the regular, canonical, but not necessarily infinitesimal, transformation $(q, p) \to (\bar{q}, \bar{p})$, and if $(q(t), p(t))$ is a solution to the equations of motion, so is the transformed (translated, rotated, etc.) trajectory, $(\bar{q}(t), \bar{p}(t))$.

Let us now analyze these two consequences.

Consequence I. Let us first verify that g is indeed conserved if $\mathscr{H}$ is invariant under the transformation it generates. Working to first order in ε, if we equate the change in $\mathscr{H}$ under the change of its arguments to zero, we get

$$\delta\mathscr{H} = \sum_i \frac{\partial\mathscr{H}}{\partial q_i}\left(\varepsilon\frac{\partial g}{\partial p_i}\right) + \frac{\partial\mathscr{H}}{\partial p_i}\left(-\varepsilon\frac{\partial g}{\partial q_i}\right) = \varepsilon\{\mathscr{H}, g\} = 0 \tag{2.8.4}$$

But according to Eq. (2.7.2),

$$\{g, \mathscr{H}\} = 0 \to g \text{ is conserved} \tag{2.8.5}$$

(More generally, the response of any variable ω to the transformation is

$$\delta\omega = \varepsilon\{\omega, g\} \tag{2.8.6}$$

Note that δp and δq in Eq. (2.8.3) may also be written as PBs.) Consider as an example, a particle in one dimension and the case $g=p$. From Eq. (2.8.3),

$$\delta x=\varepsilon\frac{\partial p}{\partial p}=\varepsilon$$
$$\delta p=-\varepsilon\frac{\partial p}{\partial x}=0 \tag{2.8.7}$$

which we recognize to be an infinitesimal translation. Thus the linear momentum p is the generator of spatial translations and is conserved in a translationally invariant problem. The physics behind this result is clear. Since p is unchanged in a translation, so is $T=p^2/2m$. Consequently $V(x+\varepsilon)=V(x)$. But if the potential doesn't vary from point to point, there is no force and p is conserved.

Next consider an example from two dimensions with $g=l_z=xp_y-yp_x$. Here,

$$\delta x=-y\varepsilon\left(=\varepsilon\frac{\partial l_z}{\partial p_x}\right)$$
$$\delta y=x\varepsilon\left(=\varepsilon\frac{\partial l_z}{\partial p_y}\right)$$
$$\delta p_x=-p_y\varepsilon\left(=-\varepsilon\frac{\partial l_z}{\partial x}\right) \tag{2.8.8}$$
$$\delta p_y=p_x\varepsilon\left(=-\varepsilon\frac{\partial l_z}{\partial y}\right)$$

which we recognize to be an infinitesimal rotation around the z axis, [Eq. (2.8.2)]. Thus the angular momentum around the z axis is the generator of rotations around that axis, and is conserved if $\mathscr{H}$ is invariant under rotations of the state around that axis. The relation between the symmetry and the conservation law may be understood in the following familiar terms. Under the rotation of the coordinates and the momenta, $|\mathbf{p}|$ doesn't change and so neither does $T=|\mathbf{p}|^2/2m$. Consequently, V is a constant as we go along any circle centered at the origin. This in turn means that there is no force in the tangential direction and so no torque around the z axis. The conservation of l_z then follows.

Exercise 2.8.1. Show that $p=p_1+p_2$, the total momentum, is the generator of infintesimal translations for a two-particle system.

*Exercise 2.8.2.** Verify that the infinitesimal transformation generated by any dynamical variable g is a canonical transformation. (Hint: Work, as usual, to first order in ε.)

Exercise 2.8.3. Consider

$$\mathscr{H}=\frac{p_x^2+p_y^2}{2m}+\frac{1}{2}m\omega^2(x^2+y^2)$$

whose invariance under the rotation of the coordinates *and* momenta leads to the conservation of l_z. But $\mathscr{H}$ is also invariant under the rotation of *just the coordinates*. Verify that this is a *noncanonical* transformation. Convince yourself that in this case it is not possible to write $\delta\mathscr{H}$ as $\varepsilon\{\mathscr{H}, g\}$ for any g, i.e., that no conservation law follows.

*Exercise 2.8.4.** Consider $\mathscr{H}=\frac{1}{2}p^2+\frac{1}{2}x^2$, which is invariant under infinitesimal rotations in *phase space* (the x–p plane). Find the generator of this transformation (after verifying that it is canonical). (You could have guessed the answer based on Exercise 2.5.2.).

The preceding analysis yields, as a by-product, a way to generate infinitesimal canonical transformations. We take any function $g(q, p)$ and obtain the transformation given by Eq. (2.8.6). (Recall that although we defined a canonical transformation earlier, until now we had no means of generating one.) Given an infinitesimal canonical transformation, we can get a finite one by "integrating" it. The following examples should convince you that this is possible. Consider the transformation generated by $g=\mathscr{H}$. We have

$$\begin{aligned}\delta q_i&=\varepsilon\{q_i, \mathscr{H}\}\\ \delta p_i&=\varepsilon\{p_i, \mathscr{H}\}\end{aligned}\tag{2.8.9}$$

But we know from the equations of motion that $\dot{q}_i=\{q_i, \mathscr{H}\}$ etc. So

$$\begin{aligned}\delta q_i&=\varepsilon\dot{q}_i\\ \delta p_i&=\varepsilon\dot{p}_i\end{aligned}\tag{2.8.10}$$

Thus the new point in phase space $(\bar{q}, \bar{p})=(q+\delta q, p+\delta p)$ obtained by this canonical transformation of (q, p) is just the point to which (q, p) would move in an infinitesimal time interval ε. In other words, the motion of points in phase space under the time evolution generated by $\mathscr{H}$ is an active canonical transformation. Now, you know that by integrating the equations of motion, we can find $(\bar{q}, \bar{p})$ at any future time, i.e., get the finite canonical transformation. Consider now a general case of $g\neq\mathscr{H}$. We still have

$$\begin{aligned}\delta q_i&=\varepsilon\{q_i, g\}\\ \delta p_i&=\varepsilon\{p_i, g\}\end{aligned}\tag{2.8.11}$$

Mathematically, these equations are identical to Eq. (2.8.9), with g playing the role of the Hamiltonian. Clearly there should be no problem integrating these equations for the evolution of the phase space points under the "fake" Hamiltonian g, and fake "time" ε. Let us consider for instance the case $g=l_z$ which has units erg sec and the corresponding fake time $\varepsilon=\delta\theta$, an angle. The transformation of the coordinates is

$$\begin{aligned}\delta x&=\varepsilon\{x, l_z\}=-\varepsilon y\equiv(-\delta\theta)y\\ \delta y&=(\delta\theta)x\end{aligned}\tag{2.8.12}$$

The fake equations of motion are

$$\frac{dx}{d\theta}=-y, \qquad \frac{dy}{d\theta}=x \tag{2.8.13}$$

Differentiating first with respect to θ, and using the second, we get

$$\frac{d^2x}{d\theta^2}+x=0$$

and likewise,

$$\frac{d^2y}{d\theta^2}+y=0$$

So

$$x=A\cos\theta+B\sin\theta$$

$$y=C\sin\theta+D\cos\theta$$

We find the constants from the "initial" ($\theta=0$) coordinates and "velocities": $A=x_0$, $D=y_0$, $B=(\partial x/\partial\theta)_0=-y_0$, $C=(\partial y/\partial\theta)_0=x_0$. Reverting to the standard notation in which (x, y), rather than (x_0, y_0), labels the initial point and $(\bar{x}, \bar{y})$, rather than (x, y), denotes the transformed one, we may write the finite canonical transformation (a finite rotation) as

$$\begin{aligned}\bar{x}&=x\cos\theta-y\sin\theta\\ \bar{y}&=x\sin\theta+y\cos\theta\end{aligned} \tag{2.8.14}$$

Similar equations may be derived for $\bar{p}_x$ and $\bar{p}_y$ in terms of p_x and p_y.

Although a wide class of canonical transformations is now open to us, there are many that aren't. For instance, $(q, p)\to(-q, -p)$ is a discrete canonical transformation that has no infinitesimal version. There are also the transformations that are not regular, such as the change from Cartesian to spherical coordinates, which have neither infintesimal forms, nor an active interpretation. We do not consider ways of generating these.‡

Consequence II. Let us understand the content of this result through an example before turning to the proof. Consider a two-particle system whose Hamiltonian is invariant under the translation of the entire system, i.e., both particles. Let an observer S_A prepare, at $t=0$, a state $(x_1^0, x_2^0; p_1^0, p_2^0)$ which evolves as $(x_1(t), x_2(t); p_1(t), p_2(t))$ for some time and ends up in the state $(x_1^T, x_2^T; p_1^T, p_2^T)$ at time T. Let

‡ For an excellent and lucid treatment of this question and many other topics in advanced classical mechanics, see H. Goldstein, *Classical Mechanics*, Addison-Wesley, Reading, Massachusetts (1950); E. C. G. Sudharshan and N. Mukunda, *Classical Dynamics: A Modern Perspective*, Wiley, New York (1974).

us call the final state the outcome of the experiment conducted by S_A. We are told that as a result of the translational invariance of $\mathscr{H}$, any other trajectory that is related to this by an arbitrary translation a is also a solution to the equations of motion. In this case, the initial state, for example, is $(x_1^0+a, x_2^0+a; p_1^0, p_2^0)$. *The final state and all intermediate states are likewise displaced by the same amount.* To an observer S_B, displaced relative to S_A by an amount a, the evolution of the second system will appear to be identical to what S_A saw in the first. Assuming for the sake of this argument that S_B had in fact prepared the second system, we may say that a given experiment and its translated version will give the same result (as seen by the observers who conducted them) if $\mathscr{H}$ is translationally invariant.

The physical idea is the following. For the usual reasons, translational invariance of $\mathscr{H}$ implies the invariance of $V(x_1, x_2)$. This in turn means that $V(x_1, x_2) = V(x_1 - x_2)$. Thus each particle cares only about where the other is relative to it, and not about where the system as a whole is in space. Consequently the outcome of the experiment is not affected by an overall translation.

Consequence II is just a generalization of this result to other canonical transformations that leave $\mathscr{H}$ invariant. For instance, if $\mathscr{H}$ is rotationally invariant, a given experiment and its rotated version will give the same result (according to the observers who conducted them).

Let us now turn to the proof of the general result.

Proof. Imagine a trajectory $(q(t), p(t))$ in phase space that satisfies the equations of motion. Let us associate with it an image trajectory, $(\bar{q}(t), \bar{p}(t))$, which is obtained by transforming each point (q, p) to the image point $(\bar{q}, \bar{p})$ by means of a regular canonical transformation. We ask if the image point moves according to Hamilton's equation of motion, i.e., if

$$\dot{\bar{q}}_j = \frac{\partial \mathscr{H}(\bar{q}, \bar{p})}{\partial \bar{p}_j}, \qquad \dot{\bar{p}}_j = -\frac{\partial \mathscr{H}(\bar{q}, \bar{p})}{\partial \bar{q}_j} \tag{2.8.15}$$

if $\mathscr{H}$ is invariant under the transformation $(q, p) \to (\bar{q}, \bar{p})$. Now $\bar{q}_j(q, p)$, like any dynamical variable $\omega(q, p)$, obeys

$$\dot{\bar{q}}_j = \{\bar{q}_j, \mathscr{H}(q, p)\}_{q,p} \tag{2.8.16}$$

If $(q, p) \to (\bar{q}, \bar{p})$ were a *passive* canonical transformation, we could write, since the PB are invariant under such a transformation,

$$\dot{\bar{q}}_j = \{\bar{q}_j, \mathscr{H}(q, p)\}_{q,p} = \{\bar{q}_j, \mathscr{H}(\bar{q}, \bar{p})\}_{\bar{q},\bar{p}} = \frac{\partial \mathscr{H}(\bar{q}, \bar{p})}{\partial \bar{p}_j}$$

But it is an active transformation. However, *because of the symmetry of* $\mathscr{H}$, i.e., $\mathscr{H}(q, p) = \mathscr{H}(\bar{q}, \bar{p})$, we can go through the very same steps that led to Eq. (2.7.16) from Eq. (2.7.14) and prove the result. If you do not believe this, you may verify it by explicit computation using $\mathscr{H}(q,p)=\mathscr{H}(\bar{q},\bar{p})$. A similar argument shows that

$$\dot{\bar{p}}_j=-\frac{\partial\mathscr{H}(\bar{q},\bar{p})}{\partial\bar{q}_j} \tag{2.8.17}$$

So the image point moves according to Hamilton's equations. Q.E.D.

Exercise 2.8.5. Why is it that a *non*canonical transformation that leaves $\mathscr{H}$ invariant does not map a solution into another? Or, in view of the discussions on consequence II, why is it that an experiment and its transformed version do not give the same result when the transformation that leaves $\mathscr{H}$ invariant is not canonical? It is best to consider an example. Consider the potential given in Exercise 2.8.3. Suppose I release a particle at $(x=a, y=0)$ with $(p_x=b, p_y=0)$ and you release one in the transformed state in which $(x=0, y=a)$ and $(p_x=b, p_y=0)$, i.e., you rotate the coordinates but not the momenta. This is a noncanonical transformation that leaves $\mathscr{H}$ invariant. Convince yourself that at later times the states of the two particles are not related by the same transformation. Try to understand what goes wrong in the general case.

As you go on and learn quantum mechanics, you will see that the symmetries of the Hamiltonian have similar consequences for the dynamics of the system.

A Useful Relation Between *S* and *E*

We now prove a result that will be invoked in Chapter 16:

$$\frac{\partial S_{\rm cl}(x_f,t_f;x_i,t_i)}{\partial t_f}=-\mathscr{H}(t_f)$$

where $S_{\rm cl}(x_f,t_f;x_i,t_i)$ is the action of the classical path from x_i, t_i to x_f, t_f and $\mathscr{H}$ is the Hamiltonian at the upper end point. Since we shall be working with problems where energy is conserved we may write

$$\frac{\partial S_{\rm cl}(x_f,t_f;x_i,t_i)}{\partial t_f}=-E \tag{2.8.18}$$

where E is the conserved energy, constant on the whole trajectory.

At first sight you may think that since

$$S_{\rm cl}=\int_{t_i}^{t_f}\mathscr{L}\,dt$$

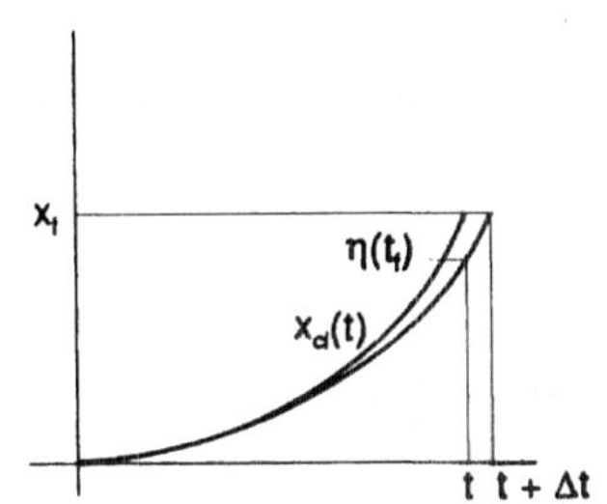

Figure 2.5. The upper trajectory takes time t while the lower takes $t+\Delta t$.

the right side must equal $\mathscr{L}$ and not $-E$. The explanation requires Fig. 2.5 wherein we have set $x_i=t_i=0$ for convenience.

The derivative we are computing is governed by the change in action of the *classical path* due to a change in travel by Δt holding the end points x_i and x_f fixed. From the figure it is clear that now the particle takes a different classical trajectory

$$x(t)=x_{\text{cl}}(t)+\eta(t) \quad \text{with} \quad \eta(0)=0.$$

so that the total change in action comes from the difference in paths between $t=0$ and $t=t_f$ as well as the entire action due to the extra travel between t_f and $t_f+\Delta t_f$. Only the latter is given $\mathscr{L}\Delta t$. The correct answer is then

$$\begin{aligned}\delta S_{\text{cl}}&=\int_0^{t_f}\left[\frac{\partial\mathscr{L}}{\partial x}\,\eta(t)+\frac{\partial\mathscr{L}}{\partial\dot{x}}\,\dot{\eta}(t)\right]dt+\mathscr{L}(t_f)\,\Delta t\\&=\int_0^{t_f}\left(-\frac{d}{dt}\frac{\partial\mathscr{L}}{\partial\dot{x}}+\frac{\partial\mathscr{L}}{\partial x}\right)_{x_{\text{cl}}}\eta(t)\,dt+\int_0^{t_f}\frac{d}{dt}\left[\frac{\partial\mathscr{L}}{\partial\dot{x}}\,\eta(t)\right]dt+\mathscr{L}(t_f)\,\Delta t\\&=0+\frac{\partial\mathscr{L}}{\partial\dot{x}}\,\eta(t)\bigg|_{t_f}+\mathscr{L}(t_f)\,\Delta t.\end{aligned}$$

It is clear from the figure that $\eta(t_f)=-\dot{x}(t_f)\,\Delta t$ so that

$$\delta S=\left[-\frac{\partial\mathscr{L}}{\partial\dot{x}}\,\dot{x}+\mathscr{L}\right]_{t_f}\Delta t=-\mathscr{H}(t_f)\,\Delta t$$

from which the result follows.

Exercise 2.8.6. Show that $\partial S_{\text{cl}}/\partial x_f=p(t_f)$.

Exercise 2.8.7. Consider the harmonic oscillator, for which the general solution is

$$x(t)=A\cos\omega t+B\sin\omega t.$$

Express the energy in terms of A and B and note that it does not depend on time. Now choose A and B such that $x(0)=x_1$ and $x(T)=x_2$. Write down the energy in terms of x_1, x_2, and T. Show that the action for the trajectory connecting x_1 and x_2 is

$$S_{cl}(x_1, x_2, T)=\frac{m\omega}{2\sin\omega T}[(x_1^2+x_2^2)\cos\omega T-2x_1x_2].$$

Verify that $\partial S_{cl}/\partial T=-E$.

3

All Is Not Well with Classical Mechanics

It was mentioned in the Prelude that as we keep expanding our domain of observations we must constantly check to see if the existing laws of physics continue to explain the new phenomena, and that, if they do not, we must try to find new laws that do. In this chapter you will get acquainted with experiments that betray the inadequacy of the classical scheme. The experiments to be described were never performed exactly as described here, but they contain the essential features of the actual experiments that were performed (in the first quarter of this century) with none of their inessential complications.

3.1. Particles and Waves in Classical Physics

There exist in classical physics two distinct entities: particles and waves. We have studied the particles in some detail in the last chapter and may summarize their essential features as follows. Particles are localized bundles of energy and momentum. They are described at any instant by the state parameters q and $\dot{q}$ (or q and p). These parameters evolve in time according to some equations of motion. Given the initial values $q(t_i)$ and $\dot{q}(t_i)$ at time t_i, the trajectory $q(t)$ may be deduced for all future times from the equations of motion. A wave, in contrast, is a disturbance spread over space. It is described by a wave function $\psi(\mathbf{r}, t)$ which characterizes the disturbance at the point $\mathbf{r}$ at time t.

In the case of sound waves, ψ is the excess air pressure above the normal, while in the case of electromagnetic waves, ψ can be any component of the electric field vector $\mathbf{E}$. The analogs of q and $\dot{q}$ for a wave are ψ and $\dot{\psi}$ at each point $\mathbf{r}$, assuming ψ obeys a second-order wave equation in time, such as

$$\nabla^2\psi = \frac{1}{c^2}\frac{\partial^2\psi}{\partial t^2}$$

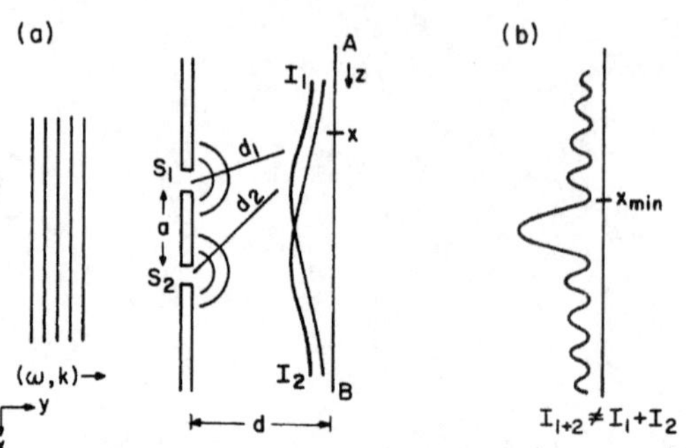

Figure 3.1. (a) When a wave $\psi = e^{i(ky-\omega t)}$ is incident on the screen with *either* slit S_1 or S_2 open, the intensity patterns I_1 and I_2, respectively, are measured by the row of detectors on AB. (b) With both slits open, the pattern I_{1+2} is observed. Note that $I_{1+2} \neq I_1 + I_2$. This is called interference.

which describes waves propagating at the speed of light, c. Given $\psi(\mathbf{r}, 0)$ and $\dot{\psi}(\mathbf{r}, 0)$ one can get the wave function $\psi(\mathbf{r}, t)$ for all future times by solving the wave equation.

Of special interest to us are waves that are periodic in space and time, called *plane waves*. In one dimension, the plane wave may be written as

$$\psi(x, t) = A \exp\left[i\left(\frac{2\pi}{\lambda}x - \frac{2\pi}{T}t\right)\right] \equiv A \exp[i\phi] \tag{3.1.1}$$

At some given time t, the wave is periodic in space with a period λ, called its *wavelength*, and likewise at a given point x, it is periodic in time, repeating itself every T seconds, T being called the *time period.* We will often use, instead of λ and T, the related quantities $k = 2\pi/\lambda$ called the *wave number* and $\omega = 2\pi/T$ called the *(angular) frequency*. In terms of the phase ϕ in Eq. (3.1.1), k measures the phase change per unit length at any fixed time t, while ω measures the phase change per unit time at any fixed point x. This wave travels at a speed $v = \omega/k$. To check this claim, note that if we start out at a point where $\phi = 0$ and move along x at a rate $x = (\omega/k)t$, ϕ remains zero. The overall scale A up front is called the *amplitude*. For any wave, the intensity is defined to be $I = |\psi|^2$. For a plane wave this is a constant equal to $|A|^2$. If ψ describes an electromagnetic wave, the intensity is a measure of the energy and momentum carried by the wave. [Since the electromagnetic field is real, only the real part of ψ describes it. However, time averages of the energy and momentum flow are still proportional to the intensity (as defined above) in the case of plane waves.]

Plane waves in three dimension are written as

$$\psi(\mathbf{r}, t) = A\, e^{i(\mathbf{k}\cdot\mathbf{r} - \omega t)}, \qquad \omega = |\mathbf{k}|v \tag{3.1.2}$$

where each component k_i gives the phase changes per unit length along the ith axis. One calls $\mathbf{k}$ the *wave vector.*‡

3.2. An Experiment with Waves and Particles (Classical)

Waves exhibit a phenomenon called *interference*, which is peculiar to them and is not exhibited by particles described by classical mechanics. This phenomenon is illustrated by the following experiment (Fig. 3.1a). Let a wave $\psi = A\, e^{i(ky-\omega t)}$ be

‡ Unfortunately we also use $\mathbf{k}$ to denote the unit vector along the z axis. It should be clear from the context what it stands for.

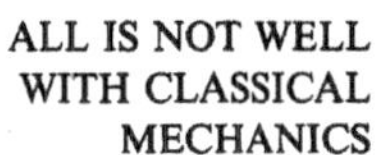

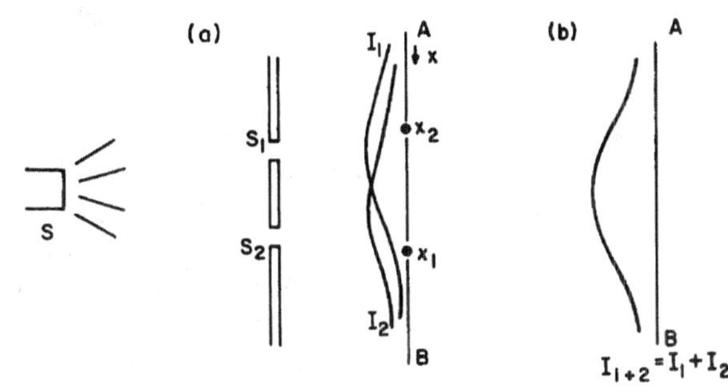

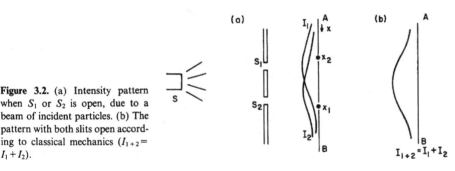

Figure 3.2. (a) Intensity pattern when S_1 or S_2 is open, due to a beam of incident particles. (b) The pattern with both slits open according to classical mechanics ($I_{1+2}=I_1+I_2$).

incident normally on a screen with slits S_1 and S_2, which are a distance a apart. At a distance d parallel to it is a row of detectors that measures the intensity as a function of the position x measured along AB.

If we first keep only S_1 open, the incident wave will come out of S_1 and propagate radially outward. One may think of S_1 as the virtual source of this wave ψ_1, which has the same frequency and wavelength as the incident wave. The intensity pattern $I_1=|\psi_1|^2$ is registered by the detectors. Similarly if S_2 is open instead of S_1, the wave ψ_2 produces the pattern $I_2=|\psi_2|^2$. In both cases the arrival of energy at the detectors is a smooth function of x and t.

Now if both S_1 and S_2 are opened, both waves ψ_1 and ψ_2 are present and produce an intensity pattern $I_{1+2}=|\psi_1+\psi_2|^2$.

The interesting thing is that $I_{1+2}\neq I_1+I_2$, but rather the interference pattern shown in Fig. 3.1b. The ups and downs are due to the fact that the waves ψ_1 and ψ_2 have to travel different distances d_1 and d_2 to arrive at some given x (see Fig. 3.1a) and thus are not always in step. In particular, the maxima correspond to the case $d_2-d_1=n\lambda$ (n is an integer), when the waves arrive exactly in step, and the minima correspond to the case $d_2-d_1=(2n+1)\lambda/2$, when the waves are exactly out of step. In terms of the phases ϕ_1 and ϕ_2, $\phi_2(x)-\phi_1(x)=2n\pi$ at a maximum and $\phi_2(x)-\phi_1(x)=(2n+1)\pi$ at a minimum. One can easily show that the spacing Δx between two adjacent maxima is $\Delta x=\lambda d/a$.

The feature to take special note of is that if $x_{\min}$ is an interference minimum, there is more energy flowing into $x_{\min}$ with just one slit open than with both. In other words, the opening of an extra slit can actually reduce the energy flow into $x_{\min}$.

Consider next the experiment with particles (Fig. 3.2a). The source of the incident plane waves is replaced by a source of particles that shoots them toward the screen with varying directions but fixed energy. Let the line AB be filled with an array of particle detectors. Let us define the intensity $I(x)$ to be the number of particles arriving per second at any given x. The patterns with S_1 or S_2 open are shown in (Fig. 3.2a). These look very much like the corresponding patterns for the wave. The only difference will be that the particles arrive not continuously, but in a staccato fashion, each particle triggering a counter at some single point x at the time of arrival. Although this fact may be obscured if the beam is dense, it can be easily detected as the incident flux is reduced.

What if both S_1 and S_2 are opened? Classical mechanics has an unambiguous prediction: $I_{1+2}=I_1+I_2$. The reasoning is as follows: each particle travels along a definite trajectory that passes via S_1 *or* S_2 to the destination x. To a particle headed for S_1, it is immaterial whether S_2 is open or closed. Being localized in space it has no way of even knowing if S_2 is open or closed, and thus cannot respond to it in any way. Thus the number coming via S_1 to x is independent of whether S_2 is open or not and vice versa. It follows that $I_{1+2}=I_1+I_2$ (Fig. 3.2b).

The following objection may be raised: although particles heading for S_1 are not aware that S_2 is open, they certainly can be deflected by those coming out of S_2, if, for instance, the former are heading for x_1 and the latter for x_2 (see Fig. 3.1a).

This objection can be silenced by sending in one particle at a time. A given particle will of course not produce a pattern like I_1 or I_2 by itself, it will go to some point x. If, however, we make a histogram, the envelope of this histogram, after many counts, will define the smooth functions I_1, I_2, and I_{1+2}. Now the conclusion $I_{1+2}=I_1+I_2$ is inevitable.

This is what classical physics predicts particles and waves will do in the double-slit experiment.

3.3. The Double-Slit Experiment with Light

Consider now what happens when we perform the following experiment to check the classical physics notion that light is an electromagnetic wave phenomenon.

We set up the double slit as in Fig. 3.1a, with a row of light-sensitive meters along AB and send a beam $\psi=A\,e^{i(ky-\omega t)}$ in a direction perpendicular to the screen. (Strictly speaking, the electromagnetic wave must be characterized by giving the orientation of the **E** and **B** vectors in addition to ω and k. However, for a plane wave, **B** is uniquely fixed by **E**. If we further assume **E** is polarized perpendicular to the page, this polarization is unaffected by the double slit. We can therefore suppress the explicit reference to this constant vector and represent the field as a scalar function ψ.) We find that with the slits open one at a time we get patterns I_1 and I_2, and with both slits open we get the interference pattern I_{1+2} as in Figs. 3.1a and 3.1b. (The interference pattern is of course what convinced classical physicists that light was a wave phenomenon.) The energy arrives at the detectors smoothly and continuously as befitting a wave.

Say we repeat the experiment with a change that is expected (in classical physics) to produce no qualitative effects. We start with S_1 open and cut down the intensity. A very strange thing happens. We find that the energy is not arriving continuously, but in sudden bursts, a burst here, a burst there, etc. We now cut down the intensity further so that only one detector gets activated at a given time and there is enough of a gap, say a millisecond, between counts. As each burst occurs at some x, we record it and plot a histogram. With enough data, the envelope of the histogram becomes, of course, the pattern I_1. We have made an important discovery: light energy is not continuous—it comes in bundles. This discrete nature is obscured in intense beams, for the bundles come in so fast and all over the line AB, that the energy flow seems continuous in space and time.

We pursue our study of these bundles, called photons, in some detail and find the following properties:

1. Each bundle carries the same energy E.
2. Each bundle carries the same momentum p.

3. $E = pc$. From the famous equation $E^2 = p^2c^2 + m^2c^4$, we deduce that these bundles are particles of zero mass.
4. If we vary the frequency of the light source we discover that

$$E = \hbar\omega \tag{3.3.1}$$

$$p = \hbar k \tag{3.3.2}$$

where $\hbar = h/2\pi$ is a constant. The constant h is called *Planck's constant*, and has the dimensions of erg sec, which is the same as that of action and angular momentum. Its value is

$$\frac{h}{2\pi} = \hbar \simeq 10^{-27} \text{ erg sec} \tag{3.3.3}$$

For those interested in history, the actual experiment that revealed the granular nature of light is called the *photoelectric effect*. The correct explanation of this experiment, in terms of photons, was given by Einstein in 1905.

That light is made of particles will, of course, surprise classical physicists but will not imply the end of classical physics, for physicists are used to the idea that phenomena that seem continuous at first sight may in reality be discrete. They will cheerfully plunge into the study of the dynamics of the photons, trying to find the equations of motion for its trajectory and so on. What really undermines classical physics is the fact that if we now open both slits, still keeping the intensity so low that only one photon is in the experimental region at a given time, and watch the histogram take shape, we won't find that I_{1+2} equals $I_1 + I_2$ as would be expected of particles, but is instead an interference pattern characteristic of wave number k. This result completely rules out the possibility that photons move in well-defined trajectories like the particles of classical mechanics—for if this were true, a photon going in via S_1 should be insensitive to whether S_2 is open or not (and vice versa), and the result $I_{1+2} = I_1 + I_2$ is inescapable! To say this another way, consider a point x_{min} which is an interference minimum. More photons arrive here with either S_1 or S_2 open than with both open. If photons followed definite trajectories, it is incomprehensible how opening an extra pathway can *reduce* the number coming to x_{min}. Since we are doing the experiment with one photon at a time, one cannot even raise the improbable hypothesis that photons coming out of S_1 collide with those coming out of S_2 to modify (miraculously) the smooth pattern $I_1 + I_2$ into the wiggly interference pattern.

From these facts Born drew the following conclusion: with *each* photon is associated a wave ψ, called the *probability amplitude* or simply *amplitude*, whose modulus squared $|\psi(x)|^2$ gives the probability of finding the particle at x. [Strictly speaking, we must not refer to $|\psi(x)|^2$ as the probability for a given x, but rather as the probability density at x since x is a continuous variable. These subtleties can, however, wait.] The entire experiment may be understood in terms of this hypothesis as follows. Every incoming photon of energy E and momentum p has a wave function ψ associated with it, which is a plane wave with $\omega = E/\hbar$ and $k = p/\hbar$. This wave interferes with itself and forms the oscillating pattern $|\psi(x)|^2$ along AB, which gives the probability that the given photon will arive at x. A given photon of course arrives at some definite x and does not reveal the probability distribution. If, however, we wait till several photons, all described by the same ψ, have arrived, the number at any x will become proportional to the probability function $|\psi(x)|^2$. Likewise, if an intense (macroscopic) monochromatic beam is incident, many photons, all described by the same wave and hence the same probability distribution, arrive at the same time and all along the line AB. The intensity distribution then assumes the shape of the probability distribution right away and the energy flow seems continuous and in agreement with the predictions of classical electromagnetic theory.

The main point to note, besides the probability interpretation, is that a wave is associated not with a beam of photons, but with *each* photon. If the beam is monochromatic, every photon is given by the same ψ and the same probability distribution. A large ensemble of such photons will reproduce the phenomena expected of a classical electromagnetic wave ψ and the probabilistic aspect will be hidden.

3.4. Matter Waves (de Broglie Waves)

That light, which one thought was a pure wave phenomenon, should consist of photons, prompted de Broglie to conjecture that entities like the electron, generally believed to be particles, should exhibit wavelike behavior. More specifically, he conjectured, in analogy with photons, that particles of momentum p will produce an interference pattern corresponding to a wave number $k=p/\hbar$ in the double-slit experiment. This prediction was verified for electrons by Davisson and Germer, shortly thereafter. It is now widely accepted that all particles are described by probability amplitudes $\psi(x)$, and that the assumption that they move in definite trajectories is ruled out by experiment.

But what about common sense, which says that billiard balls and baseballs travel along definite trajectories? How did classical mechanics survive for three centuries? The answer is that the wave nature of matter is not apparent for macroscopic phenomena since $\hbar$ is so small. The precise meaning of this explanation will become clear only after we fully master quantum mechanics. Nonetheless, the following example should be instructive. Suppose we do the double-slit experiment with pellets of mass 1 g, moving at 1 cm/sec. The wavelength associated with these particles is

$$\lambda=\frac{2\pi}{k}=\frac{h}{p}\simeq 10^{-26}\text{ cm}$$

which is 10^{-13} times smaller than the radius of the proton! For any reasonable values of the parameters a and d (see Fig. 3.1b), the interference pattern would be so dense in x that our instruments will only measure the smooth average, which will obey $I_{1+2}=I_1+I_2$ as predicted classically.

3.5. Conclusions

The main objective of this chapter was to expose the inadequacy of classical physics in explaining certain phenomena and, incidentally, to get a glimpse of what the new (quantum) physics ought to look like. We found that entities such as the electron are particles in the classical sense in that when detected they seem to carry all their energy, momentum, charge, etc. in localized form; and at the same time they are not particlelike in that assuming they move along definite trajectories leads to conflict with experiment. It appears that each particle has associated with it a wave function $\psi(x, t)$, such that $|\psi(x, t)|^2$ gives the probability of finding it at a point x at time t. This is called *wave-particle duality*.

The dynamics of the particle is then the dynamics of this function $\psi(x, t)$ or, if we think of functions as vectors in an infinite-dimensional space, of the ket $|\psi(t)\rangle$. In the next chapter the postulates of quantum theory will define the dynamics in terms of $|\psi(t)\rangle$. The postulates, which specify what sort of information is contained in $|\psi(t)\rangle$ and how $|\psi(t)\rangle$ evolves with time, summarize the results of the double-slit experiment and *many others not mentioned here*. The double-slit experiment was described here to expose the inadequacy of classical physics and not to summarize the entire body of experimental results from which all the postulates could be inferred. Fortunately, the double-slit experiment contains most of the central features of the theory, so that when the postulates are encountered in the next chapter, they will appear highly plausible.

4

The Postulates—a General Discussion

Having acquired the necessary mathematical training and physical motivation, you are now ready to get acquainted with the postulates of quantum mechanics. In this chapter the postulates will be stated and discussed in broad terms to bring out the essential features of quantum theory. The subsequent chapters will simply be applications of these postulates to the solution of a variety of physically interesting problems. Despite your preparation you may still find the postulates somewhat abstract and mystifying on this first encounter. These feelings will, however, disappear after you have worked with the subject for some time.

4.1. The Postulates‡

The following are the postulates of nonrelativistic quantum mechanics. We consider first a system with one degree of freedom, namely, a single particle in one space dimension. The straightforward generalization to more particles and higher dimensions will be discussed towards the end of the chapter. In what follows, the quantum postulates are accompanied by their classical counterparts (in the Hamiltonian formalism) to provide some perspective.

Classical Mechanics	Quantum Mechanics
I. The state of a particle at any given time is specified by the two variables $x(t)$ and $p(t)$, i.e., as a point in a two-dimensional phase space.	I. The state of the particle is represented by a vector $\|\psi(t)\rangle$ in a Hilbert space.
II. Every dynamical variable ω is a function of x and p: $\omega=\omega(x, p)$.	II. The independent variables x and p of classical mechanics are represented

‡ Recall the discussion in the Preface regarding the sense in which the word is used here.

by Hermitian operators X and P with the following matrix elements in the eigenbasis of X‡

$$\langle x|X|x'\rangle = x\delta(x-x')$$

$$\langle x|P|x'\rangle = -i\hbar\delta'(x-x')$$

The operators corresponding to dependent variables $\omega(x, p)$ are given Hermitian operators

$$\Omega(X, P) = \omega(x \to X, p \to P)\S$$

III. If the particle is in a state given by x and p, the measurement‖ of the variable ω will yield a value $\omega(x, p)$. The state will remain unaffected.

III. If the particle is in a state $|\psi\rangle$, measurement‖ of the variable (corresponding to) Ω will yield one of the eigenvalues ω with probability $P(\omega) \propto |\langle\omega|\psi\rangle|^2$. The state of the system will change from $|\psi\rangle$ to $|\omega\rangle$ as a result of the measurement.

IV. The state variables change with time according to Hamilton's equations:

$$\dot{x} = \frac{\partial \mathscr{H}}{\partial p}$$

$$\dot{p} = -\frac{\partial \mathscr{H}}{\partial x}$$

IV. The state vector $|\psi(t)\rangle$ obeys the *Schrödinger equation*

$$i\hbar \frac{d}{dt}|\psi(t)\rangle = H|\psi(t)\rangle$$

where $H(X, P) = \mathscr{H}(x \to X, p \to P)$ is the quantum Hamiltonian operator and $\mathscr{H}$ is the Hamiltonian for the corresponding classical problem.

4.2. Discussion of Postulates I–III

The postulates (of classical and quantum mechanics) fall naturally into two sets: the first three, which tell us how the system is depicted at a given time, and the last, which specifies how this picture changes with time. We will confine our attention to the first three postulates in this section, leaving the fourth for the next.

The first postulate states that a particle is described by a ket $|\psi\rangle$ in a Hilbert space which, you will recall, contains *proper vectors* normalizable to unity as well as

‡ Note that the X operator is the same one discussed at length in Section 1.10. Likewise $P = \hbar K$, where K was also discussed therein. You may wish to go over that section now to refresh your memory.

§ By this we mean that Ω is the same function of X and P as ω is of x and p.

‖ That is, in an ideal experiment consistent with the theory. It is assumed you are familiar with the ideal classical measurement which can determine the state of the system without disturbing it in any way. A discussion of ideal quantum measurements follows.

improper vectors, normalizable only to the Dirac delta functions.‡ Now, a ket in such a space has in general an infinite number of components in a given basis. One wonders why a particle, which had only two independent degrees of freedom, x and p, in classical mechanics, now needs to be specified by an infinite number of variables. What do these variables tell us about the particle? To understand this we must go on to the next two postulates, which answer exactly this question. For the present let us note that the double-slit experiment has already hinted to us that a particle such as the electron needs to be described by a wave function $\psi(x)$. We have seen in Section 1.10 that a function $f(x)$ may be viewed as a ket $|f\rangle$ in a Hilbert space. The ket $|\psi\rangle$ of quantum mechanics is none other than the vector representing the probability amplitude $\psi(x)$ introduced in the double-slit experiment.

When we say that $|\psi\rangle$ is an element of a vector space we mean that if $|\psi\rangle$ and $|\psi'\rangle$ represent possible states of a particle so does $\alpha|\psi\rangle+\beta|\psi'\rangle$. This is called the *principle of superposition*. The principle by itself is not so new: we know in classical physics, for example, that if $f(x)$ and $g(x)$ [with $f(0)=f(L)=g(0)=g(L)=0$] are two possible displacements of a string, so is the superposition $\alpha f(x)+\beta g(x)$. What *is* new is the interpretation of the superposed state $\alpha|\psi\rangle+\beta|\psi'\rangle$. In the case of the string, the state $\alpha f+\beta g$ has very different attributes from the states f and g: it will look different, have a different amount of stored elastic energy, and so on. In quantum theory, on the other hand, the state $\alpha|\psi\rangle+\beta|\psi'\rangle$ will, loosely speaking, have attributes that sometimes resemble that of $|\psi\rangle$ and at other times those of $|\psi'\rangle$. There is, however, no need to speak loosely, since we have postulates II and III to tell us exactly how the state vector $|\psi\rangle$ is to be interpreted in quantum theory. Let us find out.

In classical mechanics when a state (x, p) is given, one can say that any dynamical variable ω has a value $\omega(x, p)$, in the sense that if the variable is measured the result $\omega(x, p)$ will obtain. What is the analogous statement one can make in quantum mechanics given that the particle is in a state $|\psi\rangle$? The answer is provided by Postulates II and III, in terms of the following steps:

Step 1. Construct the corresponding quantum operator $\Omega=\omega(x\to X,\ p\to P)$, where X and P are the operators defined in postulate II.

Step 2. Find the orthonormal eigenvectors $|\omega_i\rangle$ and eigenvalues ω_i of Ω.

Step 3. Expand $|\psi\rangle$ in this basis:

$$|\psi\rangle=\sum_i |\omega_i\rangle\langle\omega_i|\psi\rangle$$

Step 4. The probability $P(\omega)$ that the result ω will obtain is proportional to the modulus squared of the projection of $|\psi\rangle$ along the eigenvector $|\omega\rangle$, that is $P(\omega)\propto|\langle\omega|\psi\rangle|^2$. In terms of the projection operator $\mathbb{P}_\omega=|\omega\rangle\langle\omega|$, $P(\omega)\propto|\langle\omega|\psi\rangle|^2=\langle\psi|\omega\rangle\langle\omega|\psi\rangle=\langle\psi|\mathbb{P}_\omega|\psi\rangle=\langle\psi|\mathbb{P}_\omega\mathbb{P}_\omega|\psi\rangle=\langle\mathbb{P}_\omega\psi|\mathbb{P}_\omega\psi\rangle$.

There is a tremendous amount of information contained in these steps. Let us note, for the present, the following salient points.

‡ The status of the two classes will be clarified later in this chapter.

(1) The theory makes only probabilistic predictions for the result of a measurement of Ω. Further, it assigns (relative) probabilities only for obtaining some eigenvalue ω of Ω. *Thus the only possible values of Ω are its eigenvalues.* Since postulate II demands that Ω be Hermitian, these eigenvalues are all real.

(2) Since we are told that $P(\omega_i) \propto |\langle \omega_i | \psi \rangle|^2$, the quantity $|\langle \omega_i | \psi \rangle|^2$ is only the relative probability. To get the absolute probability, we divide $|\langle \omega_i | \psi \rangle|^2$ by the sum of all relative probabilities:

$$P(\omega_i) = \frac{|\langle \omega_i | \psi \rangle|^2}{\sum_j |\langle \omega_j | \psi \rangle|^2} = \frac{|\langle \omega_i | \psi \rangle|^2}{\langle \psi | \psi \rangle} \tag{4.2.1}$$

It is clear that if we had started with a normalized state

$$|\psi'\rangle = \frac{|\psi\rangle}{\langle \psi | \psi \rangle^{1/2}}$$

we would have had

$$P(\omega_i) = |\langle \omega_i | \psi' \rangle|^2 \tag{4.2.2}$$

If $|\psi\rangle$ is a proper vector, such a rescaling is possible and will be assumed hereafter. The probability interpretation breaks down if $|\psi\rangle$ happens to be one of the improper vectors in the space, for in this case $\langle \psi | \psi \rangle = \delta(0)$ is the only sensible normalization. The status of such vectors will be explained in Example 4.2.2 below.

Note that the condition $\langle \psi | \psi \rangle = 1$ is a matter of convenience and not a physical restriction on the proper vectors. (In fact the set of all normalized vectors does not even form a vector space. If $|\psi\rangle$ and $|\psi'\rangle$ are normalized, then an arbitrary linear combination, $\alpha|\psi\rangle + \beta|\psi'\rangle$, is not.)

Note that the relative probability distributions corresponding to the states $|\psi\rangle$ and $\alpha|\psi\rangle$, when they are renormalized to unity, reduce to the same absolute probability distribution. Thus, corresponding to each physical state, there exists not one vector, but a *ray* or "direction" in Hilbert space. When we speak of the state of the particle, we usually mean the ket $|\psi\rangle$ with unit norm. Even with the condition $\langle \psi | \psi \rangle = 1$, we have the freedom to multiply the ket by a number of the form $e^{i\theta}$ without changing the physical state. This freedom will be exploited at times to make the components of $|\psi\rangle$ in some basis come out real.

(3) If $|\psi\rangle$ is an eigenstate $|\omega_i\rangle$, the measurement of Ω is guaranteed to yield the result ω_i. A particle in such a state may be said to have a value ω_i for Ω in the classical sense.

(4) When two states $|\omega_1\rangle$ and $|\omega_2\rangle$ are superposed to form a (normalized) state, such as

$$|\psi\rangle = \frac{\alpha|\omega_1\rangle + \beta|\omega_2\rangle}{(|\alpha|^2 + |\beta|^2)^{1/2}}$$

one gets the state, which upon measurement of Ω, can yield either ω_1 or ω_2 with probabilities $|\alpha|^2/(|\alpha|^2+|\beta|^2)$ and $|\beta|^2/(|\alpha|^2+|\beta|^2)$, respectively. This is the peculiar

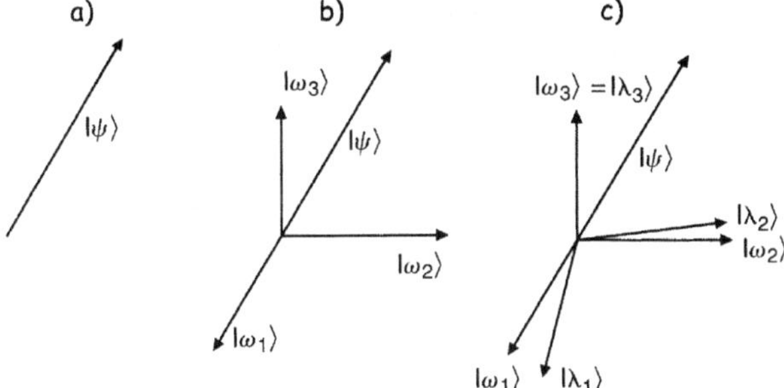

Figure 4.1. (a) The normalized ket in $\mathbb{V}^3(R)$ representing the state of the particle. (b) The Ω basis, $|\omega_1\rangle$, $|\omega_2\rangle$, and $|\omega_3\rangle$. (c) The Ω and the Λ bases. To get the statistical information on a variable, we find the eigenvectors of the corresponding operator and project $|\psi\rangle$ on that basis.

consequence of the superposition principle in quantum theory, referred to earlier. It has no analog in classical mechanics. For example, if a dynamical variable of the string in the state $\alpha f + \beta g$ is measured, one does not expect to get the value corresponding to f some of the time and that corresponding to g the rest of the time; instead, one expects a unique value generally distinct from both. Likewise, the functions f and αf (α real) describe two distinct configurations of the string and are not physically equivalent.

(5) When one wants information about another variable Λ, one repeats the whole process, finding the eigenvectors $|\lambda_i\rangle$ and the eigenvalues λ_i. Then

$$P(\lambda) \propto |\langle\lambda|\psi\rangle|^2$$

The bases of Ω and Λ will of course be different in general. In summary, we have a single ket $|\psi\rangle$ representing the state of the particle in Hilbert space, and it contains the statistical prediction for all observables. To extract this information for any observable, we must determine the eigenbasis of the corresponding operator and find the projection of $|\psi\rangle$ along all its eigenkets.

(6) As our interest switches from one variable Ω, to another, Λ, so does our interest go from the kets $|\omega\rangle$, to the kets $|\lambda\rangle$. There is, however, no need to change the basis each time. Suppose for example we are working in the Ω basis in which

$$|\psi\rangle = \sum_i |\omega_i\rangle\langle\omega_i|\psi\rangle$$

and $P(\omega_i) = |\langle\omega_i|\psi\rangle|^2$. If we want $P(\lambda_i)$ we take the operator Λ (which is some given matrix with elements $\Lambda_{ij} = \langle\omega_i|\Lambda|\omega_j\rangle$); find its eigenvectors $|\lambda_i\rangle$ (which are column vectors with components $\langle\omega_j|\lambda_i\rangle$), and take the inner product $\langle\lambda_i|\psi\rangle$ in this basis:

$$\langle\lambda_i|\psi\rangle = \sum_j \langle\lambda_i|\omega_j\rangle\langle\omega_j|\psi\rangle$$

Example 4.2.1. Consider the following example from a fictitious Hilbert space $\mathbb{V}^3(R)$ (Fig. 4.1). In Fig. 4.1a we have the normalized state $|\psi\rangle$, with no reference to any basis. To get predictions on Ω, we find its eigenbasis and express the state vector $|\psi\rangle$ in terms of the orthonormal eigenvectors $|\omega_1\rangle$, $|\omega_2\rangle$, and $|\omega_3\rangle$ (Fig. 4.1b). Let us suppose

$$|\psi\rangle = \frac{1}{2}|\omega_1\rangle + \frac{1}{2}|\omega_2\rangle + \frac{1}{2^{1/2}}|\omega_3\rangle$$

This means that the values ω_1, ω_2, and ω_3 are expected with probabilities $\frac{1}{4}$, $\frac{1}{4}$, and $\frac{1}{2}$, respectively, *and other values of ω are impossible.* If instead $|\psi\rangle$ were some eigenvector, say $|\omega_1\rangle$, then the result ω_1 would obtain with unit probability. Only a particle in a state $|\psi\rangle = |\omega_i\rangle$ has a well-defined value of Ω in the classical sense. If we want $P(\lambda_i)$ we construct the basis $|\lambda_1\rangle$, $|\lambda_2\rangle$, and $|\lambda_3\rangle$, which can in general be distinct from the Ω basis. In our example (Fig. 4.1c) there is just one common eigenvector $|\omega_3\rangle = |\lambda_3\rangle$. □

Returning to our main discussion, there are a few complications that could arise as one tries to carry out the steps 1–4. We discuss below the major ones and how they are to be surmounted.

Complication 1: The Recipe $\Omega = \omega(x \to X, p \to P)$ Is Ambiguous. If, for example, $\omega = xp$, we don't know if $\Omega = XP$ or PX since $xp = px$ classically. There is no universal recipe for resolving such ambiguities. In the present case, the rule is to use the symmetric sum: $\Omega = (XP + PX)/2$. Notice incidentally that symmetrization also renders Ω Hermitian. Symmetrization is the answer as long as Ω does not involve products of two or more powers of X with two or more powers of P. If it does, only experiment can decide the correct prescription. We will not encounter such cases in this book.

Complication 2: The Operator Ω Is Degenerate. Let us say $\omega_1 = \omega_2 = \omega$. What is $P(\omega)$ in this case? We select some orthonormal basis $|\omega, 1\rangle$ and $|\omega, 2\rangle$ in the eigenspace $\mathbb{V}_\omega$ with eigenvalue ω. Then

$$P(\omega) = |\langle \omega, 1|\psi\rangle|^2 + |\langle \omega, 2|\psi\rangle|^2$$

which is the modulus squared of the projection of $|\psi\rangle$ in the degenerate eigenspace. This is the result we will get if we assume that ω_1 and ω_2 are infintesimally distinct and ask for $P(\omega_1$ or $\omega_2)$. In terms of the projection operator for the *eigenspace*,

$$\mathbb{P}_\omega = |\omega, 1\rangle\langle \omega, 1| + |\omega, 2\rangle\langle \omega, 2| \tag{4.2.3a}$$

we have

$$P(\omega) = \langle \psi|\mathbb{P}_\omega|\psi\rangle = \langle \mathbb{P}_\omega \psi|\mathbb{P}_\omega \psi\rangle \tag{4.2.3b}$$

In general, one can replace in Postulate III

$$P(\omega) \propto \langle \psi|\mathbb{P}_\omega|\psi\rangle$$

where $\mathbb{P}_\omega$ is the projection operator for the eigenspace with eigenvalue ω. Then postulate III as stated originally would become a special case in which there is no degeneracy and each eigenspace is simply an eigenvector.

In our example from $\mathbb{V}^3(R)$, if $\omega_1 = \omega_2 = \omega$ (Fig. 4.1b) then $P(\omega)$ is the square of the component of $|\psi\rangle$ in the "$x\ y$" plane.

Complication 3: The Eigenvalue Spectrum of Ω Is Continuous. In this case one expands $|\psi\rangle$ as

$$|\psi\rangle = \int |\omega\rangle\langle\omega|\psi\rangle\, d\omega$$

One expects that as ω varies continuously, so will $\langle\omega|\psi\rangle$, that is to say, one expects $\langle\omega|\psi\rangle$ to be a smooth function $\psi(\omega)$. To visualize this function one introduces an auxiliary one-dimensional space, called the ω space, the points in which are labeled by the coordinate ω. In this space $\psi(\omega)$ will be a smooth function of ω and is called the *wave function in the ω space.* We are merely doing the converse of what we did in Section 1.10 wherein we started with a function $f(x)$ and tried to interpret it as the components of an infinite-dimensional ket $|\psi\rangle$ in the $|x\rangle$ basis. As far as the state vector $|\psi\rangle$ is concerned, there is just one space, the Hilbert space, in which it resides. The ω space, the λ space, etc. are auxiliary manifolds introduced for the purpose of visualizing the components of the infinite-dimensional vector $|\psi\rangle$ in the Ω basis, the Λ basis, and so on. The wave function $\psi(\omega)$ is also called the *probability amplitude* for finding the particle with $\Omega = \omega$.

Can we interpret $|\langle\omega|\psi\rangle|^2$ as the probability for finding the particle with a value ω for Ω? No. Since the number of possible values for ω is infinite and the total probability is unity, each single value of ω can be assigned only an infinitesimal probability. One interprets $P(\omega) = |\langle\omega|\psi\rangle|^2$ to be the *probability density* at ω, by which one means that $P(\omega)\, d\omega$ is the probability of obtaining a result between ω and $\omega + d\omega$. This definition meets the requirement that the total probability be unity, since

$$\int P(\omega)\, d\omega = \int |\langle\omega|\psi\rangle|^2\, d\omega = \int \langle\psi|\omega\rangle\langle\omega|\psi\rangle\, d\omega$$
$$= \langle\psi|I|\psi\rangle = \langle\psi|\psi\rangle = 1 \tag{4.2.4}$$

If $\langle\psi|\psi\rangle = \delta(0)$ is the only sensible normalization possible, the state cannot be normalized to unity and $P(\omega)$ must be interpreted as the *relative probability density.* We will discuss such improper states later.

An important example of a continuous spectrum is that of X, the operator corresponding to the position x. The wave function in the X basis (or the x space), $\psi(x)$, is usually referred to as just the wave function, since the X basis is almost always what one uses. In our discussions in the last chapter, $|\psi(x)|^2$ was referred to as the probability for finding the particle *at* a given x, rather than as the probability density, in order to avoid getting into details. Now the time has come to become precise!

Earlier on we were wondering why it was that a classical particle defined by just two numbers x and p now needs to be described by a ket which has an infinite number of components. The answer is now clear. A classical particle has, at any given time, a definite position. One simply has to give this value of x in specifying the state. A quantum particle, on the other hand, can take on any value of x upon measurement and one must give the relative probabilities for *all possible outcomes.* This is part of the information contained in $\psi(x)=\langle x|\psi\rangle$, the components of $|\psi\rangle$ in the X basis. Of course, in the case of the classical particle, one needs also to specify the momentum p as well. In quantum theory one again gives the odds for getting different values of momenta, but one doesn't need a new vector for specifying this; the same ket $|\psi\rangle$ when expanded in terms of the eigenkets $|p\rangle$ of the momentum operator P gives the odds through the wave function in p space, $\psi(p)=\langle p|\psi\rangle$.

Complication 4: The Quantum Variable Ω Has No Classical Counterpart. Even "point" particles such as the electron are now known to carry "spin," which is an internal angular momentum, that is to say, angular momentum unrelated to their motion through space. Since such a degree of freedom is absent in classical mechanics, our postulates do not tell us which operator is to describe this variable in quantum theory. As we will see in Chapter 14, the solution is provided by a combination of intuition and semi-classical reasoning. It is worth bearing in mind that no matter how diligently the postulates are constructed, they must often be supplemented by intuition and classical ideas.

Having discussed the four-step program for extracting statistical information from the state vector, we continue with our study of what else the postulates of quantum theory tell us.

Collapse of the State Vector

We now examine another aspect of postulate III, namely, that the measurement of the variable Ω changes the state vector, which is in general some superposition of the form

$$|\psi\rangle=\sum_{\omega}|\omega\rangle\langle\omega|\psi\rangle$$

into the eigenstate $|\omega\rangle$ corresponding to the eigenvalue ω obtained in the measurement. This phenomenon is called the *collapse or reduction of the state vector.*

Let us first note that any definitive statement about the impact of the measurement process presupposes that the measurement process is of a definite kind. For example, the classical mechanics maxim that any dynamical variable can be measured without changing the state of the particle, assumes that the measurement is an ideal measurement (consistent with the classical scheme). But one *can* think up *nonideal* measurements which *do* change the state; imagine trying to locate a chandelier in a dark room by waving a broom till one makes contact. What makes Postulate III profound is that the measurement process referred to there is an *ideal quantum measurement*, which in a sense is the best one can do. We now illustrate the notion of an ideal quantum measurement and the content of this postulate by an example.

Consider a particle in a momentum eigenstate $|p\rangle$. The postulate tells us that if the momentum in this state is measured we are assured a result p, and that the state will be the same after the measurement (since $|\psi\rangle = |p\rangle$ is already an eigenstate of the operator P in question). One way to measure the momentum of the particle is by *Compton scattering*, in which a photon of definite momentum bounces off the particle.

Let us assume the particle is forced to move along the x-axis and that we send in a right-moving photon of energy $\hbar\omega$ that bounces off the particle and returns as a left-moving photon of energy $\hbar\omega'$. (How do we know what the photon energies are? We assume we have atoms that are known to emit and absorb photons of any given energy.) Using momentum and energy conservation:

$$cp' = cp + \hbar(\omega + \omega')$$

$$E' = E + \hbar(\omega - \omega')$$

it is now possible from this data to reconstruct the initial and final momenta of the particle:

$$cp = -\frac{(\hbar\omega + \hbar\omega')}{2} + \sqrt{1 + \frac{m^2c^4}{\hbar^2\omega\omega'}}\,\frac{\hbar\omega - \hbar\omega'}{2}$$

$$cp' = \frac{(\hbar\omega + \hbar\omega')}{2} + \sqrt{1 + \frac{m^2c^4}{\hbar^2\omega\omega'}}\,\frac{\hbar\omega - \hbar\omega'}{2}$$

Solving for ω' and p' in terms of ω and p, one readily sees that for any choice of p, if $\omega \to 0$, then so does ω'. Thus one can always make the change in momentum $p' - p$ arbitrarily small. Hereafter, when we speak of a momentum measurement, this is what we will mean. We will also assume that to each dynamical variable there exists a corresponding ideal measurement. We will discuss, for example, the ideal position measurement, which, when conducted on a particle in state $|x\rangle$, will give the result x with unit probability and leave the state vector unchanged.

Suppose now that we measure the *position* of a particle in a *momentum* eigenstate $|p\rangle$. Since $|p\rangle$ is a sum of position eigenkets $|x\rangle$,

$$|p\rangle = \int |x\rangle\langle x|p\rangle\, dx$$

the measurement will force the system into some state $|x\rangle$. Thus even the *ideal* position measurement will change the state which is not a position eigenstate. Why does a position measurement alter the state $|p\rangle$, while momentum measurement does not? The answer is that an ideal position measurement uses photons of infinitely high momentum (as we will see) while an ideal momentum measurement uses photons of infinitesimally low momentum (as we have seen).

This then is the big difference between classical and quantum mechanics: an ideal measurement of any variable ω in classical mechanics leaves any state invariant, whereas the ideal measurement of Ω in quantum mechanics leaves only the eigenstates of Ω invariant.

The effect of measurement may be represented schematically as follows:

$$|\psi\rangle \xrightarrow[\Omega \text{ measured, } \omega \text{ obtained}]{} = \frac{\mathbb{P}_\omega|\psi\rangle}{\langle \mathbb{P}_\omega\psi|\mathbb{P}_\omega\psi\rangle^{1/2}}$$

where $\mathbb{P}_\omega$ is the projection operator associated with $|\omega\rangle$, and the state after measurement has been normalized. If ω is degenerate,

$$|\psi\rangle \to \frac{\mathbb{P}_\omega|\psi\rangle}{\langle \mathbb{P}_\omega\psi|\mathbb{P}_\omega\psi\rangle^{1/2}}$$

where $\mathbb{P}_\omega$ is the projection operator for the eigenspace $\mathbb{V}_\omega$. Special note should be taken of the following point: if the initial state $|\psi\rangle$ were unknown, and the measurement yielded a degenerate eigenvalue ω, we could not say what the state was after the measurement, except that it was some state in the eigenspace with eigenvalue ω. On the other hand, if the initial state $|\psi\rangle$ were known, and the measurement yielded a degenerate value ω, the state after measurement is known to be $\mathbb{P}_\omega|\psi\rangle$ (up to normalization). Consider our example from $\mathbb{V}^3(R)$ (Fig. 4.1b). Say we had $\omega_1 = \omega_2 = \omega$. Let us use an orthonormal basis $|\omega, 1\rangle$, $|\omega, 2\rangle$, $|\omega_3\rangle$, where, as usual, the extra labels 1 and 2 are needed to distinguish the basis vectors in the degenerate eigenspace. If in this basis we know, for example, that

$$|\psi\rangle = \tfrac{1}{2}|\omega, 1\rangle + \tfrac{1}{2}|\omega, 2\rangle + (\tfrac{1}{2})^{1/2}|\omega_3\rangle$$

and the measurement gives a value ω, the normalized state after measurement is known to us to be

$$|\psi\rangle = 2^{-1/2}(|\omega, 1\rangle + |\omega, 2\rangle)$$

If, on the other hand, the initial state were unknown and a measurement gave a result ω, we could only say

$$|\psi\rangle = \frac{\alpha|\omega, 1\rangle + \beta|\omega, 2\rangle}{(\alpha^2 + \beta^2)^{1/2}}$$

where α and β are arbitrary real numbers.

Note that although *we* do not know what α and β are from the measurement, they are not arbitrary. In other words, the system had a well-defined state vector $|\psi\rangle$ before the measurement, though *we* did not know $|\psi\rangle$, and has a well-defined state vector $\mathbb{P}_\omega|\psi\rangle$ after the measurement, although all *we* know is that it lies within a subspace $\mathbb{V}_\omega$.

One of the outstanding features of classical mechanics is that it makes fully deterministic predictions. It may predict for example that a particle leaving $x=x_i$ with momentum p_i in some potential $V(x)$ will arrive 2 seconds later at $x=x_f$ with momentum $p=p_f$. To test the prediction we release the particle at $x=x_i$ with $p=p_i$ at $t=0$ and wait at $x=x_f$ and see if the particle arrives there with $p=p_f$ at $t=2$ seconds.

Quantum theory, on the other hand, makes statistical predictions about a particle in a state $|\psi\rangle$ and claims that this state evolves in time according to Schrödinger's equation. To test these predictions we must be able to

(1) Create particles in a well-defined state $|\psi\rangle$.
(2) Check the probabilistic predictions at any time.

The collapse of the state vector provides us with a good way of preparing definite states: we begin with a particle in an arbitrary state $|\psi\rangle$ and meaure a variable Ω. If we get a nondegenerate eigenvalue ω, we have in our hands the state $|\omega\rangle$. (If ω is degenerate, further measurement is needed. We are not ready to discuss this problem.) Notice how in quantum theory, measurement, instead of telling us what the system was doing *before* the measurement, tells us what it is doing just *after* the measurement. (Of course it does tell us that the original state had some projection on the state $|\omega\rangle$ obtained after measurement. But this information is nothing compared to the complete specifications of the state just *after* measurement.)

Anyway, assume we have prepared a state $|\omega\rangle$. If we measure some variable Λ, immediately thereafter, so that the state could not have changed from $|\omega\rangle$, and if say,

$$|\omega\rangle=\frac{1}{3^{1/2}}|\lambda_1\rangle+\left(\frac{2}{3}\right)^{1/2}|\lambda_2\rangle+0\cdot(\text{others})$$

the theory predicts that λ_1 and λ_2 will obtain with probabilities 1/3 and 2/3, respectively. If our measurement gives a λ_i, $i\neq 1, 2$ (or worse still a $\lambda\neq$any eigenvalue!) that is the end of the theory. So let us assume we get one of the allowed values, say λ_1. This is consistent with the theory but does not fully corroborate it, since the odds for λ_1 could have been 1/30 instead of 1/3 and we could still get λ_1. Therefore, we must repeat the experiment many times. But we cannot repeat the experiment with *this* particle, since after the measurement the state of the particle is $|\lambda_1\rangle$. We must start afresh with another particle in $|\omega\rangle$. For this purpose we require a *quantum ensemble*, which consists of a large number N of particles *all in the same state* $|\omega\rangle$. If a measurement of Λ is made on every one of these particles, approximately $N/3$ will yield a value λ_1 and end up in the state $|\lambda_1\rangle$ while approximately $2N/3$ will yield a value λ_2 and end up in a state $|\lambda_2\rangle$. For sufficiently large N, the deviations from the fractions 1/3 and 2/3 will be negligible. The chief difference between a classical ensemble, of the type one encounters in, say, classical statistical mechanics, and the quantum ensemble referred to above, is the following. If in a classical ensemble of N particles $N/3$ gave a result λ_1 and $2N/3$ a result λ_2, one can think of the ensemble as having contained $N/3$ particles with $\lambda=\lambda_1$ and the others with $\lambda=\lambda_2$ *before* the measurement. In a quantum ensemble, on the other hand, every particle is assumed to be in the same state $|\omega\rangle$ prior to measurement (i.e., every particle is potentially capable of yielding *either* result λ_1 or λ_2). Only after that measurement are a third of them forced into the state $|\lambda_1\rangle$ and the rest into $|\lambda_2\rangle$.

Once we have an ensemble, we can measure any other variable and test the expectations of quantum theory. We can also prepare an ensemble, let it evolve in time, and study it at a future time to see if the final state is what the Schrödinger equation tells us it should be.

Example 4.2.2. An example of an ensemble being used to test quantum theory was encountered in the double-slit experiment, say with photons. A given photon of momentum p and energy E was expected to hit the detectors with a probability density given by the oscillating function $|\psi(x)|^2$. One could repeat the experiment N times, sending one such photon at a time to see if the final number distribution indeed was given by $|\psi(x)|^2$. One could equally well send in a macroscopic, monochromatic beam of light of frequency $\omega=E/\hbar$ and wave number $k=p/\hbar$, which consists of a large number of photons of energy E and momentum p. If one makes the assumption (correct to a high degree) that the photons are noninteracting, sending in the beam is equivalent to experimenting with the ensemble. In this case the intensity pattern will take the shape of the probability density $|\psi(x)|^2$, the instant the beam is turned on. □

Example 4.2.3. The following example is provided to illustrate the distinction between the probabilistic descriptions of systems in classical mechanics and in quantum mechanics.

We choose as our classical system a six-faced die for which the probabilities $P(n)$ of obtaining a number n have been empirically determined. As our quantum system we take a particle in a state

$$|\psi\rangle=\sum_{i=1}^{6} C_i|\omega_i\rangle$$

Suppose we close our eyes, toss the die, and cover it with a mug. Its statistical description has many analogies with the quantum description of the state $|\psi\rangle$:

(1) The state of the die is described by a probability function $P(n)$ before the mug is lifted.
(2) The only possible values of n are 1, 2, 3, 4, 5, and 6.
(3) If the mug is lifted, and some value—say $n=3$—is obtained, the function $P(n)$ collapses to δ_{n3}.
(4) If an ensemble of N such dice are thrown, $NP(n)$ of them will give the result n (as $N\to\infty$).

The corresponding statements for the particle in the state $|\psi\rangle$ are no doubt known to you. Let us now examine some of the key differences between the statistical descriptions in the two cases.

(1) It is possible, *at least in principle*, to predict exactly which face of the die will be on top, given the mass of the die, its position, orientation, velocity, and angular velocity at the time of release, the viscosity of air, the elasticity of the table top, and so on. The statistical description is, however, the only possibility in the quantum case, *even in principle*.

(2) If the result $n=3$ was obtained upon lifting the mug, it is consistent to assume that the die was in such a state *even prior to measurement*. In the quantum case, however, the state after measurement, say $|\omega_3\rangle$, is not the state before measurement, namely $|\psi\rangle$.

(3) If N such dice are tossed and covered with N mugs, there will be $NP(1)$ dice with $n=1$, $NP(2)$ dice with $n=2$, etc. in the ensemble *before and after the measurement*. In contrast, the quantum ensemble corresponding to $|\psi\rangle$ will contain N particles all of which are in the same state $|\psi\rangle$ (that is, each can yield any of the values $\omega_1, \ldots, \omega_6$) before the measurement, and $NP(\omega_i)$ particles in $|\omega_i\rangle$ *after* the measurement. Only the ensemble before the measurement represents the state $|\psi\rangle$. The ensemble after measurement is a mixture of six ensembles representing the states $|\omega_1\rangle, \ldots, |\omega_6\rangle$.‡ □

Having seen the utility of the ensemble concept in quantum theory, we now define and discuss the two statistical variables that characterize an ensemble.

Expectation Value

Given a large ensemble of N particles in a state $|\psi\rangle$, quantum theory allows us to predict what fraction will yield a value ω if the variable Ω is measured. This prediction, however, involves solving the eigenvalue problem of the operator Ω. If one is not interested in such detailed information on the state (or the corresponding ensemble) one can calculate instead an average over the ensemble, called the *expectation value*, $\langle\Omega\rangle$. The expectation value is just the mean value defined in statistics:

$$\begin{aligned}\langle\Omega\rangle &= \sum_i P(\omega_i)\omega_i = \sum_i |\langle\omega_i|\psi\rangle|^2\omega_i \\ &= \sum_i \langle\psi|\omega_i\rangle\langle\omega_i|\psi\rangle\omega_i \end{aligned} \tag{4.2.5}$$

But for the factors ω_i multiplying each projection operator $|\omega_i\rangle\langle\omega_i|$, we could have used $\sum_i |\omega_i\rangle\langle\omega_i| = I$. To get around this, note that $\omega_i|\omega_i\rangle = \Omega|\omega_i\rangle$. Feeding this in and continuing, we get

$$\langle\Omega\rangle = \sum_i \langle\psi|\Omega|\omega_i\rangle\langle\omega_i|\psi\rangle$$

Now we can use $\sum_i |\omega_i\rangle\langle\omega_i| = I$ to get

$$\langle\Omega\rangle = \langle\psi|\Omega|\psi\rangle \tag{4.2.6}$$

‡ This is an example of a *mixed* ensemble. These will be discussed in the digression on density matrices, which follows in a while.

There are a few points to note in connection with this formula.

(1) To calculate $\langle\Omega\rangle$, one need only be given the state vector and the operator Ω (say as a column vector and a matrix, respectively, in some basis). There is no need to find the eigenvectors or eigenvalues of Ω.
(2) If the particle is in an eigenstate of Ω, that is $\Omega|\psi\rangle=\omega|\psi\rangle$, then $\langle\Omega\rangle=\omega$.
(3) By the average value of Ω we mean the average over the ensemble. A given particle will of course yield only one of the eigenvalues upon measurement. The mean value will generally be an inaccessible value for a single measurement unless it accidentally equals an eigenvalue. [A familiar example of this phenomenon is that of the mean number of children per couple, which may be 2.12, although the number in a given family is restricted to be an integer.]

The Uncertainty

In any situation described probabilistically, another useful quantity to specify besides the mean is the *standard deviation*, which measures the average fluctuation around the mean. It is defined as

$$\Delta\Omega=\langle(\Omega-\langle\Omega\rangle)^2\rangle^{1/2} \tag{4.2.7}$$

and often called the root-mean-squared deviation. In quantum mechanics, it is referred to as the *uncertainty in* Ω. If Ω has a discrete spectrum

$$(\Delta\Omega)^2=\sum_i P(\omega_i)(\omega_i-\langle\Omega\rangle)^2 \tag{4.2.8}$$

and if it has a continuous spectrum,

$$(\Delta\Omega)^2=\int P(\omega)(\omega-\langle\Omega\rangle)^2\,d\omega \tag{4.2.9}$$

Notice that $\Delta\Omega$, just like $\langle\Omega\rangle$, is also calculable given just the state and the operator, for Eq. (4.2.7) means just

$$\Delta\Omega=[\langle\psi|(\Omega-\langle\Omega\rangle)^2|\psi\rangle]^{1/2} \tag{4.2.10}$$

Usually the expectation value and the uncertainty provide us with a fairly good description of the state. For example, if we are given that a particle has $\langle X\rangle=a$ and $\Delta X=\Delta$, we know that the particle is likely to be spotted near $x=a$, with deviations of order Δ.

So far, we have concentrated on the measurement of a single variable at a time. We now turn our attention to the measurement of more than one variable at a time. (Since no two independent measurements can really be performed at the same time, we really mean the measurement of two or more dynamical variables in rapid succession.)

Exercise 4.2.1 (Very Important). Consider the following operators on a Hilbert space $\mathbb{V}^3(C)$:

$$L_x = \frac{1}{2^{1/2}}\begin{bmatrix} 0 & 1 & 0 \\ 1 & 0 & 1 \\ 0 & 1 & 0 \end{bmatrix}, \qquad L_y = \frac{1}{2^{1/2}}\begin{bmatrix} 0 & -i & 0 \\ i & 0 & -i \\ 0 & i & 0 \end{bmatrix}, \qquad L_z = \begin{bmatrix} 1 & 0 & 0 \\ 0 & 0 & 0 \\ 0 & 0 & -1 \end{bmatrix}$$

(1) What are the possible values one can obtain if L_z is measured?
(2) Take the state in which $L_z = 1$. In this state what are $\langle L_x \rangle$, $\langle L_x^2 \rangle$, and ΔL_x?
(3) Find the normalized eigenstates and the eigenvalues of L_x in the L_z basis.
(4) If the particle is in the state with $L_z = -1$, and L_x is measured, what are the possible outcomes and their probabilities?
(5) Consider the state

$$|\psi\rangle = \begin{bmatrix} 1/2 \\ 1/2 \\ 1/2^{1/2} \end{bmatrix}$$

in the L_z basis. If L_z^2 is measured in this state and a result $+1$ is obtained, what is the state after the measurement? How probable was this result? If L_z is measured immediately afterwards, what are the outcomes and respective probabilities?

(6) A particle is in a state for which the probabilities are $P(L_z = 1) = 1/4$, $P(L_z = 0) = 1/2$, and $P(L_z = -1) = 1/4$. Convince yourself that the most general, normalized state with this property is

$$|\psi\rangle = \frac{e^{i\delta_1}}{2}|L_z = 1\rangle + \frac{e^{i\delta_2}}{2^{1/2}}|L_z = 0\rangle + \frac{e^{i\delta_3}}{2}|L_z = -1\rangle$$

It was stated earlier on that if $|\psi\rangle$ is a normalized state then the state $e^{i\theta}|\psi\rangle$ is a physically equivalent normalized state. Does this mean that the factors $e^{i\delta_i}$ multiplying the L_z eigenstates are irrelevant? [Calculate for example $P(L_x = 0)$.]

Compatible and Incompatible Variables

A striking feature of quantum theory is that given a particle in a state $|\psi\rangle$, one cannot say in general that the particle has a definite value for a given dynamical variable Ω: a measurement can yield any eigenvalue ω for which $\langle \omega | \psi \rangle$ is not zero. The exceptions are the states $|\omega\rangle$. A particle in one of these states can be said, as in classical mechanics, to have a value ω for Ω, since a measurement is assured to give this result. To produce such states we need only take an arbitrary state $|\psi\rangle$ and measure Ω. The measurement process acts as a filter that lets through just one component of $|\psi\rangle$, along some $|\omega\rangle$. The probability that this will happen is $P(\omega) = |\langle\omega|\psi\rangle|^2$.

We now wish to extend these ideas to more than one variable. We consider first the question of two operators. The extension to more than two will be straightforward. We ask:

Question 1. Is there some multiple filtering process by which we can take an ensemble of particles in some state $|\psi\rangle$ and produce a state with well-defined values ω and λ for two variables Ω and Λ?

Question 2. What is the probability that the filtering will give such a state if we start with the state $|\psi\rangle$?

To answer these questions, let us try to devise a multiple filtering scheme. Let us first measure Ω on the ensemble described by $|\psi\rangle$ and take the particles that yield a result ω. These are in a state that has a well-defined value for Ω. We *immediately* measure Λ and pick those particles that give a result λ. Do we have now an ensemble that is in a state with $\Omega=\omega$ and $\Lambda=\lambda$? Not generally. The reason is clear. After the first measurement, we had the system in the state $|\omega\rangle$, which assured a result ω for Ω, but nothing definite for Λ (since $|\omega\rangle$ need not be an eigenstate of Λ). Upon performing the second measurement, the state was converted to

$$|\psi'\rangle=|\lambda\rangle$$

and we are now assured a result λ for Λ, but nothing definite for Ω (since $|\lambda\rangle$ need not be an eigenstate of Ω).

In other words, the second filtering generally alters the state produced by the first. This change is just the collapse of the state vector $|\omega\rangle=\sum|\lambda\rangle\langle\lambda|\omega\rangle$ into the eigenstate $|\lambda\rangle$.

An exception occurs when the state produced after the first measurement is unaffected by the second. This in turn requires that $|\omega\rangle$ also be an eigenstate of Λ. The answer to the first question above is then in the affirmative only for the simultaneous eigenstates $|\omega\lambda\rangle$. The means for producing them are just as described above. These kets satisfy the equations

$$\Omega|\omega\lambda\rangle=\omega|\omega\lambda\rangle \tag{4.2.11}$$

$$\Lambda|\omega\lambda\rangle=\lambda|\omega\lambda\rangle \tag{4.2.12}$$

The question that arises naturally is: When will two operators admit simultaneous eigenkets? A necessary (but not sufficient) condition is obtained by operating Eq. (4.2.12) with Ω, Eq. (4.2.11) with Λ, and taking the difference:

$$(\Omega\Lambda-\Lambda\Omega)|\omega\lambda\rangle=0 \tag{4.2.13}$$

Thus $[\Omega, \Lambda]$ must have eigenkets with zero eigenvalue if simultaneous eigenkets are to exist. A pair of operators Ω and Λ will fall into one of the three classes:

A. Compatible: $[\Omega, \Lambda]=0$
B. Incompatible: $[\Omega, \Lambda]=$ something that obviously has no zero eigenvalue
C. Others

Class A. If two operators commute, we know a complete basis of simultaneous eigenkets can be found. Each element $|\omega\lambda\rangle$ of this basis has well-defined values for Ω and Λ.

Class B. The most famous example of this class is provided by the position and momentum operators X and P, which obey the *canonical commutation rule*

$$[X, P] = i\hbar \tag{4.2.14}$$

Evidently we cannot ever have $i\hbar|\psi\rangle = 0|\psi\rangle$ for any nontrivial $|\psi\rangle$. *This means there doesn't exist even a single ket for which both X and P are well defined.* Any attempt to filter X is ruined by a subsequent filtering for P and vice vesa. This is the origin of the famous *Heisenberg uncertainty principle*, which will be developed as we go along.

Class C. In this case there are *some* states that are simultaneous eigenkets. There is nothing very interesting we can say about this case except to emphasize that even if two operators don't commute, one can still find a few common eigenkets, though not a full basis. (Why?)

Let us now turn to the second question of the probability of obtaining a state $|\omega\lambda\rangle$ upon measurement of Ω and Λ in a state $|\psi\rangle$. We will consider just case A; the question doesn't arise for case B, and case C is not very interesting. (You should be able to tackle case C yourself after seeing the other two cases.)

Case A. Let us first assume there is no degeneracy. Thus, to a given eigenvalue λ, there is just one ket and this must be a simultaneous eigenket $|\omega\lambda\rangle$. Suppose we measured Ω first. We get ω with a probability $P(\omega) = |\langle\omega\lambda|\psi\rangle|^2$. After the measurement, the particle is in a state $|\omega\lambda\rangle$. The measurement of Λ is certain to yield the result λ. The probability for obtaining ω for Ω and λ for Λ is just the product of the two probabilities

$$P(\omega, \lambda) = |\langle\omega\lambda|\psi\rangle|^2 \cdot 1 = |\langle\omega\lambda|\psi\rangle|^2$$

Notice that if Λ were measured first and Ω next, the probability is the same for getting the results λ and ω. Thus if we expand $|\psi\rangle$ in the complete common eigenbasis as

$$|\psi\rangle = \sum |\omega\lambda\rangle\langle\omega\lambda|\psi\rangle \tag{4.2.15a}$$

then

$$P(\omega, \lambda) = |\langle\omega\lambda|\psi\rangle|^2 = P(\lambda, \omega) \tag{4.2.15b}$$

The reason for calling Ω and Λ compatible if $[\Omega, \Lambda] = 0$ is that the measurement of one variable followed by the other doesn't alter the *eigenvalue* obtained in the first measurement and we have in the end a state with a well-defined value for both observables. Note the emphasis on the invariance of the *eigenvalue* under the second measurement. In the non-degenerate case, this implies the invariance of the state vector as well. In the degenerate case, the state vector can change due to the second measurement, though the eigenvalue will not, as the following example will show. Consider two operators Λ and Ω on $\mathbb{V}^3(R)$. Let $|\omega_3\lambda_3\rangle$ be one common eigenvector. Let $\lambda_1=\lambda_2=\lambda$. Let $\omega_1\neq\omega_2$ be the eigenvalues of Ω in this degenerate space. Let us use as a basis $|\omega_1\lambda\rangle$, $|\omega_2\lambda\rangle$, and $|\omega_3\lambda_3\rangle$. Consider a normalized state

$$|\psi\rangle=\alpha|\omega_3\lambda_3\rangle+\beta|\omega_1\lambda\rangle+\gamma|\omega_2\lambda\rangle \tag{4.2.16}$$

Let us say we measure Ω first and get ω_3. The state becomes $|\omega_3\lambda_3\rangle$ and the subsequent measurement of Λ is assured to give a value λ_3 and to leave the state alone. Thus $P(\omega_3, \lambda_3)=|\langle\omega_3\lambda_3|\psi\rangle|^2=\alpha^2$. Evidently $P(\omega_3, \lambda_3)=P(\lambda_3, \omega_3)$.

Suppose that the measurement of Ω gave a value ω_1. The resulting state is $|\omega_1\lambda\rangle$ and the probability for this outcome is $|\langle\omega_1\lambda|\psi\rangle|^2$. The subsequent measurement of Λ will leave the state alone and yield the result λ with unit probability. Thus $P(\omega_1, \lambda)$ is the product of the probabilities:

$$P(\omega_1, \lambda)=|\langle\omega_1\lambda|\psi\rangle|^2\cdot 1=|\langle\omega_1\lambda|\psi\rangle|^2=\beta^2 \tag{4.2.17}$$

Let us now imagine the measurements carried out in reverse order. Let the result of the measurement be λ. The state $|\psi'\rangle$ after measurement is the projection of $|\psi\rangle$ in the degenerate λ eigenspace:

$$|\psi'\rangle=\frac{\mathbb{P}_\lambda|\psi\rangle}{|\langle\mathbb{P}_\lambda\psi|\mathbb{P}_\lambda\psi\rangle|^{1/2}}=\frac{\beta|\omega_1\lambda\rangle+\gamma|\omega_2\lambda\rangle}{(\beta^2+\gamma^2)^{1/2}} \tag{4.2.18}$$

where, in the expression above, the projected state has been normalized. The probability for this outcome is $P(\lambda)=\beta^2+\gamma^2$, the square of the projection of $|\psi\rangle$ in the eigenspace. If Ω is measured now, both results ω_1 and ω_2 are possible. The probability for obtaining ω_1 is $|\langle\omega_1\lambda|\psi'\rangle|^2=\beta^2/(\beta^2+\gamma^2)$. Thus, the probability for the result $\Lambda=\lambda$, $\Omega=\omega_1$, is the product of the probabilities:

$$P(\lambda, \omega_1)=(\beta^2+\gamma^2)\cdot\frac{\beta^2}{\beta^2+\gamma^2}=\beta^2=P(\omega_1, \lambda) \tag{4.2.19}$$

Thus $P(\omega_1, \lambda)=P(\lambda, \omega_1)$ independent of the degeneracy. *But this time the state suffered a change due to the second measurement* (unless by accident $|\psi'\rangle$ has no component along $|\omega_2\lambda\rangle$). Thus compatibility generally implies the invariance under the second measurement of the *eigenvalue* measured in the first. Therefore, the state can only be said to remain in the same eigenspace after the second measurement. If the first eigenvalue is non-degenerate, the eigenspace is one dimensional and the state vector itself remains invariant.

In our earlier discussion on how to produce well-defined states $|\psi\rangle$ for testing quantum theory, it was observed that the measurement process could itself be used as a preparation mechanism: if the measurement of Ω on an arbitrary, unknown initial state given a result ω, we are sure we have the state $|\psi\rangle=|\omega\rangle$. But this presumes ω is not a degenerate eigenvalue. If it is degenerate, we cannot nail down the state, except to within an eigenspace. It was therefore suggested that we stick to variables with a nondegenerate spectrum. We can now lift that restriction. Let us say a degenerate eigenvalue ω for the variable Ω was obtained. We have then some vector in the ω eigenspace. We now measure another compatible variable Λ. If we get a result λ, we have a definite state $|\omega\lambda\rangle$, unless the value (ω, λ) itself is degenerate. We must then measure a third variable Γ compatible with Ω and Λ and so on. Ultimately we will get a state that is unique, given all the simultaneous eigenvalues: $|\omega, \lambda, \gamma, \ldots\rangle$. It is presumed that such a set of compatible observables, called a *complete set of commuting observables*, exists. To prepare a state for studying quantum theory then, we take an arbitrary initial state and filter it by a sequence of compatible measurements till it is down to a unique, known vector. Any nondegenerate operator, all by itself, is a "complete set."

Incidentally, even if the operators Ω and Λ are incompatible, we can specify the probability $P(\omega, \lambda)$ that the measurement of Ω followed by that of Λ on a state $|\psi\rangle$ will give the results ω and λ, respectively. However, the following should be noted:

(1) $P(\omega, \lambda) \neq P(\lambda, \omega)$ in general.

(2) The probability $P(\omega, \lambda)$ is not the probability for producing a final state that has well-defined values ω and λ for Ω and Λ. (Such a state doesn't exist by the definition of incompatibility.) The state produced by the two measurements is just the eigenstate of the second operator with the measured eigenvalue.

The Density Matrix—a Digression‡

So far we have considered ensembles of N systems all in the same state $|\psi\rangle$. They are hard to come by in practice. More common are ensembles of N systems, n_i $(i=1, 2, \ldots, k)$ of which are in the state $|i\rangle$. (We restrict ourselves to the case where $|i\rangle$ is an element of an orthonormal basis.) Thus the ensemble is described by k kets $|1\rangle, |2\rangle, \ldots, |k\rangle$, and k *occupancy numbers* $n_1, \ldots, n_k$. A convenient way to assemble all this information is in the form of the *density matrix* (which is really an operator that becomes a matrix in some basis):

$$\rho = \sum_i p_i |i\rangle\langle i| \tag{4.2.20}$$

where $p_i = n_i/N$ is the probability that a system picked randomly out of the ensemble is in the state $|i\rangle$. The ensembles we have dealt with so far are said to be *pure*; they correspond to all $p_i=0$ except one. A general ensemble is *mixed*.

Consider now the ensemble average of Ω. It is

$$\langle\bar{\Omega}\rangle = \sum_i p_i \langle i|\Omega|i\rangle \tag{4.2.21}$$

The bar on $\langle\bar{\Omega}\rangle$ reminds us that two kinds of averaging have been carried out: a quantum average $\langle i|\Omega|i\rangle$ for each system in $|i\rangle$ and a classical average over the

‡ This digression may be omitted or postponed without loss of continuity.

systems in different states $|i\rangle$. Observe that

$$\begin{aligned}\operatorname{Tr}(\Omega\rho)&=\sum_j \langle j|\Omega\rho|j\rangle\\&=\sum_j\sum_i \langle j|\Omega|i\rangle\langle i|j\rangle p_i=\sum_i\sum_j \langle i|j\rangle\langle j|\Omega|i\rangle p_i\\&=\sum_i \langle i|\Omega|i\rangle p_i\\&=\langle\bar{\Omega}\rangle\end{aligned}\tag{4.2.22}$$

The density matrix contains all the statistical information about the ensemble. Suppose we want, not $\langle\bar{\Omega}\rangle$, but instead $\overline{P(\omega)}$, the probability of obtaining a particular value ω. We first note that, for a pure ensemble,

$$P(\omega)=|\langle\omega|\psi\rangle|^2=\langle\psi|\omega\rangle\langle\omega|\psi\rangle=\langle\psi|\mathbb{P}_\omega|\psi\rangle=\langle\mathbb{P}_\omega\rangle$$

which combined with Eq. (4.2.22) tells us that

$$\overline{P(\omega)}=\operatorname{Tr}(\mathbb{P}_\omega\rho)$$

The following results may be easily established:

$$\begin{aligned}&(1)\ \rho^\dagger=\rho\\&(2)\ \operatorname{Tr}\rho=1\\&(3)\ \rho^2=\rho &&\text{for a pure ensemble}\\&(4)\ \rho=(1/k)I &&\text{for an ensemble uniformly distributed over } k \text{ states}\\&(5)\ \operatorname{Tr}\rho^2\le 1 &&\text{(equality holds for a pure ensemble)}\end{aligned}\tag{4.2.23}$$

You are urged to convince yourself of these relations.

Example 4.2.4. To gain more familiarity with quantum theory let us consider an infinite-dimensional ket $|\psi\rangle$ expanded in the basis $|x\rangle$ of the position operator X:

$$|\psi\rangle=\int_{-\infty}^{\infty}|x\rangle\langle x|\psi\rangle\,dx=\int_{-\infty}^{\infty}|x\rangle\psi(x)\,dx$$

We call $\psi(x)$ the wave function (in the X basis). Let us assume $\psi(x)$ is a Gaussian, that is, $\psi(x)=A\exp[-(x-a)^2/2\Delta^2]$ (Fig. 4.2a). We now try to extract information about this state by using the postulates. Let us begin by normalizing the state:

$$\begin{aligned}1=\langle\psi|\psi\rangle&=\int_{-\infty}^{\infty}\langle\psi|x\rangle\langle x|\psi\rangle\,dx=\int_{-\infty}^{\infty}|\psi(x)|^2\,dx\\&=\int_{-\infty}^{\infty}A^2e^{-(x-a)^2/\Delta^2}\,dx=A^2(\pi\Delta^2)^{1/2}\quad\text{(see Appendix A.2)}\end{aligned}$$

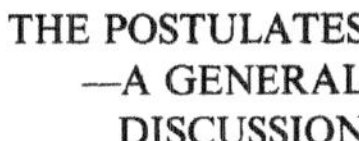

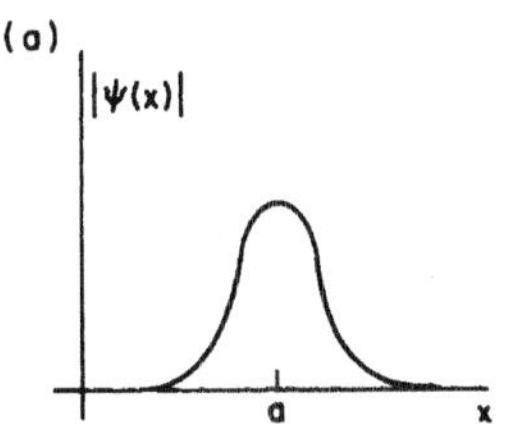

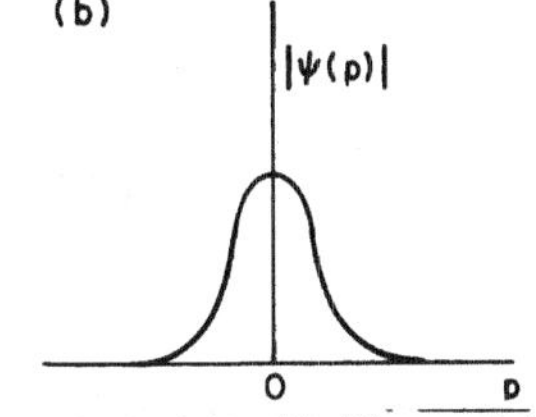

Figure 4.2. (a) The modulus of the wave function, $|\langle x|\psi\rangle| = |\psi(x)|$. (b) The modulus of the wave function, $|\langle p|\psi\rangle| = |\psi(p)|$.

So the normalized state is

$$\psi(x) = \frac{1}{(\pi\Delta^2)^{1/4}} e^{-(x-a)^2/2\Delta^2}$$

The probability for finding the particle between x and $x+dx$ is

$$P(x)\,dx = |\psi(x)|^2\,dx = \frac{1}{(\pi\Delta^2)^{1/2}} e^{-(x-a)^2/\Delta^2}\,dx$$

which looks very much like Fig. 4.2a. Thus the particle is most likely to be found around $x=a$, and chances of finding it away from this point drop rapidly beyond a distance Δ. We can quantify these statements by calculating the expectation value and uncertainty for X. Let us do so.

Now, the operator X defined in postulate II is the same one we discussed at length in Section 1.10. Its action in the X basis is simply to multiply by x, i.e., if

$$\langle x|\psi\rangle = \psi(x)$$

then,

$$\begin{aligned}\langle x|X|\psi\rangle &= \int_{-\infty}^{\infty} \langle x|X|x'\rangle\langle x'|\psi\rangle\,dx' = \int_{-\infty}^{\infty} x\delta(x-x')\psi(x')\,dx' \\ &= x\psi(x)\end{aligned}$$

Using this result, the mean or expectation value of X is

$$\begin{aligned}\langle X\rangle = \langle\psi|X|\psi\rangle &= \int_{-\infty}^{\infty} \langle\psi|x\rangle\langle x|X|\psi\rangle\,dx \\ &= \int_{-\infty}^{\infty} \psi^*(x)x\psi(x)\,dx \\ &= \frac{1}{(\pi\Delta^2)^{1/2}}\int_{-\infty}^{\infty} e^{-(x-a)^2/\Delta^2}x\,dx\end{aligned}$$

If we define $y=x-a$,

$$\langle X\rangle=\frac{1}{(\pi\Delta^2)^{1/2}}\int_{-\infty}^{\infty}(y+a)\,e^{-y^2/\Delta^2}\,dy$$
$$=a$$

We should have anticipated this result of course, since the probability density is symmetrically distributed around $x=a$.

Next, we calculate the fluctuations around $\langle X\rangle=a$, i.e., the uncertainty

$$\begin{aligned}\Delta X&=[\langle\psi|(X-\langle X\rangle)^2|\psi\rangle]^{1/2}\\&=[\langle\psi|X^2-2X\langle X\rangle+\langle X\rangle^2|\psi\rangle]^{1/2}\\&=[\langle\psi|X^2-\langle X\rangle^2|\psi\rangle]^{1/2}\qquad(\text{since }\langle\psi|X|\psi\rangle=\langle X\rangle)\\&=[\langle X^2\rangle-\langle X\rangle^2]^{1/2}\\&=[\langle X^2\rangle-a^2]^{1/2}\end{aligned}$$

Now

$$\langle X^2\rangle=\frac{1}{(\pi\Delta^2)^{1/2}}\int_{-\infty}^{\infty}e^{-(x-a)^2/2\Delta^2}\cdot x^2\cdot e^{-(x-a)^2/2\Delta^2}\,dx$$
$$=\frac{1}{(\pi\Delta^2)^{1/2}}\int_{-\infty}^{\infty}e^{-y^2/\Delta^2}\,(y^2+2ya+a^2)\,dy=\frac{\Delta^2}{2}+0+a^2$$

So

$$\Delta X=\frac{\Delta}{2^{1/2}}$$

So much for the information on the variable X. Suppose we next want to know the probability distribution for different values of another dynamical variable, say the momentum P.

(1) First we must construct the operator P in this basis.
(2) Then we must find its eigenvalues p, and eigenvectors $|p\rangle$.
(3) Finally, we must take the inner product $\langle p|\psi\rangle$.
(4) If p is discrete, $|\langle p_i|\psi\rangle|^2=P(p_i)$, and if p is continuous, $|\langle p|\psi\rangle|^2=P(p)$, the probability density.

Now, the P operator is just the K operator discussed in Section 1.10 multiplied by $\hbar$ and has the action of $-i\hbar\, d/dx$ in the X basis, for if

$$\langle x|\psi\rangle=\psi(x)$$

then

$$\langle x|P|\psi\rangle=\int_{-\infty}^{\infty}\langle x|P|x'\rangle\langle x'|\psi\rangle\,dx'$$

$$=\int_{-\infty}^{\infty}[-i\hbar\delta'(x-x')]\psi(x')\,dx' \qquad \text{(Postulate II)}$$

$$=-i\hbar\frac{d\psi}{dx}$$

Thus, if we project the eigenvalue equation

$$P|p\rangle=p|p\rangle$$

onto the X basis, we get

$$\langle x|P|p\rangle=p\langle x|p\rangle$$

or

$$-i\hbar\frac{d\psi_p(x)}{dx}=p\psi_p(x)$$

where $\psi_p(x)=\langle x|p\rangle$. The solutions, normalized to the Dirac delta function‡ are (from Section 1.10)

$$\psi_p(x)=\frac{1}{(2\pi\hbar)^{1/2}}e^{ipx/\hbar}$$

Now we can compute

$$\langle p|\psi\rangle=\int\langle p|x\rangle\langle x|\psi\rangle\,dx=\int\psi_p^*(x)\psi(x)\,dx$$

$$=\int_{-\infty}^{\infty}\frac{e^{-ipx/\hbar}}{(2\pi\hbar)^{1/2}}\frac{e^{-(x-a)^2/2\Delta^2}}{(\pi\Delta^2)^{1/4}}\,dx=\left(\frac{\Delta^2}{\pi\hbar^2}\right)^{1/4}e^{-ipa/\hbar}\,e^{-p^2\Delta^2/2\hbar^2}$$

The modulus of $\psi(p)$ is a Gaussian (Fig. 4.2b) of width $\hbar/2^{1/2}\Delta$. It follows that $\langle P\rangle=0$, and $\Delta P=\hbar/2^{1/2}\Delta$. Since $\Delta X=\Delta/2^{1/2}$; we get the relation

$$\Delta X\cdot\Delta P=\hbar/2$$

‡ Here we want $\langle p|p'\rangle=\delta(p-p')=\delta(k-k')/\hbar$, where $p=\hbar k$. This explains the $(2\pi\hbar)^{-1/2}$ normalization factor.

The Gaussian happens to saturate the lower bound of the uncertainty relation (to be formally derived in chapter 9):

$$\Delta X \cdot \Delta P \geq \hbar/2$$

The uncertainty relation is a consequence of the general fact that anything narrow in one space is wide in the transform space and vice versa. So if you are a 110-lb weakling and are taunted by a 600-lb bully, just ask him to step into momentum space! □

This is a good place to point out that the plane waves $e^{ipx/\hbar}$ (and *all improper vectors*, i.e., vectors that can't be normalized to unity but only to the Dirac delta function) are introduced into the formalism as purely mathematical entities. Our inability to normalize them to unity translates into our inability to associate with them a sensible absolute probability distribution, so essential to the physical interpretation of the wave function. In the present case we have a particle whose relative probability density is uniform in all of space. Thus the absolute probability of finding it in any finite volume, even as big as our solar system, is zero. Since any particle that we are likely to be interested in will definitely be known to exist in some finite volume of such large dimensions, it is clear that no physically interesting state will be given by a plane wave. But, since the plane waves are eigenfunctions of P, does it mean that states of well-defined momentum do not exist? Yes, in the strict sense. However, there do exist states that are both normalizable to unity (i.e., correspond to *proper* vectors) and come arbitrarily close to having a precise momentum. For example, a wave function that behaves as $e^{ip_0x/\hbar}$ over a large region of space and tapers off to zero beyond, will be normalizable to unity and will have a Fourier transform so sharply peaked at $p=p_0$ that momentum measurements will only give results practically indistinguishable from p_0. Thus there is no conflict between the fact that plane waves are unphysical, while states of well-defined momentum exist, for "well defined" never means "mathematically exact," but only "exact to any measurable accuracy." Thus a particle coming out of some accelerator with some advertised momentum, say 500 GeV/c, is in a proper normalizable state (since it is known to be located in our laboratory) and not in a plane wave state corresponding to $|p=500\ \text{GeV}/c\rangle$.

But despite all this, we will continue to use the eigenkets $|p\rangle$ as basis vectors and to speak of a particle being in the state $|p\rangle$, because these vectors are so much more convenient to handle mathematically than the proper vectors. It should, however, be borne in mind that when we say a particle is (coming out of the accelerator) in a state $|p_0\rangle$, it is really in a proper state with a momentum space wave function so sharply peaked at $p=p_0$ that it may be replaced by a delta function $\delta(p-p_0)$.

The other set of improper kets we will use in the same spirit are the position eigenkets $|x\rangle$, which also form a convenient basis. Again, when we speak of a particle being in a state $|x_0\rangle$ we shall mean that its wave function is so sharply peaked at $x=x_0$ that it may be treated as a delta function to a good accuracy.‡

‡ Thus, by the physical Hilbert space, we mean the space of interest to physicists, not one whose elements all correspond to physically realizable states.

Occasionally, the replacement of a proper wave function by its improper counterpart turns out to be a poor approximation. Here is an example from Chapter 19: Consider the probability that a particle coming out of an accelerator with a nearly exact momentum scatters off a target and enters a detector placed far away, and not in the initial direction. Intuition says that the answer must be zero if the target is absent. This reasonable condition is violated if we approximate the initial state of the particle by a plane wave (which is nonzero everywhere). So we proceed as follows. In the vicinity of the target, we use the plane wave to approximate the initial wave function, for the two are indistinguishable over the (finite and small) range of influence of the target. At the detector, however, we go back to the proper wave (which has tapered off) to represent the initial state.

*Exercise 4.2.2.** Show that for a real wave function $\psi(x)$, the expectation value of momentum $\langle P\rangle=0$. (Hint: Show that the probabilities for the momenta $\pm p$ are equal.) Generalize this result to the case $\psi=c\psi_r$, where ψ_r is real and c an arbitrary (real or complex) constant. (Recall that $|\psi\rangle$ and $\alpha|\psi\rangle$ are physically equivalent.)

*Exercise 4.2.3.** Show that if $\psi(x)$ has mean momentum $\langle P\rangle$, $e^{ip_0x/\hbar}\,\psi(x)$ has mean momentum $\langle P\rangle+p_0$.

Example 4.2.5. The collapse of the state vector and the uncertainty principle play a vital role in explaining the following extension of the double slit experiment. Suppose I say, "I don't believe that a given particle (let us say an electron) doesn't really go through one slit or the other. So I will set up a light source in between the slits to the right of the screen. Each passing electron will be exposed by the beam and I note which slit it comes out of. Then I note where it arrives on the screen. I make a table of how many electrons arrive at each x and which slit they came from. Now there is no escape from the conclusion that the number arriving at a given x is the sum of the numbers arriving via S_1 and S_2. So much for quantum theory and its interference pattern!"

But the point of course is that quantum theory no longer predicts an interference pattern! The theory says that if an electron of definite momentum p is involved, the corresponding wave function is a wave with a well-defined wave number $k=p/\hbar$, which interferes with itself and produces a nice interference pattern. This prediction is valid only as long as the state of the electron is what we say it is. But this state is necessarily altered by the light source, which upon measuring the position of the electron (as being next to S_1, say) changes its wave function from something that was extended in space to something localized near S_1. Once the state is changed, the old prediction of interference is no longer valid.

Now, once in a while some electrons will get to the detectors without being detected by the light source. We note where these arrive, but cannot classify them as coming via S_1 or S_2. When the distribution of just these electrons is plotted; sure enough we get the interference pattern. We had better, for quantum theory predicts it, the state not having been tampered with in these cases.

The above experiment can also be used to demystify to some extent the collapse of the wave function under measurement. Why is it that even the ideal measurement produces unavoidable changes in the state? The answer, as we shall see, has to do with the fact that $\hbar$ is not zero.

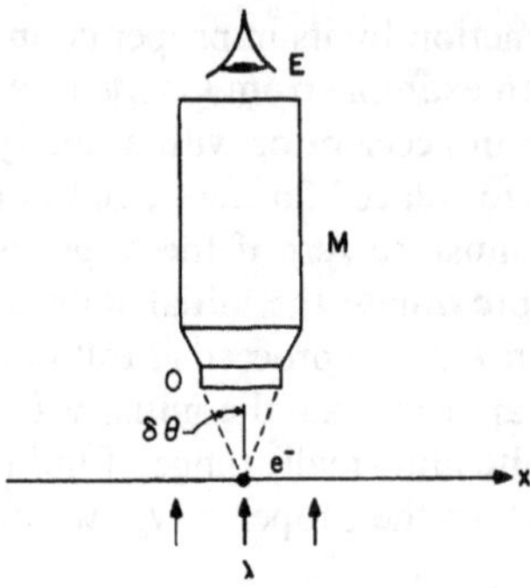

Figure 4.3. Light of frequency λ bounces off the electron, enters the objective O of the microscope, and enters the eye E of the observer.

Consider the schematic set up in Fig. 4.3. Light of wavelength λ illuminates an electron (e^-), enters the objective (O) of a microscope (M) and reaches our eye (E). If $\delta\theta$ is the opening angle of the cone of light entering the objective after interacting with the electron, classical optics limits the accuracy of the position measurement by an uncertainty

$$\Delta X \cong \lambda/\sin\delta\theta$$

Both classically and quantum mechanically, we can reduce ΔX to 0 by reducing λ to zero.‡ In the latter description however, the improved accuracy in the position measurement is at the expense of producing an increased uncertainty in the x component (p_x) of the electron momentum. The reason is that light of wavelength λ is not a continuous wave whose impact on the electron momentum may be arbitrarily reduced by a reduction of its amplitude, but rather a flux of photons of momentum $p = 2\pi\hbar/\lambda$. As λ decreases, the collisions between the electron and the photons become increasingly violent. This in itself would not lead to an uncertainty in the electron momentum, were it not for the fact that the x component of the photons entering the objective can range from 0 to $p\sin\delta\theta = 2\pi\hbar\sin\delta\theta/\lambda$. Since at least one photon must reach our eyes after bouncing off the electron for us to see it, there is a minimum uncertainty in the recoil momentum of the electron given by

$$\Delta P_x \simeq \frac{2\pi\hbar}{\lambda}\sin\delta\theta$$

Consequently, we have at the end of our measurement an electron whose position and momenta are uncertain by ΔX and ΔP_x such that

$$\Delta X \cdot \Delta P_x \simeq 2\pi\hbar \simeq \hbar$$

[The symbols ΔX and ΔP_x are not precisely the quantities defined in Eq. (4.2.7) but are of the same order of magnitude.] This is the famous *uncertainty principle*. There is no way around it. If we soften the blow of each photon by increasing λ or narrowing the objective to better constrain the final photon momentum, we lose in resolution.

‡ This would be the ideal position measurement.

More elaborate schemes, which determine the recoil of the microscope, are equally futile. Note that if $\hbar$ were 0, we could have ΔX and ΔP_x simultaneously 0. Physically, it means that we can increase our position resolution without increasing the punch carried by the photons. Of course $\hbar$ is not zero and we can't make it zero in any experiment. But what we can do is to use bigger and bigger objects for our experiment so that in the scale of these objects $\hbar$ appears to be negligible. We then regain classical mechanics. The position of a billiard ball can be determined very well by shining light on it, but this light hardly affects its momentum. This is why one imagines in classical mechanics that momentum and position can be well defined simultaneously. □

Generalization to More Degrees of Freedom

Our discussion so far has been restricted to a system with one degree of freedom—namely, a single particle in one dimension. We now extend our domain to a system with N degrees of freedom. The only modification is in postulate II, which now reads as follows.

Postulate II. Corresponding to the N Cartesian coordinates $x_1, \ldots, x_N$ describing the classical system, there exist in quantum theory N mutually commuting operators $X_1, \ldots, X_N$. In the simultaneous eigenbasis $|x_1, x_2, \ldots, x_N\rangle$ of these operators, called the *coordinate basis* and normalized as

$$\langle x_1, x_2, \ldots, x_N | x_1', x_2', \ldots, x_N'\rangle = \delta(x_1 - x_1') \ldots \delta(x_N - x_N')$$

(the product of delta functions vanishes unless all the arguments vanish) we have the following correspondence:

$$|\psi\rangle \rightarrow \langle x_1, \ldots, x_N | \psi\rangle = \psi(x_1, \ldots, x_N)$$

$$X_i |\psi\rangle \rightarrow \langle x_1, \ldots, x_N | X_i | \psi\rangle = x_i \psi(x_1, \ldots, x_N)$$

$$P_i |\psi\rangle \rightarrow \langle x_1, \ldots, x_N | P_i | \psi\rangle = -i\hbar \frac{\partial}{\partial x_i} \psi(x_1, \ldots, x_N)$$

P_i being the momentum operator corresponding to the classical momentum p_i. Dependent dynamical variables $\omega(x_i, p_j)$ are represented by operators $\Omega = \omega(x_i \rightarrow X_i, p_j \rightarrow P_j)$.

The other postulates remain the same. For example $|\psi(x_1, \ldots, x_N)|^2 \times dx_1 \ldots dx_N$ is the probability that the particle coordinates lie between $x_1, x_2, \ldots, x_N$ and $x_1 + dx_1, x_2 + dx_2, \ldots, x_N + dx_N$.

This postulate is stated in terms of Cartesian coordinates since only in terms of these can one express the operator assignments in the simple form $X_i \rightarrow x_i$, $P_i \rightarrow -i\hbar\, \partial/\partial x_i$. Once the substitutions have been made and the desired equations obtained in the coordinate basis, one can perform any desired change of variable before solving them. Suppose, for example, that we want to find the eigenvalues and eigenvectors of the operator Ω, corresponding to the classical variable

$$\omega=\frac{p_1^2+p_2^2+p_3^2}{2m}+x_1^2+x_2^2+x_3^2 \tag{4.2.24}$$

where x_1, x_2, and x_3 are the three Cartesian coordinates and p_i the corresponding momenta of a particle of mass m in three dimensions. Since the coordinates are usually called x, y, and z, let us follow this popular notation and rewrite Eq. (4.2.24) as

$$\omega=\frac{p_x^2+p_y^2+p_z^2}{2m}+x^2+y^2+z^2 \tag{4.2.25}$$

To solve the equation

$$\Omega|\omega\rangle=\omega|\omega\rangle$$

with

$$\Omega=\frac{P_x^2+P_y^2+P_z^2}{2m}+X^2+Y^2+Z^2$$

we make the substitution

$$|\omega\rangle\to\psi_\omega(x,y,z)$$

$$X\to x, \qquad P_x\to -i\hbar\frac{\partial}{\partial x}$$

etc. and get

$$\left[\frac{-\hbar^2}{2m}\left(\frac{\partial^2}{\partial x^2}+\frac{\partial^2}{\partial y^2}+\frac{\partial^2}{\partial z^2}\right)+x^2+y^2+z^2\right]\psi_\omega(x,y,z)=\omega\psi_\omega(x,y,z) \tag{4.2.26}$$

Once we have obtained this differential equation, we can switch to any other set of coordinates. In the present case the spherical coordinates r, θ, and ϕ recommend themselves. Since

$$\frac{\partial^2}{\partial x^2}+\frac{\partial^2}{\partial y^2}+\frac{\partial^2}{\partial z^2}$$

$$\equiv\nabla^2\equiv\frac{1}{r^2}\left[\frac{\partial}{\partial r}\left(r^2\frac{\partial}{\partial r}\right)+\frac{1}{\sin\theta}\frac{\partial}{\partial\theta}\left(\sin\theta\frac{\partial}{\partial\theta}\right)+\frac{1}{\sin^2\theta}\frac{\partial^2}{\partial\phi^2}\right]$$

Eq. (4.2.26) becomes

$$\frac{-\hbar^2}{2m}\left[\frac{1}{r^2}\frac{\partial}{\partial r}\left(r^2\frac{\partial\psi_\omega}{\partial r}\right)+\frac{1}{r^2\sin\theta}\frac{\partial}{\partial\theta}\left(\sin\theta\frac{\partial\psi_\omega}{\partial\theta}\right)+\frac{1}{r^2\sin^2\theta}\frac{\partial^2\psi_\omega}{\partial\phi^2}\right]$$
$$+r^2\psi_\omega=\omega\psi_\omega \qquad (4.2.27)$$

What if we wanted to go directly from ω in spherical coordinates

$$\omega=\frac{1}{2m}\left(p_r^2+\frac{p_\theta^2}{r^2}+\frac{p_\phi^2}{r^2\sin^2\theta}\right)+r^2$$

to Eq. (4.2.27)? It is clear upon inspection that there exists no simple rule [such as $p_r\to(-i\hbar\,\partial/\partial r)$] for replacing the classical momenta by differential operators in r, θ, and ϕ which generates Eq. (4.2.27) starting from the ω above. There does exist a complicated procedure for quantizing in non-Cartesian coordinates, but we will not discuss it, since the recipe eventually reproduces what the Cartesian recipe (which seems to work‡) yields so readily.

There are further generalizations, namely, to relativistic quantum mechanics and to quantum mechanics of systems in which particles are created and destroyed (so that the number of degrees of freedom changes!). Except for a brief discussion of these toward the end of the program, we will not address these matters.

4.3. The Schrödinger Equation (Dotting Your i's and Crossing Your $\hbar$'s)

Having discussed in some detail the state at a given time, we now turn our attention to postulate IV, which specifies the change of this state with time. According to this postulate, the state obeys the Schrödinger equation

$$i\hbar\frac{d}{dt}|\psi(t)\rangle=H|\psi(t)\rangle \qquad (4.3.1)$$

Our discussion of this equation is divided into three sections:

(1) Setting up the equation
(2) General approach to its solution
(3) Choosing a basis for solving the equation

Setting Up the Schrödinger Equation

To set up the Schrödinger equation one must simply make the substitution $\mathscr{H}(x\to X, p\to P)$, where $\mathscr{H}$ is the classical Hamiltonian for the same problem. Thus,

‡ In the sense that in cases where comparison with experiment is possible, as in say the hydrogen spectrum, there is agreement.

if we are describing a harmonic oscillator, which is classically described by the Hamiltonian

$$\mathscr{H}=\frac{p^2}{2m}+\frac{1}{2}m\omega^2x^2 \tag{4.3.2}$$

the Hamiltonian operator in quantum mechanics is

$$H=\frac{P^2}{2m}+\frac{1}{2}m\omega^2X^2 \tag{4.3.3}$$

In three dimensions, the Hamiltonian operator for the quantum oscillator is likewise

$$H=\frac{P_x^2+P_y^2+P_z^2}{2m}+\frac{1}{2}m\omega^2(X^2+Y^2+Z^2) \tag{4.3.4}$$

assuming the force constant is the same in all directions.

If the particle in one dimension is subject to a constant force f, then

$$\mathscr{H}=\frac{p^2}{2m}-fx$$

and

$$H=\frac{P^2}{2m}-fX \tag{4.3.5}$$

For a particle of charge q in an electromagnetic field in three dimensions,

$$\mathscr{H}=\frac{|\mathbf{p}-(q/c)\mathbf{A}(\mathbf{r},t)|^2}{2m}+q\phi(\mathbf{r},t) \tag{4.3.6}$$

In constructing the corresponding quantum Hamiltonian operator, we must use the symmetrized form

$$H=\frac{1}{2m}\left(\mathbf{P}\cdot\mathbf{P}-\frac{q}{c}\mathbf{P}\cdot\mathbf{A}-\frac{q}{c}\mathbf{A}\cdot\mathbf{P}+\frac{q^2}{c^2}\mathbf{A}\cdot\mathbf{A}\right)+q\phi \tag{4.3.7}$$

since $\mathbf{P}$ does not commute with $\mathbf{A}$, which is a function of X, Y, and Z.

In this manner one can construct the Hamiltonian H for any problem with a classical counterpart. Problems involving spin have no classical counterparts and some improvisation is called for. We will discuss this question when we study spin in some detail in Chapter 14.

Let us first assume that H has no explicit t dependence. In this case the equation

$$i\hbar|\dot{\psi}\rangle = H|\psi\rangle$$

is analogous to equations discussed in Chapter 1

$$|\ddot{x}\rangle = \Omega|x\rangle$$

and

$$|\ddot{\psi}\rangle = -K^2|\psi\rangle$$

describing the coupled masses and the vibrating string, respectively. Our approach will once again be to find the eigenvectors and eigenvalues of H and to construct the propagator $U(t)$ in terms of these. Once we have $U(t)$, we can write

$$|\psi(t)\rangle = U(t)|\psi(0)\rangle$$

There is no need to make assumptions about $|\dot{\psi}(0)\rangle$ here, since it is determined by Eq. (4.3.1):

$$|\dot{\psi}(0)\rangle = \frac{-i}{\hbar} H|\psi(0)\rangle$$

In other words, Schrödinger's equation is first order in time, and the specification of $|\psi\rangle$ at $t=0$ is a sufficient initial-value datum.

Let us now construct an explicit expression for $U(t)$ in terms of $|E\rangle$, the normalized eigenkets of H with eigenvalues E which obey

$$H|E\rangle = E|E\rangle \tag{4.3.8}$$

This is called the *time-independent Schrödinger equation*. Assume that we have solved it and found the kets $|E\rangle$. If we expand $|\psi\rangle$ as

$$|\psi(t)\rangle = \sum |E\rangle\langle E|\psi(t)\rangle \equiv \sum a_E(t)|E\rangle \tag{4.3.9}$$

the equation for $a_E(t)$ follows if we act on both sides with $(i\hbar\,\partial/\partial t - H)$:

$$0 = (i\hbar\,\partial/\partial t - H)|\psi(t)\rangle = \sum (i\hbar\dot{a}_E - Ea_E)|E\rangle \Rightarrow i\hbar\dot{a}_E = Ea_E \tag{4.3.10}$$

where we have used the linear independence of the kets $|E\rangle$. The solution to Eq. (4.3.10) is

$$a_E(t) = a_E(0)\, e^{-iEt/\hbar} \tag{4.3.11a}$$

or

$$\langle E|\psi(t)\rangle = \langle E|\psi(0)\rangle\, e^{-iEt/\hbar} \tag{4.3.11b}$$

so that

$$|\psi(t)\rangle = \sum_E |E\rangle\langle E|\psi(0)\rangle\, e^{-iEt/\hbar} \tag{4.3.12}$$

We can now extract $U(t)$:

$$U(t) = \sum_E |E\rangle\langle E|\, e^{-iEt/\hbar} \tag{4.3.13}$$

We have been assuming that the energy spectrum is discrete and nondegenerate. If E is degenerate, one must first introduce an extra label α (usually the eigenvalue of a compatible observable) to specify the states. In this case

$$U(t) = \sum_\alpha \sum_E |E, \alpha\rangle\langle E, \alpha|\, e^{-iEt/\hbar}$$

If E is continuous, the sum must be replaced by an integral. The normal modes

$$|E(t)\rangle = |E\rangle\, e^{-iEt/\hbar}$$

are also called *stationary states* for the following reason: the probability distribution $P(\omega)$ for any variable Ω is time-independent in such a state:

$$\begin{aligned} P(\omega, t) &= |\langle\omega|\psi(t)\rangle|^2 \\ &= |\langle\omega|E(t)\rangle|^2 \\ &= |\langle\omega|E\rangle\, e^{-iEt/\hbar}|^2 \\ &= |\langle\omega|E\rangle|^2 \\ &= P(\omega, 0) \end{aligned}$$

There exists another expression for $U(t)$ besides the sum, Eq. (4.3.13), and that is

$$U(t) = e^{-iHt/\hbar} \tag{4.3.14}$$

If this exponential series converges (and it sometimes does not), this form of $U(t)$ can be very useful. (Convince yourself that $|\psi(t)\rangle = e^{-iHt/\hbar}|\psi(0)\rangle$ satisfies Schrödinger's equation.)

Since H (the energy operator) is Hermitian, it follows that $U(t)$ is unitary. We may therefore think of the time evolution of a ket $|\psi(t)\rangle$ as a "rotation" in Hilbert space. One immediate consequence is that the norm $\langle\psi(t)|\psi(t)\rangle$ is invariant:

$$\langle\psi(t)|\psi(t)\rangle=\langle\psi(0)|U^{\dagger}(t)U(t)|\psi(0)\rangle=\langle\psi(0)|\psi(0)\rangle \tag{4.3.15}$$

so that a state, once normalized, stays normalized. There are other consequences of the fact that the time evolution may be viewed as a rotation. For example, one can abandon the fixed basis we have been using, and adopt one that also rotates at the same rate as the state vectors. In such a basis the vectors would appear frozen, but the operators, which were constant matrices in the fixed basis, would now appear to be time dependent. Any physical entity, such as a matrix element, would, however, come out the same as before since $\langle\phi|\Omega|\psi\rangle$, which is the dot product of $\langle\phi|$ and $|\Omega\psi\rangle$, is invariant under rotations. This view of quantum mechanics is called the *Heisenberg picture*, while the one we have been using is called the *Schrödinger picture*. Infinitely many pictures are possible, each labeled by how the basis is rotating. So if you think you were born too late to make a contribution to quantum theory fear not, for you can invent your own picture. We will take up the study of various pictures in Chapter 18.

Let us now consider the case $H=H(t)$. We no longer look for normal modes, since the operator in question is changing with time. There exists no fixed strategy for solving such problems. In the course of our study we will encounter a time-dependent problem involving spin which can be solved exactly. We will also study a systematic approximation scheme for solving problems with

$$H(t)=H^0+H^1(t)$$

where H^0 is a large time-independent piece and $H^1(t)$ is a small time-dependent piece.

What is the propagator $U(t)$ in the time-dependent case? In other words, how is $U(t)$ in $|\psi(t)\rangle=U(t)|\psi(0)\rangle$ related to $H(t)$? To find out, we divide the interval $(0-t)$ into N pieces of width $\Delta=t/N$, where N is very large and Δ is very small. By integrating the Schrödinger equation over the first interval, we can write *to first order in* Δ

$$\begin{aligned}|\psi(\Delta)\rangle&=|\psi(0)\rangle+\Delta\left.\frac{d|\psi\rangle}{dt}\right|_0\\&=|\psi(0)\rangle-\frac{i\Delta}{\hbar}H(0)|\psi(0)\rangle\\&=\left[1-\frac{i\Delta}{\hbar}H(0)\right]|\psi(0)\rangle\end{aligned}$$

which, to this order

$$=\exp\left[\frac{-i\Delta}{\hbar}H(0)\right]|\psi(0)\rangle$$

[One may wonder whether in the interval $0-\Delta$, one must use $H(0)$ or $H(\Delta)$ or $H(\Delta/2)$ and so on. The difference between these possibilities is of order Δ and hence irrelevant, since there is already one power of Δ in front of H.] Inching forth in steps of Δ, we get

$$|\psi(t)\rangle = \prod_{n=0}^{N-1} e^{-i\Delta H(n\Delta)/\hbar} |\psi(0)\rangle$$

We cannot simply add the exponents to get, in the $N \to \infty$ limit,

$$U(t) = \exp\left[-(i/\hbar) \int_0^t H(t')\, dt'\right]$$

since

$$[H(t_1), H(t_2)] \neq 0$$

in general. For example, if

$$H(t) = X^2 \cos^2 \omega t + P^2 \sin^2 \omega t$$

then

$$H(0) = X^2$$

and

$$H(\pi/2\omega) = P^2$$

and

$$[H(0), H(\pi/2\omega)] \neq 0$$

It is common to use the symbol, called the *time-ordered integral*

$$T\left\{\exp\left[-(i/\hbar) \int_0^t H(t')\, dt'\right]\right\} = \lim_{N\to\infty} \prod_{n=0}^{N-1} \exp[-(i/\hbar) H(n\Delta)\Delta]$$

in such problems. We will not make much use of this form of $U(t)$. But notice that being a product of unitary operators, $U(t)$ is unitary, and time evolution continues to be a "rotation" whether or not H is time independent.

Whether or not H depends on time, the propagator satisfies the following conditions:

$$U(t_3, t_2)U(t_2, t_1) = U(t_3, t_1)$$
$$U^\dagger(t_2, t_1) = U^{-1}(t_2, t_1) = U(t_1, t_2) \tag{4.3.16}$$

It is intuitively clear that these equations are correct. You can easily prove them by applying the U's to some arbitrary state and using the fact that U is unitary and $U(t, t) = I$.

Choosing a Basis for Solving Schrödinger's Equation

Barring a few exceptions, the Schrödinger equation is always solved in a particular basis. Although all bases are equal mathematically, some are more equal than others. First of all, since $H = H(X, P)$ the X and P bases recommend themselves, for in going to one of them the corresponding operator is rendered diagonal. Thus one can go to the X basis in which $X \to x$ and $P \to -i\hbar\, d/dx$ or to the P basis in which $P \to p$ and $X \to i\hbar\, d/dp$. The choice between the two depends on the Hamiltonian. Assuming it is of the form (in one dimension)

$$H = T + V = \frac{P^2}{2m} + V(X) \tag{4.3.17}$$

the choice is dictated by $V(X)$. Since $V(X)$ is usually a more complicated function of X than T is of P, one prefers the X basis. Thus if

$$H = \frac{P^2}{2m} + \frac{1}{\cosh^2 X} \tag{4.3.18}$$

the equation

$$H|E\rangle = E|E\rangle$$

becomes in the X basis the second-order equation

$$\left(-\frac{\hbar^2}{2m}\frac{d^2}{dx^2} + \frac{1}{\cosh^2 x}\right)\psi_E(x) = E\psi_E(x) \tag{4.3.19}$$

which can be solved. Had one gone to the P basis, one would have ended up with the equation

$$\left[\frac{p^2}{2m} + \frac{1}{\cosh^2(i\hbar\, d/dp)}\right]\psi_E(p) = E\psi_E(p) \tag{4.3.20}$$

which is quite frightening.

A problem where the P basis is preferred is that of a particle in a constant force field f, for which

$$H=\frac{P^2}{2m}-fX \qquad (4.3.21)$$

In the P basis one gets a first-order differential equation

$$\left(\frac{p^2}{2m}-i\hbar f\frac{d}{dp}\right)\psi_E(p)=E\psi_E(p) \qquad (4.3.22)$$

whereas in the X basis one gets the second-order equation

$$\left(-\frac{\hbar^2}{2m}\frac{d^2}{dx^2}-fx\right)\psi_E(x)=E\psi_E(x) \qquad (4.3.23)$$

The harmonic oscillator can be solved with equal ease in either basis since H is quadratic in X and P. It turns out to be preferable to solve it in a third basis in which neither X nor P is diagonal! You must wait till Chapter 7 before you see how this happens.

There exists a built-in bias in favor of the X basis. This has to do with the fact that the x space is the space we live in. In other words, when we speak of the probability of obtaining a value between x and $x+dx$ if the variable X is measured, we mean simply the probability of finding the particle between x and $x+dx$ *in our space*. One may thus visualize $\psi(x)$ as a function in our space, whose modulus squared gives the probability density for finding a particle near x. Such a picture is useful in thinking about the double-slit experiment or the electronic states in a hydrogen atom.

But like all pictures, it has its limits. First of all it must be borne in mind that even though $\psi(x)$ can be visualized as a wave in our space, it is not a real wave, like the electromagnetic wave, which carries energy, momentum, etc. To understand this point, consider a particle in three dimensions. The function $\psi(x, y, z)$ can be visualized as a wave in our space. But, if we consider next a two-particle system, $\psi(x_1, y_1, z_1, x_2, y_2, z_2)$ is a function in a six-dimensional *configuration space* and cannot be visualized in our space.

Thus the case of the single particle is really an exception: there is only one position operator and the space of its eigenvalues *happens to coincide* with the space in which we live and in which the drama of physics takes place.

This brings us to the end of our general discussion of the postulates. We now turn to the application of quantum theory to various physical problems. For pedagogical reasons, we will restrict ourselves to problems of a single particle in one dimension in the next few chapters.

5

Simple Problems in One Dimension

Now that the postulates have been stated and explained, it is all over but for the applications. We begin with the simplest class of problems—concerning a single particle in one dimension. Although these one-dimensional problems are somewhat artificial, they contain most of the features of three-dimensional quantum mechanics but little of its complexity. One problem we will not discuss in this chapter is that of the harmonic oscillator. This problem is so important that a separate chapter has been devoted to its study.

5.1. The Free Particle

The simplest problem in this family is of course that of the free particle. The Schrödinger equation is

$$i\hbar|\dot{\psi}\rangle = H|\psi\rangle = \frac{P^2}{2m}|\psi\rangle \tag{5.1.1}$$

The normal modes or stationary states are solutions of the form

$$|\psi\rangle = |E\rangle\, e^{-iEt/\hbar} \tag{5.1.2}$$

Feeding this into Eq. (5.1.1), we get the time-independent Schrödinger equation for $|E\rangle$:

$$H|E\rangle = \frac{P^2}{2m}|E\rangle = E|E\rangle \tag{5.1.3}$$

This problem can be solved without going to any basis. First note that any eigenstate of P is also an eigenstate of P^2. So we feed the trial solution $|p\rangle$ into Eq. (5.1.3) and find

$$\frac{P^2}{2m}|p\rangle = E|p\rangle$$

or

$$\left(\frac{p^2}{2m} - E\right)|p\rangle = 0 \tag{5.1.4}$$

Since $|p\rangle$ is not a null vector, we find that the allowed values of p are

$$p = \pm(2mE)^{1/2} \tag{5.1.5}$$

In other words, there are two orthogonal eigenstates for each eigenvalue E:

$$|E, +\rangle = |p = (2mE)^{1/2}\rangle \tag{5.1.6}$$

$$|E, -\rangle = |p = -(2mE)^{1/2}\rangle \tag{5.1.7}$$

Thus, we find that to the eigenvalue E there corresponds a degenerate two-dimensional eigenspace, spanned by the above vectors. Physically this means that a particle of energy E can be moving to the right or to the left with momentum $|p| = (2mE)^{1/2}$. Now, you might say, "This is exactly what happens in classical mechanics. So what's new?" What is new is the fact that the state

$$|E\rangle = \beta|p = (2mE)^{1/2}\rangle + \gamma|p = -(2mE)^{1/2}\rangle \tag{5.1.8}$$

is also an eigenstate of energy E and represents a *single* particle of energy E that can be caught moving either to the right or to the left with momentum $(2mE)^{1/2}$!

To construct the complete orthonormal eigenbasis of H, we must pick from each degenerate eigenspace any two orthonormal vectors. The obvious choice is given by the kets $|E, +\rangle$ and $|E, -\rangle$ themselves. In terms of the ideas discussed in the past, we are using the eigenvalue of a compatible variable P as an extra label within the space degenerate with respect to energy. Since P is a nondegenerate operator, the label p by itself is adequate. In other words, there is no need to call the state $|p, E = P^2/2m\rangle$, since the value of $E = E(p)$ follows, given p. We shall therefore drop this redundant label.

The propagator is then

$$\begin{aligned} U(t) &= \int_{-\infty}^{\infty} |p\rangle\langle p|\, e^{-iE(p)t/\hbar}\, dp \\ &= \int_{-\infty}^{\infty} |p\rangle\langle p|\, e^{-ip^2t/2m\hbar}\, dp \end{aligned} \tag{5.1.9}$$

Exercise 5.1.1. Show that Eq. (5.1.9) may be rewritten as an integral over E and a sum over the $\pm$ index as

$$U(t)=\sum_{\alpha=\pm}\int_0^\infty\left[\frac{m}{(2mE)^{1/2}}\right]|E,\alpha\rangle\langle E,\alpha|\,e^{-iEt/\hbar}\,dE$$

*Exercise 5.1.2.** By solving the eigenvalue equation (5.1.3) in the X basis, regain Eq. (5.1.8), i.e., show that the general solution of energy E is

$$\psi_E(x)=\beta\frac{\exp[i(2mE)^{1/2}x/\hbar]}{(2\pi\hbar)^{1/2}}+\gamma\frac{\exp[-i(2mE)^{1/2}x/\hbar]}{(2\pi\hbar)^{1/2}}$$

[The factor $(2\pi\hbar)^{-1/2}$ is arbitrary and may be absorbed into β and γ.] Though $\psi_E(x)$ will satisfy the equation even if $E<0$, are these functions in the Hilbert space?

The propagator $U(t)$ can be evaluated explicitly in the X basis. We start with the matrix element

$$U(x,t;x')\equiv\langle x|U(t)|x'\rangle=\int_{-\infty}^{\infty}\langle x|p\rangle\langle p|x'\rangle\,e^{-ip^2t/2m\hbar}\,dp$$

$$=\frac{1}{2\pi\hbar}\int_{-\infty}^{\infty}e^{ip(x-x')/\hbar}\cdot e^{-ip^2t/2m\hbar}\,dp$$

$$=\left(\frac{m}{2\pi\hbar it}\right)^{1/2}e^{im(x-x')^2/2\hbar t}\qquad(5.1.10)$$

using the result from Appendix A.2 on Gaussian integrals. In terms of this propagator, any initial-value problem can be solved, since

$$\psi(x,t)=\int U(x,t;x')\psi(x',0)\,dx'\qquad(5.1.11)$$

Had we chosen the initial time to be t' rather than zero, we would have gotten

$$\psi(x,t)=\int U(x,t;x',t')\psi(x',t')dx'\qquad(5.1.12)$$

where $U(x,t;x',t')=\langle x|U(t-t')|x'\rangle$, since U depends only on the time interval $t-t'$ and not the absolute values of t and t'. [Had there been a time-dependent potential such as $V(t)=V_0\,e^{-at^2}$ in H, we could have told what absolute time it was by looking at $V(t)$. In the absence of anything defining an absolute time in the problem, only time differences have physical significance.] Whenever we set $t'=0$, we will resort to our old convention and write $U(x,t;x',0)$ as simply $U(x,t;x')$.

A nice physical interpretation may be given to $U(x,t;x',t')$ by considering a special case of Eq. (5.1.12). Suppose we started off with a particle localized at

$x'=x_0'$, that is, with $\psi(x', t')=\delta(x'-x_0')$. Then

$$\psi(x, t)=U(x, t; x_0', t') \tag{5.1.13}$$

In other words, the propagator (in the X basis) is the amplitude that a particle starting out at the space-time point (x_0', t') ends with at the space–time point (x, t). [It can obviously be given such an interpretation in any basis: $\langle\omega|U(t, t')|\omega'\rangle$ is the amplitude that a particle in the state $|\omega'\rangle$ at t' ends up with in the state $|\omega\rangle$ at t.] Equation (5.1.12) then tells us that the total amplitude for the particle's arrival at (x, t) is the sum of the contributions from all points x' with a weight proportional to the initial amplitude $\psi(x', t')$ that the particle was at x' at time t'. One also refers to $U(x, t; x_0', t')$ as the "fate" of the delta function $\psi(x', t')=\delta(x'-x_0')$.

Time Evolution of the Gaussian Packet

There is an unwritten law which says that the derivation of the free-particle propagator be followed by its application to the Gaussian packet. Let us follow this tradition.

Consider as the initial wave function the wave packet

$$\psi(x', 0)=e^{ip_0x'/\hbar}\frac{e^{-x'^2/2\Delta^2}}{(\pi\Delta^2)^{1/4}} \tag{5.1.14}$$

This packet has mean position $\langle X\rangle=0$, with an uncertainty $\Delta X=\Delta/2^{1/2}$, and mean momentum p_0 with uncertainty $\hbar/2^{1/2}\Delta$. By combining Eqs. (5.1.10) and (5.1.12) we get

$$\psi(x, t)=\left[\pi^{1/2}\left(\Delta+\frac{i\hbar t}{m\Delta}\right)\right]^{-1/2}\cdot\exp\left[\frac{-(x-p_0t/m)^2}{2\Delta^2(1+i\hbar t/m\Delta^2)}\right]$$
$$\times\exp\left[\frac{ip_0}{\hbar}\left(x-\frac{p_0t}{2m}\right)\right] \tag{5.1.15}$$

The corresponding probability density is

$$P(x, t)=\frac{1}{\pi^{1/2}(\Delta^2+\hbar^2t^2/m^2\Delta^2)^{1/2}}\cdot\exp\left\{\frac{-[x-(p_0/m)t]^2}{\Delta^2+\hbar^2t^2/m^2\Delta^2}\right\} \tag{5.1.16}$$

The main features of this result are as follows:

(1) The mean position of the particles is

$$\langle X\rangle=\frac{p_0t}{m}=\frac{\langle P\rangle t}{m}$$

In other words, the classical relation $x=(p/m)t$ now holds between average quantities. This is just one of the consequences of the *Ehrenfest theorem* which states that the classical equations obeyed by dynamical variables will have counterparts in quantum mechanics as relations among expectation values. The theorem will be proved in the next chapter.

(2) The width of the packet grows as follows:

$$\Delta X(t)=\frac{\Delta(t)}{2^{1/2}}=\frac{\Delta}{2^{1/2}}\left(1+\frac{\hbar^2 t^2}{m^2\Delta^4}\right)^{1/2} \tag{5.1.17}$$

The increasing uncertainty in position is a reflection of the fact that any uncertainty in the initial velocity (that is to say, the momentum) will be reflected with passing time as a growing uncertainty in position. In the present case, since $\Delta V(0)=\Delta P(0)/m=\hbar/2^{1/2}m\Delta$, the uncertainty in X grows approximately as $\Delta X\simeq\hbar t/2^{1/2}m\Delta$ which agrees with Eq. (5.1.17) for large times. Although we are able to understand the spreading of the wave packet in classical terms, the fact that the initial spread $\Delta V(0)$ is *unavoidable* (given that we wish to specify the position to an accuracy Δ) is a purely quantum mechanical feature.

If the particle in question were macroscopic, say of mass 1 g, and we wished to fix its initial position to within a proton width, which is approximately 10^{-13} cm, the uncertainty in velocity would be

$$\Delta V(0)\simeq\frac{\hbar}{2^{1/2}m\Delta}\simeq 10^{-14}\ \text{cm/sec}$$

It would be over 300,000 years before the uncertainty $\Delta(t)$ grew to 1 millimeter! We may therefore treat a macroscopic particle classically for any reasonable length of time. This and similar questions will be taken up in greater detail in the next chapter.

Exercise 5.1.3 (*Another Way to Do the Gaussian Problem*). We have seen that there exists another formula for $U(t)$, namely, $U(t)=e^{-iHt/\hbar}$. For a free particle this becomes

$$U(t)=\exp\left[\frac{i}{\hbar}\left(\frac{\hbar^2 t}{2m}\frac{d^2}{dx^2}\right)\right]=\sum_{n=0}^{\infty}\frac{1}{n!}\left(\frac{i\hbar t}{2m}\right)^n\frac{d^{2n}}{dx^{2n}} \tag{5.1.18}$$

Consider the initial state in Eq. (5.1.14) with $p_0=0$, and set $\Delta=1$, $t'=0$:

$$\psi(x,0)=\frac{e^{-x^2/2}}{(\pi)^{1/4}}$$

Find $\psi(x,t)$ using Eq. (5.1.18) above and compare with Eq. (5.1.15).

Hints: (1) Write $\psi(x,0)$ as a power series:

$$\psi(x,0)=(\pi)^{-1/4}\sum_{n=0}^{\infty}\frac{(-1)^n x^{2n}}{n!(2)^n}$$

(2) Find the action of a few terms

$$1, \quad \left(\frac{i\hbar t}{2m}\right)\frac{d^2}{dx^2}, \quad \frac{1}{2!}\left(\frac{i\hbar t}{2m}\frac{d^2}{dx^2}\right)^2$$

etc., on this power series.

(3) Collect terms with the same power of x.

(4) Look for the following series expansion in the coefficient of x^{2n}:

$$\left(1+\frac{it\hbar}{m}\right)^{-n-1/2}=1-(n+1/2)\left(\frac{i\hbar t}{m}\right)+\frac{(n+1/2)(n+3/2)}{2!}\left(\frac{it\hbar}{m}\right)^2+\cdots$$

(5) Juggle around till you get the answer.

Exercise 5.1.4: A Famous Counterexample. Consider the wave function

$$\begin{aligned}\psi(x,0)&=\sin\left(\frac{\pi x}{L}\right), \qquad |x|\le L/2\\ &=0, \qquad |x|>L/2\end{aligned}$$

It is clear that when this function is differentiated any number of times we get another function confined to the interval $|x|\le L/2$. Consequently the action of

$$U(t)=\exp\left[\frac{i}{\hbar}\left(\frac{\hbar^2 t}{2m}\right)\frac{d^2}{dx^2}\right]$$

on this function is to give a function confined to $|x|\le L/2$. What about the spreading of the wave packet?

[Answer: Consider the derivatives at the boundary. We have here an example where the (exponential) operator power series doesn't converge. Notice that the convergence of an operator power series depends not just on the operator but also on the operand. So there is no paradox: if the function dies abruptly as above, so that there seems to be a paradox, the derivatives are singular at the boundary, while if it falls off continuously, the function will definitely leak out given enough time, no matter how rapid the falloff.]

Some General Features of Energy Eigenfunctions

Consider now the energy eigenfunctions in some potential $V(x)$. These obey

$$\psi''=-\frac{2m(E-V)}{\hbar^2}\psi$$

where each prime denotes a spatial derivative. Let us ask what the continuity of $V(x)$ implies. Let us start at some point x_0 where ψ and ψ' have the values $\psi(x_0)$ and $\psi'(x_0)$. If we pretend that x is a time variable and that ψ is a particle coordinate, the problem of finding ψ everywhere else is like finding the trajectory of a particle (for all times past and future) given its position and velocity at some time and its acceleration as a function of its position and time. It is clear that if we integrate

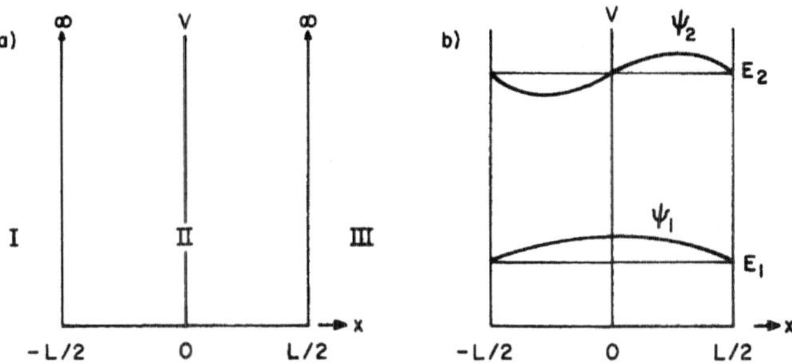

Figure 5.1. (a) The box potential. (b) The first two levels and wave functions in the box.

these equations we will get continuous $\psi'(x)$ and $\psi(x)$. This is the typical situation. There are, however, some problems where, for mathematical simplicity, we consider potentials that change abruptly at some point. This means that ψ'' jumps abruptly there. However, ψ' will still be continuous, for the area under a function is continuous even if the function jumps a bit. What if the change in V is infinitely large? It means that ψ'' is also infinitely large. This in turn means that ψ' *can* change abruptly as we cross this point, for the area under ψ'' can be finite over an infinitesimal region that surrounds this point. But whether or not ψ' is continuous, ψ, which is the area under it, will be continuous.‡

Let us turn our attention to some specific cases.

5.2. The Particle in a Box

We now consider our first problem with a potential, albeit a rather artificial one:

$$\begin{aligned} V(x) &= 0, \qquad |x| < L/2 \\ &= \infty, \qquad |x| \geq L/2 \end{aligned} \tag{5.2.1}$$

This potential (Fig. 5.1a) is called the box since there is an infinite potential barrier in the way of a particle that tries to leave the region $|x| < L/2$. The eigenvalue equation in the X basis (which is the only viable choice) is

$$\frac{d^2\psi}{dx^2} + \frac{2m}{\hbar^2}(E - V)\psi = 0 \tag{5.2.2}$$

We begin by partitioning space into three regions I, II, and III (Fig. 5.1a). The solution ψ is called ψ_{I}, ψ_{II}, and ψ_{III} in regions I, II, and III, respectively.

Consider first region III, in which $V = \infty$. It is convenient to first consider the case where V is not infinite but equal to some V_0 which is greater than E. Now

‡ We are assuming that the jump in ψ' is finite. This will be true even in the artificial potentials we will encounter. But can you think of a potential for which this is not true? (Think delta.)

Eq. (5.2.2) becomes

$$\frac{d^2\psi_{\text{III}}}{dx^2}-\frac{2m(V_0-E)}{\hbar^2}\psi_{\text{III}}=0 \tag{5.2.3}$$

which is solved by

$$\psi_{\text{III}}=A\,e^{-\kappa x}+B\,e^{\kappa x} \tag{5.2.4}$$

where $\kappa=[2m(V_0-E)/\hbar^2]^{1/2}$.

Although A and B are arbitrary coefficients from a mathematical standpoint, we must set $B=0$ on physical grounds since $B\,e^{\kappa x}$ blows up exponentially as $x\to\infty$ and such functions are not members of our Hilbert space. If we now let $V\to\infty$, we see that

$$\psi_{\text{III}}\equiv 0$$

It can similarly be shown that $\psi_{\text{I}}\equiv 0$. In region II, since $V=0$, the solutions are exactly those of a free particle:

$$\psi_{\text{II}}=A\exp[i(2mE/\hbar^2)^{1/2}x]+B\exp[-i(2mE/\hbar^2)^{1/2}x] \tag{5.2.5}$$

$$=A\,e^{ikx}+B\,e^{-ikx},\qquad k=(2mE/\hbar^2)^{1/2} \tag{5.2.6}$$

It therefore appears that the energy eigenvalues are once again continuous as in the free-particle case. This is not so, for $\psi_{\text{II}}(x)=\psi$ only in region II and not in all of space. We must require that ψ_{II} goes continuously into its counterparts ψ_{I} and ψ_{III} as we cross over to regions I and III, respectively. In other words we require that

$$\psi_{\text{I}}(-L/2)=\psi_{\text{II}}(-L/2)=0 \tag{5.2.7}$$

$$\psi_{\text{III}}(+L/2)=\psi_{\text{II}}(+L/2)=0 \tag{5.2.8}$$

(We make no such continuity demands on ψ' at the walls of the box since V jumps to infinity there.) These constraints applied to Eq. (5.2.6) take the form

$$A\,e^{-ikL/2}+B\,e^{ikL/2}=0 \tag{5.2.9a}$$

$$A\,e^{ikL/2}+B\,e^{-ikL/2}=0 \tag{5.2.9b}$$

or in matrix form

$$\begin{bmatrix} e^{-ikL/2} & e^{ikL/2} \\ e^{ikL/2} & e^{-ikL/2} \end{bmatrix}\begin{bmatrix} A \\ B \end{bmatrix}=\begin{bmatrix} 0 \\ 0 \end{bmatrix} \tag{5.2.10}$$

Such an equation has nontrivial solutions only if the determinant vanishes:

$$e^{-ikL}-e^{ikL}=-2i\sin(kL)=0 \tag{5.2.11}$$

that is, only if

$$k=\frac{n\pi}{L},\qquad n=0,\ \pm 1,\ \pm 2,\ldots \tag{5.2.12}$$

To find the corresponding eigenfunctions, we go to Eqs. (5.2.9a) and (5.2.9b). Since only one of them is independent, we study just Eq. (5.2.9a), which says

$$A\,e^{-in\pi/2}+B\,e^{in\pi/2}=0 \tag{5.2.13}$$

Multiplying by $e^{i\pi n/2}$, we get

$$A=-e^{in\pi}B \tag{5.2.14}$$

Since $e^{in\pi}=(-1)^n$, Eq. (5.2.6) generates two families of solutions (normalized to unity):

$$\psi_n(x)=\left(\frac{2}{L}\right)^{1/2}\sin\left(\frac{n\pi x}{L}\right),\qquad n\ \text{even} \tag{5.2.15}$$

$$=\left(\frac{2}{L}\right)^{1/2}\cos\left(\frac{n\pi x}{L}\right),\qquad n\ \text{odd} \tag{5.2.16}$$

Notice that the case $n=0$ is uninteresting since $\psi_0\equiv 0$. Further, since $\psi_n=\psi_{-n}$ for n odd and $\psi_n=-\psi_{-n}$ for n even, and since eigenfunctions differing by an overall factor are not considered distinct, we may restrict ourselves to positive nonzero n. In summary, we have

$$\psi_n=\left(\frac{2}{L}\right)^{1/2}\cos\left(\frac{n\pi x}{L}\right),\qquad n=1,3,5,7,\ldots \tag{5.2.17a}$$

$$=\left(\frac{2}{L}\right)^{1/2}\sin\left(\frac{n\pi x}{L}\right),\qquad n=2,4,6,\ldots \tag{5.2.17b}$$

and from Eqs. (5.2.6) and (5.2.12),

$$E_n=\frac{\hbar^2k_n^2}{2m}=\frac{\hbar^2\pi^2n^2}{2mL^2} \tag{5.2.17c}$$

[It is tacitly understood in Eqs. (5.2.17a) and (5.2.17b) that $|x|<L/2$.]

We have here our first encounter with the quantization of a dynamical variable. Both the variables considered so far, X and P, had a continuous spectrum of eigenvalues from $-\infty$ to $+\infty$, which coincided with the allowed values in classical mechanics. In fact, so did the spectrum of the Hamiltonian in the free-particle case. The particle in the box is the simplest example of a situation that will be encountered again and again, wherein Schrödinger's equation, combined with appropriate boundary conditions, leads to the quantization of energy. These solutions are also examples of *bound states*, namely, states in which a potential prevents a particle from escaping to infinity. Bound states are thus characterized by

$$\psi(x) \xrightarrow[|x|\to\infty]{} 0$$

Bound states appear in quantum mechanics exactly where we expect them classically, namely, in situations where $V(\pm\infty)$ is greater than E.

The energy levels of bound states are always quantized. Let us gain some insight into how this happens. In the problem of the particle in a box, quantization resulted from the requirement that ψ_{II} completed an integral number of half-cycles within the box so that it smoothly joined its counterparts ψ_{I} and ψ_{III} which vanished identically. Consider next a particle bound by a finite well, i.e., by a potential that jumps from 0 to V_0 at $|x|=L/2$. We have already seen [Eq. (5.2.4)] that in the classically forbidden region ($E<V_0$, $|x|\geq L/2$) ψ is a sum of rising and falling exponentials (as $|x|\to\infty$) and that we must choose the coefficient of the rising exponential to be zero to get an admissible solution. In the classically allowed region ($|x|\leq L/2$) ψ is a sum of a sine and cosine. Since V is everywhere finite, we demand that ψ and ψ' be continuous at $x=\pm L/2$. Thus we impose four conditions on ψ, which has only three free parameters. (It may seem that there are four—the coefficients of the two falling exponentials, the sine, and the cosine. However, the overall scale of ψ is irrelevant both in the eigenvalue equation and the continuity conditions, these being *linear* in ψ and ψ'. Thus if say, ψ' does not satisfy the continuity condition at $x=L/2$, an overall rescaling of ψ and ψ' will not help.) Clearly, the continuity conditions cannot be fulfilled except possibly at certain special energies. (See Exercise 5.2.6 for details). This is the origin of energy quantization here.

Consider now a general potential $V(x)$ which tends to limits $V_\pm$ as $x\to\pm\infty$ and which binds a particle of energy E (less than both $V_\pm$). We argue once again that we have one more constraint than we have parameters, as follows. Let us divide space into tiny intervals such that in each interval $V(x)$ is essentially constant. As $x\to\pm\infty$, these intervals can be made longer and longer since V is stabilizing at its asymptotic values $V_\pm$. The right- and leftmost intervals can be made infinitely wide, since by assumption V has a definite limit as $x\to\pm\infty$. Now in all the finite intervals, ψ has two parameters: these will be the coefficients of the sine/cosine if $E>V$ or growing/falling exponential if $E<V$. (The rising exponential is not disallowed, since it doesn't blow up within the finite intervals.) Only in the left- and rightmost intervals does ψ have just one parameter, for in these infinite intervals, the growing exponential can blow up. All these parameters are constrained by the continuity of ψ and ψ' at each interface between adjacent regions. To see that we have one more constraint than we have parameters, observe that every extra interval brings with it two free parameters and one new interface, i.e., two new constraints. Thus as we go from

three intervals in the finite well to the infinite number of intervals in the arbitrary potential, the constraints are always one more than the free parameters. Thus only at special energies can we expect an allowed solution.

[Later we will study the oscillator *potential*, $V=\frac{1}{2}m\omega^2x^2$, which grows without limit as $|x|\to\infty$. How do we understand energy quantization here? Clearly, any *allowed* ψ will vanish even more rapidly than before as $|x|\to\infty$, since $V-E$, instead of being a constant, grows quadratically, so that the particle is "even more forbidden than before" from escaping to infinity. If E is an allowed energy,‡ we expect ψ to fall off rapidly as we cross the classical turning points $x_0=\pm(2E/m\omega^2)^{1/2}$. To a particle in such a state, it shouldn't matter if we flatten out the potential to some constant at distances much greater than $|x_0|$, i.e., the *allowed* levels and eigenfunctions must be the same in the two potentials which differ only in a region that the particle is so strongly inhibited from going to. Since the flattened-out potential has the asymptotic behavior we discussed earlier, we can understand energy quantization as we did before.]

Let us restate the origin of energy quantization in another way. Consider the search for acceptable energy eigenfunctions, taking the finite well as an example. If we start with some arbitrary values $\psi(x_0)$ and $\psi'(x_0)$, at some point x_0 to the right of the well, we can integrate Schrödinger's equation numerically. (Recall the analogy with the problem of finding the trajectory of a particle given its initial position and velocity and the force on it.) As we integrate out to $x\to\infty$, ψ will surely blow up since $\psi_{\rm III}$ contains a growing exponential. Since $\psi(x_0)$ merely fixes the overall scale, we vary $\psi'(x_0)$ until the growing exponential is killed. [Since we can solve the problem analytically in region III, we can even say what the desired value of $\psi'(x_0)$ is: it is given by $\psi'(x_0)=-\kappa\psi(x_0)$. Verify, starting with Eq. (5.2.4), that this implies $B=0$.] We are now out of the fix as $x\to\infty$, but we are committed to whatever comes out as we integrate to the left of x_0. We will find that ψ grows exponentially till we reach the well, whereupon it will oscillate. When we cross the well, ψ will again start to grow exponentially, for $\psi_{\rm I}$ also contains a growing exponential in general. Thus there will be no acceptable solution at some randomly chosen energy. It can, however, happen that for certain values of energy, ψ will be exponentially damped in both regions I *and* III. [At any point x_0' in region I, there is a ratio $\psi'(x_0')/\psi(x_0')$ for which only the damped exponential survives. The ψ we get integrating from region III will not generally have this feature. At special energies, however, this can happen.] These are the allowed energies and the corresponding functions are the allowed eigenfunctions. Having found them, we can choose $\psi(x_0)$ such that they are normalized to unity. For a nice numerical analysis of this problem see the book by Eisberg and Resnick.§

It is clear how these arguments generalize to a particle bound by some arbitrary potential: if we try to keep ψ exponentially damped as $x\to-\infty$, it blows up as $x\to\infty$ (and vice versa), except at some special energies. It is also clear why there is no quantization of energy for unbound states: since the particle is classically allowed at infinity, ψ *oscillates* there and so we have two more parameters, one from each end (why?), and so two solutions (normalizable to $\delta(0)$) at *any* energy.

‡ We are not assuming E is quantized.

§ R. Eisberg and R. Resnick, *Quantum Physics of Atoms, Molecules, Solids, Nuclei and Particles*, Wiley, New York (1974). See Section 5.7 and Appendix F.

Let us now return to the problem of the particle in a box and discuss the fact that the lowest energy is not zero (as it would be classically, corresponding to the particle at rest inside the well) but $\hbar^2\pi^2/2mL^2$. The reason behind it is the uncertainty principle, which prevents the particle, whose position (and hence ΔX) is bounded by $|x| \le L/2$, from having a well-defined momentum of zero. This in turn leads to a lower bound on the energy, which we derive as follows. We begin with‡

$$H=\frac{P^2}{2m} \tag{5.2.18}$$

so that

$$\langle H\rangle=\frac{\langle P^2\rangle}{2m} \tag{5.2.19}$$

Now $\langle P\rangle = 0$ in any bound state for the following reason. Since a bound state is a stationary state, $\langle P\rangle$ is time independent. If this $\langle P\rangle \neq 0$, the particle must (in the average sense) drift either to the right or to the left and eventually escape to infinity, which cannot happen in a bound state.

Consequently we may rewrite Eq. (5.2.19) as

$$\langle H\rangle=\frac{\langle (P-\langle P\rangle)^2\rangle}{2m}=\frac{(\Delta P)^2}{2m}$$

If we now use the uncertainty relation

$$\Delta P\cdot\Delta X \ge \hbar/2$$

we find

$$\langle H\rangle \ge \frac{\hbar^2}{8m(\Delta X)^2}$$

Since the variable x is constrained by $-L/2 \le x \le L/2$, its standard deviation ΔX cannot exceed $L/2$. Consequently

$$\langle H\rangle \ge \hbar^2/2mL^2$$

In an energy eigenstate, $\langle H\rangle = E$ so that

$$E \ge \hbar^2/2mL^2 \tag{5.2.20}$$

The actual ground-state energy E_1 happens to be π^2 times as large as the lower

‡ We are suppressing the infinite potential due to the walls of the box. Instead we will restrict x to the range $|x| \le L/2$.

bound. The uncertainty principle is often used in this fashion to provide a quick order-of-magnitude estimate for the ground-state energy.

If we denote by $|n\rangle$ the abstract ket corresponding to $\psi_n(x)$, we can write the propagator as

$$U(t)=\sum_{n=1}^{\infty}|n\rangle\langle n|\exp\left[-\frac{i}{\hbar}\left(\frac{\hbar^2\pi^2n^2}{2mL^2}\right)t\right] \tag{5.2.21}$$

The matrix elements of $U(t)$ in the X basis are then

$$\begin{aligned}\langle x|U(t)|x'\rangle &= U(x,t;x')\\ &=\sum_{n=1}^{\infty}\psi_n(x)\psi_n^*(x')\exp\left[-\frac{i}{\hbar}\left(\frac{\hbar^2\pi^2n^2}{2mL^2}\right)t\right]\end{aligned} \tag{5.2.22}$$

Unlike in the free-particle case, there exists no simple closed expression for this sum.

*Exercise 5.2.1.** A particle is in the ground state of a box of length L. Suddenly the box expands (symmetrically) to twice its size, leaving the wave function undisturbed. Show that the probability of finding the particle in the ground state of the new box is $(8/3\pi)^2$.

*Exercise 5.2.2.** (a) Show that for any normalized $|\psi\rangle$, $\langle\psi|H|\psi\rangle\geq E_0$, where E_0 is the lowest-energy eigenvalue. (Hint: Expand $|\psi\rangle$ in the eigenbasis of H.)

(b) Prove the following theorem: Every attractive potential in one dimension has at least one bound state. Hint: Since V is attractive, if we define $V(\infty)=0$, it follows that $V(x)=-|V(x)|$ for all x. To show that there exists a bound state with $E<0$, consider

$$\psi_\alpha(x)=\left(\frac{\alpha}{\pi}\right)^{1/4}e^{-\alpha x^2/2}$$

and calculate

$$E(\alpha)=\langle\psi_\alpha|H|\psi_\alpha\rangle,\qquad H=-\frac{\hbar^2}{2m}\frac{d^2}{dx^2}-|V(x)|$$

Show that $E(\alpha)$ can be made negative by a suitable choice of α. The desired result follows from the application of the theorem proved above.

*Exercise 5.2.3.** Consider $V(x)=-aV_0\delta(x)$. Show that it admits a bound state of energy $E=-ma^2V_0^2/2\hbar^2$. Are there any other bound states? Hint: Solve Schrödinger's equation outside the potential for $E<0$, and keep only the solution that has the right behavior at infinity and is continuous at $x=0$. Draw the wave function and see how there is a cusp, or a discontinuous change of slope at $x=0$. Calculate the change in slope and equate it to

$$\int_{-\varepsilon}^{+\varepsilon}\left(\frac{d^2\psi}{dx^2}\right)dx$$

(where ε is infinitesimal) determined from Schrödinger's equation.

Exercise 5.2.4. Consider a particle of mass m in the state $|n\rangle$ of a box of length L. Find the force $F=-\partial E/\partial L$ encountered when the walls are slowly pushed in, assuming the particle remains in the nth state of the box as its size changes. Consider a classical particle of energy E_n in this box. Find its velocity, the frequency of collision on a given wall, the momentum transfer per collision, and hence the average force. Compare it to $-\partial E/\partial L$ computed above.

*Exercise 5.2.5.** If the box extends from $x=0$ to L (instead of $-L/2$ to $L/2$) show that $\psi_n(x)=(2/L)^{1/2}\sin(n\pi x/L)$, $n=1, 2, \ldots, \infty$ and $E_n=\hbar^2\pi^2n^2/2mL^2$.

*Exercise 5.2.6.** *Square Well Potential.* Consider a particle in a square well potential:

$$V(x)=\begin{cases}0, & |x|\le a\\ V_0, & |x|\ge a\end{cases}$$

Since when $V_0\to\infty$, we have a box, let us guess what the lowering of the walls does to the states. First of all, all the bound states (which alone we are interested in), will have $E\le V_0$. Second, the wave functions of the low-lying levels will look like those of the particle in a box, with the obvious difference that ψ will not vanish at the walls but instead spill out with an exponential tail. The eigenfunctions will still be even, odd, even, etc.

(1) Show that the even solutions have energies that satisfy the transcendental equation

$$k\tan ka=\kappa \tag{5.2.23}$$

while the odd ones will have energies that satisfy

$$k\cot ka=-\kappa \tag{5.2.24}$$

where k and $i\kappa$ are the real and complex wave numbers inside and outside the well, respectively. Note that k and κ are related by

$$k^2+\kappa^2=2mV_0/\hbar^2 \tag{5.2.25}$$

Verify that as V_0 tends to ∞, we regain the levels in the box.

(2) Equations (5.2.23) and (5.2.24) must be solved graphically. In the ($\alpha=ka$, $\beta=\kappa a$) plane, imagine a circle that obeys Eq. (5.2.25). The bound states are then given by the intersection of the curve $\alpha\tan\alpha=\beta$ or $\alpha\cot\alpha=-\beta$ with the circle. (Remember α and β are positive.)

(3) Show that there is always one even solution and that there is no odd solution unless $V_0\ge\hbar^2\pi^2/8ma^2$. What is E when V_0 just meets this requirement? Note that the general result from Exercise 5.2.2b holds.

5.3. The Continuity Equation for Probability

We interrupt our discussion of one-dimensional problems to get acquainted with two concepts that will be used in the subsequent discussions, namely, those of the *probability current density* and the *continuity equation* it satisfies. Since the probability current concept will also be used in three-dimensional problems, we discuss here a particle in three dimensions.

As a prelude to our study of the continuity equation in quantum mechanics, let us recall the analogous equation from electromagnetism. We know in this case that the total charge in the universe is a constant, that is

$$Q(t) = \text{const, independent of time } t \tag{5.3.1}$$

This is an example of a global conservation law, for it refers to the total charge in the universe. But charge is also conserved locally, a fact usually expressed in the form of the continuity equation

$$\frac{\partial \rho(\mathbf{r}, t)}{\partial t} = -\nabla \cdot \mathbf{j} \tag{5.3.2}$$

where ρ and $\mathbf{j}$ are the charge and current densities, respectively. By integrating this equation over a volume V bounded by a surface S_V we get, upon invoking Gauss's law,

$$\frac{d}{dt}\int_V \rho(\mathbf{r}, t)\, d^3\mathbf{r} = -\int_V \nabla \cdot \mathbf{j}\, d^3\mathbf{r} = -\int_{S_V} \mathbf{j} \cdot d\mathbf{S} \tag{5.3.3}$$

This equation states that any decrease in charge in the volume V is accounted for by the flow of charge out of it, that is to say, charge is not created or destroyed in any volume.

The continuity equation forbids certain processes that obey global conservation, such as the sudden disappearance of charge from one region of space and its immediate reappearance in another.

In quantum mechanics the quantity that is globally conserved is the total probability for finding the particle anywhere in the universe. We get this result by expressing the invariance of the norm in the coordinate basis: since

$$\langle \psi(t) | \psi(t) \rangle = \langle \psi(0) | U^\dagger(t) U(t) | \psi(0) \rangle = \langle \psi(0) | \psi(0) \rangle$$

then

$$\begin{aligned}
\text{const} = \langle \psi(t) | \psi(t) \rangle &= \iiint \langle \psi(t) | x, y, z \rangle \langle x, y, z | \psi(t) \rangle \, dx\, dy\, dz\ddagger \\
&= \iiint \langle \psi(t) | \mathbf{r} \rangle \langle \mathbf{r} | \psi(t) \rangle \, d^3\mathbf{r} \\
&= \iiint \psi^*(\mathbf{r}, t) \psi(\mathbf{r}, t) \, d^3\mathbf{r} \\
&= \iiint P(\mathbf{r}, t) \, d^3\mathbf{r}
\end{aligned} \tag{5.3.4}$$

‡ The range of integration will frequently be suppressed when obvious.

This global conservation law is the analog of Eq. (5.3.1). To get the analog of Eq. (5.3.2), we turn to the Schrödinger equation

$$i\hbar\frac{\partial\psi}{\partial t}=-\frac{\hbar^2}{2m}\nabla^2\psi+V\psi \tag{5.3.5}$$

and its conjugate

$$-i\hbar\frac{\partial\psi^*}{\partial t}=-\frac{\hbar^2}{2m}\nabla^2\psi^*+V\psi^* \tag{5.3.6}$$

Note that V has to be real if H is to be Hermitian. Multiplying the first of these equations by ψ^*, the second by ψ, and taking the difference, we get

$$i\hbar\frac{\partial}{\partial t}(\psi^*\psi)=-\frac{\hbar^2}{2m}(\psi^*\nabla^2\psi-\psi\nabla^2\psi^*)$$

$$\frac{\partial P}{\partial t}=-\frac{\hbar}{2mi}\nabla\cdot(\psi^*\nabla\psi-\psi\nabla\psi^*)$$

$$\frac{\partial P}{\partial t}=-\nabla\cdot\mathbf{j} \tag{5.3.7}$$

where

$$\mathbf{j}=\frac{\hbar}{2mi}(\psi^*\nabla\psi-\psi\nabla\psi^*) \tag{5.3.8}$$

is the *probability current density*, that is to say, the probability flow per unit time per unit area perpendicular to $\mathbf{j}$. To regain the global conservation law, we integrate Eq. (5.3.7) over all space:

$$\frac{d}{dt}\int P(\mathbf{r},t)\,d^3\mathbf{r}=-\int_{S_\infty}\mathbf{j}\cdot d\mathbf{S} \tag{5.3.9}$$

where S_∞ is the sphere at infinity. For (typical) wave functions which are normalizable to unity, $r^{3/2}\psi\to 0$ as $r\to\infty$ in order that $\int\psi^*\psi r^2\,dr\,d\Omega$ is bounded, and the surface integral of $\mathbf{j}$ on S_∞ vanishes. The case of momentum eigenfunctions that do not vanish on S_∞ is considered in one of the following exercises.

Exercise 5.3.1. Consider the case where $V=V_r-iV_i$, where the imaginary part V_i is a constant. Is the Hamiltonian Hermitian? Go through the derivation of the continuity equation and show that the total probability for finding the particle decreases exponentially as $e^{-2V_it/\hbar}$. Such complex potentials are used to describe processes in which particles are absorbed by a sink.

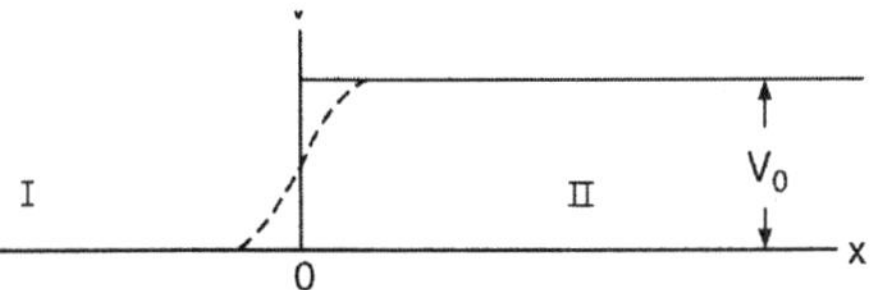

Figure 5.2. The single-step potential. The dotted line shows a more realistic potential idealized by the step, which is mathematically convenient. The total energy E and potential energy V are measured along the y axis.

Exercise 5.3.2. Convince yourself that if $\psi = c\tilde{\psi}$, where c is constant (real or complex) and $\tilde{\psi}$ is real, the corresponding $\mathbf{j}$ vanishes.

Exercise 5.3.3. Consider

$$\psi_{\mathbf{p}} = \left(\frac{1}{2\pi\hbar}\right)^{3/2} e^{i(\mathbf{p}\cdot\mathbf{r})/\hbar}$$

Find $\mathbf{j}$ and P and compare the relation between them to the electromagnetic equation $\mathbf{j} = \rho\mathbf{v}$, $\mathbf{v}$ being the velocity. Since ρ and $\mathbf{j}$ are constant, note that the continuity Eq. (5.3.7) is trivially satisfied.

*Exercise 5.3.4.** Consider $\psi = Ae^{ipx/\hbar} + Be^{-ipx/\hbar}$ in one dimension. Show that $j = (|A|^2 - |B|^2)p/m$. The absence of cross terms between the right- and left-moving pieces in ψ allows us to associate the two parts of j with corresponding parts of ψ.

Ensemble Interpretation of j

Recall that $\mathbf{j}\cdot d\mathbf{S}$ is the rate at which probability flows past the area $d\mathbf{S}$. If we consider an ensemble of N particles all in some state $\psi(\mathbf{r}, t)$, then $N\mathbf{j}\cdot d\mathbf{S}$ particles will trigger a particle detector of area $d\mathbf{S}$ per second, assuming that N tends to infinity and that $\mathbf{j}$ is the current associated with $\psi(\mathbf{r}, t)$.

5.4. The Single-Step Potential: A Problem in Scattering‡

Consider the step potential (Fig. 5.2)

$$\begin{aligned} V(x) &= 0 \qquad x < 0 \text{ (region I)} \\ &= V_0 \qquad x > 0 \text{ (region II)} \end{aligned} \tag{5.4.1}$$

Such an abrupt change in potential is rather unrealistic but mathematically convenient. A more realistic transition is shown by dotted lines in Figure 5.2.

Imagine now that a classical particle of energy E is shot in from the left (region I) toward the step. One expects that if $E > V_0$, the particle would climb the barrier and travel on to region II, while if $E < V_0$, it would get reflected. We now compare this classical situation with its quantum counterpart.

‡ This rather difficult section may be postponed till the reader has gone through Chapter 7 and gained more experience with the subject. It is for the reader or the instructor to decide which way to go.

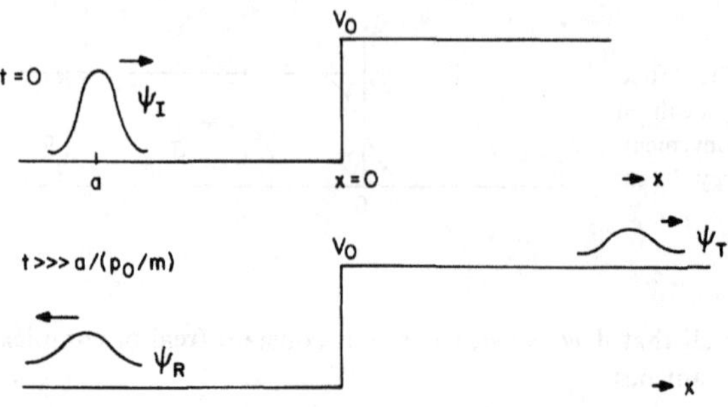

Figure 5.3. A schematic description of the wave function long before and long after it hits the step. The area under $|\psi_I|^2$ is unity. The areas under $|\psi_R|^2$ and $|\psi_T|^2$, respectively, are the probabilities for reflection and transmission.

First of all, we must consider an initial state that is compatible with quantum principles. We replace the incident particle possessing a well-defined trajectory with a wave packet.‡ Though the detailed wave function will be seen to be irrelevant in the limit we will consider, we start with a Gaussian, which is easy to handle analytically§:

$$\psi_I(x, 0) = \psi_I(x) = (\pi\Delta^2)^{-1/4} e^{ik_0(x+a)} e^{-(x+a)^2/2\Delta^2} \tag{5.4.2}$$

This packet has a mean momentum $p_0 = \hbar k_0$, a mean position $\langle X\rangle = -a$ (which we take to be far away from the step), with uncertainties

$$\Delta X = \frac{\Delta}{2^{1/2}}, \qquad \Delta P = \frac{\hbar}{2^{1/2}\Delta}$$

We shall be interested in the case of large Δ, where the particle has essentially well-defined momentum $\hbar k_0$ and energy $E_0 \simeq \hbar^2 k_0^2/2m$. *We first consider the case* $E_0 > V_0$.

After a time $t \simeq a[p_0/m]^{-1}$, the packet will hit the step and in general break into two packets: ψ_R, the reflected packet, and ψ_T, the transmitted packet (Fig. 5.3). The area under $|\psi_R|^2$ at large t is the probability of finding the particle in region I in the distant future, that is to say, the probability of reflection. Likewise the area under $|\psi_T|^2$ at large t is the probability of transmission. Our problem is to calculate the *reflection coefficient*

$$R = \int |\psi_R|^2\, dx, \qquad t \to \infty \tag{5.4.3}$$

and *transmission coefficient*

$$T = \int |\psi_T|^2\, dx, \qquad t \to \infty \tag{5.4.4}$$

Generally R and T will depend on the detailed shape of the initial wave function. If, however, we go to the limit in which the initial momentum is well defined (i.e.,

‡ A wave packet is any wave function with reasonably well-defined position and momentum.
§ This is just the wave packet in Eq. (5.1.14), displaced by an amount $-a$.

when the Gaussian in x space has infinite width), we expect the answer to depend only on the initial energy, it being the only characteristic of the state. In the following analysis we will assume that $\Delta X = \Delta/2^{1/2}$ is large and that the wave function in k space is very sharply peaked near k_0.

We follow the standard procedure for finding the fate of the incident wave packet, ψ_I:

Step 1: Solve for the *normalized* eigenfunction of the step potential Hamiltonian, $\psi_E(x)$.
Step 2: Find the projection $a(E) = \langle \psi_E | \psi_I \rangle$.
Step 3: Append to each coefficient $a(E)$ a time dependence $e^{-iEt/\hbar}$ and get $\psi(x, t)$ at any future time.
Step 4: Identify ψ_R and ψ_T in $\psi(x, t\to\infty)$ and determine R and T using Eqs. (5.4.3) and (5.4.4).

Step 1. In region I, as $V=0$, the (*unnormalized*) solution is the familiar one:

$$\psi_E(x) = A\, e^{ik_1 x} + B\, e^{-ik_1 x}, \qquad k_1 = \left(\frac{2mE}{\hbar^2}\right)^{1/2} \tag{5.4.5}$$

In region II, we simply replace E by $E - V_0$ [see Eq. (5.2.2)],

$$\psi_E(x) = C\, e^{ik_2 x} + D\, e^{-ik_2 x}, \qquad k_2 = \left[\frac{2m(E - V_0)}{\hbar^2}\right]^{1/2} \tag{5.4.6}$$

(We consider only $E > V_0$; the eigenfunction with $E < V_0$ will be orthogonal to ψ_I as will be shown on the next two pages.) Of interest to us are eigenfunctions with $D = 0$, since we want only a transmitted (right-going) wave in region II, and incident plus reflected waves in region I. If we now impose the continuity of ψ and its derivative at $x=0$; we get

$$A + B = C \tag{5.4.7}$$

$$ik_1(A - B) = ik_2 C \tag{5.4.8}$$

In anticipation of future use, we solve these equations to express B and C in terms of A:

$$B = \left(\frac{k_1 - k_2}{k_1 + k_2}\right) A = \left(\frac{E^{1/2} - (E - V_0)^{1/2}}{E^{1/2} + (E - V_0)^{1/2}}\right) A \tag{5.4.9}$$

$$C = \left(\frac{2k_1}{k_1 + k_2}\right) A = \left(\frac{2E^{1/2}}{E^{1/2} + (E - V_0)^{1/2}}\right) A \tag{5.4.10}$$

Note that if $V_0=0$, $B=0$ and $C=A$ as expected. The solution with energy E is then

$$\psi_E(x)=A\left[\left(e^{ik_1x}+\frac{B}{A}e^{-ik_1x}\right)\theta(-x)+\frac{C}{A}e^{ik_2x}\,\theta(x)\right] \tag{5.4.11}$$

where

$$\begin{aligned}\theta(x)&=1 \qquad \text{if } x>0\\ &=0 \qquad \text{if } x<0\end{aligned}$$

Since to each E there is a unique $k_1=+(2mE/\hbar^2)^{1/2}$, we can label the eigenstates by k_1 instead of E. Eliminating k_2 in favor of k_1, we get

$$\begin{aligned}\psi_{k_1}(x)=A\Bigg[&\left(\exp(ik_1x)+\frac{B}{A}\exp(-ik_1x)\right)\theta(-x)\\ &+\frac{C}{A}\exp[i(k_1^2-2mV_0/\hbar^2)^{1/2}x]\theta(x)\Bigg]\end{aligned} \tag{5.4.12}$$

Although the overall scale factor A is generally arbitrary (and the physics depends only on B/A and C/A), here we must choose $A=(2\pi)^{-1/2}$ because ψ_k has to be properly normalized in the four-step procedure outlined above. We shall verify shortly that $A=(2\pi)^{-1/2}$ is the correct normalization factor.

Step 2. Consider next

$$\begin{aligned}a(k_1)&=\langle\psi_{k_1}|\psi_I\rangle\\ &=\frac{1}{(2\pi)^{1/2}}\left\{\int_{-\infty}^{\infty}\left[e^{-ik_1x}+\left(\frac{B}{A}\right)^*e^{ik_1x}\right]\theta(-x)\psi_I(x)\,dx\right.\\ &\qquad\left.+\int_{-\infty}^{\infty}\left(\frac{C}{A}\right)^*e^{-ik_2x}\theta(x)\psi_I(x)\,dx\right\}\end{aligned} \tag{5.4.13}$$

The second integral vanishes (to an excellent approximation) since $\psi_I(x)$ is nonvanishing far to the left of $x=0$, while $\theta(x)$ is nonvanishing only for $x>0$. Similarly the second piece of the first integral also vanishes since ψ_I in k space is peaked around $k=+k_0$ and is orthogonal to (left-going) negative momentum states. [We can ignore the $\theta(-x)$ factor in Eq. (5.4.13) since it equals 1 where $\psi_I(x)\neq0$.] So

$$\begin{aligned}a(k_1)&=\left(\frac{1}{2\pi}\right)^{1/2}\int_{-\infty}^{\infty}e^{-ik_1x}\psi_I(x)\,dx\\ &=\left(\frac{\Delta^2}{\pi}\right)^{1/4}e^{-(k_1-k_0)^2\Delta^2/2}\,e^{ik_1a}\end{aligned} \tag{5.4.14}$$

is just the Fourier transform of ψ_I. Notice that for large Δ, $a(k_1)$ is very *sharply peaked at* $k_1 = k_0$. This justifies our neglect of eigenfunctions with $E < V_0$, for these correspond to k_1 not near k_0.

Step 3. The wave function at any future time t is

$$\psi(x, t) = \int_{-\infty}^{\infty} a(k_1)\, e^{-iE(k_1)t/\hbar}\, \psi_{k_1}(x)\, dk_1 \tag{5.4.15}$$

$$= \left(\frac{\Delta^2}{4\pi^3}\right)^{1/4} \int_{-\infty}^{\infty} \exp\left(\frac{-i\hbar k_1^2 t}{2m}\right) \cdot \exp\left[\frac{-(k_1 - k_0)^2 \Delta^2}{2}\right] \exp(ik_1 a)$$

$$\times \left\{ e^{ik_1 x}\theta(-x) + \left(\frac{B}{A}\right) e^{-ik_1 x}\theta(-x) \right.$$

$$\left. + \left(\frac{C}{A}\right) \exp[i(k_1^2 - 2mV_0/\hbar^2)^{1/2} x]\theta(x) \right\} dk_1 \tag{5.4.16}$$

You can convince yourself that if we set $t = 0$ above we regain $\psi_I(x)$, which corroborates our choice $A = (2\pi)^{-1/2}$.

Step 4. Consider the first of the three terms. If $\theta(-x)$ were absent, we would be propagating the original Gaussian. After replacing x by $x + a$ in Eq. (5.1.15), and inserting the $\theta(-x)$ factor, the first term of $\psi(x, t)$ is

$$\theta(-x)\pi^{-1/4}\left(\Delta + \frac{i\hbar t}{m}\right)^{-1/2} \exp\left[\frac{-(x + a - \hbar k_0 t/m)^2}{2\Delta^2(1 + i\hbar t/m\Delta^2)}\right]$$

$$\times \exp\left[ik_0\left(x + a - \frac{\hbar k_0 t}{2m}\right)\right] \equiv \theta(-x)G(-a, k_0, t) \tag{5.4.17}$$

Since the Gaussian $G(-a, k_0, t)$ is centered at $x = -a + \hbar k_0 t/m \simeq \hbar k_0 t/m$ as $t \to \infty$, and $\theta(-x)$ vanishes for $x > 0$, the product θG vanishes. Thus the initial packet has disappeared and in its place are the reflected and transmitted packets given by the next two terms. In the middle term if we replace B/A, which is a function of k_1, by its value $(B/A)_0$ at $k_1 = k_0$ (because $a(k_1)$ is very sharply peaked at $k_1 = k_0$) and pull it out of the integral, changing the dummy variable from k_1 to $-k_1$, it is easy to see that apart from the factor $(B/A)_0\theta(-x)$ up front, the middle term represents the free propagation of a normalized Gaussian packet that was originally peaked at $x = +a$ and began drifting to the *left* with mean momentum $-\hbar k_0$. Thus

$$\psi_R = \theta(-x)G(a, -k_0, t)(B/A)_0 \tag{5.4.18}$$

As $t\to\infty$, we can set $\theta(-x)$ equal to 1, since G is centered at $x=a-\hbar k_0 t/m\simeq -\hbar k_0 t/m$. Since the Gaussian G has unit norm, we get from Eqs. (5.4.3) and (5.4.9),

$$R=\int|\psi_R|^2\,dx=\left|\frac{B}{A}\right|_0^2=\left|\frac{E_0^{1/2}-(E_0-V_0)^{1/2}}{E_0^{1/2}+(E_0-V_0)^{1/2}}\right|^2$$

where

$$E_0=\frac{\hbar^2k_0^2}{2m} \tag{5.4.19}$$

This formula is exact only when the incident packet has a well-defined energy E_0, that is to say, when the width of the incident Gaussian tends to infinity. But it is an excellent approximation for *any* wave packet that is narrowly peaked in momentum space.

To find T, we can try to evaluate the third piece. But there is no need to do so, since we know that

$$R+T=1 \tag{5.4.20}$$

which follows from the global conservation of probability. It then follows that

$$T=1-R=\frac{4E_0^{1/2}(E_0-V_0)^{1/2}}{[E_0^{1/2}+(E_0-V_0)^{1/2}]^2}=\left|\frac{C}{A}\right|_0^2\frac{(E_0-V_0)^{1/2}}{E_0^{1/2}} \tag{5.4.21}$$

By inspecting Eqs. (5.4.19) and (5.4.21) we see that both R and T are readily expressed in terms of the ratios $(B/A)_0$ and $(C/A)_0$ and a kinematical factor, $(E_0-V_0)^{1/2}/E_0^{1/2}$. Is there some way by which we can directly get to Eqs. (5.4.19) and (5.4.21), which describe the dynamic phenomenon of scattering, from Eqs. (5.4.9) and (5.4.10), which describe the static solution to Schrödinger's equation? Yes.

Consider the unnormalized eigenstate

$$\psi_{k_0}(x)=[A_0\exp(ik_0x)+B_0\exp(-ik_0x)]\theta(-x) + C_0\exp\left[i\left(k_0^2-\frac{2mV_0}{\hbar^2}\right)^{1/2}x\right]\theta(x) \tag{5.4.22}$$

The incoming plane wave $A\,e^{ik_0x}$ has a probability current associated with it equal to

$$j_I=|A_0|^2\frac{\hbar k_0}{m} \tag{5.4.23}$$

while the currents associated with the reflected and transmitted pieces are

$$j_R = |B_0|^2 \frac{\hbar k_0}{m} \tag{5.4.24}$$

and

$$j_T = |C_0|^2 \frac{\hbar(k_0^2 - 2mV_0/\hbar^2)^{1/2}}{m} \tag{5.4.25}$$

(Recall Exercise 5.3.4, which provides the justification for viewing the two parts of the j in region I as being due to the incident and reflected wave functions.) In terms of these currents

$$R = \frac{j_R}{j_I} = \left|\frac{B_0}{A_0}\right|^2 \tag{5.4.26}$$

and

$$T = \frac{j_T}{j_I} = \left|\frac{C_0}{A_0}\right|^2 \frac{(k_0^2 - 2mV_0/\hbar^2)^{1/2}}{k_0} = \left|\frac{C_0}{A_0}\right|^2 \frac{(E_0 - V_0)^{1/2}}{E_0^{1/2}} \tag{5.4.27}$$

Let us now enquire as to why it is that R and T are calculable in these two ways. Recall that R and T were exact only for the incident packet whose momentum was well defined and equal to $\hbar k_0$. From Eq. (5.4.2) we see that this involves taking the width of the Gaussian to infinity. As the incident Gaussian gets wider and wider (we ignore now the $\Delta^{-1/2}$ factor up front and the normalization) the following things happen:

(1) It becomes impossible to say when it hits the step, for it has spread out to be a right-going plane wave in region I.
(2) The reflected packet also gets infinitely wide and coexists with the incident one, as a left-going plane wave.
(3) The transmitted packet becomes a plane wave with wave number $(k_0^2 - 2mV_0/\hbar^2)^{1/2}$ in region II.

In other words, the dynamic picture of an incident packet hitting the step and disintegrating into two becomes the steady-state process described by the eigenfunction Eq. (5.4.22). We cannot, however, find R and T by calculating areas under $|\psi_T|^2$ and $|\psi_R|^2$ since all the areas are infinite, the wave packets having been transformed into plane waves. We find instead that the ratios of the probability currents associated with the incident, reflected, and transmitted waves give us R and T. The equivalence between the wave packet and static descriptions that we were able to demonstrate in this simple case happens to be valid for any potential. When we come to scattering in three dimensions, we will assume that the equivalence of the two approaches holds.

Exercise 5.4.1 (Quite Hard). Evaluate the third piece in Eq. (5.4.16) and compare the resulting T with Eq. (5.4.21). [Hint: Expand the factor $(k_1^2-2mV_0/\hbar^2)^{1/2}$ near $k_1=k_0$, keeping just the first derivative in the Taylor series.]

Before we go on to examine some of the novel features of the reflection and transmission coefficients, let us ask how they are used in practice. Consider a general problem with some $V(x)$, which tends to constants V_+ and V_- as $x\to\pm\infty$. For simplicity we take $V_\pm=0$. Imagine an accelerator located to the far left ($x\to-\infty$) which shoots out a beam of nearly monoenergetic particles with $\langle P\rangle=\hbar k_0$ toward the potential. The question one asks in practice is what fraction of the particles will get transmitted and what fraction will get reflected to $x=-\infty$, respectively. In general, the question cannot be answered because we know only the mean momenta of the particles and not their individual wave functions. But the preceding analysis shows that *as long as the wave packets are localized sharply in momentum space, the reflection and transmission probabilities* (R *and* T) *depend only on the mean momentum and not the detailed shape of the wave functions.* So the answer to the question raised above is that a fraction $R(k_0)$ will get reflected and a fraction $T(k_0)=1-R(k_0)$ will get transmitted. To find R and T we solve for the time-independent eigenfunctions of $H=T+V$ with energy eigenvalue $E_0=\hbar^2k_0^2/2m$, and asymptotic behavior

$$\psi_{k_0}(x)\begin{cases}\xrightarrow[x\to-\infty]{} A\,e^{ik_0x}+B\,e^{-ik_0x}\\[2ex] \xrightarrow[x\to\infty]{} C\,e^{ik_0x}\end{cases}$$

and obtain from it $R=|B/A|^2$ and $T=|C/A|^2$. Solutions with this asymptotic behavior (namely, free-particle behavior) will always exist provided V vanishes rapidly enough as $|x|\to\infty$. [Later we will see that this means $|xV(x)|\to 0$ as $|x|\to\infty$.] The general solution will also contain a piece $D\exp(-ik_0x)$ as $x\to\infty$, but we set $D=0$ here, for if $a\exp(ik_0x)$ is to be identified with the incident wave, it must only produce a right-moving transmitted wave $C\,e^{ik_0x}$ as $x\to\infty$.

Let us turn to Eqs. (5.4.19) and (5.4.21) for R and T. These contain many nonclassical features. First of all we find that an incident particle with $E_0>V_0$ gets reflected some of the time. It can also be shown that a particle with $E_0>V_0$ *incident from the right* will also get reflected some of the time, contrary to classical expectations.

Consider next the case $E_0<V_0$. Classically one expects the particle to be reflected at $x=0$, and never to get to region II. This is not so quantum mechanically. In region II, the solution to

$$\frac{d^2\psi_{\text{II}}}{dx^2}+\frac{2m}{\hbar^2}(E_0-V_0)\psi_{\text{II}}=0$$

with $E_0<V_0$ is

$$\psi_{\text{II}}(x)=C\,e^{-\kappa x},\qquad \kappa=\left(\frac{2m|(E_0-V_0)|}{\hbar^2}\right)^{1/2} \tag{5.4.28}$$

(The growing exponential $e^{\kappa x}$ does not belong to the physical Hilbert space.) Thus there is a finite probability for finding the particle in the region where its kinetic energy $E_0 - V_0$ is negative. There is, however, no steady flow of probability current into region II, since $\psi_{II}(x) = C\tilde{\psi}$ where $\tilde{\psi}$ is real. This is also corroborated by the fact the reflection coefficient in this case is

$$R = \left|\frac{(E_0)^{1/2} - (E_0 - V_0)^{1/2}}{(E_0)^{1/2} + (E_0 - V_0)^{1/2}}\right|^2 = \left|\frac{k_0 - i\kappa}{k_0 + i\kappa}\right|^2 = 1 \tag{5.4.29}$$

The fact that the particle can penetrate into the classically forbidden region leads to an interesting quantum phenomenon called *tunneling*. Consider a modification of Fig. 5.2, in which $V = V_0$ only between $x = 0$ and L (region II) and is once again zero beyond $x = L$ (region III). If now a plane wave is incident on this barrier from the left with $E < V_0$, there is an exponentially small probability for the particle to get to region III. Once a particle gets to region III, it is free once more and described by a plane wave. An example of tunneling is that of α particles trapped in the nuclei by a barrier. Every once in a while an α particle manages to penetrate the barrier and come out. The rate for this process can be calculated given V_0 and L.

Exercise 5.4.2. (a)* Calculate R and T for scattering off a potential $V(x) = V_0 a\delta(x)$. (b) Do the same for the case $V = 0$ for $|x| > a$ and $V = V_0$ for $|x| < a$. Assume that the energy is positive but less than V_0.

Exercise 5.4.3. Consider a particle subject to a constant force f in one dimension. Solve for the propagator in momentum space and get

$$U(p, t; p', 0) = \delta(p - p' - ft)e^{i(p'^3 - p^3)/6m\hbar f} \tag{5.4.30}$$

Transform back to coordinate space and obtain

$$U(x, t; x', 0) = \left(\frac{m}{2\pi\hbar i t}\right)^{1/2} \exp\left\{\frac{i}{\hbar}\left[\frac{m(x - x')^2}{2t} + \frac{1}{2}ft(x + x') - \frac{f^2t^3}{24m}\right]\right\} \tag{5.4.31}$$

[Hint: Normalize $\psi_E(p)$ such that $\langle E|E'\rangle = \delta(E - E')$. Note that E is not restricted to be positive.]

5.5. The Double-Slit Experiment

Having learned so much quantum mechanics, it now behooves us to go back and understand the double-slit experiment (Fig. 3.1). Let us label by I and II the regions to the left and right of the screen. The incident particle, which must really be represented by a wave packet, we approximate by a plane wave of wave number $k = p/\hbar$. The impermeable screen we treat as a region with $V = \infty$, and hence the region of vanishing ψ. Standard wave theory (which we can borrow from classical electromagnetism) tells us what happens in region II: the two slits act as sources of radially outgoing waves of the same wavelength. These two waves interfere on the line AB and produce the interference pattern. We now return to quantum mechanics and interpret the intensity $|\psi|^2$ as the probability density for finding the particle.

5.6. Some Theorems

Theorem 15. There is no degeneracy in one-dimensional bound states.

Proof. Let ψ_1 and ψ_2 be two solutions with the same eigenvalue E:

$$\frac{-\hbar^2}{2m}\frac{d^2\psi_1}{dx^2}+V\psi_1=E\psi_1 \tag{5.6.1}$$

$$\frac{-\hbar^2}{2m}\frac{d^2\psi_2}{dx^2}+V\psi_2=E\psi_2 \tag{5.6.2}$$

Multiply the first by ψ_2, the second by ψ_1 and subtract, to get

$$\psi_1\frac{d^2\psi_2}{dx^2}-\psi_2\frac{d^2\psi_1}{dx^2}=0$$

or

$$\frac{d}{dx}\left(\psi_1\frac{d\psi_2}{dx}-\psi_2\frac{d\psi_1}{dx}\right)=0$$

so that

$$\psi_1\frac{d\psi_2}{dx}-\psi_2\frac{d\psi_1}{dx}=c \tag{5.6.3}$$

To find the constant c, go to $|x|\to\infty$, where ψ_1 and ψ_2 vanish, since they describe bound states by assumption.‡ It follows that $c=0$. So

$$\frac{1}{\psi_1}\,d\psi_1=\frac{1}{\psi_2}\,d\psi_2$$

$$\log\psi_1=\log\psi_2+d \qquad (d \text{ is a constant})$$

$$\psi_1=e^d\psi_2 \tag{5.6.4}$$

‡ The theorem holds even if ψ vanishes at either $+\infty$ or $-\infty$. In a bound state it vanishes at both ends. But one can think of situations where the potential confines the wave function at one end but not the other.

Thus the two eigenfunctions differ only by a scale factor and represent the same state. Q.E.D.

What about the free-particle case, where to every energy there are two degenerate solutions with $p = \pm(2mE/\hbar^2)^{1/2}$? The theorem doesn't apply here since $\psi_p(x)$ does not vanish at spatial infinity. [Calculate c in Eq. (5.6.3).]

Theorem 16. The eigenfunctions of H can always be chosen pure real in the coordinate basis.

Proof. If

$$\left[\frac{-\hbar^2}{2m}\frac{d^2}{dx^2} + V(x)\right]\psi_n = E_n\psi_n$$

then by conjugation

$$\left[\frac{-\hbar^2}{2m}\frac{d^2}{dx^2} + V(x)\right]\psi_n^* = E_n\psi_n^*$$

Thus ψ_n and ψ_n^* are eigenfunctions with the same eigenvalue. It follows that the real and imaginary parts of ψ_n,

$$\psi_r = \frac{\psi_n + \psi_n^*}{2}$$

and

$$\psi_i = \frac{\psi_n - \psi_n^*}{2i}$$

are also eigenfunctions with energy E. Q.E.D.

The theorem holds in higher dimensions as well for *Hamiltonians of the above form*, which in addition to being Hermitian, are *real*. Note, however, that while Hermiticity is preserved under a unitary change of basis, reality is not.

If the problem involves a magnetic field, the Hamiltonian is no longer real in the coordinate basis, as is clear from Eq. (4.3.7). In this case the eigenfunctions cannot be generally chosen real. This question will be explored further at the end of Chapter 11.

Returning to one dimension, due to nondegeneracy of bound states, we must have

$$\psi_i = c\psi_r, \qquad c, \text{ a constant}$$

Consequently,

$$\psi = \psi_r + i\psi_i = (1+ic)\psi_r = \tilde{c}\psi_r$$

Since the overall scale $\tilde{c}$ is irrelevant, we can ignore it, i.e., work with real eigenfunctions with no loss of generality.

This brings us to the end of our study of one-dimensional problems, except for the harmonic oscillator, which is the subject of Chapter 7.

6

The Classical Limit

It is intuitively clear that when quantum mechanics is applied to a macroscopic system it should reproduce the results of classical mechanics, very much the way that relativistic dynamics, when applied to slowly moving ($v/c \ll 1$) objects, reproduces Newtonian dynamics. In this chapter we examine how classical mechanics is regained from quantum mechanics in the appropriate domain. When we speak of regaining classical mechanics, we refer to the numerical aspects. Qualitatively we know that the deterministic world of classical mechanics does not exist. Once we have bitten the quantum apple, our loss of innocence is permanent.

We commence by examining the time evolution of the expectation values. We find

$$\begin{aligned}\frac{d}{dt}\langle\Omega\rangle &= \frac{d}{dt}\langle\psi|\Omega|\psi\rangle \\ &= \langle\dot{\psi}|\Omega|\psi\rangle + \langle\psi|\Omega|\dot{\psi}\rangle + \langle\psi|\dot{\Omega}|\psi\rangle\ddagger \end{aligned} \tag{6.1}$$

In what follows we will assume that Ω has no explicit time dependence. We will therefore drop the third term $\langle\psi|\dot{\Omega}|\psi\rangle$. From the Schrödinger equation, we get

$$|\dot{\psi}\rangle = \frac{-i}{\hbar} H|\psi\rangle$$

and from its adjoint,

$$\langle\dot{\psi}| = \frac{i}{\hbar}\langle\psi|H$$

‡ If you are uncomfortable differentiating bras and kets, work in a basis and convince yourself that this step is correct.

Feeding these into Eq. (6.1) we get the relation

$$\frac{d}{dt}\langle\Omega\rangle=\left(\frac{-i}{\hbar}\right)\langle\psi|[\Omega, H]|\psi\rangle$$
$$=\left(\frac{-i}{\hbar}\right)\langle[\Omega, H]\rangle \tag{6.2}$$

which is called *Ehrenfest's theorem.*

Notice the structural similarity between this equation and the corresponding one from classical mechanics:

$$\frac{d\omega}{dt}=\{\omega, \mathscr{H}\} \tag{6.3}$$

We continue our investigation to see how exactly the two mechanics are related. Let us, for simplicity, discuss a particle in one dimension. If we consider $\Omega=X$ we get

$$\langle\dot{X}\rangle=\left(\frac{-i}{\hbar}\right)\langle[X, H]\rangle \tag{6.4}$$

If we assume

$$H=\frac{P^2}{2m}+V(X)$$

then

$$\langle\dot{X}\rangle=\left(\frac{-i}{\hbar}\right)\langle[X, P^2/2m]\rangle$$

Now

$$[X, P^2]=P[X, P]+[X, P]P \qquad \text{[from Eq. (1.5.10)]}$$
$$=2i\hbar P$$

so that

$$\langle\dot{X}\rangle=\frac{\langle P\rangle}{m} \tag{6.5}$$

The relation $\dot{x}=p/m$ of classical mechanics now appears as a relation among the mean values. We can convert Eq. (6.5) to a more suggestive form by writing

$$\frac{P}{m}=\frac{\partial H}{\partial P}$$

where $\partial H/\partial P$ is a formal derivative of H with respect to P, calculated by pretending that H, P, and X are just c numbers. The rule for finding such derivatives is just as in calculus, as long as the function being differentiated has a power series, as in this case. We now get, in the place of Eq. (6.5),

$$\langle \dot{X} \rangle = \left\langle \frac{\partial H}{\partial P} \right\rangle \tag{6.6}$$

Consider next

$$\begin{aligned}\langle \dot{P} \rangle &= \frac{1}{i\hbar}\langle [P, H] \rangle \\ &= \frac{1}{i\hbar}\langle [P, V(X)] \rangle\end{aligned}$$

To find $[P, V(X)]$ we go to the X basis, in which

$$P \to -i\hbar\frac{d}{dx} \quad \text{and} \quad V(X) \to V(x)$$

and for any $\psi(x)$,

$$\left[-i\hbar\frac{d}{dx}, V(x)\right]\psi(x) = -i\hbar\frac{dV}{dx}\psi(x)$$

We conclude that in the abstract,

$$[P, V(X)] = -i\hbar\frac{dV}{dX} \tag{6.7}$$

where dV/dX is again a formal derivative. Since $dV/dX=\partial H/\partial X$, we get

$$\langle \dot{P} \rangle = \left\langle -\frac{\partial H}{\partial X} \right\rangle \tag{6.8}$$

The *similarity* between Eqs. (6.6) and (6.8) and Hamilton's equations is rather striking. We would like to see how the quantum equations reduce to Hamilton's equations when applied to a macroscopic particle (of mass 1 g, say).

First of all, it is clear that we must consider an initial state that resembles the states of classical mechanics, i.e., states with well-defined position and momentum. Although simultaneous eigenstates of X and P do not exist, there do exist states which we can think of as approximate eigenstates of both X and P. In these states, labeled $|x_0 p_0 \Delta\rangle$, $\langle X\rangle = x_0$ and $\langle P\rangle = p_0$, with uncertainties $\Delta X = \Delta$ and $\Delta P \simeq \hbar/\Delta$, *both of which are small* in the *macroscopic scale*. A concrete example of such a state is

$$|x_0 p_0 \Delta\rangle \to \Psi_{x_0, p_0, \Delta} = \left(\frac{1}{\pi\Delta^2}\right)^{1/4} e^{ip_0 x/\hbar}\, e^{-(x-x_0)^2/2\Delta^2} \tag{6.9}$$

If we choose $\Delta \simeq 10^{-13}$ cm, say, *which is the size of a proton*, $\Delta P \simeq 10^{-14}$ g cm/sec. For a particle of mass 1 g, this implies $\Delta V \simeq 10^{-14}$ cm/sec, an uncertainty far below the experimentally detectable range. *In the classical scale*, such a state can be said to have well-defined values for X and P, namely, x_0 and p_0, since the uncertainties (fluctuations) around these values are truly negligible. If we let such a state evolve with time, the mean values $x_0(t)$ and $p_0(t)$ will follow Hamilton's equations, once again with negligible deviations. We establish this result as follows.

Consider Eqs. (6.6) and (6.8) which govern the evolution of $\langle X\rangle = x_0$ and $\langle P\rangle = p_0$. These would reduce to Hamilton's equations *if we could replace the mean values of the functions on the right-hand side by the functions of the mean values*:

$$\dot{x}_0 = \langle \dot{X}\rangle = \left\langle \frac{\partial H(X, P)}{\partial P}\right\rangle \simeq \frac{\partial H}{\partial P}\bigg|_{(X = x_0, P = p_0)} = \frac{\partial \mathscr{H}(x_0, p_0)}{\partial p_0} \tag{6.10}$$

and

$$\dot{p}_0 = \langle \dot{P}\rangle = -\left\langle \frac{\partial H}{\partial X}\right\rangle \simeq -\frac{\partial H}{\partial X}\bigg|_{(X = x_0, P = p_0)} = -\frac{\partial \mathscr{H}(x_0, p_0)}{\partial x_0} \tag{6.11}$$

If we consider some function of X and P, we will find in the same approximation

$$\langle \Omega(X, P)\rangle \simeq \Omega(x_0, p_0) = \omega(x_0, p_0) \tag{6.12}$$

Thus we regain classical physics as a good approximation whenever it is a good approximation to replace the mean of the functions $\partial H/\partial P$, $-\partial H/\partial X$, and $\Omega(X, P)$ by the functions of the mean. This in turn requires that the fluctuations about the mean have to be small. (The result is exact *if* there are no fluctuations.) Take as a concrete example Eqs. (6.10) and (6.11). There is no approximation involved in the first equation since $\langle \partial H/\partial P\rangle$ is just $\langle P/m\rangle = p_0/m$. In the second one, we need to approximate $\langle \partial H/\partial X\rangle = \langle dV/dX\rangle = \langle V'(X)\rangle$ by $V'(X = x_0)$. To see when this is a good approximation, let us expand V' in a Taylor series around x_0. Here it is convenient to work in the coordinate basis where $V(X) = V(x)$. The series is

$$V'(x) = V'(x_0) + (x - x_0)V''(x_0) + \tfrac{1}{2}(x - x_0)^2 V'''(x_0) + \cdots$$

Let us now take the mean of both sides. The first term on the right-hand side, which alone we keep in our approximation, corresponds to the classical force at x_0, and thus reproduces Newton's second law. The second vanishes in all cases, since the mean of $x-x_0$ does. The succeeding terms, which are corrections to the classical approximation, represent the fact that unlike the classical particle, which responds only to the force $F=-V'$ *at* x_0, the quantum particle responds to the force at neighboring points as well. (Note, incidentally, that these terms are zero if the potential is at the most quadratic in the variable x.) Each of these terms is a product of two factors, one of which measures the size or nonlocality of the wave packet and the other, the variation of the force with x. (See the third term for example.) At an intuitive level, we may say that these terms are negligible if the force varies very little over the "size" of the wave packet. (There is no unique definition of "size." The uncertainty is one measure. We see above that the uncertainty squared has to be much smaller than the inverse of the second derivative of the force.) In the present case, where the size of the packet is of the order of 10^{-13} cm, it is clear that the classical approximation is good for any potential that varies appreciably only over macroscopic scales.

There is one apparent problem: although we may start the system out in a state with $\Delta\simeq 10^{-13}$ cm, which is certainly a very small uncertainty, we know that with passing time the wave packet will spread. The uncertainty in the particle's position will inevitably become macroscopic. True. But recall the arguments of Section 5.1. We saw that the spreading of the wave packet can be attributed to the fact that any initial uncertainty in velocity, however small, will eventually manifest itself as a giant uncertainty in position. But in the present case ($\Delta V\simeq 10^{-14}$ cm/sec) it would take 300,000 years before the packet is even a millimeter across! (It is here that we invoke the fact that the particle is macroscopic: but for this, a small ΔP would not imply a small ΔV.) The problem is thus of academic interest only; and besides, it exists in classical mechanics as well, since the perfect measurement of velocity is merely an idealization.

There remains yet another question. We saw that for a macroscopic particle prepared in a state $|x_0p_0\Delta\rangle$, the time evolution of x_0 and p_0 will be in accordance with Hamilton's equations. Question: While it is true that a particle in such a conveniently prepared state obeys classical mechanics, are these the only states one encounters in classical mechanics? What if the initial position of the macroscopic particle is fixed to an accuracy of 10^{-27} cm? Doesn't its velocity now have uncertainties that are classically detectable? Yes. But such states do not occur in practice. The classical physicist talks about making exact position measurements, but never does so in practice. This is clear from the fact that he uses light of a finite frequency to locate the particle's positions, while only light of infinite frequency has perfect resolution. For example light in the visible spectrum has a wavelength of $\lambda\simeq 10^{-5}$ cm and thus the minimum ΔX is $\simeq 10^{-5}$ cm. If one really went towards the classical ideal and used photons of decreasing wavelength, one would soon find that the momentum of the macroscopic particle *is* affected by the act of measuring its position. For example, by the time one gets to a wavelength of 10^{-27} cm, each photon would carry a momentum of approximately 1 g cm/sec and one would see *macroscopic* objects recoiling under their impact.

In summary then, a typical macroscopic particle, described classically as possessing a well-defined value of x and p, is in reality an approximate eigenstate $|x_0p_0\Delta\rangle$, where Δ is at least 10^{-5} cm if visible light is used to locate the particle. The quantum equations for the time evolution of these approximate eigenvalues x_0 and p_0 reduce to Hamilton's equations, up to truly negligible uncertainties. The same goes for any other dynamical variable dependent on x and p.

We conclude this chapter by repeating an earlier observation to underscore its importance. Ehrenfest's theorem does not tell us that, in general, the expectation values of quantum operators evolve as do their classical counterparts. In particular $\langle X\rangle = x_0$ and $\langle P\rangle = p_0$ do not obey Hamilton's equations in all problems. For them to obey Hamilton's equations, we must be able to replace the mean values (expectation values) of the functions $\partial H/\partial P$ and $\partial H/\partial X$ of X and P by the corresponding functions of the mean values $\langle X\rangle = x_0$ and $\langle P\rangle = p_0$. For Hamiltonians that are at the most quadratic in X and P, this replacement can be done with no error for all wave functions. In the general case, such a replacement is a poor approximation unless the fluctuations about the means x_0 and p_0 are small. Even in those cases where x_0 and p_0 obey classical equations, the expectation value of some dependent variable $\Omega(X, P)$ need not, unless we can replace $\langle\Omega(X, P)\rangle$ by $\Omega(\langle X\rangle, \langle P\rangle) = \omega(x_0, p_0)$.

Example 6.1. Consider $\langle\Omega(X)\rangle$, where $\Omega = X^2$, in a state given by $\psi(x) = A\exp[-(x-a)^2/2\Delta^2]$. Is $\langle\Omega(X)\rangle = \Omega(\langle X\rangle)$? No, for the difference between the two is $\langle X^2\rangle - \langle X\rangle^2 = (\Delta X)^2 \neq 0$.

7

The Harmonic Oscillator

7.1. Why Study the Harmonic Oscillator?

In this section I will put the harmonic oscillator in its place—on a pedestal. Not only is it a system that can be exactly solved (in classical and quantum theory) and a superb pedagogical tool (which will be repeatedly exploited in this text), but it is also a system of great physical relevance. As will be shown below, any system fluctuating by small amounts near a configuration of stable equilibrium may be described either by an oscillator or by a collection of decoupled harmonic oscillators. Since the dynamics of a collection of noninteracting oscillators is no more complicated than that of a single oscillator (apart from the obvious N-fold increase in degrees of freedom), in addressing the problem of the oscillator we are actually confronting the general problem of small oscillations near equilibrium of an arbitrary system.

A concrete example of a single harmonic oscillator is a mass m coupled to a spring of force constant k. For small deformations x, the spring will exert the force given by Hooke's law, $F=-kx$, (k being its force constant) and produce a potential $V=\frac{1}{2}kx^2$. The Hamiltonian for this system is

$$\mathscr{H}=T+V=\frac{p^2}{2m}+\frac{1}{2}m\omega^2x^2 \tag{7.1.1}$$

where $\omega=(k/m)^{1/2}$ is the classical frequency of oscillation. Any Hamiltonian of the above form, quadratic in the coordinate and momentum, will be called the *harmonic oscillator Hamiltonian*. Now, the mass-spring system is just one among the following family of systems described by the oscillator Hamiltonian. Consider a particle moving in a potential $V(x)$. If the particle is placed at one of its minima x_0, it will remain there in a state of stable, static equilibrium. (A maximum, which is a point of unstable static equilibrium, will not interest us here.) Consider now the dynamics of this particle as it fluctuates by small amounts near $x=x_0$. The potential it experiences may be expanded in a Taylor series:

$$V(x)=V(x_0)+\left.\frac{dV}{dx}\right|_{x_0}(x-x_0)+\frac{1}{2!}\left.\frac{d^2V}{dx^2}\right|_{x_0}(x-x_0)^2+\cdots \tag{7.1.2}$$

Now, the constant piece $V(x_0)$ is of no physical consequence and may be dropped. [In other words, we may choose $V(x_0)$ as the arbitrary reference point for measuring the potential.] The second term in the series also vanishes since x_0 is a minimum of $V(x)$, or equivalently, since at a point of static equilibrium, the force $-dV/dx$, vanishes. If we now shift our origin of coordinates to x_0 Eq. (7.1.2) reads

$$V(x)=\frac{1}{2!}\left.\frac{d^2V}{dx^2}\right|_0 x^2+\frac{1}{3!}\left.\frac{d^3V}{dx^3}\right|_0 x^3+\cdots \tag{7.1.3}$$

For *small* oscillations, we may neglect all but the leading term and arrive at the potential (or Hamiltonian) in Eq. (7.1.1), d^2V/dx^2 being identified with $k=m\omega^2$. (By definition, x is small if the neglected terms in the Taylor series are small compared to the leading term, which alone is retained. In the case of the mass-spring system, x is small as long as Hooke's law is a good approximation.)

As an example of a system described by a collection of independent oscillators, consider the coupled-mass system from Example 1.8.6. (It might help to refresh your memory by going back and reviewing this problem.) The Hamiltonian for this system is

$$\begin{aligned}\mathscr{H}&=\frac{p_1^2}{2m}+\frac{p_2^2}{2m}+\frac{1}{2}m\omega^2[x_1^2+x_2^2+(x_1-x_2)^2]\\&=\mathscr{H}_1+\mathscr{H}_2+\tfrac{1}{2}m\omega^2(x_1-x_2)^2\end{aligned} \tag{7.1.4}$$

Now this $\mathscr{H}$ is not of the promised form, since the oscillators corresponding to $\mathscr{H}_1$ and $\mathscr{H}_2$ (associated with the coordinates x_1 and x_2) are coupled by the $(x_1-x_2)^2$ term. But we already know of an alternate description of this system in which it can be viewed as two *decoupled* oscillators. The trick is of course the introduction of normal coordinates. We exchange x_1 and x_2 for

$$x_{\mathrm{I}}=\frac{x_1+x_2}{2^{1/2}} \tag{7.1.5a}$$

and

$$x_{\mathrm{II}}=\frac{x_1-x_2}{2^{1/2}} \tag{7.1.5b}$$

By differentiating these equations with respect to time, we get similar ones for the velocities, and hence the momenta. In terms of the normal coordinates (and the corresponding momenta),

$$\mathscr{H}=\mathscr{H}_{\mathrm{I}}+\mathscr{H}_{\mathrm{II}}=\frac{p_{\mathrm{I}}^2}{2m}+\frac{1}{2}m\omega^2x_{\mathrm{I}}^2+\frac{p_{\mathrm{II}}^2}{2m}+\frac{3}{2}m\omega^2x_{\mathrm{II}}^2 \tag{7.1.6}$$

Thus the problem of the two coupled masses reduces to that of two uncoupled oscillators of frequencies $\omega_{\mathrm{I}}=\omega=(k/m)^{1/2}$ and $\omega_{\mathrm{II}}=3^{1/2}\omega=(3k/m)^{1/2}$.

Let us rewrite Eq. (7.1.4) as

$$\mathscr{H}=\frac{1}{2m}\sum_{i=1}^{2}\sum_{j=1}^{2}p_i\delta_{ij}p_j+\frac{1}{2}\sum_{i=1}^{2}\sum_{j=1}^{2}x_iV_{ij}x_j \tag{7.1.7}$$

where V_{ij} are elements of a real symmetric (Hermitian) matrix V with the following values:

$$V_{11}=V_{22}=2m\omega^2, \qquad V_{12}=V_{21}=-m\omega^2 \tag{7.1.8}$$

In switching to the normal coordinates $x_{\rm I}$ and $x_{\rm II}$ (and $p_{\rm I}$ and $p_{\rm II}$), we are going to a basis that diagonalizes V and reduces the potential energy to a sum of decoupled terms, one for each normal mode. The kinetic energy piece remains decoupled in both bases.

Now, just as the mass–spring system was just a representative element of a family of systems described by the oscillator Hamiltonian, the coupled-mass system is also a special case of a family that can be described by a collection of coupled harmonic oscillators. Consider a system with N Cartesian degrees of freedom $x_1\ldots x_N$, with a potential energy function $V(x_1,\ldots,x_N)$. Near an equilibrium point (chosen as the origin), the expansion of V, in analogy with Eq. (7.1.3), is

$$V(x_1\ldots x_N)=\frac{1}{2}\sum_{i=1}^{N}\sum_{j=1}^{N}\left.\frac{\partial^2 V}{\partial x_i\,\partial x_j}\right|_0 x_ix_j+\cdots \tag{7.1.9}$$

For small oscillations, the Hamiltonian is

$$\mathscr{H}=\sum_{i=1}^{N}\sum_{j=1}^{N}\frac{p_i\delta_{ij}p_j}{2m}+\frac{1}{2}\sum_{i=1}^{N}\sum_{j=1}^{N}x_iV_{ij}x_j \tag{7.1.10}$$

where

$$V_{ij}=\left.\frac{\partial^2 V}{\partial x_i\,\partial x_j}\right|_0=\left.\frac{\partial^2 V}{\partial x_j\,\partial x_i}\right|_0=V_{ji} \tag{7.1.11}$$

are the elements of a *Hermitian* matrix V. (We are assuming for simplicity that the masses associated with all N degrees of freedom are equal.) From the mathematical theory of Chapter 1, we know that there exists a new basis (i.e., a new set of coordinates $x_{\rm I}$, $x_{\rm II}$, ...) which will diagonalize V and reduce $\mathscr{H}$ to a sum of N decoupled oscillator Hamiltonians, one for each normal mode. Thus the general problem of small fluctuations near equilibrium of an arbitrary system reduces to the study of a single harmonic oscillator.

This section concludes with a brief description of two important systems which are described by a collection of independent oscillators. The first is a crystal (in three dimensions), the atoms in which jiggle about their mean positions on the lattice. The second is the electromagnetic field in free space. A crystal with N_0 atoms (assumed to be point particles) has $3N_0$ degrees of freedom, these being the displacements from equilibrium points on the lattice. For small oscillations, the Hamiltonian will be quadratic in the coordinates (and of course the momenta). Hence there will exist $3N_0$ normal coordinates and their conjugate momenta, in terms of which $\mathscr{H}$ will be a decoupled sum over oscillator Hamiltonians. What are the corresponding normal modes? Recall that in the case of two coupled masses, the normal modes corresponded to collective motions of the entire system, with the two masses in step in one case, and exactly out of step in the other. Likewise, in the present case, the motion is collective in the normal modes, and corresponds to plane waves traveling across the lattice. For a given wavevector $\mathbf{k}$, the atoms can vibrate parallel to k (*longitudinal polarization*) or in any one of the two independent directions perpendicular to $\mathbf{k}$ (*transverse polarization*). Most books on solid state physics will tell you why there are only N_0 possible values for $\mathbf{k}$. (This must of course be so, for with three polarizations at each $\mathbf{k}$, we will have exactly $3N_0$ normal modes.) The modes, labeled $(\mathbf{k}, \lambda)$, where λ is the polarization index ($\lambda = 1, 2, 3$), form a complete basis for expanding any state of the system. The coefficients of the expansion, $a(\mathbf{k}, \lambda)$, are the normal coordinates. The normal frequencies are labeled $\omega(\mathbf{k}, \lambda)$.‡

In the case of the electromagnetic field, the coordinate is the potential $\mathbf{A}(\mathbf{r}, t)$ at each point in space. [$\dot{\mathbf{A}}(\mathbf{r}, t)$ is the "velocity" corresponding to the coordinate $\mathbf{A}(\mathbf{r}, t)$.] The normal modes are once again plane waves but with two differences: there is no restriction on $\mathbf{k}$, but the polarization has to be transverse. The quantum theory of the field will be discussed at length in Chapter 18.

7.2. Review of the Classical Oscillator

The equations of motion for the oscillator are, from Eq. (7.1.1),

$$\dot{x} = \frac{\partial \mathscr{H}}{\partial p} = \frac{p}{m} \tag{7.2.1}$$

$$\dot{p} = -\frac{\partial \mathscr{H}}{\partial x} = -m\omega^2 x \tag{7.2.2}$$

By eliminating $\dot{p}$, we arrive at the familiar equation

$$\ddot{x} + \omega^2 x = 0$$

with the solution

$$x(t) = A \cos \omega t + B \sin \omega t = x_0 \cos(\omega t + \phi) \tag{7.2.3}$$

where x_0 is the amplitude and ϕ the phase of oscillator. The conserved energy associated with the oscillator is

$$E = T + V = \tfrac{1}{2}m\dot{x}^2 + \tfrac{1}{2}m\omega^2 x^2 = \tfrac{1}{2}m\omega^2 x_0^2 \tag{7.2.4}$$

‡ To draw a parallel with the two-mass system, $(\mathbf{k}, \lambda)$ is like I or II, $a(\mathbf{k}, \lambda)$ is like x_{I} or x_{II} and $\omega(\mathbf{k}, \lambda)$ is like $(k/m)^{1/2}$ or $(3k/m)^{1/2}$.

Since x_0 is a continuous variable, so is the energy of the classical oscillator. The lowest value for E is zero, and corresponds to the particle remaining at rest at the origin.

By solving for $\dot{x}$ in terms of E and x from Eq. (7.2.4) we obtain

$$\dot{x}=(2E/m-\omega^2x^2)^{1/2}=\omega(x_0^2-x^2)^{1/2} \tag{7.2.5}$$

which says that the particle starts from rest at a turning point ($x=\pm x_0$), picks up speed till it reaches the origin, and slows down to rest by the time it reaches the other turning point.

You are reminded of these classical results, so that you may readily compare and contrast them with their quantum counterparts.

7.3. Quantization of the Oscillator (Coordinate Basis)

We now consider the quantum oscillator, that is to say, a particle whose state vector $|\psi\rangle$ obeys the Schrödinger equation

$$i\hbar\frac{d}{dt}|\psi\rangle=H|\psi\rangle$$

with

$$H=\mathscr{H}(x\to X, p\to P)=\frac{P^2}{2m}+\frac{1}{2}m\omega^2X^2$$

As observed repeatedly in the past, the complete dynamics is contained in the propagator $U(t)$, which in turn may be expressed in terms of the eigenvectors and eigenvalues of H. In this section and the next, we will solve the eigenvalue problem in the X basis and the H basis, respectively. In Section 7.5 the passage from the H basis to the X basis will be discussed. The solution in the P basis, trivially related to the solution in the X basis in this case, will be discussed in an exercise.

With an eye on what is to follow, let us first establish that the eigenvalues of H cannot be negative. For any $|\psi\rangle$,

$$\begin{aligned}\langle H\rangle&=\frac{1}{2m}\langle\psi|P^2|\psi\rangle+\frac{1}{2}m\omega^2\langle\psi|X^2|\psi\rangle\\&=\frac{1}{2m}\langle\psi|P^\dagger P|\psi\rangle+\frac{1}{2}m\omega^2\langle\psi|X^\dagger X|\psi\rangle\\&=\frac{1}{2m}\langle P\psi|P\psi\rangle+\frac{1}{2}m\omega^2\langle X\psi|X\psi\rangle\geq 0\end{aligned}$$

since the norms of the states $|P\psi\rangle$ and $|X\psi\rangle$ cannot be negative. If we now set $|\psi\rangle$ equal to any eigenstate of H, we get the desired result.

Armed with the above result, we are now ready to attack the problem in the X basis.

We begin by projecting the eigenvalue equation,

$$\left(\frac{P^2}{2m}+\frac{1}{2}m\omega^2X^2\right)|E\rangle = E|E\rangle \tag{7.3.1}$$

onto the X basis, using the usual substitutions

$$X\to x$$
$$P\to -i\hbar\frac{d}{dx}$$
$$|E\rangle\to\psi_E(x)$$

and obtain

$$\left(-\frac{\hbar^2}{2m}\frac{d^2}{dx^2}+\frac{1}{2}m\omega^2x^2\right)\psi = E\psi \tag{7.3.2}$$

(The argument of ψ and the subscript E are implicit.)

We can rearrange this equation to the form

$$\frac{d^2\psi}{dx^2}+\frac{2m}{\hbar^2}\left(E-\frac{1}{2}m\omega^2x^2\right)\psi=0 \tag{7.3.3}$$

We wish to find all solutions to this equation that lie in the physical Hilbert space (of functions normalizable to unity or the Dirac delta function). Follow the approach closely—it will be invoked often in the future.

The first step is to write Eq. (7.3.3) in terms of dimensionless variables. We look for a new variable y which is dimensionless and related to x by

$$x=by \tag{7.3.4}$$

where b is a scale factor with units of length. Although any length b (say the radius of the solar system) will generate a dimensionless variable y, the idea is to choose the natural length scale generated by the equation itself. By feeding Eq. (7.3.4) into Eq. (7.3.3), we arrive at

$$\frac{d^2\psi}{dy^2}+\frac{2mEb^2}{\hbar^2}\psi-\frac{m^2\omega^2b^4}{\hbar^2}y^2\psi=0 \tag{7.3.5}$$

The last terms suggests that we choose

$$b=\left(\frac{\hbar}{m\omega}\right)^{1/2} \tag{7.3.6}$$

Let us also define a dimensionless variable ε corresponding to E:

$$\varepsilon=\frac{mEb^2}{\hbar^2}=\frac{E}{\hbar\omega} \tag{7.3.7}$$

(We may equally well choose $\varepsilon = 2mEb^2/\hbar^2$. Constants of order unity are not uniquely suggested by the equation. In the present case, our choice of ε is in anticipation of the results.) In terms of the dimensionless variables, Eq. (7.3.5) becomes

$$\psi''+(2\varepsilon-y^2)\psi=0 \tag{7.3.8}$$

where the prime denotes differentiation with respect to y.

Not only do dimensionless variables lead to a more compact equation, they also provide the natural scales for the problem. By measuring x and E in units of $(\hbar/m\omega)^{1/2}$ and $\hbar\omega$, which are scales generated intrinsically by the parameters entering the problem, we develop a feeling for what the words "small" and "large" mean: for example the displacement of the oscillator is large if y is large. If we insist on using the same units for all problems ranging from atomic physics to cosmology, we will not only be dealing with extremely large or extremely small numbers, we will also have no feeling for the size of quantities in the relevant scale. (A distance of 10^{-20} parsecs, small on the cosmic scale, is enormous if one is dealing with an atomic system.)

The next step is to examine Eq. (7.3.8) at limiting values of y to learn about the solution in these limits. In the limit $y\to\infty$, we may neglect the $2\varepsilon\psi$ term and obtain

$$\psi''-y^2\psi=0 \tag{7.3.9}$$

The solution to this equation *in the same limit* is

$$\psi=Ay^me^{\pm y^2/2}$$

for

$$\psi''=Ay^{m+2}\cdot e^{\pm y^2/2}\left[1\pm\frac{2m+1}{y^2}+\frac{m(m-1)}{y^4}\right]$$

$$\xrightarrow[y\to\infty]{} Ay^{m+2}e^{\pm y^2/2}=y^2\psi$$

where we have dropped all but the leading power in y as $y \to \infty$. Of the two possibilities $y^m e^{\pm y^2/2}$, we pick $y^m e^{-y^2/2}$, for the other possibility is not a part of the physical Hilbert space since it grows exponentially as $y \to \infty$.

Consider next the $y \to 0$ limit. Equation (7.3.8) becomes, upon dropping the $y^2\psi$ term,

$$\psi'' + 2\varepsilon\psi = 0$$

which has the solution

$$\psi = A \cos[\sqrt{2\varepsilon}y] + B \sin[\sqrt{2\varepsilon}y]$$

Since we have dropped the y^2 term in the equation as being too small, consistency demands that we expand the cosine and sine and drop terms of order y^2 and beyond. We then get

$$\psi \xrightarrow[y \to 0]{} A + cy + O(y^2)$$

where c is a new constant $[=B(2\varepsilon)^{1/2}]$.

We therefore infer that ψ is of the form

$$\psi(y) = u(y)\, e^{-y^2/2} \tag{7.3.10}$$

where u approaches $A + cy$ (plus higher powers) as $y \to 0$, and y^m (plus lower powers) as $y \to \infty$. To determine $u(y)$ completely, we feed the above *ansatz* into Eq. (7.3.8) and obtain

$$u'' - 2yu' + (2\varepsilon - 1)u = 0 \tag{7.3.11}$$

This equation has the desired features (to be discussed in Exercise 7.3.1) that indicate that a power-series solution is possible, i.e., if we assume

$$u(y) = \sum_{n=0}^{\infty} C_n y^n \tag{7.3.12}$$

the equation will determine the coefficients. [The series begins with $n=0$, and not some negative n, since we know that as $y \to 0$, $u \to A + cy + O(y^2)$.] Feeding this series into Eq. (7.3.11) we find

$$\sum_{n=0}^{\infty} C_n[n(n-1)y^{n-2} - 2ny^n + (2\varepsilon - 1)y^n] = 0 \tag{7.3.13}$$

Consider the first of three pieces in the above series:

$$\sum_{n=0}^{\infty} C_n n(n-1)y^{n-2}$$

Due to the $n(n-1)$ factor, this series also equals

$$\sum_{n=2}^{\infty} C_n n(n-1)y^{n-2}$$

In terms of a new variable $m=n-2$ the series becomes

$$\sum_{m=0}^{\infty} C_{m+2}(m+2)(m+1)y^m \equiv \sum_{n=0}^{\infty} C_{n+2}(n+2)(n+1)y^n$$

since m is a dummy variable. Feeding this equivalent series back into Eq. (7.3.13) we get

$$\sum_{n=0}^{\infty} y^n[C_{n+2}(n+2)(n+1)+C_n(2\varepsilon-1-2n)]=0 \tag{7.3.14}$$

Since the functions y^n are linearly independent (you cannot express y^n as a linear combination of other powers of y) each coefficient in the linear relation above must vanish. We thus find

$$C_{n+2}=C_n\frac{(2n+1-2\varepsilon)}{(n+2)(n+1)} \tag{7.3.15}$$

Thus for any C_0 and C_1, the *recursion relation* above generates $C_2, C_4, C_6, \ldots$ and $C_3, C_5, C_7, \ldots$. The function $u(y)$ is given by

$$u(y)=C_0\left[1+\frac{(1-2\varepsilon)y^2}{(0+2)(0+1)}+\frac{(1-2\varepsilon)}{(0+2)(0+1)}\frac{(4+1-2\varepsilon)}{(2+2)(2+1)}y^4+\cdots\right]$$
$$+C_1\left[y+\frac{(2+1-2\varepsilon)y^3}{(1+2)(1+1)}+\frac{(2+1-2\varepsilon)}{(1+2)(1+1)}\frac{(6+1-2\varepsilon)}{(3+2)(3+1)}y^5+\cdots\right] \tag{7.3.16}$$

where C_0 and C_1 are arbitrary.

It appears as if the energy of the quantum oscillator is arbitrary, since ε has not been constrained in any way. But we know something is wrong, since we saw at the outset that the oscillator eigenvalues are nonnegative. The first sign of sickness in our solution, Eq. (7.3.16), is that $u(y)$ does not behave like y^m as $y\to\infty$ (as deduced at the outset) since it contains arbitrarily high powers of y. There is only one explanation. We have seen that as $y\to\infty$, there are just two possibilities

$$\psi(y)\xrightarrow[y\to\infty]{} y^m e^{\pm y^2/2}$$

If we write $\psi(y)=u(y)\,e^{-y^2/2}$, then the two possibilities for $u(y)$ are

$$u(y)\xrightarrow[y\to\infty]{} y^m \quad \text{or} \quad y^m e^{y^2}$$

Clearly $u(y)$ in Eq. (7.3.16), which is not bounded by any finite power of y as $y\to\infty$, corresponds to the latter case. We may explicitly verify this as follows.

Consider the power series for $u(y)$ as $y\to\infty$. Just as the series is controlled by C_0 (the coefficient of the lowest power of y) as $y\to 0$, it is governed by its coefficients $C_{n\to\infty}$ as $y\to\infty$. The growth of the series is characterized by the ratio [see Eq. (7.3.15)]

$$\frac{C_{n+2}}{C_n}\xrightarrow[n\to\infty]{}\frac{2}{n} \tag{7.3.17}$$

Compare this to the growth of $y^m e^{y^2}$. Since

$$y^m e^{y^2}=\sum_{k=0}^{\infty}\frac{y^{2k+m}}{k!}$$

$C_n=$coefficient of $y^n=1/k!$; with $n=2k+m$ or $k=(n-m)/2$. Likewise

$$C_{n+2}=\frac{1}{[(n+2-m)/2]!}$$

so

$$\frac{C_{n+2}}{C_n}\xrightarrow[n\to\infty]{}\frac{[(n-m)/2]!}{[(n+2-m)/2]!}=\frac{1}{(n-m+2)/2}\sim\frac{2}{n}$$

In other words, $u(y)$ in Eq. (7.3.16) grows as $y^m e^{y^2}$, so that $\psi(y)\simeq y^m e^{y^2} e^{-y^2/2}\simeq y^m e^{+y^2/2}$, which is the rejected solution raising its ugly head! Our predicament is now reversed: from finding that every ε is allowed, we are now led to conclude that no ε is allowed. Fortunately there is a way out. If ε is one of the special values

$$\varepsilon_n=\frac{2n+1}{2},\qquad n=0, 1, 2, \ldots \tag{7.3.18}$$

the coefficient C_{n+2} (and others dependent on it) vanish. If we choose $C_1=0$ when n is even (or $C_0=0$ when n is odd) we have a finite polynomial of order n which satisfies the differential equation and behaves as y^n as $y\to\infty$:

$$\psi(y)=u(y)\,e^{-y^2/2}=\begin{Bmatrix} C_0+C_2y^2+C_4y^4+\cdots+C_ny^n \\ C_1y+C_3y^3+C_5y^5+\cdots+C_ny^n\end{Bmatrix}\cdot e^{-y^2/2} \tag{7.3.19}$$

Equation (7.3.18) tells us that energy is quantized: the only allowed values for $E=\varepsilon\hbar\omega$ (i.e., values that yield solutions in the physical Hilbert space) are

$$E_n=(n+\tfrac{1}{2})\hbar\omega,\qquad n=0, 1, 2, \ldots \tag{7.3.20}$$

For each value of n, Eq. (7.3.15) determines the corresponding polynomials of nth order, called *Hermite polynomials*, $H_n(y)$:

$$\begin{aligned} H_0(y) &= 1 \\ H_1(y) &= 2y \\ H_2(y) &= -2(1-2y^2) \\ H_3(y) &= -12(y-\tfrac{2}{3}y^3) \\ H_4(y) &= 12(1-4y^2+\tfrac{4}{3}y^4) \end{aligned} \tag{7.3.21}$$

The arbitrary initial coefficients C_0 and C_1 in H_n are chosen according to a standard convention. The normalized solutions are then

$$\begin{aligned} \psi_E(x) &\equiv \psi_{(n+1/2)\hbar\omega}(x) \equiv \psi_n(x) \\ &= \left(\frac{m\omega}{\pi\hbar 2^{2n}(n!)^2}\right)^{1/4} \exp\left(-\frac{m\omega x^2}{2\hbar}\right) H_n\left[\left(\frac{m\omega}{\hbar}\right)^{1/2} x\right] \end{aligned} \tag{7.3.22}$$

The derivation of the normalization constant

$$A_n = \left[\frac{m\omega}{\pi\hbar 2^{2n}(n!)^2}\right]^{1/4} \tag{7.3.23}$$

is rather tedious and will not be discussed here in view of a shortcut to be discussed in the next section.

The following recursion relations among Hermite polynomials are very useful:

$$H_n'(y) = 2nH_{n-1} \tag{7.3.24}$$

$$H_{n+1}(y) = 2yH_n - 2nH_{n-1} \tag{7.3.25}$$

as is the integral

$$\int_{-\infty}^{\infty} H_n(y)H_{n'}(y)\, e^{-y^2}\, dy = \delta_{nn'}(\pi^{1/2}2^n n!) \tag{7.3.26}$$

which is just the orthonormality condition of the eigenfunctions $\psi_n(x)$ and $\psi_{n'}(x)$ written in terms of $y = (m\omega/\hbar)^{1/2}x$.

We can now express the propagator as

$$\begin{aligned} U(x, t; x', t') = \sum_{n=0}^{\infty} A_n \exp\left(-\frac{m\omega}{2\hbar}x^2\right) H_n(x) A_n \exp\left(-\frac{m\omega}{2\hbar}x'^2\right) \\ \times H_n(x') \exp[-i(n+1/2)\omega(t-t')] \end{aligned} \tag{7.3.27}$$

Evaluation of this sum is a highly formidable task. We will not attempt it here since we will find an extremely simple way for calculating U in Chapter 8, devoted to the path integral formalism. The result happens to be

$$U(x,t;x',t')=\left(\frac{m\omega}{2\pi i\hbar\sin\omega T}\right)^{1/2}\exp\left[\frac{im\omega}{\hbar}\frac{(x^2+x'^2)\cos\omega T-2xx'}{2\sin\omega T}\right] \tag{7.3.28}$$

where $T=t-t'$.

This concludes the solution of the eigenvalue problem. Before analyzing our results let us recapitulate our strategy.

Step 1. Introduce dimensionless variables natural to the problem.
Step 2. Extract the asymptotic ($y\to\infty$, $y\to 0$) behavior of ψ.
Step 3. Write ψ as a product of the asymptotic form and an unknown function u. The function u will usually be easier to find than ψ.
Step 4. Try a power series to see if it will yield a recursion relation of the form Eq. (7.3.15).

*Exercise 7.3.1.** Consider the question why we tried a power-series solution for Eq. (7.3.11) but not Eq. (7.3.8). By feeding in a series into the latter, verify that a three-term recursion relation between C_{n+2}, C_n, and C_{n-2} obtains, from which the solution does not follow so readily. The problem is that ψ'' has two powers of y less than $2\varepsilon\psi$, while the $-y^2$ piece has two more powers of y. In Eq. (7.3.11) on the other hand, of the three pieces u'', $-2yu'$, and $(2\varepsilon-1)u$, the last two have the same powers of y.

Exercise 7.3.2. Verify that $H_3(y)$ and $H_4(y)$ obey the recursion relation, Eq. (7.3.15).

Exercise 7.3.3. If $\psi(x)$ is even and $\phi(x)$ is odd under $x\to -x$, show that

$$\int_{-\infty}^{\infty}\psi(x)\phi(x)\,dx=0$$

Use this to show that $\psi_2(x)$ and $\psi_1(x)$ are orthogonal. Using the values of Gaussian integrals in Appendix A.2 verify that $\psi_2(x)$ and $\psi_0(x)$ are orthogonal.

Exercise 7.3.4. Using Eqs. (7.3.23)–(7.3.25), show that

$$\langle n'|X|n\rangle=\left(\frac{\hbar}{2m\omega}\right)^{1/2}[\delta_{n',n+1}(n+1)^{1/2}+\delta_{n',n-1}n^{1/2}]$$

$$\langle n'|P|n\rangle=\left(\frac{m\omega\hbar}{2}\right)^{1/2}i[\delta_{n',n+1}(n+1)^{1/2}-\delta_{n',n-1}n^{1/2}]$$

*Exercise 7.3.5.** Using the symmetry arguments from Exercise 7.3.3 show that $\langle n|X|n\rangle=\langle n|P|n\rangle=0$ and thus that $\langle X^2\rangle=(\Delta X)^2$ and $\langle P^2\rangle=(\Delta P)^2$ in these states. Show that $\langle 1|X^2|1\rangle=3\hbar/2m\omega$ and $\langle 1|P^2|1\rangle=\frac{3}{2}m\omega\hbar$. Show that $\psi_0(x)$ saturates the uncertainty bound $\Delta X\cdot\Delta P\geq\hbar/2$.

*Exercise 7.3.6.** Consider a particle in a potential

$$V(x)=\tfrac{1}{2}m\omega^2x^2, \qquad x>0$$
$$=\infty, \qquad x\leq 0$$

What are the boundary conditions on the wave functions now? Find the eigenvalues and eigenfunctions.

We now discuss the eigenvalues and eigenfunctions of the oscillator. The following are the main features:

(1) The energy is quantized. In contrast to the classical oscillator whose energy is continuous, the quantum oscillator has a discrete set of levels given by Eq. (7.3.20). Note that the quantization emerges only after we supplement Schrödinger's equation with the requirement that ψ be an element of the physical Hilbert space. In this case it meant the imposition of the boundary condition $\psi(|x|\to\infty)\to 0$ [as opposed to $\psi(|x|\to\infty)\to\infty$, which is what obtained for all but the special values of E].

Why does the classical oscillator seem to have a continuum of energy values? The answer has to do with the relative sizes of the energy gap and the total energy of the classical oscillator. Consider, for example, a mass of 2 g, oscillating at a frequency of 1 rad/sec, with an amplitude of 1 cm. Its energy is

$$E=\tfrac{1}{2}m\omega^2x_0^2=1 \text{ erg}$$

Compare this to the gap between allowed energies:

$$\Delta E=\hbar\omega\simeq 10^{-27} \text{ erg}$$

At the macroscopic level, it is practically impossible to distinguish between a system whose energy is continuous and one whose allowed energy levels are spaced 10^{-27} erg apart. Stated differently, the *quantum number* associated with this oscillator is

$$n=\frac{E}{\hbar\omega}-\frac{1}{2}\simeq 10^{27}$$

while the difference in n between adjacent levels is unity. We have here a special case of the *correspondence principle*, which states that as the quantum number tends to infinity, we regain the classical picture. (We know vaguely that when a system is big, it may be described classically. The correspondence principle tells us that the quantum number is a good measure of bigness.)

(2) The levels are spaced uniformly. The fact that the oscillator energy levels go up in steps of $\hbar\omega$ allows one to construct the following picture. We pretend that associated with an oscillator of classical frequency ω there exist fictitious particles called *quanta* each endowed with energy $\hbar\omega$. We view the $n\hbar\omega$ piece in the energy formula Eq. (7.3.20) as the energy of n such quanta. In other words, we forget about the mass and spring and think in terms of the quanta. When the quantum number n goes up (or down) by Δn, we say that Δn quanta have been created (or destroyed).

Although it seems like a matter of semantics, thinking of the oscillator in terms of these quanta has proven very useful.

In the case of the crystal, there are $3N_0$ oscillators, labeled by the $3N_0$ values of $(\mathbf{k}, \lambda)$, with frequencies $\omega(\mathbf{k}, \lambda)$. The quantum state of the crystal is specified by giving the number of quanta, called *phonons*, at each $(\mathbf{k}, \lambda)$. For a crystal whose Hamiltonian is exactly given by a sum of oscillator pieces, the introduction of the phonon concept is indeed a matter of semantics. If, however, we consider deviations from this, say to take into account nonleading terms in the Taylor expansion of the potential, or the interaction between the crystal and some external probe such as an electron shot at it, the phonon concept proves very useful. (The two effects mentioned above may be seen as phonon–phonon interactions and phonon–electron interactions, respectively.)

Similarly, the interaction of the electromagnetic field with matter may be viewed as the interaction between light quanta or *photons* and matter, which is discussed in Chapter 18.

(3) The lowest possible energy is $\hbar\omega/2$ and not 0. Unlike the classical oscillator, which can be in a state of zero energy (with $x=p=0$) the quantum oscillator has a minimum energy of $\hbar\omega/2$. This energy, called the *zero-point energy*, is a reflection of the fact that the simultaneous eigenstate $|x=0, p=0\rangle$ is precluded by the canonical commutation relation $[X, P]=i\hbar$. This result is common to all oscillators, whether they describe a mechanical system or a normal mode of the electromagnetic field, since all these problems are mathematically identical and differ only in what the coordinate and its conjugate momentum represent. Thus, a crystal has an energy $\frac{1}{2}\hbar\omega(\mathbf{k}, \lambda)$ in each mode $(\mathbf{k}, \lambda)$ even when phonons are absent, and the electromagnetic field has an energy $\frac{1}{2}\hbar\omega(\mathbf{k}, \lambda)$ in each mode of frequency ω even when photons are absent. (The zero-point fluctuation of the field has measurable consequences, which will be discussed in Chapter 18.)

In the following discussion let us restrict ourselves to the mechanical oscillator and examine more closely the zero-point energy. We saw that it is the absence of the state $|x=0, p=0\rangle$ that is responsible for this energy. Such a state, with $\Delta X=\Delta P=0$, is forbidden by the uncertainty principle. Let us therefore try to find a state that is quantum mechanically allowed and comes as close as possible (in terms of its energy) to the classical state $x=p=0$. If we choose a wave function $\psi(x)$ that is sharply peaked near $x=0$ to minimize the mean potential energy $\langle \frac{1}{2}m\omega^2X^2\rangle$, the wave function in P space spreads out and the mean kinetic energy $\langle P^2/2m\rangle$ grows. The converse happens if we pick a momentum space wave function sharply peaked near $p=0$. What we need then is a compromise $\psi_{\min}(x)$ that minimizes the *total* mean energy without violating the uncertainty principle. Let us now begin our quest for $\psi_{\min}(x)$. We start with a normalized trial state $|\psi\rangle$ and consider

$$\langle\psi|H|\psi\rangle=\langle H\rangle=\frac{\langle P^2\rangle}{2m}+\frac{1}{2}m\omega^2\langle X^2\rangle \tag{7.3.29}$$

Now

$$(\Delta P)^2=\langle P^2\rangle-\langle P\rangle^2 \tag{7.3.30}$$

and

$$(\Delta X)^2=\langle X^2\rangle-\langle X\rangle^2 \tag{7.3.31}$$

so that

$$\langle H\rangle = \frac{(\Delta P)^2 + \langle P\rangle^2}{2m} + \frac{1}{2}m\omega^2[(\Delta X)^2 + \langle X\rangle^2] \tag{7.3.32}$$

The first obvious step in minimizing $\langle H\rangle$ is to restrict ourselves to states with $\langle X\rangle = \langle P\rangle = 0$. (Since $\langle X\rangle$ and $\langle P\rangle$ are independent of each other and of $(\Delta X)^2$ and $(\Delta P)^2$, such a choice is always possible.) For these states (from which we must pick the winner)

$$\langle H\rangle = \frac{(\Delta P)^2}{2m} + \frac{1}{2}m\omega^2(\Delta X)^2 \tag{7.3.33}$$

Now we use the uncertainty relation

$$\Delta X \cdot \Delta P \geq \hbar/2 \tag{7.3.34}$$

where the equality sign holds only for a Gaussian, as will be shown in Section 9.3. We get

$$\langle H\rangle \geq \frac{\hbar^2}{8m(\Delta X)^2} + \frac{1}{2}m\omega^2(\Delta X)^2 \tag{7.3.35}$$

We minimize $\langle H\rangle$ by choosing a Gaussian wave function, for which

$$\langle H\rangle_{\text{Gaussian}} = \frac{\hbar^2}{8m(\Delta X)^2} + \frac{1}{2}m\omega^2(\Delta X)^2 \tag{7.3.36}$$

What we have found is that the mean energy associated with the trial wave function is sensitive only to the corresponding ΔX and that, of all functions with the same ΔX, the Gaussian has the lowest energy. Finally we choose, from the family of Gaussians, the one with the ΔX that minimizes $\langle H\rangle_{\text{Gaussian}}$. By requiring

$$\frac{\partial\langle H\rangle_{\text{Gaussian}}}{\partial(\Delta X)^2} = 0 = \frac{-\hbar^2}{8m(\Delta X)^4} + \frac{1}{2}m\omega^2 \tag{7.3.37}$$

we obtain

$$(\Delta X)^2 = \hbar/2m\omega \tag{7.3.38}$$

and

$$\langle H\rangle_{\text{min}} = \hbar\omega/2 \tag{7.3.39}$$

Thus, by systematically hunting in Hilbert space, we have found that the following normalized function has the lowest mean energy:

$$\psi_{\min}(x) = \left(\frac{m\omega}{\pi\hbar}\right)^{1/4} \exp\left(-\frac{m\omega x^2}{2\hbar}\right), \qquad \langle H\rangle_{\min} = \frac{\hbar\omega}{2} \tag{7.3.40}$$

If we apply the above result

$$\langle\psi_{\min}|H|\psi_{\min}\rangle \le \langle\psi|H|\psi\rangle \quad (\text{for all } |\psi\rangle)$$

to $|\psi\rangle = |\psi_0\rangle$ = ground-state vector, we get

$$\langle\psi_{\min}|H|\psi_{\min}\rangle \le \langle\psi_0|H|\psi_0\rangle = E_0 \tag{7.3.41}$$

Now compare this with the result of Exercise 5.2.2:

$$E_0 = \langle\psi_0|H|\psi_0\rangle \le \langle\psi|H|\psi\rangle \quad \text{for all } |\psi\rangle$$

If we set $|\psi\rangle = |\psi_{\min}\rangle$ we get

$$E_0 = \langle\psi_0|H|\psi_0\rangle \le \langle\psi_{\min}|H|\psi_{\min}\rangle \tag{7.3.42}$$

It follows from Eq. (7.3.41) and (7.3.42) that

$$E_0 = \langle\psi_0|H|\psi_0\rangle = \langle\psi_{\min}|H|\psi_{\min}\rangle = \frac{\hbar\omega}{2} \tag{7.3.43}$$

Also, since there was only one state, $|\psi_{\min}\rangle$, with energy $\hbar\omega/2$, it follows that

$$|\psi_0\rangle = |\psi_{\min}\rangle \tag{7.3.44}$$

We have thus managed to find the oscillator ground-state energy and state vector without solving the Schrödinger equation.

It would be a serious pedagogical omission if it were not emphasized at this juncture that the uncertainty relation has been unusually successful in the above context. Our ability here to obtain all the information about the ground state using the uncertainty relation is a consequence of the special form of the oscillator Hamiltonian [which allowed us to write $\langle H\rangle$ in terms of $(\Delta X)^2$ and $(\Delta P)^2$] and the fact that its ground-state wave function is a Gaussian (which has a privileged role with respect to the uncertainty relation). In more typical instances, the use of the uncertainty relation will have to be accompanied by some hand-waving [before $\langle H\rangle$ can be approximated by a function of $(\Delta X)^2$ and $(\Delta P)^2$] and then too will yield only an *estimate* for the ground-state energy. As for the wave function, we can only get an estimate for ΔX, the spread associated with it.

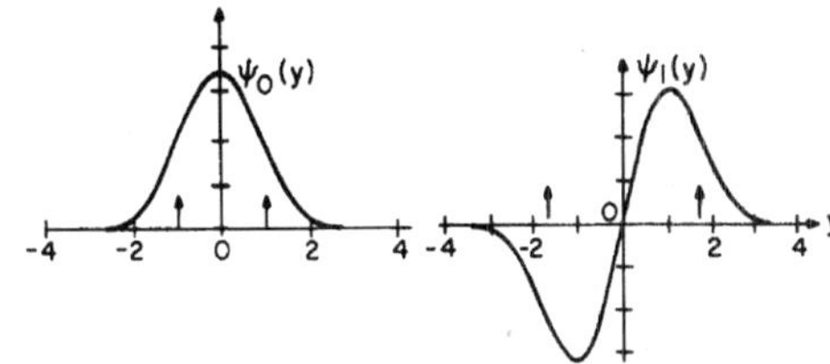

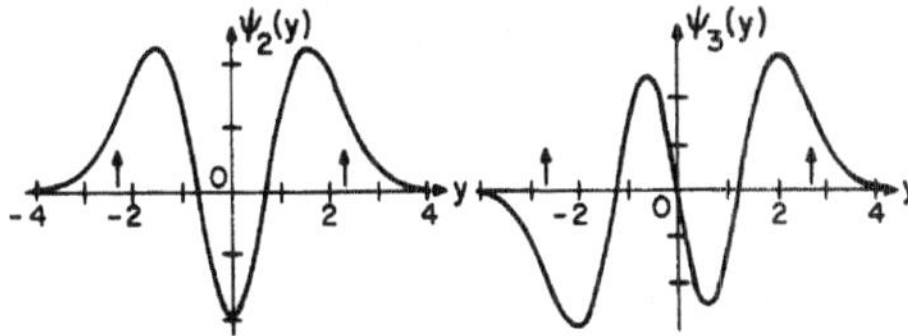

Figure 7.1. Normalized eigenfunctions for $n=0, 1, 2,$ and 3. The small arrows at $|y|=(2n+1)^{1/2}$ stand for the classical turning points. Recall that $y=(m\omega/\hbar)^{1/2}x$.

(4) The solutions (Fig. 7.1) $\psi_n(x)$ contain only even or odd powers of x, depending on whether n is even or odd. Consequently the eigenfunctions are even or odd:

$$\begin{aligned} \psi_n(-x) &= \psi_n(x), \qquad n \text{ even} \\ &= -\psi_n(x), \qquad n \text{ odd} \end{aligned}$$

In Chapter 11 on symmetries it will be shown that the eigenfunctions had to have this property.

(5) The wave function does not vanish beyond the classical turning points, but dies out exponentially as $x \to \infty$. [Verify that the classical turning points are given by $y_0 = \pm(2n+1)^{1/2}$.] Notice, however, that when n is large (Fig. 7.2) the excursions outside the turning points are small compared to the classical amplitude. This exponentially damped amplitude in the classically forbidden region was previously encountered in Chapter 5 when we studied tunneling.

(6) The probability distribution $P(x)$ is very different from the classical case. The position of a given classical oscillator is of course exactly known. But we could ask the following probabilistic question: if I suddenly walk into a room containing the oscillator, where am I likely to catch it? If the velocity of the oscillator at a point x is $v(x)$, the time it spends near the x, and hence the probability of our catching it there during a random spot check, varies inversely with $v(x)$:

$$P_{\mathrm{cl}}(x) \propto \frac{1}{v(x)} = \frac{1}{\omega(x_0^2 - x^2)^{1/2}} \tag{7.3.45}$$

which is peaked near $\pm x_0$ and has a minimum at the origin. In the quantum case, for the ground state in particular, $|\psi(x)|^2$ seems to go just the other way (Fig. 7.1). There is no contradiction here, for quantum mechanics *is* expected to differ from classical mechanics. The correspondence principle, however, tells us that for large n

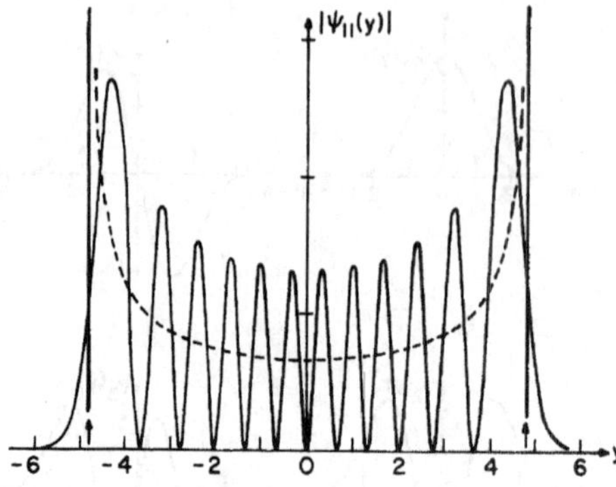

Figure 7.2. Probability density in the state $n=11$. The broken curve gives the classical probability distribution in a state with the same energy.

the two must become indistinguishable. From Fig. 7.2, which shows the situations at $n=11$, we can see how the classical limit is reached: the quantum distribution $P(x)=|\psi(x)|^2$ wiggles so rapidly (in a scale set by the classical amplitude) that only its mean can be detected at these scales, and this agrees with $P_{\text{cl}}(x)$. We are reminded here of the double-slit experiment performed with macroscopic particles: there is a dense interference pattern, whose mean is measured in practice and agrees with the classical probability curve.

A remark that was made in more general terms in Chapter 6: the classical oscillator that we often refer to, is a figment lodged in our imagination and doesn't exist. In other words, all oscillators, including the 2-g mass and spring system, are ultimately governed by the laws of quantum mechanics, and thus have discrete energies, can shoot past the "classical" turning points, and have a zero-point energy of $\frac{1}{2}\hbar\omega$ even while they play dead. Note however that what I am calling nonexistent is an oscillator that *actually* has the properties attributed to it in classical mechanics, and not one that *seems* to have them when examined at the macroscopic level.

Exercise 7.3.7. The Oscillator in Momentum Space.* By setting up an eigenvalue equation for the oscillator in the P basis and comparing it to Eq. (7.3.2), show that the momentum space eigenfunctions may be obtained from the ones in coordinate space through the substitution $x\to p$, $m\omega\to 1/m\omega$. Thus, for example,

$$\psi_0(p)=\left(\frac{1}{m\pi\hbar\omega}\right)^{1/4} e^{-p^2/2m\hbar\omega}$$

There are several other pairs, such as ΔX and ΔP in the state $|n\rangle$, which are related by the substitution $m\omega\to 1/m\omega$. You may wish to watch out for them. (Refer back to Exercise 7.3.5.)

7.4. The Oscillator in the Energy Basis

Let us orient ourselves by recalling how the eigenvalue equation

$$\left(\frac{P^2}{2m}+\frac{1}{2}m\omega^2X^2\right)|E\rangle=E|E\rangle \tag{7.4.1}$$

was solved in the coordinate basis: (1) We made the assignments $X \to x$, $P \to -i\hbar\, d/dx$. (2) We solved for the components $\langle x|E\rangle = \psi_E(x)$ and the eigenvalues.

To solve the problem in the momentum basis, we first compute the X and P operators in this basis, given their form in the coordinate basis. For instance,

$$\langle p'|X|p\rangle = \iint \underbrace{\langle p'|x\rangle}_{\frac{e^{-ip'x/\hbar}}{(2\pi\hbar)^{1/2}}} \underbrace{\langle x|X|x'\rangle}_{\substack{x\delta(x-x') \\ \text{(given)}}} \underbrace{\langle x'|p\rangle}_{\frac{e^{ipx'/\hbar}}{(2\pi\hbar)^{1/2}}} dx\, dx'$$

$$= -i\hbar\delta'(p-p')$$

We then find P and $H(X, P)$ in this basis. The eigenvalue equation, (7.4.1), will then become a differential equation that we will proceed to solve.

Now suppose that we want to work in the energy basis. We must first find the eigenfunctions of H, i.e., $\langle x|E\rangle$, so that we can carry out the change of basis. But finding $\langle x|E\rangle = \psi_E(x)$ amounts to solving the full eigenvalue problem in the coordinate basis. Once we have done this, there is not much point in setting up the problem in the E basis.

But there is a clever way due to Dirac, which allows us to work in the energy basis without having to know ahead of time the operators X and P in this basis. All we will need is the commutation relation

$$[X, P] = i\hbar I = i\hbar \tag{7.4.2}$$

which follows from $X \to x$, $P \to -i\hbar\, d/dx$, *but is basis independent*. The next few steps will seem rather mysterious and will not fit into any of the familiar schemes discussed so far. You must be patient till they begin to pay off.

Let us first introduce the operator

$$a = \left(\frac{m\omega}{2\hbar}\right)^{1/2} X + i\left(\frac{1}{2m\omega\hbar}\right)^{1/2} P \tag{7.4.3}$$

and its adjoint

$$a^\dagger = \left(\frac{m\omega}{2\hbar}\right)^{1/2} X - i\left(\frac{1}{2m\omega\hbar}\right)^{1/2} P \tag{7.4.4}$$

(Note that $m\omega \to 1/m\omega$ as $X \leftrightarrow P$.) They satisfy the commutation relation (which you should verify)

$$[a, a^\dagger] = 1 \tag{7.4.5}$$

Note next that the Hermitian operator $a^\dagger a$ is simply related to H:

$$a^\dagger a=\frac{m\omega}{2\hbar}X^2+\frac{1}{2m\omega\hbar}P^2+\frac{i}{2\hbar}[X, P]$$
$$=\frac{H}{\hbar\omega}-\frac{1}{2}$$

so that

$$H=(a^\dagger a+1/2)\hbar\omega \tag{7.4.6}$$

[This method is often called the "method of factorization" since we are expressing $H=P^2+X^2$ (ignoring constants) as a product of $(X+iP)=a$ and $(X-iP)=a^\dagger$. The extra $\hbar\omega/2$ in Eq. (7.4.6) comes from the non-commutative nature of X and P.]

Let us next define an operator $\hat{H}$,

$$\hat{H}=\frac{H}{\hbar\omega}=(a^\dagger a+1/2) \tag{7.4.7}$$

whose eigenvalues ε measure energy in units of $\hbar\omega$. We wish to solve the eigenvalue equation for $\hat{H}$:

$$\hat{H}|\varepsilon\rangle=\varepsilon|\varepsilon\rangle \tag{7.4.8}$$

where ε is the energy measured in units of $\hbar\omega$. Two relations we will use shortly are

$$[a, \hat{H}]=[a, a^\dagger a+1/2]=[a, a^\dagger a]=a \tag{7.4.9}$$

and

$$[a^\dagger, \hat{H}]=-a^\dagger \tag{7.4.10}$$

The utility of a and $a^\dagger$ stems from the fact that given an eigenstate of $\hat{H}$, they generate others. Consider

$$\begin{aligned}\hat{H}a|\varepsilon\rangle&=(a\hat{H}-[a, \hat{H}])|\varepsilon\rangle\\&=(a\hat{H}-a)|\varepsilon\rangle\\&=(\varepsilon-1)a|\varepsilon\rangle\end{aligned} \tag{7.4.11}$$

We infer from Eq. (7.4.11) that $a|\varepsilon\rangle$ is an eigenstate with eigenvalue $\varepsilon-1$, i.e.,

$$a|\varepsilon\rangle=C_\varepsilon|\varepsilon-1\rangle \tag{7.4.12}$$

where C_ε is a constant, and $|\varepsilon-1\rangle$ and $|\varepsilon\rangle$ are normalized eigenkets.‡

Similarly we see that

$$\begin{aligned}\hat{H}a^\dagger|\varepsilon\rangle&=(a^\dagger\hat{H}-[a^\dagger, H])|\varepsilon\rangle\\&=(a^\dagger\hat{H}+a^\dagger)|\varepsilon\rangle\\&=(\varepsilon+1)a^\dagger|\varepsilon\rangle\end{aligned} \tag{7.4.13}$$

so that

$$a^\dagger|\varepsilon\rangle=C_{\varepsilon+1}|\varepsilon+1\rangle \tag{7.4.14}$$

One refers to a and $a^\dagger$ as *lowering and raising operators* for obvious reasons. They are also called *destruction and creation operators* since they destroy or create quanta of energy $\hbar\omega$.

We are thus led to conclude that if ε is an eigenvalue of $\hat{H}$, so are $\varepsilon+1$, $\varepsilon+2$, $\varepsilon+3, \ldots, \varepsilon+\infty$; and $\varepsilon-1, \ldots, \varepsilon-\infty$. The latter conclusion is in conflict with the result that the eigenvalues of H are nonnegative. So, it must be that the downward chain breaks at some point: there must be a state $|\varepsilon_0\rangle$ that cannot be lowered further:

$$a|\varepsilon_0\rangle=0 \tag{7.4.15}$$

Operating with $a^\dagger$, we get

$$a^\dagger a|\varepsilon_0\rangle=0$$

or

$$(\hat{H}-1/2)|\varepsilon_0\rangle=0 \quad \text{[from Eq. (7.4.7)]}$$

or

$$\hat{H}|\varepsilon_0\rangle=\tfrac{1}{2}|\varepsilon_0\rangle$$

or

$$\varepsilon_0=\tfrac{1}{2} \tag{7.4.16}$$

‡ We are using the fact that there is no degeneracy in one dimension.

We may, however, raise the state $|\varepsilon_0\rangle$ indefinitely by the repeated application of $a^\dagger$. We thus find that the oscillator has a sequence of levels given by

$$\varepsilon_n = (n+1/2), \qquad n=0, 1, 2, \ldots$$

or

$$E_n = (n+1/2)\hbar\omega, \qquad n=0, 1, 2, \ldots \tag{7.4.17}$$

Are these the only levels? If there were another family, it too would have to have a ground state $|\varepsilon_0'\rangle$ such that

$$a|\varepsilon_0'\rangle = 0$$

or

$$a^\dagger a|\varepsilon_0'\rangle = 0$$

or

$$\hat{H}|\varepsilon_0'\rangle = \tfrac{1}{2}|\varepsilon_0'\rangle \tag{7.4.18}$$

But we know that there is no degeneracy in one dimension (Theorem 15). Consequently it follows from Eqs. (7.4.16) and (7.4.18) that $|\varepsilon_0\rangle$ and $|\varepsilon_0'\rangle$ represent the same state. The same goes for the families built from $|\varepsilon_0\rangle$ and $|\varepsilon_0'\rangle$ by the repeated action of $a^\dagger$.

We now calculate the constants C_ε and $C_{\varepsilon+1}$ appearing in Eqs. (7.4.12) and (7.4.14). Since $\varepsilon = n+1/2$, let us label the kets by the integer n. We want to determine the constant C_n appearing in the equation

$$a|n\rangle = C_n|n-1\rangle \tag{7.4.19a}$$

Consider the adjoint of this equation

$$\langle n|a^\dagger = \langle n-1|C_n^* \tag{7.4.19b}$$

By combining these equations we arrive at

$$\begin{aligned}
\langle n|a^\dagger a|n\rangle &= \langle n-1|n-1\rangle C_n^* C_n \\
\langle n|\hat{H}-\tfrac{1}{2}|n\rangle &= C_n^* C_n \quad \text{(since } |n-1\rangle \text{ is normalized)} \\
\langle n|n|n\rangle &= |C_n|^2 \quad \text{(since } \hat{H}|n\rangle = (n+1/2)|n\rangle\text{)} \\
|C_n|^2 &= n \\
C_n &= (n)^{1/2}\, e^{i\phi} \quad (\phi \text{ is arbitrary})
\end{aligned} \tag{7.4.20}$$

It is conventional to choose ϕ as zero. So we have

$$a|n\rangle = n^{1/2}|n-1\rangle \tag{7.4.21}$$

It can similarly be shown (by you) that

$$a^\dagger|n\rangle = (n+1)^{1/2}|n+1\rangle \tag{7.4.22}$$

[Note that in Eqs. (7.4.21) and (7.4.22) the larger of the n's labeling the two kets appears under the square root.] By combining these two equations we find

$$a^\dagger a|n\rangle = a^\dagger n^{1/2}|n-1\rangle = n^{1/2}n^{1/2}|n\rangle = n|n\rangle \tag{7.4.23}$$

In terms of

$$N = a^\dagger a \tag{7.4.24}$$

called the *number operator* (since it counts the quanta)

$$\hat{H} = N + \tfrac{1}{2} \tag{7.4.25}$$

Equations (7.4.21) and (7.4.22) are very important. They allow us to compute the matrix elements of all operators in the $|n\rangle$ basis. First consider a and $a^\dagger$ themselves:

$$\langle n'|a|n\rangle = n^{1/2}\langle n'|n-1\rangle = n^{1/2}\delta_{n',n-1} \tag{7.4.26}$$

$$\langle n'|a^\dagger|n\rangle = (n+1)^{1/2}\langle n'|n+1\rangle = (n+1)^{1/2}\delta_{n',n+1} \tag{7.4.27}$$

To find the matrix elements of X and P, we invert Eqs. (7.4.3) and (7.4.4) to obtain

$$X = \left(\frac{\hbar}{2m\omega}\right)^{1/2}(a+a^\dagger) \tag{7.4.28}$$

$$P = i\left(\frac{m\omega\hbar}{2}\right)^{1/2}(a^\dagger - a) \tag{7.4.29}$$

and then use Eqs. (7.4.26) and (7.4.27). The details are left as an exercise. The two basic matrices in this energy basis are

$$a^\dagger \leftrightarrow \begin{array}{c} \\ n=0 \\ n=1 \\ n=2 \\ \cdot \\ \cdot \end{array} \begin{array}{c} \begin{array}{cccc} n=0 & n=1 & n=2 & \cdots \end{array} \\ \begin{bmatrix} 0 & 0 & 0 & \cdots \\ 1^{1/2} & 0 & 0 & \\ 0 & 2^{1/2} & 0 & \\ 0 & 0 & 3^{1/2} & \\ \vdots & & & \end{bmatrix} \end{array} \tag{7.4.30}$$

and its adjoint

$$a \leftrightarrow \begin{bmatrix} 0 & 1^{1/2} & 0 & 0 & \cdots \\ 0 & 0 & 2^{1/2} & 0 & \\ 0 & 0 & 0 & 3^{1/2} & \\ \vdots & & & & \end{bmatrix} \tag{7.4.31}$$

Both matrices can be constructed either from Eqs. (7.4.26) and (7.4.27) or Eqs. (7.4.21) and (7.4.22) combined with our mnemonic involving images of the transformed vectors $a^\dagger|n\rangle$ and $a|n\rangle$. We get the matrices representing X and P by turning to Eqs. (7.4.28) and (7.4.29):

$$X \leftrightarrow \left(\frac{\hbar}{2m\omega}\right)^{1/2} \begin{bmatrix} 0 & 1^{1/2} & 0 & 0 & \cdots \\ 1^{1/2} & 0 & 2^{1/2} & 0 & \\ 0 & 2^{1/2} & 0 & 3^{1/2} & \\ 0 & 0 & 3^{1/2} & 0 & \\ \vdots & & & & \end{bmatrix} \tag{7.4.32}$$

$$P \leftrightarrow i\left(\frac{m\omega\hbar}{2}\right)^{1/2} \begin{bmatrix} 0 & -1^{1/2} & 0 & 0 & \cdots \\ 1^{1/2} & 0 & -2^{1/2} & 0 & \\ 0 & 2^{1/2} & 0 & -3^{1/2} & \\ 0 & 0 & 3^{1/2} & 0 & \\ \vdots & & & & \end{bmatrix} \tag{7.4.33}$$

The Hamiltonian is of course diagonal in its own basis:

$$H \leftrightarrow \hbar\omega \begin{bmatrix} 1/2 & 0 & 0 & 0 & \cdots \\ 0 & 3/2 & 0 & 0 & \\ 0 & 0 & 5/2 & & \\ \vdots & & & & \end{bmatrix} \tag{7.4.34}$$

Equation (7.4.22) also allows us to express all normalized eigenvectors $|n\rangle$ in terms of the ground state $|0\rangle$:

$$|n\rangle = \frac{a^\dagger}{n^{1/2}}|n-1\rangle = \frac{a^\dagger}{n^{1/2}}\frac{a^\dagger}{(n-1)^{1/2}}|n-2\rangle \cdots = \frac{(a^\dagger)^n}{(n!)^{1/2}}|0\rangle \tag{7.4.35}$$

The a and $a^\dagger$ operators greatly facilitate the calculation of the matrix of elements of other operators between oscillator eigenstates. Consider, for example, $\langle 3|X^3|2\rangle$. In the X basis one would have to carry out the following integral:

$$\langle 3|X^3|2\rangle = \left(\frac{m\omega}{\pi\hbar}\right)^{1/2}\left(\frac{1}{2^3 3!}\cdot\frac{1}{2^2 2!}\right)^{1/2}\int_{-\infty}^{\infty}\left\{\exp\left(-\frac{m\omega x^2}{2\hbar}\right)\right.$$

$$\left.\times H_3\left[\left(\frac{m\omega}{\hbar}\right)^{1/2}x\right]x^3\exp\left(-\frac{m\omega x^2}{2\hbar}\right)H_2\left[\left(\frac{m\omega}{\hbar}\right)^{1/2}x\right]\right\}dx$$

whereas in the $|n\rangle$ basis

$$\langle 3|X^3|2\rangle = \left(\frac{\hbar}{2m\omega}\right)^{3/2}\langle 3|(a+a^\dagger)^3|2\rangle$$

$$= \left(\frac{\hbar}{2m\omega}\right)^{3/2}\langle 3|(a^3+a^2a^\dagger+aa^\dagger a+aa^\dagger a^\dagger$$

$$+a^\dagger aa+a^\dagger aa^\dagger+a^\dagger a^\dagger a+a^\dagger a^\dagger a^\dagger)|2\rangle$$

Since a lowers n by one unit and $a^\dagger$ raises it by one unit and we want to go up by one unit from $n=2$ to $n=3$, the only nonzero contribution comes from $a^\dagger a^\dagger a$, $aa^\dagger a^\dagger$, and $a^\dagger aa^\dagger$. Now

$$a^\dagger a^\dagger a|2\rangle = 2^{1/2}a^\dagger a^\dagger|1\rangle = 2^{1/2}2^{1/2}a^\dagger|2\rangle = 2^{1/2}2^{1/2}3^{1/2}|3\rangle$$

$$aa^\dagger a^\dagger|2\rangle = 3^{1/2}aa^\dagger|3\rangle = 3^{1/2}4^{1/2}a|4\rangle = 3^{1/2}4^{1/2}4^{1/2}|3\rangle$$

$$a^\dagger aa^\dagger|2\rangle = 3^{1/2}a^\dagger a|3\rangle = 3^{1/2}N|3\rangle = 3^{1/2}3|3\rangle$$

so that

$$\langle 3|X^3|2\rangle = \left(\frac{\hbar}{2m\omega}\right)^{3/2}[2(3^{1/2})+4(3^{1/2})+3(3^{1/2})]$$

What if we want not some matrix element of X, but the probability of finding the particle in $|n\rangle$ at position x? We can of course fall back on Postulate III, which tells us to find the eigenvectors $|x\rangle$ of the matrix X [Eq. (7.4.32)] and evaluate the inner product $\langle x|n\rangle$. A more practical way will be developed in the next section.

Consider a remarkable feature of the above solution to the eigenvalue problem of H. Usually we work in the X basis and set up the eigenvalue problem (as a differential equation) by invoking Postulate II, which gives the action of X and P in the X basis ($X\to x$, $P\to -i\hbar\, d/dx$). In some cases (the linear potential problem), the P basis recommends itself, and then we use the Fourier-transformed version of Postulate II, namely, $X\to i\hbar\, d/dp$, $P\to p$. In the present case we could not transform this operator assignment to the energy eigenbasis, for to do so we first had to solve for the energy eigenfunctions in the X basis, which was begging the question. Instead we used just the commutation relation $[X, P]=i\hbar$, which follows from Postulate II, but is true in all bases, in particular the energy basis. Since we obtained the complete solution given just this information, it would appear that the essence of Postulate II is just the commutator. This in fact is the case. In other words, we may trade our present Postulate II for a more general version:

> *Postulate II.* The independent variables x and p of classical mechanics now become Hermitian operators X and P defined by the canonical commutator $[X, P] = i\hbar$. Dependent variables $\omega(x, p)$ are given by operators $\Omega = \omega(x \to X, p \to P)$.

To regain our old version, we go to the X basis. Clearly in its own basis $X \to x$. We must then pick P such that $[X, P] = i\hbar$. If we make the conventional choice $P = -i\hbar\, d/dx$, we meet this requirement and arrive at Postulate II as stated earlier. But the present version of Postulate II allows us some latitude in the choice of P, for we can add to $-i\hbar\, d/dx$ any function of x without altering the commutator: the assignment

$$X \xrightarrow[X \text{ basis}]{} x \tag{7.4.36a}$$

$$P \xrightarrow[X \text{ basis}]{} -i\hbar \frac{d}{dx} + f(x) \tag{7.4.36b}$$

is equally satisfactory. Now, it is not at all obvious that in every problem (and not just the harmonic oscillator) the same physics will obtain if we make this our starting point. For example if we project the eigenvalue equation

$$P|p\rangle = p|p\rangle \tag{7.4.37a}$$

onto the X basis, we now get

$$\left[-i\hbar \frac{d}{dx} + f(x)\right]\psi_p(x) = p\psi_p(x) \tag{7.4.37b}$$

from which it follows that $\psi_p(x)$ is no longer a plane wave $\propto e^{ipx/\hbar}$. How can the physics be the same as before? The answer is that the wave function is never measured directly. What we do measure are probabilities $|\langle\omega|\psi\rangle|^2$ for obtaining some result ω when Ω is measured, squares of matrix elements $|\langle\psi_1|\Omega|\psi_2\rangle|^2$, or the eigenvalue spectrum of operators such as the Hamiltonian. In one of the exercises that follows, you will be guided toward the proof that these measurable quantities are in fact left invariant under the change to the nontraditional operator assignment Eq. (7.4.36).

Dirac emphasized the close connection between the commutation rule

$$[X, P] = i\hbar$$

of the quantum operators and the Poisson brackets (PB) of their classical counterparts

$$\{x, p\} = 1$$

which allows us to write the defining relation of the quantum operators as

$$[X, P] = i\hbar\{x, p\} = i\hbar \qquad (7.4.38)$$

The virtue of this viewpoint is that its generalization to the "quantization" of a system of N degrees of freedom is apparent:

Postulate II (For N Degrees of Freedom). The Cartesian coordinates $x_1, \ldots, x_N$ and momenta $p_1, \ldots, p_N$ of the classical description of a system with N degrees of freedom now become Hermitian operators $X_1, \ldots, X_N$; $P_1, \ldots, P_N$ obeying the commutation rules

$$\begin{aligned}[]
[X_i, P_j] &= i\hbar\{x_i, p_j\} = i\hbar\delta_{ij} \\
[X_i, X_j] &= i\hbar\{x_i, x_j\} = 0 \\
[P_i, P_j] &= i\hbar\{p_i, p_j\} = 0
\end{aligned} \qquad (7.4.39)$$

Similarly $\omega(x, p) \to \omega(x \to X, p \to P) = \Omega$.

[We restrict ourselves to Cartesian coordinates to avoid certain subtleties associated with the quantization of non-Cartesian but canonical coordinates; see Exercise (7.4.10). Once the differential equations are obtained, we may abandon Cartesian coordinates in looking for the solutions.]

It is evident that the generalization provided towards the end of Section 4.2, namely,

$$X_i \xrightarrow[X \text{ basis}]{} x_i$$

$$P_i \xrightarrow[X \text{ basis}]{} -i\hbar \frac{\partial}{\partial x_i}$$

is *a* choice but not *the* choice satisfying the canonical commutation rules, Eq. (7.4.39), for the same reason as in the $N=1$ case.

Given the commutation relations between X and P, the ones among dependent operators follow from the repeated use of the relations

$$[\Omega, \Lambda\Gamma] = \Lambda[\Omega, \Gamma] + [\Omega, \Lambda]\Gamma$$

and

$$[\Omega\Lambda, \Gamma] = \Omega[\Lambda, \Gamma] + [\Omega, \Gamma]\Lambda$$

Since PB obey similar rules (Exercise 2.7.1) except for the lack of emphasis on ordering of the classical variables, it turns out that if

$$\{\omega(x, p), \lambda(x, p)\} = \gamma(x, p)$$

then

$$[\Omega(X, P), \Lambda(X, P)] = i\hbar\Gamma(X, P) \tag{7.4.40}$$

except for differences arising from ordering ambiguities; hence the formal similarity between classical and quantum mechanics, first encountered in Chapter 6.

Although the new form of postulate II provides a general, basis-independent specification of the quantum operators corresponding to classical variables, that is to say for "quantizing," in practice one typically works in the X basis and also ignores the latitude in the choice of P_i and sticks to the traditional one, $P_i = -i\hbar\, \partial/\partial x_i$, which leads to the simplest differential equations. The solution to the oscillator problem, given just the commutation relations (and a little help from Dirac) is atypical.

*Exercise 7.4.1.** Compute the matrix elements of X and P in the $|n\rangle$ basis and compare with the result from Exercise 7.3.4.

*Exercise 7.4.2.** Find $\langle X\rangle$, $\langle P\rangle$, $\langle X^2\rangle$, $\langle P^2\rangle$, $\Delta X \cdot \Delta P$ in the state $|n\rangle$.

*Exercise 7.4.3.** (*Virial Theorem*). The virial theorem in classical mechanics states that for a particle bound by a potential $V(r) = ar^k$, the average (over the orbit) kinetic and potential energies are related by

$$\bar{T} = c(k)\bar{V}$$

when $c(k)$ depends only on k. Show that $c(k) = k/2$ by considering a circular orbit. Using the results from the previous exercise show that for the oscillator ($k=2$)

$$\langle T\rangle = \langle V\rangle$$

in the quantum state $|n\rangle$.

Exercise 7.4.4. Show that $\langle n|X^4|n\rangle = (\hbar/2m\omega)^2[3 + 6n(n+1)]$.

*Exercise 7.4.5.** At $t=0$ a particle starts out in $|\psi(0)\rangle = 1/2^{1/2}(|0\rangle + |1\rangle)$. (1) Find $|\psi(t)\rangle$; (2) find $\langle X(0)\rangle = \langle\psi(0)|X|\psi(0)\rangle$, $\langle P(0)\rangle$, $\langle X(t)\rangle$, $\langle P(t)\rangle$; (3) find $\langle\dot{X}(t)\rangle$ and $\langle\dot{P}(t)\rangle$ using Ehrenfest's theorem and solve for $\langle X(t)\rangle$ and $\langle P(t)\rangle$ and compare with part (2).

*Exercise 7.4.6.** Show that $\langle a(t)\rangle = e^{-i\omega t}\langle a(0)\rangle$ and that $\langle a^\dagger(t)\rangle = e^{i\omega t}\langle a^\dagger(0)\rangle$.

Exercise 7.4.7. Verify Eq. (7.4.40) for the case

(1) $\Omega = X$, $\Lambda = X^2 + P^2$
(2) $\Omega = X^2$, $\Lambda = P^2$

The second case illustrates the ordering ambiguity.

*Exercise 7.4.8.** Consider the three angular momentum variables in classical mechanics:

$$l_x = yp_z - zp_y$$

$$l_y = zp_x - xp_z$$

$$l_z = xp_y - yp_x$$

(1) Construct L_x, L_y, and L_z, the quantum counterparts, and note that there are no ordering ambiguities.
(2) Verify that $\{l_x, l_y\} = l_z$ [see Eq. (2.7.3) for the definition of the PB].
(3) Verify that $[L_x, L_y] = i\hbar L_z$.

Exercise 7.4.9 (*Important*). Consider the unconventional (but fully acceptable) operator choice

$$X \to x$$

$$P \to -i\hbar \frac{d}{dx} + f(x)$$

in the X basis.

(1) Verify that the canonical commutation relation is satisfied.
(2) It is possible to interpret the change in the operator assignment as a result of a unitary change of the X basis:

$$|x\rangle \to |\tilde{x}\rangle = e^{ig(X)/\hbar}\,|x\rangle = e^{ig(x)/\hbar}\,|x\rangle$$

where

$$g(x) = \int^x f(x')\,dx'$$

First verify that

$$\langle \tilde{x}|X|\tilde{x}'\rangle = x\delta(x - x')$$

i.e.,

$$X \xrightarrow[\text{new } X \text{ basis}]{} x$$

Next verify that

$$\langle \tilde{x}|P|\tilde{x}'\rangle = \left[-i\hbar \frac{d}{dx} + f(x)\right]\delta(x - x')$$

i.e.,

$$P \xrightarrow[\text{new } X \text{ basis}]{} -i\hbar\frac{d}{dx}+f(x)$$

This exercise teaches us that the "X basis" is not unique; given a basis $|x\rangle$, we can get another $|\tilde{x}\rangle$, by multiplying by a phase factor which changes neither the norm nor the orthogonality. The matrix elements of P change with f, the standard choice corresponding to $f=0$. Since the presence of f is related to a change of basis, the invariance of the physics under a change in f (from zero to nonzero) follows. What is novel here is that we are changing from one X basis to another X basis rather than to some other Ω basis. Another lesson to remember is that two different differential operators $\omega(x, -i\hbar\, d/dx)$ and $\omega(x, -i\hbar\, d/dx+f)$ can have the same eigenvalues and a one-to-one correspondence between their eigenfunctions, since they both represent the same abstract operator $\Omega(X, P)$. □

*Exercise 7.4.10.** Recall that we always quantize a system by promoting the Cartesian coordinates $x_1, \ldots, x_N$; and momenta $p_1, \ldots, p_N$ to operators obeying the canonical commutation rules. If non-Cartesian coordinates seem more natural in some cases, such as the eigenvalue problem of a Hamiltonian with spherical symmetry, we first set up the differential equation in Cartesian coordinates and *then* change to spherical coordinates (Section 4.2). In Section 4.2 it was pointed out that if $\mathscr{H}$ is written in terms of non-Cartesian but canonical coordinates $q_1 \ldots q_N$; $p_1 \ldots p_N$; $\mathscr{H}(q_i \to q_i, p_i \to -i\hbar\, \partial/\partial q_i)$ does not generate the correct Hamiltonian H, even though the operator assignment satisfies the canonical commutation rules. In this section we revisit this problem in order to explain some of the subtleties arising in the direct quantization of non-Cartesian coordinates without the use of Cartesian coordinates in intermediate stages.

(1) Consider a particle in two dimensions with

$$\mathscr{H}=\frac{p_x^2+p_y^2}{2m}+a(x^2+y^2)^{1/2}$$

which leads to

$$H \to \frac{-\hbar^2}{2m}\left(\frac{\partial^2}{\partial x^2}+\frac{\partial^2}{\partial y^2}\right)+a(x^2+y^2)^{1/2}$$

in the coordinate basis. Since the problem has rotational symmetry we use polar coordinates

$$\rho=(x^2+y^2)^{1/2}, \qquad \phi=\tan^{-1}(y/x)$$

in terms of which

$$H \xrightarrow[\text{coordinate basis}]{} \frac{-\hbar^2}{2m}\left(\frac{\partial^2}{\partial \rho^2}+\frac{1}{\rho}\frac{\partial}{\partial \rho}+\frac{1}{\rho^2}\frac{\partial^2}{\partial \phi^2}\right)+a\rho \tag{7.4.41}$$

Since ρ and ϕ are not mixed up as x and y are [in the $(x^2+y^2)^{1/2}$ term] the polar version can be more readily solved.

The question we address is the following: why not *start* with $\mathscr{H}$ expressed in terms of polar coordinates and the conjugate momenta

$$p_\rho = \mathbf{e}_\rho \cdot \mathbf{p} = \frac{xp_x + yp_y}{(x^2+y^2)^{1/2}}$$

(where $\mathbf{e}_\rho$ is the unit vector in the radial direction), and

$$p_\phi = xp_y - yp_x \quad \text{(the angular momentum, also called } l_z\text{)}$$

i.e.,

$$\mathscr{H} = \frac{p_\rho^2}{2m} + \frac{p_\phi^2}{2m\rho^2} + a\rho \quad \text{(verify this)}$$

and directly promote all classical variables ρ, p_ρ, ϕ, and p_ϕ to quantum operators obeying the canonical commutations rules? Let's do it and see what happens. If we choose operators

$$P_\rho \to -i\hbar \frac{\partial}{\partial \rho}$$

$$P_\phi \to -i\hbar \frac{\partial}{\partial \phi}$$

that obey the commutation rules, we end up with

$$H \xrightarrow[\text{basis}]{\text{coordinate}} \frac{-\hbar^2}{2m}\left(\frac{\partial^2}{\partial \rho^2} + \frac{1}{\rho^2}\frac{\partial^2}{\partial \phi^2}\right) + a\rho \tag{7.4.42}$$

which disagrees with Eq. (7.4.41). Now this in itself is not serious, for as seen in the last exercise the same physics may be hidden in two different equations. In the present case this isn't true: as we will see, the Hamiltonians in Eqs. (7.4.41) and (7.4.42) do not have the same eigenvalues.‡ We know Eq. (7.4.41) is the correct one, since the quantization procedure in terms of Cartesian coordinates has empirical support. What do we do now?

(2) A way out is suggested by the fact that although the choice $P_\rho \to -i\hbar\, \partial/\partial\rho$ leads to the correct commutation rule, it is not Hermitian! Verify that

$$\begin{aligned}\langle \psi_1 | P_\rho | \psi_2 \rangle &= \int_0^\infty \int_0^{2\pi} \psi_1^* \left(-i\hbar \frac{\partial \psi_2}{\partial \rho}\right) \rho\, d\rho\, d\phi \\ &\neq \int_0^\infty \int_0^{2\pi} \left(-i\hbar \frac{\partial \psi_1}{\partial \rho}\right)^* \psi_2 \rho\, d\rho\, d\phi \\ &= \langle \mathbf{P}_\rho \psi_1 | \psi_2 \rangle\end{aligned}$$

(You may assume $\rho\psi_1^*\psi_2 \to 0$ as $\rho \to 0$ or ∞. The problem comes from the fact that $\rho\, d\rho\, d\phi$ and not $d\rho\, d\phi$ is the measure for integration.)

‡ What we will see is that $P_\rho = -i\hbar\, d/d\rho$, and hence the H constructed with it, are non-Hermitian.

Show, however, that

$$P_\rho \to -i\hbar\left(\frac{\partial}{\partial\rho}+\frac{1}{2\rho}\right) \tag{7.4.43}$$

is indeed Hermitian and also satisfies the canonical commutation rule. The angular momentum $P_\phi \to -i\hbar\,\partial/\partial\phi$ is Hermitian, as it stands, on single-valued functions: $\psi(\rho,\phi)=\psi(\rho,\phi+2\pi)$.

(3) In the Cartesian case we saw that adding an arbitrary $f(x)$ to $-i\hbar\,\partial/\partial x$ didn't have any physical effect, whereas here the addition of a function of ρ to $-i\hbar\,\partial/\partial\rho$ seems important. Why? [Is $f(x)$ completely arbitrary? Mustn't it be real? Why? Is the same true for the $-i\hbar/2\rho$ piece?]

(4) Feed in the new momentum operator P_ρ and show that

$$H \xrightarrow[\text{basis}]{\text{coordinate}} \frac{-\hbar^2}{2m}\left(\frac{\partial^2}{\partial\rho^2}+\frac{1}{\rho}\frac{\partial}{\partial\rho}-\frac{1}{4\rho^2}+\frac{1}{\rho^2}\frac{\partial^2}{\partial\phi^2}\right)+a\rho$$

which still disagrees with Eq. (7.4.41). We have satisfied the commutation rules, chosen Hermitian operators, and yet do not get the right quantum Hamiltonian. The key to the mystery lies in the fact that $\mathscr{H}$ doesn't determine H uniquely since terms of order $\hbar$ (or higher) may be present in H but absent in $\mathscr{H}$. While this ambiguity is present even in the Cartesian case, it is resolved by symmetrization in all interesting cases. With non-Cartesian coordinates the ambiguity is more severe. There *are* ways of constructing H given $\mathscr{H}$ (the path integral formulation suggests one) such that the substitution $P_\rho \to -i\hbar(\partial/\partial\rho+1/2\rho)$ leads to Eq. (7.4.41). In the present case the quantum Hamiltonian corresponding to

$$\mathscr{H}=\frac{p_\rho^2}{2m}+\frac{p_\phi^2}{2m\rho^2}+a\rho$$

is given by

$$H \xrightarrow[\text{basis}]{\text{coordinate}} \mathscr{H}\left(\rho\to\rho,\, p_\rho\to -i\hbar\left[\frac{\partial}{\partial\rho}+\frac{1}{2\rho}\right];\, \phi\to\phi,\, p_\phi\to -i\hbar\frac{\partial}{\partial\phi}\right)-\frac{\hbar^2}{8m\rho^2} \tag{7.4.44}$$

Notice that the additional term is indeed of nonzero order in $\hbar$.

We will not get into a discussion of these prescriptions for generating H since they finally reproduce results more readily available in the approach we are adopting. □

7.5. Passage from the Energy Basis to the *X* Basis

It was remarked in the last section that although the $|n\rangle$ basis was ideally suited for evaluating the matrix elements of operators between oscillator eigenstates, the amplitude for finding the particle in a state $|n\rangle$ at the point x could not be readily computed: it seemed as if one had to find the eigenkets $|x\rangle$ of the operators X [Eq. (7.4.32)] and then take the inner product $\langle x|n\rangle$. But there is a more direct way to get $\psi_n(x)=\langle x|n\rangle$.

We start by projecting the equation defining the ground state of the oscillator

$$a|0\rangle = 0 \tag{7.5.1}$$

on the X basis:

$$|0\rangle \to \langle x|0\rangle = \psi_0(x)$$

$$a = \left(\frac{m\omega}{2\hbar}\right)^{1/2} X + i\left(\frac{1}{2m\omega\hbar}\right)^{1/2} P$$

$$\to \left(\frac{m\omega}{2\hbar}\right)^{1/2} x + \left(\frac{\hbar}{2m\omega}\right)^{1/2} \frac{d}{dx} \tag{7.5.2}$$

In terms of $y = (m\omega/\hbar)^{1/2}x$,

$$a = \frac{1}{2^{1/2}}\left(y + \frac{d}{dy}\right) \tag{7.5.3}$$

For later use we also note that (since d/dy is *anti*-Hermitian),

$$a^\dagger = \frac{1}{2^{1/2}}\left(y - \frac{d}{dy}\right) \tag{7.5.4}$$

In the X basis Eq. (7.5.1) then becomes

$$\left(y + \frac{d}{dy}\right)\psi_0(y) = 0 \tag{7.5.5}$$

or

$$\frac{d\psi_0(y)}{\psi_0(y)} = -y\,dy$$

or

$$\psi_0(y) = A_0\,e^{-y^2/2}$$

or

$$\psi_0(x) = A_0 \exp\left(-\frac{m\omega x^2}{2\hbar}\right)$$

or (upon normalizing)

$$=\left(\frac{m\omega}{\pi\hbar}\right)^{1/4}\exp\left(-\frac{m\omega x^2}{2\hbar}\right) \tag{7.5.6}$$

By projecting the equation

$$|n\rangle=\frac{(a^\dagger)^n}{(n!)^{1/2}}|0\rangle$$

onto the X basis, we get the *normalized* eigenfunctions

$$\langle x|n\rangle=\psi_n\left[x=\left(\frac{\hbar}{m\omega}\right)^{1/2}y\right]=\frac{1}{(n!)^{1/2}}\left[\frac{1}{2^{1/2}}\left(y-\frac{d}{dy}\right)\right]^n\left(\frac{m\omega}{\pi\hbar}\right)^{1/4}e^{-y^2/2} \tag{7.5.7}$$

A comparison of the above result with Eq. (7.3.22) shows that

$$H_n(y)=e^{y^2/2}\left(y-\frac{d}{dy}\right)^n e^{-y^2/2} \tag{7.5.8}$$

We now conclude our rather lengthy discussion of the oscillator. If you understand this chapter thoroughly, you should have a good grasp of how quantum mechanics works.

Exercise 7.5.1. Project Eq. (7.5.1) on the P basis and obtain $\psi_0(p)$.

Exercise 7.5.2. Project the relation

$$a|n\rangle=n^{1/2}|n-1\rangle$$

on the X basis and derive the recursion relation

$$H_n'(y)=2nH_{n-1}(y)$$

using Eq. (7.3.22).

Exercise 7.5.3. Starting with

$$a+a^\dagger=2^{1/2}y$$

and

$$(a+a^\dagger)|n\rangle=n^{1/2}|n-1\rangle+(n+1)^{1/2}|n+1\rangle$$

and Eq. (7.3.22), derive the relation

$$H_{n+1}(y)=2yH_n(y)-2nH_{n-1}(y)$$

Exercise 7.5.4. Thermodynamics of Oscillators.* The Boltzman formula

$$P(i)=e^{-\beta E(i)}/Z$$

where

$$Z=\sum_i e^{-\beta E(i)}$$

gives the probability of finding a system in a state i with energy $E(i)$, when it is in thermal equilibrium with a reservoir of absolute temperature $T=1/\beta k$, $k=1.4\times 10^{-16}$ ergs/° K; being Boltzman's constant. (The "probability" referred to above is in relation to a classical ensemble of similar systems and has nothing to do with quantum mechanics.)

(1) Show that the thermal average of the system's energy is

$$\bar{E}=\sum_i E(i)P(i)=\frac{-\partial}{\partial\beta}\ln Z$$

(2) Let the system be a classical oscillator. The index i is now continuous and corresponds to the variables x and p describing the state of the oscillator, i.e.,

$$i\to x, p$$

and

$$\sum_i \to \iint dx\, dp$$

and

$$E(i)\to E(x,p)=\frac{p^2}{2m}+\frac{1}{2}m\omega^2x^2$$

Show that

$$Z_{\text{cl}}=\left(\frac{2\pi}{\beta m\omega^2}\right)^{1/2}\left(\frac{2\pi m}{\beta}\right)^{1/2}=\frac{2\pi}{\omega\beta}$$

and that

$$\bar{E}_{\text{cl}}=\frac{1}{\beta}=kT$$

Note that E_{cl} is independent of m and ω.

(3) For the quantum oscillator the quantum number n plays the role of the index i. Show that

$$Z_{qu}=e^{-\beta\hbar\omega/2}(1-e^{-\beta\hbar\omega})^{-1}$$

and

$$\bar{E}_{qu}=\hbar\omega\left(\frac{1}{2}+\frac{1}{e^{\beta\hbar\omega}-1}\right)$$

(4) It is intuitively clear that as the temperature T increases (and $\beta=1/kT$ decreases) the oscillator will get more and more excited and eventually (from the correspondence principle)

$$\bar{E}_{qu}\xrightarrow[T\to\infty]{}\bar{E}_{cl}$$

Verify that this is indeed true and show that "large T" means $T\gg\hbar\omega/k$.

(5) Consider a crystal with N_0 atoms, which, for small oscillations, is equivalent to $3N_0$ decoupled oscillators. The mean thermal energy of the crystal $\bar{E}_{crystal}$ is $\bar{E}_{cl}$ or $\bar{E}_{qu}$ summed over all the normal modes. Show that if the oscillators are treated classicaly, the specific heat per atom is

$$C_{cl}(T)=\frac{1}{N_0}\frac{\partial\bar{E}_{crystal}}{\partial T}=3k$$

which is independent of T and the parameters of the oscillators and hence the same for all crystals.‡ This agrees with experiment at high temperatures but not as $T\to 0$. Empirically,

$$\begin{aligned} C(T)&\to 3k \quad (T\text{ large})\\ &\to 0 \quad (T\to 0)\end{aligned}$$

Following Einstein, treat the oscillators quantum mechanically, asuming for simplicity that they all have the same frequency ω. Show that

$$C_{qu}(T)=3k\left(\frac{\theta_E}{T}\right)^2\frac{e^{\theta_E/T}}{(e^{\theta_E/T}-1)^2}$$

where $\theta_E=\hbar\omega/k$ is called the *Einstein temperature* and varies from crystal to crystal. Show that

$$C_{qu}(T)\xrightarrow[T\gg\theta_E]{}3k$$

$$C_{qu}(T)\xrightarrow[T\ll\theta_E]{}3k\left(\frac{\theta_E}{T}\right)^2e^{-\theta_E/T}$$

Although $C_{qu}(T)\to 0$ as $T\to 0$, the exponential falloff disagrees with the observed $C(T)\to_{T\to 0}T^3$ behavior. This discrepancy arises from assuming that the frequencies of all

‡ More precisely, for crystals whose atoms behave as point particles with no internal degrees of freedom.

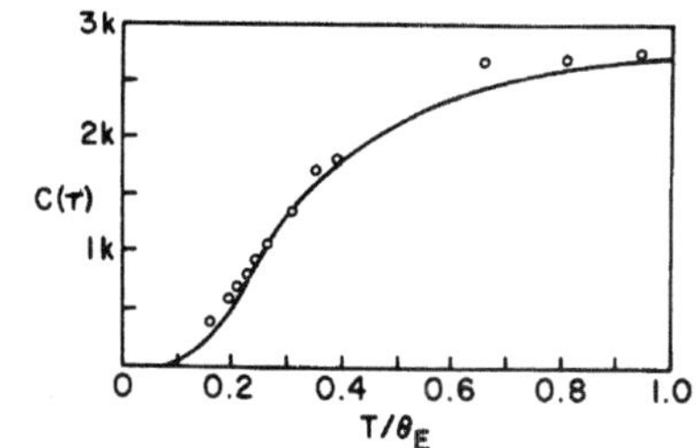

Figure 7.3. Comparison of experiment with Einstein's theory for the specific heat in the case of diamond. (θ_E is chosen to be 1320 K.)

normal modes are equal, which is of course not generally true. [Recall that in the case of two coupled masses we get $\omega_{\mathrm{I}}=(k/m)^{1/2}$ and $\omega_{\mathrm{II}}=(3k/m)^{1/2}$.] This discrepancy was eliminated by Debye.

But Einstein's simple picture by itself is remarkably successful (see Fig. 7.3).

8

The Path Integral Formulation of Quantum Theory

We consider here an alternate formulation of quantum mechanics invented by Feynman in the forties.‡ In contrast to the Schrödinger formulation, which stems from Hamiltonian mechanics, the Feynman formulation is tied to the Lagrangian formulation of mechanics. Although we are committed to the former approach, we discuss in this chapter Feynman's alternative, not only because of its aesthetic value, but also because it can, in a class of problems, give the full propagator with tremendous ease and also give valuable insight into the relation between classical and quantum mechanics.

8.1. The Path Integral Recipe

We have already seen that the quantum problem is fully solved once the propagator is known. Thus far our practice has been to first find the eigenvalues and eigenfunctions of H, and then express the propagator $U(t)$ in terms of these. In the path integral approach one computes $U(t)$ directly. For a single particle in one dimension, the procedure is the following.

To find $U(x, t; x', t')$:

(1) Draw all paths in the x–t plane connecting (x', t') and (x, t) (see Fig. 8.1).
(2) Find the action $S[x(t)]$ for each path $x(t)$.
(3) $$U(x, t; x', t') = A \sum_{\text{all paths}} e^{iS[x(t)]/\hbar} \tag{8.1.1}$$

where A is an overall normalization factor.

‡ The nineteen forties that is, and in his twenties. An interesting account of how he was influenced by Dirac's work in the same direction may be found in his Nobel lectures. See, *Nobel Lectures—Physics*, Vol. III, Elsevier Publication, New York (1972).

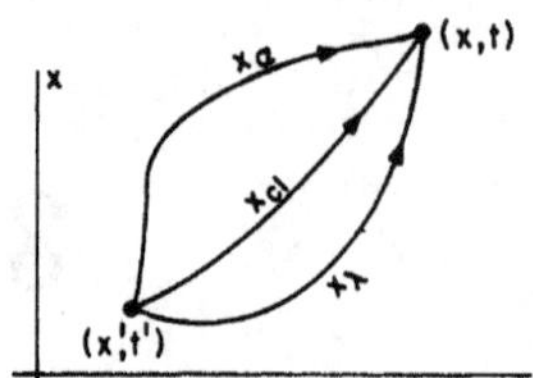

Figure 8.1. Some of the paths that contribute to the propagator. The contribution from the path $x(t)$ is $Z=\exp\{iS[x(t)]/\hbar\}$.

8.2. Analysis of the Recipe

Let us analyze the above recipe, postponing for a while the proof that it reproduces conventional quantum mechanics. The most surprising thing about it is the fact that every path, including the classical path, $x_{\text{cl}}(t)$, gets the same weight, that is to say, a number of unit modulus. How are we going to regain classical mechanics in the appropriate limit if the classical path does not seem favored in any way?

To understand this we must perform the sum in Eq. (8.1.1). Now, the correct way to sum over all the paths, that is to say, path integration, is quite complicated and we will discuss it later. For the present let us take the heuristic approach. Let us first pretend that the continuum of paths linking the end points is actually a discrete set. A few paths in the set are shown in Fig. 8.1.

We have to add the contributions $Z_\alpha=e^{iS[x_\alpha(t)]/\hbar}$ from each path $x_\alpha(t)$. This summation is done schematically in Fig. 8.2. Since each path has a different action, it contributes with a different phase, and the contributions from the paths essentially cancel each other, until we come near the classical path. Since S is stationary here, the Z's add constructively and produce a large sum. As we move away from $x_{\text{cl}}(t)$, destructive interference sets in once again. It is clear from the figure that $U(t)$ is dominated by the paths near $x_{\text{cl}}(t)$. Thus the classical path is important, not because it contributes a lot by itself, but because in its vicinity the paths contribute coherently.

How far must we deviate from x_{cl} before destructive interference sets in? One may say crudely that coherence is lost once the phase differs from the stationary value $S[x_{\text{cl}}(t)]/\hbar\equiv S_{\text{cl}}/\hbar$ by about π. This in turn means that the action for the coherence paths must be within $\hbar\pi$ of S_{cl}. For a macroscopic particle this means a very tight constraint on its path, since S_{cl} is typically $\simeq 1$ erg sec $\simeq 10^{27}\hbar$, while for an electron there is quite a bit of latitude. Consider the following example. A free particle leaves the origin at $t=0$ and arrives at $x=1$ cm at $t=1$ second. The classical path is

$$x=t \tag{8.2.1}$$

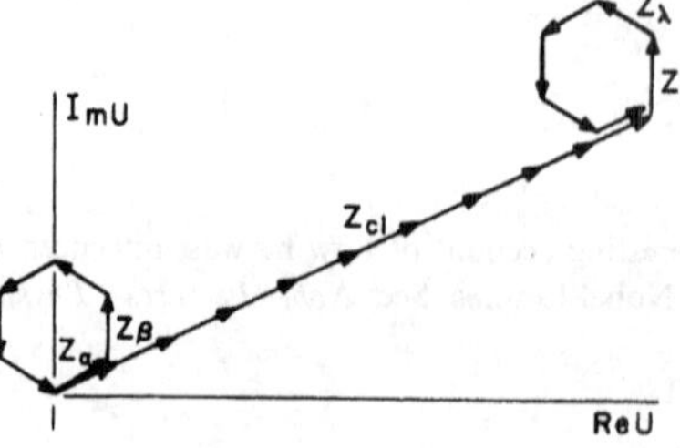

Figure 8.2. Schematic representation of the sum ΣZ_α. Paths near $x_{\text{cl}}(t)$ contribute coherently since S is stationary there, while others cancel each other out and may be ignored in the first approximation when we calculate $U(t)$.

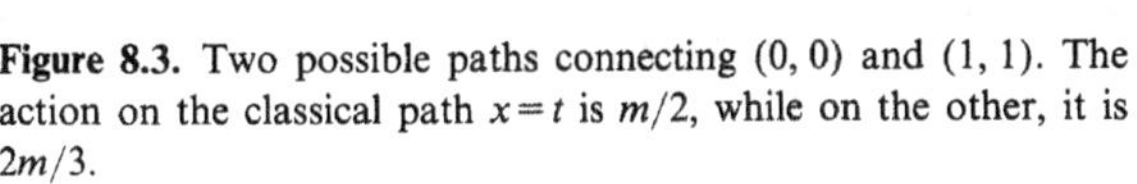

Figure 8.3. Two possible paths connecting (0, 0) and (1, 1). The action on the classical path $x=t$ is $m/2$, while on the other, it is $2m/3$.

Consider another path

$$x=t^2 \tag{8.2.2}$$

which also links the two space-time points (Fig. 8.3.)

For a classical particle, of mass, say 1 g, the action changes by roughly $1.6\times 10^{26}\hbar$, and the phase by roughly 1.6×10^{26} rad as we move from the classical path $x=t$ to the nonclassical path $x=t^2$. We may therefore completely ignore the nonclassical path. On the other hand, for an electron whose mass is $\simeq 10^{-27}$ g, $\delta S \simeq \hbar/6$ and the phase change is just around a sixth of a radian, which is well within the coherence range $\delta S/\hbar \lesssim \pi$. It is in such cases that assuming that the particle moves along a well-defined trajectory, $x_{\rm cl}(t)$, leads to conflict with experiment.

8.3. An Approximation to $U(t)$ for a Free Particle

Our previous discussions have indicated that, to an excellent approximation, we may ignore all but the classical path and its neighbors in calculating $U(t)$. Assuming that each of these paths contributes the same amount $\exp(iS_{\rm cl}/\hbar)$, since S is stationary, we get

$$U(t)=A'\, e^{iS_{\rm cl}/\hbar} \tag{8.3.1}$$

where A' is some normalizing factor which "measures" the number of paths in the coherent range. Let us find $U(t)$ for a free particle in this approximation and compare the result with the exact result, Eq. (5.1.10).

The classical path for a free particle is just a straight line in the x–t plane:

$$x_{\rm cl}(t'')=x'+\frac{x-x'}{t-t'}(t''-t') \tag{8.3.2}$$

corresponding to motion with uniform velocity $v=(x-x')/(t-t')$. Since $\mathscr{L}=mv^2/2$ is a constant,

$$S_{\rm cl}=\int_{t'}^{t}\mathscr{L}\,dt''=\frac{1}{2}m\frac{(x-x')^2}{t-t'}$$

so that

$$U(x, t; x', t') = A' \exp\left[\frac{im(x-x')^2}{2\hbar(t-t')}\right] \tag{8.3.3}$$

To find A', we use the fact that as $t-t'$ tends to 0, U must tend to $\delta(x-x')$. Comparing Eq. (8.3.3) to the representation of the delta function encountered in Section 1.10 (see footnote on page 61),

$$\delta(x-x') \equiv \lim_{\Delta\to 0} \frac{1}{(\pi\Delta^2)^{1/2}} \exp\left[-\frac{(x-x')^2}{\Delta^2}\right]$$

(valid even if Δ is imaginary) we get

$$A' = \left[\frac{m}{2\pi\hbar i(t-t')}\right]^{1/2}$$

so that

$$U(x, t; x', 0) \equiv U(x, t; x') = \left(\frac{m}{2\pi\hbar it}\right)^{1/2} \exp\left[\frac{im(x-x')^2}{2\hbar t}\right] \tag{8.3.4}$$

which is the exact answer! We have managed to get the exact answer by just computing the classical action! However, we will see in Section 8.6 that only for potentials of the form $V = a + bx + cx^2 + d\dot{x} + ex\dot{x}$ is it true that $U(t) = A(t)\, e^{iS_{\text{cl}}/\hbar}$. Furthermore, we can't generally find $A(t)$ using $U(x, 0; x') = \delta(x-x')$ since A can contain an arbitrary dimensionless function f such that $f \to 1$ as $t \to 0$. Here $f \equiv 1$ because we can't construct a nontrivial dimensionless f using just m, $\hbar$, and t (check this).

8.4. Path Integral Evaluation of the Free-Particle Propagator

Although our heuristic analysis yielded the exact free-particle propagator, we will now repeat the calculation without any approximation to illustrate path integration.

Consider $U(x_N, t_N; x_0, t_0)$. The peculiar labeling of the end points will be justified later. Our problem is to perform the path integral

$$\int_{x_0}^{x_N} e^{iS[x(t)]/\hbar} \mathscr{D}[x(t)] \tag{8.4.1}$$

where

$$\int_{x_0}^{x_N} \mathscr{D}[x(t)]$$

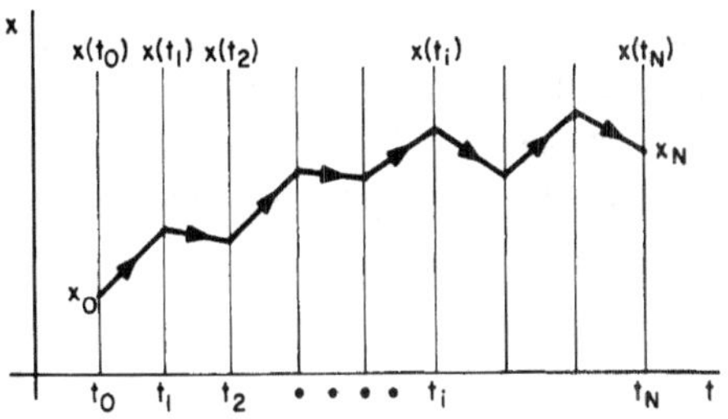

Figure 8.4. The discrete approximation to a path $x(t)$. Each path is specified by $N-1$ numbers $x(t_1), \ldots, x(t_{N-1})$. To sum over paths we must integrate each x_i from $-\infty$ to $+\infty$. Once all integrations are done, we can take the limit $N\to\infty$.

is a symbolic way of saying "integrate over all paths connecting x_0 and x_N (in the interval t_0 and t_N)." Now, a path $x(t)$ is fully specified by an infinity of numbers $x(t_0), \ldots, x(t), \ldots, x(t_N)$, namely, the values of the function $x(t)$ at every point t in the interval t_0 to t_N. To sum over all paths we must integrate over all possible values of these infinite variables, except of course $x(t_0)$ and $x(t_N)$, which will be kept fixed at x_0 and x_N, respectively. To tackle this problem, we follow the idea that was used in Section 1.10: we trade the function $x(t)$ for a discrete approximation which agrees with $x(t)$ at the $N+1$ points $t_n=t_0+n\varepsilon$, $n=0, \ldots, N$, where $\varepsilon=(t_N-t_0)/N$. In this approximation each path is specified by $N+1$ numbers $x(t_0), x(t_1), \ldots, x(t_N)$. The gaps in the discrete function are interpolated by straight lines. One such path is shown in Fig. 8.4. We hope that if we take the limit $N\to\infty$ at the end we will get a result that is insensitive to these approximations.‡ Now that the paths have been discretized, we must also do the same to the action integral. We replace the continuous path definition

$$S=\int_{t_0}^{t_N} \mathscr{L}(t)\, dt=\int_{t_0}^{t_N} \frac{1}{2} m\dot{x}^2\, dt$$

by

$$S=\sum_{i=0}^{N-1} \frac{m}{2}\left(\frac{x_{i+1}-x_i}{\varepsilon}\right)^2 \varepsilon \tag{8.4.2}$$

where $x_i=x(t_i)$. We wish to calculate

$$\begin{aligned} U(x_N, t_N; x_0, t_0) &= \int_{x_0}^{x_N} \exp\{iS[x(t)]/\hbar\}\mathscr{D}[x(t)] \\ &= \lim_{\substack{N\to\infty \\ \varepsilon\to 0}} A \int_{-\infty}^{\infty}\int_{-\infty}^{\infty}\cdots\int_{-\infty}^{\infty} \exp\left[\frac{i}{\hbar}\frac{m}{2}\sum_{i=0}^{N-1}\frac{(x_{i+1}-x_i)^2}{\varepsilon}\right] \\ &\quad \times dx_1\cdots dx_{N-1} \end{aligned} \tag{8.4.3}$$

‡ We expect that the abrupt changes in velocity at the points $t_0+n\varepsilon$ that arise due to our approximation will not matter because $\mathscr{L}$ does not depend on the acceleration or higher derivatives.

It is implicit in the above that x_0 and x_N have the values we have chosen at the outset. The factor A in the front is to be chosen at the end such that we get the correct scale for U when the limit $N\to\infty$ is taken.

Let us first switch to the variables

$$y_i=\left(\frac{m}{2\hbar\varepsilon}\right)^{1/2} x_i$$

We then want

$$\lim_{N\to\infty} A'\int_{-\infty}^{\infty}\int_{-\infty}^{\infty}\cdots\int_{-\infty}^{\infty}\exp\left[-\sum_{i=0}^{N-1}\frac{(y_{i+1}-y_i)^2}{i}\right]dy_1\cdots dy_{N-1} \qquad (8.4.4)$$

where

$$A'=A\left(\frac{2\hbar\varepsilon}{m}\right)^{(N-1)/2}$$

Although the multiple integral looks formidable, it is not. Let us begin by doing the y_1 integration. Considering just the part of the integrand that involves y_1, we get

$$\int_{-\infty}^{\infty}\exp\left\{-\frac{1}{i}[(y_2-y_1)^2+(y_1-y_0)^2]\right\}dy_1=\left(\frac{i\pi}{2}\right)^{1/2}e^{-(y_2-y_0)^2/2i} \qquad (8.4.5)$$

Consider next the integration over y_2. Bringing in the part of the integrand involving y_2 and combining it with the result above we compute next

$$\left(\frac{i\pi}{2}\right)^{1/2}\int_{-\infty}^{\infty}e^{-(y_3-y_2)^2/i}\cdot e^{-(y_2-y_0)^2/2i}\,dy_2$$

$$=\left(\frac{i\pi}{2}\right)^{1/2}e^{-(2y_3^2+y_0^2)/2i}\left(\frac{2\pi i}{3}\right)^{1/2}e^{(y_0+2y_3)^2/6i}$$

$$=\left[\frac{(i\pi)^2}{3}\right]^{1/2}e^{-(y_3-y_0)^2/3i} \qquad (8.4.6)$$

By comparing this result to the one from the y_1 integration, we deduce the pattern: if we carry out this process $N-1$ times so as to evaluate the integral in Eq. (8.4.4), it will become

$$\frac{(i\pi)^{(N-1)/2}}{N^{1/2}}e^{-(y_N-y_0)^2/Ni}$$

or

$$\frac{(i\pi)^{(N-1)/2}}{N^{1/2}}\, e^{-m(x_N-x_0)^2/2\hbar\varepsilon Ni}$$

Bringing in the factor $A(2\hbar\varepsilon/m)^{(N-1)/2}$ from up front, we get

$$U=A\left(\frac{2\pi\hbar\varepsilon i}{m}\right)^{N/2}\left(\frac{m}{2\pi\hbar iN\varepsilon}\right)^{1/2}\exp\left[\frac{im(x_N-x_0)^2}{2\hbar N\varepsilon}\right]$$

If we now let $N\to\infty$, $\varepsilon\to 0$, $N\varepsilon\to t_N-t_0$, we get the right answer provided

$$A=\left[\frac{2\pi\hbar\varepsilon i}{m}\right]^{-N/2}\equiv B^{-N} \tag{8.4.7}$$

It is conventional to associate a factor $1/B$ with each of the $N-1$ integrations and the remaining factor $1/B$ with the overall process. In other words, we have just learnt that the precise meaning of the statement "integrate over all paths" is

$$\int \mathscr{D}[x(t)]=\lim_{\substack{\varepsilon\to 0\\ N\to\infty}}\frac{1}{B}\int_{-\infty}^{\infty}\int\int\cdots\int_{-\infty}^{\infty}\frac{dx_1}{B}\cdot\frac{dx_2}{B}\cdots\frac{dx_{N-1}}{B}$$

where

$$B=\left(\frac{2\pi\hbar\varepsilon i}{m}\right)^{1/2} \tag{8.4.8}$$

8.5. Equivalence to the Schrödinger Equation

The relation between the Schrödinger and Feynman formalisms is quite similar to that between the Newtonian and the least action formalisms of mechanics, in that the former approach is local in time and deals with time evolution over infinitesimal periods while the latter is global and deals directly with propagation over finite times.

In the Schrödinger formalism, the change in the state vector $|\psi\rangle$ over an infinitesimal time ε is

$$|\psi(\varepsilon)\rangle-|\psi(0)\rangle=\frac{-i\varepsilon}{\hbar}H|\psi(0)\rangle \tag{8.5.1}$$

which becomes in the X basis

$$\psi(x,\varepsilon)-\psi(x,0)=\frac{-i\varepsilon}{\hbar}\left[\frac{-\hbar^2}{2m}\frac{\partial^2}{\partial x^2}+V(x,0)\right]\psi(x,0) \tag{8.5.2}$$

to first order in ε. To compare this result with the path integral prediction to the same order in ε, we begin with

$$\psi(x, \varepsilon)=\int_{-\infty}^{\infty} U(x, \varepsilon; x')\psi(x', 0)\, dx' \tag{8.5.3}$$

The calculation of $U(\varepsilon)$ is simplified by the fact that there is no need to do any integrations over intermediate x's since there is just one slice of time ε between the start and finish. So

$$U(x, \varepsilon; x')=\left(\frac{m}{2\pi\hbar i\varepsilon}\right)^{1/2} \exp\left\{i\left[\frac{m(x-x')^2}{2\varepsilon}-\varepsilon V\left(\frac{x+x'}{2}, 0\right)\right]\Big/\hbar\right\} \tag{8.5.4}$$

where the $(m/2\pi\hbar i\varepsilon)^{1/2}$ factor up front is just the $1/B$ factor from Eq. (8.4.8). We take the time argument of V to be zero since there is already a factor of ε before it and any variation of V with time in the interval 0 to ε will produce an effect of second order in ε. So

$$\psi(x, \varepsilon)=\left(\frac{m}{2\pi\hbar i\varepsilon}\right)^{1/2}\int_{-\infty}^{\infty} \exp\left[\frac{im(x-x')^2}{2\varepsilon\hbar}\right]\exp\left[-\frac{i\varepsilon}{\hbar}V\left(\frac{x+x'}{2}, 0\right)\right] \times \psi(x', 0)\, dx' \tag{8.5.5}$$

Consider the factor $\exp[im(x-x')^2/2\varepsilon\hbar]$. It oscillates very rapidly as $(x-x')$ varies since ε is infinitesimal and $\hbar$ is so small. When such a rapidly oscillating function multiplies a smooth function like $\psi(x', 0)$, the integral vanishes for the most part due to the random phase of the exponential. Just as in the case of the path integration, the only substantial contribution comes from the region where the phase is stationary. In this case the only stationary point is $x=x'$, where the phase has the minimum value of zero. In terms of $\eta=x'-x$, the region of coherence is, as before,

$$\frac{m\eta^2}{2\varepsilon\hbar}\lesssim\pi$$

or

$$|\eta|\lesssim\left(\frac{2\varepsilon\hbar\pi}{m}\right)^{1/2} \tag{8.5.6}$$

Consider now

$$\psi(x, \varepsilon)=\left(\frac{m}{2\pi\hbar i\varepsilon}\right)^{1/2}\int_{-\infty}^{\infty} \exp(im\eta^2/2\hbar\varepsilon)\cdot\exp\left[-\left(\frac{i}{\hbar}\right)\varepsilon V\left(x+\frac{\eta}{2}, 0\right)\right] \times \psi(x+\eta, 0)\, d\eta \tag{8.5.7}$$

We will work to first order in ε and therefore to second order in η [see Eq. (8.5.6) above]. We expand

$$\psi(x+\eta, 0)=\psi(x, 0)+\eta \frac{\partial \psi}{\partial x}+\frac{\eta^2}{2} \frac{\partial^2 \psi}{\partial x^2}+\cdots$$

$$\begin{aligned} \exp\left[-\left(\frac{i}{\hbar}\right)\varepsilon V\left(x+\frac{\eta}{2}, 0\right)\right] &= 1-\frac{i\varepsilon}{\hbar} V\left(x+\frac{\eta}{2}, 0\right)+\cdots \\ &= 1-\frac{i\varepsilon}{\hbar} V(x, 0)+\cdots \end{aligned}$$

since terms of order $\eta\varepsilon$ are to be neglected. Equation (8.5.7) now becomes

$$\begin{aligned} \psi(x, \varepsilon)=\left(\frac{m}{2\pi\hbar i\varepsilon}\right)^{1/2} \int_{-\infty}^{\infty} \exp\left(\frac{im\eta^2}{2\hbar\varepsilon}\right)\left[\psi(x, 0)-\frac{i\varepsilon}{\hbar} V(x, 0)\psi(x, 0)\right. \\ \left. +\eta \frac{\partial \psi}{\partial x}+\frac{\eta^2}{2} \frac{\partial^2 \psi}{\partial x^2}\right] d\eta \end{aligned}$$

Consulting the list of Gaussian integrals in Appendix A.2, we get

$$\begin{aligned} \psi(x, \varepsilon)=\left(\frac{m}{2\pi\hbar i\varepsilon}\right)^{1/2}\left[\psi(x, 0)\left(\frac{2\pi\hbar i\varepsilon}{m}\right)^{1/2}-\frac{\hbar\varepsilon}{2im}\left(\frac{2\pi\hbar i\varepsilon}{m}\right)^{1/2} \frac{\partial^2 \psi}{\partial x^2}\right. \\ \left. -\frac{i\varepsilon}{\hbar}\left(\frac{2\pi\hbar i\varepsilon}{m}\right)^{1/2} V(x, 0)\psi(x, 0)\right] \end{aligned}$$

or

$$\psi(x, \varepsilon)-\psi(x, 0)=\frac{-i\varepsilon}{\hbar}\left[\frac{-\hbar^2}{2m} \frac{\partial^2}{\partial x^2}+V(x, 0)\right]\psi(x, 0) \tag{8.5.8}$$

which agrees with the Schrödinger prediction, Eq. (8.5.1).

8.6. Potentials of the Form $V=a+bx+cx^2+d\dot{x}+ex\dot{x}$‡

We wish to compute

$$U(x, t; x')=\int_{x'}^{x} e^{iS[x(t'')]/\hbar}\mathscr{D}[x(t'')] \tag{8.6.1}$$

‡ This section may be omitted without loss of continuity.

Let us write every path as

$$x(t'') = x_{\mathrm{cl}}(t'') + y(t'') \tag{8.6.2}$$

It follows that

$$\dot{x}(t'') = \dot{x}_{\mathrm{cl}}(t'') + \dot{y}(t'') \tag{8.6.3}$$

Since all the paths agree at the end points, $y(0) = y(t) = 0$. When we slice up the time into N parts, we have for intermediate integration variables

$$x_i \equiv x(t_i'') = x_{\mathrm{cl}}(t_i'') + y(t_i'') \equiv x_{\mathrm{cl}}(t_i'') + y_i$$

Since $x_{\mathrm{cl}}(t_i'')$ is just some constant at t_i'',

$$dx_i = dy_i$$

and

$$\int_{x'}^{x} \mathscr{D}[x(t'')] = \int_0^0 \mathscr{D}[y(t'')] \tag{8.6.4}$$

so that Eq. (8.6.1) becomes

$$U(x, t; x') = \int_0^0 \exp\left\{\frac{i}{\hbar} S[x_{\mathrm{cl}}(t'') + y(t'')]\right\} \mathscr{D}[y(t'')] \tag{8.6.5}$$

The next step is to expand the functional S in a Taylor series about x_{cl}:

$$\begin{aligned} S[x_{\mathrm{cl}} + y] &= \int_0^t \mathscr{L}(x_{\mathrm{cl}} + y, \dot{x}_{\mathrm{cl}} + \dot{y})\, dt'' \\ &\equiv \int_0^t \left[\mathscr{L}(x_{\mathrm{cl}}, \dot{x}_{\mathrm{cl}}) + \left(\left.\frac{\partial \mathscr{L}}{\partial x}\right|_{x_{\mathrm{cl}}} y + \left.\frac{\partial \mathscr{L}}{\partial \dot{x}}\right|_{x_{\mathrm{cl}}} \dot{y} \right) \right. \\ &\quad \left. + \frac{1}{2}\left(\left.\frac{\partial^2 \mathscr{L}}{\partial x^2}\right|_{x_{\mathrm{cl}}} y^2 + 2 \left.\frac{\partial^2 \mathscr{L}}{\partial x\, \partial \dot{x}}\right|_{x_{\mathrm{cl}}} y\dot{y} + \left.\frac{\partial^2 \mathscr{L}}{\partial \dot{x}^2}\right|_{x_{\mathrm{cl}}} \dot{y}^2 \right) \right] dt'' \end{aligned} \tag{8.6.6}$$

The series terminates here since $\mathscr{L}$ is a quadratic polynominal.

The first piece $\mathscr{L}(x_{\mathrm{cl}}, \dot{x}_{\mathrm{cl}})$ integrates to give $S[x_{\mathrm{cl}}] \equiv S_{\mathrm{cl}}$. The second piece, linear in y and $\dot{y}$, vanishes due to the classical equation of motion. In the last piece, if we recall

$$\mathscr{L} = \tfrac{1}{2} m\dot{x}^2 - a - bx - cx^2 - d\dot{x} - ex\dot{x} \tag{8.6.7}$$

we get

$$\frac{1}{2}\frac{\partial^2 \mathscr{L}}{\partial x^2}=-c \tag{8.6.8}$$

$$\frac{\partial^2 \mathscr{L}}{\partial x\,\partial \dot{x}}=-e \tag{8.6.9}$$

$$\frac{1}{2}\frac{\partial^2 \mathscr{L}}{\partial \dot{x}^2}=m \tag{8.6.10}$$

Consequently Eq. (8.6.5) becomes

$$U(x,t;x')=\exp\left(\frac{iS_{\rm cl}}{\hbar}\right)\int_0^0 \exp\left[\frac{i}{\hbar}\int_0^t\left(\frac{1}{2}m\dot{y}^2-cy^2-ey\dot{y}\right)dt''\right] \times \mathscr{D}[y(t'')] \tag{8.6.11}$$

Since the path integral has no memory of $x_{\rm cl}$, it can only depend on t. So

$$U(x,t;x')=e^{iS_{\rm cl}/\hbar}A(t) \tag{8.6.12}$$

where $A(t)$ is some unknown function of t. Now if we were doing the free-particle problem, we would get Eq. (8.6.11) with $c=e=0$. In this case we know that [see Eq. (8.3.4)]

$$A(t)=\left(\frac{m}{2\pi\hbar it}\right)^{1/2} \tag{8.6.13}$$

Since the coefficient b does not figure in Eq. (8.6.11), it follows that the same value of $A(t)$ corresponds to the linear potential $V=a+bx$ as well. For the harmonic oscillator, $c=\frac{1}{2}m\omega^2$, and we have to do the integral

$$A(t)=\int_0^0 \exp\left[i/\hbar\int_0^t \frac{1}{2}m(\dot{y}^2-\omega^2y^2)\right]dt''\mathscr{D}[y(t'')] \tag{8.6.14}$$

The evaluation of this integral is discussed in the book by Feynman and Hibbs referred to at the end of this section. Note that even if the factor $A(t)$ in $\psi(x,t)$ is not known, we can extract all the probabilistic information at time t.

Notice the ease with which the Feynman formalism yields the full propagator in these cases. Consider in particular the horrendous alternative of finding the eigenfunctions of the Hamiltonian and constructing from them the harmonic oscillator propagator.

The path integral method may be extended to three dimensions without any major qualitative differences. In particular, the form of U in Eq. (8.6.12) is valid for potentials that are at most quadratic in the coordinates and the velocities. An interesting problem in this class is that of a particle in a uniform magnetic field. For further details on the subject of path integral quantum mechanics, see R. P. Feynman and A. R. Hibbs, *Path Integrals and Quantum Mechanics*, McGraw-Hill (1965), and Chapter 21.

*Exercise 8.6.1.** Verify that

$$U(x, t; x', 0) = A(t)\exp(iS_{cl}/\hbar), \; A(t) = \left(\frac{m}{2\pi\hbar it}\right)^{1/2}$$

agrees with the exact result, Eq. (5.4.31), for $V(x) = -fx$. Hint: Start with $x_{cl}(t'') = x_0 + v_0 t'' + \frac{1}{2}(f/m)t''^2$ and find the constants x_0 and v_0 from the requirement that $x_{cl}(0) = x'$ and $x_{cl}(t) = x$.

Exercise 8.6.2. Show that for the harmonic oscillator with

$$\mathscr{L} = \tfrac{1}{2}m\dot{x}^2 - \tfrac{1}{2}m\omega^2 x^2$$

$$U(x, t; x') = A(t)\exp\left\{\frac{im\omega}{2\hbar\sin\omega t}[(x^2 + x'^2)\cos\omega t - 2xx']\right\}$$

where $A(t)$ is an unknown function. (Recall Exercise 2.8.7.)

Exercise 8.6.3. We know that given the eigenfunctions and the eigenvalues we can construct the propagator:

$$U(x, t; x', t') = \sum_n \psi_n(x)\psi_n^*(x')\, e^{-iE_n(t-t')/\hbar} \tag{8.6.15}$$

Consider the reverse process (since the path integral approach gives U directly), for the case of the oscillator.

(1) Set $x = x' = t' = 0$. Assume that $A(t) = (m\omega/2\pi i\hbar\sin\omega t)^{1/2}$ for the oscillator. By expanding both sides of Eq. (8.6.15), you should find that $E = \hbar\omega/2, 5\hbar\omega/2, 9\hbar\omega/2, \ldots$, etc. What happened to the levels in between?

(2) (Optional). Now consider the extraction of the eigenfunctions. Let $x = x'$ and $t' = 0$. Find E_0, E_1, $|\psi_0(x)|^2$, and $|\psi_1(x)|^2$ by expanding in powers of $\alpha = \exp(i\omega t)$.

*Exercise 8.6.4.** Recall the derivation of the Schrödinger equation (8.5.8) starting from Eq. (8.5.4). Note that although we chose the argument of V to be the midpoint $x + x'/2$, it did not matter very much: any choice $x + \alpha\eta$, (where $\eta = x' - x$) for $0 \le \alpha \le 1$ would have given the same result since the difference between the choices is of order $\eta\varepsilon \simeq \varepsilon^{3/2}$. All this was thanks to the factor ε multiplying V in Eq. (8.5.4) and the fact that $|\eta| \simeq \varepsilon^{1/2}$, as per Eq. (8.6.5).

Consider now the case of a vector potential which will bring in a factor

$$\exp\left[\frac{iq\varepsilon}{\hbar c}\frac{x-x'}{\varepsilon}A(x+\alpha\eta)\right]\equiv\exp\left[-\frac{iq\varepsilon}{\hbar c}\frac{\eta}{\varepsilon}A(x+\alpha\eta)\right]$$

to the propagator for one time slice. (We should really be using vectors for position and the vector potential, but the one-dimensional version will suffice for making the point here.) Note that ε now gets canceled, in contrast to the scalar potential case. Thus, going to order ε to derive the Schrödinger equation means going to order η^2 in expanding the exponential. This will not only bring in an A^2 term, but will also make the answer sensitive to the argument of A in the linear term. Choose $\alpha=1/2$ and verify that you get the one-dimensional version of Eq. (4.3.7). Along the way you will see that changing α makes an order ε difference to $\psi(x, \varepsilon)$ so that we have no choice but to use $\alpha=1/2$, i.e., use the *midpoint prescription*. This point will come up in Chapter 21.

9

The Heisenberg Uncertainty Relations

9.1. Introduction

In classical mechanics a particle in a state (x_0, p_0) has associated with it well-defined values for any dynamical variable $\omega(x, p)$, namely, $\omega(x_0, p_0)$. In quantum theory, given a state $|\psi\rangle$, one can only give the probabilities $P(\omega)$ for the possible outcomes of a measurement of Ω. The probability distribution will be characterized by a mean or expectation value

$$\langle\Omega\rangle = \langle\psi|\Omega|\psi\rangle \tag{9.1.1}$$

and an uncertainty about this mean:

$$(\Delta\Omega) = [\langle\psi|(\Omega - \langle\Omega\rangle)^2|\psi\rangle]^{1/2} \tag{9.1.2}$$

There are, however, states for which $\Delta\Omega = 0$, and these are the eigenstates $|\omega\rangle$ of Ω.

If we consider two Hermitian operators Ω and Λ, they will generally have some uncertainties $\Delta\Omega$ and $\Delta\Lambda$ in an arbitrary state. In the next section we will derive the Heisenberg uncertainty relations, which will provide a lower bound on the product of uncertainties, $\Delta\Omega \cdot \Delta\Lambda$. Generally the lower bound will depend not only on the operators but also on the state. Of interest to us are those cases in which the lower bound is independent of the state. The derivation will make clear the conditions under which such a relation will exist.

9.2. Derivation of the Uncertainty Relations

Let Ω and Λ be two Hermitian operators, with a commutator

$$[\Omega, \Lambda] = i\Gamma \tag{9.2.1}$$

You may readily verify that Γ is also Hermitian. Let us start with the uncertainty product in a normalized state $|\psi\rangle$:

$$(\Delta\Omega)^2(\Delta\Lambda)^2=\langle\psi|(\Omega-\langle\Omega\rangle)^2|\psi\rangle\langle\psi|(\Lambda-\langle\Lambda\rangle)^2|\psi\rangle \tag{9.2.2}$$

where $\langle\Omega\rangle=\langle\psi|\Omega|\psi\rangle$ and $\langle\Lambda\rangle=\langle\psi|\Lambda|\psi\rangle$. Let us next define the pair

$$\begin{aligned}\hat{\Omega}&=\Omega-\langle\Omega\rangle\\ \hat{\Lambda}&=\Lambda-\langle\Lambda\rangle\end{aligned} \tag{9.2.3}$$

which has the same commutator as Ω and Λ (verify this). In terms of $\hat{\Omega}$ and $\hat{\Lambda}$

$$\begin{aligned}(\Delta\Omega)^2(\Delta\Lambda)^2&=\langle\psi|\hat{\Omega}^2|\psi\rangle\langle\psi|\hat{\Lambda}^2|\psi\rangle\\ &=\langle\hat{\Omega}\psi|\hat{\Omega}\psi\rangle\langle\hat{\Lambda}\psi|\hat{\Lambda}\psi\rangle\end{aligned} \tag{9.2.4}$$

since

$$\hat{\Omega}^2=\hat{\Omega}\hat{\Omega}=\hat{\Omega}^\dagger\hat{\Omega}$$

and

$$\hat{\Lambda}^2=\hat{\Lambda}^\dagger\hat{\Lambda} \tag{9.2.5}$$

If we apply the Schwartz inequality

$$|V_1|^2|V_2|^2\geq|\langle V_1|V_2\rangle|^2 \tag{9.2.6}$$

(where the equality sign holds only if $|V_1\rangle=c|V_2\rangle$, where c is a constant) to the states $|\hat{\Omega}\psi\rangle$ and $|\hat{\Lambda}\psi\rangle$, we get from Eq. (9.2.4),

$$(\Delta\Omega)^2(\Delta\Lambda)^2\geq|\langle\hat{\Omega}\psi|\hat{\Lambda}\psi\rangle|^2 \tag{9.2.7}$$

Let us now use the fact that

$$\langle\hat{\Omega}\psi|\hat{\Lambda}\psi\rangle=\langle\psi|\hat{\Omega}^\dagger\hat{\Lambda}|\psi\rangle=\langle\psi|\hat{\Omega}\hat{\Lambda}|\psi\rangle \tag{9.2.8}$$

to rewrite the above inequality as

$$(\Delta\Omega)^2(\Delta\Lambda)^2\geq|\langle\psi|\hat{\Omega}\hat{\Lambda}|\psi\rangle|^2 \tag{9.2.9}$$

Now, we know that the commutator has to enter the picture somewhere. This we arrange through the following identity:

$$\begin{aligned}\hat{\Omega}\hat{\Lambda}&=\frac{\hat{\Omega}\hat{\Lambda}+\hat{\Lambda}\hat{\Omega}}{2}+\frac{\hat{\Omega}\hat{\Lambda}-\hat{\Lambda}\hat{\Omega}}{2}\\ &=\tfrac{1}{2}[\hat{\Omega},\hat{\Lambda}]_++\tfrac{1}{2}[\hat{\Omega},\hat{\Lambda}]\end{aligned} \tag{9.2.10}$$

where $[\hat{\Omega}, \hat{\Lambda}]_+$ is called the *anticommutator*. Feeding Eq. (9.2.10) into the inequality (9.2.9), we get

$$(\Delta\Omega)^2(\Delta\Lambda)^2 \geq |\langle\psi|\tfrac{1}{2}[\hat{\Omega}, \hat{\Lambda}]_+ + \tfrac{1}{2}[\hat{\Omega}, \hat{\Lambda}]|\psi\rangle|^2 \qquad (9.2.11)$$

We next use the fact that

(1) since $[\hat{\Omega}, \hat{\Lambda}] = i\Gamma$, where Γ is Hermitian, the expectation value of the commutator is pure imaginary;

(2) since $[\hat{\Omega}, \hat{\Lambda}]_+$ is Hermitian, the expectation value of the anticommutator is real.

Recalling that $|a+ib|^2 = a^2 + b^2$, we get

$$\begin{aligned}(\Delta\Omega)^2(\Delta\Lambda)^2 &\geq \tfrac{1}{4}|\langle\psi|[\hat{\Omega}, \hat{\Lambda}]_+|\psi\rangle + i\langle\psi|\Gamma|\psi\rangle|^2 \\ &\geq \tfrac{1}{4}\langle\psi|[\hat{\Omega}, \hat{\Lambda}]_+|\psi\rangle^2 + \tfrac{1}{4}\langle\psi|\Gamma|\psi\rangle^2 \end{aligned} \qquad (9.2.12)$$

This is the general uncertainty relation between any two Hermitian operators and is evidently state dependent. Consider now canonically conjugate operators, for which $\Gamma = \hbar$. In this case

$$(\Delta\Omega)^2(\Delta\Lambda)^2 \geq \frac{1}{4}\langle\psi|[\hat{\Omega}, \hat{\Lambda}]_+|\psi\rangle^2 + \frac{\hbar^2}{4} \qquad (9.2.13)$$

Since the first term is positive definite, we may assert that for *any* $|\psi\rangle$

$$(\Delta\Omega)^2(\Delta\Lambda)^2 \geq \hbar^2/4$$

or

$$\Delta\Omega \cdot \Delta\Lambda \geq \hbar/2 \qquad (9.2.14)$$

which is the celebrated uncertainty relation. Let us note that the above inequality becomes an equality only if

(1) $\hat{\Omega}|\psi\rangle = c\hat{\Lambda}|\psi\rangle$

and (9.2.15)

(2) $\langle\psi|[\hat{\Omega}, \hat{\Lambda}]_+|\psi\rangle = 0$

9.3. The Minimum Uncertainty Packet

In this section we will find the wave function $\psi(x)$ which saturates the lower bound of the uncertainty relation for X and P. According to Eq. (9.2.15) such a state is characterized by

$$(P - \langle P\rangle)|\psi\rangle = c(X - \langle X\rangle)|\psi\rangle \qquad (9.3.1)$$

and

$$\langle\psi|(P-\langle P\rangle)(X-\langle X\rangle)+(X-\langle X\rangle)(P-\langle P\rangle)|\psi\rangle=0 \tag{9.3.2}$$

where $\langle P\rangle$ and $\langle X\rangle$ refer to the state $|\psi\rangle$, implicitly defined by these equations. In the X basis, Eq. (9.3.1) becomes

$$\left(-i\hbar\frac{d}{dx}-\langle P\rangle\right)\psi(x)=c(x-\langle X\rangle)\psi(x)$$

or

$$\frac{d\psi(x)}{\psi(x)}=\frac{i}{\hbar}[\langle P\rangle+c(x-\langle X\rangle)]\,dx \tag{9.3.3}$$

Now, whatever $\langle X\rangle$ may be, it is always possible to shift our origin (to $x=\langle X\rangle$) so that in the new frame of reference $\langle X\rangle=0$. In this frame, Eq. (9.3.3) has the solution

$$\psi(x)=\psi(0)\,e^{i\langle P\rangle x/\hbar}\,e^{icx^2/2\hbar} \tag{9.3.4}$$

Let us next consider the constraint, Eq. (9.3.2), which in this frame reads

$$\langle\psi|(P-\langle P\rangle)X+X(P-\langle P\rangle)|\psi\rangle=0$$

If we now exploit Eq. (9.3.1) and its adjoint, we find

$$\langle\psi|c^*X^2+cX^2|\psi\rangle=0$$

$$(c+c^*)\langle\psi|X^2|\psi\rangle=0$$

from which it follows that c is pure imaginary:

$$c=i|c| \tag{9.3.5}$$

Our solution, Eq. (9.3.4) now becomes

$$\psi(x)=\psi(0)\,e^{i\langle P\rangle x/\hbar}\,e^{-|c|x^2/2\hbar}$$

In terms of

$$\Delta^2=\hbar/|c|$$

$$\psi(x)=\psi(0)\,e^{i\langle P\rangle x/\hbar}\,e^{-x^2/2\Delta^2} \tag{9.3.6}$$

where Δ^2, like $|c|$, is arbitrary. If the origin were not chosen to make $\langle X\rangle$ zero, we would have instead

$$\psi(x)=\psi(\langle X\rangle)\, e^{i\langle P\rangle(x-\langle X\rangle)/\hbar}\, e^{-(x-\langle X\rangle)^2/2\Delta^2} \tag{9.3.7}$$

Thus the *minimum uncertainty wave function* is a Gaussian of arbitrary width and center. This result, for the special case $\langle X\rangle=\langle P\rangle=0$, was used in the quest for the state that minimized the expectation value of the oscillator Hamiltonian.

9.4. Applications of the Uncertainty Principle

I now illustrate the use of the uncertainty principle by estimating the size of the ground-state energy and the spread in the ground-state wave function. It should be clear from this example that the success we had with the oscillator was rather atypical.

We choose as our system the hydrogen atom. The Hamiltonian for this system, assuming the proton is a spectator whose only role is to provide a Coulomb potential for the electron, may be written entirely in terms of the electron's variable as

$$H=\frac{P_x^2+P_y^2+P_z^2}{2m}-\frac{e^2}{(X^2+Y^2+Z^2)^{1/2}} \tag{9.4.1‡}$$

Let us begin by mimicking the analysis we employed for the oscillator. We evaluate $\langle H\rangle$ in a normalized state $|\psi\rangle$:

$$\begin{aligned}\langle H\rangle&=\frac{\langle P_x^2+P_y^2+P_z^2\rangle}{2m}-e^2\left\langle\frac{1}{(X^2+Y^2+Z^2)^{1/2}}\right\rangle\\&=\frac{\langle P_x^2\rangle+\langle P_y^2\rangle+\langle P_z^2\rangle}{2m}-e^2\left\langle\frac{1}{(X^2+Y^2+Z^2)^{1/2}}\right\rangle\end{aligned} \tag{9.4.2}$$

Since

$$\langle P_x^2\rangle=\langle\Delta P_x\rangle^2+\langle P_x\rangle^2 \quad \text{etc.}$$

the first step in minimizing $\langle H\rangle$ is to work only with states for which $\langle P_i\rangle=0$. For such states

$$\langle H\rangle=\frac{(\Delta P_x)^2+(\Delta P_y)^2+(\Delta P_z)^2}{2m}-e^2\left\langle\frac{1}{(X^2+Y^2+Z^2)^{1/2}}\right\rangle \tag{9.4.3}$$

‡ The operator $(X^2+Y^2+Z^2)^{-1/2}$ is just $1/r$ in the coordinate basis. We will occasionally denote it by $1/r$ even while referring to it in the abstract, to simplify the notation.

We cannot exploit the uncertainty relations

$$\Delta P_x \Delta X \geq \hbar/2, \quad \text{etc.}$$

yet since $\langle H\rangle$ is not a function of ΔX and ΔP. The problem is that $\langle (X^2+Y^2+Z^2)^{-1/2}\rangle$ is not simply related to ΔX, ΔY, and ΔZ. Now the handwaving begins. We argue that (see Exercise 9.4.2),

$$\left\langle \frac{1}{(X^2+Y^2+Z^2)^{1/2}} \right\rangle \simeq \frac{1}{\langle (X^2+Y^2+Z^2)^{1/2}\rangle} \tag{9.4.4}$$

where the $\simeq$ symbol means that the two sides of Eq. (9.4.4) are not strictly equal, but of the same order of magnitude. So we write

$$\langle H\rangle \simeq \frac{(\Delta P_x)^2+(\Delta P_y)^2+(\Delta P_z)^2}{2m} - \frac{e^2}{\langle (X^2+Y^2+Z^2)^{1/2}\rangle}$$

Once again, we argue that

$$\langle (X^2+Y^2+Z^2)^{1/2}\rangle \simeq (\langle X^2\rangle + \langle Y^2\rangle + \langle Z^2\rangle)^{1/2}$$

and get‡

$$\langle H\rangle \simeq \frac{(\Delta P_x)^2+(\Delta P_y)^2+(\Delta P_z)^2}{2m} - \frac{e^2}{(\langle X^2\rangle + \langle Y^2\rangle + \langle Z^2\rangle)^{1/2}}$$

From the relations

$$\langle X^2\rangle = (\Delta X)^2 + \langle X\rangle^2 \quad \text{etc.}$$

it follows that we may confine ourselves to states for which $\langle X\rangle = \langle Y\rangle = \langle Z\rangle = 0$ in looking for the state with the lowest mean energy. For such states

$$\langle H\rangle \simeq \frac{\Delta P_x^2+\Delta P_y^2+\Delta P_z^2}{2m} - \frac{e^2}{[(\Delta X)^2+(\Delta Y)^2+(\Delta Z)^2]^{1/2}}$$

For a problem such as this, with spherical symmetry, it is intuitively clear that the configuration of least energy will have

$$(\Delta X)^2 = (\Delta Y)^2 = (\Delta Z)^2$$

‡ We are basically arguing that the mean of the functions (of X, Y, and Z) and the functions of the mean ($\langle X\rangle$, $\langle Y\rangle$, and $\langle Z\rangle$) are of the same order of magnitude. They are in fact equal if there are no fluctuations around the mean and approximately equal if the fluctuations are small (recall the discussion toward the end of Chapter 6).

and

$$(\Delta P_x)^2=(\Delta P_y)^2=(\Delta P_z)^2$$

so that

$$\langle H\rangle \simeq \frac{3(\Delta P_x)^2}{2m}-\frac{e^2}{3^{1/2}\,\Delta X} \tag{9.4.5}$$

Now we use

$$\Delta P_x \Delta X \ge \hbar/2$$

to get

$$\langle H\rangle \gtrsim \frac{3\hbar^2}{8m(\Delta X)^2}-\frac{e^2}{3^{1/2}\,\Delta X}$$

We now differentiate the right-hand side with respect to ΔX to find its minimum:

$$\frac{-6\hbar^2}{8m(\Delta X)^3}+\frac{e^2}{3^{1/2}(\Delta X)^2}=0$$

or

$$\Delta X=\frac{3(3^{1/2})\hbar^2}{4me^2}\simeq 1.3\,\frac{\hbar^2}{me^2} \tag{9.4.6}$$

Finally,

$$\langle H\rangle \gtrsim \frac{-2me^4}{9\hbar^2} \tag{9.4.7}$$

What prevents us from concluding (as we did in the case of the oscillator), that the ground-state energy is $-2me^4/9\hbar^2$ or that the ground-state wave function is a Gaussian [of width $3(3^{1/2})\hbar^2/4me^2$] is the fact that Eq. (9.4.7) is an approximate inequality. However, the exact ground-state energy

$$E_g=-me^4/2\hbar^2 \tag{9.4.8}$$

differs from our estimate, Eq. (9.4.7), only by a factor $\simeq 2$. Likewise, the true ground-state wave function is not a Gaussian but an exponential $\psi(x, y, z)= c\exp[-(x^2+y^2+z^2)^{1/2}/a_0]$, where

$$a_0=\hbar^2/me^2$$

is called the *Bohr radius*. However, the ΔX associated with this wave function is

$$\Delta X = \hbar^2/me^2 \tag{9.4.9}$$

which also is within a factor of 2 of the estimated ΔX in (9.4.6).

In conclusion, the uncertainty principle gives us a lot of information about the ground state, but not always as much as in the case of the oscillator.

*Exercise 9.4.1.** Consider the oscillator in the state $|n=1\rangle$ and verify that

$$\left\langle \frac{1}{X^2} \right\rangle \simeq \frac{1}{\langle X^2 \rangle} \simeq \frac{m\omega}{\hbar}$$

Exercise 9.4.2. (1) By referring to the table of integrals in Appendix A.2, verify that

$$\psi = \frac{1}{(\pi a_0^3)^{1/2}} e^{-r/a_0}, \qquad r = (x^2+y^2+z^2)^{1/2}$$

is a normalized wave function (of the ground state of hydrogen). Note that in three dimensions the normalization condition is

$$\langle \psi | \psi \rangle = \int \psi^*(r, \theta, \phi) \psi(r, \theta, \phi) r^2 \, dr \, d(\cos\theta) \, d\phi$$

$$= 4\pi \int \psi^*(r) \psi(r) r^2 \, dr = 1$$

for a function of just r.

(2) Calculate $(\Delta X)^2$ in this state [argue that $(\Delta X)^2 = \frac{1}{3}\langle r^2 \rangle$] and regain the result quoted in Eq. (9.4.9).

(3) Show that $\langle 1/r \rangle \simeq 1/\langle r \rangle \simeq me^2/\hbar^2$ in this state.

Exercise 9.4.3. Ignore the fact that the hydrogen atom is a three-dimensional system and pretend that

$$H = \frac{P^2}{2m} - \frac{e^2}{(R^2)^{1/2}} \qquad (P^2 = P_x^2 + P_y^2 + P_z^2,\ R^2 = X^2 + Y^2 + Z^2)$$

corresponds to a one-dimensional problem. Assuming

$$\Delta P \cdot \Delta R \geq \hbar/2$$

estimate the ground-state energy.

*Exercise 9.4.4.** Compute $\Delta T \cdot \Delta X$, where $T = P^2/2m$. Why is this relation not so famous?

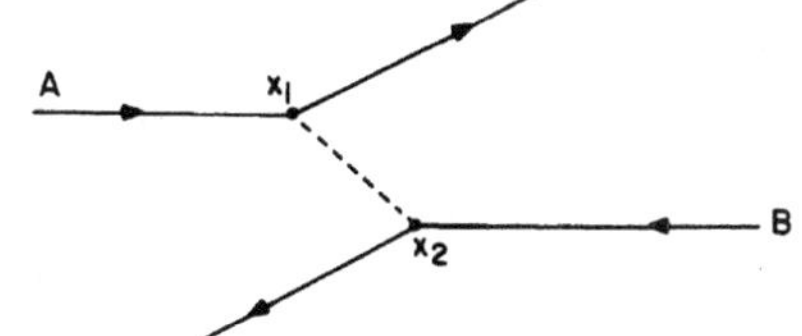

Figure 9.1. At the point x_1, skater A throws the snowball towards skater B, who catches it at the point x_2.

9.5. The Energy–Time Uncertainty Relation

There exists an uncertainty relation

$$\Delta E \cdot \Delta t \geq \hbar/2 \tag{9.5.1}$$

which does not follow from Eq. (9.2.12), since time t is not a dynamical variable but a parameter. The content of this equation is quite different from the others involving just dynamical variables. The rough meaning of this inequality is that the energy of a system that has been in existence only for a finite time Δt has a spread (or uncertainty) of at least ΔE, where ΔE and Δt are related by (9.5.1). To see how this comes about, recall that eigenstates of energy have a time dependence $e^{-iEt/\hbar}$, i.e., a definite energy is associated with a definite frequency, $\omega = E/\hbar$. Now, only a wave train that is infinitely long in time (that is to say, a system that has been in existence for infinite time) has a well-defined frequency. Thus a system that has been in existence only for a finite time, even if its time dependence goes as $e^{-iEt/\hbar}$ during this period, is not associated with a pure frequency $\omega = E/\hbar$ or definite energy E.

Consider the following example. At time $t=0$, we turn on light of frequency ω on an ensemble of hydrogen atoms all in their ground state. Since the light is supposed to consist of photons of energy $\hbar\omega$, we expect transitions to take place only to a level (if it exists) $\hbar\omega$ above the ground state. It will however be seen that initially the atoms make transitions to several levels not obeying this constraint. However, as t increases, the deviation ΔE from the expected final-state energy will decrease according to $\Delta E \simeq \hbar/t$. Only as $t \to \infty$ do we have a rigid law of conservation of energy in the classical sense. We interpret this result by saying that the light source is not associated with a definite frequency (i.e., does not emit photons of definite energy) if it has been in operation only for a finite time, even if the dial is set at a definite frequency ω during this time. [The output of the source is not just $e^{-i\omega t}$ but rather $\theta(t)\, e^{-i\omega t}$, whose transform is not a delta function peaked at ω.] Similarly when the excited atoms get deexcited and drop to the ground state, they do not emit photons of a definite energy $E = E_e - E_g$ (the subscripts e and g stand for "excited" and "ground") but rather with a spread $\Delta E \simeq \hbar/\Delta t$, Δt being the duration for which they were in the excited state. [The time dependence of the atomic wave function is not $e^{-iE_e t/\hbar}$ but rather $\theta(t)\theta(T-t)\, e^{-iE_e t/\hbar}$ assuming it abruptly got excited to this state at $t=0$ and abruptly got deexcited at $t=T$.] We shall return to this point when we discuss the interaction of atoms with radiation in a later chapter.

Another way to describe this uncertainty relation is to say that violations in the classical energy conservation law by ΔE are possible over times $\Delta t \sim \hbar/\Delta E$. The following example should clarify the meaning of this statement.

Example 9.5.1. (Range of the Nuclear Force.) Imagine two ice skaters each equipped with several snowballs, and skating toward each other on trajectories that are parallel but separated by some perpendicular distance (Fig. 9.1). When skater A reaches some point x_1

let him throw a snowball toward B. He (A) will then recoil away from B and start moving along a new straight line. Let B now catch the snowball. He too will recoil as a result, as shown in the figure. If this whole process were seen by someone who could not see the snow balls, he would conclude that there is a repulsive force between A and B. If A (or B) can throw the ball at most 10 ft, the observer would conclude that the *range of the force* is 10 ft, meaning A and B will not affect each other if the perpendicular distance between them exceeds 10 ft.

This is roughly how elementary particles interact with each other: if they throw photons at each other the force is called the electromagnetic force and the ability to throw and catch photons is called "electric charge." If the projectiles are pions the force is called the nuclear force. We would like to estimate the range of the nuclear force using the uncertainty principle. Now, unlike the two skaters endowed with snowballs, the protons and neutrons (i.e., nucleons) in the nucleus do not have a ready supply of pions, which have a mass μ and energy μc^2. A nucleon can, however, produce a pion from nowhere (violating the classical law of energy conservation by $\simeq \mu c^2$) provided it is caught by the other nucleon within a time Δt such that $\Delta t \simeq \hbar/\Delta E = \hbar/\mu c^2$. Even if the pion travels toward the receiver at the speed of light, it can only cover a distance $r = c\,\Delta t = \hbar/\mu c$, which is called the *Compton wavelength* of the pion and is a measure of the range of nuclear force. The value of r is approximately 1 Fermi $= 10^{-13}$ cm.

The picture of nuclear force given here is rather simpleminded and should be taken with a grain of salt. For example, neither is the pion the only particle that can be "exchanged" between nucleons nor is the number of exchanges limited to one per encounter. (The pion is, however, the lightest object that can be exchanged and hence responsible for the nuclear force of the longest range.) Also our analogy with snowballs does not explain any attractive interaction between particles. □

10

Systems with N Degrees of Freedom

10.1. N Particles in One Dimension

So far, we have restricted our attention (apart from minor digressions) to a system with one degree of freedom, namely, a single particle in one dimension. We now consider the quantum mechanics of systems with N degrees of freedom. The increase in degrees of freedom may be due to an increase in the number of particles, number of spatial dimensions, or both. In this section we consider N particles in one dimension, and start with the case $N=2$.

The Two-Particle Hilbert Space

Consider two particles described classically by (x_1, p_1) and (x_2, p_2). The rule for quantizing this system [Postulate II, Eq. (7.4.39)] is to promote these variables to quantum operators (X_1, P_1) and (X_2, P_2) obeying the canonical commutation relations:

$$[X_i, P_j] = i\hbar\{x_i, p_j\} = i\hbar\delta_{ij} \qquad (i=1, 2) \tag{10.1.1a}$$

$$[X_i, X_j] = i\hbar\{x_i, x_j\} = 0 \tag{10.1.1b}$$

$$[P_i, P_j] = i\hbar\{p_i, p_j\} = 0 \tag{10.1.1c}$$

It might be occasionally possible (as it was in the case of the oscillator) to extract all the physics given just the canonical commutators. In practice one works in a basis, usually the coordinate basis. This basis consists of the kets $|x_1 x_2\rangle$ which are simultaneous eigenkets of the commuting operators X_1 and X_2:

$$\begin{aligned} X_1|x_1x_2\rangle &= x_1|x_1x_2\rangle \\ X_2|x_1x_2\rangle &= x_2|x_1x_2\rangle \end{aligned} \tag{10.1.2}$$

and are normalized as‡

$$\langle x_1'x_2'|x_1x_2\rangle = \delta(x_1'-x_1)\delta(x_2'-x_2) \tag{10.1.3}$$

In this basis

$$\begin{aligned} |\psi\rangle &\to \langle x_1x_2|\psi\rangle = \psi(x_1, x_2) \\ X_i &\to x_i \\ P_i &\to -i\hbar\frac{\partial}{\partial x_i} \end{aligned} \tag{10.1.4}$$

We may interpret

$$P(x_1, x_2) = |\langle x_1x_2|\psi\rangle|^2 \tag{10.1.5}$$

as the absolute probability density for catching particle 1 near x_1 and particle 2 near x_2, provided we normalize $|\psi\rangle$ to unity

$$1 = \langle\psi|\psi\rangle = \int |\langle x_1x_2|\psi\rangle|^2\, dx_1\, dx_2 = \int P(x_1, x_2)\, dx_1\, dx_2 \tag{10.1.6}$$

There are other bases possible besides $|x_1x_2\rangle$. There is, for example, the momentum basis, consisting of the simultaneous eigenkets $|p_1p_2\rangle$ of P_1 and P_2. More generally, we can use the simultaneous eigenkets $|\omega_1\omega_2\rangle$ of two commuting operators§ $\Omega_1(X_1, P_1)$ and $\Omega_2(X_2, P_2)$ to define the Ω basis. We denote by $\mathbb{V}_{1\otimes 2}$ the two-particle Hilbert space spanned by any of these bases.

$\mathbb{V}_{1\otimes 2}$ As a Direct Product Space

There is another way to arrive at the space $\mathbb{V}_{1\otimes 2}$, and that is to build it out of two one-particle spaces. Consider a system of two particles described classically by (x_1, p_1) and (x_2, p_2). If we want the quantum theory of just particle 1, we define operators X_1 and P_1 obeying

$$[X_1, P_1] = i\hbar I \tag{10.1.7}$$

The eigenvectors $|x_1\rangle$ of X_1 form a complete (coordinate) basis for the Hilbert space

‡ Note that we denote the bra corresponding to $|x_1'x_2'\rangle$ as $\langle x_1'x_2'|$.

§ Note that any function of X_1 and P_1 commutes with any function of X_2 and P_2.

$\mathbb{V}_1$ of particle 1. Other bases, such as $|p_1\rangle$ of P_1 or in general, $|\omega_1\rangle$ of $\Omega_1(X_1, P_1)$ are also possible. Since the operators X_1, P_1, Ω_1, etc., act on $\mathbb{V}_1$, let us append a superscript (1) to all of them. Thus Eq. (10.1.7) reads

$$[X_1^{(1)}, P_1^{(1)}] = i\hbar I^{(1)} \tag{10.1.8a}$$

where $I^{(1)}$ is the identity operator on $\mathbb{V}_1$. A similar picture holds for particle 2, and in particular,

$$[X_2^{(2)}, P_2^{(2)}] = i\hbar I^{(2)} \tag{10.1.8b}$$

Let us now turn our attention to the two-particle system. What will be the coordinate basis for this system? Previously we assigned to every possible outcome x_1 of a position measurement a vector $|x_1\rangle$ in $\mathbb{V}_1$ and likewise for particle 2. Now a position measurement will yield a pair of numbers (x_1, x_2). Since after the measurement particle 1 will be in state $|x_1\rangle$ and particle 2 in $|x_2\rangle$, let us denote the corresponding ket by $|x_1\rangle \otimes |x_2\rangle$:

$$|x_1\rangle \otimes |x_2\rangle \leftrightarrow \begin{cases} \text{particle 1 at } x_1 \\ \text{particle 2 at } x_2 \end{cases} \tag{10.1.9}$$

Note that $|x_1\rangle \otimes |x_2\rangle$ is a new object, quite unlike the inner product $\langle \psi_1|\psi_2\rangle$ or the outer product $|\psi_1\rangle\langle\psi_2|$ both of which involve *two vectors from the same space*. The product $|x_1\rangle \otimes |x_2\rangle$, called the *direct product*, is the product of *vectors from two different spaces*. The direct product is a linear operation:

$$(\alpha|x_1\rangle + \alpha'|x_1'\rangle) \otimes (\beta|x_2\rangle) = \alpha\beta|x_1\rangle \otimes |x_2\rangle + \alpha'\beta|x_1'\rangle \otimes |x_2\rangle \tag{10.1.10}$$

The set of all vectors of the form $|x_1\rangle \otimes |x_2\rangle$ forms the basis for a space which we call $\mathbb{V}_1 \otimes \mathbb{V}_2$, and refer to as the *direct product of the spaces* $\mathbb{V}_1$ *and* $\mathbb{V}_2$. The dimensionality (number of possible basis vectors) of $\mathbb{V}_1 \otimes \mathbb{V}_2$ is the product of the dimensionality of $\mathbb{V}_1$ and the dimensionality of $\mathbb{V}_2$. Although all the dimensionalities are infinite here, the statement makes heuristic sense: to each basis vector $|x_1\rangle$ of $\mathbb{V}_1$ and $|x_2\rangle$ of $\mathbb{V}_2$, there is one and only one basis vector $|x_1\rangle \otimes |x_2\rangle$ of $\mathbb{V}_1 \otimes \mathbb{V}_2$. This should be compared to the direct sum (Section 1.4):

$$\mathbb{V}_{1\oplus 2} = \mathbb{V}_1 \oplus \mathbb{V}_2$$

in which case the dimensionalities of $\mathbb{V}_1$ and $\mathbb{V}_2$ add (assuming the vectors of $\mathbb{V}_1$ are linearly independent of those of $\mathbb{V}_2$).

The coordinate basis, $|x_1\rangle \otimes |x_2\rangle$, is just one possibility; we can use the momentum basis $|p_1\rangle \otimes |p_2\rangle$, or, more generally, $|\omega_1\rangle \otimes |\omega_2\rangle$. *Although these vectors span* $\mathbb{V}_1 \otimes \mathbb{V}_2$, *not every element of* $\mathbb{V}_1 \otimes \mathbb{V}_2$ *is a direct product.* For instance

$$|\psi\rangle = |x_1'\rangle \otimes |x_2'\rangle + |x_1''\rangle \otimes |x_2''\rangle$$

cannot be written as

$$|\psi\rangle = |\psi_1\rangle \otimes |\psi_2\rangle$$

where $|\psi_1\rangle$ and $|\psi_2\rangle$ are elements of $\mathbb{V}_1$ and $\mathbb{V}_2$, respectively.

The inner product of $|x_1\rangle \otimes |x_2\rangle$ and $|x_1'\rangle \otimes |x_2'\rangle$ is

$$\begin{aligned}(\langle x_1'| \otimes \langle x_2'|)(|x_1\rangle \otimes |x_2\rangle) &= \langle x_1'|x_1\rangle\langle x_2'|x_2\rangle \\ &= \delta(x_1'-x_1)\delta(x_2'-x_2) \end{aligned} \tag{10.1.11}$$

Since any vector in $\mathbb{V}_1 \otimes \mathbb{V}_2$ can be expressed in terms of the $|x_1\rangle \otimes |x_2\rangle$ basis, this defines the inner product between any two vectors in $\mathbb{V}_1 \otimes \mathbb{V}_2$.

It is intuitively clear that when two particles are amalgamated to form a single system, the position and momentum operators of each particle, $X_1^{(1)}$, $P_1^{(1)}$ and $X_2^{(2)}$, $P_2^{(2)}$, which acted on $\mathbb{V}_1$ and $\mathbb{V}_2$, respectively, must have counterparts in $\mathbb{V}_1 \otimes \mathbb{V}_2$ and have the same interpretation. Let us denote by $X_1^{(1)\otimes(2)}$ the counterpart of $X_1^{(1)}$, and refer to it also as the "X operator of particle 1." Let us define its action on $\mathbb{V}_1 \otimes \mathbb{V}_2$. Since the vectors $|x_1\rangle \otimes |x_2\rangle$ span the space, it suffices to define its action on these. Now the ket $|x_1\rangle \otimes |x_2\rangle$ denotes a state in which particle 1 is at x_1. Thus it must be an eigenket of $X_1^{(1)\otimes(2)}$ with eigenvalue x_1:

$$X_1^{(1)\otimes(2)}|x_1\rangle \otimes |x_2\rangle = x_1|x_1\rangle \otimes |x_2\rangle \tag{10.1.12}$$

Note that $X_1^{(1)\otimes(2)}$ does not really care about the second ket $|x_2\rangle$, i.e., it acts trivially (as the identity) on $|x_2\rangle$ and acts on $|x_1\rangle$ just as $X_1^{(1)}$ did. In other words

$$X_1^{(1)\otimes(2)}|x_1\rangle \otimes |x_2\rangle = |X_1^{(1)}x_1\rangle \otimes |I^{(2)}x_2\rangle \tag{10.1.13}$$

Let us define a *direct product of two operators*, $\Gamma_1^{(1)}$ and $\Lambda_2^{(2)}$ (denoted by $\Gamma_1^{(1)} \otimes \Lambda_2^{(2)}$), whose action on a direct product ket $|\omega_1\rangle \otimes |\omega_2\rangle$ is

$$(\Gamma_1^{(1)} \otimes \Lambda_2^{(2)})|\omega_1\rangle \otimes |\omega_2\rangle = |\Gamma_1^{(1)}\omega_1\rangle \otimes |\Lambda_2^{(2)}\omega_2\rangle \tag{10.1.14}$$

In this notation, we may write $X_1^{(1)\otimes(2)}$, in view of Eq. (10.1.13), as

$$X_1^{(1)\otimes(2)} = X_1^{(1)} \otimes I^{(2)} \tag{10.1.15}$$

We can similarly promote $P_2^{(2)}$, say, from $\mathbb{V}_2$ to $\mathbb{V}_1 \otimes \mathbb{V}_2$ by defining the momentum operator for particle 2, $P_2^{(1)\otimes(2)}$, as

$$P_2^{(1)\otimes(2)} = I^{(1)} \otimes P_2^{(2)} \tag{10.1.16}$$

The following properties of direct products of operators may be verified (say by acting on the basis vectors $|x_1\rangle \otimes |x_2\rangle$):

*Exercise 10.1.1.** Show the following:

(1) $$[\Omega_1^{(1)}\otimes I^{(2)}, I^{(1)}\otimes\Lambda_2^{(2)}]=0 \text{ for any } \Omega_1^{(1)} \text{ and } \Lambda_2^{(2)} \tag{10.1.17a}$$

(operators of particle 1 commute with those of particle 2).

(2) $$(\Omega_1^{(1)}\otimes\Gamma_2^{(2)})(\theta_1^{(1)}\otimes\Lambda_2^{(2)})=(\Omega\theta)_1^{(1)}\otimes(\Gamma\Lambda)_2^{(2)} \tag{10.1.17b}$$

(3) If

$$[\Omega_1^{(1)}, \Lambda_1^{(1)}]=\Gamma_1^{(1)}$$

then

$$[\Omega_1^{(1)\otimes(2)}, \Lambda_1^{(1)\otimes(2)}]=\Gamma_1^{(1)}\otimes I^{(2)} \tag{10.1.17c}$$

and similarly with $1\rightarrow 2$.

(4) $$(\Omega_1^{(1)\otimes(2)}+\Omega_2^{(1)\otimes(2)})^2=(\Omega_1^2)^{(1)}\otimes I^{(2)}+I^{(1)}\otimes(\Omega_2^2)^{(2)}+2\Omega_1^{(1)}\otimes\Omega_2^{(2)} \tag{10.1.17d}$$

The notion of direct products of vectors and operators is no doubt a difficult one, with no simple analogs in elementary vector analysis. The following exercise should give you some valuable experience. It is recommended that you reread the preceding discussion after working on the exercise.

*Exercise 10.1.2.** Imagine a fictitious world in which the single-particle Hilbert space is two-dimensional. Let us denote the basis vectors by $|+\rangle$ and $|-\rangle$. Let

$$\sigma_1^{(1)}=\begin{array}{c} \\ + \\ - \end{array}\!\!\begin{array}{cc} + & - \\ \left[\begin{array}{c} a \\ c \end{array}\right. & \left.\begin{array}{c} b \\ d \end{array}\right] \end{array} \quad\text{and}\quad \sigma_2^{(2)}=\begin{array}{c} \\ + \\ - \end{array}\!\!\begin{array}{cc} + & - \\ \left[\begin{array}{c} e \\ g \end{array}\right. & \left.\begin{array}{c} f \\ h \end{array}\right] \end{array}$$

be operators in $\mathbb{V}_1$ and $\mathbb{V}_2$, respectively (the $\pm$ signs label the basis vectors. Thus $b=\langle+|\sigma_1^{(1)}|-\rangle$ etc.) The space $\mathbb{V}_1\otimes\mathbb{V}_2$ is spanned by four vectors $|+\rangle\otimes|+\rangle$, $|+\rangle\otimes|-\rangle$, $|-\rangle\otimes|+\rangle$, $|-\rangle\otimes|-\rangle$. Show (using the method of images or otherwise) that

(1) $$\sigma_1^{(1)\otimes(2)}=\sigma_1^{(1)}\otimes I^{(2)}=\begin{array}{c} \\ ++ \\ +- \\ -+ \\ -- \end{array}\!\!\begin{array}{c} \begin{array}{cccc} ++ & +- & -+ & -- \end{array} \\ \left[\begin{array}{cccc} a & 0 & b & 0 \\ 0 & a & 0 & b \\ c & 0 & d & 0 \\ 0 & c & 0 & d \end{array}\right] \end{array}$$

(Recall that $\langle\alpha| \otimes \langle\beta|$ is the bra corresponding to $|\alpha\rangle \otimes |\beta\rangle$.)

$$(2) \qquad \sigma_2^{(1)\otimes(2)} = \begin{bmatrix} e & f & 0 & 0 \\ g & h & 0 & 0 \\ 0 & 0 & e & f \\ 0 & 0 & g & h \end{bmatrix}$$

$$(3) \qquad (\sigma_1\sigma_2)^{(1)\otimes(2)} = \sigma_1^{(1)} \otimes \sigma_2^{(2)} = \begin{bmatrix} ae & af & be & bf \\ ag & ah & bg & bh \\ ce & cf & de & df \\ cg & ch & dg & dh \end{bmatrix}$$

Do part (3) in two ways, by taking the matrix product of $\sigma_1^{(1)\otimes(2)}$ and $\sigma_2^{(1)\otimes(2)}$ and by directly computing the matrix elements of $\sigma_1^{(1)} \otimes \sigma_2^{(2)}$.

From Eqs. (10.1.17a) and (10.1.17c) it follows that the commutation relations between the position and momentum operators on $\mathbb{V}_1 \otimes \mathbb{V}_2$ are

$$\begin{aligned} &[X_i^{(1)\otimes(2)}, P_j^{(1)\otimes(2)}] = i\hbar\delta_{ij} I^{(1)\otimes I(2)} = i\hbar\delta_{ij} I^{(1)\otimes(2)} \\ &[X_i^{(1)\otimes(2)}, X_j^{(1)\otimes(2)}] = [P_i^{(1)\otimes(2)}, P_j^{(1)\otimes(2)}] = 0 \qquad i,j = 1,2 \end{aligned} \tag{10.1.18}$$

Now we are ready to assert something that may have been apparent all along: the space $\mathbb{V}_1 \otimes \mathbb{V}_2$ is just $\mathbb{V}_{1\otimes 2}$, $|x_1\rangle \otimes |x_2\rangle$ is just $|x_1x_2\rangle$, and $X_1^{(1)\otimes(2)}$ is just X_1, etc. Notice first that both spaces have the same dimensionality: the vectors $|x_1x_2\rangle$ and $|x_1\rangle \otimes |x_2\rangle$ are both in one-to-one correspondence with points in the $x_1 - x_2$ plane. Notice next that the two sets of operators $X_1, \ldots, P_2$ and $X_1^{(1)\otimes(2)}, \ldots, P_2^{(1)\otimes(2)}$ have the same connotation and commutation rules [Eqs. (10.1.1) and (10.1.18)]. Since X and P are *defined* by their commutators we can make the identification

$$\begin{aligned} X_i^{(1)\otimes(2)} &= X_i \\ P_i^{(1)\otimes(2)} &= P_i \end{aligned} \tag{10.1.19a}$$

We can also identify the simultaneous eigenkets of the position operators (since they are nondegenerate):

$$|x_1\rangle \otimes |x_2\rangle = |x_1x_2\rangle \tag{10.1.19b}$$

In the future, we shall use the more compact symbols occurring on the right-hand side of Eqs. (10.1.19). We will, however, return to the concept of direct products of vectors and operators on and off and occasionally use the symbols on the left-hand side. Although the succinct notation suppresses the label $(1 \otimes 2)$ of the space on which the operators act, it should be clear from the context. Consider, for example, the CM kinetic energy operator of the two-particle system:

$$T_{\text{CM}}=\frac{P_{\text{CM}}^2}{2(m_1+m_2)}=\frac{P_{\text{CM}}^2}{2M}=\frac{(P_1+P_2)^2}{2M}=\frac{P_1^2+P_2^2+2P_1P_2}{2M}$$

which really means

$$\begin{aligned}2MT_{\text{CM}}^{(1)\otimes(2)}&=(P_1^2)^{(1)\otimes(2)}+(P_2^2)^{(1)\otimes(2)}+2P_1^{(1)\otimes(2)}\cdot P_2^{(1)\otimes(2)}\\&=(P_1^{(1)}\otimes I^{(2)})^2+(I^{(1)}\otimes P_2^{(2)})^2+2P_1^{(1)}\otimes P_2^{(2)}\end{aligned}$$

The Direct Product Revisited

Since the notion of a direct product space is so important, we revisit the formation of $\mathbb{V}_{1\otimes 2}$ as a direct product of $\mathbb{V}_1$ and $\mathbb{V}_2$, but this time in the coordinate basis instead of in the abstract. Let $\Omega_1^{(1)}$ be an operator on $\mathbb{V}_1$ whose nondegenerate eigenfunctions $\psi_{\omega_1}(x_1)\equiv\omega_1(x_1)$ form a complete basis. Similarly let $\omega_2(x_2)$ form a basis for $\mathbb{V}_2$. Consider now a function $\psi(x_1, x_2)$, which represents the abstract ket $|\psi\rangle$ from $\mathbb{V}_{1\otimes 2}$. If we keep x_1 fixed at some value, say $\bar{x}_1$, then ψ becomes a function of x_2 alone and may be expanded as

$$\psi(\bar{x}_1, x_2)=\sum_{\omega_2} C_{\omega_2}(\bar{x}_1)\omega_2(x_2) \tag{10.1.20}$$

Notice that the coefficients of the expansion depend on the value of $\bar{x}_1$. We now expand the function $C_{\omega_2}(\bar{x}_1)$ in the basis $\omega_1(\bar{x}_1)$:

$$C_{\omega_2}(\bar{x}_1)=\sum_{\omega_1} C_{\omega_1,\omega_2}\omega_1(\bar{x}_1) \tag{10.1.21}$$

Feeding this back to the first expansion and dropping the bar on $\bar{x}_1$ we get

$$\psi(x_1, x_2)=\sum_{\omega_1}\sum_{\omega_2} C_{\omega_1,\omega_2}\omega_1(x_1)\omega_2(x_2) \tag{10.1.22a}$$

What does this expansion of an arbitrary $\psi(x_1, x_2)$ in terms of $\omega_1(x_1)\times\omega_2(x_2)$ imply? Equation (10.1.22a) is the coordinate space version of the abstract result

$$|\psi\rangle=\sum_{\omega_1}\sum_{\omega_2} C_{\omega_1,\omega_2}|\omega_1\rangle\otimes|\omega_2\rangle \tag{10.1.22b}$$

which means $\mathbb{V}_{1\otimes 2}=\mathbb{V}_1\otimes\mathbb{V}_2$, for $|\psi\rangle$ belongs to $\mathbb{V}_{1\otimes 2}$ and $|\omega_1\rangle\otimes|\omega_2\rangle$ spans $\mathbb{V}_1\otimes\mathbb{V}_2$. If we choose $\Omega=X$, we get the familiar basis $|x_1\rangle\otimes|x_2\rangle$. By dotting both sides of Eq. (10.1.22b) with these basis vectors we regain Eq. (10.1.22a). (In the coordinate basis, the direct product of the kets $|\omega_1\rangle$ and $|\omega_2\rangle$ becomes just the ordinary product of the corresponding wave functions.)

Consider next the operators. The momentum operator on $\mathbb{V}_1$, which used to be $-i\hbar\, d/dx_1$ now becomes $-i\hbar\, \partial/\partial x_1$, where the partial derivative symbol tells us it operates on x_1 as before and leaves x_2 alone. This is the coordinate space version of $P^{(1)\otimes(2)} = P_1^{(1)} \otimes I^{(2)}$. You are encouraged to pursue this analysis further.

Evolution of the Two-Particle State Vector

The state vector of the system is an element of $\mathbb{V}_{1\otimes 2}$. It evolves in time according to the equation

$$i\hbar|\dot{\psi}\rangle = \left[\frac{P_1^2}{2m_1} + \frac{P_2^2}{2m_2} + V(X_1, X_2)\right]|\psi\rangle = H|\psi\rangle \tag{10.1.23}$$

There are two classes of problems.

Class A: H is separable, i.e.,

$$H = \frac{P_1^2}{2m_1} + V_1(X_1) + \frac{P_2^2}{2m_2} + V_2(X_2) = H_1 + H_2 \tag{10.1.24}$$

Class B: H is not separable, i.e.,

$$V(X_1, X_2) \neq V_1(X_1) + V_2(X_2)$$

and

$$H \neq H_1 + H_2 \tag{10.1.25}$$

Class A corresponds to two particles interacting with external potentials V_1 and V_2 but not with each other, while in class B there is no such restriction. We now examine these two classes.

Class A: Separable Hamiltonians. Classically, the decomposition

$$\mathscr{H} = \mathscr{H}_1(x_1, p_1) + \mathscr{H}_2(x_2, p_2)$$

means that the two particles evolve independently of each other. In particular, their energies are *separately* conserved and the total energy E is $E_1 + E_2$. Let us see these results reappear in quantum theory. For a stationary state,

$$|\psi(t)\rangle = |E\rangle\, e^{-iEt/\hbar} \tag{10.1.26}$$

Eq. (10.1.23) becomes

$$[H_1(X_1, P_1) + H_2(X_2, P_2)]|E\rangle = E|E\rangle \tag{10.1.27}$$

Since $[H_1, H_2] = 0$ [Eq. (10.1.17a)] we can find their simultaneous eigenstates, which are none other than $|E_1\rangle \otimes |E_2\rangle = |E_1 E_2\rangle$, where $|E_1\rangle$ and $|E_2\rangle$ are solutions to

$$H_1^{(1)}|E_1\rangle = E_1|E_1\rangle \tag{10.1.28a}$$

and

$$H_2^{(2)}|E_2\rangle = E_2|E_2\rangle \tag{10.1.28b}$$

It should be clear that the state $|E_1\rangle \otimes |E_2\rangle$ corresponds to particle 1 being in the energy eigenstate $|E_1\rangle$ and particle 2 being in the energy eigenstate $|E_2\rangle$. Clearly

$$H|E\rangle = (H_1 + H_2)|E_1\rangle \otimes |E_2\rangle = (E_1 + E_2)|E_1\rangle \otimes |E_2\rangle = (E_1 + E_2)|E\rangle$$

so that

$$E = E_1 + E_2 \tag{10.1.28c}$$

(The basis $|E_1\rangle \otimes |E_2\rangle$ is what we would get if in forming basis vectors of the direct product $\mathbb{V}_1 \otimes \mathbb{V}_2$, we took the energy eigenvectors from each space, instead of, say, the position eigenvectors.) Finally, feeding $|E\rangle = |E_1\rangle \otimes |E_2\rangle, E = E_1 + E_2$ into Eq. (10.1.26) we get

$$|\psi(t)\rangle = |E_1\rangle e^{-iE_1t/\hbar} \otimes |E_2\rangle e^{-iE_2t/\hbar} \tag{10.1.29}$$

It is worth rederiving Eqs. (10.1.28) and (10.1.29) in the coordinate basis to illustrate a useful technique that you will find in other textbooks. By projecting the eigenvalue Eq. (10.1.27) on this basis, and making the usual operator substitutions, Eq. (10.1.4), we obtain

$$\left[\frac{-\hbar^2}{2m_1}\frac{\partial^2}{\partial x_1^2} + V_1(x_1) - \frac{\hbar^2}{2m_2}\frac{\partial^2}{\partial x_2^2} + V_2(x_2)\right]\psi_E(x_1, x_2) = E\psi_E(x_1, x_2)$$

where

$$\psi_E(x_1, x_2) = \langle x_1 x_2 | E\rangle \tag{10.1.30}$$

We solve the equation by the method of *separation of variables*. We assume

$$\psi_E(x_1, x_2) = \psi_{E_1}(x_1)\psi_{E_2}(x_2) \tag{10.1.31}$$

The subscripts E_1 and E_2 have no specific interpretation yet and merely serve as labels. Feeding this *ansatz* into Eq. (10.1.30) and *then* dividing both sides by

$\psi_{E_1}(x_1)\psi_{E_2}(x_2)$ we get

$$\frac{1}{\psi_{E_1}(x_1)}\left[\frac{-\hbar^2}{2m_1}\frac{\partial^2}{\partial x_1^2}+V_1(x_1)\right]\psi_{E_1}(x_1)$$
$$+\frac{1}{\psi_{E_2}(x_2)}\left[\frac{-\hbar^2}{2m_2}\frac{\partial^2}{\partial x_2^2}+V_2(x_2)\right]\psi_{E_2}(x_2)=E \qquad (10.1.32)$$

This equation says that a function of x_1 alone, plus one of x_2 alone, equals a constant E. Since x_1 and x_2, and hence the two functions, may be varied independently, it follows that each function separately equals a constant. We will call these constants E_1 and E_2. Thus Eq. (10.1.32) breaks down into three equations:

$$\frac{1}{\psi_{E_1}(x_1)}\left[\frac{-\hbar^2}{2m_1}\frac{\partial^2}{\partial x_1^2}+V_1(x_1)\right]\psi_{E_1}(x_1)=E_1$$
$$\frac{1}{\psi_{E_2}(x_2)}\left[\frac{-\hbar^2}{2m_2}\frac{\partial^2}{\partial x_2^2}+V_2(x_2)\right]\psi_{E_2}(x_2)=E_2 \qquad (10.1.33)$$
$$E_1+E_2=E$$

Consequently

$$\psi_E(x_1, x_2, t)=\psi_E(x_1, x_2)\, e^{-iEt/\hbar}$$
$$=\psi_{E_1}(x_1)\, e^{-iE_1t/\hbar}\psi_{E_2}(x_2)\, e^{-iE_2t/\hbar} \qquad (10.1.34)$$

where ψ_{E_1} and ψ_{E_2} are eigenfunctions of the one-particle Schrödinger equation with eigenvalues E_1 and E_2, respectively. We recognize Eqs. (10.1.33) and (10.1.34) to be the projections of Eqs. (10.1.28) and (10.1.29) on $|x_1x_2\rangle=|x_1\rangle\otimes|x_2\rangle$.

Case B: Two Interacting Particles. Consider next the more general problem of two interacting particles with

$$\mathscr{H}=\frac{p_1^2}{2m_1}+\frac{p_2^2}{2m_2}+V(x_1, x_2) \qquad (10.1.35)$$

where

$$V(x_1, x_2)\neq V_1(x_1)+V(x_2)$$

Generally this cannot be reduced to two independent single-particle problems. If, however,

$$V(x_1, x_2)=V(x_1-x_2) \qquad (10.1.36)$$

which describes two particles responding to each other but nothing external, one can always, by employing the CM coordinate

$$x_{\mathrm{CM}}=\frac{m_1x_1+m_2x_2}{m_1+m_2} \tag{10.1.37a}$$

and the relative coordinate

$$x=x_1-x_2 \tag{10.1.37b}$$

reduce the problem to that of two independent fictitious particles: one, the CM, which is free, has mass $M=m_1+m_2$ and momentum

$$p_{\mathrm{CM}}=M\dot{x}_{\mathrm{CM}}=m_1\dot{x}_1+m_2\dot{x}_2$$

and another, with the reduced mass $\mu=m_1m_2/(m_1+m_2)$, momentum $p=\mu\dot{x}$, moving under the influence of $V(x)$:

$$\begin{aligned}\mathscr{H}(x_1,p_1;x_2,p_2)&\to\mathscr{H}(x_{\mathrm{CM}},p_{\mathrm{CM}};x,p)\\&=\mathscr{H}_{\mathrm{CM}}+\mathscr{H}_{\mathrm{relative}}=\frac{p_{\mathrm{CM}}^2}{2M}+\frac{p^2}{2\mu}+V(x)\end{aligned} \tag{10.1.38}$$

which is just the result from Exercise 2.5.4 modified to one dimension. Since the new variables are also canonical (Exercise 2.7.6) and Cartesian, the quantization condition is just

$$[X_{\mathrm{CM}},P_{\mathrm{CM}}]=i\hbar \tag{10.1.39a}$$

$$[X,P]=i\hbar \tag{10.1.39b}$$

and all other commutators zero. In the quantum theory,

$$H=\frac{P_{\mathrm{CM}}^2}{2M}+\frac{P^2}{2\mu}+V(X) \tag{10.1.40}$$

and the eigenfunctions of H factorize:

$$\begin{aligned}\psi_E(x_{\mathrm{CM}},x)&=\frac{e^{ip_{\mathrm{CM}}\cdot x_{\mathrm{CM}}/\hbar}}{(2\pi\hbar)^{1/2}}\cdot\psi_{E_{\mathrm{rel}}}(x)\\E&=\frac{p_{\mathrm{CM}}^2}{2M}+E_{\mathrm{rel}}\end{aligned} \tag{10.1.41}$$

The real dynamics is contained in $\psi_{E_{\mathrm{rel}}}(x)$ which is the energy eigenfunction for a particle of mass μ in a potential $V(x)$. Since the CM drifts along as a free particle, one usually chooses to study the problem in the CM frame. In this case $E_{\mathrm{CM}}=$

$p_{\rm CM}^2/2M$ drops out of the energy, and the plane wave factor in ψ representing CM motion becomes a constant. In short, one can forget all about the CM in the quantum theory just as in the classical theory.

N Particles in One Dimension

All the results but one generalize from $N=2$ to arbitrary N. The only exception is the result from the last subsection: for $N>2$, one generally cannot, by using CM and relative coordinates (or other sets of coordinates) reduce the problem to N independent one-particle problems. There are a few exceptions, the most familiar ones being Hamiltonians quadratic in the coordinates and momenta which may be reduced to a sum over oscillator Hamiltonians by the use of normal coordinates. In such cases the oscillators become independent and their energies add both in the classical and quantum cases. This result (with respect to the quantum oscillators) was assumed in the discussion on specific heats in Chapter 7.

*Exercise 10.1.3.** Consider the Hamiltonian of the coupled mass system:

$$\mathscr{H}=\frac{p_1^2}{2m}+\frac{p_2^2}{2m}+\frac{1}{2}m\omega^2[x_1^2+x_2^2+(x_1-x_2)^2]$$

We know from Example 1.8.6 that $\mathscr{H}$ can be decoupled if we use normal coordinates

$$x_{\rm I,II}=\frac{x_1\pm x_2}{2^{1/2}}$$

and the corresponding momenta

$$p_{\rm I,II}=\frac{p_1\pm p_2}{2^{1/2}}$$

(1) Rewrite $\mathscr{H}$ in terms of normal coordinates. Verify that the normal coordinates are also canonical, i.e., that

$$\{x_i, p_j\}=\delta_{ij}\text{ etc.};\qquad i,j=\text{I, II}$$

Now quantize the system, promoting these variables to operators obeying

$$[X_i, P_j]=i\hbar\delta_{ij}\text{ etc.};\qquad i,j=\text{I, II}$$

Write the eigenvalue equation for H in the simultaneous eigenbasis of $X_{\rm I}$ and $X_{\rm II}$.

(2) Quantize the system directly, by promoting x_1, x_2, p_1, and p_2 to quantum operators. Write the eigenvalue equation for H in the simultaneous eigenbasis of X_1 and X_2. Now change from x_1, x_2 (and of course $\partial/\partial x_1$, $\partial/\partial x_2$) to $x_{\rm I}$, $x_{\rm II}$ (and $\partial/\partial x_{\rm I}$, $\partial/\partial x_{\rm II}$) *in the differential equation.* You should end up with the result from part (1).

In general, one can change coordinates and then quantize or first quantize and then change variables in the differential equation, if the change of coordinates is canonical. (We are assuming that all the variables are Cartesian. As mentioned earlier in the book, if one wants to employ non-Cartesian coordinates, it is best to first quantize the Cartesian coordinates and then change variables in the differential equation.)

10.2. More Particles in More Dimensions

Mathematically, the problem of a single particle in two dimensions (in terms of Cartesian coordinates) is equivalent to that of two particles in one dimension. It is, however, convenient to use a different notation in the two cases. We will denote the two Cartesian coordinates of the single particle by x and y rather than x_1 and x_2. Likewise the momenta will be denoted by p_x and p_y. The quantum operators will be called X and Y; and P_x, and P_y, their common eigenkets $|xy\rangle$, $|p_xp_y\rangle$, respectively, and so on. The generalization to three dimensions is obvious. We will also write a position eigenket as $|\mathbf{r}\rangle$ and the orthonormality relation $\langle xyz|x'y'z'\rangle = \delta(x-x')\delta(y-y')\delta(z-z')$ as $\langle\mathbf{r}|\mathbf{r}'\rangle = \delta^3(\mathbf{r}-\mathbf{r}')$. The same goes for the momentum eigenkets $|\mathbf{p}\rangle$ also. When several particles labeled by numbers $1, \ldots, N$ are involved, this extra label will also be used. Thus $|\mathbf{p}_1\mathbf{p}_2\rangle$ will represent a two-particle state in which particle 1 has momentum $\mathbf{p}_1$ and particle 2 has momentum $\mathbf{p}_2$ and so on.

*Exercise 10.2.1** (Particle in a Three-Dimensional Box). Recall that a particle in a one-dimensional box extending from $x=0$ to L is confined to the region $0\le x\le L$; its wave function vanishes at the edges $x=0$ and L and beyond (Exercise 5.2.5). Consider now a particle confined in a three-dimensional cubic box of volume L^3. Choosing as the origin one of its corners, and the x, y, and z axes along the three edges meeting there, show that the normalized energy eigenfunctions are

$$\psi_E(x, y, z) = \left(\frac{2}{L}\right)^{1/2} \sin\left(\frac{n_x\pi x}{L}\right)\left(\frac{2}{L}\right)^{1/2} \sin\left(\frac{n_y\pi y}{L}\right)\left(\frac{2}{L}\right)^{1/2} \sin\left(\frac{n_z\pi z}{L}\right)$$

where

$$E = \frac{\hbar^2\pi^2}{2ML^2}(n_x^2+n_y^2+n_z^2)$$

and n_i are positive integers.

*Exercise 10.2.2.** Quantize the two-dimensional oscillator for which

$$\mathscr{H} = \frac{p_x^2+p_y^2}{2m} + \frac{1}{2}m\omega_x^2x^2 + \frac{1}{2}m\omega_y^2y^2$$

(1) Show that the allowed energies are

$$E = (n_x+1/2)\hbar\omega_x + (n_y+1/2)\hbar\omega_y, \qquad n_x, n_y = 0, 1, 2, \ldots$$

(2) Write down the corresponding wave functions in terms of single oscillator wave functions. Verify that they have definite parity (even/odd) number $x\to -x$, $y\to -y$ and that the parity depends only on $n = n_x + n_y$.

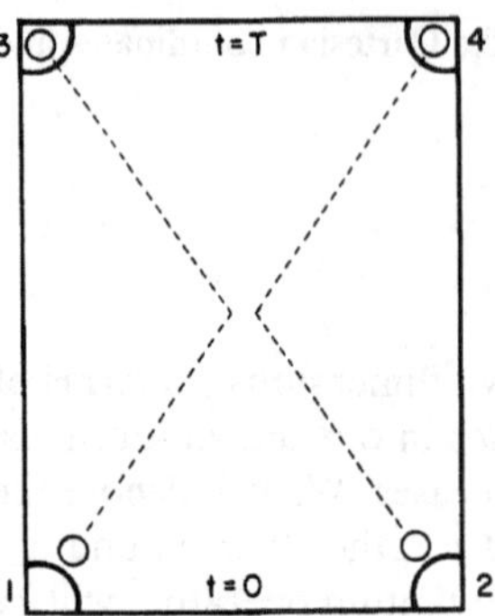

Figure 10.1. Two identical billiard balls start near holes 1 and 2 and end up in holes 3 and 4, respectively, as predicted by P_1. The prediction of P_2, that they would end up in holes 4 and 3, respectively, is wrong, even though the two final configurations would be indistinguishable to an observer who walks in at $t=T$.

(3) Consider next the *isotropic* oscillator ($\omega_x=\omega_y$). Write *explicit*, normalized eigenfunctions of the first three *states* (that is, for the cases $n=0$ and 1). Reexpress your results in terms of polar coordinates ρ and ϕ (for later use). Show that the degeneracy of a level with $E=(n+1)\hbar\omega$ is $(n+1)$.

*Exercise 10.2.3.** Quantize the three-dimensional *isotropic oscillator* for which

$$\mathscr{H}=\frac{p_x^2+p_y^2+p_z^2}{2m}+\frac{1}{2}m\omega^2(x^2+y^2+z^2)$$

(1) Show that $E=(n+3/2)\hbar\omega$; $n=n_x+n_y+n_z$; $n_x, n_y, n_z=0, 1, 2, \ldots$.

(2) Write the corresponding eigenfunctions in terms of single-oscillator wave functions and verify that the parity of the level with a given n is $(-1)^n$. Reexpress the first four states in terms of spherical coordinates. Show that the degeneracy of a level with energy $E=(n+3/2)\hbar\omega$ is $(n+1)(n+2)/2$.

10.3. Identical Particles

The formalism developed above, when properly applied to a system containing identical particles, leads to some very surprising results. We shall say two particles are identical if they are exact replicas of each other in every respect—there should be no experiment that detects any intrinsic‡ difference between them. Although the definition of identical particles is the same classically and quantum mechanically, the implications are different in the two cases.

The Classical Case

Let us first orient ourselves by recapitulating the situation in classical physics. Imagine a billiard table with four holes, numbered 1 through 4 (Fig. 10.1). Near holes 1 and 2 rest two identical billiard balls. Let us call these balls 1 and 2. The difference between the labels reflects not any intrinsic difference in the balls (for they are identical) but rather a difference in their environments, namely, the holes near which they find themselves.

‡ By intrinsic I mean properties inherent to the particle, such as its charge or mass and not its location or momentum.

Now it follows from the definition of identity, that if these two balls are exchanged, the resulting configuration would appear exactly the same. Nonetheless these two configurations are treated as distinct in classical physics. In order for this distinction to be meaningful, there must exist some experiments in which these two configurations are inequivalent. We will now discuss one such experiment.

Imagine that at time $t=0$, two players propel the balls toward the center of the table. At once two physicists P_1 and P_2 take the initial-value data and make the following predictions:

$$P_1: \quad \left.\begin{array}{l}\text{ball 1 goes to hole 3}\\ \text{ball 2 goes to hole 4}\end{array}\right\} \quad \text{at } t=\text{T}$$

$$P_2: \quad \left.\begin{array}{l}\text{ball 1 goes to hole 4}\\ \text{ball 2 goes to hole 3}\end{array}\right\} \quad \text{at } t=T$$

Say at time T we find that ball 1 ends up in hole 3 and ball 2 in hole 4. We declare that P_1 is correct and P_2 is wrong. Now, the configurations predicted by them for $t=T$ differ only by the exchange of two identical particles. If seen in isolation they would appear identical: an observer who walks in just at $t=T$ and is given the predictions of P_1 and P_2 will conclude that both are right. What do we know about the balls (that allows us to make a distinction between them and hence the two outcomes), that the newcomer does not? The answer of course is—their histories. Although both balls appear identical to the newcomer, we are able to trace the ball in hole 3 back to the vicinity of hole 1 and the one in hole 4 back to hole 2. Similarly at $t=0$, the two balls which seemed identical to us would be distinguishable to someone who had been following them from an earlier period. Now of course it is not really necessary that either we or any other observer be actually present in order for this distinction to exist. One imagines in classical physics the fictitious observer who sees everything and disturbs nothing; if he can make the distinction, the distinction exists.

To summarize, it is possible in classical mechanics to distinguish between identical particles by following their nonidentical trajectories (without disturbing them in any way). Consequently two configurations related by exchanging the identical particles are physically nonequivalent.

An immediate consequence of the above reasoning, and one that will play a dominant role in what follows, is that in quantum theory, which completely outlaws the notion of continuous trajectories for the particles, there exists no physical basis for distinguishing between identical particles. Consequently two configurations related by the exchange of identical particles must be treated as one and the same configuration and described by the same state vector. We now proceed to deduce the consequences of this restriction.

Two-Particle Systems—Symmetric and Antisymmetric States

Suppose we have a system of two *distinguishable* particles 1 and 2 and a position measurement on the system shows particle 1 to be at $x=a$ and particle 2 to be at

$x=b$. We write the state just after measurement as

$$|\psi\rangle=|x_1=a, x_2=b\rangle=|ab\rangle \tag{10.3.1}$$

where we are adopting the convention that the state of particle 1 is described by the first label (a) and that of particle 2 by the second label (b). Since the particles are distinguishable, the state obtained by exchanging them is distinguishable from the above. It is given by

$$|\psi\rangle=|ba\rangle$$

and corresponds to having found particle 1 at b and particle 2 at a.

Suppose we repeat the experiment with two *identical* particles and catch one at $x=a$ and the other at $x=b$. Is the state vector just after measurement $|ab\rangle$ or $|ba\rangle$? The answer is, neither. We have seen that in quantum theory two configurations related by the exchange of identical particles must be viewed as one and the same and be described by the same state vector. Since $|\psi\rangle$ and $\alpha|\psi\rangle$ are physically equivalent, we require that $|\psi(a, b)\rangle$, the state vector just after the measurement, satisfy the constraint

$$|\psi(a, b)\rangle=\alpha|\psi(b, a)\rangle \tag{10.3.2}$$

where α is any complex number. Since under the exchange

$$|ab\rangle \leftrightarrow |ba\rangle$$

and the two vectors are not multiples of each other‡ (i.e., are physically distinct) neither is acceptable. The problem is that our position measurement yields not an ordered pair of numbers (as in the distinguishable particle case) but just a pair of numbers: to assign them to the particles in a definite way is to go beyond what is physically meaningful in quantum theory. What our measurement *does* permit us to conclude is that the state vector is an eigenstate of X_1+X_2 with eigenvalue $a+b$, the sum of the eigenvalue being insensitive to how the values a and b are assigned to the particles. In other words, given an *unordered* pair of numbers a and b we can still define a unique sum (but not difference). Now, there are just two product vectors, $|ab\rangle$ and $|ba\rangle$ with this eigenvalue, and the state vector lies somewhere in the two-dimensional degenerate (with respect to X_1+X_2) eigenspace spanned by them. Let $|\psi(a, b)\rangle=\beta|ab\rangle+\gamma|ba\rangle$ be the allowed vector. If we impose the constraint Eq. (10.3.2):

$$\beta|ab\rangle+\gamma|ba\rangle=\alpha[\beta|ba\rangle+\gamma|ab\rangle]$$

we find, upon equating the coefficients of $|ab\rangle$ and $|ba\rangle$ that

$$\beta=\alpha\gamma, \qquad \gamma=\alpha\beta$$

‡ We are assuming $a\neq b$. If $a=b$, the state is acceptable, but the choice we are agonizing over does not arise.

so that

$$\alpha = \pm 1 \tag{10.3.3}$$

It is now easy to construct the allowed state vectors. They are

$$|ab, S\rangle = |ab\rangle + |ba\rangle \tag{10.3.4}$$

called the *symmetric* state vector ($\alpha = 1$) and

$$|ab, A\rangle = |ab\rangle - |ba\rangle \tag{10.3.5}$$

called the *antisymmetric* state vector ($\alpha = -1$). (These are *unnormalized* vectors. Their normalization will be taken up shortly.)

More generally, if some variable Ω is measured and the values ω_1 and ω_2 are obtained, the state vector immediately following the measurement is either $|\omega_1\omega_2, S\rangle$ or $|\omega_1\omega_2, A\rangle$.‡ Although we have made a lot of progress in nailing down the state vector corresponding to the measurement, we have still to find a way to choose between these two alternatives.

Bosons and Fermions

Although both S and A states seem physically acceptable (in that they respect the indistinguishability of the particles) we can go a step further and make the following assertion:

A given species of particles must choose once and for all between S and A states.

Suppose the contrary were true, and the Hilbert space of two identical particles contained both S and A vectors. Then the space also contains linear combinations such as

$$|\psi\rangle = \alpha|\omega_1\omega_2, S\rangle + \beta|\omega_1'\omega_2', A\rangle$$

which are neither symmetric nor antisymmetric. So we rule out this possibility.

Nature seems to respect the constraints we have deduced. Particles such as the pion, photon, and graviton are *always* found in symmetric states and are called *bosons*, and particles such as the electron, proton, and neutron are *always* found in antisymmetric states and are called *fermions*.

Thus if we catch two identical bosons, one at $x = a$ and the other at $x = b$, the state vector immediately following the measurement is

$$\begin{aligned} |\psi\rangle &= |x_1 = a, x_2 = b\rangle + |x_1 = b, x_2 = a\rangle \\ &= |ab\rangle + |ba\rangle = |ab, S\rangle \end{aligned}$$

‡ We are assuming Ω is nondegenerate. If not, let ω represent the eigenvalues of a complete set of commuting operators.

Had the particles been fermions, the state vector after the measurement would have been

$$|\psi\rangle = |x_1 = a, x_2 = b\rangle - |x_1 = b, x_2 = a\rangle = |ab\rangle - |ba\rangle$$
$$= |ab, A\rangle$$

Note that although we still use the labels x_1 and x_2, we do not attach them to the particles in any particular way. Thus having caught the bosons at $x = a$ and $x = b$, we need not agonize over whether $x_1 = a$ and $x_2 = b$ or vice versa. Either choice leads to the same $|\psi\rangle$ for bosons, and to state vectors differing only by an overall sign for fermions.

We are now in a position to deduce a fundamental property of fermions, which results from the antisymmetry of their state vectors. Consider a two-fermion state

$$|\omega_1\omega_2, A\rangle = |\omega_1\omega_2\rangle - |\omega_2\omega_1\rangle$$

Let us now set $\omega_1 = \omega_2 = \omega$. We find

$$|\omega\omega, A\rangle = |\omega\omega\rangle - |\omega\omega\rangle = 0 \tag{10.3.6}$$

This is the celebrated *Pauli exclusion principle: Two identical fermions cannot be in the same quantum state.* This principle has profound consequences—in statistical mechanics, in understanding chemical properties of atoms, in nuclear theory, astrophysics, etc. We will have occasion to return to it often.

With this important derivation out of our way, let us address a question that may have plagued you: our analysis has only told us that a given type of particle, say a pion, has to be either a boson or a fermion, but does not say which one. There are two ways to the answer. The first is by further cerebration, to be specific, within the framework of quantum field theory, which relates the spin of the particle to its "statistics"—which is the term physicists use to refer to its bosonic or fermionic nature. Since the relevant arguments are beyond the scope of this text I merely quote the results here. Recall that the spin of the particle is its internal angular momentum. The *magnitude* of spin happens to be an invariant for a particle (and thus serves as a label, like its mass or charge) and can have only one of the following values: 0, $\hbar/2$, $\hbar$, $3\hbar/2$, $2\hbar$, ... The spin statistics theorem, provable in quantum field theory, asserts that particles with (magnitude of spin) equal to an even multiple of $\hbar/2$ are bosons, and those with spin equal to an odd multiple of $\hbar/2$ are fermions. However, this connection, proven in three dimensions, does not apply to one dimension, where it is not possible to define spin or any form of angular momentum. (This should be clear classically.) Thus the only way to find if a particle in one dimension is a boson or fermion is to determine the symmetry of the wave function experimentally. This is the second method, to be discussed in a moment.

Before going on to this second method, let us note that the requirement that the state vector of two identical particles be symmetric or antisymmetric (under the exchange of the quantum numbers labeling them) applies in three dimensions as well, as will be clear by going through the arguments in one dimension. The only difference will be the increase in the number of labels. For example, the position eigenket of a spin-zero boson will be labeled by three numbers x, y, and z. For fermions, which have spin at least equal to $\hbar/2$, the states will be labeled by the orientation of the spin as well as the orbital labels that describe spinless bosons.‡ We shall consider just spin-$\frac{1}{2}$ particles, for which this label can take only two values, call them + and − or spin up and down (the meaning of these terms will be clear later). If we denote by ω all the orbital labels and by s the spin label, the state vector of the fermion that is antisymmetric under the exchange of the particles, i.e., under the exchange of all the labels, will be of the form

$$|\omega_1 s_1, \omega_2 s_2, A\rangle = |\omega_1 s_1, \omega_2 s_2\rangle - |\omega_2 s_2, \omega_1 s_1\rangle \tag{10.3.7}$$

We see that the state vector vanishes if

$$\omega_1 = \omega_2 \qquad \text{and} \qquad s_1 = s_2 \tag{10.3.8}$$

Thus we find once again that two fermions cannot be in the same quantum state, but we mean by a quantum state a state of definite ω and s. Thus two electrons can be in the same orbital state if their spin orientations are different.

We now turn to the second way of finding the statistics of a given species of particles, the method that works in one or three dimensions, because it appeals to a simple experiment which determines whether the two-particle state vector is symmetric or antisymmetric for the given species. As a prelude to the discussion of such an experiment, let us study in some detail the Hilbert space of bosons and fermions.

Bosonic and Fermionic Hilbert Spaces

We have seen that two identical bosons will always have symmetric state vectors and two identical fermions will always have antisymmetric state vectors. Let us call the Hilbert space of symmetric bosonic vectors $\mathbb{V}_S$ and the Hilbert space of the antisymmetric fermionic vectors $\mathbb{V}_A$. We first examine the relation between these two spaces on the one hand and the direct product space $\mathbb{V}_{1\otimes 2}$ on the other.

The space $\mathbb{V}_{1\otimes 2}$ consists of all vectors of the form $|\omega_1\omega_2\rangle = |\omega_1\rangle \otimes |\omega_2\rangle$. To each pair of vectors $|\omega_1 = a, \omega_2 = b\rangle$ and $|\omega_1 = b, \omega_2 = a\rangle$ there is one (unnormalized) bosonic vector $|\omega_1 = a, \omega_2 = b\rangle + |\omega_1 = b, \omega_2 = a\rangle$ and one fermionic vector $|\omega_1 = a, \omega_2 = b\rangle - |\omega_1 = a, \omega_2 = b\rangle$. If $a = b$; the vector $|\omega_1 = a, \omega_2 = a\rangle$ is already symmetric and we may take it to be the bosonic vector. There is no corresponding fermionic vector (the Pauli principle). Thus $\mathbb{V}_{1\otimes 2}$ has just enough basis vectors to form one bosonic Hilbert space and one fermionic Hilbert space. We express this relation as

$$\mathbb{V}_{1\otimes 2} = \mathbb{V}_S \oplus \mathbb{V}_A \tag{10.3.9}$$

‡ Since spin has no classical counterpart, the operator representing it is not a function of the coordinate and momentum operators and it commutes with any orbital operator Ω. Thus spin may be specified simultaneously with the orbital variables.

with $\mathbb{V}_S$ getting slightly more than half the dimensionality of $\mathbb{V}_{1\otimes 2}$.‡ Our analysis has shown that at any given time, the state of two bosons is an element of $\mathbb{V}_S$ and that of two fermions an element of $\mathbb{V}_A$. It can also be shown that a system that starts out in $\mathbb{V}_S(\mathbb{V}_A)$ remains in $\mathbb{V}_S(\mathbb{V}_A)$ (see Exercise 10.3.5). *Thus in studying two identical particles we need only consider* $\mathbb{V}_S$ *or* $\mathbb{V}_A$. It is however convenient, for bookkeeping purposes, to view $\mathbb{V}_S$ and $\mathbb{V}_A$ as subspaces of $\mathbb{V}_{1\otimes 2}$ and the elements of $\mathbb{V}_S$ or $\mathbb{V}_A$ as elements also of $\mathbb{V}_{1\otimes 2}$.

Let us now consider the normalization of the vectors in $\mathbb{V}_S$. Consider first the eigenkets $|\omega_1\omega_2, S\rangle$ corresponding to a variable Ω with discrete eigenvalues. The unnormalized state vector is

$$|\omega_1\omega_2, S\rangle = |\omega_1\omega_2\rangle + |\omega_2\omega_1\rangle$$

Since $|\omega_1\omega_2\rangle$ and $|\omega_2\omega_1\rangle$ are orthonormal states in $\mathbb{V}_{1\otimes 2}$, the normalization factor is just $2^{-1/2}$, i.e.,

$$|\omega_1\omega_2, S\rangle = 2^{-1/2}[|\omega_1\omega_2\rangle + |\omega_2\omega_1\rangle] \tag{10.3.10a}$$

is a normalized eigenvector. You may readily check that $\langle\omega_1\omega_2, S|\omega_1\omega_2, S\rangle = 1$. The preceding discussion assumes $\omega_1 \neq \omega_2$. If $\omega_1 = \omega_2 = \omega$ the product ket $|\omega\omega\rangle$ is itself both symmetric and normalized and we choose

$$|\omega\omega, S\rangle = |\omega\omega\rangle \tag{10.3.10b}$$

Any vector $|\psi_S\rangle$ in $\mathbb{V}_S$ may be expanded in terms of this Ω basis. As usual we identify

$$P_S(\omega_1, \omega_2) = |\langle\omega_1\omega_2, S|\psi_S\rangle|^2 \tag{10.3.11}$$

as the *absolute* probability of finding the particles in state $|\omega_1\omega_2, S\rangle$ when an Ω measurement is made on a system in state $|\psi_S\rangle$. The normalization condition of $|\psi_S\rangle$ and $P_S(\omega_1, \omega_2)$ may be written as

$$\begin{aligned} 1 = \langle\psi_S|\psi_S\rangle &= \sum_{\text{dist}} |\langle\omega_1\omega_2, S|\psi_S\rangle|^2 \\ &= \sum_{\text{dist}} P_S(\omega_1, \omega_2) \end{aligned} \tag{10.3.12a}$$

where $\sum_{\text{dist}}$ denotes a sum over all physically *distinct* states. If ω_1 and ω_2 take values between ω_{min} and ω_{max}, then

$$\sum_{\text{dist}} = \sum_{\omega_2=\omega_{\text{min}}}^{\omega_{\text{max}}} \sum_{\omega_1=\omega_{\text{min}}}^{\omega_2} \tag{10.3.12b}$$

In this manner we avoid counting *both* $|\omega_1\omega_2, S\rangle$ *and* $|\omega_2\omega_1, S\rangle$, which are physically equivalent. Another way is to count them both and then divide by 2.

‡ Since every element of $\mathbb{V}_S$ is perpendicular to every element of $\mathbb{V}_A$ (you should check this) the dimensionality of $\mathbb{V}_{1\otimes 2}$ equals the sum of the dimensionalities of $\mathbb{V}_S$ and $\mathbb{V}_A$.

What if we want the absolute probability density for some continuous variable such as X? In this case we must take the projection of $|\psi_S\rangle$ on the normalized position eigenket:

$$|x_1x_2, S\rangle = 2^{-1/2}[|x_1x_2\rangle + |x_2x_1\rangle] \tag{10.3.13}$$

to obtain

$$P_S(x_1, x_2) = |\langle x_1x_2, S|\psi_S\rangle|^2 \tag{10.3.14}$$

The normalization condition for $P_S(x_1, x_2)$ and $|\psi_S\rangle$ is

$$1 = \iint P_S(x_1, x_2)\,\frac{dx_1\,dx_2}{2} = \iint |\langle x_1x_2, S|\psi_S\rangle|^2\,\frac{dx_1\,dx_2}{2} \tag{10.3.15}$$

where the factor 1/2 makes up for the double counting done by the $dx_1\,dx_2$ integration.‡ In this case it is convenient to define the wave function as

$$\psi_S(x_1, x_2) = 2^{-1/2}\langle x_1x_2, S|\psi_S\rangle \tag{10.3.16}$$

so that the normalization of ψ_S is

$$1 = \iint |\psi_S(x_1, x_2)|^2\,dx_1\,dx_2 \tag{10.3.17}$$

However, in this case

$$P_S(x_1, x_2) = 2|\psi_S(x_1, x_2)|^2 \tag{10.3.18}$$

due to the rescaling. Now, note that

$$\begin{aligned}\psi_S(x_1, x_2) &= \frac{1}{2^{1/2}}\langle x_1x_2, S|\psi_S\rangle = \frac{1}{2}[\langle x_1x_2|\psi_S\rangle + \langle x_2x_1|\psi_S\rangle] \\ &= \langle x_1x_2|\psi_S\rangle \end{aligned} \tag{10.3.19}$$

where we have exploited the fact that $|\psi_S\rangle$ is symmetrized between the particles and has the same inner product with $\langle x_1x_2|$ and $\langle x_2x_1|$. Consequently, the normalization

‡ The points $x_1 = x_2 = x$ pose some subtle questions both with respect to the factor 1/2 and the normalization of the kets $|xx, S\rangle$. We do not get into these since the points on the line $x_1 = x_2 = x$ make only an infinitesimal contribution to the integration in the $x_1 - x_2$ plane (of any smooth function). In the following discussion you may assume that quantities such as $P_S(x, x)$, $\psi_S(x, x)$ are all given by the limits $x_1 \to x_2 \to x$ of $P_S(x_1, x_2)$, $\psi_S(x_1, x_2)$, etc.

condition Eq. (10.3.17) becomes

$$1=\langle\psi_S|\psi_S\rangle=\iint|\psi_S|^2\,dx_1\,dx_2=\iint\langle\psi_S|x_1x_2\rangle\langle x_1x_2|\psi_S\rangle\,dx_1\,dx_2$$

which makes sense, as $|\psi_S\rangle$ is an element of $\mathbb{V}_{1\otimes 2}$ as well. Note, however, that the kets $|x_1x_2\rangle$ enter the definition of the wave function Eq. (10.3.19), and the normalization integral above, only as bookkeeping devices. They are not elements of $\mathbb{V}_S$ and the inner product $\langle x_1x_2|\psi\rangle$ would be of no interest to us, were it not for the fact that the quantity that *is* of physical interest $\langle x_1x_2, S|\psi_S\rangle$, is related to it by just a scale factor of $2^{1/2}$. Let us now consider a concrete example. We measure the energy of two noninteracting bosons in a box extending from $x=0$ to $x=L$ and find them to be in the quantum states $n=3$ and $n=4$. The normalized state vector just after measurement is then

$$|\psi_S\rangle=\frac{|3,4\rangle+|4,3\rangle}{2^{1/2}} \tag{10.3.20}$$

in obvious notation. The wave function is

$$\begin{aligned}\psi_S(x_1,x_2)&=2^{-1/2}\langle x_1x_2, S|\psi_S\rangle\\&=\frac{1}{2}(\langle x_1x_2|+\langle x_2x_1|)\left(\frac{|3,4\rangle+|4,3\rangle}{2^{1/2}}\right)\\&=\frac{1}{2(2^{1/2})}[\langle x_1x_2|3,4\rangle+\langle x_1x_2|4,3\rangle+\langle x_2x_1|3,4\rangle+\langle x_2x_1|4,3\rangle]\\&=\frac{1}{2(2^{1/2})}[\psi_3(x_1)\psi_4(x_2)+\psi_4(x_1)\psi_3(x_2)+\psi_3(x_2)\psi_4(x_1)\\&\quad+\psi_4(x_2)\psi_3(x_1)]\\&=2^{-1/2}[\psi_3(x_1)\psi_4(x_2)+\psi_4(x_1)\psi_3(x_2)]\\&=\langle x_1x_2|\psi_S\rangle\end{aligned} \tag{10.3.21a}$$

where in all of the above,

$$\psi_n(x)=\left(\frac{2}{L}\right)^{1/2}\sin\left(\frac{n\pi x}{L}\right) \tag{10.3.21b}$$

These considerations apply with obvious modifications to the fermionic space $\mathbb{V}_A$. The basis vectors are of the form

$$|\omega_1\omega_2, A\rangle=2^{-1/2}[|\omega_1\omega_2\rangle-|\omega_2\omega_1\rangle] \tag{10.3.22}$$

(The case $\omega_1=\omega_2$ does not arise here.) The wave function is once again

$$\begin{aligned}\psi_A(x_1, x_2) &= 2^{-1/2}\langle x_1 x_2, A|\psi_A\rangle \\ &= \langle x_1 x_2|\psi_A\rangle\end{aligned} \tag{10.3.23}$$

and as in the bosonic case

$$P_A(x_1, x_2) = 2|\psi_A(x_1, x_2)|^2 \tag{10.3.24}$$

The normalization condition is

$$1 = \iint P_A(x_1, x_2)\frac{dx_1\, dx_2}{2} = \iint |\psi_A(x_1, x_2)|^2\, dx_1\, dx_2 \tag{10.3.25}$$

Returning to our example of two particles in a box, if we had obtained the values $n=3$ and $n=4$, then the state just after measurement would have been

$$|\psi_A\rangle = \frac{|3, 4\rangle - |4, 3\rangle}{2^{1/2}} \tag{10.3.26}$$

(We may equally well choose

$$|\psi_A\rangle = \frac{|4, 3\rangle - |3, 4\rangle}{2^{1/2}}$$

which makes no physical difference). The corresponding wave function may be written in the form of a determinant:

$$\begin{aligned}\psi_A(x_1, x_2) = \langle x_1 x_2|\psi_A\rangle &= 2^{-1/2}[\psi_3(x_1)\psi_4(x_2) - \psi_4(x_1)\psi_3(x_2)] \\ &= 2^{-1/2}\begin{vmatrix}\psi_3(x_1) & \psi_4(x_1) \\ \psi_3(x_2) & \psi_4(x_2)\end{vmatrix}\end{aligned} \tag{10.3.27}$$

Had we been considering the state $|\omega_1\omega_2, A\rangle$ [Eq. (10.3.22)],‡

$$\psi_A(x_1, x_2) = 2^{-1/2}\begin{vmatrix}\psi_{\omega_1}(x_1) & \psi_{\omega_2}(x_1) \\ \psi_{\omega_1}(x_2) & \psi_{\omega_2}(x_2)\end{vmatrix} \tag{10.3.28}$$

Determination of Particle Statistics

We are finally ready to answer the old question: how does one determine empirically the statistics of a given species, i.e., whether it is a boson or fermion, without turning to the spin statistics theorem? For concreteness, let us say we have two identical noninteracting pions and wish to find out if they are bosons or fermions.

‡ The determinantal form of ψ_A makes it clear that ψ_A vanishes if $x_1=x_2$ or $\omega_1=\omega_2$.

We proceed as follows. We put them in a one-dimensional box‡ and make an energy measurement. Say we find one in the state $n=3$ and the other in the state $n=4$. The probability distribution in x space would be, depending on their statistics,

$$\begin{aligned}P_{S/A}(x_1, x_2) &= 2|\psi_{S/A}(x_1, x_2)|^2 \\ &= 2|2^{-1/2}[\psi_3(x_1)\psi_4(x_2) \pm \psi_4(x_1)\psi_3(x_2)]|^2 \\ &= |\psi_3(x_1)|^2|\psi_4(x_2)|^2 + |\psi_4(x_1)|^2|\psi_3(x_2)|^2 \\ &\quad \pm [\psi_3^*(x_1)\psi_4(x_1)\psi_4^*(x_2)\psi_3(x_2) + \psi_4^*(x_1)\psi_3(x_1)\psi_3^*(x_2)\psi_4(x_2)] \end{aligned} \quad (10.3.29)$$

Compare this situation with two particles carrying labels 1 and 2, but otherwise identical,§ with particle 1 in state 3 and described by a probability distribution $|\psi_3(x)|^2$, and particle 2 in state 4 and described by the probability distribution $|\psi_4(x)|^2$. In this case, the first term represents the probability that particle 1 is at x_1 and particle 2 is at x_2, while the second gives the probability for the exchanged event. The sum of these two terms then gives $P_D(x_1, x_2)$, the probability for finding one at x_1 and the other at x_2, with no regard paid to their labels. (The subscript D denotes distinguishable.) The next two terms, called interference terms, remind us that there is more to identical particles in quantum theory than just their identical characteristics: they have no separate identities. Had they separate identities (as in the classical case) and we were just indifferent to which one arrives at x_1 and which one at x_2, we would get just the first two terms. There is a parallel between this situation and the double-slit experiment, where the probabilities for finding a particle at a given point x on the screen with both slits open was not the sum of the probabilities with either slit open. In both cases, the interference terms arise, because in quantum theory, when an event can take place in two (or more) indistinguishable ways, we add the corresponding amplitudes and not the corresponding probabilities.

Just as we were not allowed then to assign a definite trajectory to the particle (through slits 1 or 2), we are not allowed now to assign definite labels to the two particles.

The interference terms tell us if the pions are bosons or fermions. The difference between the two cases is most dramatic as $x_1 \to x_2 \to x$:

$$P_A(x_1 \to x, x_2 \to x) \to 0 \text{ (Pauli principle applied to state } |x\rangle) \quad (10.3.30)$$

whereas

$$P_S(x_1 \to x, x_2 \to x) = 2[|\psi_3(x)|^2|\psi_4(x)|^2 + |\psi_4(x)|^2|\psi_3(x)|^2] \quad (10.3.31)$$

which is twice as big as $P_D(x_1 \to x, x_2 \to x)$, the probability density for two distinct label carrying (but otherwise identical) particles, whose labels are disregarded in the position measurement.

One refers to the tendency of fermions to avoid each other (i.e., avoid the state $x_1 = x_2 = x$) as obeying "Fermi–Dirac statistics" and the tendency of bosons to

‡ We do this to simplify the argument. The basic idea works just as well in three dimensions.
§ The label can, for example, be the electric charge.

conglomerate as "obeying Bose–Einstein statistics," after the physicists who first explored the consequences of the antisymmetrization and symmetrization requirements on the statistical mechanics of an ensemble of fermions and bosons, respectively. (This is the reason for referring to the bosonic/fermionic nature of a particle as its statistics.)

Given the striking difference in the two distributions, we can readily imagine deciding (once and for all) whether pions are bosons or fermions by preparing an ensemble of systems (with particles in $n=3$ and 4) and measuring $P(x_1, x_2)$.

Note that $P(x_1, x_2)$ helps us decide not only whether the particles are bosons or fermions, but also whether they are identical in the first place. In other words, if particles that we think are identical differ with respect to some label that we are not aware of, the nature of the interference term will betray this fact. Imagine, for example, two bosons, call them K and $\bar{K}$, which are identical with respect to mass and charge, but different with respect to a quantum number called "hypercharge." Let us assume we are ignorant of hypercharge. In preparing an ensemble that we think contains N identical pairs, we will actually be including some (K, K) pairs, some $(\bar{K}, \bar{K})$ pairs. If we now make measurements on the ensemble and extract the distribution $P(x_1, x_2)$ (once again ignoring the hypercharge), we will find the interference term has the + sign but is not as big as it should be. If the ensemble contained only identical bosons, $P(x, x)$ should be twice as big as $P_D(x, x)$, which describes label-carrying particles; if we get a ratio less than 2, we know the ensemble is contaminated by label-carrying particles which produce no interference terms.

From the above discussions, it is also clear that one cannot hastily conclude, upon catching two electrons in the same orbital state in three dimensions that they are not fermions. In this case, the label we are ignoring is the spin orientation s. As mentioned earlier on, s can have only two values, call them + and −. If we assume that s never changes (during the course of the experiment) it can serve as a particle label that survives with time. If $s=+$ for one electron and − for the other, they are like two distinct particles and *can* be in the same orbital state. The safe thing to do here is once again to work with an ensemble rather than an isolated measurement. Since we are ignorant of spin, our ensemble will contain (+, +) pairs, (−, −) pairs, and (+, −) pairs. The (+, +) and (−, −) pairs are identical fermions and will produce a negative interference term, while the (+, −) pairs will not. Thus we will find $P(\mathbf{r}, \mathbf{r})$ is smaller than $P_D(\mathbf{r}, \mathbf{r})$ describing labeled particles, but not zero. This will tell us that our ensemble has identical fermion pairs contaminated by pairs of distinguishable particles. It will then be up to us to find the nature of the hidden degree of freedom which provides the distinction.

Systems of N Identical Particles

The case $N=2$ lacks one feature that is found at larger N. We illustrate it by considering the case of three identical particles in a box. Let us say that an energy measurement shows the quantum numbers of the particles to be n_1, n_2, and n_3. Since the particles are identical, all we can conclude from this observation is that the total energy is

$$E=\left(\frac{\hbar^2\pi^2}{2mL^2}\right)(n_1^2+n_2^2+n_3^2)$$

Now there are $3! = $ six product states with this energy: $|n_1n_2n_3\rangle$, $|n_1n_3n_2\rangle$, $|n_2n_3n_1\rangle$, $|n_2n_1n_3\rangle$, $|n_3n_2n_1\rangle$, and $|n_3n_1n_2\rangle$. The physical states are elements of the six-dimensional eigenspace spanned by these vectors and distinguished by the property that under the exchange of *any* two particle labels, the state vector changes only by a factor α. Since double exchange of the same two labels is equivalent to no exchange, we conclude as before that $\alpha = \pm 1$. There are only two states with this property:

$$|n_1n_2n_3, S\rangle = \frac{1}{(3!)^{1/2}}[|n_1n_2n_3\rangle + |n_1n_3n_2\rangle + |n_2n_3n_1\rangle + |n_2n_1n_3\rangle + |n_3n_2n_1\rangle + |n_3n_1n_2\rangle] \tag{10.3.32}$$

called the totally symmetric state,‡ for which $\alpha = +1$ for all three possible exchanges $(1 \leftrightarrow 2,\ 2 \leftrightarrow 3,\ 1 \leftrightarrow 3)$; and

$$|n_1n_2n_3, A\rangle = \frac{1}{(3!)^{1/2}}[|n_1n_2n_3\rangle - |n_1n_3n_2\rangle + |n_2n_3n_1\rangle - |n_2n_1n_3\rangle + |n_3n_1n_2\rangle - |n_3n_2n_1\rangle] \tag{10.3.33}$$

called the totally antisymmetric state, for which $\alpha = -1$ for all three possible exchanges.

Bosons will always pick the S states and fermions, the A states. It follows that no two fermions can be in the same state.

As in the $N = 2$ case, the wave function in the X basis is

$$\psi_{S/A}(x_1, x_2, x_3) = (3!)^{-1/2}\langle x_1x_2x_3, S/A|\psi_{S/A}\rangle = \langle x_1x_2x_3|\psi_{S/A}\rangle \tag{10.3.34}$$

and

$$\int_{-\infty}^{\infty}\int_{-\infty}^{\infty}\int_{-\infty}^{\infty} |\psi_{S/A}|^2\, dx_1\, dx_2\, dx_3 = 1$$

For instance, the wave function associated with $|n_1n_2n_3, S/A\rangle$, Eqs. (10.3.33) and (10.3.34), is

$$\begin{aligned}\psi_{n_1n_2n_3}(x_1, x_2, x_3, S/A) &= (3!)^{-1/2}[\psi_{n_1}(x_1)\psi_{n_2}(x_2)\psi_{n_3}(x_3) \pm \psi_{n_1}(x_1)\psi_{n_3}(x_2)\psi_{n_2}(x_3) \\ &\quad + \psi_{n_2}(x_1)\psi_{n_3}(x_2)\psi_{n_1}(x_3) \pm \psi_{n_2}(x_1)\psi_{n_1}(x_2)\psi_{n_3}(x_3) \\ &\quad + \psi_{n_3}(x_1)\psi_{n_1}(x_2)\psi_{n_2}(x_3) \pm \psi_{n_3}(x_1)\psi_{n_2}(x_2)\psi_{n_1}(x_3)]\end{aligned} \tag{10.3.35}$$

‡ The normalization factor $(3!)^{-1/2}$ is correct only if all three n's are different. If, for example, $n_1 = n_2 = n_3 = n$, then the product state $|nnn\rangle$ is normalized and symmetric and can be used as the S state. A similar question does not arise for the fermion state due to the Pauli principle.

The fermion wave function may once again be written as a determinant:

$$\psi_{n_1n_2n_3}(x_1, x_2, x_3, A) = \frac{1}{(3!)^{1/2}} \begin{vmatrix} \psi_{n_1}(x_1) & \psi_{n_2}(x_1) & \psi_{n_3}(x_1) \\ \psi_{n_1}(x_2) & \psi_{n_2}(x_2) & \psi_{n_3}(x_2) \\ \psi_{n_1}(x_3) & \psi_{n_2}(x_3) & \psi_{n_3}(x_3) \end{vmatrix} \tag{10.3.36}$$

Using the properties of the determinant, one easily sees that ψ vanishes if two of the x's or n's coincide. All these results generalize directly to any higher N.

Two questions may bother you at this point.

Question I. Consider the case $N=3$. There are three possible exchanges here: $(1 \leftrightarrow 2)$, $(1 \leftrightarrow 3)$, and $(2 \leftrightarrow 3)$. The S states pick up a factor $\alpha=+1$ for all three exchanges, while the A states pick up $\alpha=-1$ for all three exchanges. What about states for which some of the α's are $+1$ and the others -1? Such states do not exist. You may verify this by exhaustion: take the 3! product vectors and try to form such a linear combination. Since a general proof for this case and all N involves group theory, we will not discuss it here. Note that since we get only two acceptable vectors for every $N!$ product vectors, the direct product space for $N \geq 3$ is bigger (in dimensionality) than $\mathbb{V}_S \oplus \mathbb{V}_A$.

Question II. We have tacitly assumed that if *two* identical particles of a given species always pick the S (or A) state, so will three or more, i.e., we have extended our definition of bosons and fermions from $N=2$ to all N. What if two pions always pick the S state while *three* always pick the A state? While intuition revolts at such a possibility, it still needs to be formally ruled out. We do so at the end of the next subsection.

When Can We Ignore Symmetrization and Antisymmetrization?

A basic assumption physicists make before they can make any headway is that they can single out some part of the universe (the system) and study it in isolation from the rest. While no system is truly isolated, one can often get close to this ideal. For instance, when we study the oscillations of a mass coupled to a spring, we ignore the gravitational pull of Pluto.

Classically, the isolation of the system is expressed by the separability of the Hamiltonian of the universe:

$$\mathscr{H}_{\text{universe}} = \mathscr{H}_{\text{sys}} + \mathscr{H}_{\text{rest}} \tag{10.3.37}$$

where $\mathscr{H}_{\text{sys}}$ is a function of the system coordinates and momenta alone. It follows that the time evolution of the system's p's and q's are independent of what is going on in the rest of the universe. In our example, this separability is ruined (to give just one example) by the gravitational interaction between the mass and Pluto, which depends on their relative separation. If we neglect this absurdly small effect (and other such effects) we obtain separability to an excellent approximation.

Quantum mechanically, separability of H leads to the factorization of the wave function of the universe:

$$\psi_{\text{universe}} = \psi_{\text{sys}} \cdot \psi_{\text{rest}} \tag{10.3.38}$$

where ψ_{sys} is a function only of system coordinates, collectively referred to as x_s. Thus if we want the probability that the system has a certain coordinate x_s, and do not care about the rest, we find (symbolically)

$$\begin{aligned} P(x_s) &= \int |\psi_{\text{universe}}(x_s, x_{\text{rest}})|^2 \, dx_{\text{rest}} \\ &= |\psi_{\text{sys}}(x_s)|^2 \int |\psi(x_{\text{rest}})|^2 \, dx_{\text{rest}} \\ &= |\psi_{\text{sys}}(x_s)|^2 \end{aligned} \tag{10.3.39}$$

We could have obtained this result by simply ignoring ψ_{rest} from the outset.

Things get complicated when the system and the "rest" contain identical particles. Even if there is no interaction between the system and the rest, i.e., the Hamiltonian is separable, product states are not allowed and only S or A states must be used. Once the state vector fails to factorize, we will no longer have

$$P(x_s, x_{\text{rest}}) = P(x_s) P(x_{\text{rest}}) \tag{10.3.40}$$

(i.e., the systems will not be statistically independent), and we can not integrate out $P(x_{\text{rest}})$ and regain $P(x_s)$.

Now it seems reasonable that at least in certain cases it should be possible to get away with the product state and ignore the symmetrization or antisymmetrization conditions.

Suppose, for example, that at $t=0$, we find one pion in the ground state of an oscillator potential centered around a point on earth and another pion in the same state, but on the moon. It seems reasonable that we can give the particles the labels "earth pion" and "moon pion," which will survive with time. Although we cannot follow their trajectories, we can follow their wave functions: we know the first wave function is a Gaussian $G_E(x_E)$ centered at a point in the lab on earth and that the second is a Gaussian $G_M(x_M)$ centered at a point on the moon. If we catch a pion somewhere on earth at time t, the theory tells us that it is almost certainly the "earth pion" and that the chances of its being the "moon pion" are absurdly small. Thus the uncertainty in the position of each pion is compensated by a separation that is much larger. (Even in classical mechanics, it is not necessary to know the trajectories exactly to follow the particles; the band of uncertainty about each trajectory has merely to be much thinner than the minimum separation between the particles during their encounter.) We therefore believe that if we assumed

$$\psi(x_E, x_M) = G_E(x_E) G_M(x_M) \tag{10.3.41}$$

we should be making an error that is as negligible as is the chance of finding the earth pion on the moon and vice versa. Given this product form, the person on earth can compute the probability for finding the earth pion at some x by integrating out the moon pion:

$$P(x_E)=|G_E(x_E)|^2\int|G_M(x_M)|^2\,dx_M$$

$$=|G_E(x_E)|^2 \tag{10.3.42}$$

Likewise the person on the moon, who does not care about (i.e., sums over) the earth pion will obtain

$$P(x_M)=|G_M(x_M)|^2 \tag{10.3.43}$$

Let us now verify that if we took a properly symmetrized wave function it leads to essentially the same predictions (with negligible differences).

Let us start with

$$\psi_S(x_1, x_2)=2^{-1/2}[G_E(x_1)G_M(x_2)+G_M(x_1)G_E(x_2)] \tag{10.3.44}$$

We use the labels x_1 and x_2 rather than x_E and x_M to emphasize that the pions are indeed being treated as indistinguishable. Now, the probability (density) of finding one particle near x_1 and one near x_2 is

$$\begin{aligned}P(x_1, x_2)=2|\psi|^2&=|G_E(x_1)|^2|G_M(x_2)|^2+|G_M(x_1)|^2|G_E(x_2)|^2\\&+G_E^*(x_1)G_M(x_1)G_M^*(x_2)G_E(x_2)\\&+G_M^*(x_1)G_E(x_1)G_E^*(x_2)G_M(x_2)\end{aligned} \tag{10.3.45}$$

Let us ask for the probability of finding one particle near some point x_E on the earth, with no regard to the other. This is given by setting *either* one of the variables (say x_1) equal to x_E and integrating out the other [since $P(x_1, x_2)=P(x_2, x_1)$]. There is no need to divide by 2 in doing this integration (why?). We get

$$\begin{aligned}P(x_E)=|G_E(x_E)|^2\int|G_M(x_2)|^2\,dx_2&+|G_M(x_E)|^2\int|G_E(x_2)|^2\,dx_2\\&+G_E^*(x_E)G_M(x_E)\int G_M^*(x_2)G_E(x_2)\,dx_2\\&+G_M^*(x_E)G_E(x_E)\int G_E^*(x_2)G_M(x_2)\,dx_2\end{aligned} \tag{10.3.46}$$

The first term is what we would get if we begin with a product wave function Eq. (10.3.41) and integrate out x_M. The other three terms are negligible since G_M is peaked on the moon and is utterly negligible at a point x_E on the earth. Similarly if we asked for $P(x_M)$, where x_M is a point on the moon, we will again get $|G_M(x_M)|^2$.

The labels "earth pion" and "moon pion" were useful only because the two Gaussians remained well separated for all times (being stationary states). If the two Gaussians had not been bound by the oscillating wells, and were wave packets drifting toward each other, the labeling (and the factorized wave function) would have become invalid when the Gaussians begin to have a significant overlap. The point is that at the start of any experiment, one can always assign the particles some labels. These labels acquire a physical significance only if they survive for some time. Labels like "a particle of mass m and charge $+1$" survive forever, while the longevity of a label like "earth pion" is controlled by whether or not some other pion is in the vicinity.

A dramatic illustration of this point is provided by the following example. At $t=0$ we catch two pions, one at $x=a$ and the other at $x=b$. We can give them the labels a and b since the two delta functions do not overlap even if a and b are in the same room. We may describe the initial state by a product wave function. But this labeling is quite useless, since after the passage of an infinitesimal period of time, the delta functions spread out completely: the probability distributions become constants. You may verify this by examining $|U(x, t; a, 0)|^2$ (the "fate" of the delta function)‡ or by noting that $\Delta P=\infty$ for a delta function (the particle has all possible velocities from 0 to ∞) and which, therefore, spreads out in no time.

All these considerations apply with no modification to two fermions: the two cases differ in the sign of the interference term, which is irrelevant to these considerations.

What if there are three pions, two on earth and one on the moon? Since the two on the earth (assuming that their wave functions appreciably overlap) *can* be confused with each other, we must symmetrize between them, and the total wave function will be, in obvious notation,

$$\psi(x_{E_1}, x_{E_2}, x_M)=\psi_S(x_{E_1}, x_{E_2})\cdot\psi(x_M) \tag{10.3.47}$$

The extension of this result to more particles and to fermions is obvious.

At this point the answer to Question II raised at the end of the last subsection becomes apparent. Suppose three-pion systems picked the A state while two-pion systems picked the S state. Let two of the three pions be on earth and the third one on the moon. Then, by assumption, the following function should provide an excellent approximation:

$$\psi(x_{E_1}, x_{E_2}, x_M)=\psi_A(x_{E_1}, x_{E_2})\psi(x_M) \tag{10.3.48}$$

If we integrate over the moon pion we get

$$P(x_{E_1}, x_{E_2})=2|\psi_A(x_{E_1}, x_{E_2})|^2 \tag{10.3.49}$$

We are thus led to conclude that two pions on earth will have a probability distribution corresponding to two fermions if there is a third pion on the moon and a distribution expected to two bosons if there is not a third one on the moon. Such

‡ It is being assumed that the particles are free.

absurd conclusions are averted only if the statistics depend on the species and not the number of particles.

A word of caution before we conclude this long discussion. If two particles have nonoverlapping wave functions in x space, then it is only in x space that a product wave function provides a good approximation to the exact symmetrized wave function, which in our example was

$$\psi_S(x_1, x_2) = 2^{-1/2}[G_E(x_1)G_M(x_2) + G_M(x_1)G_E(x_2)] \tag{10.3.50}$$

The formal reason is that for any choice of the arguments x_1 and x_2, only one or the other of the two terms in the right-hand side is important. (For example, if x_1 is on the earth and x_2 is on the moon, only the first piece is important.) Physically it is because the chance of finding one pion in the territory of the other is negligible and interference effects can be ignored.

If, however, we wish to switch to another basis, say the P basis, we must consider the Fourier transform of the symmetric function ψ_S and not the product, so that we end up with a symmetrized wave function in p space. The physical reason for this is that the two pions have the same momentum distributions—with $\langle P\rangle = 0$ and identical Gaussian fluctuations about this mean—since the momentum content of the oscillator is independent of its location. Consequently, there are no grounds in P space for distinguishing between them. Thus when a momentum measurement (which says nothing about the positions) yields two numbers, we cannot assign them to the pions in a unique way. Formally, symmetrization is important because the p-space wave functions of the pions overlap strongly and there exist values for the two momenta (both $\simeq 0$) for which both terms in the symmetric wave function are significant.

By the same token, if there are two particles with nonoverlapping wave functions in p space, we may describe the system by a product wave function in this space (using labels like "fast" and "slow" instead of "earth" and "moon" to distinguish between them), but not in another space where the distinction between them is absent. It should be clear that these arguments apply not just to X or P but to any arbitrary variable Ω.

*Exercise 10.3.1.** Two identical bosons are found to be in states $|\phi\rangle$ and $|\psi\rangle$. Write down the normalized state vector describing the system when $\langle\phi|\psi\rangle \neq 0$.

*Exercise 10.3.2.** When an energy measurement is made on a system of three bosons in a box, the n values obtained were 3, 3, and 4. Write down a symmetrized, normalized state vector.

*Exercise 10.3.3.** Imagine a situation in which there are three particles and only three states a, b, and c available to them. Show that the total number of allowed, distinct configurations for this system is

(1) 27 if they are labeled
(2) 10 if they are bosons
(3) 1 if they are fermions

*Exercise 10.3.4.** Two identical particles of mass m are in a one-dimensional box of length L. Energy measurement of the system yields the value $E_{sys}=\hbar^2\pi^2/mL^2$. Write down the state vector of the system. Repeat for $E_{sys}=5\hbar^2\pi^2/2mL^2$. (There are two possible vectors in this case.) You are not told if they are bosons or fermions. You may assume that the only degrees of freedom are orbital.

*Exercise 10.3.5.** Consider the *exchange operator* P_{12} whose action on the X basis is

$$P_{12}|x_1, x_2\rangle = |x_2, x_1\rangle$$

(1) Show that P_{12} has eigenvalues ± 1. (It is Hermitian and unitary.)

(2) Show that its action on the basis ket $|\omega_1, \omega_2\rangle$ is also to exchange the labels 1 and 2, and hence that $\mathbb{V}_{S/A}$ are its eigenspaces with eigenvalues ± 1.

(3) Show that $P_{12}X_1P_{12}=X_2$, $P_{12}X_2P_{12}=X_1$ and similarly for P_1 and P_2. Then show that $P_{12}\Omega(X_1, P_1; X_2, P_2)P_{12}=\Omega(X_2, P_2; X_1, P_1)$. [Consider the action on $|x_1, x_2\rangle$ or $|p_1, p_2\rangle$. As for the functions of X and P, assume they are given by power series and consider any term in the series. If you need help, peek into the discussion leading to Eq. (11.2.22).]

(4) Show that the Hamiltonian and propagator for two *identical* particles are left unaffected under $H\to P_{12}HP_{12}$ and $U\to P_{12}UP_{12}$. Given this, show that any eigenstate of P_{12} continues to remain an eigenstate with the same eigenvalue as time passes, i.e., elements of $\mathbb{V}_{S/A}$ never leave the symmetric or antisymmetric subspaces they start in.

*Exercise 10.3.6.** Consider a composite object such as the hydrogen atom. Will it behave as a boson or fermion? Argue in general that objects containing an even/odd number of fermions will behave as bosons/fermions.

11

Symmetries and Their Consequences

11.1. Overview

In Chapter 2, we explored the consequences of the symmetries of the Hamiltonian. We saw the following:

(1) If $\mathscr{H}$ is invariant under the infinitesimal canonical transformation generated by a variable $g(q, p)$, then g is conserved.

(2) Any canonical transformation that leaves $\mathscr{H}$ invariant maps solutions to the equations of motion into other solutions. Equivalently, an experiment and its transformed version will give the same result if the transformation is canonical and leaves $\mathscr{H}$ invariant.

Here we address the corresponding results in quantum mechanics.‡

11.2. Translational Invariance in Quantum Theory

Consider a single particle in one dimension. How shall we define translational invariance? Since a particle in an arbitrary state has neither a well-defined position nor a well-defined energy, we cannot define translational invariance to be the invariance of the energy under an infinitesimal shift in the particle position. Our previous experience, however, suggests that in the quantum formulation the expectation values should play the role of the classical variables. We therefore make the correspondence shown in Table 11.1.

Having agreed to formulate the problem in terms of expectation values, we still have two equivalent ways to interpret the transformations:

$$\langle X\rangle \to \langle X\rangle + \varepsilon \qquad (11.2.1a)$$

$$\langle P\rangle \to \langle P\rangle \qquad (11.2.1b)$$

‡ It may be worth refreshing your memory by going through Sections 2.7 and 2.8.

Table 11.1. Correspondence between Classical and Quantum Mechanical Concepts Related to Translational Invariance

Concept	Classical mechanics	Quantum mechanics
Translation	$x \to x+\varepsilon$	$\langle X\rangle \to \langle X\rangle + \varepsilon$
	$p \to p$	$\langle P\rangle \to \langle P\rangle$
Translational invariance	$\mathscr{H} \to \mathscr{H}$	$\langle H\rangle \to \langle H\rangle$
Conservation law	$\dot{p}=0$	$\langle \dot{P}\rangle = 0$ (anticipated)

The first is to say that under the infinitesimal translation, each state $|\psi\rangle$ gets modified into a *translated state*, $|\psi_\varepsilon\rangle$ such that

$$\langle \psi_\varepsilon|X|\psi_\varepsilon\rangle = \langle \psi|X|\psi\rangle + \varepsilon \tag{11.2.2a}$$

$$\langle \psi_\varepsilon|P|\psi_\varepsilon\rangle = \langle \psi|P|\psi\rangle \tag{11.2.2b}$$

In terms of $T(\varepsilon)$, the *translation operator*, which translates the state (and which will be constructed explicitly in a while)

$$T(\varepsilon)|\psi\rangle = |\psi_\varepsilon\rangle \tag{11.2.3}$$

Eq. (11.2.2) becomes

$$\langle \psi|T^\dagger(\varepsilon)XT(\varepsilon)|\psi\rangle = \langle \psi|X|\psi\rangle + \varepsilon \tag{11.2.4a}$$

$$\langle \psi|T^\dagger(\varepsilon)PT(\varepsilon)|\psi\rangle = \langle \psi|P|\psi\rangle \tag{11.2.4b}$$

This point of view is called the active transformation picture (in the terminology of Section 1.7) and corresponds to physically displacing the particle to the *right* by ε.

The second point of view is to say that nothing happens to the state vectors; it is the operators X and P that get modified by $T(\varepsilon)$ as follows:

$$X \to T^\dagger(\varepsilon)XT(\varepsilon)$$
$$P \to T^\dagger(\varepsilon)PT(\varepsilon)$$

such that

$$T^\dagger(\varepsilon)XT(\varepsilon) = X + \varepsilon I \tag{11.2.5a}$$

$$T^\dagger(\varepsilon)PT(\varepsilon) = P \tag{11.2.5b}$$

This is called the passive transformation picture. Physically it corresponds to moving the environment (the coordinate system, sources of external field if any, etc.) to the *left* by ε.

Physically, the equivalence of the active and passive pictures is due to the fact that moving the particle one way is equivalent to moving the environment the other way by an equal amount.

Mathematically, we show the equivalence as follows. If we sandwich the operator equation (11.2.5) between $\langle\psi|$ and $|\psi\rangle$, we get Eq. (11.2.4). To go the other way, we first rewrite Eq. (11.2.4) as

$$\langle\psi|T^{\dagger}(\varepsilon)XT(\varepsilon)-X-\varepsilon I|\psi\rangle=0$$
$$\langle\psi|T^{\dagger}(\varepsilon)PT(\varepsilon)-P|\psi\rangle=0$$

We now reason as follows:

(1) The operators being sandwiched are Hermitian (verify).
(2) Since $|\psi\rangle$ is arbitrary, we can choose it to be any of the eigenvectors of these operators. It follows that all the eigenvalues vanish.
(3) The operators themselves vanish, implying Eq. (11.2.5).

In what follows, we will examine both pictures. We will find that it is possible to construct $T(\varepsilon)$ given either of Eqs. (11.2.4) or (11.2.5), and of course that the two yield the same result. The active transformation picture is nice in that we work with the quantum state $|\psi\rangle$, which now plays the role of the classical state (x, p). The passive transformation picture is nice because the response of the quantum operators X and P to a translation is formally similar to that of their classical counterparts.‡

We begin by discussing translations in terms of active transformations. Let us examine how the ket $|\psi_\varepsilon\rangle$ is related to $|\psi\rangle$ or, equivalently, the action of the Hilbert space operator $T(\varepsilon)$. The answer appears obvious if we work with kets of definite position, $|x\rangle$. In this case it is clear that

$$T(\varepsilon)|x\rangle=|x+\varepsilon\rangle \tag{11.2.6}$$

In other words, if the particle is originally at x, it must end up at $x+\varepsilon$. Notice that $T(\varepsilon)$ is unitary: it acts on an orthonormal basis $|x\rangle$, $-\infty\le x\le\infty$, and gives another, $|x+\varepsilon\rangle$, $-\infty\le x+\varepsilon\le\infty$. Once the action of $T(\varepsilon)$ on a complete basis is known, its action on any ket $|\psi\rangle$ follows:

$$\begin{aligned}|\psi_\varepsilon\rangle=T(\varepsilon)|\psi\rangle&=T(\varepsilon)\int_{-\infty}^{\infty}|x\rangle\langle x|\psi\rangle\,dx=\int_{-\infty}^{\infty}|x+\varepsilon\rangle\langle x|\psi\rangle\,dx\\&=\int_{-\infty}^{\infty}|x'\rangle\langle x'-\varepsilon|\psi\rangle\,dx'\qquad(x'=x+\varepsilon)\end{aligned} \tag{11.2.7}$$

In other words if

$$\langle x|\psi\rangle=\psi(x)$$

‡ As we shall see, it is this point of view that best exposes many formal relations between classical and quantum mechanics.

then

$$\langle x|T(\varepsilon)|\psi\rangle=\psi(x-\varepsilon) \tag{11.2.8}$$

For example, if $\psi(x)\sim e^{-x^2}$ is a Gaussian peaked at the origin, $\psi(x-\varepsilon)\sim e^{-(x-\varepsilon)^2}$ is an identical Gaussian peaked at $x=\varepsilon$. Thus the wave function $\psi_\varepsilon(x)$ is obtained by translating (without distortion) the wave function $\psi(x)$ by an amount ε to the right. You may verify that the action of $T(\varepsilon)$ defined by Eq. (11.2.8) satisfies the condition Eq. (11.2.1a). How about the condition Eq. (11.2.1b)? It is *automatically* satisfied:

$$\begin{aligned}
\langle\psi_\varepsilon|P|\psi_\varepsilon\rangle &= \int_{-\infty}^{\infty} \psi_\varepsilon^*(x)\left(-i\hbar\frac{d}{dx}\right)\psi_\varepsilon(x)\,dx \\
&= \int_{-\infty}^{\infty} \psi^*(x-\varepsilon)\left(-i\hbar\frac{d}{dx}\right)\psi(x-\varepsilon)\,dx \\
&= \int_{-\infty}^{\infty} \psi^*(x')\left(-i\hbar\frac{d}{dx'}\right)\psi(x')\,dx' \qquad (x'=x-\varepsilon) \\
&= \langle\psi|P|\psi\rangle
\end{aligned} \tag{11.2.9}$$

Now there is something odd here. Classically, translation is specified by two *independent* relations

$$x\to x+\varepsilon$$

$$p\to p$$

while in the quantum version we seem to find that in enforcing the former (on position eigenkets), the latter automatically follows. The reason is that in our derivation we have assumed more than what was explicitly stated. We reasoned earlier, on physical grounds, that since a particle initially located at x must end up at $x+\varepsilon$, it follows that

$$T(\varepsilon)|x\rangle=|x+\varepsilon\rangle$$

While our intuition was correct, our implementation was not. As seen in chapter 7, the X basis is not unique, and the general result consistent with our intuition is not Eq. (11.2.6) but rather

$$T(\varepsilon)|x\rangle=e^{i\varepsilon g(x)/\hbar}|x+\varepsilon\rangle \tag{11.2.10}$$

(Note that as $\varepsilon\to 0$, $T(\varepsilon)|x\rangle\to|x\rangle$ as it should.) In ignoring $g(x)$, we had essentially assumed the quantum analog of $p\to p$. Let us see how. If we start with Eq. (11.2.10) instead of Eq. (11.2.6), we find that

$$\langle X\rangle \xrightarrow[T(\varepsilon)]{} \langle X\rangle+\varepsilon \tag{11.2.11a}$$

$$\langle P\rangle \to \langle P\rangle+\varepsilon\langle f(X)\rangle \tag{11.2.11b}$$

where $f=g'$. Demanding now that $\langle P\rangle\to\langle P\rangle$, we eliminate f and reduce g to a harmless constant (which can be chosen to be 0).

Exercise 11.2.1. Verify Eq. (11.2.11b)

Note that there was nothing wrong with our initial choice $T|x\rangle=|x+\varepsilon\rangle$—it was too restrictive given just the requirement $\langle X\rangle\to\langle X\rangle+\varepsilon$, but not so if we also considered $\langle P\rangle\to\langle P\rangle$. This situation reappears when we go to two or three dimensions and when we consider rotations. In all those cases we will make the analog of the naive choice $T(\varepsilon)|x\rangle=|x+\varepsilon\rangle$ to shorten the derivations.

Having defined translations, let us now define *translational invariance* in the same spirit. We define it by the requirement

$$\langle\psi|H|\psi\rangle=\langle\psi_\varepsilon|H|\psi_\varepsilon\rangle \tag{11.2.12}$$

To derive the conservation law that goes with the above equation, we must first construct the operator $T(\varepsilon)$ explicitly. Since $\varepsilon=0$ corresponds to no translation, we may expand $T(\varepsilon)$ *to order* ε as

$$T(\varepsilon)=I-\frac{i\varepsilon}{\hbar}G \tag{11.2.13}$$

The operator G, called the *generator of translations*, is Hermitian (see Exercise 11.2.2 for the proof) and is to be determined. The constant $(-i/\hbar)$ is introduced in anticipation of what is to follow.

*Exercise 11.2.2.** Using $T^\dagger(\varepsilon)T(\varepsilon)=I$ to order ε, deduce that $G^\dagger=G$.

We find G by turning to Eq. (11.2.8):

$$\langle x|T(\varepsilon)|\psi\rangle=\psi(x-\varepsilon)$$

Expanding both sides to order ε, we find

$$\langle x|I|\psi\rangle-\frac{i\varepsilon}{\hbar}\langle x|G|\psi\rangle=\psi(x)-\frac{d\psi}{dx}\varepsilon$$

so that

$$\langle x|G|\psi\rangle = -i\hbar \frac{d\psi}{dx}$$

Clearly G is the momentum operator,

$$G = P$$

and

$$T(\varepsilon) = I - \frac{i\varepsilon}{\hbar} P \tag{11.2.14}$$

We see that exactly as in classical mechanics, the momentum is the generator of (infinitesimal) translations.

The momentum conservation law now follows from translational invariance, Eq. (11.2.12), if we combine it with Eq. (11.2.14):

$$\begin{aligned}\langle\psi|H|\psi\rangle &= \langle\psi_\varepsilon|H|\psi_\varepsilon\rangle \\ &= \langle T(\varepsilon)\psi|H|T(\varepsilon)\psi\rangle = \langle\psi|T^\dagger(\varepsilon)HT(\varepsilon)|\psi\rangle \\ &= \langle\psi|\left(I + \frac{i\varepsilon}{\hbar}P\right)H\left(I - \frac{i\varepsilon}{\hbar}P\right)|\psi\rangle \\ &= \langle\psi|H|\psi\rangle + \frac{i\varepsilon}{\hbar}\langle\psi|[P, H]|\psi\rangle + O(\varepsilon^2)\end{aligned}$$

so that, we get, upon equating the coefficient of ε to zero,

$$\langle\psi|[P, H]|\psi\rangle = 0 \tag{11.2.15}$$

It now follows from Ehrenfest's theorem that

$$\langle[P, H]\rangle = 0 \rightarrow \langle\dot{P}\rangle = 0 \tag{11.2.16}$$

Translation in Terms of Passive Transformations

Let us rederive $T(\varepsilon)$, given that it acts as follows on X and P:

$$T^\dagger(\varepsilon)XT(\varepsilon) = X + \varepsilon I \tag{11.2.17a}$$

$$T^\dagger(\varepsilon)PT(\varepsilon) = P \tag{11.2.17b}$$

The operator $T^\dagger(\varepsilon)XT(\varepsilon)$ is also a position operator, but it measures position from a new origin, shifted to the *left* by ε: This is the meaning of Eq. (11.2.17a).

Equation (11.2.17b) states that under the shift in the origin, the momentum is unaffected.

Writing once again

$$T(\varepsilon)=I-\frac{i\varepsilon G}{\hbar}$$

we find from Eq. (11.2.17a) (using the fact that $G^{\dagger}=G$)

$$\left(I+\frac{i\varepsilon G}{\hbar}\right)X\left(I-\frac{i\varepsilon G}{\hbar}\right)=X+\varepsilon I$$

or

$$-\frac{i\varepsilon}{\hbar}[X, G]=\varepsilon I \tag{11.2.18a}$$

$$[X, G]=i\hbar I \tag{11.2.18b}$$

This allows us to conclude that

$$G=P+f(X) \tag{11.2.19}$$

If we now turn to Eq. (11.2.17b) we find

$$-\frac{i\varepsilon}{\hbar}[P, G]=0 \tag{11.2.20a}$$

or

$$[P, G]=0 \tag{11.2.20b}$$

which eliminates $f(X)$.‡ So once again

$$T(\varepsilon)=I-\frac{i\varepsilon P}{\hbar}$$

Having derived the translation operator in the passive transformation picture, let us reexamine the notion of translational invariance.

We define translational invariance by the requirement

$$T^{\dagger}(\varepsilon)HT(\varepsilon)=H \tag{11.2.21}$$

‡ For the purists, it reduces f to a c number which commutes with X and P, which we choose to be zero.

We can rewrite Eq. (11.2.21) in a form that is closer to the classical definition of translational invariance. But first we need the following result: for any $\Omega(X, P)$ that can be expanded in a power series, and for any unitary operator U,

$$U^\dagger\Omega(X, P)U=\Omega(U^\dagger XU, U^\dagger PU)$$

For the proof, consider a typical term in the series such as PX^2P. We have, using $UU^\dagger=I$,

$$U^\dagger PX^2PU=U^\dagger PUU^\dagger XUU^\dagger XUU^\dagger PU \qquad \text{Q.E.D.}$$

Applying this result to the case $U=T(\varepsilon)$ we get the response of any dynamical variable to a translation:

$$\Omega(X, P)\to T^\dagger\Omega(X, P)T=\Omega(T^\dagger XT, T^\dagger PT)=\Omega(X+\varepsilon I, P) \tag{11.2.22}$$

Thus the transformed Ω is found by replacing X by $X+\varepsilon I$ and P by P. If we now apply this to Eq. (11.2.21) we get the following definition of translation invariance:

$$H(X+\varepsilon I, P)=H(X, P) \tag{11.2.23}$$

Not only does this condition have the same form as its classical counterpart

$$\mathscr{H}(x+\varepsilon, p)=\mathscr{H}(x, p)$$

but it is also satisfied whenever the classical counterpart is. The reason is simply that H is the same function of X and P as $\mathscr{H}$ is of x and p, and both sets of variables undergo identical changes in a translation.

The conservation of momentum follows if we write $T(\varepsilon)$ in Eq. (11.2.21) in terms of P and expand things out to first order in ε:

$$\begin{aligned}0=T^\dagger(\varepsilon)HT(\varepsilon)-H&=(I+i\varepsilon P/\hbar)H(I-i\varepsilon P/\hbar)-H\\&=\frac{-i\varepsilon}{\hbar}[H, P]\end{aligned} \tag{11.2.24}$$

which implies that $\langle\dot{P}\rangle=0$, because of the Ehrenfest's theorem.

A Digression on the Analogy with Classical Mechanics‡

The passive transformation picture has the virtue that it bears a close formal resemblance to classical mechanics, with operators Ω in place of the classical variables

‡ In a less advanced course, the reader may skip this digression.

ω [Eqs. (11.2.17), (11.2.22), (11.2.23)]. In fact, the infinitesimal unitary transformation $T(\varepsilon)$ generated by P is the quantum image of the infinitesimal canonical transformation generated by p: if we *define* the changes δX and δP by

$$\delta X = T^\dagger(\varepsilon) X T(\varepsilon) - X$$
$$\delta P = T^\dagger(\varepsilon) P T(\varepsilon) - P$$

we get, on the one hand, from Eq. (11.2.17),

$$\delta X = X + \varepsilon I - X = \varepsilon I$$
$$\delta P = P - P = 0$$

and on the other, from $T = I - i\varepsilon P/\hbar$ (working to first order in ε),

$$\delta X = (I + i\varepsilon P/\hbar) X (I - i\varepsilon P/\hbar) - X = \frac{-i\varepsilon}{\hbar}[X, P]$$

$$\delta P = (I + i\varepsilon P/\hbar) P (I - i\varepsilon P/\hbar) - P = \frac{-i\varepsilon}{\hbar}[P, P]$$

combining which we obtain

$$\delta X = \frac{-i\varepsilon}{\hbar}[X, P] = \varepsilon I$$

$$\delta P = \frac{-i\varepsilon}{\hbar}[P, P] = 0$$

More generally, upon combining, Eq. (11.2.22) and $T = I - i\varepsilon P/\hbar$, we obtain

$$\delta\Omega = \frac{-i\varepsilon}{\hbar}[\Omega, P] = \Omega(X + \varepsilon I, P) - \Omega(X, P)$$

These are the analogs of the canonical transformation generated by p:

$$\delta x = \varepsilon\{x, p\} = \varepsilon$$
$$\delta p = \varepsilon\{p, p\} = 0$$
$$\delta\omega = \varepsilon\{\omega, p\} = \omega(x + \varepsilon, p) - \omega(x, p)$$

If the problem is translationally invariant, we have

$$\delta H = \frac{-i\varepsilon}{\hbar}[H, P] = 0 \rightarrow \langle \dot{P} \rangle = 0 \quad \text{by Ehrenfest's theorem}$$

while classically

$$\delta\mathscr{H}=\varepsilon\{\mathscr{H},p\}=0\rightarrow\dot{p}=0 \quad \text{by } \dot{p}=\{p,\mathscr{H}\}$$

The correspondence is achieved through the substitution rules already familiar to us:

$$\Omega\leftrightarrow\omega$$

$$\frac{-i}{\hbar}[\Omega,\Lambda]\leftrightarrow\{\omega,\lambda\}$$

In general, the infinitesimal canonical transformation generated by $g(x,p)$,

$$\delta\omega=\varepsilon\{\omega,g\}$$

has as its image in quantum theory the infinitesimal unitary transformation $U_G(\varepsilon)=I-i\varepsilon G/\hbar$ in response to which

$$\delta\Omega=\frac{-i\varepsilon}{\hbar}[\Omega,G]$$

Now, we have seen that the transformation generated by any $g(x,p)$ is canonical, i.e., it preserves the PB between the x's and the p's. In the quantum theory, the quantities preserved are the commutation relations between the X's and the P's, for if

$$[X_i,P_j]=i\hbar\delta_{ij}I$$

then upon premultiplying by the unitary operator $U_G^\dagger(\varepsilon)$ and postmultiplying by $U_G(\varepsilon)$, we find that the transformed operators obey‡

$$[U^\dagger X_i U,\,U^\dagger P_j U]=i\hbar\delta_{ij}I$$

This completes the proof of the correspondence

$$\left\{\begin{array}{l}\text{infinitesimal canonical}\\ \quad\text{transformation generated}\\ \quad\text{by } g(x,p)\end{array}\right\}\leftrightarrow\left\{\begin{array}{l}\text{infinitesimal unitary}\\ \quad\text{transformation generated}\\ \quad\text{by } G(X,P)\end{array}\right\}$$

The correspondence holds for finite transformations as well, for these may be viewed as a sequence of infinitesimal transformations.

‡ More generally if $[\Omega,\theta]=\Gamma$, then a similar relation holds between the transformed operators $U^\dagger\Omega U$, $U^\dagger\theta U$, $U^\dagger\Gamma U$. This is the quantum version of the result that PB are invariant under canonical transformation.

The correspondence with unitary transformations also holds for *regular* canonical transformations which have no infinitesimal versions. For instance, in the coupled oscillator problem, Exercise 10.1.3, we performed a canonical transformation from x_1, x_2, p_1, p_2 to $x_{\text{I}}, x_{\text{II}}, p_{\text{I}}$, and p_{II}, where, for example, $x_{\text{I}} = (x_1 + x_2)/2$. In the quantum theory there will exist a unitary operator such that, for example, $U^\dagger X_1 U = (X_1 + X_2)/2 = X_{\text{I}}$ and so on.‡

We can see why we can either perform the canonical transformation at the classical level and then quantize, or first quantize and then perform the unitary transformation—since the quantum operators respond to the unitary transformation as do their classical counterparts to the canonical transformation, the end effect will be the same.§

Let us now return to the problem of translational invariance. Notice that in a problem with translational invariance, Eq. (11.2.24) tells us that we can find the simultaneous eigenbasis of P and H. (This agrees with our result from Chapter 5, that the energy eigenstates of a free particle could be chosen to be momentum eigenstates as well.‖) If a system starts out in such an eigenstate, its momentum eigenvalue remains constant. To prove this, first note that

$$[P, H] = 0 \rightarrow [P, U(t)] = 0 \tag{11.2.25}$$

since the propagator is a function of just H.*

Suppose at $t = 0$ we have a system in an eigenstate of P:

$$P|p\rangle = p|p\rangle \tag{11.2.26}$$

After time t, the state is $U(t)|p\rangle$ and we find

$$PU(t)|p\rangle = U(t)P|p\rangle = U(t)p|p\rangle = pU(t)|p\rangle \tag{11.2.27}$$

In other words, the state at time t is also an eigenstate of P with the same eigenvalue. For such states with well-defined momentum, the conservation law $\langle \dot{P} \rangle = 0$ reduces to the classical form $\dot{p} = 0$.

Finite Translations

What is the operator $T(a)$ corresponding to a finite translation a? We find it by the following trick. We divide the interval a into N parts of size a/N. As $N \rightarrow \infty$,

‡ If the transformation is not regular, we cannot find a unitary transformation in the quantum theory, since unitary transformations preserve the eigenvalue spectrum.

§ End of digression.

‖ Note that a single particle whose H is translationally invariant is necessarily free.

* When H is time independent, we know $U(t) = \exp(-iHt/\hbar)$. If $H = H(t)$, the result is true if P commutes with $H(t)$ for all t. (Why?)

a/N becomes infinitesimal and we know

$$T(a/N)=I-\frac{ia}{\hbar N}P \tag{11.2.28}$$

Since a translation by a equals N translations by a/N,

$$T(a)=\lim_{N\to\infty}[T(a/N)]^N=e^{-iaP/\hbar} \tag{11.2.29}$$

by virtue of the formula

$$e^{-ax}=\lim_{N\to\infty}\left(1-\frac{ax}{N}\right)^N$$

We may apply this formula, true for c numbers, to the present problem, since P is the only operator in the picture and commutes with everything in sight, i.e., behaves like a c number. Since

$$T(a)\xrightarrow[X\text{ basis}]{}e^{-a\,d/dx} \tag{11.2.30}$$

we find

$$\langle x|T(a)|\psi\rangle=\psi(x)-\frac{d\psi}{dx}a+\frac{d^2\psi}{dx^2}\frac{a^2}{2!}+\cdots \tag{11.2.31}$$

which is the full Taylor series for $\psi(x-a)$ about the point x.

A Consistency Check. A translation by a followed by a translation by b equals a translation by $a+b$. This result has nothing to do with quantum mechanics and is true whether you are talking about a quantum system or a sack of potatoes. It is merely a statement about how translations combine in space. Now, we have just built operators T, which are supposed to translate quantum states. For this interpretation to be consistent, it is necessary that the law of combination of the translation operators coincide with the law of combination of the translations they represent. Now, although we presumed this [see Eq. (11.2.29), and the line above it] in the very act of deriving the formula for $T(a)$, let us verify that our result $T(a)=\exp(-iaP/\hbar)$ satisfies

$$T(a)T(b)=T(a+b)? \tag{11.2.32}$$

We find that this is indeed so:

$$T(a)T(b)=e^{-iaP/\hbar}\cdot e^{-ibP/\hbar}=e^{-i(a+b)P/\hbar}=T(a+b) \tag{11.2.33}$$

A Digression on Finite Canonical and Unitary Transformations‡

Though it is clear that the correspondence between canonical and unitary transformations, established for the infinitesimal case in earlier discussions, must carry

‡ Optional.

over to the finite case, let us nonetheless go through the details. Consider, for definiteness, the case of translations. In the quantum theory we have

$$\Omega \to T^{\dagger}(a)\Omega T(a) = e^{iaP/\hbar}\Omega e^{-iaP/\hbar}$$

Using the identity

$$e^{-A}B\,e^{+A} = B + [B, A] + \frac{1}{2!}[[B, A], A] + \frac{1}{3!}\cdots$$

we find

$$\Omega \to \Omega + a\left(\frac{-i}{\hbar}\right)[\Omega, P] + \frac{1}{2!}a^2\left(\frac{-i}{\hbar}\right)^2[[\Omega, P], P] + \cdots \qquad (11.2.34)$$

For example, if we set $\Omega = X^2$ we get $X^2 \to (X + aI)^2$.

In the classical case, under an infinitesimal displacement δa,

$$\delta\omega = \delta a\{\omega, p\}$$

or

$$\frac{d\omega}{da} = \{\omega, p\}$$

Applying the above result to the variable $d\omega/da$, we get

$$\frac{d}{da}(d\omega/da) = d^2\omega/da^2 = \{d\omega/da, p\} = \{\{\omega, p\}, p\}$$

and so on. The response of ω to the finite translation is given by the Taylor series about the point $a = 0$:

$$\omega \to \omega + a\{\omega, p\} + \frac{a^2}{2!}\{\{\omega, p\}, p\} + \cdots \qquad (11.2.35)$$

which we see is in correspondence with Eq. (11.2.34) if we make the usual substitutions.

*Exercise 11.2.3.** Recall that we found the finite rotation transformation from the infinitesimal one, by solving differential equations (Section 2.8). Verify that if, instead, you relate the transformed coordinates $\bar{x}$ and $\bar{y}$ to x and y by the infinite string of Poisson brackets, you get the same result, $\bar{x} = x\cos\theta - y\sin\theta$, etc. (Recall the series for $\sin\theta$, etc.)

System of Particles

We will not belabor the extension of the preceding ideas to a system of N particles. Starting with the analog of Eq. (11.2.8),

$$\langle x_1, \ldots, x_N | T(\varepsilon) | \psi \rangle = \psi(x_1 - \varepsilon, \ldots, x_N - \varepsilon) \tag{11.2.36}$$

we find, on expanding both sides to order ε, that

$$\langle x_1, \ldots, x_N | I - \frac{i\varepsilon}{\hbar} P | \psi \rangle = \psi(x_1, \ldots, x_N) - \sum_{i=1}^{N} \varepsilon \frac{\partial \psi}{\partial x_i} \tag{11.2.37}$$

from which it follows that

$$T(\varepsilon) = I - \frac{i\varepsilon}{\hbar} \sum_{i=1}^{N} P_i = I - \frac{i\varepsilon}{\hbar} P \tag{11.2.38}$$

where P is the *total momentum operator*. You may verify that

$$\begin{aligned} T^\dagger(\varepsilon) X_i T(\varepsilon) &= X_i + \varepsilon I \\ T^\dagger(\varepsilon) P_i T(\varepsilon) &= P_i, \qquad i = 1, \ldots, N \end{aligned} \tag{11.2.39}$$

Translational invariance means in this case (suppressing indices),

$$H(X, P) = T^\dagger(\varepsilon) H(X, P) T(\varepsilon) = H(X + \varepsilon I, P) \tag{11.2.40}$$

Whereas in the single-particle cases this implied the particle was free, here it merely requires that H (or rather V) be a function of the coordinate *differences*. Any system whose parts interact with each other, but nothing external, will have this property.

There are some profound consequences of translational invariance besides momentum conservation. We take these up next.

Implications of Translational Invariances

Consider a system with translational invariance. Premultiplying both sides of Eq. (11.2.21) with T and using its unitarity, we get

$$[T(a), H] = 0$$

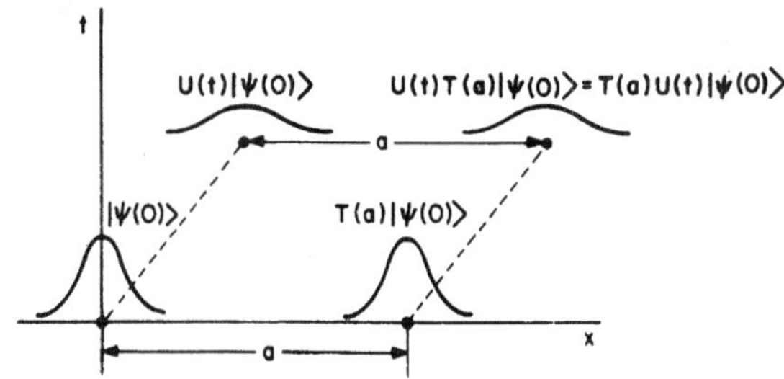

Figure 11.1. A symbolic depiction of translational invariance. The states are represented schematically by wave functions.

It follows that

$$[T(a), U(t)]=0 \quad \text{or} \quad T(a)U(t)=U(t)T(a) \tag{11.2.41}$$

The consequence of this relation is illustrated by the following example (Fig. 11.1). At $t=0$ two observers A and B prepare identical systems at $x=0$ and $x=a$, respectively. If $|\psi(0)\rangle$ is the state vector of the system prepared by A, then $T(a)|\psi(0)\rangle$ is the state vector of the system prepared by B. The two systems look identical to the observers who prepared them. After time t, the state vectors evolve into $U(t)|\psi(0)\rangle$ and $U(t)T(a)|\psi(0)\rangle$. Using Eq. (11.2.41) the latter may be rewritten as $T(a)U(t)|\psi(0)\rangle$, which is just the translated version of A's system at time t. Therefore the two systems, which differed only by a translation at $t=0$, differ only by the same translation at future times. In other words, the time evolution of each system appears the same to the observer who prepared it. Translational invariance of H implies that the same experiment repeated at two different places will give the same result (as seen by the local observers). We have already seen this result in the classical framework. We pursue it further now.

Now it turns out that every known interaction—gravitational, weak, electromagnetic, and strong (e.g., nuclear)—is translationally invariant, in that every experiment, if repeated at a new site, will give the same result. Consider the following illustrative example, which clarifies the meaning of this remark. A hydrogen atom is placed between the plates of a charged condenser. The Hamiltonian is

$$H=\frac{|\mathbf{P}_1|^2}{2m_1}+\frac{|\mathbf{P}_2|^2}{2m_2}+\frac{e_1e_2}{|\mathbf{R}_1-\mathbf{R}_2|}+e_1V(\mathbf{R}_1)+e_2V(\mathbf{R}_2) \tag{11.2.42}$$

where the subscripts 1 and 2 refer to the electron and the proton and $V(\mathbf{R})$‡ to the potential due to the plates. Now this problem has no translation invariance, i.e.,

$$H(\mathbf{R}_1+\boldsymbol{\varepsilon}, \mathbf{P}_1; \mathbf{R}_2+\boldsymbol{\varepsilon}, \mathbf{P}_2)\neq H(\mathbf{R}_1, \mathbf{P}_1; \mathbf{R}_2, \mathbf{P}_2)$$

which in turn means that if the atom alone is translated (away from the condenser) it will behave differently. But this does not correspond to repeating the *same* experiment and getting a different result, since the condenser, which materially affects the

‡ Remember that $\mathbf{R}$ is the operator corresponding to the classical variable $\mathbf{r}$.

dynamics, is left behind. To incorporate it in what is translated, we redefine our system to include the (N) charges on the condenser and write

$$H=\sum_{i=1}^{N+2}\frac{|\mathbf{P}_i|^2}{2m_i}+\frac{1}{2}\sum_{i=1}^{N+2}\sum_{\substack{j\neq i\\ j=1}}^{N+2}\frac{e_ie_j}{|\mathbf{R}_i-\mathbf{R}_j|} \tag{11.2.43}$$

Now the charges on the condenser enter H, not via the external field which breaks translational invariance, but through the Coulomb interaction, which does not. Now it is true that (dropping indices),

$$H(\mathbf{R}+\varepsilon, \mathbf{P})=H(\mathbf{R}, \mathbf{P})$$

which implies that if the atom *and* the condenser are moved to a new site, the behavior of the composite system will be unaffected. This result should be viewed not as obvious or self-evident, but rather as a profound statement about the Coulomb interaction.

The content of the assertion made above is that every known interaction has translational invariance at the fundamental level—if we expand our system to include all degrees of freedom that affect the outcome of an experiment (so that there are not external fields, only interactions between parts of the system) the total H is translationally invariant. This is why we apply momentum conservation to every problem whatever be the underlying interaction. The translational invariance of natural laws reflects the uniformity or homogeneity of space. The fact that the dynamics of an isolated‡ system (the condenser plus atom in our example) depends only on where the parts of the system are relative to each other and not on where the system is as a whole, represents the fact that one part of free space is as good as another.

It is translational invariance that allows experimentalists in different parts of the earth to claim they all did the "same" experiment, and to corroborate, correct, and complement each other. It is the invariance of the natural laws under translations that allows us to describe a hydrogen atom in some distant star as we do one on earth and to apply to its dynamics the quantum mechanical laws deduced on earth. We will examine further consequences of translational invariance toward the end of the next section.

11.3. Time Translational Invariance

Just as the homogeneity of space ensures that the same experiment performed at two different places gives the same result, homogeneity in time ensures that the

‡ To be exact, no system is truly "isolated" except the whole universe (and only its momentum is exactly conserved). But in practice one draws a line somewhere, between what constitutes the system and what is irrelevant (for practical purposes) to its evolution. I use the term "isolated" in this practical sense. The real utility of the concepts of translational invariance and momentum conservation lies in these approximate situations. Who cares if the universe as a whole is translationally invariant and its momentum is conserved? What matters to me is that I can take my equipment to another town and get the same results and that the momentum of my system is conserved (to a good accuracy).

same experiment repeated at two different times gives the same result. Let us see what feature of the Hamiltonian ensures this and what conservation law follows.

Let us prepare at time t_1 a system in state $|\psi_0\rangle$ and let it evolve for an infinitesimal time ε. The state at time $t_1+\varepsilon$, to first order in ε, will be

$$|\psi(t_1+\varepsilon)\rangle=\left[I-\frac{i\varepsilon}{\hbar}H(t_1)\right]|\psi_0\rangle \tag{11.3.1}$$

If we repeat the experiment at time t_2, beginning with the same initial state, the state at time $t_2+\varepsilon$ will be

$$|\psi(t_2+\varepsilon)\rangle=\left[I-\frac{i\varepsilon}{\hbar}H(t_2)\right]|\psi_0\rangle \tag{11.3.2}$$

The outcome will be the same in both cases if

$$\begin{aligned}0&=|\psi(t_2+\varepsilon)\rangle-|\psi(t_1+\varepsilon)\rangle\\&=\left(-\frac{i\varepsilon}{\hbar}\right)[H(t_2)-H(t_1)]|\psi_0\rangle\end{aligned} \tag{11.3.3}$$

Since $|\psi_0\rangle$ is arbitrary, it follows that

$$H(t_2)=H(t_1) \tag{11.3.4}$$

Since t_2 and t_1 are arbitrary, it follows that H is time-independent:

$$\frac{dH}{dt}=0 \tag{11.3.5}$$

Thus time translational invariance requires that H have no t dependence. Now Ehrenfest's theorem for an operator Ω *that has no time dependence*‡ is

$$i\hbar\langle\dot{\Omega}\rangle=\langle[\Omega, H]\rangle$$

Applying it to $\Omega=H$ in a problem with time translational invariance, we find

$$\langle\dot{H}\rangle=0 \tag{11.3.6}$$

which is the law of conservation of energy.

‡ If $d\Omega/dt\neq 0$ there will be an extra piece $i\hbar\langle d\Omega/dt\rangle$ on the right-hand side.

An important simplification that arises if $dH/dt=0$ is one we have repeatedly exploited in the past: Schrödinger's equation

$$i\hbar\frac{\partial|\psi\rangle}{\partial t}=H|\psi\rangle \tag{11.3.7}$$

admits solutions of the form

$$|\psi(t)\rangle=|E\rangle e^{-iEt/\hbar} \tag{11.3.8}$$

where the time-independent ket $|E\rangle$ satisfies

$$H|E\rangle=E|E\rangle \tag{11.3.9}$$

The entire dynamics, i.e., the determination of the propagator $U(t)$, boils down to the solution of the time-independent Schrödinger equation (11.3.9).

The considerations that applied to space translation invariance apply here as well. In particular, all known interactions—from gravitational to strong—are time translational invariant. Consequently, if we suitably define the system (to include the sources of external fields that affect the experiment) the total H will be independent of t. Consider, for example, a hydrogen atom between the plates of a *discharging* condenser. If the system includes just the electron and the proton, H will depend on time—it will have the form of Eq. (11.2.42), with $V=V(\mathbf{R},t)$. This simply means that repeating the experiment *without recharging the condenser*, will lead to a different result. If, however, we enlarge the system to include the N charges on the condenser, we end up with the H in Eq. (11.2.43), which has no t dependence.

The space–time invariance of natural laws has a profound impact on our quest for understanding nature. The very cycle of physics—of deducing laws from some phenomena studied at some time and place and then applying them to other phenomena at a different time and place—rests on the assumption that natural laws are space–time invariant. If nature were not to follow the same rules over space–time, there would be no rules to find, just a sequence of haphazard events with no rhyme or reason. By repeating the natural laws over and over through all of space–time, nature gives tiny earthlings, who probe just a miniscule region of space for a fleeting moment (in the cosmic scale), a chance of comprehending the universe at large. Should we at times be despondent over the fact that we know so few of nature's laws, let us find solace in these symmetry principles, which tell us that what little we know is universal and eternal.‡

‡ The invariance of the laws of nature is not to be confused with our awareness of them, which does change with time. For example, Einstein showed that Newtonian mechanics and gravitation are approximations to relativistic mechanics and gravitation. But this is not to say that the Newtonian scheme worked till Einstein came along. In other words, the relation of Newton's scheme to Einstein's (as a good approximation in a certain limit) has always been the same, before and after we learned of it.

11.4. Parity Invariance

Unlike space–time translations, and rotations, (which we will study in the next chapter), parity is a discrete transformation. Classically, the parity operation corresponds to reflecting the state of the particle through the origin

$$x \xrightarrow[\text{parity}]{} -x \qquad p \xrightarrow[\text{parity}]{} -p \tag{11.4.1}$$

In quantum theory, we define the action of the parity operator on the X basis as follows

$$\Pi|x\rangle = |-x\rangle \tag{11.4.2}$$

in analogy with the classical case. Given this,

$$\Pi|p\rangle = |-p\rangle \tag{11.4.3}$$

follows, as you will see in a moment.

Given the action of Π on a complete (X) basis, its action on an arbitrary ket follows:

$$\begin{aligned}\Pi|\psi\rangle &= \Pi\int_{-\infty}^{\infty} |x\rangle\langle x|\psi\rangle\,dx \\ &= \int_{-\infty}^{\infty} |-x\rangle\langle x|\psi\rangle\,dx \\ &= \int_{-\infty}^{\infty} |x'\rangle\langle -x'|\psi\rangle\,dx' \qquad (\text{where } x' = -x)\end{aligned} \tag{11.4.4}$$

It follows that if

$$\langle x|\psi\rangle = \psi(x)$$

$$\langle x|\Pi|\psi\rangle = \psi(-x) \tag{11.4.5}$$

The function $\psi(-x)$ is the mirror image of $\psi(x)$ about the origin. Applying Eq. (11.4.5) to a momentum eigenstate, it will be readily found that $\Pi|p\rangle = |-p\rangle$.

The eigenvalues of Π are just ± 1. A moment's "reflection" will prove this. Since

$$\Pi|x\rangle = |-x\rangle$$
$$\Pi^2|x\rangle = |-(-x)\rangle = |x\rangle$$

Since this is true for an entire basis,

$$\Pi^2 = I \tag{11.4.6}$$

Please note that

(1) $\Pi = \Pi^{-1}$
(2) The eigenvalues of Π are ± 1.
(3) Π is Hermitian and unitary.
(4) Or $\Pi^{-1} = \Pi^\dagger = \Pi$.

The eigenvectors with eigenvalue ± 1 are said to have even/odd parity. In the X basis, where

$$\psi(x) \xrightarrow[\Pi]{} \psi(-x)$$

even-parity vectors have even wave functions and odd-parity vectors have odd wave functions. The same goes for the P basis since

$$\psi(p) \xrightarrow[\Pi]{} \psi(-p)$$

In an arbitrary Ω basis, $\psi(\omega)$ need not be even or odd even if $|\psi\rangle$ is a parity eigenstate (check this).

Rather than define Π in terms of its action on the kets, we may also define it through its action on the operators:

$$\begin{aligned} \Pi^\dagger X \Pi &= -X \\ \Pi^\dagger P \Pi &= -P \end{aligned} \tag{11.4.7}$$

We say $H(X, P)$ is parity invariant if

$$\Pi^\dagger H(X, P)\Pi = H(-X, -P) = H(X, P) \tag{11.4.8}$$

In this case

$$[\Pi, H] = 0$$

and a common eigenbasis of Π and H can be found. In particular, if we consider just bound states in one dimension (which we saw are nondegenerate), every eigenvector of H is necessarily an eigenvector of Π. For example, the oscillator Hamiltonian satisfies Eq. (11.4.8) and its eigenfunctions have definite parity equal to $(-1)^n$, n being the quantum number of the state. The particle in a box has a parity-invariant Hamiltonian if the box extends from $-L/2$ to $L/2$. In this case the eigenfunctions have parity $(-1)^{n+1}$, n being the quantum number. If the box extends from 0 to L, $V(x)$ is not parity invariant and the eigenfunctions

$$\psi_n(x)=\left(\frac{2}{L}\right)^{1/2}\sin\left(\frac{n\pi x}{L}\right)$$

have no definite parity. (When $x\to -x$ they vanish, since ψ_n is given by the sine function *only between* 0 *and L, and vanishes outside.*)

If H is parity invariant, then

$$\Pi U(t)=U(t)\Pi \qquad (11.4.9)$$

This means that if at $t=0$ I start with a system in a state $|\psi(0)\rangle$, and someone else starts with a system in the parity operated state $\Pi|\psi(0)\rangle$, then at a later time the state of his system will be related to mine by the parity transformation.

Whereas all natural laws are invariant under space-time translations (and rotations) some are not invariant under parity. These are the laws of weak interactions, which are responsible for nuclear β decay (among other things). This means formally that the Hamiltonian cannot be made parity invariant by any redefinition of the system if weak interactions are involved. Physically this means that if two observers prepare initial states $|\psi(0)\rangle$ and $\Pi|\psi(0)\rangle$ which are mirror images of each other, the final states $U(t)|\psi(0)\rangle$ and $U(t)\Pi|\psi(0)\rangle$ will not be mirror images of each other (since $\Pi U\neq U\Pi$).‡ Consider the following concrete example of a β decay:

$$^{60}\text{Co}\to{}^{60}\text{Ni}+e^-+\bar{\nu}$$

where e^- is an electron and $\bar{\nu}$ is an antineutrino. Now it turns out that the electron likes to come flying out in a direction opposite to the spin of ^{60}Co—and this implies parity noninvariance. Let us see how. At $t=0$ I prepare a system that consists of a ^{60}Co nucleus with its spin up along the z axis (Fig. 11.2) (experiment A). Although you are not yet familiar with spin, you may pretend here that ^{60}Co is spinning in the sense shown. Let another observer set up another system which is just the mirror image of mine (experiment B). Let M denote a *fictitious* experiment, which is what I see in a mirror in front of me. Notice how the spin **S** gets reversed under a mirror reflection. Let the β decay take place. My electron comes out *down* the z axis. Of course the mirror also shows an electron coming down the z axis. In the other *real* experiment (B), the dynamics forces the electron to come *up* the z axis, since the initial **S** was *down.* Thus B starts out as the mirror image of A but ends up different. Consequently, what I see in the mirror (experiment M) does not correspond to what can happen in real life, i.e., is not a solution to the equations of motion.

‡ See Exercise 11.4.4 for a discussion of why the parity transformation is essentially a mirror reflection in three dimensions.

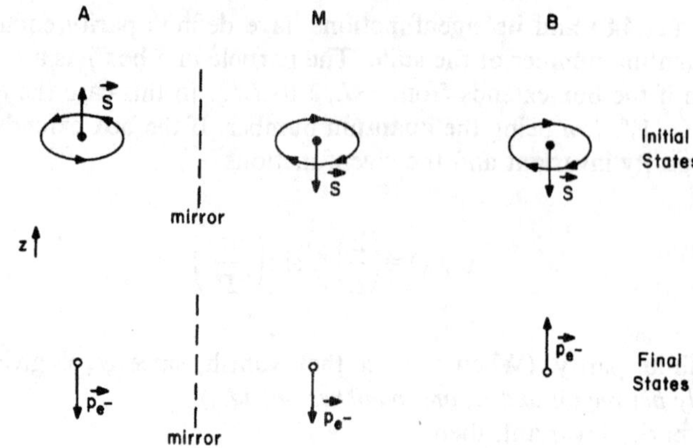

Figure 11.2. An example of parity noninvariance. In experiment A, which I perform, the spin of the nucleus points up the z axis. In its actual mirror image, it points down (experiment M). In experiment B, which is a real experiment, the spin is chosen to be down, i.e., B starts out as the mirror image of A. After the decay, the momentum of my electron, $\mathbf{p}_e$, is down the z axis. The mirror image of course also shows the electron coming down. But in the actual experiment B, the dynamics forces the electron to come up the z axis, i.e., antiparallel to the initial nuclear spin **S**.

This then is the big difference between parity and other transformations such as space-time translations and rotations. If a certain phenomenon can happen, its translated or rotated version can also happen, but not its mirror-reflected version, if the phenomenon involves weak interactions. In terms of conservation laws, if an isolated system starts out in a state of definite parity, it need not end in a state of same parity if weak interactions are at work. The possibility that weak interactions could be parity noninvariant was discussed in detail by Lee and Yang in 1956 and confirmed shortly thereafter by the experiments of C. S. Wu and collaborators.‡

*Exercise 11.4.1.** Prove that if $[\Pi, H]=0$, a system that starts out in a state of even/odd parity maintains its parity. (Note that since parity is a discrete operation, it has no associated conservation law in classical mechanics.)

*Exercise 11.4.2.** A particle is in a potential

$$V(x)=V_0 \sin(2\pi x/a)$$

which is invariant under the translations $x \to x+ma$, where m is an integer. Is momentum conserved? Why not?

*Exercise 11.4.3.** You are told that in a certain reaction, the electron comes out with its spin always parallel to its momentum. Argue that parity is violated.

*Exercise 11.4.4.** We have treated parity as a mirror reflection. This is certainly true in one dimension, where $x \to -x$ may be viewed as the effect of reflecting through a (point) mirror at the origin. In higher dimensions when we use a plane mirror (say lying on the $x-y$

‡ T. D. Lee and C. N. Yang, *Phys. Rev.*, **104**, 254 (1956); C. S. Wu, E. Ambler, R. W. Hayward, and R. P. Hudson, *Phys. Rev.*, **105**, 1413 (1957).

plane), only one (z) coordinate gets reversed, whereas the parity transformation reverses all three coordinates.

Verify that reflection on a mirror in the $x-y$ plane is the same as parity followed by 180° rotation about the z axis. Since rotational invariance holds for weak interactions, noninvariance under mirror reflection implies noninvariance under parity.

11.5. Time-Reversal Symmetry

This is a discrete symmetry like parity. Let us first understand what it means in classical physics. Consider a planet that is on a circular orbit around the sun. At $t=0$ it starts at $\theta=0$ and has a velocity in the direction of increasing θ. In other words, the orbit is running counterclockwise. Let us call the initial position and momentum $x(0), p(0)$. (We should really be using vectors, but ignore this fact for this discussion.)

We now define the *time-reversed state* as one in which the position is the same but the momentum is reversed:

$$x_r(t)=x(t) \qquad p_r(t)=-p(t).$$

In general, any quantity like position or kinetic energy, which involves an even power of t in its definition is left invariant and any quantity like momentum or angular momentum is reversed in sign under the time-reversal operation.

Say that after time T the planet has come to a final state $x(T)$, $p(T)$ at $\theta=\pi/2$ after doing a quarter of a revolution. Now Superman (for reasons best known to him) stops it dead in its tracks, reverses its speed, and lets it go. What will it do? We know it will retrace its path and at time $2T$ end up in the time-reversed state of the initial state:

$$x(2T)=x(0) \qquad p(2T)=-p(0) \tag{11.5.1}$$

The above equation defines time-reversal invariance (TRI).

We can describe TRI more graphically as follows. Suppose we take a movie of the planet from $t=0$ to $t=T$. At $t=T$, we start playing the film backward. The backward motion of the planet will bring it back to the time-reversal initial state at $t=2T$. What we see in the movie can really happen, indeed, it was shown how Superman could make it happen even as you are watching the movie. More generally, if you see a movie of some planetary motion you will have no way of knowing if the projector is running forwards or backward. In some movies they get a big laugh out of the audience by showing cars and people zooming in reverse. As a serious physics student you should not laugh when you see this since these motions obey Newton's laws. In other words, it is perfectly possible for a set of people and cars to execute this motion. On the other hand, when a cartoon character falling under gravity suddenly starts clawing his way upwards in thin air using sheer will power, you may laugh since this is a gross violation of Newton's laws.

While the correctness of Eq.(11.5.1) is intuitively clear, we will now prove it with the help of Newton's Second Law using the fact that it is invariant under $t\to-t$: the acceleration is even in time and the potential or force has no reference to t. Here are the details. Just for this discussion let us use a new clock that has its zero at the point of time-reversal, so that $t=0$ defines the point when the motion is time-reversed. When the movie is run backward we see the trajectory

$$x_r(t)=x(-t)$$

In other words, 5 seconds after the reversal, the object is where it was 5 seconds before the reversal. The reversal of velocities follows from this:

$$\dot{x}_r(t)=\frac{dx(-t)}{dt}=-\frac{dx(-t)}{d(-t)}=-\dot{x}(-t)$$

and does not have to be additionally encoded. The question is this: Does this orbit $x_r(t)$ obey Newton's Second Law

$$m\frac{d^2x_r(t)}{dt^2}=F(x_r)$$

given that $x(t)$ does? We find it does:

$$m\frac{d^2x_r(t)}{dt^2}=m\frac{d^2x(-t)}{dt^2}=m\frac{d^2x(-t)}{d(-t)^2}=F(x(-t))=F(x_r(t))$$

Not all problems are time-reversal invariant. Consider a positively charged particle in the x–y plane moving under a magnetic field *down* the z-axis. Let us say it is released at $t=0$ just like the planet, with its velocity in the direction of increasing θ. Due to the $\mathbf{v}\times\mathbf{B}$ force it will go in a counterclockwise orbit. Let us wait till it has gone around by $\pi/2$ and at this time, $t=T$, time-reverse its state. Will it return to the time-reserved initial state at $t=2T$? No, it is readily seen that starting from $t=T$ it will once again go on a counterclockwise circular orbit tangential to the first at the point of reversal. We blame the magnetic interaction for this failure of TRI: the force now involves the velocity which is odd under time-reversal.

We now ask how all this appears in quantum mechanics. The ideas will be illustrated in the simplest context. Let us consider a particle in one dimension with a time-independent Hamiltonian H. In the x-representation the wave equation is

$$i\hbar\frac{\partial\psi(x,t)}{\partial t}=H(x)\psi(x,t)$$

Let us first note that

$$\psi\to\psi^*$$

performs time-reversal. This is clear from the fact that the detailed probability distribution in x is unaffected by this change. On the other hand, it is clear from looking at plane waves (or the momentum operator $-i\hbar(\partial/\partial x)$) that $p\to -p$ under complex conjugation.

If the system has TRI, we must find the analog of Eq. (11.5.1). So let us prepare a state $\psi(x, 0)$, let it evolve for time T, complex conjugate it, let *that* evolve for another time T and see if we end up with the complex conjugate of the initial state. We find the following happens at each stage:

$$\psi(x,0)\to e^{-iH(x)T/\hbar}\psi(x,0)\to e^{iH^*(x)T/\hbar}\psi^*(x,0)\to e^{-iH(x)T/\hbar}e^{iH^*(x)T/\hbar}\psi^*(x,0)$$

It is clear that in order for the end result, which is $\psi(x, 2T)$, to obey

$$\psi(x,2T)=\psi^*(x,0)$$

we require that

$$H(x)=H^*(x) \tag{11.5.2}$$

i.e., that the Hamiltonian be real. For $H=P^2/2m+V(x)$ this is the case, even in higher dimensions. On the other hand, if we have a magnetic field, P enters linearly and $H(x)\neq H^*(x)$.

If H has TRI, i.e., is real, we have seen at the end of Chapter 6 that every eigenfunction implies a degenerate one which is its complex conjugate.

Notice that the failure of TRI in the presence of a magnetic field does not represent any *fundamental* asymmetry under time-reversal in electrodynamics. The laws of electrodynamics are invariant under $t\to -t$. The asymmetry in our example arose due to our treating the magnetic field as external to the system and hence not to be time-reversed. If we had included in our system the currents producing the magnetic field, and reversed them also, the entire system would have followed the time-reversed trajectory. Indeed, if you had taken a movie of the experiment and played it back, and you could have seen the charges in the wire, you would have found them running backward, the field would have been reversed at $t=T$, and the charge we chose to focus on would have followed the time-reversed trajectory.

On the other hand, certain experiments together with general arguments from quantum field theory suggest that there exist interactions in this universe which do not have this symmetry at the fundamental level.

There are ways to formulate TRI in a basis-independent way but we will not do so here. For most problems where the coordinate basis is the natural choice the above discussion will do. There will be a minor twist when the problem involves spin which has no classical counterpart. This can be handled by treating spin as we would treat orbital angular momentum.

12

Rotational Invariance and Angular Momentum

In the last chapter on symmetries, rotational invariance was not discussed, not because it is unimportant, but because it is all too important and deserves a chapter on its own. The reason is that most of the problems we discuss involve a single particle (which may be the reduced mass) in an external potential, and whereas translational invariance of H implies that the particle is free, rotational invariance of H leaves enough room for interesting dynamics. We first consider two dimensions and then move on to three.

12.1. Translations in Two Dimensions

Although we are concerned mainly with rotations, let us quickly review translations in two dimensions. By a straightforward extension of the arguments that led to Eq. (11.2.14) from Eq. (11.2.13), we may deduce that the generators of infinitesimal translations along the x and y directions are, respectively,

$$P_x \xrightarrow[\text{basis}]{\text{coordinate}} -i\hbar\frac{\partial}{\partial x} \tag{12.1.1}$$

$$P_y \xrightarrow[\text{basis}]{\text{coordinate}} -i\hbar\frac{\partial}{\partial y} \tag{12.1.2}$$

In terms of the *vector operator* **P**, which represents momentum,

$$\mathbf{P} = P_x\mathbf{i} + P_y\mathbf{j} \tag{12.1.3}$$

P_x and P_y are the dot products of **P** with the unit vector (**i** or **j**) in the direction of the translation. Since there is nothing special about these two directions, we conclude that in general,

$$\hat{n}\cdot\mathbf{P}\equiv P_{\hat{n}} \tag{12.1.4}$$

is the generator of translations in the direction of the unit vector $\hat{n}$. Finite translation operators are found by exponentiation. Thus $T(\mathbf{a})$, which translates by $\mathbf{a}$, is given by

$$T(\mathbf{a})=e^{-iaP_{\hat{a}}/\hbar}=e^{-ia\hat{a}\cdot\mathbf{P}/\hbar}=e^{-i\mathbf{a}\cdot\mathbf{P}/\hbar} \tag{12.1.5}$$

where $\hat{a}=\mathbf{a}/a$.

The Consistency Test. Let us now ask if the translation operators we have constructed have the right laws of combination, i.e., if

$$T(\mathbf{b})T(\mathbf{a})=T(\mathbf{a}+\mathbf{b}) \tag{12.1.6}$$

or equivalently if

$$e^{-i\mathbf{b}\cdot\mathbf{P}/\hbar}e^{-i\mathbf{a}\cdot\mathbf{P}/\hbar}=e^{-i(\mathbf{a}+\mathbf{b})\cdot\mathbf{P}/\hbar} \tag{12.1.7}$$

This amounts to asking if P_x and P_y may be treated as c numbers in manipulating the exponentials. The answer is yes, since in view of Eqs. (12.1.1) and (12.1.2), the operators commute

$$[P_x, P_y]=0 \tag{12.1.8}$$

and their q number nature does not surface here. The commutativity of P_x and P_y reflects the commutativity of translations in the x and y directions.

*Exercise 12.1.1.** Verify that $\hat{a}\cdot\mathbf{P}$ is the generator of infinitesimal translations along $\mathbf{a}$ by considering the relation

$$\langle x, y|I-\frac{i}{\hbar}\delta a\cdot\mathbf{P}|\psi\rangle=\psi(x-\delta a_x, y-\delta a_y)$$

12.2. Rotations in Two Dimensions

Classically, the effect of a rotation $\phi_0\mathbf{k}$, i.e., by an angle ϕ_0 about the z axis (counterclockwise in the x–y plane) has the following effect on the state of a particle:

$$\begin{bmatrix} x \\ y \end{bmatrix}\to\begin{bmatrix} \bar{x} \\ \bar{y} \end{bmatrix}=\begin{bmatrix} \cos\phi_0 & -\sin\phi_0 \\ \sin\phi_0 & \cos\phi_0 \end{bmatrix}\begin{bmatrix} x \\ y \end{bmatrix} \tag{12.2.1}$$

$$\begin{bmatrix} p_x \\ p_y \end{bmatrix}\to\begin{bmatrix} \bar{p}_x \\ \bar{p}_y \end{bmatrix}=\begin{bmatrix} \cos\phi_0 & -\sin\phi_0 \\ \sin\phi_0 & \cos\phi_0 \end{bmatrix}\begin{bmatrix} p_x \\ p_y \end{bmatrix} \tag{12.2.2}$$

Let us denote the operator that rotates these two-dimensional vectors by $R(\phi_0\mathbf{k})$. It is represented by the 2×2 matrix in Eqs. (12.2.1) and (12.2.2). Just as $T(\mathbf{a})$ is the operator in Hilbert space associated with the translation $\mathbf{a}$, let $U[R(\phi_0\mathbf{k})]$ be the operator associated with the rotation $R(\phi_0\mathbf{k})$. In the active transformation picture‡

$$|\psi\rangle \xrightarrow[U[R]]{} |\psi_R\rangle = U[R]|\psi\rangle \tag{12.2.3}$$

The rotated state $|\psi_R\rangle$ must be such that

$$\langle X\rangle_R = \langle X\rangle \cos\phi_0 - \langle Y\rangle \sin\phi_0 \tag{12.2.4a}$$

$$\langle Y\rangle_R = \langle X\rangle \sin\phi_0 + \langle Y\rangle \cos\phi_0 \tag{12.2.4b}$$

$$\langle P_x\rangle_R = \langle P_x\rangle \cos\phi_0 - \langle P_y\rangle \sin\phi_0 \tag{12.2.5a}$$

$$\langle P_y\rangle_R = \langle P_x\rangle \sin\phi_0 + \langle P_y\rangle \cos\phi_0 \tag{12.2.5b}$$

where

$$\langle X\rangle_R = \langle \psi_R|X|\psi_R\rangle$$

and

$$\langle X\rangle = \langle \psi|X|\psi\rangle, \text{ etc.}$$

In analogy with the translation problem, we define the action of $U[R]$ on position eigenkets:

$$U[R]|x, y\rangle = |x\cos\phi_0 - y\sin\phi_0,\ x\sin\phi_0 + y\cos\phi_0\rangle \tag{12.2.6}$$

As in the case of translations, this equation is guided by more than just Eq. (12.2.4), which specifies how $\langle X\rangle$ and $\langle Y\rangle$ transform: in omitting a possible phase factor $g(x, y)$, we are also ensuring that $\langle P_x\rangle$ and $\langle P_y\rangle$ transform as in Eq. (12.2.5).

One way to show this is to keep the phase factor and use Eqs. (12.2.5a) and (12.2.5b) to eliminate it. We will take the simpler route of dropping it from the outset and proving at the end that $\langle P_x\rangle$ and $\langle P_y\rangle$ transform according to Eq. (12.2.5).

Explicit Construction of $U[R]$

Let us now construct $U[R]$. Consider first an infinitesimal rotation $\varepsilon_z\mathbf{k}$. In this case we set

$$U[R(\varepsilon_z\mathbf{k})] = I - \frac{i\varepsilon_z L_z}{\hbar} \tag{12.2.7}$$

‡ We will suppress the rotation angle when it is either irrelevant or obvious.

where L_z, *the generator of infinitesimal rotations*, is to be determined. Starting with Eq. (12.2.6), which becomes to first order in ε_z

$$U[R]|x, y\rangle = |x - y\varepsilon_z, x\varepsilon_z + y\rangle \tag{12.2.8}$$

it can be shown that

$$\langle x, y|I - \frac{i\varepsilon_z L_z}{\hbar}|\psi\rangle = \psi(x + y\varepsilon_z, y - x\varepsilon_z) \tag{12.2.9}$$

*Exercise 12.2.1.** Provide the steps linking Eq. (12.2.8) to Eq. (12.2.9). [Hint: Recall the derivation of Eq. (11.2.8) from Eq. (11.2.6).]

Expanding both sides to order ε_z

$$\langle x, y|I|\psi\rangle - \frac{i\varepsilon_z}{\hbar}\langle x, y|L_z|\psi\rangle = \psi(x, y) + \frac{\partial\psi}{\partial x}(y\varepsilon_z) + \frac{\partial\psi}{\partial y}(-x\varepsilon_z)$$

$$\langle x, y|L_z|\psi\rangle = \left[x\left(-i\hbar\frac{\partial}{\partial y}\right) - y\left(-i\hbar\frac{\partial}{\partial x}\right)\right]\psi(x, y)$$

So

$$L_z \xrightarrow[\text{basis}]{\text{coordinate}} x\left(-i\hbar\frac{\partial}{\partial y}\right) - y\left(-i\hbar\frac{\partial}{\partial x}\right) \tag{12.2.10}$$

or in the abstract

$$L_z = XP_y - YP_x \tag{12.2.11}$$

Let us verify that $\langle P_x\rangle$ and $\langle P_y\rangle$ transform according to Eq. (12.2.5). Since

$$L_z \xrightarrow[\text{basis}]{\text{momentum}} \left(i\hbar\frac{\partial}{\partial p_x}p_y - i\hbar\frac{\partial}{\partial p_y}p_x\right) \tag{12.2.12}$$

it is clear that

$$\frac{-i\varepsilon_z}{\hbar}\langle p_x, p_y|L_z|\psi\rangle = \frac{\partial\psi}{\partial p_x}(p_y\varepsilon_z) + \frac{\partial\psi}{\partial p_y}(-p_x\varepsilon_z) \tag{12.2.13}$$

Thus $I - i\varepsilon_z L_z/\hbar$ rotates the momentum space wave function $\psi(p_x, p_y)$ by ε_z in momentum space, and as a result $\langle P_x\rangle$ and $\langle P_y\rangle$ transform just as $\langle X\rangle$ and $\langle Y\rangle$ do, i.e., in accordance with Eq. (12.2.5).

We could have also derived Eq. (12.2.11) for L_z by starting with the passive transformation equations for an infinitesimal rotation:

$$U^\dagger[R]XU[R]=X-Y\varepsilon_z \tag{12.2.14a}$$

$$U^\dagger[R]YU[R]=X\varepsilon_z+Y \tag{12.2.14b}$$

$$U^\dagger[R]P_xU[R]=P_x-P_y\varepsilon_z \tag{12.2.15a}$$

$$U^\dagger[R]P_yU[R]=P_x\varepsilon_z+P_y \tag{12.2.15b}$$

By feeding Eq. (12.2.7) into the above we can deduce that

$$[X, L_z]=-i\hbar Y \tag{12.2.16a}$$

$$[Y, L_z]=i\hbar X \tag{12.2.16b}$$

$$[P_x, L_z]=-i\hbar P_y \tag{12.2.17a}$$

$$[P_y, L_z]=i\hbar P_x \tag{12.2.17b}$$

These commutation relations suffice to fix L_z as XP_y-YP_x.

Exercise 12.2.2. Using these commutation relations (and your keen hindsight) derive $L_z=XP_y-YP_x$. At least show that Eqs. (12.2.16) and (12.2.17) are consistent with $L_z=XP_y-YP_x$.

The finite rotation operator $U[R(\phi_0\mathbf{k})]$ is

$$U[R(\phi_0\mathbf{k})]=\lim_{N\to\infty}\left(I-\frac{i}{\hbar}\frac{\phi_0}{N}L_z\right)^N=\exp(-i\phi_0L_z/\hbar) \tag{12.2.18}$$

Given

$$L_z\xrightarrow[\text{basis}]{\text{coordinate}}x\left(-i\hbar\frac{\partial}{\partial y}\right)-y\left(-i\hbar\frac{\partial}{\partial x}\right)$$

it is hard to see that $e^{-i\phi_0L_z/\hbar}$ indeed rotates the state by the angle ϕ_0. For one thing, expanding the exponential is complicated by the fact that $x(-i\hbar\partial/\partial y)$ and $y(-i\hbar\partial/\partial x)$ do not commute. So let us consider an alternative form for L_z. It can be shown, by changing to polar coordinates, that

$$L_z\xrightarrow[\text{basis}]{\text{coordinate}}-i\hbar\frac{\partial}{\partial\phi} \tag{12.2.19}$$

This result can also be derived more directly by starting with the requirement that under an infinitesimal rotation $\varepsilon_z\mathbf{k}$, $\psi(x, y)=\psi(\rho, \phi)$ becomes $\psi(\rho, \phi-\varepsilon_z)$.

*Exercise 12.2.3.** Derive Eq. (12.2.19) by doing a coordinate transformation on Eq. (12.2.10), and also by the direct method mentioned above.

Now it is obvious that

$$\exp(-i\phi_0 L_z/\hbar) \xrightarrow[\text{basis}]{\text{coordinate}} \exp\left(-\phi_0 \frac{\partial}{\partial\phi}\right) \tag{12.2.20}$$

rotates the state by an angle ϕ_0 about the z axis, for

$$\exp(-\phi_0\partial/\partial\phi)\psi(\rho, \phi)=\psi(\rho, \phi-\phi_0)$$

by Taylor's theorem. It is also obvious that $U[R(\phi_0'\mathbf{k})]U[R(\phi_0\mathbf{k})]= U[R((\phi_0+\phi_0')\mathbf{k})]$. Thus the rotation operators have the right law of combination.

Physical Interpretation of L_z. We identify L_z as the angular momentum operator, since (i) it is obtained from $l_z=xp_y-yp_x$ by the usual substitution rule (Postulate II), and (ii) it is the generator of infinitesimal rotations about the z axis. L_z is conserved in a problem with rotational invariance: if

$$U^\dagger[R]H(X, P_x;\, Y, P_y)U[R]=H(X, P_x;\, Y, P_y) \tag{12.2.21}$$

it follows (by choosing an infinitesimal rotation) that

$$[L_z, H]=0 \tag{12.2.22}$$

Since X, P_x, Y, and P_y respond to the rotation as do their classical counterparts [Eqs. (12.2.14) and (12.2.15)] and H is the same function of these operators as $\mathscr{H}$ is of the corresponding classical variables, H is rotationally invariant whenever $\mathscr{H}$ is.

Besides the conservation of $\langle L_z\rangle$, Eq. (12.2.22) also implies the following:

(1) An experiment and its rotated version will give the same result if H is rotationally invariant.
(2) There exists a common basis for L_z and H. (We will spend a lot of time discussing this basis as we go along.)

The Consistency Check. Let us now verify that our rotation and translation operators combine as they should. In contrast to pure translations or rotations, which have a simple law of composition, the combined effect of translations and rotations is nothing very simple. We seem to be facing the prospect of considering every possible combination of rotations and translations, finding their net effect, and then verifying that the product of the corresponding quantum operators equals the operator corresponding to the result of all the transformations. Let us take one small step in this direction, which will prove to be a giant step toward our goal.

Consider the following product of four infinitesimal operations:

$$U[R(-\varepsilon_z\mathbf{k})]T(-\boldsymbol{\varepsilon})U[R(\varepsilon_z\mathbf{k})]T(\boldsymbol{\varepsilon})$$

where $\boldsymbol{\varepsilon}=\varepsilon_x\mathbf{i}+\varepsilon_y\mathbf{j}$. By subjecting a point in the x–y plane to these four operations we find

$$\begin{bmatrix} x \\ y \end{bmatrix} \xrightarrow[\boldsymbol{\varepsilon}]{} \begin{bmatrix} x+\varepsilon_x \\ y+\varepsilon_y \end{bmatrix} \xrightarrow[R(\varepsilon_z\mathbf{k})]{} \begin{bmatrix} (x+\varepsilon_x)-(y+\varepsilon_y)\varepsilon_z \\ (x+\varepsilon_x)\varepsilon_z+(y+\varepsilon_y) \end{bmatrix}$$

$$\xrightarrow[-\boldsymbol{\varepsilon}]{} \begin{bmatrix} x-(y+\varepsilon_y)\varepsilon_z \\ (x+\varepsilon_x)\varepsilon_z+y \end{bmatrix} \xrightarrow[R(-\varepsilon_z\mathbf{k})]{} \begin{bmatrix} x-\varepsilon_y\varepsilon_z \\ y+\varepsilon_x\varepsilon_z \end{bmatrix} \tag{12.2.23}$$

i.e., that the net effect is a translation by $-\varepsilon_y\varepsilon_z\mathbf{i}+\varepsilon_x\varepsilon_z\mathbf{j}$.‡ In the above, we have ignored terms involving ε_x^2, ε_y^2, ε_z^2, and beyond. We do, however, retain the $\varepsilon_x\varepsilon_z$ and $\varepsilon_y\varepsilon_z$ terms since they contain the first germ of noncommutativity. Note that although these are second-order terms, they are fully determined in our approximation, i.e. unaffected by the second-order terms that we have ignored. Equation (12.2.23) imposes the following restriction on the quantum operators:

$$U[R(-\varepsilon_z\mathbf{k})]T(-\boldsymbol{\varepsilon})U[R(\varepsilon_z\mathbf{k})]T(\boldsymbol{\varepsilon})=T(-\varepsilon_y\varepsilon_z\mathbf{i}+\varepsilon_x\varepsilon_z\mathbf{j}) \tag{12.2.24}$$

or

$$\left(I+\frac{i}{\hbar}\varepsilon_z L_z\right)\left[I+\frac{i}{\hbar}(\varepsilon_x P_x+\varepsilon_y P_y)\right]\left(I-\frac{i}{\hbar}\varepsilon_z L_z\right)\left[I-\frac{i}{\hbar}(\varepsilon_x P_x+\varepsilon_y P_y)\right]$$

$$=I+\frac{i}{\hbar}\varepsilon_y\varepsilon_z P_x-\frac{i}{\hbar}\varepsilon_x\varepsilon_z P_y \tag{12.2.25}$$

By matching coefficients (you should do this) we can deduce the following constraints:

$$[P_x, L_z]=-i\hbar P_y$$

$$[P_y, L_z]=i\hbar P_x$$

which are indeed satisfied by the generators [Eq. (12.2.17)].

So our operators have passed this test. But many other tests are possible. How about the coefficients of terms such as $\varepsilon_x\varepsilon_z^2$, or more generally, how about finite

‡ Note that if rotations and translations commuted, the fourfold product would equal I, as can be seen by rearranging the factors so that the two opposite rotations and the two opposite translations cancel each other. The deviation from this result of I is a measure of noncommutativity. Given two symmetry operations that do not commute, the fourfold product provides a nice characterization of their noncommutavity. As we shall see, this characterization is complete.

rotations? How about tests other than the fourfold product, such as one involving 14 translations and six rotations interlaced?

There is a single answer to all these equations: there is no need to conduct any further tests. Although it is beyond the scope of this book to explain why this is so, it is not hard to explain when it is time to stop testing. We can stop the tests when all possible commutators between the generators have been considered. In the present case, given the generators P_x, P_y, and L_z, the possible commutators are $[P_x, L_z]$, $[P_y, L_z]$, and $[P_x, P_y]$. We have just finished testing the first two. Although the third was tested implicitly in the past, let us do it explicitly again. If we convert the law of combination

$$\begin{bmatrix} x \\ y \end{bmatrix} \xrightarrow[\varepsilon_x \mathbf{i}]{} \begin{bmatrix} x+\varepsilon_x \\ y \end{bmatrix} \xrightarrow[\varepsilon_y \mathbf{j}]{} \begin{bmatrix} x+\varepsilon_x \\ y+\varepsilon_y \end{bmatrix} \xrightarrow[-\varepsilon_x \mathbf{i}]{} \begin{bmatrix} x \\ y+\varepsilon_y \end{bmatrix} \xrightarrow[-\varepsilon_y \mathbf{j}]{} \begin{bmatrix} x \\ y \end{bmatrix} \tag{12.2.26}$$

into the operator constraint

$$T(-\varepsilon_y \mathbf{j})T(-\varepsilon_x \mathbf{i})T(\varepsilon_y \mathbf{j})T(\varepsilon_x \mathbf{i}) = I \tag{12.2.27}$$

we deduce that

$$[P_x, P_y] = 0$$

which of course is satisfied by the generators P_x and P_y. [Although earlier on, we did not consider the fourfold product, Eq. (12.2.27), we did verify that the arguments of the T operators combined according to the laws of vector analysis. Equation (12.2.26) is just a special case which brings out the commutativity of P_x and P_y.]

When I say that there are no further tests to be conducted, I mean the following:

(1) Every consistency test will reduce to just another relation between the *commutators* of the generators.

(2) This relation will be automatically satisfied if the generators pass the tests we have finished conducting. The following exercise should illustrate this point.

*Exercise 12.2.4.** Rederive the equivalent of Eq. (12.2.23) keeping terms of order $\varepsilon_x \varepsilon_z^2$. (You may assume $\varepsilon_y = 0$.) Use this information to rewrite Eq. (12.2.24) to order $\varepsilon_x \varepsilon_z^2$. By equating coefficients of this term deduce the constraint

$$-2L_z P_x L_z + P_x L_z^2 + L_z^2 P_x = \hbar^2 P_x$$

This seems to conflict with statement (1) made above, but not really, in view of the identity

$$-2\Lambda\Omega\Lambda + \Omega\Lambda^2 + \Lambda^2\Omega \equiv [\Lambda, [\Lambda, \Omega]]$$

Using the identify, verify that the new constraint coming from the $\varepsilon_x \varepsilon_z^2$ term is satisfied given the commutation relations between P_x, P_y, and L_z.

We call $\mathbf{V}=V_x\mathbf{i}+V_y\mathbf{j}$ a vector operator if V_x and V_y transform as components of a vector under a passive transformation generated by $U[R]$:

$$U^\dagger[R]V_iU[R]=\sum_j R_{ij}V_j$$

where R_{ij} is the 2×2 rotation matrix appearing in Eq. (12.2.1). Examples of $\mathbf{V}$ are $\mathbf{P}=P_x\mathbf{i}+P_y\mathbf{j}$ and $\mathbf{R}=X\mathbf{i}+Y\mathbf{j}$ [see Eqs. (12.2.14) and (12.2.15)]. Note the twofold character of a vector operator such as $\mathbf{P}$: on the one hand, its components are operators in Hilbert space, and on the other, it transforms as a vector in $\mathbb{V}^2(R)$.

The same definition of a vector operator holds in three dimensions as well, with the obvious difference that R_{ij} is a 3×3 matrix.

12.3. The Eigenvalue Problem of L_z

We have seen that in a rotationally invariant problem, H and L_z share a common basis. In order to exploit this fact we must first find the eigenfunctions of L_z. We begin by writing

$$L_z|l_z\rangle=l_z|l_z\rangle \tag{12.3.1}$$

in the coordinate basis:

$$-i\hbar\frac{\partial\psi_{l_z}(\rho,\phi)}{\partial\phi}=l_z\psi_{l_z}(\rho,\phi) \tag{12.3.2}$$

The solution to this equation is

$$\psi_{l_z}(\rho,\phi)=R(\rho)\,e^{il_z\phi/\hbar} \tag{12.3.3}$$

where $R(\rho)$ is an arbitrary function normalizable with respect to $\int_0^\infty \rho\,d\rho$.‡ We shall have more to say about $R(\rho)$ in a moment. But first note that l_z seems to be arbitrary: it can even be complex since ϕ goes only from 0 to 2π. (Compare this to the eigenfunctions $e^{ipx/\hbar}$ of linear momentum, where we could argue that p had to be real to keep $|\psi|$ bounded as $|x|\to\infty$.) The fact that complex eigenvalues enter the answer, signals that we are overlooking the Hermiticity constraint. Let us impose it. The condition

$$\langle\psi_1|L_z|\psi_2\rangle=\langle\psi_2|L_z|\psi_1\rangle^* \tag{12.3.4}$$

‡ This will ensure that ψ is normalizable with respect to

$$\iint dx\,dy=\int_0^\infty\int_0^{2\pi}\rho\,d\rho\,d\phi$$

becomes in the coordinate basis

$$\int_0^\infty \int_0^{2\pi} \psi_1^*\left(-i\hbar\frac{\partial}{\partial\phi}\right)\psi_2 \rho\, d\rho\, d\phi = \left[\int_0^\infty \int_0^{2\pi} \psi_2^*\left(-i\hbar\frac{\partial}{\partial\phi}\right)\psi_1 \rho\, d\rho\, d\phi\right]^* \tag{12.3.5}$$

If this requirement is to be satisfied for all ψ_1 and ψ_2, one can show (upon integrating by parts) that it is enough if each ψ obeys

$$\psi(\rho, 0) = \psi(\rho, 2\pi) \tag{12.3.6}$$

If we impose this constraint on the L_z eigenfunctions, Eq. (12.3.3), we find

$$1 = e^{2\pi i l_z/\hbar} \tag{12.3.7}$$

This forces l_z not merely to be real, but also to be an integral multiple of $\hbar$:

$$l_z = m\hbar, \qquad m = 0, \pm 1, \pm 2, \ldots \tag{12.3.8}$$

One calls m the *magnetic quantum number*. Notice that $l_z = m\hbar$ implies that ψ is a single-valued function of ϕ. (However, see Exercise 12.3.2.)

Exercise 12.3.1. Provide the steps linking Eq. (12.3.5) to Eq. (12.3.6).

Exercise 12.3.2. Let us try to deduce the restriction on l_z from another angle. Consider a superposition of two allowed l_z eigenstates:

$$\psi(\rho, \phi) = A(\rho)\, e^{i\phi l_z/\hbar} + B(\rho)\, e^{i\phi l'_z/\hbar}$$

By demanding that upon a 2π rotation we get the same physical state (not necessarily the same state vector), show that $l_z - l'_z = m\hbar$, where m is an integer. By arguing on the grounds of symmetry that the allowed values of l_z must be symmetric about zero, show that these values are *either* $\ldots, 3\hbar/2, \hbar/2, -\hbar/2, -3\hbar/2, \ldots$ or $\ldots, 2\hbar, \hbar, 0, -\hbar, -2\hbar, \ldots$. It is not possible to restrict l_z any further this way. □

Let us now return to the arbitrary function $R(\rho)$ that accompanies the eigenfunctions of L_z. Its presence implies that the eigenvalue $l_z = m\hbar$ does not nail down a unique state in Hilbert space but only a subspace $\mathbb{V}_m$. The dimensionality of this space is clearly infinite, for the space of all normalizable functions R is infinite dimensional. The natural thing to do at this point is to introduce some operator that commutes with L_z and whose simultaneous eigenfunctions with L_z pick out a unique basis in each $\mathbb{V}_m$. We shall see in a moment that the Hamiltonian in a rotationally invariant problem does just this. Physically this means that a state is not uniquely specified by just its angular momentum (which only fixes the angular part of the wave function), but it can be specified by its energy and angular momentum in a rotationally invariant problem.

It proves convenient to introduce the functions

$$\Phi_m(\phi)=(2\pi)^{-1/2}\, e^{im\phi} \tag{12.3.9}$$

which would have been nondegenerate eigenfunctions of L_z if the ρ coordinate had not existed. These obey the orthonormality condition

$$\int_0^{2\pi} \Phi_m^*(\phi)\Phi_{m'}(\phi)\, d\phi=\delta_{mm'} \tag{12.3.10}$$

It will be seen that these functions play an important role in problems with rotational invariance.

*Exercise 12.3.3.** A particle is described by a wave function

$$\psi(\rho, \phi)=A\, e^{-\rho^2/2\Delta^2} \cos^2 \phi$$

Show (by expressing $\cos^2 \phi$ in terms of Φ_m) that

$$P(l_z=0)=2/3$$

$$P(l_z=2\hbar)=1/6$$

$$P(l_z=-2\hbar)=1/6$$

(Hint: Argue that the radial part $e^{-\rho^2/2\Delta^2}$ is irrelevant here.)

*Exercise 12.3.4.** A particle is described by a wave function

$$\psi(\rho, \phi)=A\, e^{-\rho^2/2\Delta^2}\left(\frac{\rho}{\Delta} \cos \phi+\sin \phi\right)$$

Show that

$$P(l_z=\hbar)=P(l_z=-\hbar)=\tfrac{1}{2}$$

Solutions to Rotationally Invariant Problems

Consider a problem where $V(\rho, \phi)=V(\rho)$. The eigenvalue equation for H is

$$\left[\frac{-\hbar^2}{2\mu}\left(\frac{\partial^2}{\partial\rho^2}+\frac{1}{\rho}\frac{\partial}{\partial\rho}+\frac{1}{\rho^2}\frac{\partial^2}{\partial\phi^2}\right)+V(\rho)\right]\psi_E(\rho, \phi)=E\psi_E(\rho, \phi) \tag{12.3.11}$$

(We shall use μ to denote the mass, since m will denote the angular momentum quantum number.) Since $[H, L_z]=0$ in this problem, we seek simultaneous eigenfunctions of H and L_z. We have seen that the most general eigenfunction of L_z with eigenvalue $m\hbar$ is of the form

$$\psi_m(\rho, \phi) = R(\rho)(2\pi)^{-1/2} e^{im\phi} = R(\rho)\Phi_m(\phi)$$

where $R(\rho)$ is undetermined. In the present case R is determined by the requirement that

$$\psi_{Em}(\rho, \phi) = R_{Em}(\rho)\Phi_m(\phi) \tag{12.3.12}$$

be an eigenfunction of H as well, with eigenvalue E, i.e., that ψ_{Em} satisfy Eq. (12.3.11). Feeding the above form into Eq. (12.3.11), we get the *radial equation* that determines $R_{Em}(\rho)$ and the allowed values for E:

$$\left[\frac{-\hbar^2}{2\mu}\left(\frac{d^2}{d\rho^2} + \frac{1}{\rho}\frac{d}{d\rho} - \frac{m^2}{\rho^2}\right) + V(\rho)\right] R_{Em}(\rho) = E R_{Em}(\rho) \tag{12.3.13}$$

As we change the potential, only the *radial part* of the wave function, R, changes; the angular part Φ_m is unchanged. Thus the functions $\Phi_m(\phi)$, which were obtained by pretending ρ does not exist, provide the angular part of the wave function in the eigenvalue problem of any rotationally invariant Hamiltonian.

*Exercise 12.3.5**. Note that the angular momentum seems to generate a repulsive potential in Eq. (12.3.13). Calculate its gradient and identify it as the centrifugal force.

Exercise 12.3.6. Consider a particle of mass μ constrained to move on a circle of radius a. Show that $H = L_z^2/2\mu a^2$. Solve the eigenvalue problem of H and interpret the degeneracy.

Exercise 12.3.7. (The Isotropic Oscillator)*. Consider the Hamiltonian

$$H = \frac{P_x^2 + P_y^2}{2\mu} + \frac{1}{2}\mu\omega^2(X^2 + Y^2)$$

(1) Convince yourself $[H, L_z] = 0$ and reduce the eigenvalue problem of H to the radial differential equation for $R_{Em}(\rho)$.

(2) Examine the equation as $\rho \to 0$ and show that

$$R_{Em}(\rho) \underset{\rho \to 0}{\longrightarrow} \rho^{|m|}$$

(3) Show likewise that up to powers of ρ

$$R_{Em}(\rho) \underset{\rho \to \infty}{\longrightarrow} e^{-\mu\omega\rho^2/2\hbar}$$

So assume that $R_{Em}(\rho) = \rho^{|m|} e^{-\mu\omega\rho^2/2\hbar} U_{Em}(\rho)$.

(4) Switch to dimensionless variables $\varepsilon = E/\hbar\omega, y = (\mu\omega/\hbar)^{1/2}\rho$.

(5) Convert the equation for R into an equation for U. (I suggest proceeding in two stages: $R = y^{|m|}f, f = e^{-y^2/2}U$.) You should end up with

$$U'' + \left[\left(\frac{2|m|+1}{y}\right) - 2y\right]U' + (2\varepsilon - 2|m| - 2)U = 0$$

(6) Argue that a power series for U of the form

$$U(y) = \sum_{r=0}^{\infty} C_r y^r$$

will lead to a *two-term* recursion relation.

(7) Find the relation between C_{r+2} and C_r. Argue that the series must terminate at some finite r if the $y \to \infty$ behavior of the solution is to be acceptable. Show $\varepsilon = r + |m| + 1$ leads to termination after r terms. Now argue that r is necessarily even—i.e., $r = 2k$. (Show that if r is odd, the behavior of R as $\rho \to 0$ is not $\rho^{|m|}$.) So finally you must end up with

$$E = (2k + |m| + 1)\hbar\omega, \qquad k = 0, 1, 2, \ldots$$

Define $n = 2k + |m|$, so that

$$E_n = (n+1)\hbar\omega$$

(8) For a given n, what are the allowed values of $|m|$? Given this information show that for a given n, the degeneracy is $n+1$. Compare this to what you found in Cartesian coordinates (Exercise 10.2.2).

(9) Write down all the normalized eigenfunctions corresponding to $n = 0, 1$.

(10) Argue that the $n = 0$ function *must* equal the corresponding one found in Cartesian coordinates. Show that the two $n = 1$ solutions are linear combinations of their counterparts in Cartesian coordinates. Verify that the parity of the states is $(-1)^n$ as you found in Cartesian coordinates.

*Exercise 12.3.8.** Consider a particle of mass μ and charge q in a vector potential

$$\mathbf{A} = \frac{B}{2}(-y\mathbf{i} + x\mathbf{j})$$

(1) Show that the magnetic field is $\mathbf{B} = B\mathbf{k}$.

(2) Show that a classical particle in this potential will move in circles at an angular frequency $\omega_0 = qB/\mu c$.

(3) Consider the Hamiltonian for the corresponding quantum problem:

$$H = \frac{[P_x + qYB/2c]^2}{2\mu} + \frac{[P_y - qXB/2c]^2}{2\mu}$$

Show that $Q = (cP_x + qYB/2)/qB$ and $P = (P_y - qXB/2c)$ are canonical. Write H in terms of P and Q and show that allowed levels are $E = (n + 1/2)\hbar\omega_0$.

(4) Expand H out in terms of the original variables and show

$$H=H\left(\frac{\omega_0}{2}, \mu\right)-\frac{\omega_0}{2} L_z$$

where $H(\omega_0/2, \mu)$ is the Hamiltonian for an isotropic two-dimensional harmonic oscillator of mass μ and frequency $\omega_0/2$. Argue that the same basis that diagonalized $H(\omega_0/2, \mu)$ will diagonalize H. By thinking in terms of this basis, show that the allowed levels for H are $E=(k+\frac{1}{2}|m|-\frac{1}{2}m+\frac{1}{2})\hbar\omega_0$, where k is any integer and m is the angular momentum. Convince yourself that you get the same levels from this formula as from the earlier one $[E=(n+1/2)\hbar\omega_0]$. We shall return to this problem in Chapter 21.

12.4. Angular Momentum in Three Dimensions

It is evident that as we pass from two to three dimensions, the operator L_z picks up two companions L_x and L_y which generate infinitesimal rotations about the x and y axes, respectively. So we have

$$L_x=YP_z-ZP_y \tag{12.4.1a}$$

$$L_y=ZP_x-XP_z \tag{12.4.1b}$$

$$L_z=XP_y-YP_x \tag{12.4.1c}$$

As usual, we subject these to the consistency test. It may be verified, (Exercise 12.4.2), that if we take a point in three-dimensional space and subject it to the following rotations: $R(\varepsilon_x\mathbf{i})$, $R(\varepsilon_y\mathbf{j})$, $R(-\varepsilon_x\mathbf{i})$ and lastly $R(-\varepsilon_y\mathbf{j})$, it ends up rotated by $-\varepsilon_x\varepsilon_y\mathbf{k}$. In other words

$$R(-\varepsilon_y\mathbf{j})R(-\varepsilon_x\mathbf{i})R(\varepsilon_y\mathbf{j})R(\varepsilon_x\mathbf{i})=R(-\varepsilon_x\varepsilon_y\mathbf{k}) \tag{12.4.2}$$

It follows that the quantum operators $U[R]$ must satisfy

$$U[R(-\varepsilon_y\mathbf{j})]U[R(-\varepsilon_x\mathbf{i})]U[R(\varepsilon_y\mathbf{j})]U[R(\varepsilon_x\mathbf{i})]=U[R(-\varepsilon_x\varepsilon_y\mathbf{k})] \tag{12.4.3}$$

If we write each U to order ε and match coefficients of $\varepsilon_x\varepsilon_y$, we will find

$$[L_x, L_y]=i\hbar L_z \tag{12.4.4a}$$

By considering two similar tests involving $\varepsilon_y\varepsilon_z$ and $\varepsilon_z\varepsilon_x$, we can deduce the constraints

$$[L_y, L_z]=i\hbar L_x \tag{12.4.4b}$$

$$[L_z, L_x]=i\hbar L_y \tag{12.4.4c}$$

You may verify that the operators in Eq. (12.4.1) satisfy these constraints. So they are guaranteed to generate finite rotation operators that obey the right laws of combination.

The three relations above may be expressed compactly as one vector equation

$$\mathbf{L}\times\mathbf{L}=i\hbar\mathbf{L} \tag{12.4.5}$$

Yet another way to write the commutation relations is

$$[L_i, L_j]=i\hbar\sum_{k=1}^{3}\varepsilon_{ijk}L_k \tag{12.4.6}$$

In this equation, i and j run from 1 to 3, L_1, L_2, and L_3 stand for L_x, L_y, and L_z, respectively,‡ and ε_{ijk} are the components of an antisymmetric tensor of rank 3, with the following properties:

(1) They change sign when any two indices are exchanged. Consequently no two indices can be equal.
(2) $\varepsilon_{123}=1$.

This fixes all other components. For example,

$$\varepsilon_{132}=-1, \qquad \varepsilon_{312}=(-1)(-1)=+1 \tag{12.4.7}$$

and so on. In short, ε_{ijk} is $+1$ for any cyclic permutation of the indices in ε_{123} and -1 for the others. (The relation

$$\mathbf{c}=\mathbf{a}\times\mathbf{b} \tag{12.4.8}$$

between three vectors from $\mathbb{V}^3(R)$ may be written in component form as

$$c_i=\sum_{j=1}^{3}\sum_{k=1}^{3}\varepsilon_{ijk}a_jb_k \tag{12.4.9}$$

Of course $\mathbf{a}\times\mathbf{a}$ is zero if $\mathbf{a}$ is a vector whose components are numbers, but not zero if it is an operator such as $\mathbf{L}$.)

*Exercise 12.4.1.** (1) Verify that Eqs. (12.4.9) and Eq. (12.4.8) are equivalent, given the definition of ε_{ijk}.

(2) Let U_1, U_2, and U_3 be three energy eigenfunctions of a single particle in some potential. Construct the wave function $\psi_A(x_1, x_2, x_3)$ of three fermions in this potential, one of which is in U_1, one in U_2, and one in U_3, using the ε_{ijk} tensor.

*Exercise 12.4.2.** (1) Verify Eq. (12.4.2) by first constructing the 3×3 matrices corresponding to $R(\varepsilon_x\mathbf{i})$ and $R(\varepsilon_y\mathbf{j})$, to order ε.

(2) Provide the steps connecting Eqs. (12.4.3) and (12.4.4a).

‡ We will frequently let the indices run over 1, 2, and 3 instead of x, y, and z.

(3) Verify that L_x and L_y defined in Eq. (12.4.1) satisfy Eq. (12.4.4a). The proof for other commutators follows by cyclic permutation.

We next define the total angular momentum operator squared

$$L^2 = L_x^2 + L_y^2 + L_z^2 \tag{12.4.10}$$

It may be verified (by you) that

$$[L^2, L_i] = 0, \quad i = x, y, \text{ or } z \tag{12.4.11}$$

Finite Rotation Operators. Rotations about a given axis commute. So a finite rotation may be viewed as a sequence of infinitesimal rotations about the same axis. What is the operator that rotates by angle $\boldsymbol{\theta}$, i.e., by an amount θ about an axis parallel to $\hat{\theta}$? If $\boldsymbol{\theta} = \theta_x \mathbf{i}$, then clearly

$$U[R(\theta_x \mathbf{i})] = e^{-i\theta_x L_x/\hbar}$$

The same goes for $\boldsymbol{\theta}$ along the unit vectors $\mathbf{j}$ and $\mathbf{k}$. What if $\boldsymbol{\theta}$ has some arbitrary direction? We conjecture that $L_{\hat{\theta}} \equiv \hat{\theta} \cdot \mathbf{L}$ (where $\hat{\theta} = \boldsymbol{\theta}/\theta$) is the generator of infinitesimal rotations about that axis and that

$$\begin{aligned} U[R(\boldsymbol{\theta})] = \lim_{N\to\infty} \left(I - \frac{i}{\hbar}\frac{\theta}{N}\hat{\theta} \cdot \mathbf{L}\right)^N &= e^{-i\theta\hat{\theta}\cdot\mathbf{L}/\hbar} \\ &= e^{-i\mathbf{L}\cdot\boldsymbol{\theta}} \end{aligned} \tag{12.4.12}$$

Our conjecture is verified in the following exercise.

*Exercise 12.4.3.** We would like to show that $\hat{\theta} \cdot \mathbf{L}$ generates rotations about the axis parallel to $\hat{\theta}$. Let $\delta\boldsymbol{\theta}$ be an infinitesimal rotation parallel to $\boldsymbol{\theta}$.

(1) Show that when a vector $\mathbf{r}$ is rotated by an angle $\delta\boldsymbol{\theta}$, it changes to $\mathbf{r} + \delta\boldsymbol{\theta} \times \mathbf{r}$. (It might help to start with $\mathbf{r} \perp \delta\boldsymbol{\theta}$ and then generalize.)

(2) We therefore demand that (to first order, as usual)

$$\psi(\mathbf{r}) \underset{U[R(\delta\boldsymbol{\theta})]}{\longrightarrow} \psi(\mathbf{r} - \delta\boldsymbol{\theta} \times \mathbf{r}) = \psi(\mathbf{r}) - (\delta\boldsymbol{\theta} \times \mathbf{r}) \cdot \nabla\psi$$

Comparing to $U[R(\delta\boldsymbol{\theta})] = I - (i\delta\theta/\hbar)L_{\hat{\theta}}$, show that $L_{\hat{\theta}} = \hat{\theta} \cdot \mathbf{L}$.

*Exercise 12.4.4.** Recall that $\mathbf{V}$ is a vector operator if its components V_i transform as

$$U^\dagger[R] V_i U[R] = \sum_j R_{ij} V_j \tag{12.4.13}$$

(1) For an infinitesimal rotation $\delta\boldsymbol{\theta}$, show, on the basis of the previous exercise, that

$$\sum_j R_{ij} V_j = V_i + (\delta\boldsymbol{\theta} \times \mathbf{V})_i = V_i + \sum_j \sum_k \varepsilon_{ijk} (\delta\theta)_j V_k$$

(2) Feed in $U[R] = 1 - (i/\hbar)\delta\boldsymbol{\theta} \cdot \mathbf{L}$ into the left-hand side of Eq. (12.4.13) and deduce that

$$[V_i, L_j] = i\hbar \sum_k \varepsilon_{ijk} V_k \tag{12.4.14}$$

This is as good a definition of a vector operator as Eq. (12.4.13). By setting $\mathbf{V} = \mathbf{L}$, we can obtain the commutation rules among the L's.

If the Hamiltonian is invariant under arbitrary rotations,

$$U^\dagger[R] H U[R] = H \tag{12.4.15}$$

it follows (upon considering infinitesimal rotations around the x, y, and z axes) that

$$[H, L_i] = 0 \tag{12.4.16}$$

and from it

$$[H, L^2] = 0 \tag{12.4.17}$$

Thus L^2 and all three components of $\mathbf{L}$ are conserved. It does not, however, follow that there exists a basis common to H and all three L's. This is because the L's do not commute with each other. So the best one can do is find a basis common to H, L^2, and one of the L's, usually chosen to be L_z.

We now examine the eigenvalue problem of the commuting operators L^2 and L_z. When this is solved, we will turn to the eigenvalue problem of H, L^2, and L_z.

12.5. The Eigenvalue Problem of L^2 and L_z

There is a close parallel between our approach to this problem and that of the harmonic oscillator. Recall that in that case we (1) solved the eigenvalue problem of H in the coordinate basis; (2) solved the problem in the energy basis directly, using the a and $a^\dagger$ operators, the commutation rules, and the positivity of H; (3) obtained the coordinate wave function $\psi_n(y)$ given the results of part (2), by the following trick. We wrote

$$a|0\rangle = 0$$

in the coordinate basis as

$$\left(y + \frac{\partial}{\partial y}\right)\psi_0(y) = 0$$

which immediately gave us $\psi_0(y) \sim e^{-y^2/2}$, up to a normalization that could be easily determined.

Given the normalized eigenfunction $\psi_0(y)$, we got $\psi_n(y)$ by the application of the (differential) operator $(a^\dagger)^n/(n!)^{1/2} \to (y-\partial/\partial y)^n/(2^n n!)^{1/2}$.

In the present case we omit part (1), which involves just one more bout with differential equations and is not particularly enlightening.

Let us now consider part (2). It too has many similarities with part (2) of the oscillator problem.‡ We begin by assuming that there exists a basis $|\alpha, \beta\rangle$ common to L^2 and L_z:

$$L^2|\alpha\beta\rangle = \alpha|\alpha\beta\rangle \tag{12.5.1}$$

$$L_z|\alpha\beta\rangle = \beta|\alpha\beta\rangle \tag{12.5.2}$$

We now define *raising and lower operators*

$$L_\pm = L_x \pm iL_y \tag{12.5.3}$$

which satisfy

$$[L_z, L_\pm] = \pm\hbar L_\pm \tag{12.5.4}$$

and of course (since L^2 commutes with L_x and L_y)

$$[L^2, L_\pm] = 0 \tag{12.5.5}$$

Equations (12.5.4) and (12.5.5) imply that $L_\pm$ raise/lower the eigenvalue of L_z by $\hbar$, while leaving the eigenvalue of L^2 alone. For example,

$$\begin{aligned} L_z(L_+|\alpha\beta\rangle) &= (L_+L_z + \hbar L_+)|\alpha\beta\rangle \\ &= (L_+\beta + \hbar L_+)|\alpha\beta\rangle \\ &= (\beta+\hbar)(L_+|\alpha\beta\rangle) \end{aligned} \tag{12.5.6}$$

and

$$L^2L_+|\alpha\beta\rangle = L_+L^2|\alpha\beta\rangle = \alpha L_+|\alpha\beta\rangle \tag{12.5.7}$$

From Eqs. (12.5.6) and (12.5.7) it is clear that $L_+|\alpha\beta\rangle$ is proportional to the normalized eigenket $|\alpha, \beta+\hbar\rangle$:

$$L_+|\alpha\beta\rangle = C_+(\alpha, \beta)|\alpha, \beta+\hbar\rangle \tag{12.5.8a}$$

‡ If you have forgotten the latter, you are urged to refresh your memory at this point.

It can similarly be shown that

$$L_-|\alpha\beta\rangle = C_-(\alpha, \beta)|\alpha, \beta-\hbar\rangle \tag{12.5.8b}$$

The existence of $L_\pm$ implies that given an eigenstate $|\alpha\beta\rangle$ there also exist eigenstates $|\alpha, \beta+\hbar\rangle$, $|\alpha, \beta+2\hbar\rangle, \ldots$; and $|\alpha, \beta-\hbar\rangle$, $|\alpha, \beta-2\hbar\rangle, \ldots$. This clearly signals trouble, for classical intuition tells us that the z component of angular momentum cannot take arbitrarily large positive or negative values for a given value of the square of the total angular momentum; in fact classically $|l_z| \le (l^2)^{1/2}$.

Quantum mechanically we have

$$\langle\alpha\beta|L^2 - L_z^2|\alpha\beta\rangle = \langle\alpha\beta|L_x^2 + L_y^2|\alpha\beta\rangle \tag{12.5.9}$$

which implies

$$\alpha - \beta^2 \ge 0$$

(since $L_x^2 + L_y^2$ is positive definite) or

$$\alpha \ge \beta^2 \tag{12.5.10}$$

Since β^2 is bounded by α, it follows that there must exist a state $|\alpha\beta_{\max}\rangle$ such that it cannot be raised:

$$L_+|\alpha\beta_{\max}\rangle = 0 \tag{12.5.11}$$

Operating with L_- and using $L_-L_+ = L^2 - L_z^2 - \hbar L_z$, we get

$$(L^2 - L_z^2 - \hbar L_z)|\alpha\beta_{\max}\rangle = 0$$

$$(\alpha - \beta_{\max}^2 - \hbar\beta_{\max})|\alpha\beta_{\max}\rangle = 0$$

$$\alpha = \beta_{\max}(\beta_{\max} + \hbar) \tag{12.5.12}$$

Starting with $|\alpha\beta_{\max}\rangle$ let us operate k times with L_-, till we reach a state $|\alpha\beta_{\min}\rangle$ that cannot be lowered further without violating the inequality (12.5.10):

$$L_-|\alpha\beta_{\min}\rangle = 0$$

$$L_+L_-|\alpha\beta_{\min}\rangle = 0$$

$$(L^2 - L_z^2 + \hbar L_z)|\alpha\beta_{\min}\rangle = 0$$

$$\alpha = \beta_{\min}(\beta_{\min} - \hbar) \tag{12.5.13}$$

A comparison of Eqs. (12.5.12) and (12.5.13) shows (as is to be expected)

$$\beta_{\min} = -\beta_{\max} \tag{12.5.14}$$

Table 12.1. Some Low-Angular-Momentum States

(Angular momentum) $k/2$	β_{max}	α	$\lvert\alpha\beta\rangle$
0	0	0	$\lvert 0, 0\rangle$
1/2	$\hbar/2$	$(1/2)(3/2)\hbar^2$	$\lvert(3/4)\hbar^2, \hbar/2\rangle$ $\lvert(3/4)\hbar^2, -\hbar/2\rangle$
1	$\hbar$	$(1)(2)\hbar^2$	$\lvert 2\hbar^2, \hbar\rangle$ $\lvert 2\hbar^2, 0\rangle$ $\lvert 2\hbar^2, -\hbar\rangle$
3/2			
⋮	⋮	⋮	⋮

Since we got to $|\alpha\beta_{min}\rangle$ from $|\alpha\beta_{max}\rangle$ in k steps of $\hbar$ each, it follows that

$$\beta_{max}-\beta_{min}=2\beta_{max}=\hbar k$$

$$\beta_{max}=\frac{\hbar k}{2}, \qquad k=0, 1, 2, \ldots \tag{12.5.15a}$$

$$\alpha=(\beta_{max})(\beta_{max}+\hbar)=\hbar^2\left(\frac{k}{2}\right)\left(\frac{k}{2}+1\right) \tag{12.5.15b}$$

We shall refer to $(k/2)=(\beta_{max}/\hbar)$ as the *angular momentum of the state*. Notice that unlike in classical physics, β^2_{max} is less than α, the square of the magnitude of angular momentum, except when $\alpha=\beta_{max}=0$, i.e., in a state of zero angular momentum.

Let us now take a look at a few of the low-angular-momentum states listed in Table 12.1.

At this point the astute reader raises the following objection.

A.R.: I am disturbed by your results for odd k. You seem to find that L_z can have half-integral eigenvalues (in units of $\hbar$). But you just convinced us in Section 12.3 that L_z has only integral eigenvalues m (in units of $\hbar$). Where did you go wrong?

R.S.: Nowhere, but your point is well taken. The extra (half-integral) eigenvalues arise because we have solved a more general problem than that of L_x, L_y, L_z, and L^2 (although we didn't intend to). Notice that nowhere in the derivation did we use the explicit expressions for the L's [Eq. (12.4.1)] and in particular $L_z\to -i\hbar\partial/\partial\phi$. (Had we done so, we would have gotten only integral eigenvalues as you expect.) We relied instead on just the commutation relations, $\mathbf{L}\times\mathbf{L}=i\hbar\mathbf{L}$. Now, these commutation relations reflect the law of combinations of infinitesimal rotations in three dimensions and must be satisfied by the three generators of rotations whatever the nature of the wave functions they rotate. We have so far considered just scalar wave functions $\psi(x, y, z)$, which assign a complex number (scalar) to each point. Now, there can be particles in nature for which the wave function is more complicated, say a vector field $\mathbf{\Psi}(x, y, z)=\psi_x(x, y, z)\mathbf{i}+\psi_y(x, y, z)\mathbf{j}+\psi_z(x, y, x)\mathbf{k}$. The response of such a wave function to rotations is more involved. Whereas in the scalar case the effect of rotation by $\delta\boldsymbol{\theta}$ is to take the number assigned to each point (x, y, z)

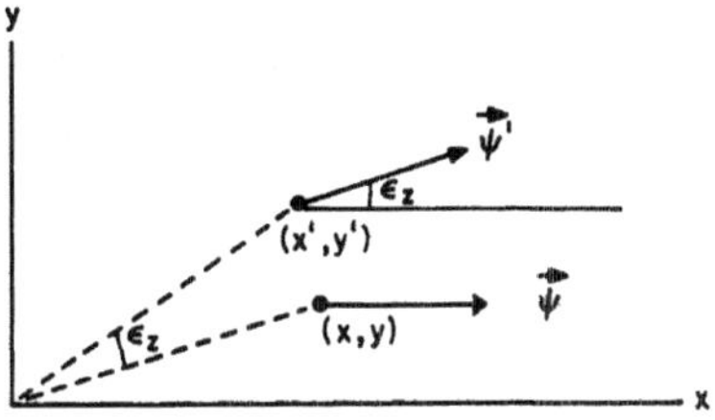

Figure 12.1. The effect of the infinitesimal rotations by ε_z on a vector ψ in two dimensions is to (1) first reassign it to the rotated point (x', y') (2) and then rotate the vector itself by the infinitesimal angle. The differential operator L_z does the first part while a 2×2 spin matrix S_z does the second.

and reassign it to the rotated point (x', y', z'), in the vector case the vector at (x, y, z) (i) must itself be rotated by $\delta\boldsymbol{\theta}$ and (ii) then reassigned to (x', y', z'). (A simple example from two dimensions is given in Fig. 12.1.) The differential operators L_x, L_y, and L_z will only do part (ii) but not part (i), which has to be done by 3×3 matrices S_x, S_y, and S_z which shuffle the components ψ_x, ψ_y, ψ_z of $\boldsymbol{\Psi}$. In such cases, the generators of infinitesimal rotations will be of the form

$$J_i = L_i + S_i$$

where L_i does part (2) and S_i does part (1) (see Exercise 12.5.1 for a concrete example). One refers to L_i as the *orbital angular* momentum, S_i as the *spin angular momentum* (or simply spin), and J_i as the *total angular momentum*. We do not yet know what J_i or S_i look like in these general cases, but we do know this: the J_i's must obey the same commutation rules as the L_i's, for the commutation rules reflect the law of combination of rotations and must be obeyed by any triplet of generators (the consistency condition), whatever be the nature of wave function they rotate. So in general we have

$$\mathbf{J}\times\mathbf{J} = i\hbar\mathbf{J} \qquad (12.5.16)$$

with $\mathbf{L}$ as a special case when the wave function is a scalar. So our result, which followed from just the commutation relations, applies to the problem of arbitrary $\mathbf{J}$ and not just $\mathbf{L}$. Thus the answer to the question raised earlier is that unlike L_z, J_z is not restricted to have integral eigenvalues. But our analysis tells us, who know very little about spin, that S_z can have only integral or half-integral eigenvalues if the commutation relations are to be satisfied. Of course, our analysis doesn't imply that there *must* exist particles with spin integral or half integral—but merely reveals the possible variety in wave functions. But the old maxim—if something can happen, it will—is true here and nature does provide us with particles that possess spin—i.e., particles whose wave functions are more complicated than scalars. We will study them in Chapter 14 on spin.

*Exercise 12.5.1.** Consider a vector field $\boldsymbol{\Psi}(x, y)$ in two dimensions. From Fig. 12.1 it follows that under an infinitesimal rotation $\varepsilon_z\mathbf{k}$,

$$\psi_x \to \psi'_x(x, y) = \psi_x(x + y\varepsilon_z, y - x\varepsilon_z) - \psi_y(x + y\varepsilon_z, y - x\varepsilon_z)\varepsilon_z$$

$$\psi_y \to \psi'_y(x, y) = \psi_x(x + y\varepsilon_z, y - x\varepsilon_z)\varepsilon_z + \psi_y(x + y\varepsilon_z, y - x\varepsilon_z)$$

Show that (to order ε_z)

$$\begin{bmatrix}\psi'_x\\ \psi'_y\end{bmatrix}=\left(\begin{bmatrix}1 & 0\\ 0 & 1\end{bmatrix}-\frac{i\varepsilon_z}{\hbar}\begin{bmatrix}L_z & 0\\ 0 & L_z\end{bmatrix}-\frac{i\varepsilon_z}{\hbar}\begin{bmatrix}0 & -i\hbar\\ i\hbar & 0\end{bmatrix}\right)\begin{bmatrix}\psi_x\\ \psi_y\end{bmatrix}$$

so that

$$\begin{aligned}J_z&=L_z^{(1)}\otimes I^{(2)}+I^{(1)}\otimes S_z^{(2)}\\ &=L_z+S_z\end{aligned}$$

where $I^{(2)}$ is a 2×2 identity matrix with respect to the vector components, $I^{(1)}$ is the identity operator with respect to the argument (x, y) of $\Psi(x, y)$. This example only illustrates the fact that $J_z=L_z+S_z$ if the wave function is not a scalar. An example of half-integral eigenvalues will be provided when we consider spin in a later chapter. (In the present example, S_z has eigenvalues $\pm\hbar$.)

Let us return to our main discussion. To emphasize the generality of the results we have found, we will express them in terms of J's rather than L's and also switch to a more common notation. Here is a summary of what we have found. The eigenvectors of the operators J^2 and J_z are given by

$$J^2|jm\rangle=j(j+1)\hbar^2|jm\rangle, \qquad j=0, 1/2, 1, 3/2, \ldots \tag{12.5.17a}$$

$$J_z|jm\rangle=m\hbar|jm\rangle, \qquad m=j, j-1, j-2, \ldots, -j \tag{12.5.17b}$$

We shall call j the angular momentum of the state. Note that in the above m can be an integer or half-integer depending on j.

The results for the restricted problem $\mathbf{J}=\mathbf{L}$ that we originally set out to solve are contained in Eq. (12.5.17): we simply ignore the states with half-integral m and j. To remind us in these cases that we are dealing with $\mathbf{J}=\mathbf{L}$, we will denote these states by $|lm\rangle$. They obey

$$L^2|lm\rangle=l(l+1)\hbar^2|lm\rangle, \qquad l=0, 1, 2, \ldots \tag{12.5.18a}$$

$$L_z|lm\rangle=m\hbar|lm\rangle, \qquad m=l, l-1, \ldots, -l \tag{12.5.18b}$$

Our problem has not been fully solved: we have only found the eigenvalues—the eigenvectors aren't fully determined yet. (As in the oscillator problem, finding the eigenvectors means finding the matrices corresponding to the basic operators whose commutation relations are given.) Let us continue our analysis in terms of the J's. If we rewrite Eq. (12.5.8) in terms of $J_\pm$, j, and m (instead of $L_\pm$, α, and β), we get

$$J_\pm|jm\rangle=C_\pm(j, m)|j, m\pm 1\rangle \tag{12.5.19}$$

where $C_\pm(j, m)$ are yet to be determined. We will determine them now.

If we take the adjoint of

$$J_+|jm\rangle = C_+(j, m)|j, m+1\rangle$$

we get

$$\langle jm|\, J_- = C_+^*(j, m)\langle j, m+1|$$

Equating the inner product of the objects on the left-hand side to the product of the objects on the right-hand side, we obtain

$$\begin{aligned}\langle jm|\, J_-J_+|jm\rangle &= |C_+(j, m)|^2\langle j, m+1|j, m+1\rangle\\ &= |C_+(j, m)|^2\\ \langle jm|\, J^2 - J_z^2 - \hbar J_z|jm\rangle &= |C_+(j, m)|^2\end{aligned}$$

or

$$\begin{aligned}|C_+(j, m)|^2 &= j(j+1)\hbar^2 - m^2\hbar^2 - m\hbar^2\\ &= \hbar^2(j-m)(j+m+1)\end{aligned}$$

or‡

$$C_+(j, m) = \hbar[(j-m)(j+m+1)]^{1/2}$$

It can likewise be shown that

$$C_-(j, m) = \hbar[(j+m)(j-m+1)]^{1/2}$$

so that finally

$$J_\pm|jm\rangle = \hbar[(j\mp m)(j\pm m+1)]^{1/2}|j, m\pm 1\rangle \tag{12.5.20}$$

Notice that when $J_\pm$ act on $|j, \pm j\rangle$ they kill the state, so that each family with a given angular momentum j has only $2j+1$ states with eigenvalues $j\hbar$, $(j-1)\hbar, \ldots, -(j\hbar)$ for J_z.

Equation (12.5.20) brings us to the end of our calculation, for we can write down the matrix elements of J_x and J_y in this basis:

$$\begin{aligned}\langle j'm'|\, J_x|jm\rangle &= \langle j'm'|\,\frac{J_+ + J_-}{2}\,|jm\rangle\\ &= \frac{\hbar}{2}\{\delta_{jj'}\,\delta_{m',m+1}[(j-m)(j+m+1)]^{1/2} + \delta_{jj'}\,\delta_{m',m-1}\\ &\quad\times[(j+m)(j-m+1)]^{1/2}\}\end{aligned} \tag{12.5.21a}$$

‡ There can be an overall phase factor in front of C_+. We choose it to be unity according to standard convention.

$$\langle j'm'|J_y|jm\rangle = \langle j'm'|\frac{J_+ - J_-}{2i}|jm\rangle$$

$$= \frac{\hbar}{2i}\{\delta_{jj'}\delta_{m',m+1}[(j-m)(j+m+1)]^{1/2} - \delta_{jj'}\delta_{m',m-1} \times [(j+m)(j-m+1)]^{1/2}\} \quad (12.5.21b)$$

Using these (or our mnemonic based on images) we can write down the matrices corresponding to J^2, J_z, J_x, and J_y in the $|jm\rangle$ basis‡:

$$J^2 \rightarrow \begin{array}{c|ccccccc} j'm' \backslash jm & (0,0) & (\frac{1}{2},\frac{1}{2}) & (\frac{1}{2},-\frac{1}{2}) & (1,1) & (1,0) & (1,-1) & \cdots \\ \hline (0,0) & 0 & 0 & 0 & 0 & 0 & 0 & \cdots \\ (\frac{1}{2},\frac{1}{2}) & 0 & \frac{3}{4}\hbar^2 & 0 & 0 & 0 & 0 & \\ (\frac{1}{2},-\frac{1}{2}) & 0 & 0 & \frac{3}{4}\hbar^2 & 0 & 0 & 0 & \\ (1,1) & 0 & 0 & 0 & 2\hbar^2 & 0 & 0 & \\ (1,0) & 0 & 0 & 0 & 0 & 2\hbar^2 & 0 & \\ (1,-1) & 0 & 0 & 0 & 0 & 0 & 2\hbar^2 & \\ & \vdots & & & & & & \ddots \end{array} \quad (12.5.22)$$

J_z is also diagonal with elements $m\hbar$.

$$J_x \rightarrow \begin{bmatrix} 0 & 0 & 0 & 0 & 0 & 0 & \cdots \\ 0 & 0 & \hbar/2 & 0 & 0 & 0 & \\ 0 & \hbar/2 & 0 & 0 & 0 & 0 & \\ 0 & 0 & 0 & 0 & \hbar/2^{1/2} & 0 & \\ 0 & 0 & 0 & \hbar/2^{1/2} & 0 & \hbar/2^{1/2} & \\ 0 & 0 & 0 & 0 & \hbar/2^{1/2} & 0 & \\ \vdots & & & & & & \ddots \end{bmatrix} \quad (12.5.23)$$

$$J_y \rightarrow \begin{bmatrix} 0 & 0 & 0 & 0 & 0 & 0 & \cdots \\ 0 & 0 & -i\hbar/2 & 0 & 0 & 0 & \\ 0 & i\hbar/2 & 0 & 0 & 0 & 0 & \\ 0 & 0 & 0 & 0 & -i\hbar/2^{1/2} & 0 & \\ 0 & 0 & 0 & i\hbar/2^{1/2} & 0 & -i\hbar/2^{1/2} & \\ 0 & 0 & 0 & 0 & i\hbar/2^{1/2} & 0 & \\ \vdots & & & & & & \ddots \end{bmatrix} \quad (12.5.24)$$

Notice that although J_x and J_y are not diagonal in the $|jm\rangle$ basis, they are *block diagonal*: they have no matrix elements between one value of j and another. This is

‡ The quantum numbers j and m do not fully label a state; a state is labeled by $|\alpha jm\rangle$, where α represents the remaining labels. In what follows, we suppress α but assume it is the same throughout.

because $J_\pm$ (out of which they are built) do not change j when they act on $|jm\rangle$. Since the J's are all block diagonal, the blocks do not mix when we multiply them. *In particular when we consider a commutation relation such as* $[J_x, J_y]=i\hbar J_z$, *it will be satisfied within each block.* If we denote the $(2j+1)\times(2j+1)$ block in J_i, corresponding to a certain j, by $J_i^{(j)}$, then we have

$$[J_x^{(j)}, J_y^{(j)}]=i\hbar J_z^{(j)}, \qquad j=0, \tfrac{1}{2}, 1, \ldots \tag{12.5.25}$$

Exercise 12.5.2. (1) Verify that the 2×2 matrices $J_x^{(1/2)}$, $J_y^{(1/2)}$, and $J_z^{(1/2)}$ obey the commutation rule $[J_x^{(1/2)}, J_y^{(1/2)}]=i\hbar J_z^{(1/2)}$.

(2) Do the same for the 3×3 matrices $J_i^{(1)}$.

(3) Construct the 4×4 matrices and verify that

$$[J_x^{(3/2)}, J_y^{(3/2)}]=i\hbar J_z^{(3/2)}$$

*Exercise 12.5.3.** (1) Show that $\langle J_x\rangle=\langle J_y\rangle=0$ in a state $|jm\rangle$.

(2) Show that in these states

$$\langle J_x^2\rangle=\langle J_y^2\rangle=\tfrac{1}{2}\hbar^2[j(j+1)-m^2]$$

(use symmetry arguments to relate $\langle J_x^2\rangle$ to $\langle J_y^2\rangle$).

(3) Check that $\Delta J_x\cdot\Delta J_y$ from part (2) satisfies the inequality imposed by the uncertainty principle [Eq. (9.2.9)].

(4) Show that the uncertainty bound is saturated in the state $|j, \pm j\rangle$.

Finite Rotations‡

Now that we have explicit matrices for the generators of rotations, J_x, J_y, and J_z, we can construct the matrices representing $U[R]$ by exponentiating $(-i\boldsymbol{\theta}\cdot\mathbf{J}/\hbar)$. But this is easier said than done. The matrices J_i are infinite dimensional and exponentiating them is not practically possible. But the situation is not as bleak as it sounds for the following reason. First note that since J_i are block diagonal, so is the linear combination $\boldsymbol{\theta}\cdot\mathbf{J}$, and so is its exponential. Consequently, *all* rotation operators $U[R]$ will be represented by block diagonal matrices. The $(2j+1)$-dimensional block at a given j is denoted by $D^{(j)}[R]$. The block diagonal form of the rotation matrices implies (recall the mnemonic of images) that any vector $|\psi_j\rangle$ in the subspace $\mathbb{V}_j$ spanned by the $(2j+1)$ vectors $|jj\rangle, \ldots, |j-j\rangle$ goes into another element $|\psi_j'\rangle$ of $\mathbb{V}_j$. Thus to rotate $|\psi_j\rangle$, we just need the matrix $D^{(j)}$. More generally, if $|\psi\rangle$ has components only in $\mathbb{V}_0, \mathbb{V}_1, \mathbb{V}_2, \ldots, \mathbb{V}_j$, we need just the first $(j+1)$ matrices $D^{(j)}$. What makes the situation hopeful is that it is possible, in practice, to evaluate these if j is small. Let us see why. Consider the series representing $D^{(j)}$:

$$D^{(j)}[R(\boldsymbol{\theta})]=\exp\left[-\frac{i\boldsymbol{\theta}\cdot\mathbf{J}^{(j)}}{\hbar}\right]=\sum_0^\infty\left(\frac{-i\theta}{\hbar}\right)^n(\hat{\theta}\cdot\mathbf{J}^{(j)})^n\frac{1}{n!}$$

‡ The material from here to the end of Exercise 12.5.7 may be skimmed over in a less advanced course.

It can be shown (Exercise 12.5.4) that $(\hat{\theta}\cdot\mathbf{J}^{(j)})^n$ for $n>2j$ can be written as a linear combination of the first $2j$ powers of $\hat{\theta}\cdot\mathbf{J}^{(j)}$. Consequently the series representing $D^{(j)}$ may be reduced to

$$D^{(j)}=\sum_{0}^{2j} f_n(\theta)(\hat{\theta}\cdot\mathbf{J}^{(j)})^n$$

It is possible, in practice, to find closed expressions for $f_n(\theta)$ in terms of trigonometric functions, for modest values of j (see Exercise 12.5.5). For example,

$$D^{(1/2)}[R]=\cos\left(\frac{\theta}{2}\right)-\frac{2i}{\hbar}\,\hat{\theta}\cdot\mathbf{J}^{(1/2)}\sin\left(\frac{\theta}{2}\right)$$

Let us return to the subspaces $\mathbb{V}_j$. Since they go into themselves under arbitrary rotations, they are called *invariant subspaces.* The physics behind the invariance is simple: each subspace contains states of a definite magnitude of angular momentum squared $j(j+1)\hbar^2$, and a rotation cannot change this. Formally it is because $[J^2, U[R]]=0$ and so $U[R]$ cannot change the eigenvalue of J^2.

The invariant subspaces have another feature: they are *irreducible.* This means that $\mathbb{V}_j$ itself does not contain *invariant subspaces.* We prove this by showing that any invariant subspace $\bar{\mathbb{V}}_j$ of $\mathbb{V}_j$ is as big as the latter. Let $|\psi\rangle$ be an element of $\bar{\mathbb{V}}_j$. Since we haven't chosen a basis yet, let us choose one such that $|\psi\rangle$ is one of the basis vectors, and furthermore, such that it is the basis vector $|jj\rangle$, up to a normalization factor, which is irrelevant in what follows. (What if we had already chosen a basis $|jj\rangle,\ldots,|j,-j\rangle$ generated by the operators J_i? Consider any unitary transformation U which converts $|jj\rangle$ into $|\psi\rangle$ and a different triplet of operators J_i' defined by $J_i'=UJ_iU^\dagger$. The primed operators have the same commutation rules and hence eigenvalues as the J_i. The eigenvectors are just $|jm\rangle'=U|jm\rangle$, with $|jj\rangle'=|\psi\rangle$. In the following analysis we drop all primes.)

Let us apply an infinitesimal rotation $\delta\boldsymbol{\theta}$ to $|\psi\rangle$. This gives

$$\begin{aligned}|\psi'\rangle&=U[R(\delta\boldsymbol{\theta})]|jj\rangle\\&=[I-(i/\hbar)(\delta\boldsymbol{\theta}\cdot\mathbf{J})]|jj\rangle\\&=[I-(i/2\hbar)(\delta\theta_+J_-+\delta\theta_-J_++2\delta\theta_zJ_z)]|jj\rangle\end{aligned}$$

where

$$\delta\theta_\pm=(\delta\theta_x\pm i\delta\theta_y)$$

Since $J_+|jj\rangle=0$, $J_z|jj\rangle=j\hbar|jj\rangle$, and $J_-|jj\rangle=\hbar(2j)^{1/2}|j,j-1\rangle$, we get

$$|\psi'\rangle=(1-ij\delta\theta_z)|jj\rangle-\tfrac{1}{2}i(2j)^{1/2}\delta\theta_+|j,j-1\rangle$$

Since $\bar{\mathbb{V}}_j$ is assumed to be invariant under any rotation, $|\psi'\rangle$ also belongs to $\bar{\mathbb{V}}_j$. Subtracting $(1-ij\delta\theta_z)|jj\rangle$, which also belongs to $\bar{\mathbb{V}}_j$, from $|\psi'\rangle$, we find that $|j,j-1\rangle$ also belongs to $\bar{\mathbb{V}}_j$. By considering more of such rotations, we can easily establish that the $(2j+1)$ orthonormal vectors, $|jj\rangle$, $|j,j-1\rangle,\ldots,|j,-j\rangle$ all belong to $\bar{\mathbb{V}}_j$.

Thus $\bar{\mathbb{V}}_j$ has the same dimensionality as $\mathbb{V}_j$. Thus $\mathbb{V}_j$ has no invariant subspaces. (In a technical sense, $\mathbb{V}_j$ is its own subspace and is invariant. We are concerned here with subspaces of smaller dimensionality.)

The irreducibility of $\mathbb{V}_j$ means that we cannot, by a change of basis within $\mathbb{V}_j$, further block diagonalize all the $D^{(j)}$. We show that if this were not true, then a contradiction would arise. Let it be possible to block diagonalize *all* the $D^{(j)}$, say, as follows:

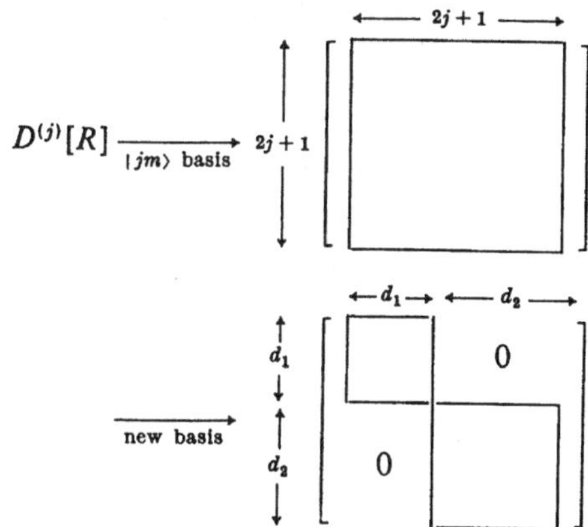

(The boxed regions are generally nonzero). It follows that $\mathbb{V}_j$ contains two invariant subspaces of dimensionalities d_1 and d_2, respectively. (For example, any vector with just the first d_1 components nonzero will get rotated into another such vector. Such vectors form a d_1-dimensional subspace.) We have seen this is impossible.

The block diagonal matrices representing the rotation operators $U[R]$ are said to provide an *irreducible* (*matrix*) *representation* of these operators. For the set of all rotation operators, the elements of which do not generally commute with each other, this irreducible form is the closest one can come to simultaneous diagonalization. All this is summarized schematically in the sketch below, where the boxed regions represent the blocks, $D^{(0)}$, $D^{(1)}$, . . . etc. The unboxed regions contain zeros.

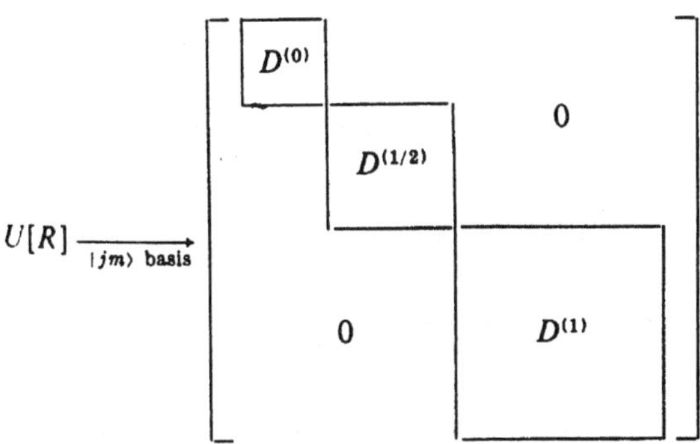

Consider next the matrix representing a rotationally invariant Hamiltonian in this basis. Since $[H, \mathbf{J}]=0$, H has the same form as J^2, which also commutes with all the generators, namely,

(1) H is diagonal, since $[H, J^2]=0$, $[H, J_z]=0$.
(2) Within each block, H has the same eigenvalue E_j, since $[H, J_\pm]=0$.

It follows from (2) that $\mathbb{V}_j$ is an eigenspace of H with eigenvalue E_j, i.e., all states of a given j are degenerate in a rotationally invariant problem. Although the same result is true classically, the relation between degeneracy and rotational invariance is different in the two cases. Classically, if we are given two states with the same magnitude of angular momentum but different orientation, we argue that they are degenerate because

(1) One may be rotated into the other.
(2) This rotation does not change the energy.

Quantum mechanically, given two elements of $\mathbb{V}_j$, it is not always true that they may be rotated into each other (Exercise 12.5.6). However, we argue as follows:

(1) One may be reached from the other (in general) by the combined action of $J_\pm$ and $U[R]$.
(2) These operators commute with H.

In short, rotational invariance is the cause of degeneracy in both cases, but the degenerate states are not always rotated versions of each other in the quantum case (Exercises 12.5.6 and 12.5.7).

*Exercise 12.5.4.** (1) Argue that the eigenvalues of $J_x^{(j)}$ and $J_y^{(j)}$ are the same as those of $J_z^{(j)}$, namely, $j\hbar$, $(j-1)\hbar, \ldots, (-j\hbar)$. Generalize the result to $\hat{\theta}\cdot\mathbf{J}^{(j)}$.

(2) Show that

$$(J-j\hbar)[J-(j-1)\hbar][J-(j-2)\hbar]\cdots(J+j\hbar)=0$$

where $J\equiv\hat{\theta}\cdot\mathbf{J}^{(j)}$. (Hint: In the case $J=J_z$ what happens when both sides are applied to an arbitrary eigenket $|jm\rangle$? What about an arbitrary superpositions of such kets?)

(3) It follows from (2) that J^{2j+1} is a linear combination of $J^0, J^1, \ldots, J^{2j}$. Argue that the same goes for J^{2j+k}, $k=1, 2, \ldots$.

Exercise 12.5.5. (Hard). Using results from the previous exercise and Eq. (12.5.23), show that

(1) $D^{(1/2)}[R]=\exp(-i\hat{\theta}\cdot\mathbf{J}^{(1/2)}/\hbar)=\cos(\theta/2)I^{(1/2)}-(2i/\hbar)\sin(\theta/2)\hat{\theta}\cdot\mathbf{J}^{(1/2)}$

(2) $D^{(1)}[R]=\exp(-i\theta_x J_x^{(1)}/\hbar)=(\cos\theta_x-1)\left(\dfrac{J_x^{(1)}}{\hbar}\right)^2-i\sin\theta_x\left(\dfrac{J_x^{(1)}}{\hbar}\right)+I^{(1)}$

Exercise 12.5.6. Consider the family of states $|jj\rangle, \ldots, |jm\rangle, \ldots, |j, -j\rangle$. One refers to them as states of the same magnitude but different orientation of angular momentum. If ones takes this remark literally, i.e., in the classical sense, one is led to believe that one may rotate these into each other, as is the case for classical states with these properties. Consider, for instance, the family $|1, 1\rangle$, $|1, 0\rangle$, $|1, -1\rangle$. It may seem, for example, that the state with zero angular momentum along the z axis, $|1, 0\rangle$, may be obtained by rotating $|1, 1\rangle$ by some suitable ($\frac{1}{2}\pi$?) angle about the x axis. Using $D^{(1)}[R(\theta_x \mathbf{i})]$ from part (2) in the last exercise show that

$$|1, 0\rangle \neq D^{(1)}[R(\theta_x \mathbf{i})]|1, 1\rangle \quad \text{for any } \theta_x$$

The error stems from the fact that classical reasoning should be applied to $\langle \mathbf{J} \rangle$, which responds to rotations like an ordinary vector, and not direcly to $|jm\rangle$, which is a vector in Hilbert space. Verify that $\langle \mathbf{J} \rangle$ responds to rotations like its classical counterpart, by showing that $\langle \mathbf{J} \rangle$ in the state $D^{(1)}[R(\theta_x \mathbf{i})]|1, 1\rangle$ is $\hbar[-\sin\theta_x \mathbf{j} + \cos\theta_x \mathbf{k}]$.

It is not too hard to see why we can't always satisfy

$$|jm'\rangle = D^{(j)}[R]|jm\rangle$$

or more generally, for two normalized kets $|\psi_j'\rangle$ and $|\psi_j\rangle$, satisfy

$$|\psi_j'\rangle = D^{(j)}[R]|\psi_j\rangle$$

by any choice of R. These abstract equations imply $(2j+1)$ linear, complex relations between the components of $|\psi_j'\rangle$ and $|\psi_j\rangle$ that can't be satisfied by varying R, which depends on only three parameters, θ_x, θ_y, and θ_z. (Of course one can find a unitary matrix in $\mathbb{V}_j$ that takes $|jm\rangle$ into $|jm'\rangle$ or $|\psi_j\rangle$ into $|\psi_j'\rangle$, but it will not be a *rotation* matrix corresponding to $U[R]$.)

Exercise 12.5.7: Euler Angles. Rather than parametrize an arbitrary rotation by the angle $\boldsymbol{\theta}$, which describes a *single* rotation by θ about an axis parallel to $\boldsymbol{\theta}$, we may parametrize it by three angles, γ, β, and α called *Euler angles*, which define three successive rotations:

$$U[R(\alpha, \beta, \gamma)] = e^{-i\alpha J_z/\hbar}\, e^{-i\beta J_y/\hbar}\, e^{-i\gamma J_z/\hbar}$$

(1) Construct $D^{(1)}[R(\alpha, \beta, \gamma)]$ explicitly as a product of three 3×3 matrices. (Use the result from Exercise 12.5.5 with $J_x \to J_y$.)

(2) Let it act on $|1, 1\rangle$ and show that $\langle \mathbf{J} \rangle$ in the resulting state is

$$\langle \mathbf{J} \rangle = \hbar(\sin\beta \cos\alpha \mathbf{i} + \sin\beta \sin\alpha \mathbf{j} + \cos\beta \mathbf{k})$$

(3) Show that for no value of α, β, and γ can one rotate $|1, 1\rangle$ into just $|1, 0\rangle$.

(4) Show that one can always rotate any $|1, m\rangle$ into a linear combination that involves $|1, m'\rangle$, i.e.,

$$\langle 1, m'|D^{(1)}[R(\alpha, \beta, \gamma)]|1, m\rangle \neq 0$$

for some α, β, γ and any m, m'.

(5) To see that one can occasionally rotate $|jm\rangle$ into $|jm'\rangle$, verify that a 180° rotation about the y axis applied to $|1, 1\rangle$ turns it into $|1, -1\rangle$.

Angular Momentum Eigenfunctions in the Coordinate Basis

We now turn to step (3) outlined at the beginning of this section, namely, the construction of the eigenfunctions of L^2 and L_z in the coordinate basis, given the information on the kets $|lm\rangle$.

Consider the states corresponding to a given l. The "topmost" state $|ll\rangle$ satisfies

$$L_+|ll\rangle = 0 \tag{12.5.26}$$

If we write the operator $L_\pm = L_x \pm iL_y$ in spherical coordinates we find

$$L_\pm \xrightarrow[\text{basis}]{\text{coordinate}} \pm \hbar e^{\pm i\phi}\left(\frac{\partial}{\partial\theta} \pm i\cot\theta\,\frac{\partial}{\partial\phi}\right) \tag{12.5.27}$$

Exercise 12.5.8 (Optional). Verify that

$$L_x \xrightarrow[\text{basis}]{\text{coordinate}} i\hbar\left(\sin\phi\,\frac{\partial}{\partial\theta} + \cos\phi\cot\theta\,\frac{\partial}{\partial\phi}\right)$$

$$L_y \xrightarrow[\text{basis}]{\text{coordinate}} i\hbar\left(-\cos\phi\,\frac{\partial}{\partial\theta} + \sin\phi\cot\theta\,\frac{\partial}{\partial\phi}\right)$$

If we denote by $\psi_l^l(r, \theta, \phi)$ the eigenfunction corresponding to $|ll\rangle$, we find that it satisfies

$$\left(\frac{\partial}{\partial\theta} + i\cot\theta\,\frac{\partial}{\partial\phi}\right)\psi_l^l(r, \theta, \phi) = 0 \tag{12.5.28}$$

Since ψ_l^l is an eigenfunction of L_z with eigenvalue $l\hbar$, we let

$$\psi_l^l(r, \theta, \phi) = U_l^l(r, \theta)\, e^{il\phi} \tag{12.5.29}$$

and find that

$$\left(\frac{\partial}{\partial\theta} - l\cot\theta\right)U_l^l = 0 \tag{12.5.30}$$

$$\frac{dU_l^l}{U_l^l} = l\,\frac{d(\sin\theta)}{\sin\theta}$$

or

$$U_l^l(r, \theta) = R(r)(\sin\theta)^l \tag{12.5.31}$$

where $R(r)$ is an arbitrary (normalizable) function of r. When we address the eigenvalue problem of rotationally invariant Hamiltonians, we will see that H will nail down R if we seek simultaneous eigenfunctions of H, L^2, and L_z. But first let us introduce, as we did in the study of L_z in two dimensions, the function that would have been the unique, nondegenerate solution in the absence of the radial coordinate:

$$Y_l^l(\theta, \phi) = (-1)^l \left[\frac{(2l+1)!}{4\pi}\right]^{1/2} \frac{1}{2^l l!} (\sin\theta)^l e^{il\phi} \tag{12.5.32}$$

Whereas the phase factor $(-1)^l$ reflects our convention, the others ensure that

$$\int |Y_l^l|^2 \, d\Omega \equiv \int_{-1}^{1} \int_0^{2\pi} |Y_l^l|^2 \, d(\cos\theta) \, d\phi = 1 \tag{12.5.33}$$

We may obtain Y_l^{l-1} by using the lowering operator. Since

$$L_-|ll\rangle = \hbar[(l+l)(1)]^{1/2}|l, l-1\rangle = \hbar(2l)^{1/2}|l, l-1\rangle$$

$$Y_l^{l-1}(\theta, \phi) = \frac{1}{(2l)^{1/2}} \frac{(-1)}{\hbar} \left[\hbar e^{-i\phi}\left(\frac{\partial}{\partial\theta} - i\cot\theta \frac{\partial}{\partial\phi}\right)\right] Y_l^l \tag{12.5.34}$$

We can keep going in this manner until we reach Y_l^{-1}. The result is, for $m \ge 0$,

$$Y_l^m(\theta, \phi) = (-1)^l \left[\frac{(2l+1)!}{4\pi}\right]^{1/2} \frac{1}{2^l l!} \left[\frac{(l+m)!}{(2l)!(l-m)!}\right]^{1/2} e^{im\phi} (\sin\theta)^{-m}$$
$$\times \frac{d^{l-m}}{d(\cos\theta)^{l-m}} (\sin\theta)^{2l} \tag{12.5.35}$$

For $m<0$, see Eq. (12.5.40). These functions are called *spherical harmonics* and satisfy the orthonormality condition

$$\int Y_l^{m*}(\theta, \phi) Y_{l'}^{m'}(\theta, \phi) \, d\Omega = \delta_{ll'} \delta_{mm'}$$

Another route to the Y_l^m is the direct solution of the L^2, L_z eigenvalue problem in the coordinate basis where

$$L^2 \to (-\hbar^2)\left(\frac{1}{\sin\theta} \frac{\partial}{\partial\theta} \sin\theta \frac{\partial}{\partial\theta} + \frac{1}{\sin^2\theta} \frac{\partial^2}{\partial\phi^2}\right) \tag{12.5.36}$$

and of course

$$L_z \to -i\hbar \frac{\partial}{\partial\phi}$$

If we seek common eigenfunctions of the form‡ $f(\theta) e^{im\phi}$, which are regular between $\theta = 0$ and π, we will find that L^2 has eigenvalues of the form $l(l+1)\hbar^2$, $l = 0, 1, 2, \ldots,$

‡ We neglect the function $R(r)$ that can tag along as a spectator.

where $l \geq |m|$. The Y_l^m functions are mutually orthogonal because they are *nondegenerate* eigenfunctions of L^2 *and* L_z, which are Hermitian on single-valued functions of θ and ϕ.

Exercise 12.5.9. Show that L^2 above is Hermitian in the sense

$$\int \psi_1^*(L^2\psi_2)d\Omega = \left[\int \psi_2^*(L^2\psi_1)d\Omega\right]^*$$

The same goes for L_z, which is insensitive to θ and is Hermitian with respect to the ϕ integration.

We may expand any $\psi(r,\theta,\phi)$ in terms of $Y_l^m(\theta,\phi)$ *using r-dependent coefficients* [consult Eq. (10.1.20) for a similar expansion]:

$$\psi(r,\theta,\phi) = \sum_{l=0}^{\infty}\sum_{m=-l}^{l} C_l^m(r)Y_l^m(\theta,\phi) \tag{12.5.37a}$$

where

$$C_l^m(r) = \int Y_l^{m*}(\theta,\phi)\psi(r,\theta,\phi)d\Omega \tag{12.5.37b}$$

If we compute $\langle\psi|L^2|\psi\rangle$ and interpret the result as a weighted average, we can readily see (assuming ψ is normalized to unity) that

$$P(L^2 = l(l+1)\hbar^2, L_z = m\hbar) = \int_0^{\infty} |C_l^m(r)|^2 r^2 dr \tag{12.5.38}$$

It is clear from the above that C_l^m is the amplitude to find the particle at a radial distance r with angular momentum (l, m).‡ The expansion Eq. (12.5.37a) tells us how to rotate any $\psi(r,\theta,\phi)$ by an angle $\boldsymbol{\theta}$ (in principle):

(1) We construct the block diagonal matrices, $\exp(-i\boldsymbol{\theta}\cdot\mathbf{L}^{(l)}/\hbar)$.

(2) Each block will rotate the C_l^m into linear combinations of each other, i.e., under the action of $U[R]$, the coefficients $C_l^m(r), m = 1, l-1, \ldots, -l$; will get mixed with each other by $D_{m'm}^{(l)}$.

In practice, one can explicitly carry out these steps only if ψ contains only Y_l^m's with small l. A concrete example will be provided in one of the exercises.

‡ Note that r is just the eigenvalue of the operator $(X^2+Y^2+Z^2)^{1/2}$ which commutes with L^2 and L_z.

Here are the first few Y_l^m functions:

$$\begin{aligned} Y_0^0 &= (4\pi)^{-1/2} \\ Y_1^{\pm 1} &= \mp(3/8\pi)^{1/2} \sin\theta e^{\pm i\phi} \\ Y_1^0 &= (3/4\pi)^{1/2} \cos\theta \\ Y_2^{\pm 2} &= (15/32\pi)^{1/2} \sin^2\theta e^{\pm 2i\phi} \\ Y_2^{\pm 1} &= \mp(15/8\pi)^{1/2} \sin\theta\cos\theta e^{\pm i\phi} \\ Y_2^0 &= (5/16\pi)^{1/2}(3\cos^2\theta - 1) \end{aligned} \tag{12.5.39}$$

Note that

$$Y_l^{-m} = (-1)^m (Y_l^m)^* \tag{12.5.40}$$

Closely related to the spherical harmonics are the *associated Legendre polynomials* P_l^m (with $0 \le m \le l$) defined by

$$Y_l^m(\theta, \phi) \doteq \left[\frac{(2l+1)(l-m)!}{4\pi(l+m)!}\right]^{1/2} (-1)^m e^{im\phi} P_l^m(\cos\theta) \tag{12.5.41}$$

If $m = 0$, $P_l^0(\cos\theta) \equiv P_l(\cos\theta)$ is called a *Legendre Polynomial.*

The Shape of the Y_l^m Functions. For large l, the functions $|Y_l^m|$ exhibit many classical features. For example, $|Y_l^l| \propto |\sin^l\theta|$, is almost entirely confined to the $x-y$ plane, as one would expect of a classical particle with all its angular momentum pointing along the z axis. Likewise, $|Y_l^0|$ is, for large l, almost entirely confined to the z axis. Polar plots of these functions may be found in many textbooks.

Exercise 12.5.10. Write the differential equation corresponding to

$$L^2|\alpha\beta\rangle = \alpha|\alpha\beta\rangle$$

in the coordinate basis, using the L^2 operator given in Eq. (12.5.36). We already know $\beta = m\hbar$ from the analysis of $-i\hbar(\partial/\partial\phi)$. So assume that the simultaneous eigenfunctions have the form

$$\psi_{\alpha m}(\theta, \phi) = P_\alpha^m(\theta) e^{im\phi}$$

and show that P_α^m satisfies the equation

$$\left(\frac{1}{\sin\theta}\frac{\partial}{\partial\theta}\sin\theta\frac{\partial}{\partial\theta} + \frac{\alpha}{\hbar^2} - \frac{m^2}{\sin^2\theta}\right) P_\alpha^m(\theta) = 0$$

We need to show that

$$(1)\ \frac{\alpha}{\hbar^2}=l(l+1), \qquad l=0, 1, 2, , \ldots$$

$$(2)\ |m|\le l$$

We will consider only part (1) and that too for the case $m=0$. By rewriting the equation in terms of $u=\cos\theta$, show that P_α^0 satisfies

$$(1-u^2)\frac{d^2P_\alpha^0}{du^2}-2u\frac{dP_\alpha^0}{du}+\left(\frac{\alpha}{\hbar^2}\right)P_\alpha^0=0$$

Convince yourself that a power series solution

$$P_\alpha^0=\sum_{n=0}^{\infty} C_n u^n$$

will lead to a two-term recursion relation. Show that $(C_{n+2}/C_n)\to 1$ as $n\to\infty$. Thus the series diverges when $|u|\to 1$ ($\theta\to 0$ or π). Show that if $\alpha/\hbar^2=(l)(l+1)$; $l=0, 1, 2, \ldots$, the series will terminate and be either an even or odd function of u. The functions $P_\alpha^0(u)=P_{l(l+1)\hbar^2}^0(u)\equiv P_l^0(u)\equiv P_l(u)$ are just the Legendre polynomials up to a scale factor. Determine P_0, P_1, and P_2 and compare (ignoring overall scales) with the Y_l^0 functions.

Exercise 12.5.11. Derive Y_1^1 starting from Eq.(12.5.28) and normalize it yourself. [Remember the $(-1)^l$ factor from Eq. (12.5.32).] Lower it to get Y_1^0 and Y_1^{-1} and compare it with Eq. (12.5.39).

*Exercise 12.5.12.** Since L^2 and L_z commute with Π, they should share a basis with it. Verify that under parity $Y_l^m \to (-1)^l Y_l^m$. (First show that $\theta\to\pi-\theta$, $\phi\to\phi+\pi$ under parity. Prove the result for Y_l^l. Verify that L_- does not alter the parity, thereby proving the result for all Y_l^m.)

*Exercise 12.5.13.** Consider a particle in a state described by

$$\psi=N(x+y+2z)\,e^{-\alpha r}$$

where N is a normalization factor.

(1) Show, by rewriting the $Y_1^{\pm 1,0}$ functions in terms of x, y, z, and r, that

$$Y_1^{\pm 1}=\mp\left(\frac{3}{4\pi}\right)^{1/2}\frac{x\pm iy}{2^{1/2}r}$$
$$Y_1^0=\left(\frac{3}{4\pi}\right)^{1/2}\frac{z}{r} \qquad (12.5.42)$$

(2) Using this result, show that for a particle described by ψ above, $P(l_z=0)=2/3$; $P(l_z=+\hbar)=1/6=P(l_z=-\hbar)$.

Exercise 12.5.14. Consider a rotation $\theta_x \mathbf{i}$. Under this

$$x \to x$$
$$y \to y\cos\theta_x - z\sin\theta_x$$
$$z \to z\cos\theta_z + y\sin\theta_x$$

Therefore we must have

$$\psi(x,y,z) \xrightarrow[U[R(\theta_x \mathbf{i})]]{} \psi_R = \psi(x, y\cos\theta_x + z\sin\theta_x, z\cos\theta_x - y\sin\theta_x)$$

Let us verify this prediction for a special case

$$\psi = Aze^{-r^2/a^2}$$

which must go into

$$\psi_R = A(z\cos\theta_x - y\sin\theta_x)e^{-r^2/a^2}$$

(1) Expand ψ in terms of Y_1^1, Y_1^0, Y_1^{-1}.

(2) Use the matrix $e^{-i\theta_x L_x/\hbar}$ to find the fate of ψ under this rotation.‡ Check your result against that anticipated above. [Hint: (1) $\psi \sim Y_1^0$, which corresponds to

$$\begin{bmatrix} 0 \\ 1 \\ 0 \end{bmatrix}$$

(3) Use Eq. (12.5.42).]

12.6. Solution of Rotationally Invariant Problems

We now consider a class of problems of great practical interest: problems where $V(r,\theta,\phi) = V(r)$. The Schrödinger equation in spherical coordinates becomes

$$\left[\frac{-\hbar^2}{2\mu}\left(\frac{1}{r^2}\frac{\partial}{\partial r}r^2\frac{\partial}{\partial r} + \frac{1}{r^2\sin\theta}\frac{\partial}{\partial\theta}\sin\theta\frac{\partial}{\partial\theta} + \frac{1}{r^2\sin^2\theta}\frac{\partial^2}{\partial\phi^2}\right) + V(r)\right] \times \psi_E(r,\theta,\phi) = E\psi_E(r,\theta,\phi) \tag{12.6.1}$$

Since $[H, \mathbf{L}] = 0$ for a spherically symmetric potential, we seek simultaneous eigenfunctions of H, L^2, and L_z:

$$\psi_{Elm}(r,\theta,\phi) = R_{Elm}(r)Y_l^m(\theta,\phi) \tag{12.6.2}$$

Feeding in this form, and bearing in mind that the angular part of ∇^2 is just the L^2 operator in the coordinate basis [up to a factor $(-\hbar^2 r^2)^{-1}$, see Eq. (12.5.36)], we get

‡ See Exercise 12.5.5.

the *radial equation*

$$\left\{-\frac{\hbar^2}{2\mu}\left[\frac{1}{r^2}\frac{\partial}{\partial r}r^2\frac{\partial}{\partial r}-\frac{l(l+1)}{r^2}\right]+V(r)\right\}R_{El}=ER_{El} \tag{12.6.3}$$

Notice that the subscript m has been dropped: neither the energy nor the radial function depends on it. We find, as anticipated earlier, the $(2l+1)$-fold degeneracy of H.

*Exercise 12.6.1.** A particle is described by the wave function

$$\psi_E(r,\theta,\phi)=A\,e^{-r/a_0} \qquad (a_0=\text{const})$$

(1) What is the angular momentum content of the state?
(2) Assuming ψ_E is an eigenstate in a potential that vanishes as $r\to\infty$, find E. (Match leading terms in Schrödinger's equation.)
(3) Having found E, consider finite r and find $V(r)$.

At this point it becomes fruitful to introduce an auxiliary function U_{El} defined as follows:

$$R_{El}=U_{El}/r \tag{12.6.4}$$

and which obeys the equation

$$\left\{\frac{d^2}{dr^2}+\frac{2\mu}{\hbar^2}\left[E-V(r)-\frac{l(l+1)\hbar^2}{2\mu r^2}\right]\right\}U_{El}=0 \tag{12.6.5}$$

*Exercise 12.6.2.** Provide the steps connecting Eq. (12.6.3) and Eq. (12.6.5).

The equation is the same as the one-dimensional Schrödinger equation except for the following differences:

(1) The independent variable (r) goes from 0 to ∞ and not from $-\infty$ to ∞.
(2) In addition to the actual potential $V(r)$, there is the *repulsive centrifugal barrier*, $l(l+1)\hbar^2/2\mu r^2$, in all but the $l=0$ states.
(3) The boundary conditions on U are different from the one-dimensional case. We find these by rewriting Eq. (12.6.5) as an eigenvalue equation

$$\left[-\frac{\hbar^2}{2\mu}\frac{d^2}{dr^2}+V(r)+\frac{l(l+1)\hbar^2}{2\mu r^2}\right]U_{El}\equiv D_l(r)U_{El}=EU_{El} \tag{12.6.6}$$

and demanding that the functions U_{El} be such that D_l is Hermitian with respect to them. In other words, if U_1 and U_2 are two such functions, then we demand that

$$\int_0^\infty U_1^*(D_l U_2)\,dr = \left[\int_0^\infty U_2^*(D_l U_1)\,dr\right]^* \equiv \int_0^\infty (D_l U_1)^* U_2\,dr \qquad (12.6.7a)$$

This reduces to the requirement

$$\left(U_1^* \frac{dU_2}{dr} - U_2 \frac{dU_1^*}{dr}\right)\Bigg|_0^\infty = 0 \qquad (12.6.7b)$$

Exercise 12.6.3. Show that Eq. (12.6.7b) follows from Eq. (12.6.7a).

Now, a necessary condition for

$$\int_0^\infty |R_{El}|^2 r^2\,dr = \int_0^\infty |U_{El}|^2\,dr$$

to be normalizable to unity or the Dirac delta function is that

$$U_{El} \xrightarrow[r\to\infty]{} 0 \qquad (12.6.8a)$$

or

$$U_{El} \xrightarrow[r\to\infty]{} e^{ikr} \qquad (12.6.8b)$$

the first corresponding to bound states and the second to unbound states. In either case, the expression in the brackets in Eq. (12.6.7b) vanishes at the upper limit‡ and the Hermiticity of D_l hinges on whether or not

$$\left[U_1^* \frac{dU_2}{dr} - U_2 \frac{dU_1^*}{dr}\right]_0 = 0 \qquad (12.6.9)$$

Now this condition is satisfied if

$$U \xrightarrow[r\to 0]{} c, \qquad c = \text{const} \qquad (12.6.10)$$

‡ For the oscillating case, we must use the limiting scheme described in Section 1.10.

If c is nonzero, then

$$R \sim \frac{U}{r} \sim \frac{c}{r}$$

diverges at the origin. This in itself is not a disqualification, for R is still square integrable. The problem with $c \neq 0$ is that the corresponding total wave function‡

$$\psi \sim \frac{c}{r} Y_0^0$$

does not satisfy Schrödinger's equation at the origin. This is because of the relation

$$\nabla^2(1/r) = -4\pi\delta^3(\mathbf{r}) \tag{12.6.11}$$

the proof of which is taken up in Exercise 12.6.4. Thus unless $V(r)$ contains a delta function at the origin (which we assume it does not) the choice $c \neq 0$ is untenable. Thus we deduce that

$$U_{El} \xrightarrow[r \to 0]{} 0 \tag{12.6.12}$$

*Exercise 12.6.4.** (1) Show that

$$\delta^3(\mathbf{r}-\mathbf{r}') \equiv \delta(x-x')\delta(y-y')\delta(z-z') = \frac{1}{r^2 \sin\theta}\,\delta(r-r')\delta(\theta-\theta')\delta(\phi-\phi')$$

(consider a test function).

(2) Show that

$$\nabla^2(1/r) = -4\pi\delta^3(\mathbf{r})$$

(Hint: First show that $\nabla^2(1/r) = 0$ if $r \neq 0$. To see what happens at $r = 0$, consider a small sphere centered at the origin and use Gauss's law and the identity $\nabla^2\phi = \nabla \cdot \nabla\phi$).§

General Properties of U_{El}

We have already discussed some of the properties of U_{El} as $r \to 0$ or ∞. We shall try to extract further information on U_{El} by analyzing the equation governing it in these limits, without making detailed assumptions about $V(r)$. Consider first the limit $r \to 0$. Assuming $V(r)$ is less singular than r^{-2}, the equation is dominated by the

‡ As we will see in a moment, $l \neq 0$ is incompatible with the requirement that $\psi(\mathbf{r}) \to r^{-1}$ as $r \to 0$. Thus the angular part of ψ has to be $Y_0^0 = (4\pi)^{-1/2}$.

§ Or compare this equation to Poisson's equation in electrostatics $\nabla^2\phi = -4\pi\rho$. Here $\rho = \delta^3(\mathbf{r})$, which represents a unit point charge at the origin. In this case we know from Coulomb's law that $\phi = 1/r$.

centrifugal barrier:

$$U_l'' \simeq \frac{l(l+1)}{r^2} U_l \tag{12.6.13}$$

We have dropped the subscript E, since E becomes inconsequential in this limit. If we try a solution of the form

$$U_l \sim r^\alpha$$

we find

$$\alpha(\alpha-1) = l(l+1)$$

or

$$\alpha = l+1 \quad \text{or } (-l)$$

and

$$U_l \sim \begin{cases} r^{l+1} & \text{(regular)} \\ r^{-l} & \text{(irregular)} \end{cases} \tag{12.6.14}$$

We reject the irregular solution since it does not meet the boundary condition $U(0)=0$. The behavior of the regular solutions near the origin is in accord with our expectation that as the angular momentum increases the particle should avoid the origin more and more.

The above arguments are clearly true only if $l \neq 0$. If $l=0$, the centrifugal barrier is absent, and the answer may be sensitive to the potential. In the problems we will consider, $U_{l=0}$ will also behave as r^{l+1} with $l=0$. Although $U_0(r) \to 0$ as $r \to 0$, note that a particle in the $l=0$ state has a nonzero amplitude to be at the origin, since $R_0(r) = U_0(r)/r \neq 0$ at $r=0$.

Consider now the behavior of U_{El} as $r \to \infty$. If $V(r)$ does not vanish as $r \to \infty$, it will dominate the result (as in the case of the isotropic oscillator, for which $V(r) \propto r^2$) and we cannot say anything in general. So let us consider the case where $rV(r) \to 0$ as $r \to \infty$. At large r the equation becomes

$$\frac{d^2 U_E}{dr^2} = -\frac{2\mu E}{\hbar^2} U_E \tag{12.6.15}$$

(We have dropped the subscript l since the answer doesn't depend on l.) There are now two cases:

1. $E>0$: the particle is allowed to escape to infinity classically. We expect U_E to oscillate as $r \to \infty$.
2. $E<0$: The particle is bound. The region $r \to \infty$ is classically forbidden and we expect U_E to fall exponentially there.

Consider the first case. The solutions to Eq. (12.6.15) are of the form

$$U_E = A\,e^{ikr} + B\,e^{-ikr}, \qquad k = (2\mu E/\hbar^2)^{1/2}$$

that is to say, the particle behaves as a free particle far from the origin.‡ Now, you might wonder why we demanded that $rV(r)\to 0$ and not simply $V(r)\to 0$ as $r\to\infty$. To answer this question, let us write

$$U_E = f(r)\,e^{\pm ikr}$$

and see if $f(r)$ tends to a constant as $r\to\infty$. Feeding in this form of U_E into Eq. (12.6.5) we find (ignoring the centrifugal barrier)

$$f'' \pm (2ik)\,f' - \frac{2\mu V(r)}{\hbar^2}\,f = 0$$

Since we expect $f(r)$ to be slowly varying as $r\to\infty$, we ignore f'' and find

$$\frac{df}{f} = \mp\frac{i}{k}\frac{\mu}{\hbar^2}\,V(r)\,dr$$

$$f(r) = f(r_0)\cdot\exp\mp\left[\frac{i\mu}{k\hbar^2}\int_{r_0}^{r} V(r')\,dr'\right] \tag{12.6.16}$$

where r_0 is some constant. If $V(r)$ falls faster than r^{-1}, i.e., $rV(r)\to 0$ as $r\to\infty$, we can take the limit as $r\to\infty$ in the integral and $f(r)$ approaches a constant as $r\to\infty$. If instead

$$V(r) = -\frac{e^2}{r}$$

as in the Coulomb problem,§ then

$$f(r) = f(r_0)\,\exp\pm\left[\frac{i\mu e^2}{k\hbar^2}\ln\left(\frac{r}{r_0}\right)\right]$$

and

$$U_E(r) \sim \exp\pm\left[i\left(kr + \frac{\mu e^2}{k\hbar^2}\ln r\right)\right] \tag{12.6.17}$$

This means that no matter how far away the particle is from the origin, it is never completely free of the Coulomb potential. If $V(r)$ falls even slower than a Coulomb potential, this problem only gets worse.

‡ Although A and B are arbitrary in this asymptotic form, their ratio is determined by the requirement that if U_E is continued inward to $r=0$, it must vanish. That there is just one free parameter in the solution (the overall scale), and not two, is because D_l is nondegenerate even for $E>0$, which in turn is due to the constraint $U_{El}(r=0)=0$; see Exercise 12.6.5.

§ We are considering the case of equal and opposite charges with an eye on the next chapter.

Consider now the case $E<0$. All the results from the $E>0$ case carry over with the change

$$k \to i\kappa, \qquad \kappa = (2\mu|E|/\hbar^2)^{1/2}$$

Thus

$$U_E \xrightarrow[r\to\infty]{} A\, e^{-\kappa r} + B\, e^{+\kappa r} \tag{12.6.18}$$

Again B/A is not arbitrary if we demand that U_E continued inward vanish at $r=0$. Now, the growing exponential is disallowed. For arbitrary $E<0$, both $e^{\kappa r}$ and $e^{-\kappa r}$ will be present in U_E. Only for certain discrete values of E will the $e^{\kappa r}$ piece be absent; these will be the allowed bound state levels. (If A/B were arbitrary, we could choose $B=0$ and get a normalizable bound state for every $E<0$.)

As before, Eq. (12.6.18) is true only if $rV(r)\to 0$. In the Coulomb case we expect [from Eq. (12.6.17) with $k\to i\kappa$]

$$\begin{aligned} U_E &\sim \exp\left(\pm\frac{\mu e^2}{\kappa\hbar^2}\ln r\right)e^{\mp\kappa r} \\ &= (r)^{\pm\mu e^2/\kappa\hbar^2}\, e^{\mp\kappa r} \end{aligned} \tag{12.6.19}$$

When we solve the problem of the hydrogen atom, we will find that this is indeed the case.

When $E<0$, the energy eigenfunctions are normalizable to unity. As the operator $D_l(r)$ is nondegenerate (Exercise 12.6.5), we have

$$\int_0^\infty U_{E'l}(r)\,U_{El}(r)\,dr = \delta_{EE'}$$

and

$$\psi_{Elm}(r,\theta,\phi) = R_{El}(r)\,Y_l^m(\theta,\phi)$$

obeys

$$\iiint \psi^*_{Elm}(r,\theta,\phi)\,\psi_{E'l'm'}(r,\theta,\phi)\,r^2\,dr\,d\Omega = \delta_{EE'}\,\delta_{ll'}\,\delta_{mm'}$$

We will consider the case $E>0$ in a moment.

Exercise 12.6.5. Show that D_l is nondegenerate in the space of functions U that vanish as $r\to 0$. (Recall the proof of Theorem 15, Section 5.6.) Note that U_{El} is nondegenerate even for $E>0$. This means that E, l and m, label a state fully in three dimensions.

The Free Particle in Spherical Coordinates‡

If we begin as usual with

$$\psi_{Elm}(r, \theta, \phi) = R_{El}(r) Y_l^m(\theta, \phi)$$

and switch to U_{El}, we end up with

$$\left[\frac{d^2}{dr^2} + k^2 - \frac{l(l+1)}{r^2}\right] U_{El} = 0, \qquad k^2 = \frac{2\mu E}{\hbar^2}$$

Dividing both sides by k^2, and changing to $\rho = kr$, we obtain

$$\left[-\frac{d^2}{d\rho^2} + \frac{l(l+1)}{\rho^2}\right] U_l = U_l \tag{12.6.20}$$

The variable k, which has disappeared, will reappear when we rewrite the answer in terms of $r = \rho/k$. This problem looks a lot like the harmonic oscillator except for the fact that we have a potential $1/\rho^2$ instead of ρ^2. So we define operators analogous to the raising and lowering operators. These are

$$d_l = \frac{d}{d\rho} + \frac{l+1}{\rho} \tag{12.6.21a}$$

and its adjoint

$$d_l^\dagger = -\frac{d}{d\rho} + \frac{l+1}{\rho} \tag{12.6.21b}$$

(Note that $d/d\rho$ is anti-Hermitian.) In terms of these, Eq. (12.6.20) becomes

$$(d_l d_l^\dagger) U_l = U_l \tag{12.6.22}$$

Now we premultiply both sides by $d_l^\dagger$ to get

$$d_l^\dagger d_l (d_l^\dagger U_l) = d_l^\dagger U_l \tag{12.6.23}$$

You may verify that

$$d_l^\dagger d_l = d_{l+1} d_{l+1}^\dagger \tag{12.6.24}$$

so that

$$d_{l+1} d_{l+1}^\dagger (d_l^\dagger U_l) = d_l^\dagger U_l \tag{12.6.25}$$

‡ The present analysis is a simplified version of the work of L. Infeld, *Phys. Rev.*, **59**, 737 (1941).

It follows that

$$d_l^\dagger U_l = c_l U_{l+1} \tag{12.6.26}$$

where c_l is a constant. We choose it to be unity, for it can always be absorbed in the normalization. We see that $d_l^\dagger$ serves as a "raising operator" in the index l. Given U_0, we can find the others.‡ From Eq. (12.6.20) it is clear that if $l=0$ there are two independent solutions:

$$U_0^A(\rho) = \sin\rho, \qquad U_0^B = -\cos\rho \tag{12.6.27}$$

The constants in front are chosen according to a popular convention. Now U_0^B is unacceptable at $\rho=0$ since it violates Eq. (12.6.12). If, however, one is considering the equation in a region that excludes the origin, U_0^B must be included. Consider now the tower of solutions built out of U_0^A and U_0^B. Let us begin with the equation

$$U_{l+1} = d_l^\dagger U_l \tag{12.6.28}$$

Now, we are really interested in the functions $R_l = U_l/\rho$.§ These obey (from the above)

$$\begin{aligned}
\rho R_{l+1} &= d_l^\dagger(\rho R_l) \\
&= \left(-\frac{d}{d\rho} + \frac{l+1}{\rho}\right)(\rho R_l) \\
R_{l+1} &= \left(-\frac{d}{d\rho} + \frac{l}{\rho}\right) R_l \\
&= \rho^l \left(-\frac{d}{d\rho}\right) \frac{R_l}{\rho^l}
\end{aligned}$$

or

$$\begin{aligned}
\frac{R_{l+1}}{\rho^{l+1}} &= \left(-\frac{1}{\rho}\frac{d}{d\rho}\right)\frac{R_l}{\rho^l} \\
&= \left(-\frac{1}{\rho}\frac{d}{d\rho}\right)^2 \frac{R_{l-1}}{\rho^{l-1}} \\
&= \left(-\frac{1}{\rho}\frac{d}{d\rho}\right)^{l+1} \frac{R_0}{\rho^0}
\end{aligned}$$

‡ In Chapter 15, we will gain some insight into the origin of such a ladder of solutions.

§ Actually we want $R_l = U_l/r = kU_l/\rho$. But the factor k may be absorbed in the normalization factors of U and R.

so that finally we have

$$R_l=(-\rho)^l\left(\frac{1}{\rho}\frac{\partial}{\partial\rho}\right)^l R_0 \tag{12.6.29}$$

Now there are two possibilities for R_0:

$$R_0^A=\frac{\sin\rho}{\rho}$$

$$R_0^B=\frac{-\cos\rho}{\rho}$$

These generate the functions

$$R_l^A\equiv j_l=(-\rho)^l\left(\frac{1}{\rho}\frac{d}{d\rho}\right)^l\left(\frac{\sin\rho}{\rho}\right) \tag{12.6.30a}$$

called the *spherical Bessel functions* of order l, and

$$R_l^B\equiv n_l=(-\rho)^l\left(\frac{1}{\rho}\frac{d}{d\rho}\right)^l\left(\frac{-\cos\rho}{\rho}\right) \tag{12.6.30b}$$

called *spherical Neumann functions* of order l.‡ Here are a few of these functions:

$$\begin{aligned}
j_0(\rho)&=\frac{\sin\rho}{\rho}, & n_0(\rho)&=\frac{-\cos\rho}{\rho}\\
j_1(\rho)&=\frac{\sin\rho}{\rho^2}-\frac{\cos\rho}{\rho}, & n_1(\rho)&=\frac{-\cos\rho}{\rho^2}-\frac{\sin\rho}{\rho}\\
j_2(\rho)&=\left(\frac{3}{\rho^3}-\frac{1}{\rho}\right)\sin\rho-\frac{3\cos\rho}{\rho^2}, & n_2(\rho)&=-\left(\frac{3}{\rho^3}-\frac{1}{\rho}\right)\cos\rho-\frac{3\sin\rho}{\rho^2}
\end{aligned} \tag{12.6.31}$$

As $\rho\to\infty$, these functions behave as

$$j_l\xrightarrow[\rho\to\infty]{}\frac{1}{\rho}\sin\left(\rho-\frac{l\pi}{2}\right) \tag{12.6.32}$$

$$n_l\xrightarrow[\rho\to\infty]{}-\frac{1}{\rho}\cos\left(\rho-\frac{l\pi}{2}\right) \tag{12.6.32}$$

Despite the apparent singularities as $\rho\to 0$, the $j_l(l)$ functions are finite and in fact

$$j_l(\rho)\xrightarrow[\rho\to 0]{}\frac{\rho^l}{(2l+1)!!} \tag{12.6.33}$$

‡ One also encounters *spherical Hankel functions* $h_l=j_l+in_l$ in some problems.

where $(2l+1)!!=(2l+1)(2l-1)(2l-3)\ldots(5)(3)(1)$. These are just the regular solutions listed in Eq. (12.6.14). The Neumann functions, on the other hand, are singular

$$n_l(\rho)\xrightarrow[\rho\to 0]{}-\frac{(2l-1)!!}{\rho^{l+1}} \tag{12.6.34}$$

and correspond to the irregular solutions listed in Eq. (12.6.14).

Free-particle solutions that are *regular in all space* are then

$$\psi_{Elm}(r,\theta,\phi)=j_l(kr)Y_l^m(\theta,\phi),\qquad E=\frac{\hbar^2k^2}{2\mu} \tag{12.6.35}$$

These satisfy

$$\iiint \psi^*_{Elm}\psi_{E'l'm'}r^2\,dr\,d\Omega=\frac{2}{\pi k^2}\delta(k-k')\delta_{ll'}\delta_{mm'} \tag{12.6.36}$$

We are using here the fact that

$$\int_0^\infty j_l(kr)j_l(k'r)r^2\,dr=\frac{\pi}{2k^2}\delta(k-k') \tag{12.6.37}$$

Exercise 12.6.6. (1)* Verify that Eqs. (12.6.21) and (12.6.22) are equivalent to Eq. (12.6.20)

(2) Verify Eq. (12.6.24).

Exercise 12.6.7. Verify that j_0 and j_1 have the limits given by Eq. (12.6.33).

*Exercise 12.6.8.** Find the energy levels of a particle in a spherical box of radius r_0 in the $l=0$ sector.

*Exercise 12.6.9.** Show that the quantization condition for $l=0$ bound states in a spherical well of depth $-V_0$ and radius r_0 is

$$k'/\kappa=-\tan k'r_0$$

where k' is the wave number inside the well and $i\kappa$ is the complex wave number for the exponential tail outside. Show that there are no bound states for $V_0<\pi^2\hbar^2/8\mu r_0^2$. (Recall Exercise 5.2.6.)

Connection with the Solution in Cartesian Coordinates

If we had attacked the free-particle problem in Cartesian coordinates, we would have readily obtained

$$\psi_E(x,y,z)=\frac{1}{(2\pi\hbar)^{3/2}}e^{i\mathbf{p}\cdot\mathbf{r}/\hbar},\qquad E=\frac{p^2}{2\mu}=\frac{\hbar^2k^2}{2\mu} \tag{12.6.38}$$

Consider now the case which corresponds to a particle moving along the z axis with momentum p. As

$$\mathbf{p}\cdot\mathbf{r}/\hbar = pr\cos\theta/\hbar = kr\cos\theta$$

we get

$$\psi_E(r,\theta,\phi) = \frac{e^{ikr\cos\theta}}{(2\pi\hbar)^{3/2}}, \quad E = \frac{\hbar^2k^2}{2\mu} \tag{12.6.39}$$

It should be possible to express this solution, describing a particle moving in the z direction with energy $E = \hbar^2k^2/2\mu$, as a linear combination of the functions ψ_{Elm} which have the same energy, or equivalently, the same k:

$$e^{ikr\cos\theta} = \sum_{l=0}^{\infty}\sum_{m=-l}^{l} C_l^m j_l(kr)Y_l^m(\theta,\phi) \tag{12.6.40}$$

Now, only terms with $m = 0$ are relevant since the left-hand side is independent of ϕ. Physically this means that a particle moving along the z axis has no angular momentum in that direction. Since we have

$$Y_l^0(\theta) = \left(\frac{2l+1}{4\pi}\right)^{1/2} P_l(\cos\theta)$$

$$e^{ikr\cos\theta} = \sum_{l=0}^{\infty} C_l j_l(kr)P_l(\cos\theta), \quad C_l = C_l^0\cdot\left(\frac{2l+1}{4\pi}\right)^{1/2}$$

It can be show that

$$C_l = i^l(2l+1)$$

so that

$$e^{ikr\cos\theta} = \sum_{l=0}^{\infty} i^l(2l+1)j_l(kr)P_l(\cos\theta) \tag{12.6.41}$$

This relation will come in handy when we study scattering. This concludes our study of the free particle.

Exercise 12.6.10 (Optional). Verify Eq. (12.6.41) given that

(1) $\int_{-1}^{1} P_l(\cos\theta)P_{l'}(\cos\theta)d(\cos\theta) = [2/(2l+1)]\delta_{ll'}$

(2) $P_l(x) = \dfrac{1}{2^l l!}\dfrac{d^l(x^2-1)^l}{dx^l}$

(3) $\int_0^1 (1-x^2)^m dx = \dfrac{(2m)!!}{(2m+1)!!}$

Hint: Consider the limit $kr\to 0$ after projecting out C_l.

We close this section on rotationally invariant problems with a brief study of the isotropic oscillator. The most celebrated member of this class, the hydrogen atom, will be discussed in detail in the next chapter.

The Isotropic Oscillator

The isotropic oscillator is described by the Hamiltonian

$$H=\frac{P_x^2+P_y^2+P_z^2}{2\mu}+\frac{1}{2}\mu\omega^2(X^2+Y^2+Z^2) \tag{12.6.42}$$

If we write as usual

$$\psi_{Elm}=\frac{U_{El}(r)}{r}\,Y_l^m(\theta,\phi) \tag{12.6.43}$$

we obtain the radial equation

$$\left\{\frac{d^2}{dr^2}+\frac{2\mu}{\hbar^2}\left[E-\frac{1}{2}\mu\omega^2r^2-\frac{l(l+1)\hbar^2}{2\mu r^2}\right]\right\}U_{El}=0 \tag{12.6.44}$$

As $r\to\infty$, we find

$$U\sim e^{-y^2/2} \tag{12.6.45}$$

where

$$y=\left(\frac{\mu\omega}{\hbar}\right)^{1/2}r \tag{12.6.46}$$

is dimensionless. So we let

$$U(y)=e^{-y^2/2}v(y) \tag{12.6.47}$$

and obtain the following equation for $v(y)$:

$$v''-2yv'+\left[2\lambda-1-\frac{l(l+1)}{y^2}\right]v=0,\qquad \lambda=\frac{E}{\hbar\omega} \tag{12.6.48}$$

It is clear upon inspection that a two-term recursion relation will obtain if a power-series solution is plugged in. We set

$$v(y)=y^{l+1}\sum_{n=0}^{\infty}C_ny^n \tag{12.6.49}$$

where we have incorporated the known behavior [Eq. (12.6.14)] near the origin.

By going through the usual steps (left as an exercise) we can arrive at the following quantization condition:

$$E=(2k+l+3/2)\hbar\omega, \qquad k=0, 1, 2, \ldots \tag{12.6.50}$$

If we define the principal quantum number (which controls the energy)

$$n=2k+l \tag{12.6.51}$$

we get

$$E=(n+3/2)\hbar\omega \tag{12.6.52}$$

At each n, the allowed l values are

$$l=n-2k=n, n-2, \ldots, 1 \text{ or } 0 \tag{12.6.53}$$

Here are the first few eigenstates:

$$\begin{array}{lll} n=0 & l=0 & m=0 \\ n=1 & l=1 & m=\pm 1, 0 \\ n=2 & l=0, 2 & m=0;\ \pm 2,\ \pm 1, 0 \\ n=3 & l=1, 3 & m=\pm 1, 0;\ \pm 3,\ \pm 2,\ \pm 1, 0 \\ \vdots & \vdots & \vdots \end{array}$$

Of particular interest to us is the fact that states of different l are degenerate. The degeneracy in m at each l we understand in terms of rotational invariance. The degeneracy of the different l states (which are not related by rotation operators or the generators) appears mysterious. For this reason it is occasionally termed *accidental degeneracy*. This is, however, a misnomer, for the degeneracy in l can be attributed to additional invariance properties of H. Exactly what these extra invariances or symmetries of H are, and how they explain the degeneracy in l, we will see in Chapter 15.

*Exercise 12.6.11.** (1) By combining Eqs. (12.6.48) and (12.6.49) derive the two-term recursion relation. Argue that $C_0 \neq 0$ if U is to have the right properties near $y=0$. Derive the quantizations condition, Eq. (12.6.50).

(2) Calculate the degeneracy and parity at each n and compare with Exercise 10.2.3, where the problem was solved in Cartesian coordinates.

(3) Construct the normalized eigenfunction ψ_{nlm} for $n=0$ and 1. Write them as linear combinations of the $n=0$ and $n=1$ eigenfunctions obtained in Cartesian coordinates.

13

The Hydrogen Atom

13.1. The Eigenvalue Problem

We have here a two-body problem, of an electron of charge $-e$ and mass m, and a proton of charge $+e$ and mass M. By using CM and relative coordinates and working in the CM frame, we can reduce the problem to the dynamics of a single particle whose mass $\mu = mM/(m+M)$ is the reduced mass and whose coordinate $\mathbf{r}$ is the relative coordinate of the two particles. However, since $m/M \simeq 1/2000$, as a result of which the relative coordinate is essentially the electron's coordinate and the reduced mass is essentially m, let us first solve the problem in the limit $M \to \infty$. In this case we have just the electron moving in the field of the immobile proton. At a later stage, when we compare the theory with experiment, we will see how we can easily take into account the finiteness of the proton mass.

Since the potential energy of the electron in the Coulomb potential

$$\phi = e/r \tag{13.1.1}$$

due to the proton is $V = -e^2/r$, the Schrödinger equation

$$\left\{\frac{d^2}{dr^2} + \frac{2m}{\hbar^2}\left[E + \frac{e^2}{r} - \frac{l(l+1)\hbar^2}{2mr^2}\right]\right\} U_{El} = 0 \tag{13.1.2}$$

determines the energy levels in the rest frame of the atom, as well as the wave functions‡

$$\psi_{Elm}(r, \theta, \phi) = R_{El}(r) Y_l^m(\theta, \phi) = \frac{U_{El}(r)}{r} Y_l^m(\theta, \phi) \tag{13.1.3}$$

It is clear upon inspection of Eq. (13.1.2) that a power series ansatz will lead to a three-term recursion relation. So we try to factor out the asymptotic behavior.

‡ It should be clear from the context whether m stands for the electron mass or the z component of angular momentum.

We already know from Section 12.6 that up to (possibly fractional) powers of r [Eq. (12.6.19)],

$$U_{El} \underset{r\to\infty}{\sim} \exp[-(2mW/\hbar^2)^{1/2} r] \tag{13.1.4}$$

where

$$W = -E$$

is the *binding energy* (which is the energy it would take to liberate the electron) and that

$$U_{El} \underset{r\to 0}{\sim} r^{l+1} \tag{13.1.5}$$

Equation (13.1.4) suggests the introduction of the dimensionless variable

$$\rho = (2mW/\hbar^2)^{1/2} r \tag{13.1.6}$$

and the auxiliary function v_{El} defined by

$$U_{El} = e^{-\rho}\, v_{El} \tag{13.1.7}$$

The equation for v is then

$$\frac{d^2v}{d\rho^2} - 2\frac{dv}{d\rho} + \left[\frac{e^2\lambda}{\rho} - \frac{l(l+1)}{\rho^2}\right] v = 0 \tag{13.1.8}$$

where

$$\lambda = (2m/\hbar^2 W)^{1/2} \tag{13.1.9}$$

and the subscripts on v are suppressed. You may verify that if we feed in a series into Eq. (13.1.8), a two-term recursion relation will obtain. Taking into account the behavior near $\rho = 0$ [Eq. (13.1.5)] we try

$$v_{El} = \rho^{l+1} \sum_{k=0}^{\infty} C_k \rho^k \tag{13.1.10}$$

and obtain the following recursion relation between *successive* coefficients:

$$\frac{C_{k+1}}{C_k} = \frac{-e^2\lambda + 2(k+l+1)}{(k+l+2)(k+l+1) - l(l+1)} \tag{13.1.11}$$

The Energy Levels

Since

$$\frac{C_{k+1}}{C_k} \underset{k\to\infty}{\longrightarrow} \frac{2}{k} \tag{13.1.12}$$

is the behavior of the series $\rho^m e^{2\rho}$, and would lead to $U \sim e^{-\rho} v \sim \rho^m e^{-\rho} e^{2\rho} \sim \rho^m e^\rho$ as $\rho \to \infty$, we demand that the series terminate at some k. This will happen if

$$e^2\lambda = 2(k+l+1) \tag{13.1.13}$$

or [from Eq. (13.1.9)]

$$E = -W = \frac{-me^4}{2\hbar^2(k+l+1)^2}, \qquad k=0, 1, 2, \ldots : \quad l=0, 1, 2, \ldots \tag{13.1.14}$$

In terms of the *principal quantum number*

$$n = k+l+1 \tag{13.1.15}$$

the allowed energies are

$$E_n = \frac{-me^4}{2\hbar^2 n^2}, \qquad n=1, 2, 3, \ldots \tag{13.1.16}$$

and at each n the allowed values of l are, according to Eq. (13.1.15),

$$l = n-k-1 = n-1, n-2, \ldots, 1, 0 \tag{13.1.17}$$

That states of different l should be degenerate indicates that H contains more symmetries besides rotational invariance. We discuss these later. For the present, let us note that the degeneracy at each n is

$$\sum_{l=0}^{n-1} (2l+1) = n^2 \tag{13.1.18}$$

It is common to refer to the states with $l=0, 1, 2, 3, 4, \ldots$ as $s, p, d, f, g, h, \ldots$ states. In this *spectroscopic notation*, $1s$ denotes the state ($n=1$, $l=0$); $2s$ and $2p$ the $l=0$ and $l=1$ states at $n=2$; $3s$, $3p$, and $3d$ the $l=0$, 1, and 2 states at $n=3$, and so on. No attempt is made to keep track of m.

It is convenient to employ a natural unit of energy, called a *Rydberg* (Ry), for measuring the energy levels of hydrogen:

$$\text{Ry} = \frac{me^4}{2\hbar^2} \tag{13.1.19}$$

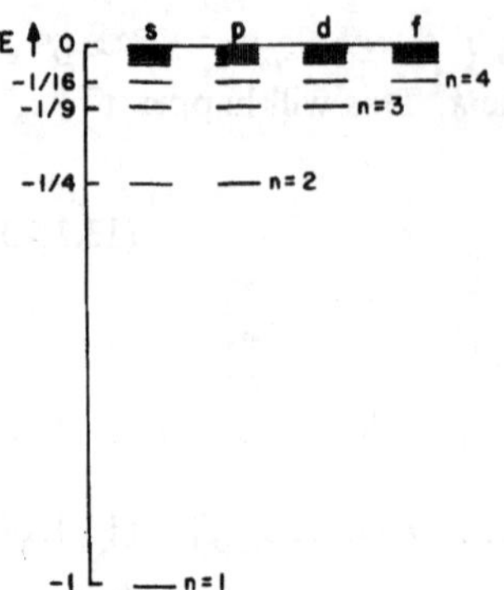

Figure 13.1. The first few eigenstates of hydrogen. The energy is measured in Rydbergs and the states are labelled in the spectroscopic notation.

in terms of which

$$E_n = \frac{-\text{Ry}}{n^2} \tag{13.1.20}$$

Figure 13.1 shows some of the lowest-energy states of hydrogen.

The Wave Functions

Given the recursion relations, it is a straightforward matter to determine the wave functions and to normalize them. Consider a given n and l. Since the series in Eq. (13.1.10) terminates at

$$k = n - l - 1 \tag{13.1.21}$$

the corresponding function v_l is ρ^{l+1} times a polynomial of degree $n-l-1$. This polynomial is called the *associated Laguerre polynomial*, $L_{n-l-1}^{2l+1}(2\rho)$.‡ The corresponding radial function is

$$R_{nl}(\rho) \sim e^{-\rho}\, \rho^l L_{n-l-1}^{2l+1}(2\rho) \tag{13.1.22}$$

Recall that

$$\rho = \left(\frac{2mW}{\hbar^2}\right)^{1/2} r = \left[\frac{2m}{\hbar^2}\left(\frac{me^4}{2\hbar^2 n^2}\right)\right]^{1/2} r$$

$$= \frac{me^2}{\hbar^2 n} r \tag{13.1.23}$$

‡ $L_p^k(x) = (-1)^k (d^k/dx^k) L_{p+k}^0$, $\quad L_p^0 = e^x (d^p/dx^p)(e^{-x} x^p)$.

In terms of the length

$$a_0=\frac{\hbar^2}{me^2} \tag{13.1.24}$$

called the *Bohr radius*, which provides the natural distance scale for the hydrogen atom,

$$R_{nl}(r)\sim e^{-r/na_0}\left(\frac{r}{na_0}\right)^l L_{n-l-1}^{2l+1}\left(\frac{2r}{na_0}\right) \tag{13.1.25}$$

As $r\to\infty$, L will be dominated by the highest power, r^{n-l-1}, and

$$R_{nl} \underset{r\to\infty}{\sim} (r)^{n-1}\, e^{-r/na_0} \quad \text{(independent of } l\text{)} \tag{13.1.26}$$

(If $l=n-1$, this form is valid at all r since L_0^{2l+1} is a constant.) Equation (13.1.26) was anticipated in the last chapter when we considered the behavior of U_{El} as $r\to\infty$, in a Coulomb potential (see Exercise 13.1.4).

The following are the first few normalized eigenfunctions, $\psi_{Elm}\equiv\psi_{nlm}$:

$$\begin{aligned}
\psi_{1,0,0}&=\left(\frac{1}{\pi a_0^3}\right)^{1/2} e^{-r/a_0}\\
\psi_{2,0,0}&=\left(\frac{1}{32\pi a_0^3}\right)^{1/2}\left(2-\frac{r}{a_0}\right)e^{-r/2a_0}\\
\psi_{2,1,0}&=\left(\frac{1}{32\pi a_0^3}\right)^{1/2}\frac{r}{a_0}\,e^{-r/2a_0}\cos\theta\\
\psi_{2,1,\pm1}&=\mp\left(\frac{1}{64\pi a_0^3}\right)^{1/2}\frac{r}{a_0}\,e^{-r/2a_0}\sin\theta\, e^{\pm i\phi}
\end{aligned} \tag{13.1.27}$$

Exercise 13.1.1. Derive Eqs. (13.1.11) and (13.1.14) starting from Eqs. (13.1.8)–(13.1.10).

Exercise 13.1.2. Derive the degeneracy formula, Eq. (13.1.18).

Exercise 13.1.3. Starting from the recursion relation, obtain ψ_{210} (normalized).

Exercise 13.1.4. Recall from the last chapter [Eq. (12.6.19)] that as $r\to\infty$, $U_E\sim(r)^{me^2/\kappa\hbar^2}\,e^{-\kappa r}$ in a Coulomb potential $V=-e^2/r$ $[\kappa=(2mW/\hbar^2)^{1/2}]$. Show that this agrees with Eq. (13.1.26).

Let us explore the statement that a_0 provides a natural length scale for the hydrogen atom. Consider the state described by

$$\psi_{n,n-1,m} \propto e^{-r/na_0} r^{n-1} Y_{n-1}^m(\theta, \phi) \tag{13.1.28}$$

Let us ask for the probability of finding the electron in a spherical shell of radius r and thickness dr:

$$\int_\Omega P(\mathbf{r}) r^2 \, dr \, d\Omega \propto e^{-2r/na_0} r^{2n} \, dr \tag{13.1.29}$$

The probability density in r reaches a maximum when

$$\frac{d}{dr}(e^{-2r/na_0} r^{2n}) = 0$$

or

$$r = n^2 a_0 \tag{13.1.30}$$

When $n=1$, this equals a_0. Thus the Bohr radius gives the most probable value of r in the ground state and this defines the "size" of the atom (to the extent one may speak of it in quantum theory). If $n>1$ we see that the size grows as n^2, at least in the state of $l=n-1$. If $l \neq n-1$, the radial function has $n-l-1$ zeros and the density in r has several bumps. In this case, we may define the size by $\langle r \rangle$.‡ It can be shown, by using properties of L_{n-l-1}^{2l+1} that

$$\langle r \rangle_{nlm} = \frac{a_0}{2}[3n^2 - l(l+1)] \tag{13.1.31}$$

Rather than go through the lengthy derivation of this formula let us consider the following argument, which indicates that the size grows as $n^2 a_0$. In any eigenstate

$$\langle H \rangle = E = \langle T \rangle + \langle V \rangle = \langle P^2/2m \rangle - \langle e^2/r \rangle \tag{13.1.32}$$

It can be shown (Exercise 13.1.5) that

$$\langle T \rangle = -\tfrac{1}{2}\langle V \rangle \tag{13.1.33}$$

which is just the quantum version of the classical virial theorem, which states that if $V = cr^k$, then the averages $\bar{T}$ and $\bar{U}$ are related by

$$\bar{T} = \frac{k}{2}\bar{V}$$

‡ Even though r represents the abstract operator $(X^2 + Y^2 + Z^2)^{1/2}$ only in the coordinate basis, we shall use the same symbol to refer to it in the abstract, so as to keep the notation simple.

It follows that

$$E=\frac{1}{2}\langle V\rangle=-\frac{1}{2}\langle e^2/r\rangle \tag{13.1.34}$$

Now, in the state labeled by n,

$$E_n=\frac{-me^4}{2\hbar^2n^2}=\frac{-e^2}{2a_0n^2} \tag{13.1.35}$$

from which it follows that

$$\left\langle\frac{1}{r}\right\rangle_n=\frac{1}{a_0n^2} \tag{13.1.36}$$

Although

$$\frac{1}{\langle r\rangle}\neq\left\langle\frac{1}{r}\right\rangle$$

the two are of the same order of magnitude (see Exercise 9.4.2) and we infer that

$$\langle r\rangle_n\sim n^2a_0 \tag{13.1.37}$$

which agrees with the result Eq. (13.1.31). (One must be somewhat cautious with statements like $\langle 1/r\rangle\simeq 1/\langle r\rangle$. For example, it is not true in an s state that $\langle 1/r^4\rangle\simeq 1/\langle r^4\rangle$, since $\langle 1/r^4\rangle$ is divergent while $1/\langle r^4\rangle$ is not. In the present case, however, $\langle 1/r\rangle$ is well defined in all states and indeed $\langle 1/r\rangle$ and $1/\langle r\rangle$ are of the same order of magnitude.)

This completes our analysis of the hydrogen spectrum and wave functions. Several questions need to be answered, such as (1) What are the numerical values of E_n, a_0, etc.? (2) How does one compare the energy levels and wave functions deduced here with experiment?

These questions will be taken up in Section 13.3. But first let us address a question raised earlier: what is the source of the degeneracy in l at each n?

*Exercise 13.1.5.** (*Virial Theorem*). Since $|n, l, m\rangle$ is a stationary state, $\langle\dot{\Omega}\rangle=0$ for any Ω. Consider $\Omega=\mathbf{R}\cdot\mathbf{P}$ and use Ehrenfest's theorem to show that $\langle T\rangle=(-1/2)\langle V\rangle$ in the state $|n, l, m\rangle$.

13.2. The Degeneracy of the Hydrogen Spectrum

The hydrogen atom, like the oscillator, exhibits "accidental degeneracy." Quotation marks are used once again, because, as in the case of the oscillator, the degeneracy can be explained in terms of other symmetries the Hamiltonian has besides rotational invariance. Now, we have seen that the symmetries of H imply the conservation of the generators of the symmetries. Consequently, if there is an extra symmetry (besides rotational invariance) there must be some extra conserved quantities (besides angular momentum). Now it is well known classically that the Coulomb‡ potential is special (among rotationally invariant potentials) in that it conserves the *Runge-Lenz vector*

$$\mathbf{n}=\frac{\mathbf{p}\times\mathbf{l}}{m}-\frac{e^2}{r}\mathbf{r} \tag{13.2.1}$$

The conservation of **n** implies that not only is the orbit confined to a plane perpendicular to **l** (as in any rotationally invariant problem) it is also *closed* (Exercise 13.2.1).

In quantum theory then, there will be an operator **N** which commutes with H:

$$[\mathbf{N}, H]=0 \tag{13.2.2}$$

and is given by§

$$\mathbf{N}=\frac{1}{2m}[\mathbf{P}\times\mathbf{L}-\mathbf{L}\times\mathbf{P}]-\frac{e^2\mathbf{R}}{(X^2+Y^2+Z^2)^{1/2}} \tag{13.2.3}$$

We have seen that the conservation of **L** implies that $[L_\pm, H]=0$, which means that we can raise and lower the m values at a given l without changing the energy. This is how the degeneracy in m is "explained" by rotational invariance.

So it must be that since $[\mathbf{N}, H]=0$, we must be able to build some operator out of the components of **N**, which commutes with H and which raises l by one unit. This would then explain the degeneracy in l at each n. Precisely what this operator is and how it manages to raise l by one unit will be explained in Section 15.4, devoted to the study of "accidental" degeneracy. You will also find therein the explanation of the degeneracy of the oscillator.

Exercise 13.2.1. Let us see why the conservation of the Runge–Lenz vector **n** implies closed orbits.

(1) Express **n** in terms of **r** and **p** alone (get rid of **l**).

(2) Since the particle is bound, it cannot escape to infinity. So, as we follow it from some arbitrary time onward, it must reach a point $\mathbf{r}_{\max}$ where its distance from the origin stops growing. Show that

$$\mathbf{n}=\mathbf{r}_{\max}\left(2E+\frac{e^2}{r_{\max}}\right)$$

‡ Or generally any $1/r$ potential, say, gravitational.

§ Since $[\mathbf{P}, \mathbf{L}]\neq 0$, we have used the symmetrization rule to construct **N** from **n**, i.e., $\mathbf{p}\times\mathbf{l}\to\frac{1}{2}[(\mathbf{P}\times\mathbf{L})+(\mathbf{P}\times\mathbf{L})^\dagger]=\frac{1}{2}[\mathbf{P}\times\mathbf{L}-\mathbf{L}\times\mathbf{P}]$ (verify this).

at this point. (Use the law of conservation of energy to eliminate p^2.) Show that, for similar reasons, if we wait some more, it will come to $\mathbf{r}_{\min}$, where

$$\mathbf{n}=\mathbf{r}_{\min}\left(2E+\frac{e^2}{r_{\min}}\right)$$

Thus $\mathbf{r}_{\max}$ and $\mathbf{r}_{\min}$ are parallel to each other and to $\mathbf{n}$. The conservation or constancy of $\mathbf{n}$ implies that the maximum (minimum) separation is always reached at the same point $\mathbf{r}_{\max}(\mathbf{r}_{\min})$, i.e., the orbit is closed. In fact, all three vectors $\mathbf{r}_{\max}$, $\mathbf{r}_{\min}$, and $\mathbf{n}$ are aligned with the major axis of the ellipse along which the particle moves; $\mathbf{n}$ and $\mathbf{r}_{\min}$ are parallel, while $\mathbf{n}$ and $\mathbf{r}_{\max}$ are antiparallel. (Why?) Convince yourself that for a circular orbit, $\mathbf{n}$ must and does vanish.

13.3. Numerical Estimates and Comparison with Experiment

In this section we (1) obtain numerical estimates for various quantities such as the Bohr radius, energy levels, etc.; (2) ask how the predictions of the theory are actually compared with experiment.

Numerical Estimates

Consider first the particle masses. We will express the rest energies of the particles in million-electron volts or MeV:

$$mc^2 \simeq 0.5 \text{ Mev} \qquad (0.511 \text{ is a more exact value}) \tag{13.3.1}$$

$$Mc^2 = 1000 \text{ MeV} \qquad (938.3)\ddagger \tag{13.3.2}$$

$$m/M \simeq 1/2000 \qquad (1/1836)\ddagger \tag{13.3.3}$$

Consequently the reduced mass μ and electron mass m are almost equal:

$$\mu=\frac{mM}{m+M}\simeq\frac{mM}{M}=m \tag{13.3.4}$$

as are the relative coordinate and the electron coordinate.

Consider now an estimate of the Bohr radius

$$a_0=\hbar^2/me^2 \tag{13.3.5}$$

‡ A more exact value.

To find this we need the values of $\hbar$ and e. It was mentioned earlier that

$$\hbar = 1.054 \times 10^{-27} \text{ erg sec}$$

A more useful thing to remember for performing quick estimates is†

$$\hbar c \simeq 2000 \text{ eV Å} \qquad (1973.3) \tag{13.3.6}$$

where 1 angstrom (Å) $= 10^{-8}$ cm. The best way to remember e^2 is through the *fine-structure constant*:

$$\alpha = \frac{e^2}{\hbar c} \simeq \frac{1}{137} \left(\frac{1}{137.04} \right) \tag{13.3.7}$$

This constant plays a fundamental role in quantum mechanical problems involving electrodynamics. Since it is dimensionless, its numerical value has an absolute significance: no matter what units we use for length, mass, and time, α will be $1/137$. Thus, although no one tries to explain why $c = 3 \times 10^{10}$ cm/sec, several attempts have been made to arrive at the magic figure of $1/137$. Since it is a God-given number (independent of mortal choice of units) one tries to relate it to fundamental numbers such as π, e, e^{π}, π^{e}, the number of space-time dimensions, etc.

Anyway, returning to our main problem, we can now estimate a_0:

$$a_0 \simeq \frac{\hbar^2}{me^2} = \frac{\hbar c}{mc^2} \left(\frac{\hbar c}{e^2} \right) = \frac{(2000)(137)}{0.5 \times 10^6} \text{ Å} \simeq 0.55 \text{ Å} \qquad (0.53)$$

Consider next the energy levels

$$E_n = -\text{Ry}/n^2$$

We estimate

$$\text{Ry} = \frac{me^4}{2\hbar^2} = \frac{mc^2}{2} \left(\frac{e^2}{\hbar c} \right)^2$$

$$\simeq \frac{0.25 \times 10^6}{(137)^2} \text{ eV} \simeq 13.3 \text{ eV} \qquad (13.6)$$

So, using the more accurate value of Ry,

$$E_n = \frac{-13.6}{n^2} \text{ eV}$$

† Many of the tricks used here were learned from Professor A. Rosenfeld at the University of California, Berkeley.

The electron in the ground state needs 13.6 eV to be liberated or ionized. One may imagine that it is 13.6 eV down the infinitely deep Coulomb potential.

Let us digress to consider two length scales related to a_0. The first

$$a_0\alpha=\frac{\hbar^2}{me^2}\cdot\frac{e^2}{\hbar c}=\frac{\hbar}{mc}\equiv\lambda_e \tag{13.3.8}$$

is called the *Compton wavelength* of the electron and is 137 times smaller than the Bohr radius. What does λ_e represent? In discussing the nuclear force, it was pointed out that the Compton wavelength of the pion was the distance over which it could be exchanged. It can also be defined as the lower limit on how well a particle can be localized. In the nonrelativistic theory we are considering, the lower limit is zero, since we admit position eigenkets $|x\rangle$. But in reality, as we try to locate the particle better and better, we use more and more energetic probes, say photons to be specific. To locate it to some ΔX, we need a photon of momentum

$$\Delta P\sim\frac{\hbar}{\Delta X}$$

Since the photon is massless, the corresponding energy is

$$\Delta E\sim\frac{\hbar c}{\Delta X}$$

in view of Einstein's formula $E^2=c^2p^2+m^2c^4$.

If this energy exceeds twice the rest energy of the particle, relativity allows the production of a particle–antiparticle pair in the measurement process. So we demand

$$\Delta E\lesssim 2mc^2$$

$$\frac{\hbar c}{\Delta X}\lesssim 2mc^2$$

or

$$\Delta X\gtrsim\frac{\hbar}{2mc}\sim\frac{\hbar}{mc}$$

If we attempt to localize the particle any better, we will see pair creation and we will have three (or more) particles instead of the one we started to locate.

In our analysis of the hydrogen atom, we treated the electron as a localized point particle. The preceding analysis shows that this is not strictly correct, but it also shows that it is a fair approximation, since the "fuzziness" or "size" of the electron is α times smaller than the size of the atom, a_0

$$\frac{\hbar/mc}{a_0}=\alpha\simeq\frac{1}{137}$$

Had the electric charge been 10 times as big, α would have been of order unity, and the size of the electron and the size of its orbit would have been of the same order and the point particle approximation would have been untenable. Let us note that

$$\lambda\!\!\!^{-}_e=\alpha\cdot a_0\simeq 0.5\times\frac{1}{137}\text{ Å}\simeq\frac{1}{250}\text{ Å}\simeq 4\times 10^{-3}\text{ Å}$$

If we multiply $\lambda\!\!\!^{-}_e$ by α we get another length, called the *classical radius of the electron*:

$$r_e=\alpha\lambda\!\!\!^{-}_e=\frac{\hbar}{mc}\cdot\frac{e^2}{\hbar c}=\frac{e^2}{mc^2}\simeq 3\times 10^{-5}\text{ Å} \tag{13.3.9}$$

If we imagine the electron to be a spherical charge distribution, the Coulomb energy of the distribution (the energy it takes to assemble it) will be of the order e^2/r_e, where r_e is the radius of the sphere. If we attribute the rest energy of the electron to this Coulomb energy, we arrive at the classical radius. In summary,

$$\begin{array}{ccccc} a_0 & \underset{\alpha}{\rightarrow} & \lambda\!\!\!^{-}_e & \underset{\alpha}{\rightarrow} & r_0 \\ \left(\frac{1}{2}\text{Å}\right) & & \left(\frac{\alpha}{2}\text{Å}\right) & & \left(\frac{\alpha^2}{2}\text{Å}\right) \end{array}$$

Let us now return to the hydrogen atom. The mnemonics discussed so far are concerned only with the numbers. Let us now consider mnemonics that help us remember the dynamics. These must be used with caution, for they are phrased in terms not allowed in quantum theory.

The source of these mnemonics is the *Bohr model* of the hydrogen atom. About a decade or so prior to the formulation of quantum mechanics as described in this text, Bohr proposed a model of the atom along the following lines. Consider a particle of mass m in $V(r)=-e^2/r$, moving in a circular orbit of radius r. The dynamical equation is

$$\frac{mv^2}{r}=\frac{e^2}{r^2} \tag{13.3.10}$$

or

$$mv^2=\frac{e^2}{r} \tag{13.3.11}$$

Thus any radius is allowed if r satisfies this equation. It also follows that any energy is allowed since

$$E=\frac{1}{2}mv^2-\frac{e^2}{r}=-\frac{e^2}{2r}=-\frac{1}{2}mv^2 \tag{13.3.12}$$

Bohr conjectured that the only allowed orbits were those that had integral angular momentum in units of $\hbar$:

$$mvr=n\hbar \tag{13.3.13}$$

Feeding this into Eq. (13.3.11) we get

$$m\cdot\frac{n^2\hbar^2}{m^2r^2}=\frac{e^2}{r}$$

or

$$r=n^2\frac{\hbar^2}{me^2}=n^2a_0 \tag{13.3.14}$$

and

$$E_n=-\frac{e^2}{2r}=-\frac{e^2}{2a_0}\cdot\left(\frac{1}{n^2}\right) \tag{13.3.15}$$

Thus, if you ever forget the formula for a_0 or E_n, you can go back to this model for the formulas (though not for the physics, since it is perched on the fence between classical and quantum mechanics; it speaks of orbits, but quantizes angular momentum and so on). The most succinct way to remember the Bohr atom (i.e., a mnemonic for the mnemonic) is the equation

$$\alpha=\beta \tag{13.3.16}$$

where β is the velocity of the electron in the ground state of hydrogen measured in units of velocity of light ($\beta=v/c$). Given this, we get the ground state energy as

$$\begin{aligned}E_1&=-\frac{1}{2}mv^2=-\frac{1}{2}mc^2(v/c)^2=-\frac{1}{2}mc^2\beta^2=-\frac{1}{2}mc^2\alpha^2\\&=-\frac{1}{2}mc^2\left(\frac{e^2}{\hbar c}\right)^2=-\frac{me^4}{2\hbar^2}\end{aligned} \tag{13.3.17}$$

Given this, how could one forget that the levels go as n^{-2}, i.e.,

$$E_n = -\frac{E_1}{n^2}?$$

If we rewrite E_1 as $-e^2/2a_0$, we can get the formula for a_0. The equation $\alpha = \beta$ also justifies the use of nonrelativistic quantum mechanics. An equivalent way (which avoids the use of velocity) is Eq. (13.3.17), which states that the binding energy is $\simeq (1/137)^2$ times the rest energy of the electron.

*Exercise 13.3.1.** The pion has a range of 1 Fermi $= 10^{-5}$ Å as a mediator of nuclear force. Estimate its rest energy.

*Exercise 13.3.2.** Estimate the de Broglie wavelength of an electron of kinetic energy 200 eV. (Recall $\lambda = 2\pi\hbar/p$.)

Comparison with Experiment

Quantum theory makes very detailed predictions for the hydrogen atom. Let us ask how these are to be compared with experiment. Let us consider first the energy levels and then the wave functions. In principle, one can measure the energy levels by simply weighing the atom. In practice, one measures the *differences in energy levels* as follows. If we start with the atom in an eigenstate $|nlm\rangle$, it will stay that way forever. However, if we perturb it for a time T, by turning on some external field (i.e., change the Hamiltonian from H^0, the Coulomb Hamiltonian, to $H^0 + H^1$) its state vector can start moving around in Hilbert space, since $|nlm\rangle$ is not a stationary state of $H^0 + H^1$. If we measure the energy at time $t > T$, we may find it corresponds to another state with $n' \neq n$. One measures the energy by detecting the photon emitted by the atom. The frequency of the detected photon will be

$$\omega_{nn'} = \frac{E_n - E_{n'}}{\hbar} \tag{13.3.18}$$

Thus the frequency of light coming out of hydrogen will be

$$\begin{aligned} \omega_{nn'} &= \frac{\text{Ry}}{\hbar}\left(-\frac{1}{n^2} + \frac{1}{n'^2}\right) \\ &= \frac{\text{Ry}}{\hbar}\left(\frac{1}{n'^2} - \frac{1}{n^2}\right) \end{aligned} \tag{13.3.19}$$

For a fixed value $n' = 1, 2, 3, \ldots$, we obtain a family of lines as we vary n. These families have in fact been seen, at least for several values of n'. The $n' = 1$ family is called the *Lyman series* (it corresponds to transitions to the ground state from the upper ones):

$$\omega_{n1}=\frac{\text{Ry}}{\hbar}\left(\frac{1}{1}-\frac{1}{n^2}\right) \tag{13.3.20}$$

The $n'=2$ family is called the *Balmer series* and corresponds to transitions to the states $|2lm\rangle$ from $n=3, 4, \ldots,$ etc. The $n'=3$ family called the *Paschen* series, etc. Let us estimate the wavelength of a typical line in the Lyman series, say the one corresponding to the transition $n=2\rightarrow n'=1$:

$$\omega_{21}=\frac{13.5\text{ eV}}{\hbar}\left(1-\frac{1}{4}\right)$$

$$\simeq\frac{10}{\hbar}\text{ eV}$$

The wavelength is estimated to be

$$\lambda=\frac{2\pi c}{\omega}=\frac{2\pi}{10}(\hbar c)\simeq 1200\ \AA$$

A more refined estimate gives a value of 1216 Å, in very good agreement with experiment. Equally good is the agreement for all other observed lines. However, there are, in all cases, small discrepancies. Much of these may be explained by corrections that are calculable in theory. First we must correct for the fact that the proton is not really immobile; that we have here a two-body problem. As explained in Chapter 10, this is done by writing Schrödinger's equation for the relative (and not electron) coordinate and working in the CM frame. This equation would differ from Eq. (13.1.2) only in that m would be replaced by μ. This in fact would be the only change in all the formulas that follow, in particular Eq. (13.1.16) for the energy levels. This would simply rescale the entire spectrum by a factor $\mu/m=M/(M+m)$, which differs from 1 by less than a tenth of a percent. This difference is, however, observable in practice: one sees it in the difference between the levels of hydrogen and deuterium (whose nucleus has a proton and a neutron).

Then there is the correction due to the fact that the kinetic energy of the electron is not $\frac{1}{2}mv^2=p^2/2m$ in Einstein's theory, but instead $mc^2[(1-v^2/c^2)^{-1/2}-1]$, which is the difference between the energy at velocity v and the energy at rest. The $\frac{1}{2}mv^2$ term is just the first in the power series expansion of the above, in the variable v^2/c^2. In Chapter 17 we will take into account the effect of the next term, which is $-3mv^4/8c^2$, or in terms of the momentum, $-3p^4/8m^3c^2$. This is a correction of order v^2/c^2 relative to the $p^2/2m$ piece we included, or since $v/c\simeq\alpha$, a correction of order α^2 relative to main piece. There are other corrections of the same order, and these go by the name of *fine-structure corrections*. They will be included (in some approximation) in Chapter 17. The Dirac equation, which we will not solve in this book, takes into account the relativistic corrections to all orders in v/c. However, it too doesn't give the full story; there are tiny corrections due to quantum fluctuations of the electromagnetic field (which we have treated classically so far). These corrections are calculable in theory and measurable experimentally. The agreement between theory and experiment is spectacular. It is, however, important to bear in mind that all these corrections are icing on the cake; that the simple nonrelativistic Schrödinger equation by itself provides an excellent description of the hydrogen spectrum. (Much of the present speculation on what the correct theory of elementary particles is will be put to rest if one can come up with a description of these particles that is half as good as the description of the hydrogen atom by Schrödinger's equation.)

Consider next the wave functions. To test the predictions, one once again relies on perturbing the system. The following example should give you a feeling for how this is done. Suppose we apply an external perturbation H^1 for a short time ε. During this time, the system goes from $|nlm\rangle$ to

$$|\psi(\varepsilon)\rangle=\left[I-\frac{i\varepsilon}{\hbar}(H^0+H^1)\right]|nlm\rangle$$
$$=|nlm\rangle-\left(\frac{i\varepsilon E_n}{\hbar}+\frac{i\varepsilon H^1}{\hbar}\right)|nlm\rangle$$

The probability of it being in a state $|n'l'm'\rangle$ (assuming $|n'l'm'\rangle$ is different from $|nlm\rangle$) is

$$|\langle n'l'm'|\psi(\varepsilon)\rangle|^2=\left|-\frac{i\varepsilon}{\hbar}\langle n'l'm'|H^1|nlm\rangle\right|^2$$

Thus quantum theory can also determine for us the rate of transition to the state $|n'l'm'\rangle$. This rate is controlled by the matrix element $\langle n'l'm'|H^1|nlm\rangle$, which in coordinate space, will be some integral over $\psi^*_{n'l'm'}$ and ψ_{nlm} with H^1 sandwiched between them. The evaluation of the integrals entails detailed knowledge of the wave functions, and conversely, agreement of the calculated rates with experiment is a check on the predicted wave functions. We shall see a concrete example of this when we discuss the interaction of radiation with matter in Chapter 18.

Exercise 13.3.3. Instead of looking at the emission spectrum, we can also look at the *absorption* spectrum of hydrogen. Say some hydrogen atoms are sitting at the surface of the sun. From the interior of the sun, white light tries to come out and the atoms at the surface absorb what they can. The atoms in the ground state will now *absorb* the Lyman series and this will lead to dark lines if we analyze the light coming from the sun. The presence of these lines will tell us that there is hydrogen at the surface of the sun. We can also estimate the surface temperature as follows. Let T be the surface temperature. The probabilities $P(n=1)$ and $P(n=2)$ of an atom being at $n=1$ and $n=2$, respectively, are related by Boltzmann's formula

$$\frac{P(n=2)}{P(n=1)}=4\,e^{-(E_2-E_1)/kT}$$

where the factor 4 is due to the degeneracy of the $n=2$ level. Now only atoms in $n=2$ can produce the Balmer lines in the absorption spectrum. The relative strength of the Balmer and

Lyman lines will tell us $P(n=2)/P(n=1)$, from which we may infer T. Show that for $T=$ 6000 K, $P(n=2)/P(n=1)$ is negligible and that it becomes significant only for $T \simeq 10^5$ K. (The Boltzmann constant is $k \simeq 9 \times 10^{-5}$ eV/K. A mnemonic is $kT \simeq \frac{1}{40}$ eV at room temperature, $T=300$ K.)

13.4. Multielectron Atoms and the Periodic Table

It is not possible to treat multielectron atoms analytically even if we treat the nucleus as immobile. Although it is possible, in principle, to treat an arbitrarily complex atom by solving the exact Schrödinger equation numerically, a more practical method is to follow some approximation scheme. Consider the one due to Hartree. Here one assumes that each electron obeys a one-particle Schrödinger equation wherein the potential energy $V=-e\phi(r)$ is due to the nucleus and the other electrons. In computing the electronic contribution to $\phi(r)$, each electron is assigned a charge distribution which is $(-e)$ times the probability density associated with its wave function. And what are the wave functions? They are the eigenstates in the potential $\phi(r)$! To break the vicious circle, one begins with a reasonable guess for the potential, call it $\phi_0(r)$, and computes the allowed energy eigenstates. One then fills them up in the order of increasing energy, putting in just two electrons in each orbital state, with opposite spins (the Pauli principle will not allow any more)‡ until all the electrons have been used up. One then computes the potential $\phi_1(r)$ due to this electronic configuration.§ If it coincides with $\phi_0(r)$ (to some desired accuracy) one stops here and takes the configuration one got to be the ground state of the atom. If not, one goes through one more round, this time starting with $\phi_1(r)$. The fact that, in practice, one soon finds a potential that reproduces itself, signals the soundness of this scheme.

What do the eigenstates look like? They are still labeled by (nlm) as in hydrogen, with states of different m degenerate at a given n and l. [This is because $\phi(r)$ is rotationally invariant.] The degeneracy in l is, however, lost. Formally this is because the potential is no longer $1/r$ and physically this is because states with lower angular momentum have a larger amplitude to be near the origin and hence sample more of the nuclear charge, while states of high angular momentum, which are suppressed at the origin, see the nuclear charge shielded by the electrons in the inner orbits. As a result, at each n the energy goes up with l. The "radius" of each state grows with n, with a slight dependence on l. States of a given n are thus said to form a *shell* (for, in a semiclassical sense, they may be viewed as moving on a sphere of radius equal to the most probable value of r). States of a given l and n are said to form a *subshell*.

Let us now consider the electronic configurations of some low Z (Z is the nuclear charge) atoms. Hydrogen (^{1}H) has just one electron, which is in the $1s$ state. This configuration is denoted by $1s^1$. Helium (^{2}He) has two electrons in the $1s$ state with opposite spins, a configuration denoted by $1s^2$. ^{2}He has its $n=1$ shell filled. Lithium (^{3}Li) has its third electron in the $2s$ state, i.e., it is in the configuration $1s^2 2s^1$. (Recall

‡ In this discussion electron spin is viewed as a spectator variable whose only role is to double the states. This is a fairly good approximation.

§ If necessary, one averages over angles to get a spherically symmetric ϕ.

that the s state is lower than the p state.) We keep going this way through beryllium (^{4}Be), boron (^{5}B), carbon (^{6}C), nitrogen (^{7}N), oxygen (^{8}O), and fluorine (^{9}F), till neon (^{10}Ne). Neon is in the configuration $1s^2 2s^2 2p^6$, i.e., has its $n=2$ shell filled. The next element, sodium (^{11}Na), has a solitary electron in the $3s$ state. The $3s$ and $3p$ subshells are filled when we get to argon (^{18}Ar). The next one, potassium (^{19}K) has its 19th electron in the $4s$ and not $3d$ state. This is because the growth in energy due to a change in n from 3 to 4 is less than the growth due to change in l from 1 to 2 at $n = 3$. This phenomenon occurs often as we move up in Z. For example, in the "rare earth" elements, the $6s$ shell is filled before the $4f$ shell.

Given the electronic configurations, one can anticipate many of the chemical properties of the elements. Consider an element such as ^{10}Ne, which has a closed outer shell. Since the total electronic charge is spherically symmetric ($|R_{nl}|^2 \sum_{m=-l}^{l} |Y_l^m|^2$ is independent of θ and ϕ), it shields the nuclear charge very effectively and the atom has no significant electrostatic affinity for electrons in other atoms. If one of the electrons in the outer shell could be excited to a higher level, this would change, but there is a large gap in energy to cross. Thus the atom is rarely excited and is chemically inert. On the other hand, consider an element like ^{11}Na, which has one more electron, which occupies the $3s$ state. This electron sees a charge of $+e$ when it looks inward (the nuclear charge of 11 shielded by the 10 electrons in the $n=1$ and 2 shells) and is thus very loosely bound. Its binding energy is 5.1 eV compared to an $n=2$ electron in Ne, which has a binding energy of 21.6 eV. If ^{11}Na could get rid of this electron, it could reach a stable configuration with a closed $n=2$ shell. If we look one place to the left (in Z) of ^{10}Ne, we see a perfect acceptor for this electron: we have here ^{9}F, whose $n=2$ shell is all full except for one electron. So when ^{11}Na and ^{9}F get together, Na passes on its electron to F and the system as a whole lowers its energy, since the binding energy in F is 17.4 eV. Having carried out the transfer, the atoms cannot part company, for they have now become charged ions, Na^+ and F^-, which are held together by electrostatic attraction, called the *ionic bond* and form the NaF molecule.

Once we grasp that the chemical behavior is dictated by what is happening in the outermost shell, we can see that several elements will have similar chemical properties because they have similar outer shells. For example, we expect all elements with filled outer shells to be chemically inert. This is true. It is also true that some elements with filled *sub*shells are also inert, such as ^{18}Ar, in which just the $3s$ and $3p$ subshells are filled. The origin of this inertness is the same as in the case with filled shells: a spherically symmetric electronic charge distribution and a large excitation energy. If we move one place to the right of the inert elements, we meet those that behave like Na, i.e., eager to give up an electron, while if we move one place to the left, we meet the likes of F, eager to accept an electron. If we move two places to the left, we see the likes of oxygen, which want two electrons, while two places to the right we have elements like magnesium, which want to get rid of two electrons. It follows that as we move in Z, we see a certain chemical tendency over and over again. This quasiperiodic behavior was emphasized in 1869 by Mendeleev, who organized the elements into a *periodic table*, in which the elements are arranged into a matrix, with all similar elements in the same column. As we go down the first column, for example, we see H, Li, Na, etc., i.e., elements with one electron to spare. In the last column we see the inert elements, He, Ne, etc. Given the maxim that happiness is a filled outer shell, we can guess who will interact with whom. For instance, not only can Na give its electron to F, it can give to Cl, which is one shy of a filled $3p$ subshell. Likewise F can get its electron from K as well, which has a lone electron in the $4s$ state. More involved things can happen, such as the formation of H_2O when two H atoms get together with an oxygen atom, forming the *covalent bond*, in which each hydrogen atom shares an electron with the oxygen atom. This way all three atoms get to fill their outer shells at least part of the time.

There are many more properties of elements that follow from the configuration of the outer electrons. Consider the rare earth elements, ^{58}Ce through ^{71}Lu, which have very similar chemical properties. Why doesn't the chemical behavior change with Z in this range? The answer is that in these elements the $6s$ subshell *is* filled and the $4f$ subshell, deep in the interior (but of a higher energy), is *being* filled. Since what happens in the interior does not affect the chemical properties, they all behave alike. The same goes for the *actinides*, ^{90}Th to ^{103}Lw, which have a filled $7s$ subshell and a $5f$ subshell that is getting filled up.

Since we must stop somewhere, let us stop here. If you want to know more, you must consult books devoted to the subject.‡

*Exercise 13.4.1.** Show that if we ignore interelectron interactions, the energy levels of a multielectron atom go as Z^2. Since the Coulomb potential is Ze/r, why is the energy $\propto Z^2$?

*Exercise 13.4.2.** Compare (roughly) the sizes of the uranium atom and the hydrogen atom. Assume levels fill in the order of increasing n, and that the nonrelativistic description holds. Ignore interelectron effects.

*Exercise 13.4.3.** Visible light has a wavelength of approximately 5000 Å. Which of the series—Lyman, Balmer, Paschen—do you think was discovered first?

‡ See, for a nice trip through the periodic table, U. Fano and L. Fano, *Basic Physics of Atoms and Molecules*, Chapter 18, Wiley, New York (1959).

14

Spin

14.1. Introduction

In this chapter we consider a class of quantum phenomena that cannot be handled by a straightforward application of the four postulates. The reason is that these phenomena involve a quantum degree of freedom called *spin*, which has no classical counterpart. Consequently, neither can we obtain the spin operator by turning to Postulate II, nor can we immediately write down the quantum Hamiltonian that governs its time evolution. The problem is very important, for most particles—electrons, protons, neutrons, photons—have the spin degree of freedom. Fortunately the problem can be solved by a shrewd mixture of classical intuition and reasoning by analogy. In this chapter we study just electron spin. The treatment of the spins of other particles is quite similar, with the exception of the photon, which moves at speed c and can't be treated nonrelativistically. Photon spin will be discussed in Chapter 18.

In the next three sections we address the following questions:

(1) What is the nature of this new spin degree of freedom?

(2) How is the Hilbert space modified to take this new degree of freedom into account? What do the spin operators look like in this space (kinematics of spin)?

(3) How does spin evolve with time, i.e., how does it enter the Hamiltonian (dynamics of spin)?

14.2. What is the Nature of Spin?

The best way to characterize spin is as a form of angular momentum. It is, however, not the angular momentum associated with the operator **L**, as the following experiment shows. An electron is prepared in a state of zero linear momentum, i.e., in a state with a constant (space-independent) wave functión. As the operators L_x, L_y, and L_z will give zero when acting on it, our existing formalism predicts that if the angular momentum along, say the z direction, is measured, a result of zero will obtain. The actual experiment, however, shows that this is wrong, that the result is

$\pm\hbar/2$.‡ It follows that the electron has "intrinsic" angular momentum, not associated with its orbital motion. This angular momentum is called *spin*, for it was imagined in the early days that if the electron has angular momentum without moving through space, then it must be spinning like a top. We adopt this nomenclature, but not the mechanical model that goes with it, for a consistent mechanical model doesn't exist. Fortunately one can describe spin and its dynamics without appealing to any model, starting with just the observed fact that it is a form of angular momentum. Let us now develop the formalism that deals with spin and, in particular, allows us to understand the above experiment.

14.3. Kinematics of Spin

The discussion following the general solution to the eigenvalue problem of angular momentum (Section 12.5) suggests the way for treating particles with intrinsic angular momentum or spin. Recall that if a particle is described by a wave function with many (n) components, the generator of infinitesimal rotation is not just **L** but something more. The reason is that under an infinitesimal rotation *two* things happen to the wave function: (1) the values at each spatial point are re-assigned to the rotated point, and (2) the components of the wave function get transformed into linear combinations of each other.

The differential operator **L** does part (1), while an $n\times n$ matrix **S** is responsible for part (2).

By generalizing our findings from Exercise 12.5.1 to an n component wave function in three dimensions, we can say that under an infinitesimal rotation around the z axis, the wave function is transformed as follows:

$$\begin{bmatrix}\psi_1'\\ \vdots \\ \psi_n'\end{bmatrix}=\left(\begin{bmatrix}1 & & \\ & \ddots & \\ & & 1\end{bmatrix}-\frac{i\varepsilon}{\hbar}\begin{bmatrix}-i\hbar\,\partial/\partial\phi & & 0\\ & \ddots & \\ 0 & & -i\hbar\,\partial/\partial\phi\end{bmatrix}-\frac{i\varepsilon}{\hbar}S_z\right)\begin{bmatrix}\psi_1\\ \vdots \\ \psi_n\end{bmatrix}\qquad(14.3.1)$$

where S_z is an $n\times n$ matrix. In abstract form, this equation reads§

$$\begin{aligned}|\psi'\rangle&=\left[I-\frac{i\varepsilon}{\hbar}(L_z+S_z)\right]|\psi\rangle\\ &=\left[I-\frac{i\varepsilon}{\hbar}J_z\right]|\psi\rangle\end{aligned}\qquad(14.3.2)$$

We identify J_z, the generator of infinitesimal rotations about the z axis, as the z component of angular momentum. We see it has two parts:

$$J_z=L_z+S_z$$

‡ In practice one measures not the angular momentum, but a related quantity called magnetic moment. More on this later. Also spin was first discovered on the basis of spectroscopic evidence and not from an experiment of the above type.

§ The spin operators will be denoted by the same symbol (S) whether they are referred to in the abstract or as matrices in some basis.

and more generally

$$\mathbf{J}=\mathbf{L}+\mathbf{S} \tag{14.3.3}$$

Our problem is to find the number (n) of components appropriate to the electron and the three spin matrices that rotate its components. We proceed as follows.

Since J_i are generators of rotations, they must obey the consistency condition

$$[J_i, J_j]=i\hbar\sum_k \varepsilon_{ijk}J_k \tag{14.3.4}$$

Since $\mathbf{L}$ and $\mathbf{S}$ act on different parts of the wave function (the former on x, y, z, the latter on the indices $i=1, \ldots, n$) they commute, and we may infer from Eq. (14.3.4) that

$$[L_i, L_j]+[S_i, S_j]=i\hbar\left[\sum_k \varepsilon_{ijk}L_k+\sum_k \varepsilon_{ijk}S_k\right] \tag{14.3.5}$$

Using the known commutation rules of the L_i, we deduce

$$[S_i, S_j]=i\hbar\sum_k \varepsilon_{ijk}S_k \tag{14.3.6}$$

Now recall that in Chapter 12 we found matrices J_x, J_y, and J_z [Eqs. (12.5.22)–(12.5.24)] that obey precisely these commutation relations. But these matrices were infinite dimensional. However, the infinite-dimensional matrices were built out of $(2j+1)\times(2j+1)$ blocks, with $j=0, 1/2, 1, 3/2, \ldots$, and the commutation relations were satisfied block by block. So which block shall we pick for the electron spin operators? The answer is given by the empirical fact that S_z has only the eigenvalues $\pm\hbar/2$. This singles out the 2×2 blocks in Eqs. (12.5.22)–(12.5.24):

$$S_x=\frac{\hbar}{2}\begin{bmatrix}0 & 1\\ 1 & 0\end{bmatrix}, \qquad S_y=\frac{\hbar}{2}\begin{bmatrix}0 & -i\\ i & 0\end{bmatrix}, \qquad S_z=\frac{\hbar}{2}\begin{bmatrix}1 & 0\\ 0 & -1\end{bmatrix} \tag{14.3.7}$$

Thus, the way to describe the electron is through a two-component wave function called a *spinor*:

$$\psi=\begin{bmatrix}\psi_+(x, y, z)\\ \psi_-(x, y, z)\end{bmatrix} \tag{14.3.8a}$$

$$\equiv\psi_+\begin{bmatrix}1\\ 0\end{bmatrix}+\psi_-\begin{bmatrix}0\\ 1\end{bmatrix} \tag{14.3.8b}$$

If $\psi_-=0$, $\psi_+\neq 0$, we have an eigenstate of S_z with eigenvalue $\hbar/2$; if $\psi_-\neq 0$, $\psi_+=0$, the S_z eigenvalue is $(-\hbar/2)$.

Let us now proceed to interpret the experiment mentioned earlier. Since we prepared a state of zero momentum, we want the operator **P** to give zero when acting on ψ. The operator **P** simply differentiates both components of ψ:

$$\mathbf{P}\rightarrow\begin{bmatrix} -i\hbar\nabla & 0 \\ 0 & -i\hbar\nabla \end{bmatrix} \tag{14.3.9}$$

We deduce from $\mathbf{P}|\psi\rangle = 0$, i.e.,

$$\begin{bmatrix} -i\hbar\nabla\psi_+ \\ -i\hbar\nabla\psi_- \end{bmatrix} = \begin{bmatrix} 0 \\ 0 \end{bmatrix} \tag{14.3.10}$$

that ψ_+ and ψ_- are independent of x, y, and z. It follows that L_z acting on ψ gives zero. However, S_z doesn't: there is an amplitude $\psi_\pm$ for obtaining $\pm\hbar/2$.

The electron spinor is a *two*-component object, which puts it between a scalar, which has one component, and a vector, which has three. However, the components of the spinor are complex.

A significant difference between spin and orbital angular momentum is this: we can change the magnitude of orbital angular momentum of a particle (by applying external fields) but not the magnitude of its spin. The S^2 operator is

$$S^2 = \hbar^2 \begin{bmatrix} (\frac{1}{2})(\frac{1}{2}+1) & 0 \\ 0 & (\frac{1}{2})(\frac{1}{2}+1) \end{bmatrix} = \frac{3}{4}\hbar^2 \begin{bmatrix} 1 & 0 \\ 0 & 1 \end{bmatrix} \tag{14.3.11}$$

and yields a value $\frac{3}{4}\hbar^2$ on any state ψ. [For any particle, the magnitude of spin is decided by the number of components in the wave function and is an invariant. Thus the spin of the electron is always 1/2 (in units of $\hbar$) and serves as an invariant label of the particle, like its charge or rest mass.]

We have deduced that the electron is to be described by a two-component wave function in the coordinate basis.‡ Let us restate this result in Hilbert space. First, it is easy to see that the introduction of spin has doubled the size of Hilbert space; if it was ∞ dimensional before, now it is 2∞ dimensional, if you know what I mean. The basis vectors $|xyzs_z\rangle$ diagonalize the mutually commuting operators X, Y, Z, and S_z (one can also think of other bases such as $|\mathbf{p}s_z\rangle$ or $|\mathbf{p}s_x\rangle$ etc.). The state vector

‡ We made the deduction given the empirical input from experiment. When we come to the Dirac equation, we will see that incorporating relativistic kinematics will automatically lead to a multicomponent wave function, i.e., lead to spin, if we demand that the equation be first order in time and space.

$|\psi\rangle$ is a 2∞-dimensional column vector in this basis:

$$|\psi\rangle \xrightarrow[\mathbf{R},S_z \text{ basis}]{} \langle xyzs_z|\psi\rangle = \left[\begin{array}{l} \psi(x=-\infty, y=-\infty, z=-\infty, s_z=+\hbar/2) \\ \vdots \\ \psi(x, y, z, s_z=+\hbar/2) \\ \vdots \\ \psi(x=\infty, y=\infty, z=\infty, s_z=+\hbar/2) \\ \hdashline \psi(x=-\infty, y=-\infty, z=-\infty, s_z=-\hbar/2) \\ \vdots \\ \psi(x, y, z, s_z=-\hbar/2) \\ \vdots \\ \psi(x=\infty, y=\infty, z=\infty, s_z=-\hbar/2) \end{array}\right] \tag{14.3.12}$$

Clearly $\psi(\mathbf{r}, \pm\hbar/2)$ gives the amplitude to find the electron at $\mathbf{r}$ with $s_z=\pm\hbar/2$. The horizontal dashed line separates the components with $s_z=\hbar/2$ from those with $s_z=-\hbar/2$. Now if s_z is fixed at $\hbar/2$ and we vary x, y, z from $-\infty$ to ∞, the component of $|\psi\rangle$ will vary smoothly, i.e., define a continuous function $\psi_+(x, y, z)$. Likewise the components below the dotted line define a function $\psi_-(x, y, z)$. In terms of these functions, we may compactify Eq. (14.3.12) to the form

$$|\psi\rangle \xrightarrow[\mathbf{R},S_z \text{ basis}]{} \begin{bmatrix} \psi_+(x, y, z) \\ \psi_-(x, y, z) \end{bmatrix} \tag{14.3.13}$$

This notation blends two notations we have used so far: if the vector has components labeled by a discrete index i ($i=1, \ldots, n$) we denote it as a column vector, while if it is labeled by a continuous index such as x, we denote it by a function $\psi(x)$; but here, since it depends on discrete (s_z) as well as continuous (x, y, z) indices, we write it as a column vector whose components are functions. The normalization condition is

$$\begin{aligned} 1 = \langle\psi|\psi\rangle &= \sum_{s_z} \int \langle\psi|xyzs_z\rangle\langle xyzs_z|\psi\rangle\, dx\, dy\, dz \\ &= \int (|\psi_+|^2 + |\psi_-|^2)\, dx\, dy\, dz \end{aligned} \tag{14.3.14}$$

In the compact notation, S_z is a 2×2 matrix:

$$S_z|\psi\rangle \xrightarrow[\mathbf{R},S_z \text{ basis}]{} \frac{\hbar}{2}\begin{bmatrix} 1 & 0 \\ 0 & -1 \end{bmatrix}\begin{bmatrix} \psi_+(x, y, z) \\ \psi_-(x, y, z) \end{bmatrix} \tag{14.3.15a}$$

whereas in its full glory, it is a 2∞-dimensional matrix:

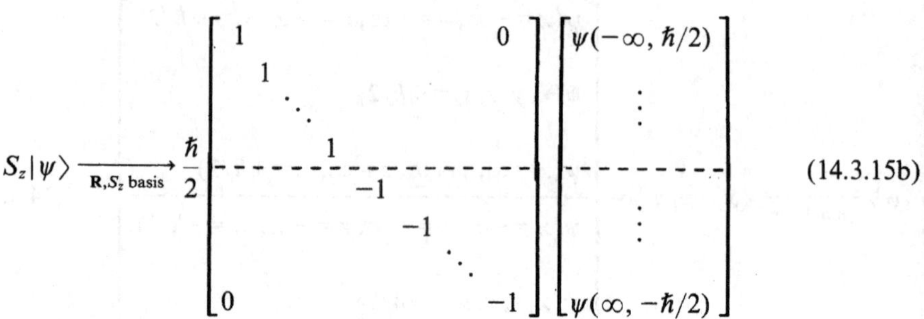

$$S_z|\psi\rangle \xrightarrow[\mathbf{R},S_z\text{ basis}]{} \frac{\hbar}{2}\left[\begin{array}{cccccc} 1 & & & & & 0 \\ & 1 & & & & \\ & & \ddots & & & \\ & & & 1 & & \\ \hline & & & -1 & & \\ & & & & -1 & \\ & & & & & \ddots \\ 0 & & & & & -1 \end{array}\right]\left[\begin{array}{c} \psi(-\infty, \hbar/2) \\ \vdots \\ \hline \vdots \\ \psi(\infty, -\hbar/2) \end{array}\right] \tag{14.3.15b}$$

What about the familiar operators $\Omega(\mathbf{R}, \mathbf{P})$? Equation (14.3.9) gives $\mathbf{P}$ in the compact notation. Likewise, L_z becomes

$$L_z|\psi\rangle \xrightarrow[\mathbf{R},S_z\text{ basis}]{} \begin{bmatrix} -i\hbar\,\partial/\partial\phi & 0 \\ 0 & -i\hbar\,\partial/\partial\phi \end{bmatrix}\begin{bmatrix} \psi_+(x, y, z) \\ \psi_-(x, y, z) \end{bmatrix} \tag{14.3.16}$$

The forms of these operators are consistent with the requirement that operators built out of $\mathbf{R}$ and $\mathbf{P}$ commute with the spin operators. Observe that the Hilbert space $\mathbb{V}_e$ of the electron may be viewed as a direct product of an infinite-dimensional space $\mathbb{V}_0$, which describes a particle with just orbital degrees of freedom, and a two-dimensional space $\mathbb{V}_s$, which describes a particle with just spin degrees of freedom:

$$\mathbb{V}_e = \mathbb{V}_0 \otimes \mathbb{V}_s \tag{14.3.17}$$

The basis vector $|x, y, z, s_z\rangle$ of $\mathbb{V}_e$ is just a direct product

$$|x, y, z, s_z\rangle = |xyz\rangle \otimes |s=1/2, s_z\rangle \tag{14.3.18}$$

Of course $\mathbb{V}_0$ and $\mathbb{V}_s$ do not describe two particles which are amalgamated into a single system, but, rather, two independent degrees of freedom of the electron.

Since we already know how to handle the orbital degrees of freedom, let us pretend from now on that only the spin degree of freedom exists. Or, to be more precise, let us assume the orbital degree of freedom exists but evolves independently. Formally this means that the Hamiltonian is separable:

$$H = H_0 + H_s \tag{14.3.19}$$

where H_0 and H_s depend on just the orbital and spin operators, respectively. Consequently the state vector factorizes into‡

$$|\psi(t)\rangle = |\psi_0(t)\rangle \otimes |\chi_s(t)\rangle \tag{14.3.20}$$

where $|\psi_0\rangle$ and $|\chi_s\rangle$ are elements of $\mathbb{V}_0$ and $\mathbb{V}_s$, respectively. Now $|\psi_0(t)\rangle$ evolves in response to H_0, while the evolution of $|\chi_s(t)\rangle$ is dictated by H_s. We will follow just the evolution of $|\chi_s\rangle$. The product form of $|\psi\rangle$ ensures that the spin and orbital degrees of freedom are statistically independent. Of course, there are many interesting cases in which H is not separable, and the orbital and spin degrees are coupled in their evolution. We will tackle them in a later chapter.

With this assumption, we have just a (complex) two-dimensional Hilbert space $\mathbb{V}_s$ to work with. A complete basis is provided by the vectors $|s, s_z\rangle = |s, m\hbar\rangle \equiv |s, m\rangle$. They are

$$|s, m\rangle = |1/2, 1/2\rangle \xrightarrow[S_z \text{ basis}]{} \begin{bmatrix} 1 \\ 0 \end{bmatrix} \tag{14.3.21a}$$

$$|s, m\rangle = |1/2, -1/2\rangle \xrightarrow[S_z \text{ basis}]{} \begin{bmatrix} 0 \\ 1 \end{bmatrix} \tag{14.3.21b}$$

Any ket $|\chi\rangle$ in $\mathbb{V}_s$ may be expanded as

$$|\chi\rangle = \alpha|1/2, 1/2\rangle + \beta|1/2, -1/2\rangle \xrightarrow[S_z \text{ basis}]{} \begin{bmatrix} \alpha \\ \beta \end{bmatrix} \tag{14.3.22}$$

The normalization condition is

$$1 = \langle\chi|\chi\rangle \xrightarrow[S_z \text{ basis}]{} 1 = [\alpha^*, \beta^*]\begin{bmatrix} \alpha \\ \beta \end{bmatrix} = |\alpha|^2 + |\beta|^2 \tag{14.3.23}$$

If one calculates $\langle \mathbf{S} \rangle$ in the eigenstates of S_z, one finds

$$\langle 1/2, \pm 1/2|\mathbf{S}|1/2, \pm 1/2\rangle = \pm(\hbar/2)\mathbf{k} \tag{14.3.24}$$

One refers to these as states with spin pointing up/down the z axis. More generally, the eigenstates $|\hat{n}, \pm\rangle$ of $\hat{n}\cdot\mathbf{S}$ with eigenvalues $\pm\hbar/2$, in which

$$\langle \hat{n}, \pm|\mathbf{S}|\hat{n}, \pm\rangle = \pm(\hbar/2)\hat{n} \tag{14.3.25}$$

are said to be states with spin up/down the direction of the unit vector $\hat{n}$. Let us address the determination (in the S_z basis) of the components of $|\hat{n}, \pm\rangle$ and the verification of Eq. (14.3.25).

‡ In the $\mathbf{R}$, S_z basis, this means $\psi(x, y, z, s_z, t) = \psi_0(x, y, z, t)\chi(t)$ where χ is a two-component spinor independent of x, y, and z.

Let us say $\hat{n}$ points in the direction (θ, ϕ), i.e., that

$$\begin{aligned}\hat{n}_z &= \cos\theta \\ \hat{n}_x &= \sin\theta\cos\phi \\ \hat{n}_y &= \sin\theta\sin\phi\end{aligned} \tag{14.3.26}$$

The kets $|\hat{n}, \pm\rangle$ are eigenvectors of

$$\begin{aligned}\hat{n}\cdot\mathbf{S} &= n_x S_x + n_y S_y + n_z S_z \\ &= \frac{\hbar}{2}\begin{bmatrix} n_z & n_x - in_y \\ n_x + in_y & -n_z \end{bmatrix} \\ &= \frac{\hbar}{2}\begin{bmatrix} \cos\theta & \sin\theta\, e^{-i\phi} \\ \sin\theta\, e^{i\phi} & -\cos\theta \end{bmatrix}\end{aligned} \tag{14.3.27}$$

It is a simple matter to solve the eigenvalue problem (Exercise 14.3.2) and to find

$$|\hat{n}\ \text{up}\rangle \equiv |\hat{n}+\rangle = \begin{bmatrix} \cos(\theta/2)\, e^{-i\phi/2} \\ \sin(\theta/2)\, e^{i\phi/2} \end{bmatrix} \tag{14.3.28a}$$

$$|\hat{n}\ \text{down}\rangle \equiv |\hat{n}-\rangle = \begin{bmatrix} -\sin(\theta/2)\, e^{-i\phi/2} \\ \cos(\theta/2)\, e^{i\phi/2} \end{bmatrix} \tag{14.3.28b}$$

You may verify that as claimed

$$\begin{aligned}\langle\hat{n}\pm|\mathbf{S}|\hat{n}\pm\rangle &= \pm(\hbar/2)(\mathbf{i}\sin\theta\cos\phi + \mathbf{j}\sin\theta\sin\phi + \mathbf{k}\cos\theta) \\ &= \pm(\hbar/2)\hat{n}\end{aligned} \tag{14.3.29}$$

An interesting feature of $\mathbb{V}_s$ is that not only can we calculate $\langle\mathbf{S}\rangle$ given a state, but we can also go the other way, i.e, deduce the state vector given $\langle\mathbf{S}\rangle$. This has to do with the fact that any element of $\mathbb{V}_s$ has only two (complex) components α and β, constrained by the normalization requirement $|\alpha|^2+|\beta|^2=1$, i.e., three real degrees of freedom, and $\langle\mathbf{S}\rangle$ contains exactly three pieces of information. If we write $\langle\mathbf{S}\rangle$ as $(\hbar/2)\hat{n}$, then the corresponding ket is $|\hat{n}, +\rangle$ or if you want $|-\hat{n}, -\rangle$. Another way to state this result is as follows. Instead of specifying a state by α and β, we can give the operator $\hat{n}\cdot\mathbf{S}$ of which it is an eigenvector with eigenvalue $\hbar/2$. An interesting corollary is that every spinor in $\mathbb{V}_s$ is an eigenket of some spin operator $\hat{n}\cdot\mathbf{S}$ with eigenvalue $\hbar/2$.

Exercise 14.3.1. Let us verify the above corollary explicitly. Take some spinor with components $\alpha=\rho_1 e^{i\phi_1}$ and $\beta=\rho_2 e^{i\phi_2}$. From $\langle\chi|\chi\rangle=1$, deduce that we can write $\rho_1=\cos(\theta/2)$ and $\rho_2=\sin(\theta/2)$ for some θ. Next pull out a common phase factor so that the spinor takes the form in Eq. (14.3.28a). This verifies the corollary and also fixes $\hat{n}$.

So much for the state vectors in $\mathbb{V}_s$. How about the operators on this space? Let us commence with S_x, S_y, and S_z. It is convenient to introduce the *Pauli matrices* $\boldsymbol{\sigma}$, defined by

$$\mathbf{S}=\frac{\hbar}{2}\boldsymbol{\sigma} \tag{14.3.30}$$

so that

$$\sigma_x=\begin{bmatrix}0 & 1\\ 1 & 0\end{bmatrix}, \quad \sigma_y=\begin{bmatrix}0 & -i\\ i & 0\end{bmatrix}, \quad \sigma_z=\begin{bmatrix}1 & 0\\ 0 & -1\end{bmatrix} \tag{14.3.31}$$

It is worth memorizing these matrices. Here are some of their important properties.

(1) They *anti*commute with each other:

$$[\sigma_i, \sigma_j]_+=0 \quad \text{or} \quad \sigma_i\sigma_j=-\sigma_j\sigma_i \qquad (i\neq j) \tag{14.3.32}$$

(2) From the commutation rules for the spin operators **S**, we get, upon using the anticommutativity of the Pauli matrices,

$$\sigma_x\sigma_y=i\sigma_z \quad \text{and cyclic permutations} \tag{14.3.33}$$

(3) They are traceless

$$\text{Tr}\,\sigma_i=0, \qquad i=x, y, z \tag{14.3.34}$$

(See Exercise 14.3.3 for the proof.)

(4) The square of any Pauli matrix equals I:

$$\sigma_i^2=I \tag{14.3.35}$$

or more generally,

$$(\hat{n}\cdot\boldsymbol{\sigma})^2=I \tag{14.3.36}$$

Proof. Since S_z has eigenvalues $\pm\hbar/2$, it follows that

$$\left(S_z+\frac{I\hbar}{2}\right)\left(S_z-\frac{I\hbar}{2}\right)=0$$

in this Hilbert space.‡ But since what we call the z axis is arbitrary, it must be true that

$$\left(\hat{n}\cdot\mathbf{S}+\frac{I\hbar}{2}\right)\left(\hat{n}\cdot\mathbf{S}-\frac{I\hbar}{2}\right)=0$$

or

$$(\hat{n}\cdot\mathbf{S})^2=\frac{\hbar^2}{4}I$$

or

$$(\hat{n}\cdot\boldsymbol{\sigma})^2=I \qquad \text{Q.E.D.}$$

(5) We can combine Eqs. (14.3.32) and (14.3.35) into

$$[\sigma_i, \sigma_j]_+=2\delta_{ij}I \tag{14.3.37}$$

(6) Combining this relation with the commutation rules

$$[\sigma_x, \sigma_y]=2i\sigma_z \quad \text{and cyclic permutations} \tag{14.3.38}$$

we may establish a very useful identity (Exercise 14.3.4):

$$(\mathbf{A}\cdot\boldsymbol{\sigma})(\mathbf{B}\cdot\boldsymbol{\sigma})=\mathbf{A}\cdot\mathbf{B}I+i(\mathbf{A}\times\mathbf{B})\cdot\boldsymbol{\sigma} \tag{14.3.39}$$

where **A** and **B** are vectors or vector operators that commute with $\boldsymbol{\sigma}$.

(7) Combining Eqs. (14.3.33), (14.3.34), and (14.3.35) we find that

$$\operatorname{Tr}(\sigma_i\sigma_j)=2\delta_{ij}, \qquad i,j=x,y,z \tag{14.3.40a}$$

Let us view the identity, I, as the fourth Pauli matrix. If we call it σ_0, then

$$\operatorname{Tr}(\sigma_\alpha\sigma_\beta)=2\delta_{\alpha\beta} \qquad (\alpha,\beta=x,y,z,0)\S \tag{14.3.40b}$$

This equation implies that the σ_α matrices are *linearly independent*. By this I mean as usual that

$$\sum_\alpha c_\alpha\sigma_\alpha=0\rightarrow c_\alpha=0 \qquad \text{for all } \alpha \tag{14.3.41}$$

To prove this for say c_β, multiply both sides by σ_β and take the trace.

‡ See Exercise 12.5.4.

§ From now on α, β will run over four values $x, y, z, 0$; while i, j will run over just x, y, and z.

Since any 2×2 matrix M has only four independent (complex) degrees of freedom, it may be written as

$$M=\sum m_\alpha \sigma_\alpha \tag{14.3.42}$$

To find m_β, we multiply by σ_β and take the trace, to find

$$m_\beta=\tfrac{1}{2}\,\mathrm{Tr}(M\sigma_\beta) \tag{14.3.43}$$

(The coefficients m_α will be complex in general, and real if M is Hermitian.)

Thus, any operator in $\mathbb{V}_s$ may be expressed in terms of the σ_α, which form a basis that is orthonormal with respect to the inner product $\frac{1}{2}\,\mathrm{Tr}(\sigma_\alpha\sigma_\rho)$.‡

Explicit Forms of Rotation Operators

The fact that $(\hat{n}\cdot\boldsymbol{\sigma})^2=I$ greatly simplifies many computations and allows us to compute in closed form several operators such as $U(t)=\exp(-iHt/\hbar)$, $U[R(\boldsymbol{\theta})]=\exp(-i\boldsymbol{\theta}\cdot\mathbf{S}/\hbar)$, which are intractable in infinite-dimensional spaces. In this section we consider the rotation operators, and in the next, the propagator.

Consider

$$\begin{aligned}
U[R(\boldsymbol{\theta})]&=\exp(-i\boldsymbol{\theta}\cdot\mathbf{S}/\hbar)=\exp(-i\boldsymbol{\theta}\cdot\boldsymbol{\sigma}/2)\\
&=\exp\left[-i\left(\frac{\theta}{2}\right)\hat{\theta}\cdot\boldsymbol{\sigma}\right]\\
&=\sum_{n=0}^{\infty}\left(-\frac{i\theta}{2}\right)^n\frac{1}{n!}(\hat{\theta}\cdot\boldsymbol{\sigma})^n\\
&=I+\left(-\frac{i\theta}{2}\right)\hat{\theta}\cdot\boldsymbol{\sigma}+\frac{1}{2!}\left(-\frac{i\theta}{2}\right)^2 I+\frac{1}{3!}\left(-\frac{i\theta}{2}\right)^3(\hat{\theta}\cdot\boldsymbol{\sigma})+\cdots
\end{aligned}$$

Grouping together the coefficients of I and $\hat{\theta}\cdot\boldsymbol{\sigma}$, we get

$$U[R(\boldsymbol{\theta})]=\cos(\theta/2)I-i\sin(\theta/2)\hat{\theta}\cdot\boldsymbol{\sigma} \tag{14.3.44}$$

Let us put this operator to a test. Suppose we have a particle with spin up along the z direction, i.e., in the state $\left[\begin{smallmatrix}1\\0\end{smallmatrix}\right]$. If we want to get from this a particle in the state $|\hat{n},+\rangle$, it is clear that we must rotate $\left[\begin{smallmatrix}1\\0\end{smallmatrix}\right]$ by an angle θ about an axis perpendicular to the z axis and the $\hat{n}$ axis. Thus the rotation angle is

$$\boldsymbol{\theta}=\theta\hat{\theta}=\theta\,\frac{\mathbf{k}\times\hat{n}}{|\mathbf{k}\times\hat{n}|} \tag{14.3.45}$$

‡ The inner product between two matrices M and M' acting on $\mathbb{V}_s$ is actually $\mathrm{Tr}(MM'^{\dagger})$. However, the dagger is irrelevant for the Hermitian σ's. It is an interesting exercise to check that this inner product obeys the three axioms.

where **k** is the unit vector along the z axis. Since $\hat{n}=(\sin\theta\cos\phi, \sin\theta\sin\phi, \cos\theta)$, it follows that

$$\hat{\theta}=\frac{1}{\sin\theta}(-\sin\theta\sin\phi, \sin\theta\cos\phi, 0)=(-\sin\phi, \cos\phi, 0) \qquad (14.3.46)$$

The rotation matrix is, from Eq. (14.3.44),

$$\exp\left(-\frac{i\theta}{2}\hat{\theta}\cdot\boldsymbol{\sigma}\right)=\begin{bmatrix}\cos(\theta/2) & -\sin(\theta/2)\,e^{-i\phi} \\ \sin(\theta/2)\,e^{i\phi} & \cos(\theta/2)\end{bmatrix} \qquad (14.3.47)$$

According to our mnemonic, the first column gives the rotated version of $\begin{bmatrix}1\\0\end{bmatrix}$. We see that it agrees with $|\hat{n}, +\rangle$ given in Eq. (14.3.28) up to an overall phase.

Here is a summary of useful formulas that were derived or simply stated:

$$\mathbf{S}=\frac{\hbar}{2}\boldsymbol{\sigma}$$

$$[\sigma_i, \sigma_j]_+=2I\delta_{ij}$$

$$[\sigma_i, \sigma_j]=2i\sum_k \varepsilon_{ijk}\sigma_k$$

$$(\hat{n}\cdot\boldsymbol{\sigma})^2=I$$

$$\text{Tr}\,\sigma_i=0$$

$$\text{Tr}(\sigma_\alpha\sigma_\beta)=2\delta_{\alpha\beta} \qquad (\alpha, \beta=x, y, z, 0)$$

$$\exp\left(-i\frac{\theta}{2}\hat{\theta}\cdot\boldsymbol{\sigma}\right)=\cos\left(\frac{\theta}{2}\right)I-i\sin\left(\frac{\theta}{2}\right)\hat{\theta}\cdot\boldsymbol{\sigma}$$

$$(\mathbf{A}\cdot\boldsymbol{\sigma})(\mathbf{B}\cdot\boldsymbol{\sigma})=(\mathbf{A}\cdot\mathbf{B})I+i(\mathbf{A}\times\mathbf{B})\cdot\boldsymbol{\sigma}$$

*Exercise 14.3.2.** (1) Show that the eigenvectors of $\boldsymbol{\sigma}\cdot\hat{n}$ are given by Eq. (14.3.28).
(2) Verify Eq. (14.3.29).

*Exercise 14.3.3.** Using Eqs. (14.3.32) and (14.3.33) show that the Pauli matrices are traceless.

*Exercise 14.3.4.** Derive Eq. (14.3.39) in two different ways.
(1) Write $\sigma_i\sigma_j$ in terms of $[\sigma_i, \sigma_j]_+$ and $[\sigma_i, \sigma_j]$.
(2) Use Eqs. (14.3.42) and (14.3.43).

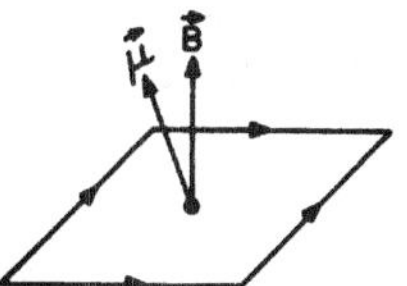

Figure 14.1. In the figure, **B** is the magnetic field and **μ** is the magnetic moment of the loop. The direction of the arrows in the loop is that of the current.

Exercise 14.3.5. Express the following matrix M in terms of the Pauli matrices:

$$M=\begin{bmatrix}\alpha & \beta \\ \gamma & \delta\end{bmatrix}$$

Exercise 14.3.6. (1) Argue that $|\hat{n}, +\rangle = U[R(\phi\mathbf{k})]U[R(\theta\mathbf{j})]|s_z=\hbar/2\rangle$. (2) Verify by explicit calculation.

Exercise 14.3.7. Express the following as linear combinations of the Pauli matrices and I:

(1) $(I+i\sigma_x)^{1/2}$. (Relate it to half a certain rotation.)
(2) $(2I+\sigma_x)^{-1}$.
(3) σ_x^{-1}.

*Exercise 14.3.8.** (1) Show that any matrix that commutes with $\boldsymbol{\sigma}$ is a multiple of the unit matrix.

(2) Show that we cannot find a matrix that anticommutes with all three Pauli matrices. (If such a matrix exists, it must equal zero.)

14.4. Spin Dynamics

Since the quest for the spin Hamiltonian is based on classical analogy, let us recall some basic ideas from classical magnetostatics. Consider a square loop (Fig. 14.1) carrying a current I, in a magnetic field **B**. From standard magnetostatics (force per unit length on a current-carrying conductor etc.) one can show that the torque on the loop is

$$\mathbf{T}=\boldsymbol{\mu}\times\mathbf{B} \tag{14.4.1}$$

where **μ**, the *magnetic moment*, is given by

$$\boldsymbol{\mu}=\frac{I\cdot A}{c}\mathbf{e}_\perp \tag{14.4.2}$$

where A is the area of the loop, c is the velocity of light, and $\mathbf{e}_\perp$ is a unit vector perpendicular to the plane of the loop.‡ The effect of **T** will be to rotate the loop until **μ** and **B** are parallel.

Since we finally wish to address a quantum mechanical problem, it is preferable to summarize the interaction between the loop and the magnetic field in terms of

‡ The sense of $\mathbf{e}_\perp$ is related to the current flow by the right-hand rule.

the potential energy associated with the torque: If θ is the angle between $\boldsymbol{\mu}$ and $\mathbf{B}$, the interaction energy is‡

$$\mathscr{H}_{\text{int}} = \int T(\theta)\, d\theta = \int \mu B \sin\theta\, d\theta = -\mu B \cos\theta = -\boldsymbol{\mu}\cdot\mathbf{B} \tag{14.4.3}$$

As we would expect, this energy is minimized, i.e., a stable configuration obtains, when $\boldsymbol{\mu}$ and $\mathbf{B}$ are parallel.

Although we derived the above equations for a square loop, they are true for any tiny planar loop, over whose extent $\mathbf{B}$ is constant. So we may apply it to the following problem. Imagine a particle of mass m, charge q, moving in a circular orbit of radius r. The current associated with this charge is

$$\begin{aligned} I &= \text{charge flow past any point in the circle per second} \\ &= \frac{qv}{2\pi r} \end{aligned} \tag{14.4.4}$$

and the magnetic moment has a magnitude

$$\mu = \frac{qv}{2\pi r}\cdot\frac{\pi r^2}{c} = \frac{qvr}{2c} = \left(\frac{q}{2mc}\right) mvr = \frac{q}{2mc}\cdot l \tag{14.4.5}$$

where l is the magnitude of the angular momentum. Since $\boldsymbol{\mu}$ and $\mathbf{l}$ are parallel,

$$\boldsymbol{\mu} = \left(\frac{q}{2mc}\right)\mathbf{l} \tag{14.4.6}$$

The ratio of $\boldsymbol{\mu}$ to $\mathbf{l}$ is called the *gyromagnetic ratio* γ. For the particle considered above,

$$\gamma = \frac{q}{2mc} \tag{14.4.7}$$

In the case of the current loop, it was stated that the effect of the torque $\mathbf{T}$ is to cause $\boldsymbol{\mu}$ to align with $\mathbf{B}$. This picture changes when $\boldsymbol{\mu}$ has its origin in angular momentum, as is the case for the particle in question. In this case, $\mathbf{T}$ causes a

‡ This is not the full Hamiltonian (for it does not include the kinetic energy of the loop) but just the potential energy of interaction with the magnetic field.

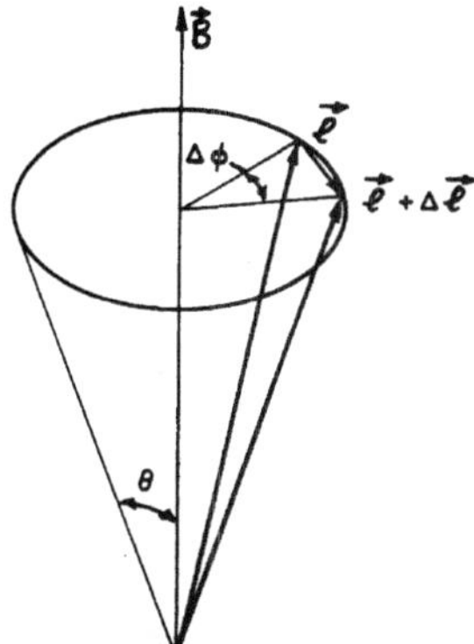

Figure 14.2. In a small time Δt, the tip of the $\mathbf{l}$ vector precesses by an angle $\Delta\phi$ around the magnetic field vector.

precession of $\boldsymbol{\mu}$ around $\mathbf{B}$. We may see this as follows (see Fig. 14.2). The equation of motion is

$$\mathbf{T}=\frac{d\mathbf{l}}{dt}=\boldsymbol{\mu}\times\mathbf{B}=\gamma(\mathbf{l}\times\mathbf{B}) \tag{14.4.8}$$

So in a small time Δt,

$$\Delta\mathbf{l}=\gamma(\mathbf{l}\times\mathbf{B})\Delta t$$

or

$$\Delta l=\gamma lB\sin\theta\,\Delta t$$

Since $\Delta\mathbf{l}$ is perpendicular to $\mathbf{l}$, the tip of the $\mathbf{l}$ vector moves by an angle

$$\Delta\phi=\left(\frac{-\Delta l}{l\sin\theta}\right)=(-\gamma B)\,\Delta t \tag{14.4.9}$$

i.e., precesses at a frequency

$$\boldsymbol{\omega}_0=-\gamma\mathbf{B} \tag{14.4.10}$$

Orbital Magnetic Moment in Quantum Theory

These ideas reemerge in the quantum theory. The Hamiltonian for a particle of mass m and charge q in a magnetic field is

$$H=\frac{(\mathbf{P}-q\mathbf{A}/c)^2}{2m}=\frac{|\mathbf{P}|^2}{2m}-\frac{q}{2mc}(\mathbf{P}\cdot\mathbf{A}+\mathbf{A}\cdot\mathbf{P})+\frac{q^2|\mathbf{A}|^2}{2mc^2} \tag{14.4.11}$$

Let

$$\mathbf{A}=\frac{B}{2}(-y\mathbf{i}+x\mathbf{j}) \tag{14.4.12}$$

so that

$$\nabla\times\mathbf{A}=\mathbf{B}=B\mathbf{k} \tag{14.4.13}$$

is constant and along the z axis. We will assume B is small and drop the last term in H, quadratic in B. When the middle term acts on any $|\psi\rangle$,

$$\begin{aligned}(\mathbf{P}\cdot\mathbf{A})|\psi\rangle &\to -i\hbar\nabla\cdot(\mathbf{A}\psi)\\ &=-i\hbar[(\nabla\cdot\mathbf{A})\psi+\mathbf{A}\cdot\nabla\psi]\\ &=(-i\hbar\mathbf{A}\cdot\nabla)\psi\to(\mathbf{A}\cdot\mathbf{P})|\psi\rangle\end{aligned}$$

since $\nabla\cdot\mathbf{A}=0$ here.‡ Thus the interaction Hamiltonian is

$$\begin{aligned}H_{\text{int}} &= -\frac{q}{2mc}(2\mathbf{A}\cdot\mathbf{P})\\ &=-\frac{q}{mc}\frac{B}{2}(-YP_x+XP_y)\\ &=-\frac{q}{2mc}\mathbf{L}\cdot\mathbf{B}\equiv-\boldsymbol{\mu}\cdot\mathbf{B}\end{aligned} \tag{14.4.14}$$

so that

$$\boldsymbol{\mu}=\frac{q}{2mc}\mathbf{L} \tag{14.4.15}$$

exactly as in the classical case. (We use the same symbol $\boldsymbol{\mu}$ to denote the classical variable and the quantum operator. We will occasionally violate our convention in this manner, so as to follow other widely used conventions.)

If we project this relation along the z axis, we get

$$\mu_z=\frac{q}{2mc}L_z=\frac{q\hbar}{2mc}(0,\pm1,\pm2,\ldots)$$

‡ It is shown in Section 18.4 that **A** corresponding to a given **B** can always be chosen divergenceless.

The quantity $q\hbar/2mc$ is called the *Bohr magneton* of the particle. The *electron Bohr magneton*, simply called the *Bohr magneton*, has a magnitude

$$\frac{e\hbar}{2mc} \simeq 0.6 \times 10^{-8}\ \text{eV/G} \tag{14.4.16}$$

where m is the mass of the electron and G stands for gauss. The *nucleon Bohr magneton* is about 2000 times smaller:

$$\frac{e\hbar}{2Mc} \simeq 0.3 \times 10^{-11}\ \text{eV/G} \tag{14.4.17}$$

where M is the nucleon (proton or neutron)‡ mass. (The nucleon Bohr magneton is also called the *nuclear Bohr magneton.*)

It may be verified, by the use of Ehrenfest's theorem, that $\langle \mathbf{L} \rangle$ precesses around the constant field $\mathbf{B}$ just as $\mathbf{l}$ would (Exercise 14.4.1).

Spin Magnetic Moment

Armed with all this knowledge, we now address the problem of how the electron interacts with an external magnetic field. We assume once again that there is a magnetic moment operator $\boldsymbol{\mu}$ associated with the spin angular momentum. Since any operator on $\mathbb{V}_s$ is a linear combination of the identity and the spin operators, and since $\boldsymbol{\mu}$ is a vector operator, we conclude that

$$\boldsymbol{\mu} = \gamma \mathbf{S} \tag{14.4.18a}$$

where γ is a constant. Since $\gamma = -e/2mc$ for the orbital case, let us write

$$\boldsymbol{\mu} = g(-e/2mc)\mathbf{S} \tag{14.4.18b}$$

where g is a constant. We also assume that

$$\begin{aligned} H_{\text{int}} &= -\boldsymbol{\mu} \cdot \mathbf{B} = \frac{ge}{2mc} \mathbf{S} \cdot \mathbf{B} \\ &= \frac{ge\hbar}{4mc} \boldsymbol{\sigma} \cdot \mathbf{B} \end{aligned} \tag{14.4.19}$$

The intrinsic magnetic moment due to spin is $g/2$ magnetons. Our present formalism does not tell us what g is; to find it we must confront the above H with experiment and hope that for some value of g it gives the right physics. This happens to be the case, and the experimental value for g is very close to 2. We assume

‡ Recall that these two are nearly equal: $M_p c^2 = 938.28$ MeV, while $M_n c^2 = 939.57$ MeV.

hereafter that

$$g=2 \tag{14.4.20}$$

Thus the gyromagnetic ratio for spin is *twice as big* as for orbital angular momentum.

Why is $g \simeq 2$? And why isn't it exactly equal to 2, which would be much prettier? Our formalism doesn't tell us. But it is irresistible to digress and mention that the Dirac equation, which we will discuss in Chapter 20, predicts that $g=2$ exactly. Quantum electrodynamics, which we will not discuss in this book, predicts that the Dirac result will receive corrections that can be calculated in a power series in α, the fine-structure constant. The physics behind the corrections is the following. Recall that the interaction between the electron and other charged particles is mediated by the exchange of photons. Occasionally, an electron will recapture the photon it emitted. Between the emission and reabsorption, the system that originally contained just the electron will contain an electron and the photon. If the magnetic moment of the system is probed at this time, we can get a result that corresponds to $g \neq 2$, since the electron in the two-particle system has both spin and orbital angular momentum. In fact, quantum electrodynamics predicts that what we call the electron is really a superposition of states that contain one Dirac electron, a Dirac electron and a photon, a Dirac electron, several photons, and several electron–positron pairs, etc.‡ The reason the observed value of g is so close to the Dirac value of 2 is that configurations of increasing complexity are suppressed by increasing powers of the fine-structure constant in the superposition. Thus the simplest configuration, with just the Dirac electron, will dominate the picture and the complicated states will provide smaller and smaller corrections to the result $g=2$. The corrections may be calculated in a power series in α:

$$g=2\left[1+\frac{1}{2\pi}\cdot\alpha+O(\alpha^2)+\cdots\right]$$

which has been evaluated to order α^3. The result is§

$$g_{\text{theory}}=2[1.001159652140(\pm 28)]$$

where the error ± 28 in the last two digits is mostly due to uncertainties in the value of α itself and in the numerical evaluation of some of the integrals in the calculation.

In addition to higher-order corrections, this result also receives corrections due to other interactions of the electron, i.e., due to its ability to exchange other quanta such as the graviton. But these effects are negligible to the accuracy considered above. The experimental value of g is‖

$$g_{\text{exp}}=2[1.0011596521884(\pm 43)]$$

‡ The time-energy uncertainty relation allows the production of these particles for short times.
§ T. Kinoshita and W. B. Lindquist, *Phys. Rev.* **D42**, 636, 1990.
‖ R. S. Van Dyck, P. B. Schwinberg, and H. G. Dehmelt, *Phys. Rev. Lett.* **59**, 26, 1987.

in splendid agreement with theory. Feynman has pointed out that this is equivalent to predicting and measuring the distance between New York and Los Angeles to within the width of a human hair!

The theoretical situation is bad for the nucleons. The reason is that these participate in strong interactions as well, i.e., can emit and absorb pions etc., and the counterpart of α is large ($\simeq 15$). In other words, the state with just the Dirac particle no longer dominates, and the corrections are no longer tiny. We can of course measure g experimentally, and the result is (to two places)

$$\gamma_{\text{proton}} = 5.6\ (e/2Mc)$$

$$\gamma_{\text{neutron}} = -3.8\ (e/2Mc)$$

Dirac theory predicts $\gamma = e/Mc$ or $g = 2$ for the proton and $\gamma = 0$ for the neutral neutron. The nonzero γ of the neutron reflects the fact that the neutron can be in a state that has particles with compensating electrical charges but not necessarily compensating magnetic moments.

Because of their large masses, the magnetic moments of the nucleons are negligible compared to that of the electron.‡

Let us now return to the dynamics of spin in a magnetic field $\mathbf{B}$. All we need from now on is the Hamiltonian

$$H = -\boldsymbol{\mu}\cdot\mathbf{B} = -\gamma\mathbf{S}\cdot\mathbf{B} \tag{14.4.21}$$

where

$$\gamma = \frac{-e\cdot 2}{2mc} = \frac{-e}{mc} \tag{14.4.22}$$

Let $|\psi(0)\rangle$ be the initial state of the electron. The state at a later time is

$$|\psi(t)\rangle = U(t)|\psi(0)\rangle$$

where

$$U(t) = e^{-iHt/\hbar} = e^{+i\gamma t(\mathbf{S}\cdot\mathbf{B})/\hbar} \tag{14.4.23}$$

Since $\exp(-i\boldsymbol{\theta}\cdot\mathbf{S}/\hbar)$ is the operator that rotates by $\boldsymbol{\theta}$, *the effect of $U(t)$ is clearly to rotate the state by an angle*

$$\boldsymbol{\theta}(t) = -\gamma\mathbf{B}t \tag{14.4.24}$$

‡ End of digression.

It follows that $\langle \mathbf{S} \rangle$ will precess around $\mathbf{B}$ at a frequency $\boldsymbol{\omega}_0 = -\gamma \mathbf{B}$. If this seems too abstract, let us consider a concrete example. Let $\mathbf{B}$ be along the z axis: $\mathbf{B} = B\mathbf{k}$. In this case

$$\begin{aligned} U(t) &= \exp(i\gamma t S_z B/\hbar) \\ &= \exp(i\omega_0 t \sigma_z/2) \qquad (\omega_0 = \gamma B) \end{aligned}$$

Since σ_z is diagonal,

$$U(t) \rightarrow \begin{bmatrix} e^{i\omega_0 t/2} & 0 \\ 0 & e^{-i\omega_0 t/2} \end{bmatrix}$$

Consider an electron that starts out in the state $|\hat{n}, +\rangle$:

$$|\psi(0)\rangle = |\hat{n}, +\rangle \rightarrow \begin{bmatrix} \cos(\theta/2)\, e^{-i\phi/2} \\ \sin(\theta/2)\, e^{i\phi/2} \end{bmatrix}$$

in which case

$$|\psi(t)\rangle = U(t)|\psi(0)\rangle \rightarrow \begin{bmatrix} \cos(\theta/2)\, e^{-i(\phi - \omega_0 t)/2} \\ \sin(\theta/2)\, e^{i(\phi - \omega_0 t)/2} \end{bmatrix}$$

i.e., since $\omega_0 < 0$ for an electron, ϕ increases at a rate $|\omega_0|$.

Paramagnetic Resonance

Consider a classical magnetic moment $\boldsymbol{\mu}$ in a field $\mathbf{B}_0 = B_0\mathbf{k}$. It will precess around $\mathbf{B}_0$ at a frequency

$$\boldsymbol{\omega}_0 = -\gamma \mathbf{B}_0$$

Suppose we view this process in a frame that is rotating at a frequency $\boldsymbol{\omega}$ parallel to $\mathbf{B}_0$. In this rotating frame, the precession frequency will be

$$\boldsymbol{\omega}_r = \boldsymbol{\omega}_0 - \boldsymbol{\omega} = -\gamma \mathbf{B}_0 - \boldsymbol{\omega} = -\gamma(\mathbf{B}_0 + \boldsymbol{\omega}/\gamma) \tag{14.4.25}$$

Thus the effective field in this rotating frame will be

$$\mathbf{B}_r = \mathbf{B}_0 + \boldsymbol{\omega}/\gamma \tag{14.4.26}$$

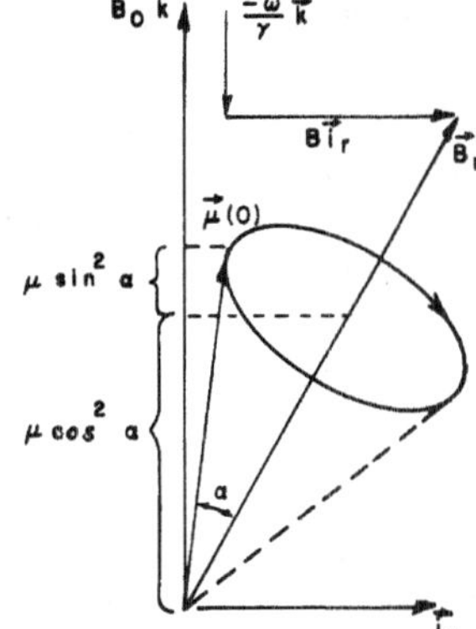

Figure 14.3. The situation in the rotating frame. The effective magnetic field is $\mathbf{B}_r$. The magnetic moment starts out along the z axis (but is slightly displaced in the figure for clarity) and precesses around $\mathbf{B}_r$. The z component of the moment oscillates with an amplitude $\mu \sin^2 \alpha$, where α is the opening angle of the cone. At resonance, $\mathbf{B}_r$ lies along the x axis and $\boldsymbol{\mu}$ precesses in the plane normal to it. The amplitude of the μ_z oscillation is then at its maximum value of μ.

This result is valid even if $\boldsymbol{\omega}$ and $\mathbf{B}_0$ are not parallel (Exercise 14.4.5). Consider now the problem of interest, where, in a nonrotating (lab) frame

$$\mathbf{B}=B\cos\omega t\mathbf{i}-B\sin\omega t\mathbf{j}+B_0\mathbf{k} \qquad (B\lll B_0) \tag{14.4.27}$$

and at $t=0$,

$$\boldsymbol{\mu}(0)=\mu\mathbf{k} \tag{14.4.28}$$

We would like to find out the evolution of $\boldsymbol{\mu}(t)$. Since $\mathbf{B}$ depends on time, it proves advantageous to first consider the problem in a frame that rotates at the same frequency $\boldsymbol{\omega}=-\omega\mathbf{k}$ as the tiny clockwise rotating field B. In this frame, the rotating component of $\mathbf{B}$ gets frozen (say along the x axis) and the constant component $B_0\mathbf{k}$ gets reduced as per Eq. (14.4.26) so that the effective, time-independent field is

$$\mathbf{B}_r=B\mathbf{i}_r+(B_0-\omega/\gamma)\mathbf{k} \tag{14.4.29}$$

where $\mathbf{i}_r$ is the unit vector in the x direction in the rotating frame. ($\mathbf{k}=\mathbf{k}_r$ of course.) In this frame, $\boldsymbol{\mu}$ will precess around $\mathbf{B}_r$ at a frequency

$$\boldsymbol{\omega}_r=-\gamma\mathbf{B}_r \tag{14.4.30a}$$

where

$$|\boldsymbol{\omega}_r|=\omega_r=\gamma[B^2+(B_0-\omega/\gamma)^2]^{1/2} \tag{14.4.30b}$$

It is a simple matter to deduce from Fig. 14.3 that μ_z oscillates as follows:

$$\begin{aligned}\mu_z(t)&=\mu\cos^2\alpha+\mu\sin^2\alpha\cos\omega_r t\\ &=\mu_z(0)\left[\frac{(\omega_0-\omega)^2}{(\omega_0-\omega)^2+\gamma^2B^2}+\frac{\gamma^2B^2\cos\omega_r t}{(\omega_0-\omega)^2+\gamma^2B^2}\right]\end{aligned} \tag{14.4.31}$$

This formula for $\mu_z(t)$ applies in the lab frame as well, since μ_z is invariant under z rotations. As ω increases from 0, the z component of $\mathbf{B}_r$ steadily decreases; α, the opening angle of the cone, increases, and the amplitude of oscillation, $\mu \sin^2 \alpha$, grows. At *paramagnetic resonance*, $\omega = \omega_0$, $\mathbf{B}_r = B\mathbf{i}_r$, $\alpha = \pi/2$, the cone becomes a circle in the $y-z$ plane, and μ_z oscillates with the largest amplitude μ at a frequency γB. The behavior for $\omega > \omega_0$ is obvious.

What if we apply the rotating field at the resonance frequency, but for a time τ such that

$$\gamma B \tau = \pi/2?$$

Such a pulse, called a 90° *pulse*, will swing the magnetic moment into the $x-y$ plane (in either frame). Thereafter $\boldsymbol{\mu}$ will precess around $B_0\mathbf{k}$ at the frequency ω_0 in the lab frame. If we apply a 180° pulse, i.e., choose τ such that

$$\gamma B \tau = \pi$$

the pulse will reverse the sign of $\boldsymbol{\mu}$ and leave it pointing down the z axis, where it will stay (in either frame).

These results for the classical moment $\boldsymbol{\mu}$ apply to the expectation value $\langle \boldsymbol{\mu} \rangle$ in the quantum problem, as you may verify by doing Exercise 14.4.1, where it is proved in general, and Exercise 14.4.3, where the explicit verification in this case is discussed.

Negative Absolute Temperature (Optional Digression)

The absolute zero of temperature, 0 K, ($\simeq -273°$ C) is defined so that nothing can be colder, yet here we speak of negative absolute temperatures! There is no conflict, however, since we will see that negative temperatures are hotter than positive temperatures! Before you give up all faith, let us quickly sort this thing out.

The absolute temperature T is defined as follows:

$$\beta = \frac{1}{kT} = \frac{1}{k}\frac{\partial S}{\partial E} = \frac{\partial \ln \Omega(E)}{\partial E} \tag{14.4.32}$$

where β is the *thermodynamic temperature*, k is Boltzmann's constant, $S = k \ln \Omega$ is the entropy and $\Omega(E)$ is the number of states available to the system as a function of its energy. (Ω depends on other variables, assumed to be fixed.) In most systems, β is positive because adding energy only opens up more states and increases Ω. For instance, if we have a box of gas molecules, they all stay in the ground state at $T = 0$. So, $S = k \ln \Omega = k \ln 1 = 0$. As we pump in energy, they can occupy higher states, and S and Ω can increase without limit.

Consider now a collection of N spin-half particles sitting on some crystal lattice which is immersed in a field $\mathbf{B} = B_0\mathbf{k}$. Each magnetic moment (or spin) has two states only, with energies $E = \pm\mu B_0$, where μ is the magnitude of the magnetic moment. At $T = 0$ K, all are in the ground state ($\boldsymbol{\mu}$ parallel to $\mathbf{B}$); $\Omega = 1$, and $S = 0$. The system has a magnetic moment $\mathbf{M} = N\mu\mathbf{k}$. If we pump in energy $2\mu B_0$, one of the moments can move to the upper energy state; there are N ways to pick the one that moves up, so that $\Omega = N$ and $S = k \ln N$. Clearly β and T are positive. As we pump in more and more energy, S keeps growing until half are up and half are down. At this point, S reaches a maximum, $\beta = \partial S/\partial E = 0$, and $T = +\infty$. The system has no mean magnetic moment along the z axis. Pumping in more energy only reduces S, with more and more particles piling up in the *upper* state. So β and T become negative. Finally, when $E = N\mu B_0$, all moments are in the upper energy state (antiparallel to $\mathbf{B}$), $\mathbf{M} = -N\mu\mathbf{k}$, there is only one such state; $\Omega = 1$ and $S = 0$. This corresponds to $\beta = -\infty$, $T = 0^-$. Thus the sequence of temperatures is $T = 0^+, \ldots, 300, \ldots, \infty, -\infty, \ldots, -300, \ldots, 0^-$. In terms of β, there is more continuity: $\beta = \infty, \ldots, 0^+, 0^-, \ldots, -\infty$. (We should have chosen $-\beta$ as the temperature, for it rises monotonically from $-\infty$ to $+\infty$ as we heat the system.) It should be clear that negative temperatures are hotter than positive temperatures since we go from the latter to the former by pumping in energy. We can also see this by imagining a system at $T = -300$ K brought in contact with identical system at $T = +300$ K. Since the populations of the two systems are identical, except for the change, parallel $\leftrightarrow$ antiparallel, they can increase their entropies by moving toward the state with equal numbers up and down. In this process energy clearly flows from the negative temperature system to the positive temperature system, i.e., the former is hotter. Also note that the final equilibrium temperature is not 0 K but ∞ K.

How does one prepare negative temperatures in the lab? One takes a sample at room temperature, say at $T = 300$ K. It will have more moments parallel than antiparallel:

$$\frac{N(\text{parallel})}{N(\text{antiparallel})} = \frac{e^{-(-B\mu B_0)}}{e^{-\beta\mu B_0}} = e^{2\beta\mu B_0} > 1 \tag{14.4.33}$$

and a net magnetic moment $\mathbf{M}$ along the z axis. If one applies a 180° pulse, there will be *population inversion* (parallel$\leftrightarrow$antiparallel), which amounts to a change in the sign of β and T [see Eq. (14.4.33)]. The spin system cannot stay in this hot state ($T = -300$ K) forever, because it is in contact with the lattice, which will eventually cool it down to room temperature.

The return to thermal equilibrium is easier to observe if one applies a 90° pulse which swings $\mathbf{M}$ into the x–y plane. The temperature now is $T = \infty$ K, since $M_z = 0 \rightarrow N(\text{parallel}) = N(\text{antiparallel}) \rightarrow T = \infty$. Thus $\mathbf{M}$, which will initially begin to precess around $\mathbf{B} = B_0\mathbf{k}$, will eventually realign itself with $\mathbf{B}$. The decay of its rotating components in the x–y plane may be observed as follows. Suppose the specimen is a long cylinder whose axis lies in the x–y plane. If one winds a coil around it, the transverse (x–y) components of $\mathbf{M}$, which simulate a bar magnet rotating in the x–y plane, will induce an oscillating voltage in the coil. The frequency of the (damped) oscillation will be ω_0 and the half-life will be a time τ, called the *transverse relaxation time*.‡

*Exercise 14.4.1.** Show that if $H = -\gamma \mathbf{L} \cdot \mathbf{B}$, and $\mathbf{B}$ is position independent,

$$\frac{d\langle \mathbf{L} \rangle}{dt} = \langle \boldsymbol{\mu} \times \mathbf{B} \rangle = \langle \boldsymbol{\mu} \rangle \times \mathbf{B}$$

‡ The transverse components of **M** decay for other reasons, besides restoration of thermal equilibrium. See R. Schumacher, *Magnetic Resonance*, W. A. Benjamin, New York (1970).

Comparing this to Eq. (14.4.8), we see that $\langle \boldsymbol{\mu} \rangle$ evolves exactly like $\boldsymbol{\mu}$. Notice that this conclusion is valid even if **B** depends on time and also if we are talking about spin instead of orbital angular momentum. A more explicit verification follows in Exercise 14.4.3.

Exercise 14.4.2. Derive (14.4.31) by studying Fig. 14.3.

*Exercise 14.4.3.** We would like to study here the evolution of a state that starts out as $\binom{1}{0}$ and is subject to the **B** field given in Eq. (14.4.27). This state obeys

$$i\hbar \frac{d}{dt}|\psi(t)\rangle = H|\psi\rangle \tag{14.4.34}$$

where $H=-\gamma\mathbf{S}\cdot\mathbf{B}$, and **B** is time dependent. Since classical reasoning suggests that in a frame rotating at frequency $(-\omega\mathbf{k})$ the Hamiltonian should be time independent and governed by $\mathbf{B}_r$ [Eq. (14.4.29)], consider the ket in the rotating frame, $|\psi_r(t)\rangle$, related to $|\psi(t)\rangle$ by a rotation angle ωt:

$$|\psi_r(t)\rangle = e^{-i\omega t S_z/\hbar}|\psi(t)\rangle \tag{14.4.35}$$

Combine Eqs. (14.4.34) and (14.4.35) to derive Schrödinger's equation for $|\psi_r(t)\rangle$ in the S_z basis and verify that the classical expectation is borne out. Solve for $|\psi_r(t)\rangle = U_r(t)|\psi_r(0)\rangle$ by computing $U_r(t)$, the propagator in the rotating frame. Rotate back to the lab and show that

$$|\psi(t)\rangle \xrightarrow[S_z \text{ basis}]{} \begin{bmatrix} \left[\cos\left(\dfrac{\omega_r t}{2}\right) + i\dfrac{\omega_0-\omega}{\omega_r}\sin\left(\dfrac{\omega_r t}{2}\right)\right] e^{+i\omega t/2} \\ \dfrac{i\gamma B}{\omega_r}\sin\left(\dfrac{\omega_r t}{2}\right) e^{-i\omega t/2} \end{bmatrix} \tag{14.4.36}$$

Compare this to the state $|\hat{\mathbf{n}}, +\rangle$ and see what is happening to the spin for the case $\omega_0=\omega$.
Calculate $\langle \mu_z(t)\rangle$ and verify that it agrees with Eq. (14.4.31).

Exercise 14.4.4. At $t=0$, an electron is in the state with $s_z=\hbar/2$. A steady field $\mathbf{B}=B\mathbf{i}$, $B=100$ G, is turned on. How many seconds will it take for the spin to flip?

Exercise 14.4.5. We would like to establish the validity of Eq. (14.4.26) when $\boldsymbol{\omega}$ and $\mathbf{B}_0$ are not parallel.

(1) Consider a vector **V** in the inertial (nonrotating) frame which changes by $\Delta\mathbf{V}$ in a time Δt. Argue, using the results from Exercise 12.4.3, that the change as seen in a frame rotating at an angular velocity $\boldsymbol{\omega}$, is $\Delta\mathbf{V}-\boldsymbol{\omega}\times\mathbf{V}\Delta t$. Obtain a relation between the time derivatives of **V** in the two frames.

(2) Apply this result to the case of **l** [Eq. (14.4.8)], and deduce the formula for the effective field in the rotating frame.

Exercise 14.4.6 (A Density Matrix Problem). (1) Show that the density matrix for an ensemble of spin-1/2 particles may be written as

$$\rho=\tfrac{1}{2}(I+\mathbf{a}\cdot\boldsymbol{\sigma})$$

where $\mathbf{a}$ is a c-number vector.

(2) Show that $\mathbf{a}$ is the mean polarization, $\langle\bar{\boldsymbol{\sigma}}\rangle$.

(3) An ensemble of electrons in a magnetic field $\mathbf{B}=B\mathbf{k}$, is in thermal equilibrium at temperature T. Construct the density matrix for this ensemble. Calculate $\langle\bar{\boldsymbol{\mu}}\rangle$.

14.5. Return of Orbital Degrees of Freedom

Let us now put back the orbital degrees of freedom. The simplest case is when H is separable:

$$H=H_0+H_s \tag{14.5.1}$$

so that the energy eigenstates factorize

$$|\psi\rangle=|\psi_0\rangle\otimes|\chi_s\rangle$$

An example is provided by the hydrogen atom, where the Coulomb interaction is independent of spin:

$$H=H_0 \tag{14.5.2}$$

Here the spin is a constant in time, and all that happens is that we attach a constant spinor χ to the wave functions we found in Chapter 13. If we choose χ to be an eigenstate of S_z, we have‡

$$\begin{aligned}|nlmm_s=1/2\rangle&\to\psi_{nlm}(r,\theta,\phi)\chi_+ &&\left[\chi_+=\begin{bmatrix}1\\0\end{bmatrix}\right]\\ |nlmm_s=-1/2\rangle&\to\psi_{nlm}(r,\theta,\phi)\chi_- &&\left[\chi_-=\begin{bmatrix}0\\1\end{bmatrix}\right]\end{aligned} \tag{14.5.3}$$

The energy levels are of course unaffected. All we have is a doubling of states, with the electron spin being up or down (the z axis) in each of the orbital states (nlm).

Consider next the problem of the hydrogen atom in a weak magnetic field $\mathbf{B}=B\mathbf{k}$. Although both the proton and the electron couple to $\mathbf{B}$, the smallness of the ratio m/M allows us to ignore, in the first approximation, the coupling of the proton's intrinsic and orbital magnetic moments [these are of order m/M and $(m/M)^2$ relative

‡ We use the subscript s on m_s to remind us that it measures the *spin* projection: $s_z=m_s\hbar$. It will be dropped whenever it is obvious that we are dealing with spin.

to that of the electron; see Exercise 14.5.1]. Thus we have, from Eqs. (14.4.14) and (14.4.19),

$$H = H_{\text{Coulomb}} - \left(\frac{-eB}{2mc}\right) L_z - \left(\frac{-eB}{mc}\right) S_z \tag{14.5.4}$$

Since the additional terms in H commute with H_{Coulomb}, L^2, L_z, and S_z, this H is diagonalized by the same states as before, namely, $|nlmm_s\rangle$. The eigenvalues are, however, different:

$$H|nlmm_s\rangle = \left[\frac{-\text{Ry}}{n^2} + \frac{eB\hbar}{2mc}(m+2m_s)\right]|nlmm_s\rangle \tag{14.5.5}$$

The degeneracy is greatly reduced by the **B** field. The ground state, which was twofold degenerate, splits into two levels:

$$E_{n=1} = -\text{Ry} \pm \frac{e\hbar B}{2mc} \tag{14.5.6}$$

The second, which was eightfold degenerate, splits into five levels:

$$E_{n=2} = -\frac{\text{Ry}}{4} + \frac{eB\hbar}{2mc} \times \begin{bmatrix} 2(m=1, m_s=1/2) \\ 1(m=0, m_s=1/2)(l=0 \text{ or } 1) \\ 0(m=1, m_s=-1/2, \text{ or } m=-1, m_s=1/2) \\ -1(m=0, m_s=-1/2)\ (l=0 \text{ or } 1) \\ -2(m=-1, m_s=-1/2) \end{bmatrix} \tag{14.5.7}$$

and so on. In a multielectron atom, one simply adds the contributions from all the electrons. The splitting of levels leads to an increase in the number of spectral lines; where there was one, there will now be several, and the spacing between them may be varied by varying B. This phenomenon is called the *Zeeman effect*.

Consider lastly the Hamiltonian

$$H = H_{\text{Coulomb}} + a\mathbf{L}\cdot\mathbf{S} \tag{14.5.8}$$

whose origin will be explained in a later chapter. For the present, we note that it is not separable, and consequently the spin and orbital degrees of freedom are coupled in their time evolution. The eigenstates of H will not be simply products of orbital and spin parts, but instead superpositions of such states that diagonalize $\mathbf{L}\cdot\mathbf{S}$. The details will be explained in the next chapter.

*Exercise 14.5.1.** (1) Why is the coupling of the proton's intrinsic moment to **B** an order m/M correction to Eq. (14.5.4)?

(2) Why is the coupling of its orbital motion an order $(m/M)^2$ correction? (You may reason classically in both parts.)

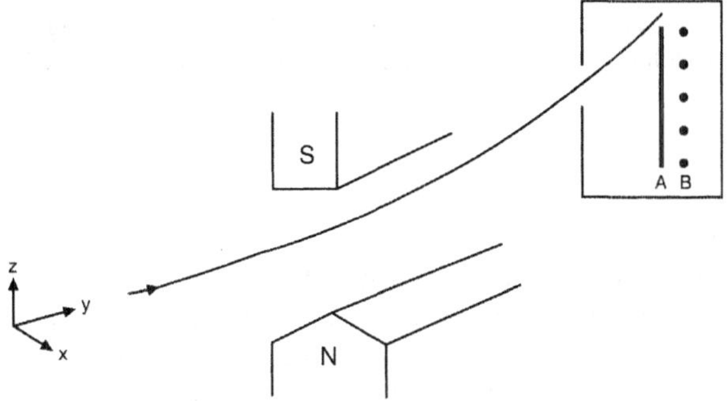

Figure 14.4. The Stern–Gerlach experiment. A beam of particles endowed with magnetic moments enters the inhomogeneous field. Classically the beam is expected to fan out and produce a continuous trace (A) on the screen. What one observes is a set of discrete dots (B). This implies the quantization of magnetic moment and angular momentum.

*Exercise 14.5.11.** (1) Estimate the relative size of the level splitting in the $n=1$ state to the unperturbed energy of the $n=1$ state, when a field $\mathbf{B}=1000\,\mathbf{k}G$ is applied.

(2) Recall that we have been neglecting the order B^2 term in H. Estimate its contribution in the $n=1$ state relative to the linear $(-\boldsymbol{\mu}\cdot\mathbf{B})$ term we have kept, by assuming the electron moves on a classical orbit of radius a_0. Above what $|\mathbf{B}|$ does it begin to be a poor approximation?

The Stern–Gerlach (SG) Experiment

We now consider (in simplified form) the SG experiment, which clearly displays the quantization of angular momentum (along any direction). The apparatus (Fig. 12.6) consists of north and south pole pieces, between which is an *inhomogeneous* magnetic field. A beam of (particles with) magnetic moments, traveling along the y axis, enters the apparatus in a region where $\mathbf{B}$ is predominantly along the z axis and $\partial B_z/\partial z<0$. What do we expect will happen classically? If we pretend that the magnetic moment is due to a pair of equal and opposite (fictitious) magnetic charges, it is clear that any inhomogeneity in $\mathbf{B}$ can lead to a net force on the dipole. This is confirmed if we calculate the force associated with the gradient of the interaction energy

$$\mathbf{F}=-\nabla\mathscr{H}=\nabla(\boldsymbol{\mu}\cdot\mathbf{B})=(\boldsymbol{\mu}\cdot\nabla)\mathbf{B}=\mu_z\frac{\partial B_z}{\partial z}\mathbf{k} \tag{14.5.9}$$

[We have used the identity $\nabla(\boldsymbol{\mu}\cdot\mathbf{B})=(\boldsymbol{\mu}\cdot\nabla)\mathbf{B}+(\mathbf{B}\cdot\nabla)\boldsymbol{\mu}+\boldsymbol{\mu}\times(\nabla\times\mathbf{B})+\mathbf{B}\times(\nabla\times\boldsymbol{\mu})$. In the present case, $\boldsymbol{\mu}$ is not a function of $\mathbf{r}$, and by Maxwell's equations, $\nabla\times\mathbf{B}=0$. Both F_x and F_y vanish on average due to the precession of spin in the $x-y$ plane.] Classically, since μ_z is continuous, the beam is expected to fan out and produce a continuous trace (A in figure) on a screen placed behind the magnet. The actual experiment performed with atoms reveals a series of discrete dots (B in figure). We understand this in semiclassical terms, by saying that μ_z in Eq. (14.5.9) is discrete and therefore so is the angular momentum along the z axis.

This experiment can also be used to reveal the existence of electron spin. For example, if we send in a beam of hydrogen atoms in their ground state, the beam will split into two parts.

Let us describe the above-mentioned hydrogen atom experiment in quantum mechanical terms. Suppose the initial state of a hydrogen atom is

$$\psi_{\text{initial}} = \psi_y(\mathbf{r}_{\text{CM}})\psi_{100}(\mathbf{r})\begin{bmatrix}1\\0\end{bmatrix} \tag{14.5.10}$$

where ψ_y is a wave packet drifting along the y axis that describes the CM motion, ψ_{100} is the ground state wave function, and $[\begin{smallmatrix}1\\0\end{smallmatrix}]$ is the electron spinor. (The proton spin is ignored, for the associated magnetic moment is too small to affect the dynamics.) Since the electron spin is up, its μ_z is down. Since $\partial B_z/\partial z < 0$, the classical force on the atom is up. So by Ehrenfest's theorem‡ we expect the atom to emerge from the apparatus in a state (up to a phase factor)

$$\psi_{\text{out}} = \psi_{y,+z}(\mathbf{r}_{\text{CM}})\psi_{100}(\mathbf{r})\begin{bmatrix}1\\0\end{bmatrix} \tag{14.5.11}$$

where $\psi_{y,+z}$ describes a wave packet that is displaced (relative to the incoming one) along the positive z axis and has also a small velocity in the same direction. Likewise, if the electron spinor had initially been $[\begin{smallmatrix}0\\1\end{smallmatrix}]$, the CM would have emerged in the state $\psi_{y,-z}$ (in the same notation). More generally, if

$$\psi_{\text{initial}} = \psi_y\psi_{100}\begin{bmatrix}\alpha\\\beta\end{bmatrix} = \psi_y\psi_{100}\begin{bmatrix}\alpha\\0\end{bmatrix} + \psi_y\psi_{100}\begin{bmatrix}0\\\beta\end{bmatrix} \tag{14.5.12}$$

then, by the linearity of Schrödinger's equation

$$\psi_{\text{out}} = \psi_{y,+z}\psi_{100}\begin{bmatrix}\alpha\\0\end{bmatrix} + \psi_{y,-z}\psi_{100}\begin{bmatrix}0\\\beta\end{bmatrix} \tag{14.5.13}$$

Assuming $\psi_{y,\pm z}$ are narrow packets with no overlap, we see that the SG apparatus has introduced a correlation between the spin and orbital coordinates: if we catch (by placing a screen) the outgoing atom above the original line of flight (i.e., in a region where $\psi_{y,+z}$ is peaked) it will have spin up, while if we catch it below, the spin is down.

The SG apparatus can be used to prepare a state of definite spin orientation: to get a pure spin up/down beam we simply block the lower/upper beam. But note that the filtering process changes the average z component of linear momentum. This can be undone and the particle restored its original momentum (but filtered with respect to spin) if we place some more magnets (with **B** along the z axis) behind this apparatus. With this modification (which is assumed in the following exercises) the

‡ Recall the warning at the end of Chapter 6. In the present case, the system follows the classical trajectory (approximately) thanks to the massive proton. If we send in just the electron, quantum fluctuations would wipe out the effect. See, for example, pages 324–330 of G. Baym, *Lectures on Quantum Mechanics*, Benjamin, New York (1969).

only effect of the SG apparatus with one or the other beams blocked is to filter the spin without affecting the orbital motion.

*Exercise 14.5.3.** A beam of spin-$1/2$ particles moving along the y axis goes through two collinear SG apparatuses, both with lower beams blocked. The first has its $\mathbf{B}$ field along the z axis and the second has its $\mathbf{B}$ field along the x axis (i.e., is obtained by rotating the first by an angle $\pi/2$ about the y axis). What fraction of particles leaving the first will exit the second? If a third filter that transmits only spin up along the z axis is introduced, what fraction of particles leaving the first will exit the third? If the middle filter transmits both spins up and down (no blocking) the x axis, but the last one transmits only spin down the z axis, what fraction of particles leaving the first will leave the last?

Exercise 14.5.4. A beam of spin-1 particles, moving along the y axis, is incident on two collinear SG apparatuses, the first with $\mathbf{B}$ along the z axis and the second with $\mathbf{B}$ along the z' axis, which lies in the x–z plane at an angle θ relative to the z axis. Both apparatuses transmit only the uppermost beams. What fraction leaving the first will pass the second?

15

Addition of Angular Momenta

15.1. A Simple Example

Consider a system of two spin-1/2 particles (whose orbital degrees of freedom we ignore). If $\mathbf{S}_1$ and $\mathbf{S}_2$‡ are their spin operators, the two-particle Hilbert space $\mathbb{V}_{1\otimes 2}$ is spanned by the four vectors

$$|s_1 m_1\rangle \otimes |s_2 m_2\rangle \equiv |s_1 m_1, s_2 m_2\rangle \tag{15.1.1}$$

which obey

$$S_i^2 |s_1 m_1, s_2 m_2\rangle = \hbar^2 s_i(s_i+1)|s_1 m_1, s_2 m_2\rangle \tag{15.1.2a}$$

$$S_{iz}|s_1 m_1, s_2 m_2\rangle = \hbar m_i |s_1 m_1, s_2 m_2\rangle \qquad (i=1, 2) \tag{15.1.2b}$$

Since $s_i = 1/2$, and $m_i = \pm 1/2$ has freedom only in sign, let us use the compact notation $|++\rangle$, $|+-\rangle$, $|-+\rangle$, $|--\rangle$ to denote the states. For instance,

$$|+-\rangle = |s_1 = \tfrac{1}{2} m_1 = \tfrac{1}{2}, s_2 = \tfrac{1}{2} m_2 = -\tfrac{1}{2}\rangle \tag{15.1.3}$$

and so on. These four vectors form the *product* basis. They represent states that have well-defined values for the magnitude and z component of the individual spins.

Suppose now that we choose not to look at the individual spins but the system as a whole. What are the possible values for the magnitude and z component of the system spin, and what are the states that go with these values? This is a problem in *addition of angular momenta*, which is the topic of this chapter.

‡ In terms of the operators $\mathbf{S}_1^{(1)}$ and $\mathbf{S}_2^{(2)}$ which act on the one-particle spaces, $\mathbf{S}_1 = \mathbf{S}_1^{(1)} \otimes I^{(2)}$ and $\mathbf{S}_2 = I^{(1)} \otimes \mathbf{S}_2^{(2)}$.

Consider the operator

$$\mathbf{S}=\mathbf{S}_1+\mathbf{S}_2 \tag{15.1.4}$$

which we call the *total angular momentum operator*. That $\mathbf{S}$ is indeed the total angular momentum operator is supported by (1) our intuition; (2) the fact that it is the generator of rotation for the product kets, i.e., rotations of the whole system; (3) the fact that it obeys the commutation rules expected of a generator of rotations, namely,

$$[S_i, S_j]=\sum_k i\hbar\varepsilon_{ijk}S_k \tag{15.1.5}$$

as may be readily verified. *Our problem is to find the eigenvalues and eigenvectors of S^2 and S_z*. Consider first

$$S_z=S_{1z}+S_{2z} \tag{15.1.6}$$

which commutes with S_1^2, S_2^2, S_{1z}, and S_{2z}. We expect it to be diagonal in the product basis. This is readily verified:

$$\begin{aligned}
S_z|++\rangle &= (S_{1z}+S_{2z})|++\rangle=\left(\frac{\hbar}{2}+\frac{\hbar}{2}\right)|++\rangle\\
S_z|+-\rangle &= 0|+-\rangle\\
S_z|-+\rangle &= 0|-+\rangle\\
S_z|--\rangle &= -\hbar|--\rangle
\end{aligned} \tag{15.1.7}$$

Thus the allowed values for the total z component are $\hbar$, 0, and $-\hbar$.

By the method of images (or any other method)

$$S_z \xrightarrow[\text{basis}]{\text{product}} \hbar \begin{array}{c} \begin{matrix} ++ & +- & -+ & -- \end{matrix} \\ \begin{bmatrix} 1 & 0 & 0 & 0\\ 0 & 0 & 0 & 0\\ 0 & 0 & 0 & 0\\ 0 & 0 & 0 & -1 \end{bmatrix} \end{array} \tag{15.1.8}$$

Note that the eigenvalue $s_z=0$ is twofold degenerate, and the eigenspace is spanned by the vectors $|+-\rangle$ and $|-+\rangle$. If we form some linear combination, $\alpha|+-\rangle+\beta|-+\rangle$, we still get an eigenstate with $s_z=0$, but this state will not have definite values for S_{1z} and S_{2z} (unless α or $\beta=0$).

Consider next the operator

$$S^2=(\mathbf{S}_1+\mathbf{S}_2)\cdot(\mathbf{S}_1+\mathbf{S}_2)=S_1^2+S_2^2+2\mathbf{S}_1\cdot\mathbf{S}_2 \tag{15.1.9}$$

Although S^2 commutes with S_1^2 and S_2^2, it does not commute with S_{1z} and S_{2z} because of the $\mathbf{S}_1 \cdot \mathbf{S}_2$ term, which has S_{1x}, S_{1y}, etc. in it. By explicit computation,

$$S^2 \xrightarrow[\text{basis}]{\text{product}} \hbar^2 \begin{array}{c} \begin{array}{cccc} ++ & +- & -+ & -- \end{array} \\ \begin{bmatrix} 2 & 0 & 0 & 0 \\ 0 & 1 & 1 & 0 \\ 0 & 1 & 1 & 0 \\ 0 & 0 & 0 & 2 \end{bmatrix} \end{array} \tag{15.1.10}$$

Thus we see that although $|++\rangle$ and $|--\rangle$ are eigenstates of $S^2[s(s+1)=2]$, the states of zero S_z, namely, $|+-\rangle$ and $|-+\rangle$, are not. However, the following linear combinations are:

$$\begin{aligned} &\frac{|+-\rangle+|-+\rangle}{2^{1/2}} \qquad (s=1) \\ &\frac{|+-\rangle-|-+\rangle}{2^{1/2}} \qquad (s=0) \end{aligned} \tag{15.1.11}$$

*Exercise 15.1.1.** Derive Eqs. (15.1.10) and (15.1.11). It might help to use

$$\mathbf{S}_1 \cdot \mathbf{S}_2 = S_{1z}S_{2z} + \tfrac{1}{2}(S_{1+}S_{2-} + S_{1-}S_{2+}) \tag{15.1.12}$$

This completes the solution to the problem we undertook. The allowed values for total spin are $s=1$ and 0, while the allowed values of s_z are $\hbar$, 0, and $-\hbar$. The corresponding eigenstates in the product basis are

$$\begin{aligned} &|s=1\, m=1, \quad s_1=1/2\; s_2=1/2\rangle = |++\rangle \\ &|s=1\, m=0, \quad s_1=1/2\; s_2=1/2\rangle = 2^{-1/2}[|+-\rangle+|-+\rangle] \\ &|s=1\, m=-1, s_1=1/2\; s_2=1/2\rangle = |--\rangle \\ &|s=0\, m=0, \quad s_1=1/2\; s_2=1/2\rangle = 2^{-1/2}[|+-\rangle-|-+\rangle] \end{aligned} \tag{15.1.13}$$

These vectors represent states with well-defined total angular momentum; they form the *total-s basis*. The three spin-1 states are called *triplets* and the solitary spin-0 state is called the *singlet*. The problem of adding angular momenta is essentially a change of basis, from one that diagonalizes $(S_1^2, S_2^2, S_{1z}, S_{2z})$ to one that diagonalizes (S^2, S_z, S_1^2, S_2^2). We can describe our findings symbolically as

$$1/2 \otimes 1/2 = 1 \oplus 0 \tag{15.1.14}$$

which means that the direct product of two spin-1/2 Hilbert spaces is a direct sum of a spin-1 space and a spin-0 space. The way the dimensionalities work out in

Eq. (15.1.14) is as follows:

$$\begin{aligned}&\text{left-hand side:} \quad (2s_1+1)(2s_2+1)=(2\times 1/2+1)(2\times 1/2+1)=4\\&\text{right-hand side:} \quad \sum_{s=0}^{1}(2s+1)=1+3=4\end{aligned} \tag{15.1.15}$$

The decomposition of the direct product space into a sum over spaces with well-defined total spin can also be viewed this way. The rotation operators for the entire system will be 4×4 matrices in the product basis. These matrices are, however, reducible: by changing to the total-s basis, they may be block diagonalized into a 3×3 block (spin-1 sector) and a 1×1 block (spin-0 sector). The total-s basis is, however, irreducible; we cannot further subdivide the spin-1 space into parts that do not mix under rotations.

The total-s states have another property: they have definite symmetry under the exchange of the two particles. The triplets are symmetric and the singlet is antisymmetric. Now, the state vector for two identical spin-1/2 particles must be antisymmetric under the exchange of particle labels, i.e., under the exchange of their spin *and* orbital degrees of freedom. We already know that if Ω is some orbital operator (built out of coordinates and momenta), then

$$|\omega_1\omega_2, S\rangle = 2^{-1/2}[|\omega_1\omega_2\rangle + |\omega_2\omega_1\rangle]$$

and

$$|\omega_1\omega_2, A\rangle = 2^{-1/2}[|\omega_1\omega_2\rangle - |\omega_2\omega_1\rangle]$$

are symmetric and antisymmetric, respectively, under the exchange of the orbital variable. To form the complete state vector, we simply multiply orbital and spin states of *opposite* symmetry:

$$|\omega_1 m_1, \omega_2 m_2, A\rangle = \begin{cases} |\omega_1\omega_2, S\rangle \otimes \dfrac{|+-\rangle - |-+\rangle}{2^{1/2}} \\ |\omega_1\omega_2, A\rangle \otimes \begin{cases} |++\rangle \\ \dfrac{|+-\rangle + |-+\rangle}{2^{1/2}} \\ |--\rangle \end{cases} \end{cases} \tag{15.1.16}$$

These vectors provide a complete basis for the Hilbert space of two identical spin-1/2 particles. As an example, consider the ground state of the He atom, which has two electrons. In connection with the periodic table it was said that in this state of lowest energy, both electrons are in the lowest orbital state $|n=1, l=0, m=0\rangle$‡ and

‡ If we neglect interelectron forces, the states allowed to the electrons are hydrogenlike, in that they are labeled $|n, l, m\rangle$. But the energies and wave functions are obtained upon making the replacement $e^2 \to Ze^2 = 2e^2$.

have opposite spins. We can sharpen that statement now. The orbital part of the ground-state ket is just the direct product,

$$|\psi_o\rangle = |100\rangle \otimes |100\rangle \tag{15.1.17}$$

which is already symmetric. So the spin part must be

$$|\chi_s\rangle = 2^{-1/2}(|+-\rangle - |-+\rangle) \tag{15.1.18}$$

and so

$$|\psi_{\text{ground}}\rangle = |\psi_o\rangle \otimes |\chi_s\rangle \tag{15.1.19}$$

In this state, both the orbital and spin angular momenta are zero.

Let us now return to the problem of just the two spins (and no orbital coordinates). Now that we have two bases, which one should we use? The answer depends on the Hamiltonian. For instance, if the two spins only interact with an external field $\mathbf{B} = B_0\mathbf{k}$,

$$H = -(\gamma_1 \mathbf{S}_1 + \gamma_2 \mathbf{S}_2) \cdot \mathbf{B} = -B_0(\gamma_1 S_{1z} + \gamma_2 S_{2z}) \tag{15.1.20}$$

the product basis, which diagonalizes S_{1z} and S_{2z} is the obvious choice. (If, however, $\gamma_1 = \gamma_2$, then $H \propto S_z$, and we can use the total-s basis as well.) On the other hand, if the spins are mutually interacting and, say,

$$H = A\mathbf{S}_1 \cdot \mathbf{S}_2 = \tfrac{1}{2}A(S^2 - S_1^2 - S_2^2) \tag{15.1.21}$$

the total-s basis diagonalizes H.

*Exercise 15.1.2.** In addition to the Coulomb interaction, there exists another, called the *hyperfine interaction*, between the electron and proton in the hydrogen atom. The Hamiltonian describing this interaction, which is due to the magnetic moments of the two particles is,

$$H_{hf} = A\mathbf{S}_1 \cdot \mathbf{S}_2 \qquad (A > 0) \tag{15.1.22}$$

(This formula assumes the orbital state of the electron is $|1, 0, 0\rangle$.) The total Hamiltonian is thus the Coulomb Hamiltonian plus H_{hf}.

(1) Show that H_{hf} splits the ground state into two levels:

$$\begin{aligned} E_+ &= -\text{Ry} + \frac{\hbar^2 A}{4} \\ E_- &= -\text{Ry} - \frac{3\hbar^2 A}{4} \end{aligned} \tag{15.1.23}$$

and that corresponding states are triplets and singlet, respectively.

(2) Try to estimate the frequency of the emitted radiation as the atom jumps from the triplet to the singlet. To do so, you may assume that the electron and proton are two dipoles μ_e and μ_p separated by a distance a_0, with an interaction energy of the order‡

$$\mathscr{H}_{hf} \cong \frac{\mu_e \cdot \mu_p}{a_0^3}$$

Show that this implies that the constant in Eq. (15.1.22) is

$$A \sim \frac{2e}{2mc}\frac{(5.6)e}{2Mc}\frac{1}{a_0^3}$$

(where 5.6 is the g factor for the proton), and that

$$\Delta E = E_+ - E_- = A\hbar^2$$

is a correction of order $(m/M)\alpha^2$ relative to the ground-state energy. Estimate that the wavelength of emitted radiation is a few tens of centimeters, using the mnemonics from Chapter 13. The measured value is 21.4 cm. This radiation, called the *21-cm line*, is a way to detect hydrogen in other parts of the universe.

(3) Estimate the probability ratio P(triplet)/P(singlet) of hydrogen atoms in thermal equilibrium at room temperature.

15.2. The General Problem

Consider now the general problem of adding two angular momenta $\mathbf{J}_1$ and $\mathbf{J}_2$. What are the eigenvalues and eigenkets of J^2 and J_z, where $\mathbf{J} = \mathbf{J}_1 + \mathbf{J}_2$? One way to find out is to mimic the last section: construct the $(2j_1+1)\cdot(2j_2+1)$-dimensional matrices J^2 and J_z and diagonalize them. Now, J_z will be diagonal in the product basis itself, for

$$J_z|j_1m_1, j_2m_2\rangle = \hbar(m_1+m_2)|j_1m_1, j_2m_2\rangle \tag{15.2.1}$$

It will be a degenerate operator, for there are many ways to build up a total $m = m_1 + m_2$, except when $m = \pm(j_1 + j_2)$ when both angular momenta have maximal projections up/down the z axis. For instance, if $m = j_1 + j_2 - 2$, there are three product kets: $(m_1 = j_1, m_2 = j_2 - 2)$, $(m_1 = j_1 - 1, m_2 = j_2 - 1)$, and $(m_1 = j_1 - 2, m_2 = j_2)$. In each of the degenerate eigenspaces of J_z, we must choose a basis that diagonalizes J^2 (and undiagonalizes J_{1z} and J_{2z}). We can do this by constructing the matrix J^2 and then diagonalizing it. But this can be a tedious business. (If you have done Exercise 15.1.1 you will know that the construction of S^2 is quite tedious even in this four-dimensional case.) There is, however, a more efficient alternative to be described now.

As a first step, we need to know the allowed values for j. Our intuition and our experience from the last section suggest that j can take on values $j_1 + j_2$,

‡ The description here is oversimplified; both $\mathscr{H}_{hf}$ and H_{hf} are rather tricky to derive. Our aim is just to estimate $|A|$ and not to get into its precise origin.

$j_1+j_2-1, \ldots, j_1-j_2$ (assuming $j_1 \geq j_2$).‡ Let us check this. The number of product kets is $(2j_1+1)\cdot(2j_2+1)$. This must equal the number of total-j kets. According to our conjecture, this number is

$$\sum_{j=j_1-j_2}^{j_1+j_2}(2j+1)=\sum_{j=0}^{j_1+j_2}(2j+1)-\sum_{j=0}^{j_1-j_2-1}(2j+1)=(2j_1+1)(2j_2+1) \qquad (15.2.2)$$

using the formula

$$\sum_{n=0}^{N} n=\frac{N(N+1)}{2}$$

We take this to be proof of our conjecture:

$$j_1 \otimes j_2=(j_1+j_2)\oplus(j_1+j_2-1)\oplus\cdots\oplus(j_1-j_2) \qquad (15.2.3)$$

In other words, the total-j kets are

$$|jm, j_1 j_2\rangle \quad \text{with} \quad j_1+j_2 \geq j \geq j_1-j_2, \qquad j \geq m \geq -j \qquad (15.2.4)$$

Let us write them in the form of an array:

$$\begin{array}{llll}
j \rightarrow & & & \\
m \downarrow \quad j_1+j_2 & j_1+j_2-1 & \cdots & j_1-j_2 \\
|j_1+j_2, j_1+j_2\rangle & & & \\
|j_1+j_2, j_1+j_2-1\rangle & |j_1+j_2-1, j_1+j_2-1\rangle & & \\
|j_1+j_2, j_1+j_2-2\rangle & |j_1+j_2-1, j_1+j_2-2\rangle & & |j_1-j_2, j_1-j_2\rangle \\
\vdots & \vdots & \cdots & \vdots \\
|j_1+j_2, -(j_1+j_2-2)\rangle & |j_1+j_2-1, -(j_1+j_2-2)\rangle & & |j_1-j_2, -(j_1-j_2)\rangle \\
|j_1+j_2, -(j_1+j_2-1)\rangle & |j_1+j_2-1, -(j_1+j_2-1)\rangle & & \\
|j_1+j_2, -(j_1+j_2)\rangle & & &
\end{array} \qquad (15.2.5)$$

(Note that the labels $j_1 j_2$ are suppressed on the total-j kets. We shall do so frequently to simplify the notation.)

Our problem is to express each of these kets as a linear combination of product kets. To get an idea of how one goes about doing this, let us consider the problem

‡ There is no loss of generality, for we can always call the larger one j_1.

solved in the last section ($j_1 = j_2 = 1/2$). In this case the states are

$$
\begin{array}{ll}
\multicolumn{2}{c}{j \rightarrow} \\
1 & 0 \\
|1,1\rangle & \\
|1,0\rangle & |0,0\rangle \\
|1,-1\rangle &
\end{array}
\quad (m \downarrow)
$$

Consider the *top state* in the first column, $|1,1\rangle$, which has the largest possible z component. There is only one product state with the right value of m, namely, both spins up. So by inspection,

$$|1,1\rangle = |++\rangle$$

We can multiply the right-hand side by a phase factor, but we follow the convention, called the *Condon–Shortley convention*, in which the coefficient of this top state is chosen to be unity. Consider next the state below this one, namely, $|1,0\rangle$. There are two product states with $m=0$, namely, $|+-\rangle$ and $|-+\rangle$; and $|1,0\rangle$ must be a linear combination of these. We find the combination as follows. We know that‡

$$S_-|1,1\rangle = 2^{1/2}\hbar|1,0\rangle$$

so that

$$|1,0\rangle = \frac{1}{2^{1/2}\hbar} S_-|1,1\rangle$$

But we do not want $|1,0\rangle$ in terms of $|1,1\rangle$, we want it in terms of the product kets. So we rewrite the right-hand side as

$$= \frac{1}{2^{1/2}\hbar}(S_{1-} + S_{2-})|++\rangle = \frac{1}{2^{1/2}\hbar}(\hbar|-+\rangle + \hbar|+-\rangle)$$

so that

$$|1,0\rangle = 2^{-1/2}(|+-\rangle + |-+\rangle)$$

in accordance with our earlier result.

The next state $|1,-1\rangle$ can be obtained by lowering this one more step in the above sense, or more simply by noting that there is only one ket with m maximally negative, namely, $|--\rangle$. So

$$|1,-1\rangle = |--\rangle$$

Our phase convention is such that this is what you would get if you lowered $|1,0\rangle$.

‡ Recall $J_\pm|j,m\rangle = \hbar[(j \mp m)(j \pm m + 1)]^{1/2}|j, m \pm 1\rangle$.

This takes care of the $j=1$ states. Consider next $j=0$. The state $|0,0\rangle$ has $m=0$ and is also a linear combination of $|+-\rangle$ and $|-+\rangle$. We find the combination using two constraints: (1) The combination must be orthogonal to the one that forms the other state with $m=0$, namely, $|1,0\rangle$ and have real coefficients.‡ (2) The combination is normalized to unity. If we call the combination $\alpha|+-\rangle+\beta|-+\rangle$, these constraints tell us that

$$\alpha+\beta=0$$

$$\alpha^2+\beta^2=1$$

It follows that

$$|0,0\rangle=2^{-1/2}(|+-\rangle-|-+\rangle)$$

Note that we could still have multiplied the state by (-1). Our convention is as follows: in each column in Eq. (15.2.5) the top state is given the overall sign which makes the coefficient of the product ket with $m_1=j_1$ positive.

Let us now turn to the general problem, Eq. (15.2.5). Once again the top state in the first column, with m equal to its maximum value of j_1+j_2, can be built out of only one product ket, the one in which both angular momenta take on maximum possible projections along the z axis:

$$|j_1+j_2, j_1+j_2\rangle=|j_1j_1, j_2j_2\rangle \tag{15.2.6}$$

The other m states at this value of j are obtained by lowering. Let us consider going down just one step. Since

$$J_-|j_1+j_2, j_1+j_2\rangle=\hbar[2(j_1+j_2)]^{1/2}|j_1+j_2, j_1+j_2-1\rangle$$

we have, as in the spin-$(1/2\otimes 1/2)$ problem

$$\begin{aligned}
&|j_1+j_2, j_1+j_2-1\rangle\\
&\quad=\frac{1}{[2(j_1+j_2)]^{1/2}\hbar}\cdot(J_{1-}+J_{2-})|j_1j_1, j_2j_2\rangle\\
&\quad=\frac{1}{[2(j_1+j_2)]^{1/2}\hbar}[\hbar(2j_1)^{1/2}|j_1(j_1-1), j_2j_2\rangle+\hbar(2j_2)^{1/2}|j_1j_1, j_2(j_2-1)\rangle]\\
&\quad=\left(\frac{j_1}{j_1+j_2}\right)^{1/2}|j_1(j_1-1), j_2j_2\rangle+\left(\frac{j_2}{j_1+j_2}\right)^{1/2}|j_1j_1, j_2(j_2-1)\rangle
\end{aligned} \tag{15.2.7}$$

Proceeding in this manner we can get to the bottom state in the first column.§

Now for the top state in the second column. Since it has $m=j_1+j_2-1$, there are two product kets that are eligible to enter the linear combination; they are

‡ This is a matter of convention.

§ In practice one goes only to $m=0$. The states of negative m can be found using special properties of the expansion, to be discussed shortly.

$|j_1j_1, j_2(j_2-1)\rangle$ and $|j_1(j_1-1), j_2j_2\rangle$. The combination must be normalized to unity, be orthogonal to the other state formed out of these kets, namely, $|j_1+j_2, j_1+j_2-1\rangle$ [see Eq. (15.2.7)], and by convention have real coefficients. The answer is, by inspection,

$$|j_1+j_2-1, j_1+j_2-1\rangle = \left(\frac{j_1}{j_1+j_2}\right)^{1/2} |j_1j_1, j_2(j_2-1)\rangle - \left(\frac{j_2}{j_1+j_2}\right)^{1/2} |j_1(j_1-1), j_2j_2\rangle \tag{15.2.8}$$

The overall sign is fixed by requirement that the coefficient of the product ket with $m_1=j_1$ be positive. Given the top state, the rest of the second column may be obtained by lowering. Let us go just one more column. The top state in the third column, $|j_1+j_2-2, j_1+j_2-2\rangle$, can be a superposition of three product kets. The three (real) coefficients are determined by these three requirements: orthogonality to the two preceding total-j kets of the same m, and unit normalization. It is clear that there are always enough constraints to determine the top states of each column, and once the top states are known, the rest follow by lowering.

Exercise 15.2.1. (1) Verify that $|j_1j_1, j_2j_2\rangle$ is indeed a state of $j=j_1+j_2$ by letting $J^2=J_1^2+J_2^2+2J_{1z}J_{2z}+J_{1+}J_{2-}+J_{1-}J_{2+}$ act on it.

(2) (optional) Verify that the right-hand side of Eq. (15.2.8) indeed has angular momentum $j=j_1+j_2-1$.

Clebsch–Gordan (CG) Coefficients

The completeness of the product kets allows us to write the total-j kets as

$$|jm, j_1j_2\rangle = \sum_{m_1,m_2}\sum |j_1m_1, j_2m_2\rangle\langle j_1m_1, j_2m_2|jm, j_1j_2\rangle$$

The coefficients of the expansion

$$\langle j_1m_1, j_2m_2|jm, j_1j_2\rangle \equiv \langle j_1m_1, j_2m_2|jm\rangle$$

are called *Clebsch–Gordan coefficients* or *vector addition coefficients*. (Since the labels j_1j_2 appear in the bra, we suppress them in the ket.) Here are some properties of these coefficients:

$$\text{(1)}\ \langle j_1m_1, j_2m_2|jm\rangle \neq 0 \quad \text{only if} \quad j_1-j_2 \leq j \leq j_1+j_2 \tag{15.2.9}$$

(This is called the *triangle inequality*, for geometrically it means that we must be able to form a triangle with sides j_1, j_2, and j).

$$\text{(2)}\ \langle j_1m_1, j_2m_2|jm\rangle \neq 0 \quad \text{only if} \quad m_1+m_2=m \tag{15.2.10}$$

(3) they are real (conventional)

(4) $\langle j_1 j_1, j_2(j-j_1)|jj\rangle$ is positive (conventional)

(This condition fixes the overall sign in the expansion of each top state and was invoked in the preceding discussion.)

(5) $\langle j_1 m_1, j_2 m_2| jm\rangle = (-1)^{j_1+j_2-j}\langle j_1(-m_1), j_2(-m_2)|j(-m)\rangle$ (15.2.11)

This relation halves the work we have to do: we start at the top state and work our way down to $m=0$ (or 1/2 if j is half-integral). The coefficients for the negative m states are then determined by this relation.

*Exercise 15.2.2.** Find the CG coefficients of

(1) $\frac{1}{2}\otimes 1 = \frac{3}{2}\oplus\frac{1}{2}$

(2) $1\otimes 1 = 2\oplus 1\oplus 0$

Exercise 15.2.3. Argue that $\frac{1}{2}\otimes\frac{1}{2}\otimes\frac{1}{2} = \frac{3}{2}\oplus\frac{1}{2}\oplus\frac{1}{2}$.

If we assemble the CG coefficients into a matrix, we find it is orthogonal (real and unitary). This follows from the fact that it relates one orthonormal basis to another. If we invert the matrix, we can write the product kets in terms of total-j kets. The coefficients in this expansion are also CG coefficients:

$$\langle jm| j_1 m_1, j_2 m_2\rangle = \langle j_1 m_1, j_2 m_2| jm\rangle^* = \langle j_1 m_1, j_2 m_2| jm\rangle$$

because the CG coefficients are real. As an example, consider the $\frac{1}{2}\otimes\frac{1}{2}$ problem. There we have

$$\begin{array}{c}|jm\rangle\\ \begin{bmatrix}|1,1\rangle\\|1,0\rangle\\|1,-1\rangle\\|0,0\rangle\end{bmatrix}\end{array} = \begin{bmatrix}1 & 0 & 0 & 0\\ 0 & 1/2^{1/2} & 1/2^{1/2} & 0\\ 0 & 0 & 0 & 1\\ 0 & 1/2^{1/2} & -1/2^{1/2} & 0\end{bmatrix}\begin{array}{c}|m_1 m_2\rangle\\ \begin{bmatrix}|++\rangle\\|+-\rangle\\|-+\rangle\\|--\rangle\end{bmatrix}\end{array}$$

(Notice that the columns contain not the components of vectors, but the basis vectors themselves.) We can invert this relation to get

$$\begin{bmatrix}|++\rangle\\|+-\rangle\\|-+\rangle\\|--\rangle\end{bmatrix} = \begin{bmatrix}1 & 0 & 0 & 0\\ 0 & 1/2^{1/2} & 0 & 1/2^{1/2}\\ 0 & 1/2^{1/2} & 0 & -1/2^{1/2}\\ 0 & 0 & 1 & 0\end{bmatrix}\begin{bmatrix}|1,1\rangle\\|1,0\rangle\\|1,-1\rangle\\|0,0\rangle\end{bmatrix}$$

Thus we can write

$$|+-\rangle = 2^{-1/2}(|1,0\rangle + |0,0\rangle)$$

etc. In practice one uses CG coefficients to go both ways, from the product to the total-j basis and vice versa.

Addition of L and S

Consider an electron bound to a proton in a state of orbital angular momentum l. Since the electron has spin 1/2, its total angular momentum $\mathbf{J} = \mathbf{L} + \mathbf{S}$ can have values of $j = l \pm 1/2$. We wish to express the total-j states in terms of product states $|lm_o, sm_s\rangle$.‡ Since $m_s = \pm 1/2$, at each m there will be at the most two eligible product kets.§ Let

$$|j = l + 1/2, m\rangle = \alpha|l, m - 1/2; 1/2, 1/2\rangle + \beta|l, m + 1/2; 1/2, -1/2\rangle \tag{15.2.12}$$

$$|j = l - 1/2, m\rangle = \alpha'|l, m - 1/2; 1/2, 1/2\rangle + \beta'|l, m + 1/2; 1/2, -1/2\rangle \tag{15.2.13}$$

The requirement that these states be orthonormal tells us that

$$\alpha^2 + \beta^2 = 1 \tag{15.2.14}$$

$$\alpha'^2 + \beta'^2 = 1 \tag{15.2.15}$$

$$\alpha\alpha' + \beta\beta' = 0 \tag{15.2.16}$$

So we only need one more constraint, say the ratio α/β. We find it by demanding that

$$J^2|j = l + 1/2, m\rangle = \hbar^2(l + 1/2)(l + 3/2)|j = l + 1/2, m\rangle \tag{15.2.17}$$

Writing

$$J^2 = L^2 + S^2 + 2L_zS_z + L_-S_+ + L_+S_- \tag{15.2.18}$$

we can dedudce that

$$\frac{\beta}{\alpha} = \left(\frac{l + 1/2 - m}{l + 1/2 + m}\right)^{1/2} \tag{15.2.19}$$

‡ Here, m_o, m_s, and m stand for orbital, spin, and total projections along the z axis.

§ It might help to construct the table as in Eq. (15.2.5). It will contain just two columns, one for $j = l + \frac{1}{2}$ and one for $j = l - 1/2$.

Given this, and our convention for the overall sign,

$$|j = l \pm 1/2, m\rangle = \frac{1}{(2l+1)^{1/2}}[\pm(l + 1/2 \pm m)^{1/2}|l, m - 1/2; 1/2, 1/2\rangle$$
$$+ (l + 1/2 \mp m)^{1/2}|l, m + 1/2; 1/2, -1/2\rangle] \qquad (15.2.20)$$

[Notice that if $j = l + 1/2, m = \pm(l + 1/2)$; only one term survives with unit coefficient.] If the Hamiltonian contains just the Coulomb interaction, or, in addition, an interaction with a weak constant magnetic field, the product basis is adequate. The total-j basis will come in handy when we study the spin–orbit interaction [which involves the operator $\mathbf{L}\cdot\mathbf{S} = \frac{1}{2}(J^2 - L^2 - S^2)$] in Chapter 17.

Exercise 15.2.4. Derive Eqs. (15.2.19) and (15.2.20).

*Exercise 15.2.5.** (1) Show that $\mathbb{P}_1 = \frac{3}{4}I + (\mathbf{S}_1\cdot\mathbf{S}_2)/\hbar^2$ and $\mathbb{P}_0 = \frac{1}{4}I - (\mathbf{S}_1\cdot\mathbf{S}_2)/\hbar^2$ are projection operators, i.e., obey $\mathbb{P}_i\mathbb{P}_j = \delta_{ij}\mathbb{P}_j$ [use Eq. (14.3.39)].

(2) Show that these project into the spin-1 and spin-0 spaces in $\frac{1}{2}\otimes\frac{1}{2} = 1\oplus 0$.

Exercise 15.2.6. Construct the projection operators $\mathbb{P}_\pm$ for the $j = l \pm 1/2$ subspaces in the addition $\mathbf{L}+\mathbf{S}=\mathbf{J}$.

Exercise 15.2.7. Show that when we add j_1 to j_1, the states with $j = 2j_1$ are symmetric. Show that the states with $j = 2j_1 - 1$ are antisymmetric. (Argue for the symmetry of the top states and show that lowering does not change symmetry.) This pattern of alternating symmetry continues as j decreases, but is harder to prove.

The Modified Spectroscopic Notation

In the absence of spin, it is sufficient to use a single letter such as $s, p, d, \ldots$ to denote the (orbital) angular momentum of a particle. In the presence of spin one changes the notation as follows:

(1) Use capital letters $S, P, D, \ldots$ (let us call a typical letter L), to indicate the value of the orbital angular momentum.
(2) Append a *sub*script J to the right of L to indicate the j value.
(3) Append a *super*script $2S+1$ to the left of L to indicate the multiplicity due to spin projections.

Thus, for example

$$^{2S+1}L_J = {}^2P_{3/2}$$

denotes a state with $l = 1, s = 1/2, j = 3/2$. For a single electron the $2S+1$ label is redundant and always equals 2. For a multielectron system, S and L stand for total spin and total orbital angular momentum, and J for their sum. Thus in the ground state of He,

$$^{2S+1}L_J = {}^1S_0$$

15.3. Irreducible Tensor Operators

We have already discussed scalar and vector operators. A scalar operator S transforms like a scalar under rotations, i.e., remains invariant:

$$S \to S' = U^\dagger[R]SU[R] = S \tag{15.3.1}$$

By considering arbitrary infinitesimal rotations we may deduce that

$$[J_i, S] = 0$$

or in a form that will be used later

$$\begin{aligned} [J_\pm, S] &= 0 \\ [J_z, S] &= 0 \end{aligned} \tag{15.3.2}$$

Examples of S are rotationally invariant Hamiltonians such as the Coulomb or isotropic oscillator Hamiltonian. A vector operator $\mathbf{V}$ was defined as a collection of three operators (V_x, V_y, V_z) which transform as the components of a vector in $\mathbb{V}^3(R)$:

$$V_i \to V_i' = U^\dagger[R]V_iU[R] = \sum_j R_{ij}V_j \tag{15.3.3}$$

where R is the usual 3×3 rotation matrix. By considering infinitesimal rotations, we may deduce that [Eq. (12.4.14)]:

$$[V_i, J_j] = i\hbar \sum_k \varepsilon_{ijk} V_k \tag{15.3.4}$$

Let us rewrite Eq. (15.3.3) in an equivalent form. Replace R by $R^{-1} = R^T$ everywhere to get

$$U[R]V_iU^\dagger[R] = \sum_j R_{ji}V_j \tag{15.3.5}$$

Notice that we are summing now over the first index of R. This seems peculiar, for we are accustomed to the likes of Eq. (15.3.3) where the sum is over the second index. The relation of Eq. (15.3.3) to Eq. (15.3.5) is the following. Let $|1\rangle$, $|2\rangle$, and $|3\rangle$ be basis kets in $\mathbb{V}^3(R)$ and R a rotation operator on it. If $|V\rangle$ is some vector with components $v_i = \langle i|V\rangle$, its rotated version $|V'\rangle = R|V\rangle$ has components

$$v_i' = \langle i|R|V\rangle = \sum_j \langle i|R|j\rangle\langle j|V\rangle = \sum_j R_{ij}v_j \tag{15.3.6}$$

If instead we ask what R does to the basis, we find $|i\rangle \to |i'\rangle = R|i\rangle$ where

$$|i'\rangle = R|i\rangle = \sum_j |j\rangle\langle j|R|i\rangle = \sum_j R_{ji}|j\rangle \tag{15.3.7}$$

Since $R_{ji} = (R^{-1})_{ij}$, we see that vector components and the basis vectors transform in "opposite" ways. Equation (15.3.3) defines a vector operator as one whose components transform under $V_i \to U^\dagger V_i U$ as do *components* of a vector $|V\rangle$ under $|V\rangle \to R|V\rangle$, while Eq. (15.3.5) defines it as one whose components V_i transform under $V_i \to UV_iU^\dagger$ as do the *kets* $|i\rangle$ under $|i\rangle \to R|i\rangle$. Both definitions are of course equivalent. The first played a prominent role in the past and the second will play a prominent role in what follows.

Tensor Operators

We know that a vector $|V\rangle$ is an element of $\mathbb{V}^3(R)$, i.e., may be written as

$$|V\rangle = \sum_{i=1}^{3} v_i|i\rangle \tag{15.3.8}$$

in terms of its components v_i and the basis kets $|i\rangle$. A *second-rank tensor* $|T^{(2)}\rangle$ is an element of the direct product space $\mathbb{V}^3(R) \otimes \mathbb{V}^3(R)$, spanned by the nine kets $|i\rangle \otimes |j\rangle$:

$$|T^{(2)}\rangle = \sum_{i=1}^{3}\sum_{j=1}^{3} t_{ij}|i\rangle \otimes |j\rangle \tag{15.3.9}$$

One refers to t_{ij} as the components of $|T^{(2)}\rangle$ in the basis $|i\rangle \otimes |j\rangle$.

As in the case of vectors, a tensor operator of rank 2 is a collection of nine operators T_{ij} which, under $T_{ij} \to U^\dagger T_{ij} U$, respond as do the tensor components t_{ij}, or, equivalently, under $T_{ij} \to UT_{ij}U^\dagger$, respond as do the basis kets $|i\rangle \otimes |j\rangle$. Tensors and tensor operators of rank $n > 2$ are defined in a similar way. (Note that a vector may be viewed as a tensor of rank 1.) We shall call these tensors *Cartesian tensors.*

Of greater interest to us are objects called *spherical tensor operators. A spherical tensor operator of rank* k has $2k+1$ components T_k^q, $q = +k, (k-1), \ldots, -k$, which, under $T_k^q \to UT_k^q U^\dagger$ respond like the angular momentum eigenkets $|j=k, m=q\rangle = |kq\rangle$‡:

$$U[R]T_k^q U^\dagger[R] = \sum_{q'} D_{q'q}^{(k)} T_k^{q'} \tag{15.3.10}$$

Since the $2k+1$ kets $|kq\rangle$ transform irreducibly, so do the operators T_k^q. For this reason, they are also called *irreducible tensor operators.*

By considering infinitesimal rotations, we may deduce from Eq. (15.3.10) that (Exercise 15.3.1):

$$\begin{aligned}[][J_\pm, T_k^q] &= \hbar[(k \mp q)(k \pm q + 1)]^{1/2} T_k^{q\pm 1} \\ [J_z, T_k^q] &= \hbar q T_k^q \end{aligned} \tag{15.3.11}$$

Notice that commuting a J with T_k^q is like letting J act on the ket $|kq\rangle$.

Why are irreducible tensor operators interesting? Consider the effect of acting on a state $|\alpha lm\rangle$ with T_k^q. (Here α denotes labels besides angular momentum.) Let us rotate the resulting state and see what happens:

$$\begin{aligned} U[R]T_k^q|jm\rangle &= U[R]T_k^q U^\dagger[R] U[R]|jm\rangle \\ &= \sum_{q'} D_{q'q}^{(k)} T_k^{q'} \sum_{m'} D_{m'm}^{(j)} |jm'\rangle \\ &= \sum_{q'}\sum_{m'} D_{q'q}^{(k)} D_{m'm}^{(j)} T_k^{q'} |jm'\rangle \end{aligned} \tag{15.3.12}$$

We find that $T_k^q|jm\rangle$ responds to rotations like the product ket $|kq\rangle \otimes |jm\rangle$. *Thus, when we act on a state with T_k^q, we add angular momentum (k,q) to the state.* In other words, an irreducible tensor operator T_k^q imparts a definite amount of angular momentum (k,q) to the state it acts on. This allows us to say the following about matrix elements of T_k^q between angular momentum eigenstates:

$$\langle \alpha' j' m' | T_k^q | \alpha j m\rangle = 0 \quad \text{unless} \quad k+j \geq j' \geq |k-j|, \quad m' = m+q \tag{15.3.13}$$

This is because $T_k^q|\alpha jm\rangle$ contains only those angular momenta that can be obtained by adding (k,q) and (j,m); so $|\alpha' j' m'\rangle$ is orthogonal to $T_k^q|\alpha jm\rangle$ unless (j', m') is one of the possible results of adding (k,q) and (j,m). Equation (15.3.13) is an example of a *selection rule.*

Let us consider some examples, starting with the tensor operator of rank 0. It has only one component T_0^0, which transforms like $|00\rangle$, i.e., remains invariant.

‡ Recall that

$$\begin{aligned} |kq\rangle \to U[R]|kq\rangle &= \sum_{k'}\sum_{q'} |k'q'\rangle\langle k'q'|U[R]|kq\rangle \\ &= \sum_{q'} D_{q'q}^{(k)} |kq'\rangle \end{aligned}$$

Thus T_0^0 is just a scalar operator S, discussed earlier. Our selection rule tells us that

$$\langle\alpha' j' m'|T_0^0|\alpha jm\rangle = 0 \quad \text{unless} \quad j = j', \qquad m = m' \tag{15.3.14}$$

Consider next $T_k^q(q = 1, 0, -1)$. Here we have three objects that go into each other under rotations. Since a vector operator **V** also has three components that transform irreducibly (why?) into each other, we conjecture that some linear combinations of the vector operator components should equal each T_k^q. In fact

$$T_1^{\pm 1} = \mp\frac{V_x \pm iV_y}{2^{1/2}} \equiv V_1^{\pm 1}$$
$$T_1^0 = V_z \equiv V_1^0 \tag{15.3.15}‡$$

Given Eq. (15.3.4) and the above definitions, it may be readily verified that $V_1^{\pm 1}$ and V_1^0 obey Eq. (15.3.11) with $k = 1, q = \pm 1, 0$. The selection rule for, say, V_x is

$$\langle\alpha' j' m'|V_x|\alpha jm\rangle = \langle\alpha' j' m'|\frac{V_1^{-1} - V_1^1}{2^{1/2}}|\alpha jm\rangle$$
$$= 0 \quad \text{unless} \quad j+1 \geq j' \geq |j-1|, \qquad m' = m \pm 1 \tag{15.3.16a}$$

and likewise

$$\langle\alpha' j' m'|V_z|\alpha jm\rangle = \langle\alpha' j' m'|V_1^0|\alpha jm\rangle$$
$$= 0 \quad \text{unless} \quad j+1 \geq j' \geq |j-1|, \qquad m' = m \tag{15.3.16b}$$

Once we go beyond rank 1, it is no longer possible to express Cartesian and spherical tensors *of the same rank* in terms of each other. A Cartesian tensor of rank n has 3^n components, whereas a spherical tensor of rank k has $(2k + 1)$ components. For $n = 0$ and $n = 1$, the Cartesian tensors happened to have the same number of components as spherical tensors of rank $k = 0$ and 1, respectively, and *also transformed irreducibly*. But consider higher ranks, say rank 2. The tensor T_2^q has *five* components that transform irreducibly. The tensor T_{ij} has *nine* components which transform reducibly, i.e., it is possible to form combinations of T_{ij} such that some of them never mix with others under rotations. There is one combination that is invariant, i.e., transforms like T_0^0; there are three combinations that transform like a vector or in light of Eq. (15.3.15) like T_1^q; and finally there are five that transform like T_2^q. We will see what these combinations are when we study the degeneracy of the isotropic oscillator of a few pages hence. Cartesian tensors of higher rank are likewise reducible. Let us now return to the selection rule, Eq. (15.3.13).

We can go a step further and relate the nonvanishing matrix elements. Consider the concrete example of R_1^q, the position operator in spherical form. We have

‡ In the special case $\mathbf{V} = \mathbf{J}$, $J_1^{\pm 1} = \mp(J_x \pm iJ_y)/2^{1/2} = \mp J_\pm/2^{1/2}$ and $J_1^0 = J_z$.

(assuming no spin, so $\mathbf{J} = \mathbf{L}$)

$$\langle \alpha_2 l_2 m_2 | R_1^q | \alpha_1 l_1 m_1 \rangle$$

$$= \int R_{\alpha_2 l_2}^*(r) Y_{l_2}^{m_2 *}(\theta, \phi) r \left(\frac{4\pi}{3}\right)^{1/2} Y_1^q R_{\alpha_1 l_1}(r) Y_{l_1}^{m_1}(\theta, \phi) r^2 \, dr \, d\Omega$$

$$= \left(\frac{4\pi}{3}\right)^{1/2} \int R_{\alpha_2 l_2}^* r R_{\alpha_1 l_1} r^2 \, dr \cdot \int Y_{l_2}^{m_2 *} Y_1^q Y_{l_1}^{m_1} \, d\Omega$$

$$= \langle \alpha_2 l_2 || R_1 || \alpha_1 l_1 \rangle \cdot \langle l_2 m_2 | 1q, l_1 m_1 \rangle \tag{15.3.17‡}$$

where $\langle \alpha_2 l_2 || R_1 || \alpha_1 l_1 \rangle$, the *reduced matrix element*, is independent of m_1, m_2, and q, which appear only in the CG coefficient, which is essentially the angular integral (up to a factor independent of m_1, m_2, and q).

This example illustrates a general result (not proven here):

$$\langle \alpha_2 j_2 m_2 | T_k^q | \alpha_1 j_1 m_1 \rangle = \langle \alpha_2 j_2 || T_k || \alpha_1 j_1 \rangle \cdot \langle j_2 m_2 | kq, j_1 m_1 \rangle \tag{15.3.18}$$

This is called the *Wigner–Eckart* theorem. It separates the dependence of the matrix element on spatial orientation (on m_2, m_1, and q) from the rest. The former is expressed entirely in terms of the CG coefficients.

Exercise 15.3.1. (1) Show that Eq. (15.3.11) follows from Eq. (15.3.10) when one considers infinitesimal rotations. (Hint: $D_{q'q}^{(k)} = \langle kq' | I - (i\delta\boldsymbol{\theta} \cdot \mathbf{J})/\hbar | kq \rangle$. Pick $\delta\boldsymbol{\theta}$ along, say, the x direction and then generalize the result to the other directions.)

(2) Verify that the spherical tensor V_1^q constructed out of $\mathbf{V}$ as in Eq. (15.3.15) obeys Eq. (15.3.11).

Exercise 15.3.2. It is claimed that $\Sigma_q (-1)^q S_k^q T_k^{(-q)}$ is a scalar operator.

(1) For $k = 1$, verify that this is just $\mathbf{S} \cdot \mathbf{T}$.

(2) Prove it in general by considering its response to a rotation. [Hint: $D_{-m,-m'}^{(j)} = (-1)^{m-m'} (D_{m,m'}^{(j)})^*$.]

Exercise 15.3.3. (1) Using $\langle jj | jj, 10 \rangle = [j/(j+1)]^{1/2}$ show that

$$\langle \alpha j || J_1 || \alpha' j' \rangle = \delta_{\alpha\alpha'} \delta_{jj'} \hbar [j(j+1)]^{1/2}$$

(2) Using $\mathbf{J} \cdot \mathbf{A} = J_z A_z + \frac{1}{2}(J_- A_+ + J_+ A_-)$ (where $A_\pm = A_x \pm i A_y$) argue that

$$\langle \alpha' j m' | \mathbf{J} \cdot \mathbf{A} | \alpha j m \rangle = c \langle \alpha' j || A || \alpha j \rangle$$

where c is a constant independent of α, α' and $\mathbf{A}$. Show that $c = \hbar [j(j+1)]^{1/2} \delta_{m,m'}$.

(3) Using the above, show that

$$\langle \alpha' j m' | A^q | \alpha j m \rangle = \frac{\langle \alpha' j m | \mathbf{J} \cdot \mathbf{A} | \alpha j m \rangle}{\hbar^2 j(j+1)} \langle j m' | J^q | j m \rangle \tag{15.3.19}$$

‡ Note that R_1^q is the tensor operator and $R_{\alpha l}(r)$ is the radial part of the wave function. We have also used Eq. (12.5.42) to obtain R_1^q.

*Exercise 15.3.4.** (1) Consider a system whose angular momentum consists of two parts $\mathbf{J}_1$ and $\mathbf{J}_2$ and whose magnetic moment is

$$\boldsymbol{\mu}=\gamma_1\mathbf{J}_1+\gamma_2\mathbf{J}_2$$

In a state $|jm, j_1j_2\rangle$ show, using Eq. (15.3.19), that

$$\langle\mu_x\rangle=\langle\mu_y\rangle=0$$

$$\langle\mu_z\rangle=m\hbar\left[\frac{\gamma_1+\gamma_2}{2}+\frac{(\gamma_1-\gamma_2)}{2}\frac{j_1(j_1+1)-j_2(j_2+1)}{j(j+1)}\right]$$

(2) Apply this to the problem of a proton ($g=5.6$) in a ${}^2P_{1/2}$ state and show that $\langle\mu_z\rangle=\pm 0.26$ nuclear magnetons.

(3) For an electron in a ${}^2P_{1/2}$ state show that $\langle\mu_z\rangle=\pm\frac{1}{3}$ Bohr magnetons.

*Exercise 15.3.5.** Show that $\langle jm|T_k^q|jm\rangle=0$ if $k>2j$.

15.4. Explanation of Some "Accidental" Degeneracies

In this section the degeneracy of states of different l at a given value of n in the hydrogen atom and the isotropic oscillator (see Section 12.6) will be explained. But first let us decide what it means to explain any degeneracy. Consider for example the $(2l+1)$-fold degeneracy of the different m states at a given l in both these problems. We explain it in terms of the rotational invariance of the Hamiltonian as follows:

(1) For every rotation $R(\boldsymbol{\theta})$ on $\mathbb{V}^3(R)$ there exists a unitary operator $U[R]$ which rotates the vector operators

$$U^\dagger V_i U=\sum_j R_{ij}V_j \tag{15.4.1}$$

If the Hamiltonian depends only on the "lengths" of various vector operators like $\mathbf{P}$, $\mathbf{R}$, $\mathbf{L}$ etc., then it is rotationally invariant:

$$U^\dagger HU=H \tag{15.4.2}$$

i.e., rotations are symmetries of H. This is the case for the two problems in question.

(2) If we write this relation in infinitesimal form, we find

$$[H, L_i]=0, \qquad i=1, 2, 3 \tag{15.4.3}$$

where L_i are the generators of rotation. For every free parameter that defines a rotation (θ_x, θ_y, and θ_z) there is a corresponding generator. They are all conserved.

(3) From the three generators we construct the operator

$$L_- = L_x - iL_y \tag{15.4.4}$$

which lowers the m value:

$$L_-|l, m\rangle = c|l, m-1\rangle \tag{15.4.5}$$

Since $[L, H]=0$, the lowering operation does not change the energy.

This explains the degeneracy in m, for, starting with the state of highest m at a given l, we can go down all the way to the lowest m without changing the energy. (We can equally well work with L_+.)

Let us try to do the same for the two problems in question. We follow these steps:

Step (1): Identify symmetries of H besides rotational invariance.
Step (2): Find the generators of the symmetry transformations.
Step (3): Construct an operator from these generators that can change l by one unit in the case of hydrogen and two units in the case of the oscillator.

Hydrogen

Steps (1) and (2). Unfortunately the only obvious symmetry of the Coulomb Hamiltonian is rotational invariance. The additional symmetry, the one we are after, is very subtle and clearest in momentum space. We will not discuss it. But how then do we go to step (2)? The answer lies in the fact that the generators of the symmetry are conserved quantities. Now we have seen that the Coulomb problem admits an extra conserved quantity, the Runge–Lenz vector. Thus the three components of

$$\mathbf{N} = \frac{1}{2m}(\mathbf{P}\times\mathbf{L} - \mathbf{L}\times\mathbf{P}) - \frac{e^2\mathbf{R}}{(X^2+Y^2+Z^2)^{1/2}} \tag{15.4.6}$$

must be the generators of the additional symmetry transformations (or linear combinations thereof).

Step (3). Since we wish to talk about angular momentum let us write $\mathbf{N}$ in spherical form:

$$N_1^{\pm 1} = \mp \frac{N_x \pm iN_y}{2^{1/2}}$$
$$N_1^0 = N_z \tag{15.4.7}$$

Consider the state $|nll\rangle$ of the H-atom. Acting on it with N_1^1, we get another state of the same energy or same n (since $[H, N_1^1]=0$) but with higher angular momentum: $N_1^1|nll\rangle$ behaves as $|11\rangle\otimes|ll\rangle = |l+1, l+1\rangle$. So

$$N_1^1|n, l, l\rangle = c|n, l+1, l+1\rangle \tag{15.4.8}$$

(It will turn out that c vanishes when $l=l_{\max}=n-1$.) Using N_1^1 we can connect all the different l states at a given n, and using L_- we can connect all the m states at a given l. For example, at $n=3$ the network that connects degenerate states is as follows:

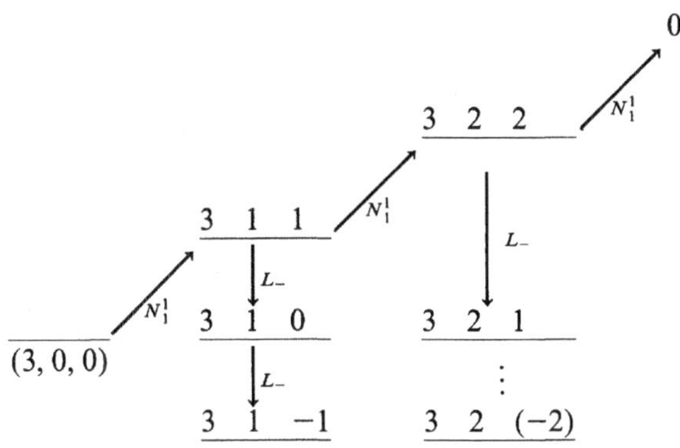

The Oscillator

Step (1). To find the extra symmetry of H, let us look at it again:

$$H=\frac{P_x^2+P_y^2+P_z^2}{2\mu}+\frac{1}{2}\mu\omega^2(X^2+Y^2+Z^2) \tag{15.4.9}$$

We say H is rotationally invariant because it depends only on the lengths squared of the (real) vectors $\mathbf{P}$ and $\mathbf{R}$. Let us now rewrite H in a way that reveals the extra symmetry. Define a *complex vector* (operator) whose real and imaginary parts are proportional to $\mathbf{R}$ and $\mathbf{P}$:

$$\mathbf{a}=\frac{1}{(2\mu\omega\hbar)^{1/2}}(\mu\omega\mathbf{R}+i\mathbf{P}) \tag{15.4.10}$$

and its adjoint, whose components are complex conjugates of those of $\mathbf{a}$:

$$\mathbf{a}^\dagger=\frac{1}{(2\mu\omega\hbar)^{1/2}}(\mu\omega\mathbf{R}-i\mathbf{P}) \tag{15.4.11}$$

The components of $\mathbf{a}$ and $\mathbf{a}^\dagger$ are just the lowering and raising operators for the x, y, and z oscillators. They obey

$$[a_i, a_j^\dagger]=\delta_{ij}$$

In terms of **a** and $\mathbf{a}^\dagger$,

$$H=\hbar\omega(\mathbf{a}^\dagger\cdot\mathbf{a}+3/2) \tag{15.4.12}$$

Thus we find that H is a function of the length squared of a *complex* three-dimensional vector **a**. So it is invariant under "rotations" in $\mathbb{V}^3(C)$, i.e., under unitary transformations in $\mathbb{V}^3(C)$. Just as we denoted the rotations in $\mathbb{V}^3(R)$ by R, let us call these C.‡ For every "rotation" C (unitary transformation) in $\mathbb{V}^3(C)$, there will exist Hilbert space operators $U[C]$ which rotate the complex vector operator **a**:

$$a_i \to a_i' = U^\dagger[C]a_iU[C]=\sum_j C_{ij}a_j \tag{15.4.13}$$

where C_{ij} are matrix elements of the unitary operator C in $\mathbb{V}^3(C)$. Since H depends only on the norm squared of **a**,

$$U^\dagger[C]HU[C]=H \tag{15.4.14}$$

Step (2). How many generators of $U[C]$ are there and what are they? The answer to the first part is the number of parameters that define a rotation in $\mathbb{V}^3(C)$, i.e., the number of independent parameters in a 3×3 unitary matrix C. Now any such matrix can be written as

$$C=e^{i\Omega} \tag{15.4.15}$$

where Ω is a 3×3 Hermitian matrix. It is easy to see that Ω has three real diagonal elements and three independent complex off-diagonal elements. Thus it depends on nine real parameters. So there are nine conserved generators. What are they? Rather than deduce them (as we did the L's by considering the effect of infinitesimal rotations on ψ) we write down the nine conserved quantities by inspection. It is clear that in the oscillator case, the nine operators

$$T_{ij}=a_i^\dagger a_j \qquad (i, j=x, y, \text{ or } z) \tag{15.4.16}$$

are conserved. The proof is simple: a_j destroys a j quantum and $a_i^\dagger$ creates an i quantum and this leaves the energy invariant since the x, y, and z oscillators have the same ω (isotropy). To see what impact T_{ij} has on l degeneracy, we must decompose T_{ij} into its irreducible parts.

Consider first the combination

$$\operatorname{Tr} T=T_{xx}+T_{yy}+T_{zz}=a_x^\dagger a_x+a_y^\dagger a_y+a_z^\dagger a_z=\mathbf{a}^\dagger\cdot\mathbf{a} \tag{15.4.17}$$

This is clearly a scalar, i.e., transforms like T_0^0. The fact that it commutes with H does not explain the degeneracy in l because it "carries" no angular momentum. In fact $\mathbf{a}^\dagger\cdot\mathbf{a}$ is just H up to a scale factor and an additive constant.

‡ We should really be calling these U. But that will complicate the notation.

Consider next the three antisymmetric combinations

$$
\begin{aligned}
T_{xy}-T_{yx}&=a_x^\dagger a_y-a_y^\dagger a_x=(\mathbf{a}^\dagger\times\mathbf{a})_z\\
T_{yz}-T_{zy}&=(\mathbf{a}^\dagger\times\mathbf{a})_x\\
T_{zx}-T_{xz}&=(\mathbf{a}^\dagger\times\mathbf{a})_y
\end{aligned}
\tag{15.4.18}
$$

These clearly transform as a vector $\mathbf{V}=\mathbf{a}^\dagger\times\mathbf{a}$. There seems to be a problem here. Suppose we form the operator $V_1^1=-(V_x+iV_y)/2^{1/2}$. Then we expect

$$
V_1^1|nll\rangle=c|n,l+1,l+1\rangle \tag{15.4.19}
$$

as in Eq. (15.4.8). This would mean that states differing by *one* unit in l are degenerate. But we know from Section 12.6 that states differing by *two* units in l are degenerate. So how do we get out of the fix? To find out, you must work out any one of the components of the operator $\mathbf{V}=\mathbf{a}^\dagger\times\mathbf{a}$ in terms of $\mathbf{R}$ and $\mathbf{P}$. If you do, you will see that c in Eq. (15.4.19) is really zero, and the paradox will be resolved.

We are now left with $9-1-3=5$ degrees of freedom out of the original nine T_{ij}'s. We argue that these must transform *irreducibly*. Why? Suppose the contrary is true. Then it must be possible to form irreducible tensors with fewer than five components out of these residual degrees of freedom. The only possibilities are tensors with 1 or 3 components, that is to say, scalars or vectors. But we know that given two vectors $\mathbf{a}^\dagger$ and $\mathbf{a}$ we can form only one scalar, $\mathbf{a}^\dagger\cdot\mathbf{a}$ and only one vector $\mathbf{a}^\dagger\times\mathbf{a}$, both of which we have already used up. So we are driven to the conclusion that the five residual degrees of freedom are linear combinations of some T_2^q. One usually refers to this object as the *quadrupole tensor* Q_2^q. All we need here is the component Q_2^2, since

$$
Q_2^2|nll\rangle=c|n,l+2,l+2\rangle \tag{15.4.20}
$$

which explains the degeneracy in l at each n. (When $l=n=l_{\max}$, c vanishes.)

Let us explicitly construct the operator Q_2^2 in terms of $a_i^\dagger a_j$ to gain some experience. Now $\mathbf{a}$ and $\mathbf{a}^\dagger$ are vector operators from which we can form the tensor operators a_1^q and $(a^\dagger)_1^q$ which behave like $|1,q\rangle$. The product $a_i^\dagger a_j$ then behaves like the direct product of (linear combinations) of two spin-1 objects. Since Q_2^2 behaves like $|22\rangle$ and since $|22\rangle=|11\rangle\otimes|11\rangle$, we deduce that

$$
\begin{aligned}
Q_2^2&=(a^\dagger)_1^1(a_1^1)\\
&=\left(\frac{a_x^\dagger+ia_y^\dagger}{2^{1/2}}\right)\left(\frac{a_x+ia_y}{2^{1/2}}\right)\\
&=\tfrac{1}{2}[a_x^\dagger a_x-a_y^\dagger a_y+i(a_x^\dagger a_y+a_y^\dagger a_x)]
\end{aligned}
\tag{15.4.21}
$$

Other components of Q_2^q may be constructed by similar techniques. (It is just a matter of adding angular momenta $1\otimes 1$ to get 2.) Starting with the smallest value of l at each n (namely, 0 or 1), we can move up in steps of 2 until we reach $l=n$, at which point c in Eq. (15.4.20) will vanish. The network for $n=4$ is shown below:

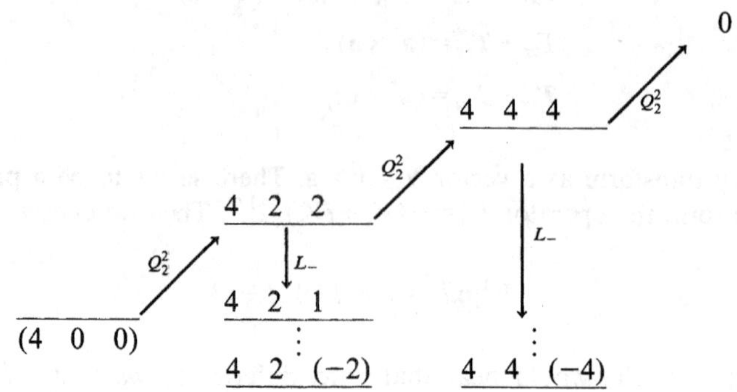

This completes the explanation of the degeneracy of the oscillator.

The Free-Particle Solutions

We examine the free-particle solutions from Section 12.6 in the light of the preceding discussion. Here again we have a case where states with different l, in fact an infinite number of them, are degenerate at each energy $E=\hbar^2k^2/2\mu$. This degeneracy is, however, not "accidental," since the extra symmetry of the free-particle Hamiltonian, namely, translational invariance, is obvious. We therefore have a conserved vector operator **P** from which we can form P_+,‡ which can raise l and m by one unit. Thus, given the state with $l=m=0$, we can move up in l using

$$|kll\rangle = c(P_+)^l|k00\rangle \tag{15.4.22}$$

where c is some normalization constant.

Recall that in the coordinate basis it was easy to find

$$|k00\rangle \to \psi_{k00} = \frac{U_0(\rho)}{\rho}\, Y_0^0 \tag{15.4.23}$$

where $\rho=kr$, and $U_0(\rho)$ is $\sin\rho$ or $-\cos\rho$ (regular or irregular solutions). It is easy to verify that

$$P_+|k00\rangle \xrightarrow[\text{basis}]{\text{coordinate}} -i\hbar(x+iy)\,\frac{1}{r}\cdot\frac{d}{dr}\left[\frac{U_0(\rho)}{\rho}\right] Y_0^0$$

$$= C_1(x+iy)\,\frac{1}{\rho}\,\frac{d}{d\rho}\left[\frac{U_0(\rho)}{\rho}\right] \tag{15.4.24}$$

‡ $P_+=P_x+iP_y$ is, up to a scale factor $(-2^{1/2})$ which does not change its rotational properties, just P_1^1.

where C_1 has absorbed all the factors that have no ρ dependence. If we operate once again with P_+ and use $[P_+, R_+]=0$ (where $R_+=R_x+iR_y \propto R_1^1$), we get

$$(P_+)^2|k00\rangle \to C_2(x+iy)^2\left(\frac{1}{\rho}\frac{\partial}{\partial\rho}\right)^2\frac{U_0(\rho)}{\rho} \tag{15.4.25}$$

and so finally

$$\begin{aligned}(P_+)^l|k00\rangle \to \psi_{kll} &= C_l\,(x+iy)^l\left(\frac{1}{\rho}\frac{d}{d\rho}\right)^l\frac{U_0(\rho)}{\rho}\\ &= \tilde{C}_l\,(\sin\theta)^l\,e^{il\phi}\rho^l\left(\frac{1}{\rho}\frac{d}{d\rho}\right)^l\frac{U_0(\rho)}{\rho}\\ &= \tilde{\tilde{C}}_l Y_l^l\rho^l\left(\frac{1}{\rho}\frac{d}{d\rho}\right)^l\frac{U_0(\rho)}{\rho}\\ &= R_l Y_l^l\end{aligned} \tag{15.4.26}$$

where

$$R_l=\tilde{\tilde{C}}_l\rho^l\left(\frac{1}{\rho}\frac{d}{d\rho}\right)^l\frac{U_0(\rho)}{\rho}=\tilde{\tilde{C}}_l\rho^l\left(\frac{1}{\rho}\frac{d}{d\rho}\right)^l R_0(\rho) \tag{15.4.27}$$

This agrees with Eq. (12.6.29) if we set $\tilde{\tilde{C}}_l=(-1)^l$.

16

The Variational and WKB Methods

16.1. The Variational Method

More often than not, it is impossible to find exact solutions to the eigenvalue problem of the Hamiltonian. One then turns to approximation methods, some of which will be described in this and the following chapters. In this section we consider a few examples that illustrate the *variational method.*

Our starting point is the inequality

$$E[\psi] \equiv \frac{\langle\psi|H|\psi\rangle}{\langle\psi|\psi\rangle} \geq E_0 \tag{16.1.1}$$

where E_0 is the lowest eigenvalue of H, i.e., the ground-state energy. Although this result was proved earlier, let us recall the idea behind it. $E[\psi]$ is just the mean value of the energy in the state $|\psi\rangle$. The inequality states that the mean value cannot be less than the lowest value that enters the average. More formally, if $|\psi\rangle$ is expanded in terms of the eigenfunctions $|E_n\rangle$ of H,

$$E[\psi] = \frac{\Sigma E_n|\langle E_n|\psi\rangle|^2}{\Sigma|\langle E_n|\psi\rangle|^2} \geq \frac{E_0\Sigma|\langle E_n|\psi\rangle|^2}{\Sigma|\langle E_n|\psi\rangle|^2} = E_0 \tag{16.1.2}$$

This inequality suggests a way (at least in principle) of determining the ground-state energy and eigenket. We take all the kets in the Hilbert space one by one and make a table of the corresponding $E[\psi]$. At the end we read off the lowest entry and the ket that goes with it. Clearly this is not a practical algorithm. What one does in practice is to consider just a subset (not necessarily a subspace) of vectors which are parametrized by some variables $(\alpha, \beta, \gamma, \ldots)$ and which have the general features one expects of the true ground-state ket. In this limited search $E[\psi]$ reduces to a function of the parameters, $E(\alpha, \beta, \ldots)$. We then find the values $(\alpha_0, \beta_0, \ldots)$ which minimize E. This minimum $E(\alpha_0, \beta_0, \ldots)$ provides an upper bound on E_0.

The name of the game is finding the lowest upper bound for a given amount of work. If H happens to be positive definite, $E_0 \geq 0$, and we will be able to restrict E_0 to the range $E(\alpha_0, \beta_0, \ldots) \geq E_0 \geq 0$.

As an example, consider the problem of a particle in a potential $V(x) = \lambda x^4$. Here are the features we expect of the ground state. It will have definite parity, and, since the ground-state function will have no nodes (more nodes→more wiggles→more kinetic energy), it will have even parity. It will be peaked at $x = 0$ so as to minimize $\langle V \rangle$. And of course it will vanish as $|x| \to \infty$. A trial function that has all these features (and is also easy to differentiate and integrate) is

$$\psi(x, \alpha) = e^{-\alpha x^2/2} \tag{16.1.3}$$

where α is a free parameter that determines the width of the Gaussian. The energy as a function of α is

$$E(\alpha) = \int e^{-\alpha x^2/2} \left(-\frac{\hbar^2}{2m} \frac{d^2}{dx^2} + \lambda x^4 \right) e^{-\alpha x^2/2} \, dx \Big/ \int e^{-\alpha x^2} \, dx = \frac{\hbar^2 \alpha}{4m} + \frac{3\lambda}{4\alpha^2}$$

We see here the familiar struggle between the kinetic and potential energy terms. The former would like to see $\alpha \to 0$ so that the wave function is wide and has only large wavelength (small momentum) components, while the latter would like to see $\alpha \to \infty$, so that the wave function is a narrow spike near $x = 0$, where the potential is a minimum. The optimal point, both effects included, is

$$\alpha_0 = \left(\frac{6m\lambda}{\hbar^2} \right)^{1/3} \tag{16.1.4}$$

The corresponding energy is

$$E(\alpha_0) = \frac{3}{8} \left(\frac{6\hbar^4 \lambda}{m^2} \right)^{1/3} \tag{16.1.5}$$

Since H is positive definite, we conclude

$$0 \leq E_0 \leq E(\alpha_0) \tag{16.1.6}$$

The best approximation to the wave function of the ground state (among all Gaussians) is $\psi(x, \alpha_0) = \exp\left(-\frac{1}{2}\alpha_0 x^2\right)$.

The inequality (16.1.6) is of course rigorous, but its utility depends on how close $E(\alpha_0)$ is to E_0. Our calculation does not tell us this. All we know is that since we paid attention to parity, nodes, etc., our upper bound $E(\alpha_0)$ is lower than that obtained by someone whose test functions had odd parity and 15 nodes. For instance, if $V(x)$ had been $\frac{1}{2}m\omega^2 x^2$ instead of λx^4, we would have found $\alpha_0 = (m\omega/\hbar)^{1/2}$ and $E(\alpha_0) = \hbar\omega/2$. Although this is the exact answer, our calculation would not tell us this. The way to estimate the quality of the bound obtained is to try to lower it further by considering a trial function with more parameters. If this produces substantial lowering, we keep going. On the other hand, if we begin to feel a "resistance" to the lowering of the bound as we try more elaborate test functions, we may suspect that E_0 is not too far below. In the case of $V(x)=\frac{1}{2}m\omega^2x^2$, it will eventually be found that there is no way of going below $E(\alpha_0)=\hbar\omega/2$.

Our faith in the variational method stems from its performance in cases where the exact answer is known either analytically or experimentally. Let us consider two examples. The first is that of the electron in a Coulomb potential $V=-e^2/r$. We expect the ground-state wave function to have no angular momentum, no nodes, behave like r^0 as $r\to 0$, and vanish as $r\to\infty$. So we choose $\psi(r, \theta, \phi, \alpha)=\exp(-\alpha r^2)$.‡ We find (upon ignoring the irrelevant angular variables throughout)

$$E(\alpha)=\int\left[e^{-\alpha r^2}\left(-\frac{\hbar^2}{2m}\frac{1}{r^2}\frac{d}{dr}r^2\frac{d}{dr}-\frac{e^2}{r}\right)e^{-\alpha r^2}\right]r^2\,dr\bigg/\int e^{-2\alpha r^2}r^2\,dr \qquad (16.1.7)$$

$$=\frac{3\hbar^2\alpha}{2m}-\left(\frac{2}{\pi}\right)^{1/2}2e^2\alpha^{1/2} \qquad (16.1.8)$$

which is minimized by

$$\alpha_0=\left(\frac{me^2}{\hbar^2}\right)^2\cdot\frac{8}{9\pi} \qquad (16.1.9)$$

The upper bound is then

$$E(\alpha_0)=-\frac{me^4}{2\hbar^2}\frac{8}{3\pi}=-0.85\text{ Ry} \qquad (16.1.10)$$

which is slightly above§ the true energy. The true wave function is of course not a Gaussian, but the general features are the same. For example $\psi(r,\alpha_0)=e^{-\alpha_0 r^2}$ predicts an uncertainty $\Delta X=(9\pi/32)^{1/2}a_0=0.94a_0$, while the exact result is $\Delta X=a_0$ (the Bohr radius).‖

The second example deals with the ground state of He. Ignoring nuclear motion ($m/M\to 0$), the Hamiltonian in the coordinate basis is

$$H\to-\frac{\hbar^2}{2m}(\nabla_1^2+\nabla_2^2)-\frac{2e^2}{r_1}-\frac{2e^2}{r_2}+\frac{e^2}{r_{12}} \qquad (16.1.11)$$

where r_1 and r_2 are the radial coordinates of the two electrons and r_{12} is the radial separation between them. We have already seen that if the mutual repulsion (e^2/r_{12}) is ignored, the ground-state wave function is just

$$\psi=\psi_{100}(\mathbf{r}_1)\psi_{100}(\mathbf{r}_2) \qquad (16.1.12)$$

‡ We could also choose $e^{-\alpha r}$, which would give the exact answer. But let us not.

§ Remember that we are dealing with negative energies here.

‖ This agreement is rather fortuitous. In general, the variational method provides much better approximations to the energies than to the wave functions. The reason follows shortly.

where the singlet spin wave function is suppressed, and ψ_{100} is the hydrogen-like wave function with $e^2 \to Ze^2$:

$$\psi_{100} = \left(\frac{Z^3}{\pi a_0^3}\right)^{1/2} e^{-Zr/a_0} \qquad (Z=2) \tag{16.1.13}$$

Consequently

$$\psi = \frac{Z^3}{\pi a_0^3} e^{-Z(r_1+r_2)/a_0} \qquad (Z=2) \tag{16.1.14}$$

The energy that goes with this simplified picture is

$$E = 2\left(-\frac{m(2e^2)^2}{2\hbar^2}\right) = -8 \text{ Ry} \simeq -108.8 \text{ eV}$$

which is far below the measured value of -78.6 eV.‡ So we find that omitting the Coulomb repulsion between the electrons is a bad approximation. But if we include the e^2/r_{12} term, the problem cannot be solved analytically. So we apply to it the variational method. For a trial wave function, we use just the product function in Eq. (16.1.14) but treat Z as a variable rather than setting it equal to 2. The idea is that as each electron shields the nuclear charge seen by the other, the effective Z is less than 2. This is borne out by the calculation of

$$\begin{aligned} E(Z) &= \frac{\left[\begin{array}{l} \displaystyle\int \psi(r_1 r_2 Z)\left[-\frac{\hbar^2}{2m}(\nabla_1^2+\nabla_2^2) - 2e^2\left(\frac{1}{r_1}+\frac{1}{r_2}\right) + \frac{e^2}{r_{12}}\right] \\ \times\, \psi(r_1 r_2 Z)\, d^3\mathbf{r}_1\, d^3\mathbf{r}_2 \end{array}\right]}{\displaystyle\int |\psi(r_1 r_2 Z)|^2\, d^3\mathbf{r}_1\, d^3\mathbf{r}_2} \\ &= -2 \text{ Ry}[4Z - Z^2 - \tfrac{5}{8}Z] \end{aligned} \tag{16.1.15}$$

whose minimum lies not at $Z=2$ but at $Z = 2 - 5/16$. The corresponding energy is

$$E(2-5/16) = -2(2-5/16)^2 \text{ Ry} \simeq -77.5 \text{ eV} \tag{16.1.16}$$

which is much closer to the real answer. Notice also that it lies above it, as demanded by the inequality (16.1.1). By considering trial functions with more parameters, one can get closer to the exact answer, and one can also feel the "resistance" to further lowering.

‡ This is not in contradiction with Eq. (16.1.1) since we are using the wrong Hamiltonian when we neglect the Coulomb repulsion between the electrons.

A virtue of the variational method is that even a poor approximation to the actual wave function can yield an excellent approximation to the actual energy. The reason is the following. Suppose we had chosen a trial function

$$|\psi\rangle = |E_0\rangle + \tfrac{1}{10}|E_1\rangle$$

which contains a 10% contamination from the state $|E_1\rangle$. The estimated energy would have been

$$E(\psi) = \frac{\langle E_0|H|E_0\rangle + \frac{1}{100}\langle E_1|H|E_1\rangle}{1+\frac{1}{100}} = \frac{E_0 + 0.01E_1}{1.01}$$
$$\simeq 0.99E_0 + 0.01E_1$$

which is off by just 1%. (We are assuming that E_1 is not anomalously large.)

More generally, let

$$|\psi\rangle = |E_0\rangle + |\delta\psi\rangle \tag{16.1.17a}$$

be a trial ket. Let us decompose $|\delta\psi\rangle$ into parts parallel to and perpendicular to $|E_0\rangle$:

$$\begin{aligned}|\delta\psi\rangle &= |\delta\psi_\parallel\rangle + |\delta\psi_\perp\rangle \\ &= \alpha|E_0\rangle + |\delta\psi_\perp\rangle\end{aligned} \tag{16.1.17b}$$

In this state

$$\begin{aligned}E[\psi] &= \frac{E_0|1+\alpha|^2 + \langle\delta\psi_\perp|H|\delta\psi_\perp\rangle}{|1+\alpha|^2 + \langle\delta\psi_\perp|\delta\psi_\perp\rangle} \\ &= E_0 + O(\delta\psi_\perp)^2\end{aligned} \tag{16.1.18}$$

Thus the error in energy is of the second order in the error in the state vector. Notice that $|\delta\psi_\parallel\rangle$ produces no error in energy. This is because rescaling the normalized eigenket does not change the mean energy.

All these results are true for any eigenket of H. If

$$|\psi_n\rangle = |E_n\rangle + |\delta\psi_n\rangle$$

is an approximation to $|E_n\rangle$, then by similar reasoning

$$E[\psi_n] = E_n + O[(\delta\psi_n)^2] \tag{16.1.19}$$

Thus the eigenkets of H are characterized by the fact that when they are changed to first order, there is no energy change to first order: *the eigenkets of H are stationary points of $E[\psi]$*. (The ground state happens, in addition, to be an absolute minimum.) If we could carry out the impossible task of tabulating all the $E[\psi]$ we can then read off *all* the eigenstates by looking for the stationary points. This is of course not a practical proposition. In practice, we use the following trick for finding the higher eigenvalues and eigenkets. Consider the case $V = \lambda x^4$. Since H is parity invariant, the states will occur with alternating parity. Suppose we take a trial state with odd parity. Then in the expansion $|\psi\rangle = \Sigma C_n |n\rangle$, $C_n = \langle n|\psi\rangle = 0$ for all even n, because the integral of an even and an odd function is zero. Consequently the lowest energy that enters the averaging is E_1 and we have the inequality

$$E[\psi] \geq E_1 \tag{16.1.20}$$

So we expect that if we take a trial state with odd parity, one node (in one dimension, there is one extra node for each upward step in energy), and the usual behavior as $|x| \to \infty$, we can get a good estimate E_1 and a rough picture of the corresponding wave function. What if we want to get a bound on E_2? The general idea is of course the same, to consider trial states in whose expansion $|E_0\rangle$ and $|E_1\rangle$ do not appear. But this cannot be done simply by choosing trial states of some definite parity. What we can do is the following. We have approximate wave functions for the first two levels from the variational energy estimates. We can choose our trial states to be orthogonal to these. The corresponding bounds are not rigorous, for we do not know $|E_0\rangle$ and $|E_1\rangle$ exactly, but they may still be useful.

This general idea is easier to implement in three dimensions if H is rotationally invariant. In this case the energy eigenstates have definite angular momentum. The ground state will have $l = 0$. By varying spherically symmetric trial functions we can estimate the ground-state energy. If we next choose $l = 1$ trial functions $[\psi = R(r)\, Y_1^m]$, $E[\psi]$ will obey

$$E[\psi] \geq E_{l=1}$$

where $E_{l=1}$ is the lowest energy level with $l = 1$. We can clearly keep going up in l. Suppose we do this for the Coulomb problem. We know that at each l, the lowest energy corresponds to $n = l+1$. The variational method applied to $l = 0, 1, 2, \ldots$ will yield energies close to those of the $n = 1, 2, \ldots$ levels. Of course we must pay attention to the radial part of ψ as well. For instance $R(r)$ must behave like r^l as $r \to 0$ in the angular momentum l sector. It must have the least number of nodes, namely zero, if it is to have the lowest energy for the given l. With these features built in, both the energy and wave function will come close to $\psi_{n,n-1,m}$.

We can also use any other operator that commutes with H in choosing trial functions. The angular momentum is especially convenient because its eigenfunctions are easy to write down and its eigenvalues are correlated (grow) with energy.

*Exercise 16.1.1.** Try $\psi = \exp(-\alpha x^2)$ for $V = \frac{1}{2} m\omega^2 x^2$ and find α_0 and $E(\alpha_0)$.

*Exercise 16.1.2.** For a particle in a box that extends from $-a$ to a, try (within the box) $\psi = (x-a)(x+a)$ and calculate E. There is no parameter to vary, but you still get an upper bound. Compare it to the true energy, E_0. (Convince yourself that the singularities in ψ'' at $x = \pm a$ do not contribute to the energy.)

*Exercise 16.1.3.** For the attractive delta function potential $V = -aV_0\delta(x)$ use a Gaussian trial function. Calculate the upper bound on E_0 and compare it to the exact answer $(-ma^2V_0^2/2\hbar^2)$ (Exercise 5.2.3).

Exercise 16.1.4 (Optional). For the oscillator choose

$$\begin{aligned} \psi &= (x-a)^2(x+a)^2, & |x| &\le a \\ &= 0, & |x| &> a \end{aligned}$$

Calculate $E(a)$, minimize it and compare to $\hbar\omega/2$.

*Exercise 16.1.5.** Solve the variational problem for the $l=1$ states of the electron in a potential $V=-e^2/r$. In your trial function incorporate (i) correct behavior as $r\to 0$, appropriate to $l=1$, (ii) correct number of nodes to minimize energy, (iii) correct behavior of wave function as $r\to\infty$ in a Coulomb potential (i.e., exponential instead of Gaussian damping). Does it matter what m you choose for Y_l^m? Comment on the relation of the energy bound you obtain to the exact answer.

16.2. The Wentzel–Kramers–Brillouin Method

Consider a particle of energy E in one dimension moving in a constant potential V. The energy eigenfunctions are

$$\psi(x) = \psi(0)\, e^{\pm ipx/\hbar}, \qquad p = [2m(E-V)]^{1/2} \tag{16.2.1}$$

where the $\pm$ signs correspond to right- and left-moving plane waves. The general solution is a combination of both waves. The real and imaginary parts of ψ oscillate in space with a wavelength $\lambda = 2\pi\hbar/p$ or equivalently, the phase change per unit length is a constant, $p/\hbar$. Suppose now that V, instead of being a constant, varies very slowly. We then expect that over a small region [small compared to the distance over which $V(x)$ varies appreciably] ψ will still behave like a plane wave, with the local value of the wavelength:

$$\lambda(x) = \frac{2\pi\hbar}{p(x)} = \frac{2\pi\hbar}{\{2m[E-V(x)]\}^{1/2}} \tag{16.2.2}$$

Since λ varies with x, the accumulated phase shift between $x=0$ and $x=x$ is given by an integral, so that

$$\psi(x) = \psi(0) \exp\left[\pm (i/\hbar) \int_0^x p(x')\, dx'\right]$$

or more generally

$$\psi(x) = \psi(x_0) \exp\left[\pm (i/\hbar) \int_{x_0}^x p(x')\, dx'\right] \tag{16.2.3}$$

Once again the $\pm$ stand for right- and left-moving waves, and the general solution is formed by taking an arbitrary linear combination of both. As mentioned above, we trust this formula only if the wavelength varies slowly. How slow is slow enough? Note that although there is a well defined function $\lambda(x)$ at each x, it makes no sense to speak of the wavelength *at* a point. The wavelength is a characteristic of a repetitive phenomenon and consequently defined only over a region that contains many repetitions. Thus the statement "a position-dependent wavelength $\lambda(x)$" makes sense only if $\delta\lambda$ over a length λ is negligible compared to λ:

$$\left|\frac{\delta\lambda}{\lambda}\right| = \left|\frac{(d\lambda/dx)\cdot\lambda}{\lambda}\right| = \left|\frac{d\lambda}{dx}\right| \ll 1 \tag{16.2.4}$$

Let us now derive all of the above results more formally. The derivation will also provide corrections to this result and clarify the nature of the approximation. Our problem is to solve the equation

$$\left\{\frac{d^2}{dx^2}+\frac{2m}{\hbar^2}[E-V(x)]\right\}\psi(x)=0$$

or

$$\left[\frac{d^2}{dx^2}+\frac{1}{\hbar^2}p^2(x)\right]\psi(x)=0$$

Let us write

$$\psi(x)=\exp[i\phi(x)/\hbar] \tag{16.2.5}$$

Since $\phi(x)$ is not assumed to be real, there is no loss of generality.‡ Feeding this form into the equation, we get

$$-\left(\frac{\phi'}{\hbar}\right)^2+\frac{i\phi''}{\hbar}+\frac{p^2(x)}{\hbar^2}=0 \tag{16.2.6}$$

We now expand ϕ in a power series in $\hbar$:

$$\phi=\phi_0+\hbar\phi_1+\hbar^2\phi_2+\cdots \tag{16.2.7}$$

The logic is the following. If $\hbar\to 0$, the wavelength $\lambda=2\pi\hbar/p$ tends to zero. Consequently any potential can be considered slowly varying in this limit and our approximation Eq. (16.2.3) should become increasingly reliable. Conversely, any corrections to this formula can be traced to the fact that $\hbar$ isn't really zero. In situations where $\hbar$ may be treated as a small number, we hope that corrections may be computed in powers of $\hbar$.

The WKB approximation (also called the *semiclassical approximation*) consists of keeping just the first two terms in Eq. (16.2.7). If we feed the truncated expansion

‡ In other words, any complex number $\psi=\rho\, e^{i\tilde{\phi}}=e^{i\tilde{\phi}+\ln\rho}=e^{i\phi}$, where $\phi=\tilde{\phi}-i\ln\rho$.

into Eq. (16.2.6) and group terms with the same $\hbar$ dependence, we get

$$\frac{-(\phi_0')^2+p^2(x)}{\hbar^2}+\frac{i\phi_0''-2\phi_1'\phi_0'}{\hbar}+O(\hbar^0)=0 \tag{16.2.8}$$

In the first approximation we concentrate on just the $\hbar^{-2}$ term. This gives

$$\phi_0'=\pm p(x)$$

or

$$\phi_0(x)=\pm\int^x p(x')\,dx' \tag{16.2.9}$$

and

$$\begin{aligned}\psi(x)&=A\exp\left[\pm(i/\hbar)\int^x p(x')\,dx'\right]\\&=\psi(x_0)\exp\left[\pm(i/\hbar)\int_{x_0}^x p(x')\,dx'\right]\end{aligned} \tag{16.2.10}$$

where A was found by setting $x=x_0$ in the first equation. All this agrees with our previous result. But we can go a step further and include the $\hbar^{-1}$ term in Eq. (16.2.8). We still choose ϕ_0' so that the $\hbar^{-2}$ term continues to vanish. For the $\hbar^{-1}$ term to vanish, we need

$$\begin{aligned}i\phi_0''&=2\phi_1'\phi_0'\\\frac{\phi_0''}{\phi_0'}&=-2i\phi_1'\\\ln\phi_0'&=-2i\phi_1+c\\\phi_1&=+i\ln(\phi_0')^{1/2}+c/2i=i\ln p^{1/2}+\tilde{c}\end{aligned} \tag{16.2.11}$$

To this order in $\hbar$,

$$\begin{aligned}\psi(x)=e^{i\phi(x)/\hbar}&=A\,e^{-\ln[p(x)]^{1/2}}\exp\left[\pm\left(\frac{i}{\hbar}\right)\int^x p(x')\,dx'\right]\\&=\frac{A}{[p(x)]^{1/2}}\exp\left[\pm\left(\frac{i}{\hbar}\right)\int^x p(x')\,dx'\right]\end{aligned} \tag{16.2.12}$$

or

$$\psi(x)=\psi(x_0)\left[\frac{p(x_0)}{p(x)}\right]^{1/2}\exp\left[\pm\frac{i}{\hbar}\int_{x_0}^{x}p(x')\,dx'\right] \tag{16.2.13}$$

The probability density associated with $\psi(x)$ behaves as $[p(x)]^{-1}$. This inverse dependence on the classical velocity is familiar to us from the classical probability distribution $P_{\text{cl}}(x)$ we studied in connection with the oscillator. Conversely, we could have written down Eq. (16.2.13) by combining classical and quantum reasoning, i.e., by semiclassical reasoning. We know classically that if a particle starts at x_0 with momentum $p(x_0)=\{2m[E-V(x_0)]\}^{1/2}$, it will have a momentum $p(x)=\{2m[E-V(x)]\}^{1/2}$ when it gets to x. Now we argue that since $p/\hbar$ is the phase change per unit length of the quantum wave function, its phase must be $(1/\hbar)\int p(x')\,dx'$. As for its amplitude, we argue that since the probability $P_{\text{cl}}(x)\sim 1/v(x)$, $|\psi|\simeq 1/[v(x)]^{1/2}\simeq 1/[p(x)]^{1/2}$.

Whenever we do an approximate calculation, it behooves us to verify at the end that the solution we obtain is consistent with the assumptions that went into its derivation. Our fundamental assumption in the recursive approach to Eq. (16.2.8) has been that the part of the equation with less powers of $\hbar$ is more important than the part with more powers, because $\hbar$ is so small. This is fine as long as the coefficients of the various powers of $\hbar$ are not anomalously big or small. For instance, if the coefficient of the $\hbar^{-2}$ term $[-(\phi_0')^2+p^2(x)]$ is very small, of the order of, say, $\hbar$, then it makes no sense to ignore the $\hbar^{-1}$ term in comparison. The same is true if the coefficient of the $\hbar^{-1}$ is as large as $\hbar^{-1}$. So we demand that the absolute magnitude of the first term be much bigger than that of the second. Since in the solution, $\phi_0'=p(x)$, we choose $(\phi_0'/\hbar)^2$ as a measure of the first term and for similar reasons $\phi_0''/\hbar$ as a measure of the second. The condition for the validity of the WKB approximation (to this order) is

$$\left|\frac{\phi_0''}{\hbar}\right|\ll\left|\frac{\phi_0'}{\hbar}\right|^2 \tag{16.2.14}$$

or

$$\hbar\left|\frac{d}{dx}\left(\frac{1}{\phi_0'}\right)\right|=\left|\frac{d}{dx}\left(\frac{\hbar}{p(x)}\right)\right|=\frac{1}{2\pi}\left|\frac{d\lambda}{dx}\right|\ll 1 \tag{16.2.15}$$

which agrees with our heuristic expectation, Eq. (16.2.4).

Connection with the Path Integral Formalism

Let us now rederive the semiclassical wave functions of Eq. (16.2.10) in the path integral approach, in the semiclassical approximation, in which one writes

$$U_{\text{cl}}(xt,x'0)=A\,e^{(i/\hbar)S_{\text{cl}}[xt;x'0]} \tag{16.2.16}$$

where $S_{cl}[xt; x'0]$ is the action for the classical path connecting the end points $(x'0)$ and (xt). We have chosen the initial time $t' = 0$, assuming the problem has time translation invariance. The prefactor A has no dependence on x, x', or t in our approximation in which the "area" under the functional integral is replaced by the value of the integrand at the stationary point times some constant. We will illustrate the procedure in the following situation.

(1) The potential is always negative and goes to a constant, chosen for convenience to be zero, as $|x| \to \infty$.
(2) The particle is in a quantum state with energy $E > 0$. This means that at the classical level there are no turning points, a fact which will prove significant.

Our strategy will be as follows. We will first show how to project out the exact wave functions from the exact propagator by performing some integrals. Then we will compute the propagator in the semiclassical approximation and project out the corresponding (approximate) wave functions of Eq. (16.2.10)).

How are the wave functions to be extracted from the propagator? In general, we have

$$U(xt; x') \equiv U(xt; x'0) = \Sigma \psi_n(x)\psi_n^*(x')e^{-iE_nt/\hbar}$$

where the sum may have to be an integral if the spectrum is continuous. Indeed, in the present problem this is the case. In the asymptotic region $|x| \to \infty$, the solutions must be plane waves with momentum

$$p_\infty = \pm\sqrt{2mE}$$

So we can use p as a label for the $E > 0$ eigenstates. (Hereafter the subscript on p_∞ will be dropped.) At each energy E, there will be two solutions just as in the free particle case, with $E = p^2/2m$. The wave functions will, however, not be plane waves in the interior and these are what we are after. So let us begin with

$$U(xt; x') = \int_{-\infty}^{\infty} \frac{dp}{2\pi\hbar}\psi_p(x)\psi_p^*(x')e^{-ip^2t/2m\hbar} + BS \qquad (16.2.17)$$

where BS stands for the sum over bound states. These will be dropped since we will soon be projecting out the states at some positive energy. Factors like π in the p-integration do not matter too much, we just want to get the right x-dependence of the wave functions and not their normalization.

Remember that $U(t)$ has been constructed to propagate the system forward in time, i.e., it is to be used for $t > 0$. Let us now define its transform

$$U(x, x', z) = \int_0^\infty dt\, U(xt; x')e^{izt/\hbar} \qquad z = E + i\varepsilon \qquad (16.2.18a)$$

where ε is a positive infinitesimal introduced to ensure convergence as $t\to\infty$. In what follows we will not distinguish between ε and a finite positive factor times ε and will ignore terms of higher order in ε.

It is readily seen by combining the two equations above that

$$U(x, x', z)=2m\int_{-\infty}^{\infty}\frac{dp}{2\pi i}\frac{\psi_p(x)\psi_p^*(x')}{p^2-2mE-i\varepsilon} \tag{16.2.18b}$$

Writing (to order ε)

$$p^2-2mE-i\varepsilon=p^2-(\sqrt{2mE}+i\varepsilon)^2$$

we factorize as follows:

$$\frac{1}{p^2-2mE-i\varepsilon}=\frac{1}{2\sqrt{2mE}}\left[\frac{1}{p-\sqrt{2mE}-i\varepsilon}-\frac{1}{p+\sqrt{2mE}+i\varepsilon}\right]$$

and use the formula (derived and explained in Appendix A4)

$$\frac{1}{(x-a)\mp i\varepsilon}=\mathscr{P}\frac{1}{x-a}\pm i\pi\delta(x-a)$$

to obtain

$$\begin{aligned}U(x, x', z)=&\sqrt{\frac{m}{2E}}\int_{-\infty}^{\infty}\frac{dp}{2\pi i}(\psi_p(x)\psi_p^*(x'))\\&\times\left[\mathscr{P}\left(\frac{1}{p-\sqrt{2mE}}\right)+i\pi\delta(p-\sqrt{2mE})\right.\\&\left.-\mathscr{P}\left(\frac{1}{p+\sqrt{2mE}}\right)+i\pi\delta(p+\sqrt{2mE})\right]\end{aligned}$$

where $\mathscr{P}$ means the principal value integral is to be evaluated. As it stands, U depends on the eigenfunctions not just at the energy E, but nearby energies also, due to the principal value integral. So we form the combination which singles out the eigenfunctions at just one energy:

$$\begin{aligned}&U(x, x', z)+[U(x', x, z)]^*\\&=\sqrt{\frac{m}{2E}}\int_{-\infty}^{\infty}dp(\psi_p(x)\psi_p^*(x'))\left[\delta(p-\sqrt{2mE})+\delta(p+\sqrt{2mE})\right]\\&=\sqrt{\frac{m}{2E}}[\psi_{\sqrt{2mE}}(x)\psi^*_{\sqrt{2mE}}(x')+\psi_{-\sqrt{2mE}}(x)\psi^*_{-\sqrt{2mE}}(x')]\end{aligned} \tag{16.2.19}$$

We now compare this with $U_{\text{cl}}(x, x', z)$:

$$U_{\text{cl}}(x, x', z) = \int_0^\infty dt\, U_{\text{cl}}(x, x', t)\, e^{(i/\hbar)(E+i\varepsilon)t} = \int_0^\infty dt\, e^{(i/\hbar)S_{\text{cl}}[x,x',t]}\, e^{(i/\hbar)(E+i\varepsilon)t}$$

Since U_{cl} was itself evaluated in the stationary point approximation of a functional integral (on the basis of the smallness of $\hbar$), to be consistent, we must evaluate the ordinary integral in t also by stationary phase, by setting the argument of the exponent to its value at the point t^* defined by

$$\frac{\partial S}{\partial t} + E = -E_{\text{cl}} + E = 0 \tag{16.2.20}$$

and equating the integral to the integrand at this point times some constant. For the stationary point, we have recalled Eq. (2.8.18) and dropped the convergence factor ε since it is not needed. What the stationary point does is to choose from all trajectories connecting the two given end points (with various travel times) the one whose classical energy equals the energy E of the quantum state [and whose travel time is $t^* = t^*(E)$].

Note that previously we were interested in trajectories that connected x' to x *in a fixed time t*. Given the equations of motion are second order in time, there will be just one trajectory that can meet these requirements. But now we are asking for a trajectory with a *fixed energy* connecting the two points x' and x with no restriction on travel time. This can generally have many solutions. For example, in a confining potential like the oscillator (or in any bound state) a particle leaving x' with some energy E will hit any other point x (within the turning points) an infinite number of times as it rattles back and forth. Each such orbit will have a different travel time and $U_{\text{cl}}(x, x', E)$ will receive contributions from an infinite number of stationary points obeying Eq. (16.2.20).

In the present problem with no turning points, there will be just two solutions—call them R and L—which are moving to the right or to the left. The right mover can go from x' to x if $x' < x$ and the left mover if $x' > x$.

Let us proceed with our calculation bearing all this in mind, and start with

$$U_{\text{cl}}(x, x', E) = A' \sum_{R,\,L} e^{(i/\hbar)[S_{\text{cl}}(x,x',t^*) + Et^*]}$$

where A' is a new constant. We now manipulate as follows bearing in mind that $E_{\text{cl}} = E$ is conserved on the classical path:

$$\begin{aligned} S_{\text{cl}}[x, x', t^*] &= \int_0^{t^*} (T - V)\, dt = \int_0^{t^*} 2T\, dt - Et^* \\ &= \int_0^{t^*} p(x(t))\,\frac{dx}{dt}\, dt - Et^* = \int_{x'}^{x} p(x'')\, dx'' - Et^* \\ &\equiv W[x, x', E] - Et^* \end{aligned} \tag{16.2.21a}$$

It follows that

$$U_{\text{cl}}(x, x', E) = A' \sum_{R,L} \exp\left[\frac{i}{\hbar}\int_{x'}^{x} p(x'')\,dx''\right]$$

$$= A' \sum_{R,L} \exp\left(\frac{i}{\hbar} W[x, x', E]\right) \tag{16.2.21b}$$

If $x > x'$, the classical momentum has to be positive and we set $p(x) = \sqrt{2m(E - V(x))}$ whereas if $x < x'$ we make the other choice for the square root. Thus

$$U_{\text{cl}}(x, x', E) = \theta(x - x')A' \exp\left(\frac{i}{\hbar}\left[\int_{x'}^{x} \sqrt{2m(E - V(x''))}\,dx''\right]\right)$$

$$+ \theta(x' - x)A' \exp\left(-\frac{i}{\hbar}\left[\int_{x'}^{x} \sqrt{2m(E - V(x''))}\,dx''\right]\right) \tag{16.2.22}$$

We now find

$$U_{\text{cl}}(x, x', E) + U_{\text{cl}}^{*}(x'x, E) = A' \exp\left(\frac{i}{\hbar}\left[\int_{x'}^{x} \sqrt{2m(E - V(x''))}\,dx''\right]\right)$$

$$+ A' \exp\left(-\frac{i}{\hbar}\left[\int_{x'}^{x} \sqrt{2m(E - V(x''))}\,dx''\right]\right) \tag{16.2.23}$$

Comparing this to Eq. (16.2.19) we find

$$\psi_{\pm\sqrt{2mE}}(x)\psi^{*}_{\pm\sqrt{2mE}}(x') \simeq \exp\left(\pm\frac{i}{\hbar}\left[\int_{x'}^{x} \sqrt{2m(E - V(x''))}\,dx''\right]\right)$$

It is easily seen that this may be written as

$$\psi_{\pm}(x) = \psi(x_0) \exp\left(\pm\frac{i}{\hbar}\left[\int_{x_0}^{x} \sqrt{2m(E - V(x''))}\,dx''\right]\right)$$

Several comments are in order.

First, if we want to get Eq. (16.2.13), with the factor $p^{-1/2}$, we need to do a more accurate calculation. So far we have evaluated the functional integral and the ordinary t integral by setting the integral equal to the value of the integrand at the stationary point, times some constant to represent the "area" around this point. To get Eq. (16.2.13) we must approximate the integrands at the stationary point by Gaussians and do the Gaussian integrals. If you are interested in the details, consult one of the works in the Bibliography at the end of Chapter 21.

Second, note that in going from $U(t)$ to $U(E)$, (at fixed x', x), we shifted our attention from paths with a fixed time to paths with a fixed energy. Since

$E = -\partial S_{cl}/\partial t$, one is trading t for the derivative of S_{cl} with respect to t. This clearly requires a Legendre transformation, as explained in Section 2.5. It is clear from Eq. (16.2.21a) that $W(E)$ is the Legendre transform in question. This idea is pursued further in Exercise (16.2.1).

Finally, let us look at the combination $U(x, x', E) + U^*(x', x, E)$ we were led to in our attempt to filter out one energy. From our discussion of time reversal you should expect that the complex conjugated version of U (with initial and final points exchanged) stands for the time-reversed propagator. If so, the integral of this combination over positive times is the integral of U for all times. It is now clear that such an integration done on Eq. (16.2.17) will indeed produce a factor $\delta(E - p^2/2m)$ which projects out states at the energy E. This point is pursued in Exercise (16.2.3).

Exercise 16.2.1. Consider the function $W(E)$ introduced in Eq. (16.2.21a). Since $-E$ is the t derivative of $S(t)$, it follows that $W(E)$, must be the Legendre transform of S. In this case t must emerge as the derivative of $W(E)$ with respect to E. Verify that differentiation of the formula

$$W(E) = \int_{x'}^{x} \sqrt{2m(E - V(x''))}\, dx''$$

gives the time t taken to go from start to finish at this preassigned energy.

Exercise 16.2.2. Consider the free particle problem using the approach given above. Now that you know the wave functions explicitly, evaluate the integral in Eq. (16.2.18b) by contour integration to obtain

$$U(x, x', t) = \theta(x - x')\, e^{(i/\hbar)\sqrt{2mE}(x - x')} + \theta(x' - x)\, e^{(-i/\hbar)\sqrt{2mE}(x - x')}.$$

In doing the contour integrals, ask in which half-plane you can close the contour for a given sign of $x - x'$. Compare the above result to the semiclassical result and make sure it all works out. Note that there is no need to form the combination $U(x, x', t) + U^*(x, x', t)$; we can calculate both the principal part and the delta function contributions explicitly because we know the p-dependence of the integrand explicitly. To see this more clearly, avoid the contour integral approach and use instead the formula for $(x \pm i\varepsilon)^{-1}$ given above. Evaluate the principal value integral and the contribution from the delta function and see how they add up to the result of contour integration. Although both routes are possible here, in the problem with $V \neq 0$ contour integration is not possible since the p-dependence of the wave function in the complex p plane is not known. The advantage of the $U + U^*$ approach is that it only refers to quantities on the real axis and at just one energy.

Exercise 16.2.3. Let us take a second look at our derivation. We worked quite hard to isolate the eigenfunctions at one energy: we formed the combination $U(x, x', z) + U^*(x, x', z)$ to get rid of the principal part and filter out the delta function. Now it is clear that if in Eq. (16.2.18a) we could integrate in the range $-\infty \leq t \leq \infty$, we would get the δ function we want. What kept us from doing that was the fact the $U(t)$ was constructed to be used for $t > 0$. However, if we use the time evolution operator $e^{-(i/\hbar)Ht}$ for negative times, it will simply tell us what the system was doing at earlier times, assuming the same Hamiltonian. Thus we can make sense of the operator for negative times as well and define a transform that extends over all times. This is true in classical mechanics as well. For instance, if a stone is thrown straight up from a tall building and we ask when it will be at ground level, we get two answers,

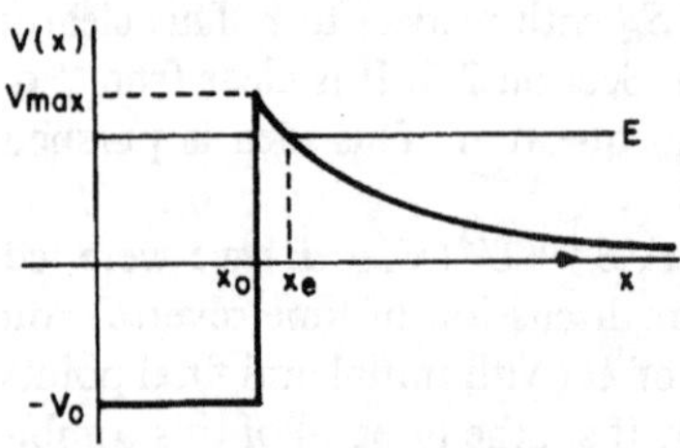

Figure 16.1. A typical barrier penetration problem. The particle has energy $E>0$, but is restrained by the barrier since $V_{\max}>E$. But there is an amplitude for it to tunnel through the barrier and escape to infinity as a free particle.

one with negative t corresponding to the extrapolation of the given initial conditions to earlier times. Verify that $U(x, x', z)+U^*(x, x', z)$ is indeed the transform of $U(t)$ for all times by seeing what happens to $\langle x|e^{-(i/\hbar)Ht}|x'\rangle$ under complex conjugation and the exchange $x \leftrightarrow x'$. Likewise, ask what happens to U_{cl} when we transform it for all times. Now you will find that no matter what the sign of $x-x'$ a single right moving trajectory can contribute—it is just that the time of the stationary point, t^*, will change sign with the sign of $x-x'$. (The same goes for a left moving trajectory.)

Tunneling Amplitudes

The WKB formula can also be used to calculate tunneling amplitudes provided $\chi(x)=\{2m[V(x)-E]\}^{1/2}$ varies slowly. As an example, consider a particle trapped in a potential shown in Fig. 16.1. If its energy is positive, there exists some probability for it to penetrate the barrier and escape to infinity as a free particle. In the first approximation, the ratio of ψ at the point of escape, x_e and at the outer wall of the well, x_0, is

$$\psi(x_e)=\psi(x_0)\exp\left(\frac{i}{\hbar}\int_{x_0}^{x_e} i\{2m[V(x)-E]\}^{1/2}\,dx\right) \tag{16.2.24}$$

$$\equiv \psi(x_0)\,e^{-\gamma/2} \tag{16.2.25}$$

The mean lifetime of the particle inside the well may be estimated by the following semiclassical computation. Since the particle inside the well has a kinetic energy $T=E-V=E+V_0$, its velocity is $v=[2m(E+V_0)]^{1/2}/m$ and it bangs against the outer wall at a frequency $f=v/2x_0$. Upon collision, there is a probability of escape $e^{-\gamma}$. Consequently probability of escape in 1 second is

$$R=\frac{[2m(E+V_0)]^{1/2}}{2mx_0}\,e^{-\gamma} \tag{16.2.26}$$

The mean lifetime is then $\tau=1/R$.

Note that just as a particle inside can tunnel out a particle with $E<V_{\max}$ can tunnel in from outside and get captured. A standard example of tunneling and capture is provided by α decay in which a nucleus emits an α particle.‡ At short distances, the force between the α particle and the nucleus is attractive (the nuclear force beats the Coulomb repulsion), while at larger distances, the Coulomb repulsion

‡ The α particle has two protons and two neutrons in it.

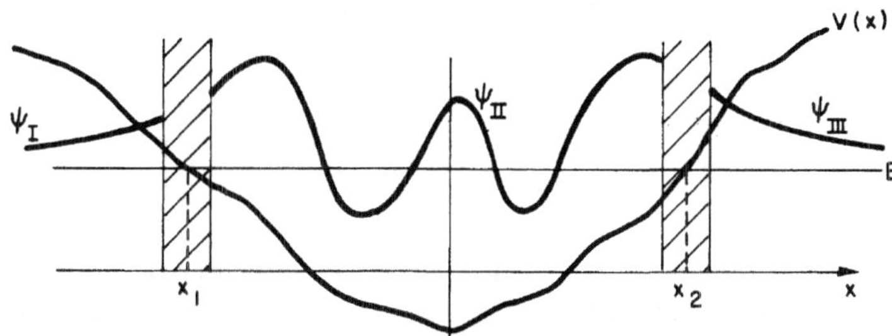

Figure 16.2. A typical bound state problem. The WKB approximation to the wave function works nicely except in the shaded bands near the classical turning points x_1 and x_2. To join ψ_{II} to ψ_I and ψ_{III}, one solves the Schrödinger equation in these bands after approximating the potential by a linear function within each band.

dominates. The sum of the two potentials is roughly described by a potential of the type shown in Fig. 16.1, with x playing the role of the radial coordinate. (The centrifugal barrier must be added onto the two potentials if the α particle comes out with nonzero orbital angular momentum.) Thus α particles with $E < V_{max}$ can tunnel out from within, or get captured from without.

Exercise 16.2.4. Alpha particles of kinetic energy 4.2 MeV come tunneling out of a nucleus of charge $Z=90$ (after emission). Assume that $V_0=0$, $x_0=10^{-12}$ cm, and $V(x)$ is just Coulombic for $x \geq x_0$. (See Fig. 16.1.) Estimate the mean lifetime. {Hint: Show $\gamma = (8Ze^2/\hbar v)[\cos^{-1} y^{1/2} - y^{1/2}(1-y)^{1/2}]$, where $y = x_0 | x_e$. Show that $y \ll 1$ and use $\cos^{-1} y^{1/2} \simeq \frac{1}{2}\pi - y^{1/2}$ before calculating numbers.}

The derivation of the tunneling amplitude, Eq. (16.2.24), is not straightforward in the path integral formalism due to the fact that there exists no classical path that can take the particle across the barrier. There is, however, such a path in the "imaginary time" formalism. This will be detailed in Chapter 21.

Bound States

The WKB method can be applied to approximate bound state energies and wave functions. Consider a particle (Fig. 16.2) bound by a potential $V(x)$. In the figure, x_1 and x_2 are the classical turning points for the energy E. Let us see how the quantization of E emerges in this formalism. We know that in the classically forbidden regions I and III, the wave function will be a damped exponential. For instance

$$\psi_{III}(x) \sim \frac{1}{\{2m[V(x)-E]\}^{1/2}} \exp\left(-\frac{1}{\hbar}\int^{x} \{2m[V(x')-E]\}^{1/2}\, dx'\right) \qquad (16.2.27)$$

In the classically allowed region II, it will be oscillating. We assume it is a real function‡ with two free parameters A and B:

$$\psi_{\text{II}}(x) = \frac{A}{[p(x)]^{1/2}} \cos\left[\frac{1}{\hbar}\int^x p(x')dx' + B\right] \tag{16.2.28}$$

[The two real parameters A and B replace the one complex parameter $\psi(x_0)$ used previously.] Unfortunately, neither Eq. (16.2.27) nor Eq. (16.2.28) is applicable near the turning points. Formally this is because $[p(x)]^{-1/2}$ and $\{2m[V(z) - E]\}^{-1/2}$ blow up there. Physically it is because the wavelength tends to infinity there and the requirement that $V(x)$ varies little over a wavelength becomes impossible to meet. It is therefore impossible to match the approximate ψ_{I}, ψ_{II}, and ψ_{III} and thereby determine the allowed energies as one does in the simple examples with piecewise constant potentials. The problem is surmounted as follows. Near each turning point, we define a *transition region* (shaded bands in the figure) inside which we solve the problem by going back to Schrödinger's equation. If $V(x)$ is slowly varying, it may be approximated by a linear function in these regions. For instance near x_1

$$\begin{aligned} V(x) &\simeq V(x_1) + V' \cdot (x - x_1) \\ &= E + V' \cdot (x - x_1) \end{aligned} \tag{16.2.29}$$

The exact solutions with this $V(x)$ are then matched to the WKB solutions outside the shaded region, that is to say, to the damped exponentials on one side and the oscillating cosine on the other.

The analysis§ near x_1 would yield the following function in region II:

$$\psi_{\text{II}}(x) = \frac{A}{[p(x)]^{1/2}} \cos\left[\frac{1}{\hbar}\int_{x_1}^x p(x')dx' - \frac{\pi}{4}\right] \tag{16.2.30}$$

while the one near x_2 would yield

$$\psi_{\text{II}}(x) = \frac{A'}{[p(x)]^{1/2}} \cos\left[\frac{1}{\hbar}\int_{x_2}^x p(x')dx' + \frac{\pi}{4}\right] \tag{16.2.31}$$

For the two solutions to coincide, A and A' must have the same magnitude and the difference in phase between the two cosines must be a multiple of π:

$$\frac{1}{\hbar}\int_{x_1}^x p(x')dx' - \frac{1}{\hbar}\int_{x_2}^x p(x')dx' - \frac{\pi}{2} = n\pi, \qquad n = 0, 1, 2, \ldots$$

‡ Recall Theorem 16, Sec. 5.6.
§ The details are omitted.

or

$$\int_{x_1}^{x_2} p(x)\,dx=(n+\tfrac{1}{2})\pi\hbar \tag{16.2.32}$$

or

$$\oint p(x)\,dx=(n+\tfrac{1}{2})2\pi\hbar \tag{16.2.33}$$

where $\oint$ denotes the integral over a full cycle, from x_1 to x_2 and back. If n is even, $A=A'$, if odd $A=-A'$.

Equation (16.2.32) expresses the quantization of energy since the integral and limits on the left-hand side are functions of energy and the other parameters such as particle mass.

As an example, consider a particle in a linear potential $V(x)=k|x|$. The turning points are

$$x_{1,2}=\mp E/k \tag{16.2.34}$$

and the quantization condition is

$$\int_{-E/k}^{E/k} [2m(E-k|x|)]^{1/2}\,dx=2\int_0^{E/k} [2m(E-kx)]^{1/2}\,dx=(n+\tfrac{1}{2})\hbar\pi \tag{16.2.35}$$

The n, k, m, $\hbar$ dependence of E can be found by a scaling argument. Let us define a variable y by

$$x=(E/k)y \tag{16.2.36}$$

in terms of which we have

$$2\int_0^1 (2mE)^{1/2}(1-y)^{1/2}\left(\frac{E}{k}\right)dy=(n+\tfrac{1}{2})\hbar\pi$$

or

$$E\propto(k)^{2/3}(m)^{-1/3}(n+\tfrac{1}{2})^{2/3}\hbar^{2/3} \tag{16.2.37}$$

The constant of proportionality may be found by carrying out the y integral. The result is

$$E_n=\left[\frac{3k\hbar\pi}{4(2m)^{1/2}}\left(n+\frac{1}{2}\right)\right]^{2/3} \tag{16.2.38}$$

If the method is applied to the potential $V(x)=\lambda x^4$, we would get, by the scaling argument,

$$E_n=\left[\frac{c\lambda^{1/4}\hbar}{m^{1/2}}\left(n+\frac{1}{2}\right)\right]^{4/3} \tag{16.2.39}$$

where c is a constant that may be found by carrying out a dimensionless integral. If the WKB energy levels are compared to the actual ones (obtained either by analytic solution or by accurate numerical integration of Schrödinger's equation) we would find that the agreement is excellent for all but very small n. In the λx^4 case, for example,

$$\frac{E_0(\text{WKB})}{E_0(\text{numerical})}=0.716$$

$$\frac{E_1(\text{WKB})}{E_1(\text{numerical})}=0.992$$

The agreement gets even better as we move up in n. Thus the WKB method complements the variational method, which works best for the lowest levels. The improved accuracy with increasing n is reasonable in view of the fact that as we move up in energy, the transitional region near the turning points (where the approximation breaks down) plays a decreasingly important role.‡

What about the WKB wave functions? They too get better at large n, except of course near the turning points, where they blow up due to the $[p(x)]^{-1/2}$ factor. If, however, we actually solve the Schrödinger equation near the turning points after approximating the potential by a linear function, the blowup can be avoided and the agreement with the true eigenfunctions greatly enhanced.

The WKB wave function has another feature that agrees with the exact answer: the wave function has n nodes ($n=0, 1, 2, \ldots$) in the nth level. We see this analytically from Eq. (16.2.30) for $\psi_{\text{II}}(x)$, and Eq. (16.2.32) for the phase integral $(1/\hbar)\int_{x_1}^{x} p(x')\,dx'$. As x goes from x_1 to x_2, the phase ϕ goes from $-\pi/4$ to $n\pi+\pi/4$ and $\cos(\phi)$ vanishes n times. We can in fact understand the quantization rule, Eq. (16.2.32), as follows. If we assume, in the first approximation, that ψ must vanish in the classically forbidden region, it follows that it must undergo an integral number of half-cycles (half-wavelengths) in the interval $x_1\le x\le x_2$. This leads to the *Bohr–Sommerfeld quantization rule.*

$$\int_{x_1}^{x_2} p(x)\,dx=(n+1)\hbar\pi, \qquad n=0, 1, 2, \ldots \tag{16.2.40}$$

But we know that ψ doesn't vanish at the turning points and has an exponential tail in the classically forbidden region. Consequently the number of half-cycles completed *between* x_1 *and* x_2 is somewhat less than $n+1$. The connection procedure tells us that

‡ There are some exceptional cases, such as the harmonic oscillator, where the method gives the exact energies for all n.

it is in fact $n+\frac{1}{2}$ and hence the usual quantization rule, Eq. (16.2.32). If, however, ψ actually vanishes at x_1 and x_2 because the potential barrier there is infinite (as in the case of a particle in a box), Eq. (16.2.40) [and not Eq. (16.3.32)] is relevant.‡ One can also consider an intermediate case where the barrier is infinite at one turning point and not at the other. In this case the quantization rule has an $(n+3/4)$ factor in it.

The WKB method may also be applied in three dimensions to solve the radial equation in a rotationally invariant problem. In the $l=0$ state, there is no centrifugal barrier, and the WKB wave function has the form

$$U(r)\sim\frac{1}{[p(r)]^{1/2}}\sin\left[\frac{1}{\hbar}\int_0^r p(r')\,dr'\right], \qquad p=\{2m[E-V(r)]\}^{1/2} \qquad (16.2.41)$$

where the lower limit in the phase integral is chosen to be 0, so that $U(0)=0$. The quantization condition, bearing in mind that the barrier at $r=0$ is infinite, is

$$\int_0^{r_{max}} p(r)\,dr=\left(n+\frac{3}{4}\right)\hbar\pi, \qquad n=0, 1, 2, \ldots \qquad (16.2.42)$$

where $r_{\max}$ is the turning point. This formula is valid only if $V(r)$ is regular at the origin. If it blows up there, the constant we add to n is not 3/4 but something else. Also if $l\neq 0$, the centrifugal barrier will alter the behavior near $r=0$ and change both the wave function and this constant.

*Exercise 16.2.5.** In 1974 two new particles called the ψ and ψ' were discovered, with rest energies 3.1 and 3.7 GeV, respectively (1 GeV$=10^9$ eV). These are believed to be nonrelativistic bound states of a "charmed" quark of mass $m=1.5$ GeV$/c^2$ (i.e., $mc^2=1.5$ GeV) and an antiquark of the same mass, in a linear potential $V(r)=V_0+kr$. By assuming that these are the $n=0$ and $n=1$ bound states of zero orbital angular momentum, calculate V_0 using the WKB formula. What do you predict for the rest mass of ψ'', the $n=2$ state? (The measured value is $\simeq 4.2$ GeV$/c^2$.) [Hints: (1) Work with GeV instead of eV. (2) There is no need to determine k explicitly.]

Exercise 16.2.6. Obtain Eq. (16.2.39) for the λx^4 potential by the scaling trick.

*Exercise 16.2.7** Find the allowed levels of the harmonic oscillator by the WKB method.

Exercise 16.2.8. Consider the $l=0$ radial equation for the Coulomb problem. Since $V(r)$ is singular at the turning point $r=0$, we can't use $(n+3/4)$.

(1) Will the additive constant be more or less than 3/4?

(2) By analyzing the exact equation near $r=0$, it can be shown that the constant equals 1. Using this constant show that the WKB energy levels agree with the exact results.

‡ The assumption that $V(x)$ may be linearized near the turning point breaks down and this invalidates Eq. (16.2.29).

17

Time-Independent Perturbation Theory

17.1. The Formalism

Time-independent perturbation theory is an approximation scheme that applies in the following context: we know the solution to the eigenvalue problem of the Hamiltonian H^0, and we want the solution to $H=H^0+H^1$, where H^1 is small compared to H^0 in a sense to be made precise shortly. For instance, H^0 can be the Coulomb Hamiltonian for an electron bound to a proton, and H^1 the addition due to an external electric field that is weak compared to the proton's field at the (average) location of the electron. One refers to H^0 as the *unperturbed Hamiltonian* and H^1 as the *perturbing Hamiltonian* or *perturbation*.

We proceed as follows. We assume that to every eigenket $|E_n^0\rangle \equiv |n^0\rangle$ of H^0 with eigenvalue E_n^0, there is an eigenket $|n\rangle$ of H with eigenvalue E_n.‡ We then assume that the eigenkets and eigenvalues of H may be expanded in a *perturbation series*§:

$$|n\rangle = |n^0\rangle + |n^1\rangle + |n^2\rangle + \cdots \tag{17.1.1}$$

$$E_n = E_n^0 + E_n^1 + E_n^2 + \cdots \tag{17.1.2}$$

The superscript k on each term gives the power of (the matrix element of) H^1 that it is expected to come out proportional to. A term with superscript equal to k is called a *kth-order term*. (Clearly a product like $E_n^k|n^{\bar{k}}\rangle$ is a term of order $k+\bar{k}$.) We hope that as the order increases, the terms get systematically smaller; this is when we can say that H^1 is small. When we find explicit formulas for $|n^k\rangle$ and E_n^k, these ideas will be sharpened.

To find the terms in the expansions for $|n\rangle$ and E_n, we start with the eigenvalue equation

$$H|n\rangle = E_n|n\rangle \tag{17.1.3}$$

‡ We are assuming that $|n^0\rangle$ is *nondegenerate*. The degenerate case follows.

§ We assume $|n^0\rangle$ is normalized (to unity). The norm of $|n\rangle$ will be discussed shortly.

or

$$(H^0+H^1)[|n^0\rangle+|n^1\rangle+\cdots]$$
$$=(E_n^0+E_n^1+\cdots)[|n^0\rangle+|n^1\rangle+\cdots] \tag{17.1.4}$$

We approach these equations as we did the differential equation in the WKB approximation. Recall that there we had an equation with terms of order $\hbar^{-2}, \hbar^{-1}, \ldots$, etc. We first ignored all but the $\hbar^{-2}$ terms and solved for ϕ_0. We then fed this into the $\hbar^{-1}$ part to determine ϕ_1. (We could have gone on this way, though we chose to stop there.) Likewise, in the present case, we first consider the zeroth-order terms of Eq. (17.1.4). We get the equation

$$H^0|n^0\rangle=E_n^0|n^0\rangle \tag{17.1.5}$$

Notice that the zeroth-order quantities $|n^0\rangle$ and E_n^0 are indeed independent of H^1 (or, equivalently, they depend on the zeroth power of H^1). By assumption, this equation may be solved and the eigenvectors $|n^0\rangle$ and eigenvalues E_n^0 determined. So we move on to the first-order terms. We get the equation

$$H^0|n^1\rangle+H^1|n^0\rangle=E_n^0|n^1\rangle+E_n^1|n^0\rangle \tag{17.1.6}$$

Let us dot both sides with $\langle n^0|$. Using $\langle n^0|H^0=\langle n^0|E_n^0$ and $\langle n^0|n^0\rangle=1$, we get

$$E_n^1=\langle n^0|H^1|n^0\rangle \tag{17.1.7}$$

i.e., *the first-order change in energy is the expectation value of H^1 in the unperturbed state*. Notice that E_n^1 is indeed proportional to the first power of H^1. Let us next dot both sides of Eq. (17.1.6) with $\langle m^0|$, $m\neq n$, to get

$$\langle m^0|H^0|n^1\rangle+\langle m^0|H^1|n^0\rangle=E_n^0\langle m^0|n^1\rangle$$

or

$$\langle m^0|n^1\rangle=\frac{\langle m^0|H^1|n^0\rangle}{E_n^0-E_m^0} \tag{17.1.8}$$

Since $m\neq n$, this equation determines all the components of $|n^1\rangle$ in the eigenbasis of H^0, except for the component parallel to $|n^0\rangle$, let's call it $|n_\parallel^1\rangle$. We determine it by the requirement that $|n\rangle$ is normalized to this order.‡ In obvious notation, we have

$$1=\langle n|n\rangle=(\langle n^0|+\langle n_\perp^1|+\langle n_\parallel^1|)(|n^0\rangle+|n_\perp^1\rangle+|n_\parallel^1\rangle) \tag{17.1.9}$$

‡ Recall that even in eigenvalue problems that can be solved exactly, there is the arbitrariness in the norm of the vector. To this order, only $|n_\parallel^1\rangle$ alters the length of $|n^0\rangle$. [See Eq. (17.1.10).]

which leads to

$$1=\langle n^0|n^0\rangle+\langle n_{\parallel}^1|n^0\rangle+\langle n^0|n_{\parallel}^1\rangle+\text{higher order}$$

or

$$0=\langle n_{\parallel}^1|n^0\rangle+\langle n^0|n_{\parallel}^1\rangle+\text{higher order} \tag{17.1.10}$$

This means that

$$\langle n^0|n_{\parallel}^1\rangle=i\alpha, \qquad \alpha \text{ real} \tag{17.1.11}$$

Using

$$1+i\alpha=e^{i\alpha} \quad \text{(to this order)} \tag{17.1.12}$$

we get

$$|n\rangle=|n^0\rangle\, e^{i\alpha}+\sum_m{}' \frac{|m^0\rangle\langle m^0|H^1|n^0\rangle}{E_n^0-E_m^0} \tag{17.1.13}$$

where the prime on $\sum'$ means that $m\neq n$. Since $|n\rangle$ has an arbitrariness in its overall phase, even after it is normalized, let us change its phase by the factor $e^{-i\alpha}$ in Eq. (17.1.13). This gets rid of the phase factor multiplying $|n^0\rangle$ and does nothing to the first-order piece, *to this order*. Calling the perturbed eigenket with the new phase also $|n\rangle$, we get the result to first order:

$$|n\rangle=|n^0\rangle+\sum_m{}' \frac{|m^0\rangle\langle m^0|H^1|n^0\rangle}{E_n^0-E_m^0}=|n^0\rangle+|n^1\rangle \tag{17.1.14}$$

Notice that $|n^1\rangle$ is orthogonal to $|n^0\rangle$ and proportional to the first power of H^1 (as anticipated). We determine E_n^2 from the second-order part of Eq. (17.1.4):

$$H^0|n^2\rangle+H^1|n^1\rangle=E_n^0|n^2\rangle+E_n^1|n^1\rangle+E_n^2|n^0\rangle \tag{17.1.15}$$

Dotting with $\langle n^0|$ and using the results from lower order ($|n^1\rangle=|n_{\perp}^1\rangle$) we obtain

$$E_n^2=\langle n^0|H^1|n^1\rangle \tag{17.1.16}$$

$$=\sum_m{}' \frac{\langle n^0|H^1|m^0\rangle\langle m^0|H^1|n^0\rangle}{E_n^0-E_m^0}$$

$$=\sum_m{}' \frac{|\langle n^0|H^1|m^0\rangle|^2}{E_n^0-E_m^0} \tag{17.1.17}$$

We can go on to higher orders, but we choose to stop here.

Before we turn to examples, let us consider some general features of our results. First we note that the energy to a given order is determined by the state vector to the next lower order, see Eqs. (17.1.7) and (17.1.16). This is in accord with the remarks made in the study of the variational method. The physics behind this phenomenon will become clear when we consider a few examples. Next we ask under what conditions the perturbation expansion is good, namely, when the correction terms are small compared to the zeroth-order (unperturbed) results. The answer follows from Eq. (17.1.14). A necessary condition for $|n^1\rangle$ to be small compared to $|n^0\rangle$ is that

$$\left|\frac{\langle m^0|H^1|n^0\rangle}{E_n^0-E_m^0}\right| \ll 1 \tag{17.1.18}$$

Thus we see that the condition depends on (1) the absolute size of H^1 (i.e., if it is due to some external field, the magnitude of the field); (2) the matrix elements of H^1 between unperturbed states; and (3) the energy difference between the levels. If the unperturbed eigenstate is $|n^0\rangle$, the perturbation mixes in orthogonal states $|m^0\rangle$; this mixing is directly proportional to the matrix element $\langle m^0|H^1|n^0\rangle$ and inversely proportional to the energy difference between the two levels, which measures the "rigidity" of the system. If for any reason the above inequality is not fulfilled (say due to degeneracy, $E_n^0=E_m^0$) we must turn to an alternate formalism called *degenerate perturbation theory* to be described later in this chapter.

17.2. Some Examples

Consider a particle of charge q and mass m in a harmonic oscillator potential $V=\frac{1}{2}m\omega^2x^2$. Suppose we apply an external electric field of magnitude f along the positive x direction. This corresponds to an electrostatic potential $\phi=-fx$ and a potential energy $V=-qfx$. Thus

$$H=H^0+H^1=\frac{P^2}{2m}+\frac{1}{2}m\omega^2X^2-qfX \tag{17.2.1}$$

We wish to handle H^1 by perturbation theory. Let us first calculate the first-order shift in energy, given by

$$E_n^1=\langle n^0|H^1|n^0\rangle=-qf\langle n^0|X|n^0\rangle \tag{17.2.2}$$

where $|n^0\rangle$ is just the nth state of the unperturbed oscillator. We can see that E^1 vanishes in many ways. At a formal level, since

$$X=\left(\frac{\hbar}{2m\omega}\right)^{1/2}(a+a^\dagger) \tag{17.2.3}$$

it has no *diagonal* matrix elements. The physics of what is happening is more transparent in the coordinate basis, where

$$E_n^1 = -qf \int (\psi_n^0)^* x \psi_n^0 \, dx$$
$$= -qf \int |\psi_n^0|^2 x \, dx \tag{17.2.4}$$

Now $\psi_n^0(x)$, being the unperturbed eigenfunction, has definite parity $(-1)^n$. Consequently $|\psi_n^0|^2$ is an even function, while the external potential is an odd function. Thus the average interaction with the external field is zero, for the particle is as likely to be found in the region of potential ϕ as in the region of potential $-\phi$. Notice that E_n^1 is the energy of interaction of the *unperturbed configuration* $|n^0\rangle$, with the applied field. Consequently this is not the whole story, for the configuration itself will get modified by the external field to $|n^0\rangle + |n^1\rangle + \cdots$, and we should really be considering the energy of interaction of the perturbed configurations and the applied field. But this is a distinction that is at least a second-order effect, for the change $\delta|n\rangle \equiv |n\rangle - |n^0\rangle$ in the configuration is at least a first-order effect and the interaction of $\delta|n\rangle$ with the applied field involves another order of H^1. So let us calculate the perturbed eigenket to first order and then energy levels to second order. From Eq. (17.1.14).

$$|n\rangle = |n^0\rangle + \sum_m{}' \frac{|m^0\rangle\langle m^0| - qf(\hbar/2m\omega)^{1/2}(a+a^\dagger)|n^0\rangle}{E_n^0 - E_m^0}$$
$$= |n^0\rangle + qf\left(\frac{1}{2m\hbar\omega^3}\right)^{1/2} [(n+1)^{1/2}|(n+1)^0\rangle - n^{1/2}|(n-1)^0\rangle] \tag{17.2.5}$$

Thus to first order, the perturbation mixes the state $|n^0\rangle$ with the states immediately above and below it. It was stated earlier that $E_n^0 - E_m^0$ measures the "rigidity" of the system. We find in this example that this quantity is proportional to ω, which in the mass–spring case measures the force constant. How does the wave function of the perturbed state look? This is not transparent from the above equation, but we expect that it will represent a probability distribution that is no longer centered at and symmetric about $x=0$, but instead is biased toward positive x (for that is the direction of the external field). We will return to confirm this picture quantitatively.

The second-order energy shift (which reflects the fact that the configuration of the system is not $|n^0\rangle$ but $|n^0\rangle + |n^1\rangle$), is

$$E_n^2 = \langle n^0|H^1|n^1\rangle = \sum_m{}' \frac{|\langle m^0|H^1|n^0\rangle|^2}{E_n^0 - E_m^0}$$
$$= q^2 \cdot f^2 \cdot \frac{\hbar}{2m\omega}\left(\frac{n+1}{-\hbar\omega} + \frac{n}{\hbar\omega}\right) = \frac{-q^2 f^2}{2m\omega^2} \tag{17.2.6}$$

The present problem is a nice testing ground for perturbation theory because it may be exactly solved. This is because H may be written as

$$H=\frac{P^2}{2m}+\frac{1}{2}m\omega^2X^2-qfX$$
$$=\frac{P^2}{2m}+\frac{1}{2}m\omega^2\left(X-\frac{qf}{m\omega^2}\right)^2-\frac{1}{2}\frac{q^2f^2}{m\omega^2} \tag{17.2.7}$$

This Hamiltonian also describes an oscillator of frequency ω, but is different in that (i) the oscillator is centered at $x=qf/m\omega^2$; (ii) each state has a constant energy $(-q^2f^2/2m\omega^2)$ added to it. Thus the eigenfunctions of H are just the eigenfunctions of H^0 shifted ay $qf/m\omega^2$ and the eigenvalues are $E_n=E_n^0-q^2f^2/2m\omega^2$. The classical picture that goes with Eq. (17.2.7) is clear: the effect of a constant force qf on a mass coupled to a spring of force constant $m\omega^2$ is to shift the equilibrium point to $x=qf/m\omega^2$. (Imagine a mass hanging from a spring attached to the ceiling and ask what gravity does to its dynamics.) Let us compare these exact results with the perturbation theory. Consider the energy

$$E_n=E_n^0-q^2f^2/2m\omega^2 \tag{17.2.8}$$

Since H^1 is proportional to qf, the power of qf gives the order of the term. According to Eq. (17.2.8), there is no first-order shift in energy, and the second-order shift is $-q^2f^2/2m\omega^2$, which agrees with Eq. (17.2.6). Had we tried to go to higher orders, we would have found nothing more.

Now consider the state vectors. The exact result is

$$|n\rangle=T(qf/m\omega^2)|n^0\rangle \tag{17.2.9}$$

where $T(a)$ is the operator that translates the system by an amount a. Since we are working to first order in qf,

$$T(qf/m\omega^2)=e^{-i(qf/m\omega^2\hbar)P}\simeq I-i\left(\frac{qf}{m\omega^2\hbar}\right)P$$
$$=I-i\left(\frac{qf}{m\omega^2\hbar}\right)\left(\frac{\hbar m\omega}{2}\right)^{1/2}\cdot\frac{a-a^\dagger}{i} \tag{17.2.10}$$

so that

$$|n\rangle=\left[I-\left(\frac{qf}{m\omega^2\hbar}\right)\left(\frac{\hbar m\omega}{2}\right)^{1/2}(a-a^\dagger)\right]|n^0\rangle$$
$$=|n^0\rangle+qf\left(\frac{1}{2m\hbar\omega^3}\right)^{1/2}[(n+1)^{1/2}|(n+1)^0\rangle-n^{1/2}|(n-1)^0\rangle] \tag{17.2.11}$$

which agrees with Eq. (17.2.5). It is clear that computing $|n\rangle$ to higher order in perturbation theory will be equivalent to expanding T to higher orders in qf.

*Exercise 17.2.1.** Consider $H^1=\lambda x^4$ for the oscillator problem.
(1) Show that

$$E_n^1=\frac{3\hbar^2\lambda}{4m^2\omega^2}[1+2n+2n^2]$$

(2) Argue that no matter how small λ is, the perturbation expansion will break down for some large enough n. What is the physical reason?

*Exercise 17.2.2.** Consider a spin-1/2 particle with gyromagnetic ratio γ in a magnetic field $\mathbf{B}=B\mathbf{i}+B_0\mathbf{k}$. Treating B as a perturbation, calculate the first- and second-order shifts in energy and first-order shift in wave function for the ground state. Then compare the exact answers expanded to the corresponding orders.

Exercise 17.2.3. In our study of the H atom, we assumed that the proton is a point charge e. This leads to the familiar Coulomb interaction $(-e^2/r)$ with the electron. (1) Show that if the proton is a uniformly dense charge distribution of radius R, the interaction is

$$\begin{aligned} V(r) &= -\frac{3e^2}{2R}+\frac{e^2r^2}{2R^3}, & r\le R \\ &= -\frac{e^2}{r}, & r>R \end{aligned}$$

(2) Calculate the first-order shift in the ground-state energy of hydrogen due to this modification. You may assume $e^{-R/a_0}\simeq 1$. You should find $E^1=2e^2R^2/5a_0^3$.

*Exercise 17.2.4.** (1) Prove the Thomas–Reiche–Kuhn sum rule

$$\sum_{n'}(E_{n'}-E_n)|\langle n'|X|n\rangle|^2=\sum_{n'}(E_{n'}-E_n)\langle n|X|n'\rangle\langle n'|X|n\rangle=\frac{\hbar^2}{2m}$$

where $|n\rangle$ and $|n'\rangle$ are eigenstates of $H=P^2/2m+V(X)$. (Hint: Eliminate the $E_{n'}-E_n$ factor in favor of H.)
(2) Test the sum rule on the nth state of the oscillator.

Exercise 17.2.5 (Hard). We have seen that if we neglect the repulsion e^2/r_{12} between the two electrons in the ground state of He, the energy is -8 Ry$=-108.8$ eV. Treating e^2/r_{12} as a perturbation, show that

$$\langle 100,100|H^1|100,100\rangle=\tfrac{5}{2}\text{ Ry}$$

so that $E_0^0+E_0^1=-5.5$. Ry$=-74.8$ eV. Recall that the measured value is -78.6 eV and the variational estimate is -77.5 eV. [Hint: $\langle H^1\rangle$ can be viewed as the interaction between two concentric, spherically symmetric exponentially falling charge distributions. Find the potential $\phi(r)$ due to one distribution and calculate the interaction energy between this potential and the other charge distribution.]

Selection Rules

The labor involved in perturbation theory calculations is greatly reduced by the use of selection rules, which allow us to conclude that certain matrix elements of H^1 are zero without explicitly calculating them. They are based on the idea that if

$$[\Omega, H^1]=0$$

then

$$\langle \alpha_2\omega_2|H^1|\alpha_1\omega_1\rangle=0 \quad \text{unless} \quad \omega_1=\omega_2 \tag{17.2.12}‡$$

Proof.

$$0=\langle \alpha_2\omega_2|\Omega H^1-H^1\Omega|\alpha_1\omega_1\rangle=(\omega_2-\omega_1)\langle \alpha_2\omega_2|H^1|\alpha_1\omega_1\rangle \qquad \text{Q.E.D.}$$

Consider for example $H^1=\lambda Z$, which is invariant under rotations around the z axis. Then $[L_z, H^1]=0$ and

$$\langle \alpha_2 m_2|H^1|\alpha_1 m_1\rangle=0 \quad \text{unless} \quad m_2=m_1 \tag{17.2.13}$$

(This result also follows from the Wigner–Eckart theorem.) Or if H^1 is parity invariant, say $H^1=\lambda Z^2$, then its matrix element between states of opposite parity is zero.

There is a simple way to understand Eq. (17.2.12). To say that $[\Omega, H^1]=0$ is to say that H^1 "carries no Ω"; in other words, when it acts on a state it imparts no Ω to it. We see this as follows. Consider $|\omega_1\rangle$, which carries a definite amount of the variable Ω, namely, ω_1:

$$\Omega|\omega_1\rangle=\omega_1|\omega_1\rangle \tag{17.2.14}$$

Let us measure Ω in the state after H^1 acts on it:

$$\Omega(H^1|\omega_1\rangle)=H^1\Omega|\omega_1\rangle=H^1\omega_1|\omega_1\rangle=\omega_1(H^1|\omega_1\rangle) \tag{17.2.15}$$

We find it is the same as before, namely, ω_1. The selection rule then merely reflects the orthogonality of eigenstates with different ω.

This discussion paves the way for an extension of the selection rule to a case where H^1 carries a definite amount of Ω. For instance, if H^1 is a tensor operator T_k^q, it carries angular momentum (k, q) and we know from the Wigner–Eckart theorem that

$$\langle \alpha_2 j_2 m_2|T_k^q|\alpha_1 j_1 m_1\rangle=0 \quad \text{unless} \quad \begin{cases} j_1+k\geq j_2\geq |j_1-k| \\ m_2=m_1+q \end{cases} \tag{17.2.16}$$

‡ α stands for other quantum numbers that label the state.

i.e., that the matrix element vanishes unless $|\alpha_2 j_2 m_2\rangle$ has the angular momentum that obtains when we add to $(j_1 m_1)$ the angular momentum (kq) imparted by the operator. For instance, if $H^1 = \lambda Z \sim T_1^0$,

$$\langle \alpha_2 j_2 m_2 | Z | \alpha_1 j_1 m_1 \rangle = 0 \quad \text{unless} \quad \begin{cases} j_2 = j_1 + 1, j_1, j_1 - 1 \\ m_2 = m_1 + 0 \end{cases} \tag{17.2.17}$$

while if $H^1 = \lambda X$ or $\lambda Y \, (\sim T_1^{\pm 1})$, we have

$$\langle \alpha_2 j_2 m_2 | X \text{ or } Y | \alpha_1 j_1 m_1 \rangle = 0 \quad \text{unless} \quad \begin{cases} j_2 = j_1 + 1, j_1, j_1 - 1 \\ m_2 = m_1 \pm 1 \end{cases} \tag{17.2.18}$$

Another example of this type is an operator that is not parity invariant, but parity odd. An example is X, which obeys

$$\Pi^\dagger X \Pi = -X \tag{17.2.19}$$

You can verify that if X acts on a state of definite parity, it changes the parity of the state. Thus the matrix element of X between eigenstates of parity vanishes unless they have opposite parity. More generally, if

$$\Pi^\dagger \Omega \Pi = -\Omega \tag{17.2.20}$$

then the matrix element of Ω between two parity eigenstates vanishes unless they have opposite parity.

We get more selection rules by combining these selection rules. For instance, we can combine the angular momentum and parity selection rules for the vector operators $\mathbf{R}$ to get (in the case of no spin, $\mathbf{J} = \mathbf{L}$),

$$\langle \alpha_2 l_2 m_2 | Z | \alpha_1 l_1 m_1 \rangle = 0 \quad \text{unless} \quad \begin{cases} l_2 = l_1 \pm 1 \\ m_2 = m_1 \end{cases}$$

$$\langle \alpha_2 l_2 m_2 | X \text{ or } Y | \alpha_1 l_1 m_1 \rangle = 0 \quad \text{unless} \quad \begin{cases} l_2 = l_1 \pm 1 \\ m_2 = m_1 \pm 1 \end{cases} \tag{17.2.21}$$

We rule out the possibility $l_2 = l_1$ by the parity selection rule, for states of orbital angular momentum l have definite parity $(-1)^l$. Equation (17.2.21) is called the *dipole selection rule.*

We now consider an example that illustrates the use of these tricks and a few more. The problem is to determine the response of the hydrogen atom in the ground state to a constant external electric field $\mathbf{E} = \mathscr{E}\mathbf{k}$. This is called the *Stark effect.* Let us first calculate H^1. We do this by determining $\mathscr{H}^1$, its classical counterpart and then making the operator substitution. If $\mathbf{r}_1$ and $\mathbf{r}_2$ are the position vectors of the electron and proton, respectively, and $\phi(\mathbf{r})$ is the electrostatic potential due to $\mathbf{E}$, then

$$\begin{aligned}\mathscr{H}^1&=-e\phi(\mathbf{r}_1)+e\phi(\mathbf{r}_2)\\&=e[\phi(\mathbf{r}_2)-\phi(\mathbf{r}_1)]\\&=e(\mathbf{r}_1-\mathbf{r}_2)\cdot\mathbf{E}\quad(\text{recall }\mathbf{E}=-\nabla\phi)\\&=e\mathbf{r}\cdot\mathbf{E}\end{aligned}\tag{17.2.22}$$

where $\mathbf{r}$ is the relative coordinate or equivalently the position vector of the electron in the CM frame in the limit $m/M=0$. $\mathscr{H}^1$ is called the *dipole interaction*, for in terms of

$$\boldsymbol{\mu}_e=e(\mathbf{r}_2-\mathbf{r}_1)=-e\mathbf{r}\tag{17.2.23}$$

the *electric dipole moment* of the system,

$$\mathscr{H}^1=-\boldsymbol{\mu}_e\cdot\mathbf{E}\tag{17.2.24}$$

(This is the electric analog of $\mathscr{H}=-\boldsymbol{\mu}\cdot\mathbf{B}$.)‡ Thus, for the given electric field

$$H^1=eZ\mathscr{E}\tag{17.2.25}$$

Let us now calculate the first-order shift in the energy of the ground state $|100\rangle$§:

$$E^1_{100}=\langle 100|eZ\mathscr{E}|100\rangle\tag{17.2.26}$$

We can argue that $E^1_{100}=0$ either on the grounds of parity or the Wigner–Eckart theorem. More physically, E^1_{100} vanishes because in the unperturbed state, the electron probability distribution is spherically symmetric and the electron samples $\phi(\mathbf{r})$ and $\phi(-\mathbf{r})=-\phi(\mathbf{r})$ equally. Another way to say this is that the unperturbed atom has no mean electric dipole moment $\langle\boldsymbol{\mu}\rangle$ (by parity or the Wigner–Eckart theorem) so that

$$E^1_{100}=\langle 100|-\boldsymbol{\mu}\cdot\mathbf{E}|100\rangle=-\langle 100|\boldsymbol{\mu}|100\rangle\cdot\mathbf{E}=0\tag{17.2.27}$$

But we expect the second-order energy shift to be nonzero, for the external field will shift the electron distribution downward and induce a dipole moment which can interact with $\mathbf{E}$. So let us calculate

$$E^2_{100}=\sum_{nlm}{}' \frac{e^2\mathscr{E}^2|\langle nlm|Z|100\rangle|^2}{E^0_{100}-E^0_{nlm}}\tag{17.2.28}$$

‡ In the rest of this chapter we will omit the subscript e on $\boldsymbol{\mu}_e$.

§ When we discuss hydrogen, we will use the symbol $|nlm\rangle$, rather than $|(nlm)^0\rangle$ to denote the *unperturbed* state.

where

$$E^0_{100} - E^0_{nlm} = -\text{Ry}\left(1 - \frac{1}{n^2}\right) = \text{Ry}\left(\frac{1-n^2}{n^2}\right) \tag{17.2.29}$$

Unlike in the case of the oscillator, the sum now involves an infinite number of terms. Although we can use dipole selection rules to reduce the sum to

$$E^2_{100} = \sum_{n=2}^{\infty} \frac{e^2\mathscr{E}^2|\langle n10|Z|100\rangle|^2}{E^0_1 - E^0_n} \tag{17.2.30}$$

let us keep the form in Eq. (17.2.28) for a while. There are several ways to proceed.

Method 1. Since the magnitude of the energy denominator grows with n, we have the inequality

$$|E^2_{100}| \le \frac{e^2\mathscr{E}^2}{|E^0_1 - E^0_2|} \sum_{nlm}{}' |\langle nlm|Z|100\rangle|^2$$

But since

$$\begin{aligned}
&\sum_{nlm}{}' |\langle nlm|Z|100\rangle|^2 \\
&\quad = \sum_{nlm}{}' \langle 100|Z|nlm\rangle\langle nlm|Z|100\rangle \\
&\quad = \sum_{nlm} \langle 100|Z|nlm\rangle\langle nlm|Z|100\rangle - \langle 100|Z|100\rangle^2 \\
&\quad = \langle 100|Z^2|100\rangle - \langle 100|Z|100\rangle^2 \\
&\quad = a_0^2 - 0 = a_0^2
\end{aligned} \tag{17.2.31}$$

we get

$$\begin{aligned}
|E^2_{100}| &\le \frac{e^2\mathscr{E}^2}{|(e^2/2a_0)(1-\frac{1}{4})|} a_0^2 \\
&= \frac{8a_0^3\mathscr{E}^2}{3}
\end{aligned} \tag{17.2.32}$$

We can also get a lower bound on $|E^2_{100}|$ by keeping just the first term in Eq. (17.2.30) (since all terms have the same sign):

$$|E^2_{100}| \ge \frac{e^2\mathscr{E}^2}{3e^2/8a_0} |\langle 210|Z|100\rangle|^2 \tag{17.2.33}$$

Now,

$$|\langle 210|Z|100\rangle|^2=\frac{2^{15}a_0^2}{3^{10}}\simeq 0.55a_0^2 \tag{17.2.34}$$

so that

$$|E_{100}^2|\geq(0.55)\tfrac{8}{3}\mathscr{E}^2a_0^3 \tag{17.2.35}$$

We thus manage to restrict $|E_{100}^2|$ to the interval

$$\tfrac{8}{3}\mathscr{E}^2a_0^3\geq|E_{100}^2|\geq 0.55(\tfrac{8}{3})\mathscr{E}^2a_0^3 \tag{17.2.36}$$

Method 2. Consider the general problem of evaluating

$$E_n^2=\sum_m{}'\frac{\langle n^0|H^1|m^0\rangle\langle m^0|H^1|n^0\rangle}{E_n^0-E_m^0} \tag{17.2.37}$$

If it weren't for the energy denominator, we could use the completeness relation to eliminate the sum (after adding and subtracting the $m=n$ term). There exists a way to eliminate the energy denominator.‡ Suppose we can find an operator Ω such that

$$H^1=[\Omega, H^0] \tag{17.2.38}$$

then

$$\begin{aligned}E_n^2&=\sum_m{}'\frac{\langle n^0|H^1|m^0\rangle\langle m^0|\Omega H^0-H^0\Omega|n^0\rangle}{E_n^0-E_m^0}\\&=\sum_m{}'\langle n^0|H^1|m^0\rangle\langle m^0|\Omega|n^0\rangle\\&=\langle n^0|H^1\Omega|n^0\rangle-\langle n^0|H^1|n^0\rangle\langle n^0|\Omega|n^0\rangle\end{aligned} \tag{17.2.39}$$

which calls for computing just three matrix elements. But it is not an easy problem to find the Ω that satisfies Eq. (17.2.38). (There are, however, exceptions, see Exercise 17.2.7.) A more modest proposal is to find Ω such that

$$H^1|n^0\rangle=[\Omega, H^0]|n^0\rangle \tag{17.2.40}$$

for a given $|n^0\rangle$. You can verify that this is all it takes to derive Eq. (17.2.39) *for this value of n.* In the problem we are interested in, we need to solve

$$H^1|100\rangle=[\Omega, H^0]|100\rangle \tag{17.2.41}$$

‡ See A. Dalgarnó and J. T. Lewis, *Proceedings of the Royal Society*, **A233**, 70 (1955).

By writing this equation in the coordinate basis and assuming Ω is a function of coordinates and not momenta, we can show that

$$\Omega \xrightarrow[\text{coordinate basis}]{} -\frac{ma_0 e\mathscr{E}}{\hbar^2}\left(\frac{r^2\cos\theta}{2}+a_0 r\cos\theta\right) \tag{17.2.42}$$

The exact second-order shift is then

$$\begin{aligned} |E^2_{100}| &= |\langle 100|H^1\Omega|100\rangle - 0| \\ &= |\langle 100|eZ\mathscr{E}\Omega|100\rangle| \\ &= \tfrac{9}{4}a_0^3\mathscr{E}^2 = \tfrac{8}{3}a_0^3\mathscr{E}^2\cdot\left(\tfrac{27}{32}\right) \\ &= (0.84)\tfrac{8}{3}a_0^3\mathscr{E}^2 \end{aligned} \tag{17.2.43}$$

which is roughly in the middle of the interval we restricted it to by Method 1.

Exercise 17.2.6. Verify Eq. (17.2.34).

*Exercise 17.2.7.** For the oscillator, consider $H^1 = -qfX$. Find an Ω that satisfies Eq. (17.2.38). Feed it into Eq. (17.2.39) for E_n^2 and compare with the earlier calculation.

Exercise 17.2.8. Fill in the steps connecting Eqs. (17.2.41) and (17.2.43). Try to use symmetry arguments to reduce the labor involved in evaluating the integrals.

We argued earlier that E^2_{100} represents the interaction of the induced dipole moment with the applied field. How big is the induced moment μ? One way to find out is to calculate $\langle\boldsymbol{\mu}\rangle$ in the perturbed ground state. An easier way is to extract it from E^2_{100}. Suppose we take a system that has no intrinsic dipole moment and turn on an external electric field that starts at 0 and grows to the full value of $\mathbf{E}$. During this time the dipole moment grows from 0 to μ. If you imagine charges $\pm q$ separated by a distance x along $\mathbf{E}$, you can see that the work done on the system as x changes by dx is

$$\begin{aligned} dW &= -q\mathscr{E}\,dx \\ &= -\mathscr{E}\,d\mu \end{aligned} \tag{17.2.44}$$

If we assume that the induced moment is proportional to $\mathbf{E}$:

$$\boldsymbol{\mu} = \alpha\mathbf{E} \tag{17.2.45}$$

(where α is called the *polarizability*), then

$$dW = -\alpha\mathscr{E}\,d\mathscr{E}$$

or

$$W = -\tfrac{1}{2}\alpha\varepsilon^2 \tag{17.2.46}$$

We identify W with E^2_{100} and determine the polarizability

$$\alpha = \frac{18}{4}a_0^3 \simeq \frac{18}{4}(0.5\,\text{Å})^3 \simeq 0.56\,\text{Å}^3 \tag{17.2.47}$$

If we use a more accurate value $a_0 = 0.53$ Å, we get $\alpha = 0.67$ Å^3, which is in excellent agreement with the measured value of 0.68 Å^3. For a given **E**, we can get $\boldsymbol{\mu}$ from Eq. (17.2.45).

Finally note that E^2_{100} is negative. From Eq. (17.1.17) it is clear that the second-order shift in the ground-state energy is always negative (unless it vanishes). Since E^2_0 measures the energy shift due to the first-order change in the ground-state state vector, we conclude that the system changes its configuration so as to lower its energy of interaction with the external field.

17.3. Degenerate Perturbation Theory

In the face of degeneracy ($E^0_n = E^0_m$) the condition for the validity of the perturbation expansion,

$$\left|\frac{\langle m^0|H^1|n^0\rangle}{E^0_n - E^0_m}\right| \ll 1 \tag{17.3.1}$$

is impossible to fulfill. The breakdown of the method may be understood in less formal terms as follows.

Let us consider the case when neither H^0 nor $H^0 + H^1$ is degenerate. For the purposes of this argument imagine that H^1 is due to some external field that can be continuously varied from zero to its full value. As the total Hamiltonian grows from H^0 to $H^0 + H^1$, the corresponding eigenbasis changes continuously from $|n^0\rangle$ to $|n\rangle$. It is this continuous or analytic variation of the eigenbasis with the perturbation that makes it possible to find $|n\rangle$ starting with $|n^0\rangle$, the way one finds the value of some analytic function at the point $x + a$ starting at the point x and using a Taylor series. Consider now the case when H^0 has a degenerate subspace and $H^0 + H^1$ is nondegenerate in this subspace. (More general cases can be handled the same way.) Imagine starting with the basis $|n\rangle$ and slowly turning *off* the perturbation. We will end up with a basis $|\bar{n}^0\rangle$ of H^0. If we now turn on the perturbation, we can retrace our path back to $|n\rangle$. It is clear that if we start with this basis, $|\bar{n}^0\rangle$, we can evaluate $|n\rangle$ perturbatively. But since H^0 is degenerate, we needn't have started with this basis; we could have started with some other basis $|n^0\rangle$, chosen randomly. But if we start with any basis except $|\bar{n}^0\rangle$, and turn on the external field of infinitesimal size, the change in the basis will not be infinitesimal. It is this nonanalytic behavior that is signaled by the divergence in the first-order matrix element. [This can be compared to the divergence of the first derivative in the Taylor series where $f(x)$ is discontinu-

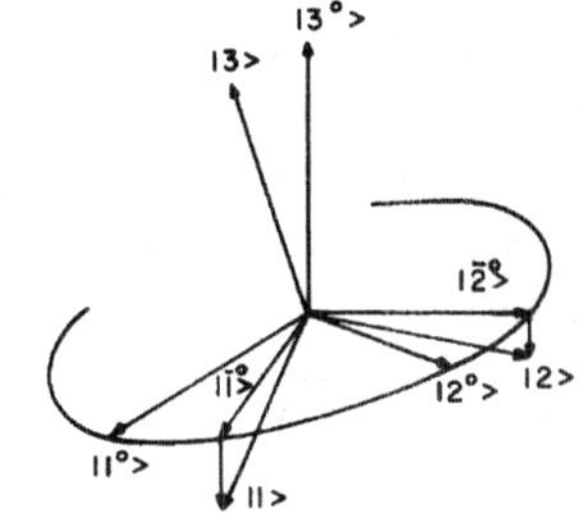

Figure 17.1. An example of the degenerate problem from $\mathbb{V}^3(R)$. In the x–y plane, which is the degenerate subspace, we must start not with some arbitrarily chosen pair of basis vectors $|1^0\rangle$ and $|2^0\rangle$, but with the pair $|\bar{1}^0\rangle$ and $|\bar{2}^0\rangle$ which diagonalizes H^1.

ous.] So, we must start with the right basis in the degenerate space. We have already characterized this basis as one we get if we start with $|n\rangle$ and slowly turn off H^1. A more useful characterization is the following: it is a basis that diagonalizes H^1 within the degenerate space. Why? Because, if we start with this basis, the first-order perturbation coefficient [Eq. (17.1.8)] does not blow up, for the (off-diagonal) matrix element in the numerator vanishes along with energy denominator whenever $|n^0\rangle$ and $|m^0\rangle$ belong to the degenerate space. Figure 17.1 depicts a simple example from $\mathbb{V}^3(R)$, where the x–y plane is the degenerate space and $|1^0\rangle$ and $|2^0\rangle$ are randomly chosen basis vectors in that subspace. The proper starting point is the pair $|\bar{1}^0\rangle$, $|\bar{2}^0\rangle$, which diagonalizes H^1 in the x–y plane.

It is worth noting that to find the proper starting point, we need to find the basis that diagonalizes H^1 only within the degenerate space and not the full Hilbert space. Thus even if we work with infinite-dimensional spaces, the exact diagonalization will usually have to be carried out only in some small, finite-dimensional subspace.

Let us consider, as a concrete example, the Stark effect in the $n=2$ level of hydrogen. (We ignore spin, which is a spectator variable.) Are we to conclude that there is no first-order shift because

$$\langle 2lm|e\mathscr{E}Z|2lm\rangle = 0 \tag{17.3.2}$$

by parity invariance, or equivalently, because the atom in these states has no intrinsic dipole moment? No, because these states need not provide the correct starting points for a perturbative calculation in view of the degeneracy. We must first find the basis in the $n=2$ sector which diagonalizes H^1. Using the selection rules, which tell us that only two of the 16 matrix elements are nonzero, we get

$$H^1 \to \begin{array}{c|cccc} nlm & 200 & 210 & 211 & 21-1 \\ \hline 200 & 0 & \Delta & 0 & 0 \\ 210 & \Delta & 0 & 0 & 0 \\ 211 & 0 & 0 & 0 & 0 \\ 21-1 & 0 & 0 & 0 & 0 \end{array} \tag{17.3.3}$$

where

$$\Delta = \langle 200|e\mathscr{E}Z|210\rangle = -3e\mathscr{E}a_0 \tag{17.3.4}$$

*Exercise 17.3.1.** Use the dipole selection rules to show that H^1 has the above form and carry out the evaluation of Δ.

Since H^1 is just Δ times the Pauli matrix σ_x in the $m=0$ sector, we infer that its eigenvalues are $\pm\Delta$ and that its eigenstates are $[|200\rangle \pm |210\rangle]/2^{1/2}$. In the $|m|=1$ sector the old states $|2, 1, \pm 1\rangle$ diagonalize H^1. Our calculation tells us the following.

(1) The zeroth-order states stable under the perturbation are $|2, 1, \pm 1\rangle$ and $[|200\rangle \pm |210\rangle]/2^{1/2}$.

(2) The first-order shift E^1 is zero for the first two states and $\pm\Delta$ for the next two. (Note that Δ is negative.)

Notice that the stable eigenstates for which $E^1 \neq 0$ are mixtures of $l=0$ and $l=1$. Thus they have indefinite parity and can have a nonzero intrinsic dipole moment which can interact with E and produce a first-order energy shift. From the energy shift, we infer that the size of the dipole moment is $3ea_0$.

Degenerate perturbation theory is relevant not only when the levels are exactly degenerate but also when they are close, that is to say, when the inequality (17.3.1) is not respected. In that case one must diagonalize $H^0 + H^1$ exactly in the almost degenerate subspace.

Exercise 17.3.2. Consider a spin-1 particle (with no orbital degrees of freedom). Let $H = AS_z^2 + B(S_x^2 - S_y^2)$, where S_i are 3×3 spin matrices, and $A \gg B$. Treating the B term as a perturbation, find the eigenstates of $H^0 = AS_z^2$ that are stable under the perturbation. Calculate the energy shifts to first order in B. How are these related to the exact answers?

Fine Structure

The Coulomb potential $(-e^2/r)$ does not give the complete interaction between the electron and the proton, though it does provide an excellent first approximation.‡ There are "fine-structure" corrections to this basic interaction, which produce energy shifts of the order of α^2 times the binding energy due to the Coulomb potential. Since the electron velocity (in a semiclassical picture) is typically $\beta = v/c \simeq O(\alpha)$, these are corrections of the order $(v/c)^2$ *relative* to binding energy, which is itself proportional to $(v/c)^2$. Thus these are relativistic in origin. There are two parts to this effect.

The first reflects the fact that to order $(v/c)^4$ the kinetic energy of the electron is not $p^2/2m$ but

$$T = (c^2p^2 + m^2c^4)^{1/2} - mc^2 = \frac{p^2}{2m} - \frac{p^4}{8m^3c^2} + O(p^6 \text{ or } v^6) \tag{17.3.5}$$

We now wish to calculate the effect of this extra term

$$H_T = -P^4/8m^3c^2 \tag{17.3.6}$$

‡ We consider here just the fine structure of hydrogen. The analysis may be extended directly to hydrogen-like atoms. We also ignore the difference between the reduced mass and the electron mass.

treating it as a perturbation. Since H_T is rotationally invariant, it is diagonal in the $|nlm\rangle$ basis. (In other words, the $|nlm\rangle$ basis is stable under this perturbation.) So we can forget about the fact that the levels at each n are degenerate and determine E_T^1 simply from

$$E_T^1 = -\frac{1}{8m^3c^2}\langle nlm|P^4|nlm\rangle \tag{17.3.7}$$

We evaluate the matrix element by noting that

$$P^4 = 4m^2\left(\frac{P^2}{2m}\right)^2 = 4m^2\left(H^0 + \frac{e^2}{r}\right)^2 \tag{17.3.8}$$

so that

$$E_T^1 = -\frac{1}{2mc^2}\left[(E_n^0)^2 + 2E_n^0 e^2\left\langle\frac{1}{r}\right\rangle_{nlm} + e^4\left\langle\frac{1}{r^2}\right\rangle_{nlm}\right] \tag{17.3.9}$$

From the virial theorem [Eq. (13.1.34)]

$$-\left\langle\frac{e^2}{r}\right\rangle_{nlm} = 2E_n^0 \tag{17.3.10}$$

while from Exercise (17.3.4)

$$\left\langle\frac{e^4}{r^2}\right\rangle_{nlm} = \frac{e^4}{a_0^2 n^3(l+1/2)} = \frac{4(E_n^0)^2 n}{l+1/2} \tag{17.3.11}$$

so that

$$\begin{aligned} E_T^1 &= -\frac{(E_n^0)^2}{2mc^2}\left(-3 + \frac{4n}{l+1/2}\right) \\ &= -\frac{1}{2}(mc^2)\alpha^4\left[-\frac{3}{4n^4} + \frac{1}{n^3(l+1/2)}\right] \end{aligned} \tag{17.3.12}$$

The other relativistic effect is called the *spin–orbit interaction*. Its origin may be understood as follows. The Coulomb interaction $(-e^2/r)$ is the whole story only if the electron is at rest. If it moves at a velocity $\mathbf{v}$, there is an extra term which we find as follows. In the electron rest frame, the proton will be moving at a velocity

$(-\mathbf{v})$ and will produce a magnetic field

$$\mathbf{B} = -\frac{e}{c}\frac{\mathbf{v}\times\mathbf{r}}{r^3} \tag{17.3.13}$$

The interaction of the magnetic moment of the electron with this field leads to the *spin–orbit energy*

$$\begin{aligned}\mathscr{H}_{\text{s.o.}} &= -\boldsymbol{\mu}\cdot\mathbf{B} = \frac{e}{mcr^3}\boldsymbol{\mu}\cdot(\mathbf{p}\times\mathbf{r}) \\ &= -\frac{e}{mc}\frac{\boldsymbol{\mu}\cdot\mathbf{l}}{r^3}\end{aligned} \tag{17.3.14}$$

So we expect that in the quantum theory there will be a perturbation

$$\begin{aligned}H_{\text{s.o.}} &= \left(-\frac{e}{mc}\right)\left(-\frac{e}{mc}\right)\frac{\mathbf{S}\cdot\mathbf{L}}{r^3} \\ &= \frac{e^2}{m^2c^2r^3}\mathbf{S}\cdot\mathbf{L}\end{aligned} \tag{17.3.15}$$

However, the correct answer is half as big:

$$H_{\text{s.o.}} = \frac{e^2}{2m^2c^2r^3}\mathbf{S}\cdot\mathbf{L} \tag{17.3.16}$$

The reason is that the "rest frame of the electron" doesn't have a fixed velocity relative to the CM of the atom since the motion of the electron is not rectilinear. Thus $\mathscr{H}_{\text{s.o.}}$ deduced in the comoving frame does not directly translate into what must be used in the CM frame. The transformation and the factor of 1/2 were found by Thomas.‡ In Chapter 20 we will derive Eq. (17.3.16) from the Dirac equation, which has relativistic kinematics built into it. The *Thomas factor* of 1/2 will drop out automatically.

Since $H_{\text{s.o.}}$ involves the spin, we must now reinstate it. Since the states at a given n are degenerate, we must start with a basis that diagonalizes $H_{\text{s.o.}}$. Since we can rewrite $H_{\text{s.o.}}$ as

$$H_{\text{s.o.}} = \frac{e^2}{4m^2c^2r^3}[J^2 - L^2 - S^2] \tag{17.3.17}$$

‡ L. H. Thomas *Nature* **117**, 574 (1926).

the states of total angular momentum suggest themselves. In this basis,

$$\langle j', m'; l', 1/2|H_{\text{s.o.}}|j, m; l, 1/2\rangle$$

$$= \delta_{jj'}\,\delta_{mm'}\,\delta_{ll'}\,\frac{e^2}{4m^2c^2}\left\langle\frac{1}{r^3}\right\rangle_{nl}\hbar^2[j(j+1)-l(l+1)-3/4] \qquad (17.3.18)$$

(Note that two states with the same total jm, but built from different l's, are orthogonal because of the orthogonality of the spherical harmonics. Thus, for example, at $n=2$, we can build $j=1/2$ either from $l=0$ or $l=1$. The states $|j=1/2, m; 0, 1/2\rangle$ and $|j=1/2, m; 1, 1/2\rangle$ are orthogonal.) Feeding $j=l\pm 1/2$ into Eq. (17.3.18) we get

$$E^1_{\text{s.o.}} = \frac{\hbar^2 e^2}{4m^2c^2}\left\langle\frac{1}{r^3}\right\rangle_{nl}\begin{Bmatrix} l \\ -(l+1)\end{Bmatrix} \qquad (17.3.19)$$

where the upper and lower values correspond to $j=l\pm 1/2$. Using the result from Exercise 17.3.4

$$\left\langle\frac{1}{r^3}\right\rangle_{nl} = \frac{1}{a_0^3}\,\frac{1}{n^3 l(l+1/2)(l+1)} \qquad (17.3.20)$$

we get

$$E^1_{\text{s.o.}} = \frac{1}{4}mc^2\alpha^4\,\frac{\begin{Bmatrix} l \\ -(l+1)\end{Bmatrix}}{n^3(l)(l+1/2)(l+1)} \qquad (17.3.21)$$

This formula has been derived for $l\neq 0$. When $l=0$, $\langle 1/r^3\rangle$ diverges and $\langle \mathbf{L}\cdot\mathbf{S}\rangle$ vanishes. But if we set $l=0$ in Eq. (17.3.21) we get a finite limit, which in fact happens to give the correct level shift for $l=0$ states. This will be demonstrated when we study the Dirac equation in Chapter 20. The physical origin behind this shift (which is clearly not the spin–orbit interaction) will be discussed then. Since $E^1_{\text{s.o.}}$ and E^1_T are both α^4 effects, we combine them to get the *total fine-structure energy shift*

$$E^1_{\text{f.s.}} = E^1_T + E^1_{\text{s.o.}} = -\frac{mc^2\alpha^2}{2n^2}\cdot\frac{\alpha^2}{n}\left(\frac{1}{j+1/2}-\frac{3}{4n}\right) \qquad (17.3.22)$$

for both $j=l\pm 1/2$.

The fine-structure formula can be extended to other atoms as well, provided we make the following change in Eq. (17.3.19):

$$\left\langle \frac{e^2}{r^3} \right\rangle \rightarrow \left\langle \frac{1}{r}\frac{dV}{dr} \right\rangle$$

where V is the potential energy of the electron in question. Consider, for example, the $n=4$ states of potassium. We have seen in the prespin treatment that due to penetration and shielding effects the $4s$ level lies below the $4p$ level. If we add spin to this picture, the s state can only become $^2S_{1/2}$ while the p state can generate both $^2P_{3/2}$ and $^2P_{1/2}$. The last two are split by the fine-structure effect‡ by an amount $(3\hbar^2/4m^2c^2)\langle(1/r)(dV/dr)\rangle$, where V is the potential seen by the $n=4$, $l=1$ electron. In the $4p\rightarrow 4s$ transition, the fine-structure interaction generates two lines in the place of one, with wavelengths 7644.9 Å and 7699.0 Å.

Exercise 17.3.3. Consider the case where H^0 includes the Coulomb plus spin–orbit interaction and H^1 is the effect of a weak magnetic field $\mathbf{B}=B\mathbf{k}$. Using the appropriate basis, show that the first-order level shift is related to j_z by

$$E^1 = \left(\frac{eB}{2mc}\right)\left(1 \pm \frac{1}{2l+1}\right) j_z, \qquad j=l\pm 1/2$$

Sketch the levels for the $n=2$ level assuming that $E^1 \ll E^1_{\text{f.s.}}$.

*Exercise 17.3.4.** We discuss here some tricks for evaluating the expectation values of certain operators in the eigenstates of hydrogen.

(1) Suppose we want $\langle 1/r\rangle_{nlm}$. Consider first $\langle \lambda/r\rangle$. We can interpret $\langle \lambda/r\rangle$ as the first-order correction due to a perturbation λ/r. Now this problem can be solved exactly; we just replace e^2 by $e^2-\lambda$ everywhere. (Why?) So the exact energy, from Eq. (13.1.16) is $E(\lambda)=-(e^2-\lambda)^2 m/2n^2\hbar^2$. The first-order correction is the term linear in λ, that is, $E^1=me^2\lambda/n^2\hbar^2=\langle\lambda/r\rangle$, from which we get $\langle 1/r\rangle = 1/n^2 a_0$, in agreement with Eq. (13.1.36). For later use, let us observe that as $E(\lambda)=E^0+E^1+\cdots=E(\lambda=0)+\lambda\,(dE/d\lambda)_{\lambda=0}+\cdots$, one way to extract E^1 from the exact answer is to calculate $\lambda\,(dE/d\lambda)_{\lambda=0}$.

(2) Consider now $\langle\lambda/r^2\rangle$. In this case, an exact solution is possible since the perturbation just modifies the centrifugal term as follows:

$$\frac{\hbar^2 l(l+1)}{2mr^2}+\frac{\lambda}{r^2}=\frac{\hbar^2 l'(l'+1)}{2mr^2} \tag{17.3.23}$$

where l' is a function of λ. Now the dependence of E on $l'(\lambda)$ is, from Eq. (13.1.14),

$$E(l')=\frac{-me^4}{2\hbar^2(k+l'+1)^2}=E(\lambda)=E^0+E^1+\cdots$$

‡ Actually the splitting at a given l is solely due to the spin–orbit interaction. The kinetic energy correction depends only on l and does not contribute to the splitting between the $P_{3/2}$ and $P_{1/2}$ levels.

Show that

$$\left\langle \frac{\lambda}{r^2} \right\rangle = E^1 = \lambda \left. \frac{dE}{d\lambda} \right|_{\lambda=0} = \left(\frac{dE}{dl'} \right)_{l'=l} \cdot \left(\frac{dl'}{d\lambda} \right)_{l'=l} \cdot \lambda = \frac{\lambda}{n^3 a_0^2 (l + \frac{1}{2})}$$

Canceling λ on both sides, we get Eq. (17.3.11).

(3) Consider finally $\langle l/r^3 \rangle$. Since there is no such term in the Coulomb Hamiltonian, we resort to another trick. Consider the *radial* momentum operator, $p_r = -i\hbar(\partial/\partial r + 1/r)$, in terms of which we may write the radial part of the Hamiltonian

$$\left(\frac{-\hbar^2}{2m} \right) \left(\frac{1}{r^2} \frac{\partial}{\partial r} r^2 \frac{\partial}{\partial r} \right)$$

as $p_r^2/2m$. (Verify this.) Using the fact that $\langle [H, p_r] \rangle = 0$ in the energy eigenstates, and by explictly evaluating the commutator, show that

$$\left\langle \frac{1}{r^3} \right\rangle = \frac{1}{a_0 (l)(l+1)} \left\langle \frac{1}{r^2} \right\rangle$$

combining which with the result from part (2) we get Eq. (17.3.20).

(4) Find the mean kinetic energy using the trick from part (1), this time rescaling the mass. Regain the virial theorem.

18

Time-Dependent Perturbation Theory

18.1. The Problem

Except for the problem of magnetic resonance, we have avoided studying phenomena governed by a time-dependent Hamiltonian. Whereas in the time-independent case the problem of solving the equation

$$i\hbar|\dot{\psi}\rangle = H|\psi\rangle \tag{18.1.1}$$

reduced to solving the eigenvalue problem of H, in the time-dependent case a frontal attack on the full time-dependent Schrödinger equation becomes inevitable.

In this chapter we consider the perturbative solution to a class of phenomena described by

$$H(t) = H^0 + H^1(t) \tag{18.1.2}$$

where H^0 is a time-independent piece whose eigenvalue problem has been solved and H^1 is a small time-dependent perturbation. For instance, H^0 could be the hydrogen atom Hamiltonian and H^1 the addition due to a weak external electromagnetic field. Whereas in the time-independent case one is interested in the eigenvectors and eigenvalues of H, the typical question one asks here is the following. If at $t=0$ the system is in the eigenstate $|i^0\rangle$ of H^0, what is the amplitude for it to be in the eigenstate $|f^0\rangle$ $(f\neq i)$ at a later time t? Our goal is to set up a scheme in which the answer may be computed in a perturbation series in powers of H^1. To zeroth order, the answer to the question raised is clearly zero, for the only effect of H^0 is to multiply $|i^0\rangle$ by a phase factor $\exp(-iE_i^0t/\hbar)$, which does not alter its orthogonality to $|f^0\rangle$. But as soon as we let H^1 enter the picture, i.e., work to nonzero order, the eigenstates of H^0 cease to be stationary and $|i^0\rangle$ can evolve into a state with a projection along $|f^0\rangle$.

The next section begins with a simple derivation of the first-order transition amplitude for the process $i \to f$ and is followed by several applications and discussions of special types of perturbations (sudden, adiabatic, periodic, etc.). In Section 3 the expressions for the transition amplitude to any order are derived, following a scheme more abstract than the one used in Section 2. Sections 4 and 5 are concerned with electromagnetic interactions. Section 4 contains a brief summary of relevant concepts from classical electrodynamics, followed by a general discussion of several fine points of the electromagnetic interaction at the classical and quantum levels. It therefore has little to do with perturbation theory. However, it paves the way for the last section, in which first-order perturbation theory is applied to the study of the interaction of atoms with the electromagnetic field. Two illustrative problems are considered, one in which the field is treated classically and the other in which it is treated quantum mechanically.

18.2. First-Order Perturbation Theory

Our problem is to solve Eq. (18.1.1) to first order in H^1. Since the eigenkets $|n^0\rangle$ of H^0 form a complete basis, we can always expand

$$|\psi(t)\rangle = \sum_n c_n(t)|n^0\rangle \tag{18.2.1}$$

To find $c_n(t)$ given $c_n(0)$ is equivalent to finding $|\psi(t)\rangle$ given $|\psi(0)\rangle$. Now $c_n(t)$ changes with time because of H^0 and H^1. Had H^1 been absent, we would know

$$c_n(t) = c_n(0)\, e^{-iE_n^0 t/\hbar} \tag{18.2.2}$$

Let us use this information and write

$$|\psi(t)\rangle = \sum_n d_n(t)\, e^{-iE_n^0 t/\hbar}|n^0\rangle \tag{18.2.3}$$

If d_n changes with time, it is because of H^1. So we expect that the time evolution of d_n can be written in a nice power series in H^1. The equation of motion for $d_f(t)$ is found by operating both sides of Eq. (18.2.3) with $(i\hbar\partial/\partial t - H^0 - H^1)$ to get

$$0 = \sum_n [i\hbar\dot{d}_n - H^1(t)d_n]\, e^{-iE_n^0 t/\hbar}|n^0\rangle \tag{18.2.4}$$

and then dotting with $\langle f^0|\exp(iE_f^0 t/\hbar)$:

$$i\hbar\dot{d}_f = \sum_n \langle f^0|H^1(t)|n^0\rangle\, e^{i\omega_{fn}t} d_n(t) \tag{18.2.5a}$$

where

$$\omega_{fn}=\frac{E_f^0-E_n^0}{\hbar} \tag{18.2.5b}$$

Notice that H^0 has been eliminated in Eq. (18.2.5), which is exact and fully equivalent to Eq. (18.1.1). Let us now consider the case where at $t=0$, the system is in the state $|i^0\rangle$, i.e.,

$$d_n(0)=\delta_{ni} \tag{18.2.6}$$

and ask what $d_f(t)$ is. To *zeroth order*, we ignore the right-hand side of Eq. (18.2.5a) completely, because of the explicit H^1, and get

$$\dot{d}_f=0 \tag{18.2.7}$$

in accordance with our expectations. To *first order*, we use the zeroth-order d_n in the right-hand side because H^1 is itself of first order. This gives us the first-order equation

$$\dot{d}_f(t)=\frac{-i}{\hbar}\langle f^0|H^1(t)|i^0\rangle\, e^{i\omega_{fi}t} \tag{18.2.8}$$

the solution to which, with the right initial conditions, is

$$d_f(t)=\delta_{fi}-\frac{i}{\hbar}\int_0^t \langle f^0|H^1(t')|i^0\rangle\, e^{i\omega_{fi}t'}\,dt' \tag{18.2.9}$$

Since we now know d to first order, we can feed it into the right-hand side of Eq. (18.2.5a) to get an equation for d that is good to second order. Although we can keep going to any desired order in this manner, we stop with the first, since a more compact scheme for calculating transition amplitudes to any desired order will be set up in the next section. At this point we merely note that the first-order calculation is reliable if $|d_f(t)|\ll 1\,(f\neq i)$. If this condition is violated, our calculation becomes internally inconsistent, for we can no longer approximate $d_n(t)$ by δ_{ni} in the right-hand side of Eq. (18.2.5a).

Let us apply our first-order result to a simple problem. Consider a one-dimensional harmonic oscillator in the ground state $|0\rangle$‡ of the unperturbed Hamiltonian at $t=-\infty$. Let a perturbation

$$H^1(t)=-e\mathscr{E}X\, e^{-t^2/\tau^2} \tag{18.2.10}$$

‡ We shall denote the nth unperturbed state by $|n\rangle$ and not $|n^0\rangle$ in this discussion.

be applied between $t=-\infty$ and $+\infty$. What is the probability that the oscillator is in the state $|n\rangle$ at $t=\infty$? According to Eq. (18.2.9), for $n\neq 0$,

$$d_n(\infty)=\frac{-i}{\hbar}\int_{-\infty}^{\infty}(-e\mathscr{E})\langle n|X|0\rangle\, e^{-t^2/\tau^2}\, e^{in\omega t}\, dt \tag{18.2.11}$$

Since

$$X=\left(\frac{\hbar}{2m\omega}\right)^{1/2}(a+a^\dagger)$$

only $d_1(\infty)\neq 0$. We find that it is (using $a^\dagger|0\rangle=|1\rangle$)

$$\begin{aligned}d_1(\infty)&=\frac{ie\mathscr{E}}{\hbar}\left(\frac{\hbar}{2m\omega}\right)^{1/2}\int_{-\infty}^{\infty}e^{-t^2/\tau^2}\, e^{i\omega t}\, dt\\&=\frac{ie\mathscr{E}}{\hbar}\left(\frac{\hbar}{2m\omega}\right)^{1/2}\cdot(\pi\tau^2)^{1/2}\, e^{-\omega^2\tau^2/4}\end{aligned} \tag{18.2.12}$$

Thus the probability of the transition $0\to 1$ is‡

$$P_{0\to 1}=|d_1|^2=\frac{e^2\mathscr{E}^2\pi\tau^2}{2m\omega\hbar}\, e^{-\omega^2\tau^2/2} \tag{18.2.13}$$

This result will be used shortly.

Exercise 18.2.1. Show that if $H^1(t)=-e\mathscr{E}X/[1+(t/\tau)^2]$, then, to first order,

$$P_{0\to 1}=\frac{e^2\mathscr{E}^2\pi^2\tau^2}{2m\omega\hbar}\, e^{-2\omega\tau}$$

*Exercise 18.2.2.** A hydrogen atom is in the ground state at $t=-\infty$. An electric field $\mathbf{E}(t)=(\mathbf{k}\mathscr{E})\, e^{-t^2/\tau^2}$ is applied until $t=\infty$. Show that the probability that the atom ends up in any of the $n=2$ states is, to first order,

$$P(n=2)=\left(\frac{e\mathscr{E}}{\hbar}\right)^2\left(\frac{2^{15}a_0^2}{3^{10}}\right)\pi\tau^2\, e^{-\omega^2\tau^2/2}$$

where $\omega=(E_{2lm}-E_{100})/\hbar$. Does the answer depend on whether or not we incorporate spin in the picture?

We now turn our attention to different types of perturbations.

‡ Since $d_n(t)$ and $c_n(t)$ differ only by a phase factor, $P(n)=|c_n|^2=|d_n|^2$.

The Sudden Perturbation

Consider a system whose Hamiltonian changes abruptly over a small time interval ε. What is the change in the state vector as $\varepsilon\to 0$? We can find the answer without resorting to perturbation theory. Assuming that the change occurred around $t=0$, we get, upon integrating Schrödinger's equation between $t=-\varepsilon/2$ and $\varepsilon/2$,

$$\begin{aligned}|\psi(\varepsilon/2)\rangle-|\psi(-\varepsilon/2)\rangle &= |\psi_{\text{after}}\rangle-|\psi_{\text{before}}\rangle \\ &= \frac{-i}{\hbar}\int_{-\varepsilon/2}^{\varepsilon/2} H(t)|\psi(t)\rangle\, dt \end{aligned} \qquad (18.2.14)$$

Since the integrand on the right-hand side is finite, the integral is of order ε. In the limit $\varepsilon\to 0$, we get

$$|\psi_{\text{after}}\rangle=|\psi_{\text{before}}\rangle \qquad (18.2.15)$$

An instantaneous change in H produces no instantaneous change in $|\psi\rangle$.‡ Now the limit $\varepsilon\to 0$ is unphysical. The utility of the above result lies in the fact that it is an excellent approximation if H changes over a time that is very small compared to the natural time scale of the system. The latter may be estimated semiclassically; several examples follow in a moment. For the present, let us consider the case of an oscillator to which is applied the perturbation in Eq. (18.2.10). It is clear that whatever be the time scale of this system, the change in the state vector must vanish as τ, the width of the Gaussian pulse, vanishes. This means in particular that the system initially in the ground state must remain there after the pulse, i.e., the $0\to 1$ transition probability must vanish. This being an exact result, we expect that if the transition probability is calculated perturbatively, it must vanish to any given order. (This is like saying that if an analytic function vanishes identically, then so does every term in its Taylor expansion.) Turning to the first-order probability for $0\to 1$ in Eq. (18.2.13), we see that indeed it vanishes as τ tends to zero.

A more realistic problem, where ε is fixed, involves a $1s$ electron bound to a nucleus of charge Z which undergoes β decay by emitting a *relativistic* electron and changing its charge to $(Z+1)$. The time the emitted electron takes to get out of the $n=1$ shell is

$$\tau\simeq a_0/Zc \qquad (18.2.16)$$

whereas the characteristic time for the $1s$ electron is

$$T\simeq\frac{\text{size of state}}{\text{velocity of } e^-}\simeq\frac{a_0}{Z}\bigg/ Z\alpha c=\frac{a_0}{Z^2\alpha c} \qquad (18.2.17)$$

so that

$$\tau/T=Z\alpha$$

‡ We are assuming H is finite in the integral $(-\varepsilon/2, \varepsilon/2)$. If it has a delta function spike, it can produce a change in $|\psi\rangle$, see Exercise 18.2.6.

For Z small, we may apply the sudden approximation and conclude that the state of the atomic electron is the same just before and just after β decay. Of course, this state is not an eigenstate of the charge $(Z+1)$ ion, but rather a superposition of such states (see Exercise 18.2.4).

*Exercise 18.2.3.** Consider a particle in the ground state of a box of length L. Argue on semiclassical grounds that the natural time period associated with it is $T \simeq mL^2/\hbar\pi$. If the box expands symmetrically to double its size in time $\tau \ll T$ what is the probability of catching the particle in the ground state of the new box? (See Exercise (5.2.1).)

*Exercise 18.2.4.** In the β decay H^3 (two neutrons + one proton in the nucleus) $\to (He^3)^+$ (two protons + one neuron in the nucleus), the emitted electron has a kinetic energy of 16 keV. Argue that the sudden approximation may be used to describe the response of an electron that is initially in the $1s$ state of H^3. Show that the amplitude for it to be in the ground state of $(He^3)^+$ is $16(2)^{1/2}/27$. What is the probability for it to be in the state

$$|n=16, l=3, m=0\rangle \text{ of } (\mathrm{He}^3)^+?$$

Exercise 18.2.5. An oscillator is in the ground state of $H = H^0 + H^1$, where the time-independent perturbation H^1 is the linear potential $(-fx)$. If at $t=0$, H^1 is abruptly turned off, show that the probability that the system is in the nth eigenstate of H^0 is given by the Poisson distribution

$$P(n) = \frac{e^{-\lambda}\lambda^n}{n!}, \quad \text{where} \quad \lambda = \frac{f^2}{2m\omega^3\hbar}$$

Hint: Use the formula

$$\exp[A+B] = \exp[A]\exp[B]\exp[-\tfrac{1}{2}[A, B]]$$

where $[A, B]$ is a c number.

*Exercise 18.2.6.** Consider a system subject to a perturbation $H^1(t) = H^1\delta(t)$. Show that if at $t=0^-$ the system is in the state $|i^0\rangle$, the amplitude to be in a state $|f^0\rangle$ at $t=0^+$ is, to first order,

$$d_f = \frac{-i}{\hbar}\langle f^0|H^1|i^0\rangle \qquad (f \neq i)$$

Notice that (1) the state of the system *does* change instantaneously; (2) Even though the perturbation is "infinite" at $t=0$, we can still use first-order perturbation theory if the "area under it" is small enough.

The Adiabatic Perturbation

We now turn to the other extreme and consider a system whose Hamiltonian $H(t)$ changes very slowly from $H(0)$ to $H(\tau)$ in a time τ. If the system starts out at $t=0$ in an eigenstate $|n(0)\rangle$ of $H(0)$, where will it end at time τ? The *adiabatic theorem* asserts that if the rate of change of H is slow enough, the system will end up in the corresponding eigenket $|n(\tau)\rangle$ of $H(\tau)$.‡ Rather than derive the theorem and the precise definition of "slow enough" we consider a few illustrative examples.

Consider a particle in a box of length $L(0)$. If the box expands slowly to a length $L(\tau)$, the theorem tells us that a particle that was in the nth state of the box of length $L(0)$ will now be in the nth state of the box of length $L(\tau)$. But how slow is slow enough?

There are two ways to estimate this. The first is a semiclassical method and goes as follows. The momentum of the particle is of the order (dropping factors of order unity like π, n, etc.)

$$p \simeq \frac{\hbar}{L} \tag{18.2.18}$$

and the time it takes to finish one full oscillation is of the order

$$T \simeq \frac{L}{v} = \frac{mL}{p} \simeq \frac{mL^2}{\hbar} \tag{18.2.19}$$

We can say the expansion or contraction is slow if the fractional change in the length of the box per cycle is much smaller than unity:

$$\frac{|\Delta L|_{\text{per cycle}}}{L} \simeq \frac{|dL/dt|mL^2/\hbar}{L} = \frac{mL}{\hbar}\left|\frac{dL}{dt}\right| \ll 1 \tag{18.2.20}$$

This can also be written as

$$\frac{v_{\text{walls}}}{v_{\text{particle}}} \ll 1 \tag{18.2.21}$$

The second approach is less intuitive§ and it estimates T as

$$T \sim \frac{1}{\omega_{\min}} \tag{18.2.22}$$

‡ This is again a result that is true to any given order in perturbation theory. We shall exploit this fact in a moment.

§ The logic behind this approach and its superiority over the intuitive one will become apparent shortly in an example where we recover the results of time-independent perturbation theory from the time-dependent one.

where $\omega_{\min}$ is the smallest of the transition frequencies between the initial state i and any *accessible* final state f‡; it is the smallest of

$$\omega_{fi}=\frac{E_f^0-E_i^0}{\hbar} \tag{18.2.23}$$

In the present case, since $E_n^0=(n^2\hbar^2\pi^2/2mL^2)$, energy differences are of the order $\hbar^2/mL^2$ and

$$T\sim\frac{1}{\omega_{\min}}\simeq\frac{mL^2}{\hbar} \tag{18.2.24}$$

which coincides with Eq. (18.2.19). This is not surprising, for we can also write T in Eq. (18.2.19) as

$$T\simeq\frac{mL^2}{\hbar}\simeq\frac{1}{E_i^0/\hbar}\sim\frac{1}{\omega_i} \tag{18.2.25}$$

Thus T in Eq. (18.2.19) is $\sim\hbar/E_i^0$, while T in eq. (18.2.24) is $\sim\hbar/|E_j^0-E_i^0|_{\min}$. Since the energy levels of a quantum system are all of the same order of magnitude (say a Rydberg or $\hbar\omega$), energies and energy differences are of the same order of magnitude and the two estimates for T are equivalent, unless *the levels are degenerate or nearly so*. In this case, it is $T\sim 1/\omega_{\min}$ that is to be trusted, for it exposes the instability of a degenerate or nearly degenerate system. An explicit example that follows later will illustrate this.

Let us consider one more example of the adiabatic theorem, an oscillator subject to the perturbation

$$H^1(t)=-e\mathscr{E}X\,e^{-t^2/\tau^2} \tag{18.2.26}$$

between $-\infty\le t\le\infty$. We expect that if τ, which measures the time over which H^1 grows from 0 to its peak, tends to infinity, the change in the system will be adiabatic. Thus, if a system starts in the ground state of $H(-\infty)=H^0$ at $t=-\infty$, it will end up in the ground state of $H(\infty)=H(-\infty)=H^0$. Our first-order formula, Eq. (18.2.13), for $P_{0\to 1}$ conforms with this expectation and vanishes exponentially as $\omega\tau\to\infty$. Our formula also tells us what large τ means: it means

$$\omega\tau\gg 1, \qquad \tau\gg 1/\omega \tag{18.2.27}$$

This is what we would expect from the semiclassical estimate or the estimate $T\sim 1/\omega_{\min}$ and the condition $\tau\gg T$.

The adiabatic theorem suggests a way of recovering the results of time-independent perturbation theory from time-dependent theory. Consider a Hamiltonian $H(t)$

‡ This is a state for which $\langle f^0|H^1|i^0\rangle\neq 0$.

which changes continuously from H^0 at $t=-\infty$ to H^0+H^1 at $t=0$:

$$H(t)=H^0+e^{t/\tau}H^1, \qquad -\infty\leq t\leq 0 \tag{18.2.28}$$

As τ, the rise time of the exponential, goes to infinity, the adiabatic theorem assures us that an eigenstate $|n^0\rangle$ of H^0 at $t=-\infty$ will evolve into the eigenstate $|n\rangle$ of H at $t=0$. If we calculate the state at $t=0$ to a *given order* in *time-dependent* theory and let $\tau\to\infty$, we should get the *time-independent* formula for the state $|n\rangle$ *to that order*. To first order, we know that the projection of the state at $t=0$ along $|m^0\rangle$ $(m\neq n)$ is

$$\begin{aligned} d_m(0)&=\frac{-i}{\hbar}\int_{-\infty}^{0}\langle m^0|H^1|n^0\rangle\, e^{t/\tau}\, e^{i\omega_{mn}t}\,dt \\ &=\frac{(-i/\hbar)\langle m^0|H^1|n^0\rangle}{1/\tau+i\omega_{mn}} \end{aligned} \tag{18.2.29}$$

If we now let $\tau\to\infty$, we regain the familiar result

$$\langle m^0|n\rangle=\frac{\langle m^0|H^1|n^0\rangle}{E_n^0-E_m^0} \tag{18.2.30}$$

In practice, $\tau\to\infty$ is replaced by some large τ. Equation (18.2.29) tells us what large τ means: it is defined by

$$|1/\tau|\ll|\omega_{\min}|$$

or

$$\tau\gg 1/\omega_{\min} \tag{18.2.31}$$

Thus we see that $T\simeq 1/\omega_{\min}$ is indeed the reliable measure of the natural time scale of the system. In particular, if the system is degenerate (or nearly so), $T\to\infty$ and it becomes impossible, in practice, to change the state of the system adiabatically.

Let us wind up the discussion on the adiabatic approximation by observing its similarity to the WKB approximation. The former tells us that if the Hamiltonian changes *in time* from H^0 to H^0+H^1, the eigenstate $|n^0\rangle$ evolves smoothly into its counterpart $|n\rangle$ in the limit $\tau/T\to\infty$, where τ is the duration over which the Hamiltonian changes and T is the *natural time scale* for the system. The latter tells us that if the potential changes *in space* from V^0 to V^1, a plane wave of momentum $p^0=[2m(E-V^0)]^{1/2}$ evolves smoothly into a plane wave of momentum $p^1=[2m(E-V^1)]^{1/2}$ in the limit $L/\lambda\to\infty$, where L is the *length* over which V changes and $\lambda=2\pi\hbar/p$ is *natural length scale* for the system.

We shall return to adiabatic evolutions in Chapter 21.

The Periodic Perturbation

Consider a system that is subject to a periodic perturbation, say an atom placed between the plates of a condenser connected to an alternating current (ac) source or in the way of a monochromatic light beam. While in reality these perturbations vary as sines and cosines, we consider here the case

$$H^1(t)=H^1\,e^{-i\omega t} \tag{18.2.32}$$

Which is easier to handle mathematically. The sines and cosines can be handled by expressing them in terms of exponentials.

Let us say the system comes into contact with this perturbation at $t=0$. The amplitude for transition from $|i^0\rangle$ to $|f^0\rangle$ in time t $(i\neq f)$ is

$$d_f(t)=\left(\frac{-i}{\hbar}\right)\int_0^t \langle f^0|H^1|i^0\rangle\, e^{i(\omega_{fi}-\omega)t'}\,dt' \tag{18.2.33}$$

$$=\frac{-i}{\hbar}\langle f^0|H^1|i^0\rangle\frac{e^{i(\omega_{fi}-\omega)t}-1}{i(\omega_{fi}-\omega)} \tag{18.2.34}$$

The probability for the transition $i\to f$ is

$$P_{i\to f}=|d_f|^2=\frac{1}{\hbar^2}|\langle f^0|H^1|i^0\rangle|^2\left\{\frac{\sin[(\omega_{fi}-\omega)t/2]}{(\omega_{fi}-\omega)\frac{1}{2}t}\right\}^2 t^2 \tag{18.2.35}$$

Since the function $(\sin^2 x)/x^2$ is peaked at the origin and has a width $\Delta x\simeq\pi$, we find that the system likes to go to states f such that

$$|(\omega_{fi}-\omega)t/2|\lesssim\pi$$

or

$$E_f^0 t=(E_i^0 t+\hbar\omega t)\pm 2\hbar\pi$$

or

$$E_f^0-E_i^0=\hbar\omega\pm\frac{2\hbar\pi}{t}=\hbar\omega\left(1\pm\frac{2\pi}{\omega t}\right) \tag{18.2.36}$$

For small t, the system shows no particular preference for the level with $E_f^0=E_i^0+\hbar\omega$. Only when $\omega t\gg 2\pi$ does it begin to favor $E_f^0=E_i^0+\hbar\omega$. The reason is simple. You and I know the perturbation has a frequency ω, say, because we set the dial on the ac source or tuned our laser to frequency ω. But the system goes by what it knows, starting from the time it made contact with the perturbation. In the beginning, it will not even know it is dealing with a periodic perturbation; it must wait a few cycles to get the message. Thus it can become selective only after a few cycles, i.e., after $\omega t\gg 2\pi$. What does it do meanwhile? It Fourier-analyzes the pulse into its frequency components and its transition amplitude to a state with $E_f^0=E_i^0+\hbar\omega_{fi}$ is proportional to the Fourier component at $\omega=\omega_{fi}$. The t' integral in Eq. (18.2.33) is precisely this Fourier transform.‡

What happens if we wait a long time? To find out, we consider the case of a system exposed to the perturbation from $t=-T/2$ to $T/2$ and let $T\to\infty$. Equation (18.2.33) becomes

$$d_f=\lim_{T\to\infty}\frac{-i}{\hbar}\int_{-T/2}^{T/2} H_{fi}^1\, e^{i(\omega_{fi}-\omega)t'}\,dt' \tag{18.2.37}$$

$$=\frac{-2\pi i}{\hbar}H_{fi}^1\delta(\omega_{fi}-\omega) \tag{18.2.38}$$

and

$$P_{i\to f}=\frac{4\pi^2}{\hbar^2}|H_{fi}^1|^2\delta(\omega_{fi}-\omega)\delta(\omega_{fi}-\omega) \tag{18.2.39}$$

We handle the product of δ functions as follows:

$$\delta\delta=\lim_{T\to\infty}\delta(\omega_{fi}-\omega)\frac{1}{2\pi}\int_{-T/2}^{T/2} e^{i(\omega_{fi}-\omega)t}\,dt \tag{18.2.40}$$

Since the δ function in front of the integral vanishes unless $\omega_{fi}=\omega$, we may set $\omega_{fi}=\omega$ in the integral to obtain

$$\delta\delta=\delta(\omega_{fi}-\omega)\lim_{T\to\infty}\frac{T}{2\pi} \tag{18.2.41}$$

Feeding this into Eq. (18.2.39) for $P_{i\to f}$, and dividing by T, we get the *average transition rate*:

$$R_{i\to f}=\frac{P_{i\to f}}{T}=\frac{2\pi}{\hbar}|\langle f^0|H^1|i^0\rangle|^2\delta(E_f^0-E_i^0-\hbar\omega) \tag{18.2.42}$$

This is called *Fermi's golden rule* and has numerous applications, some of which will be discussed later in this chapter and in the next chapter. You may be worried about the δ function in $R_{i\to f}$ and in particular whether first-order perturbation theory is to be trusted when the rate comes out infinite! As we will see, in all practical applications the δ function will get integrated over for one reason or another. The validity of the first-order formula will then depend only on the area under the δ function. (Recall Exercise 18.2.6.)

‡ The inability of a system to assign a definite frequency to an external perturbation until many cycles have elapsed is a purely classical effect. The quantum mechanics comes in when we relate frequency to energy.

18.3. Higher Orders in Perturbation Theory‡

In Section 18.2 we derived a formula for the transition amplitude from $|i^0\rangle$ to $|f^0\rangle$ to first order in perturbation theory. The procedure for going to higher orders was indicated but not pursued. We address that problem here, using a more abstract formalism, desirable for its compactness and the insight it gives us into the anatomy of the perturbation series.

The basic idea behind the approach is the same as in Section 18.2: we want to isolate the time evolution generated by H^1, for H^0 by itself causes no transitions between its own eigenstates $|i^0\rangle$ and $|f^0\rangle$. To do this, we must get acquainted with other *equivalent* descriptions of quantum dynamics besides the one we have used so far. The description we are familiar with is called the *Schrödinger picture*. In this picture the state of the particle is described by a vector $|\psi_S(t)\rangle$. (We append a subscript S to all quantities that appear in the Schrödinger picture to distinguish them from their counterparts in other pictures.) The physics is contained in the inner products $\langle\omega_S|\psi_S(t)\rangle$ which give the probabilities

$$P(\omega, t)=|\langle\omega_S|\psi_S(t)\rangle|^2 \tag{18.3.1}$$

for obtaining the result ω when Ω is measured. Here $|\omega_S\rangle$ is the normalized eigenket of the operator $\Omega_S(X_S, P_S)$ with eigenvalue ω. Since X_S and P_S are time independent so are Ω_S and $|\omega_S\rangle$. Thus the physics is contained in the dot product of the moving ket $|\psi_S(t)\rangle$ with the stationary kets $|\omega_S\rangle$.

The time evolution of $|\psi_S(t)\rangle$ is given in general by

$$i\hbar\frac{d}{dt}|\psi_S(t)\rangle=H_S|\psi_S(t)\rangle \tag{18.3.2a}$$

and in our problem by

$$i\hbar\frac{d}{dt}|\psi_S(t)\rangle=[H_S^0+H_S^1(t)]|\psi_S(t)\rangle \tag{18.3.2b}$$

The expectation values change according to

$$i\hbar\frac{d}{dt}\langle\Omega_S\rangle=\langle[\Omega_S, H_S]\rangle \tag{18.3.3}$$

If we define a propagator $U_S(t, t_0)$ by

$$|\psi_S(t)\rangle=U_S(t, t_0)|\psi_S(t_0)\rangle \tag{18.3.4}$$

‡ This section may be skimmed through by a reader pressured for time.

it follows from Eq. (18.3.2) [because $|\psi_S(t_0)\rangle$ is arbitrary] that

$$i\hbar\frac{dU_S}{dt}=H_S U_S \tag{18.3.5}$$

Here are some formulas (true for all propagators U) that will be useful in what follows (recall Eq. (4.3.16)):

$$\begin{aligned} U^\dagger U &= I \\ U(t_3,t_2)U(t_2,t_1) &= U(t_3,t_1) \\ U(t_1,t_1) &= I \\ U^\dagger(t_1,t_2) &= U(t_2,t_1) \end{aligned} \tag{18.3.6}$$

The Interaction Picture

Since $U_S(t,t_0)$ is a unitary operator, which is the generalization of the rotation operator to complex spaces, we may describe the time evolution of state vectors as "rotations" in Hilbert space.‡ The rotation is generated by $U_S(t,t_0)$ or equivalently, by $H_S(t)=H_S^0+H_S^1(t)$. Imagine for a moment that H_S^1 is absent. Then the rotation will be generated by $U_S^0(t)$, which obeys

$$i\hbar\frac{dU_S^0}{dt}=H_S^0U_S^0 \tag{18.3.7}$$

the formal solution to which is $U_S^0(t,t_0)=e^{-iH_S^0(t-t_0)/\hbar}$. If $H_S^1(t)$ is put back in, both H_S^0 and $H_S^1(t)$ jointly produce the rotation U_S.

These pictorial arguments suggest a way to freeze out the time evolution generated by H_S^0. Suppose we switch to a frame that rotates at a rate that U_S^0 (or H_S^0) by itself generates. In this frame the state vector moves because $H_S^1\neq 0$. Let us verify this conjecture. To neutralize the rotation induced by U_S^0, i.e., to see things from the rotating frame, we multiply $|\psi_S(t)\rangle$ by $(U_S^0)^\dagger$ to get

$$|\psi_I(t)\rangle=[U_S^0(t,t_0)]^\dagger|\psi_S(t)\rangle \tag{18.3.8a}$$

The ket $|\psi_I(t)\rangle$ is the state vector in the rotating frame, or in the *interaction picture.* If we set $t=t_0$ in the above equation, we get

$$|\psi_I(t_0)\rangle=|\psi_S(t_0)\rangle \tag{18.3.8b}$$

‡ In this section we use the word "rotation" in this generalized sense, and not in the sense of a spatial rotation.

i.e., the interaction and Schrödinger kets coincide at $t=t_0$, which is that instant we switch to the moving frame. The time evolution of $|\psi_I(t)\rangle$ is as follows‡:

$$
\begin{aligned}
i\hbar \frac{d}{dt}|\psi_I(t)\rangle &= i\hbar \frac{dU_S^{0\dagger}}{dt}|\psi_S\rangle + U_S^{0\dagger} i\hbar \frac{d|\psi_S\rangle}{dt} \\
&= -U_S^{0\dagger} H_S^0 |\psi_S\rangle + U_S^{0\dagger}(H_S^0 + H_S^1)|\psi_S\rangle \\
&= U_S^{0\dagger} H_S^1 |\psi_S\rangle \\
&= U_S^{0\dagger} H_S^1 U_S^0 U_S^{0\dagger} |\psi_S\rangle \\
&= U_S^{0\dagger} H_S^1 U_S^0 |\psi_I(t)\rangle
\end{aligned}
$$

Now

$$
(U_S^0)^\dagger H_S^1(t) U_S^0 = H_I^1(t) \tag{18.3.9}
$$

is the perturbing Hamiltonian as seen in the rotating frame. So we can write

$$
i\hbar \frac{d}{dt}|\psi_I(t)\rangle = H_I^1(t)|\psi_I(t)\rangle \tag{18.3.10}
$$

So, as we anticipated, the time evolution of the state vector in the interaction picture is determined by the perturbing Hamiltonian, H_I^1. Despite the fact that the state vector now rotates at a different rate, the physical predictions are the same as in the Schrödinger picture. This is because $P(\omega, t)$ depends only on the inner product between the state vector and the eigenket of Ω with eigenvalue ω, and the inner product between two vectors is unaffected by going to a rotating frame. However, both the state vector and the eigenket appear different in the interaction picture. Just as

$$
|\psi_S(t)\rangle \to U_S^{0\dagger}(t, t_0)|\psi_S(t)\rangle = |\psi_I(t)\rangle
$$

so does

$$
|\omega_S\rangle \to U_S^{0\dagger}(t, t_0)|\omega_S\rangle = |\omega_I(t)\rangle \tag{18.3.11}
$$

However,

$$
\langle \omega_S|\psi_S(t)\rangle = \langle \omega_I(t)|\psi_I(t)\rangle \tag{18.3.12}
$$

The time-dependent ket $|\omega_I(t)\rangle$ is just the eigenket of the time-dependent operator

$$
\Omega_I(t) = U_S^{0\dagger}\Omega_S U_S^0 \tag{18.3.13}
$$

which is just Ω as seen in the rotating frame:

$$
\Omega_I(t)|\omega_I(t)\rangle = U_S^{0\dagger}\Omega_S U_S^0 U_S^{0\dagger}|\omega_S\rangle = U_S^{0\dagger}\Omega_S|\omega_S\rangle = \omega|\omega_I(t)\rangle \tag{18.3.14}
$$

‡ Whenever the argument of any U is suppressed, it may be assumed to be (t, t_0).

The time dependence of Ω_I may be calculated by combining Eq. (18.3.13), which defines it, and Eq. (18.3.7), which gives the time evolution of U_S^0:

$$i\hbar \frac{d\Omega_I}{dt} = i\hbar \frac{dU_S^{0\dagger}}{dt} \Omega_S U_S^0 + U_S^{0\dagger} \Omega_S i\hbar \frac{dU_S^0}{dt}$$
$$= U_S^{0\dagger}[\Omega_S, H_S^0] U_S^0 = [\Omega_I, H_I^0] \qquad (18.3.15)$$

In the interaction picture, the operators evolve in response to the unperturbed Hamiltonian H_I^0.‡ Whereas in the Schrödinger picture, the entire burden of time evolution lies with the state vectors, in this picture it is shared by the state vectors and the operators (in such a way that the physics is the same).

Let us now address the original problem, of obtaining a perturbation series for the transition amplitude. We define a propagator $U_I(t, t_0)$ in the interaction picture:

$$|\psi_I(t)\rangle = U_I(t, t_0)|\psi_I(t_0)\rangle \qquad (18.3.16)$$

which, because of Eq. (18.3.10), obeys

$$i\hbar \frac{dU_I}{dt} = H_I^1 U_I \qquad (18.3.17)$$

Once we find $U_I(t)$, we can always go back to $U_S(t)$ by using

$$U_S(t, t_0) = U_S^0(t, t_0) U_I(t, t_0) \qquad (18.3.18)$$

which follows from Eqs. (18.3.8) and (18.3.16).

Since H_I^1 depends on time, the solution to Eq. (18.3.17) is not $U_I = \exp(-iH_I^1(t-t_0)/\hbar)$. A formal solution, with the right initial condition, is

$$U_I(t, t_0) = I - \frac{i}{\hbar} \int_{t_0}^{t} H_I(t') U_I(t', t_0)\, dt' \qquad (18.3.19)$$

as may be readily verified by feeding it into the differential equation. Since U_I occurs on both sides, this is not really a solution, but an *integral equation*, equivalent to the differential equation (18.3.17), with the right initial condition built in. So we have not got anywhere in terms of the exact solution. But the integral equation provides a nice way to carry out the perturbation expansion. Suppose we want U_I to zeroth order. We drop anything with an H_I^1 in Eq. (18.3.19):

$$U_I(t, t_0) = I + O(H_I^1) \qquad (18.3.20)$$

‡ Actually, $H_I^0 = U_S^{0\dagger} H_S^0 U_S^0 = H_S^0$ since $[H_S^0, U_S^0] = 0$ in this problem.

This is to be expected, for if we ignore H_I^1, the state vectors do not move in the interaction picture.

To first order, we keep only the first power of H_I^1. So we use the zeroth-order value for U_I in the right-hand side of Eq. (18.3.20) to get

$$U_I(t,t_0) = I - \frac{i}{\hbar}\int_{t_0}^{t} H_I^1(t')dt' + O(H_I^2) \tag{18.3.21}$$

Before going to the next order, let us compare this with Eq. (18.2.9) for the transition amplitude $d_f(t)$, computed to first order. Recall the definition of $d_f(t)$: it is the projection along $\langle f_S^0|\exp[iE_f^0(t-t_0)/\hbar]$ at time t, of a state that was initially (at $t = t_0$) $|i_S^0\rangle$‡:

$$d_f(t) = \langle f_S^0|e^{iE_f^0(t-t_0)/\hbar}U_S(t,t_0)|i_S^0\rangle \tag{18.3.22}$$

$$= \langle f_S^0|U_S^{0\dagger}(t,t_0)U_S(t,t_0)|i_S^0\rangle$$

$$= \langle f_S^0|U_I(t,t_0)|i_S^0\rangle \tag{18.3.23}$$

If we feed into this our first-order propagator given in Eq. (18.3.21), we get

$$\begin{aligned} d_f(t) &= \langle f_S^0|U_I(t,t_0)|i_S^0\rangle \\ &= \delta_{fi} - \frac{i}{\hbar}\int_{t_0}^{t}\langle f_S^0|H_I^1(t')|i_S^0\rangle dt' \\ &= \delta_{fi} - \frac{i}{\hbar}\int_{t_0}^{t}\langle f_S^0|U_S^{0\dagger}(t',t_0)H_S^1U_S^0(t',t_0)|i_S^0\rangle dt' \\ &= \delta_{fi} - \frac{i}{\hbar}\int_{t_0}^{t}(H_S^1)_{fi}e^{i\omega_{fi}(t'-t_0)}dt' \end{aligned} \tag{18.3.24}$$

which agrees with Eq. (18.2.9) if we set $t_0 = 0$.

Let us now turn to higher orders. By repeatedly feeding into the right-hand side of Eq. (18.3.19) the result for U_I to a known order, we can get U_I to higher orders:

$$\begin{aligned} U_I(t,t_0) = I &- \frac{i}{\hbar}\int_{t_0}^{t} H_I^1(t')\,dt' + (-i/\hbar)^2\int_{t_0}^{t}\int_{t_0}^{t'} H_I^1(t')H_I^1(t'')\,dt'\,dt'' \\ &+ (-i/\hbar)^3\int_{t_0}^{t}\int_{t_0}^{t'}\int_{t_0}^{t''} H_I^1(t')H_I^1(t'')H_I^1(t''')dt'\,dt''\,dt''' + \cdots \end{aligned} \tag{18.3.25}$$

‡ We have set $t_0 = 0$ in Section 18.2.

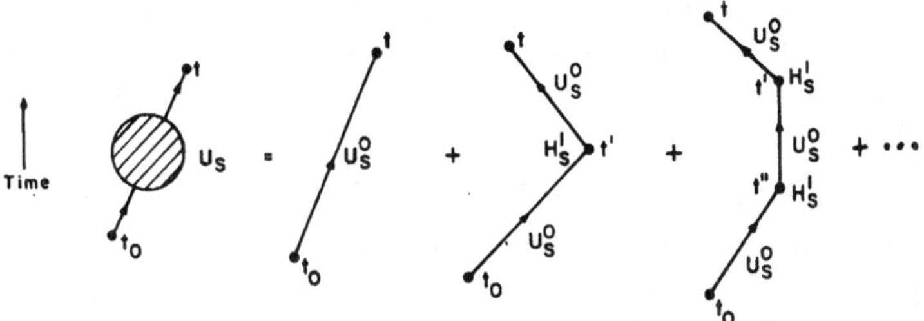

Figure 18.1. A pictorial representation of the perturbation series. The hatched circle represents the full propagator between times t_0 and t. The hatched circle is a sum of many terms, each of which corresponds to a different number of interactions with the perturbation, H_S^1. Between such interactions, the particle evolves in response to just H_S^0, i.e., is propagated by U_S^0.

Premultiplying by $U_S^0(t, t_0)$ and expressing H_I^1 in terms of H_S^1, we get the Schrödinger picture propagator

$$\begin{aligned} U_S(t, t_0) = U_S^0(t, t_0) &- \frac{i}{\hbar}\int_{t_0}^{t} U_S^0(t, t_0)U_S^{0\dagger}(t', t_0)H_S^1 U_S^0(t', t_0)\, dt' \\ &+ (-i/\hbar)^2 \int_{t_0}^{t}\int_{t_0}^{t'} U_S^0(t, t_0)U_S^{0\dagger}(t', t_0)H_S^1 U_S^0(t', t_0)U_S^{0\dagger}(t'', t_0) \\ &\times H_S^1 U_S^0(t'', t_0)\, dt'\, dt'' + \cdots \end{aligned} \tag{18.3.26}$$

$$\begin{aligned} U_S(t, t_0) = U_S^0(t, t_0) &- \frac{i}{\hbar}\int_{t_0}^{t} U_S^0(t, t')H_S^1 U_S^0(t', t_0)\, dt' \\ &+ (-i/\hbar)^2 \int_{t_0}^{t}\int_{t_0}^{t'} U_S^0(t, t')H_S^1 U_S^0(t', t'')H_S^1 U_S^0(t'', t_0)\, dt'\, dt'' + \cdots \end{aligned}$$

The above series could be described by the following words. On the left-hand side we have the complete Schrödinger picture propagator and on the right-hand side a series expansion for it. The first term says the system evolves from t_0 to t in response to just U_S^0, i.e., in response to H_S^0. The second term, if we read it from right to left (imagine it acting on some initial state) says the following: the system evolves from t_0 to t': in response to U_S^0, there it interacts once with the perturbation and thereafter responds to U_S^0 alone until time t. The integral over t' sums over the possible times at which the single encounter with H_S^1 could have taken place. The meaning of the next and higher terms is obvious. These are represented schematically in Fig. 18.1.

If we consider specifically the transition from the state $|i^0\rangle$ to $|f^0\rangle$ (we drop the subscript S everywhere) we get

$$\begin{aligned} \langle f^0|U(t, t_0)|i^0\rangle = \delta_{fi}\, e^{-iE_f^0(t-t_0)/\hbar} &+ \frac{-i}{\hbar}\int_{t_0}^{t} e^{-iE_f^0(t-t')/\hbar}\langle f^0|H^1|i^0\rangle\, e^{-iE_i^0(t'-t_0)/\hbar}\, dt' \\ &+ \left(\frac{-i}{\hbar}\right)^2 \int_{t_0}^{t}\int_{t_0}^{t'} \sum_n e^{-iE_f^0(t-t')/\hbar}\langle f^0|H^1|n^0\rangle \\ &\times e^{-iE_n^0(t'-t'')/\hbar}\langle n^0|H^1|i^0\rangle\, e^{-iE_i^0(t''-t_0)/\hbar}\, dt'\, dt'' + \cdots \end{aligned} \tag{18.3.27}$$

upon introducing a complete set of eigenstates of H^0 in the second-order term. The meaning of the first term is obvious. The second (reading right to left) says that between t_0 and t' the eigenstate $|i^0\rangle$ picks up just a phase (i.e., responds to H_S^0 alone). At t' it meets the perturbation, which has an amplitude $\langle f^0|H^1|i^0\rangle$ of converting it to the state $|f^0\rangle$. Thereafter it evolves as the eigenstate $|f^0\rangle$ until time t. The total amplitude to end up in $|f^0\rangle$ is found by integrating over the times at which the conversion could have taken place. Thus the first-order transition corresponds to a one-step process $i\to f$. At the second order, we see a sum over a complete set of states $|n^0\rangle$. It means the system can go from $|i^0\rangle$ to $|f^0\rangle$ via any *intermediate or virtual* state $|n^0\rangle$ that H^1 can knock $|i^0\rangle$ into. Thus the second-order amplitude describes a two-step process, $i\to n\to f$. Higher-order amplitudes have a similar interpretation.

The Heisenberg Picture

It should be evident that there exist not just two, but an infinite number of pictures, for one can go to frames rotating at various speeds. Not all these are worthy of study, however. We conclude this section with one picture that is very important, namely, the *Heisenberg picture*. In this picture, one freezes out the *complete time dependence* of the state vector. The Heisenberg state vector is

$$|\psi_H(t)\rangle = U_S^\dagger(t, t_0)|\psi_S(t)\rangle = |\psi_S(t_0)\rangle \tag{18.3.28}$$

The operators in this picture are

$$\Omega_H(t) = U_S^\dagger \Omega_S U_S \tag{18.3.29}$$

and obey

$$i\hbar\frac{d\Omega_H}{dt} = [\Omega_H, H_H] \tag{18.3.30}$$

*Exercise 18.3.1.** Derive Eq. (18.3.30).

Thus in the Heisenberg picture, the state vectors are fixed and the operators carry the full time dependence. (Since the interaction picture lies between this Heisenberg picture and the Schrödinger picture, in that the operators and the state vectors share the time dependence, it is also called the *intermediate picture*. Another name for it is the *Dirac picture*.)

Notice the similarity between Eq. (18.3.30) and the classical equation

$$\frac{d\omega}{dt} = \{\omega, \mathscr{H}\} \tag{18.3.31}$$

The Heisenberg picture displays the close formal similarity between quantum and classical mechanics: to every classical variable ω there is a quantum operator Ω_H, which obeys similar equations; all we need to do is make the usual substitution $\omega\to\Omega$, $\{\ \}\to(-i/\hbar)[\ ,\ ]$. The similarity between Eqs. (18.3.30) and (18.3.31) is even more striking if we actually evaluate the commutators and Poisson bracket (PB). Consider, for example, the problem of the oscillator for which

$$H_H = \frac{P_H^2}{2m} + \frac{1}{2} m\omega^2 X_H^2 \tag{18.3.32}$$

Since X_H, P_H are obtained from X_S, P_S by a unitary transformation, they satisfy the same commutation rules

$$[X_H(t), P_H(t)] = U_S^\dagger(t, t_0)[X_S, P_S]U_S(t, t_0) = U_S^\dagger i\hbar I U_S = i\hbar I \tag{18.3.33}$$

Note that the time arguments must be equal in X_H and P_H. Hence Eq. (18.3.33) is called the *equal-time commutation relation*. From Eq. (18.3.30),

$$\dot{X}_H = -\frac{i}{\hbar}[X_H, H_H] = \left(-\frac{i}{\hbar}\right)\frac{i\hbar P_H}{m} = \frac{P_H}{m} \tag{18.3.34a}$$

and likewise

$$\dot{P}_H = -m\omega^2 X_H \tag{18.3.34b}$$

which are identical in form to the classical equations

$$\begin{aligned} \dot{x} &= \frac{\partial \mathscr{H}}{\partial p} = \frac{p}{m} \\ \dot{p} &= -\frac{\partial \mathscr{H}}{\partial x} = -m\omega^2 x \end{aligned} \tag{18.3.35}$$

This is to be expected, because the recipe for quantizing is such that commutators and PB always obey the correspondence [recall Eq. (7.4.40)]

$$\{\omega, \lambda\} = \gamma \rightarrow -\frac{i}{\hbar}[\Omega, \Lambda] = \Gamma \tag{18.3.36}$$

Although the Heisenberg picture is not often used in nonrelativistic quantum mechanics, it is greatly favored in relativistic quantum field theory.

Exercise 18.3.2. In the paramagnetic resonance problem Exercise 14.4.3 we moved to a frame rotating in *real space*. Show that this is also equivalent to a Hilbert space rotation, but that it takes us neither to the interaction nor the Heisenberg picture, except at resonance. What picture is it at resonance? (If $\mathbf{B} = B_0\mathbf{k} + B\cos\omega t\,\mathbf{i} - B\sin\omega t\,\mathbf{j}$, associate B_0 with H_S^0 and B with H_S^1.)

18.4. A General Discussion of Electromagnetic Interactions

This section contains a summary of several concepts from electro-dynamics that are relevant for the next section. It also deals with certain subtle questions of basic interest, not directly linked to the rest of this chapter.

Classical Electrodynamics

Let us begin with an extremely concise review of this subject.‡ The response of matter to the electromagnetic field is given by the Lorentz force on a charge q:

$$\mathbf{F}=q\left(\mathbf{E}+\frac{\mathbf{v}}{c}\times\mathbf{B}\right) \tag{18.4.1}$$

The response of the fields to the charges is given by Maxwell's equations:

$$\nabla\cdot\mathbf{E}=4\pi\rho \tag{18.4.2}$$

$$\nabla\times\mathbf{E}+\frac{1}{c}\frac{\partial\mathbf{B}}{\partial t}=0 \tag{18.4.3}$$

$$\nabla\cdot\mathbf{B}=0 \tag{18.4.4}$$

$$\nabla\times\mathbf{B}-\frac{1}{c}\frac{\partial\mathbf{E}}{\partial t}=\frac{4\pi}{c}\mathbf{j} \tag{18.4.5}$$

where ρ and $\mathbf{j}$ are the charge and current densities bound by the continuity equation

$$\nabla\cdot\mathbf{j}+\frac{\partial\rho}{\partial t}=0 \tag{18.4.6}$$

Exercise 18.4.1. By taking the divergence of Eq. (18.4.5) show that the continuity equation must be obeyed if Maxwell's equations are to be mutually consistent.

The potentials $\mathbf{A}$ and ϕ are now introduced as follows. Equation (18.4.4), combined with the identity $\nabla\cdot\nabla\times\mathbf{A}\equiv0$, tells us that $\mathbf{B}$ can be written as a curl

$$\mathbf{B}=\nabla\times\mathbf{A} \tag{18.4.7}$$

‡ For any further information see the classic, *Classical Electrodynamics* by J. D. Jackson, Wiley, New York (1975).

Feeding this into Eq. (18.4.3), we find that

$$\nabla\times\left(\mathbf{E}+\frac{1}{c}\frac{\partial\mathbf{A}}{\partial t}\right)=0 \tag{18.4.8}$$

Based on the identity $\nabla\times\nabla\phi\equiv 0$, we deduce that $\mathbf{E}+(1/c)\partial\mathbf{A}/\partial t$ can be written as a gradient, or that

$$\mathbf{E}=-\frac{1}{c}\frac{\partial\mathbf{A}}{\partial t}-\nabla\phi \tag{18.4.9}$$

If we replace **E** and **B** by the potentials in the other two Maxwell equations and use the identity $\nabla\times\nabla\times\mathbf{A}\equiv\nabla(\nabla\cdot\mathbf{A})-\nabla^2\mathbf{A}$ (true in Cartesian coordinates) we get the equations giving the response of **A** and ϕ to the charges and currents:

$$\nabla^2\phi+\frac{1}{c}\frac{\partial}{\partial t}(\nabla\cdot\mathbf{A})=-4\pi\rho \tag{18.4.10}$$

$$\nabla^2\mathbf{A}-\frac{1}{c^2}\frac{\partial^2\mathbf{A}}{\partial t^2}-\nabla\left(\nabla\cdot\mathbf{A}+\frac{1}{c}\frac{\partial\phi}{\partial t}\right)=-\frac{4\pi\mathbf{j}}{c} \tag{18.4.11}$$

Before attacking these equations, let us note that there exists a certain arbitrariness in the potentials **A** and ϕ, in that it is possible to change them (in a certain way) without changing anything physical. It may be readily verified that **A** and ϕ and

$$\mathbf{A}'=\mathbf{A}-\nabla\Lambda \tag{18.4.12}$$

$$\phi'=\phi+\frac{1}{c}\frac{\partial\Lambda}{\partial t} \tag{18.4.13}$$

where Λ is an arbitrary function, lead to the same fields **E** and **B**.

*Exercise 18.4.2.** Calculate **E** and **B** corresponding to (**A**, ϕ) and (**A**′, ϕ') using Eqs. (18.4.7) and (18.4.9) and verify the above claim.

Since the physics, i.e., the force law and Maxwell's equations, is sensitive only to **E** and **B**, the transformation of the potentials, called a *gauge transformation*, does not affect it. This is known as *gauge invariance*, Λ is called the *gauge parameter*, and (**A**, ϕ) and (**A**′, ϕ') are called *gauge transforms* of each other, or said to be *gauge equivalent*.

Gauge invariance may be exploited to simplify Eqs. (18.4.10) and (18.4.11). We consider the case of the free electromagnetic field ($\rho=\mathbf{j}=0$), which will be of interest in the next section. In this case the gauge freedom allows us (see following exercise) to choose **A** and ϕ such that

$$\nabla\cdot\mathbf{A}=0 \tag{18.4.14}$$

$$\phi=0 \tag{18.4.15}$$

This is called the *Coulomb gauge* and will be used hereafter. There is no residual gauge freedom if we impose the above Coulomb gauge conditions and the requirement that $|\mathbf{A}|\to 0$ at spatial infinity. The potential in the Coulomb gauge is thus unique and "physical" in the sense that for a given **E** and **B** there is a unique **A**.

*Exercise 18.4.3.** Suppose we are given some **A** and ϕ that do not obey the Coulomb gauge conditions. Let us see how they can be transformed to the Coulomb gauge.

(1) Show that if we choose

$$\Lambda(\mathbf{r}, t)=-c\int_{-\infty}^{t}\phi(\mathbf{r}, t')\,dt'$$

and transform to $(\mathbf{A}', \phi')$ then $\phi'=0$. $\mathbf{A}'$ is just $\mathbf{A}-\nabla\Lambda$, with $\nabla\cdot\mathbf{A}'$ not necessarily zero.

(2) Show that if we gauge transform once more to $(\mathbf{A}'', \phi'')$ via

$$\Lambda'=-\frac{1}{4\pi}\int\frac{\nabla\cdot\mathbf{A}'(\mathbf{r}', t)\,d^3\mathbf{r}'}{|\mathbf{r}-\mathbf{r}'|}$$

then $\nabla\cdot\mathbf{A}''=0$. [Hint: Recall $\nabla^2(1/|\mathbf{r}-\mathbf{r}'|)=-4\pi\delta^3(\mathbf{r}-\mathbf{r}')$.]

(3) Verify that ϕ'' is also zero by using $\nabla\cdot\mathbf{E}=0$.

(4) Show that if we want to make any further gauge transformations *within* the Coulomb gauge, Λ must be time independent and obey $\nabla^2\Lambda=0$. If we demand that $|\mathbf{A}|\to 0$ at spatial infinity, **A** becomes unique.

In the Coulomb gauge, the equations of motion for the electromagnetic field (away from charges) simplify to

$$\nabla^2\mathbf{A}-\frac{1}{c^2}\frac{\partial^2\mathbf{A}}{\partial t^2}=0 \tag{18.4.16a}$$

$$\nabla\cdot\dot{\mathbf{A}}=0 \tag{18.4.16b}$$

$$\nabla\cdot\mathbf{A}=0 \tag{18.4.16c}$$

The first equation tells us that electromagnetic waves travel at the speed c. Of special interest to us are solutions to these equations of the form‡

$$\mathbf{A}=\mathbf{A}_0\cos(\mathbf{k}\cdot\mathbf{r}-\omega t) \tag{18.4.17}$$

‡ Here **k** denotes the wave vector and not the unit vector along the z axis.

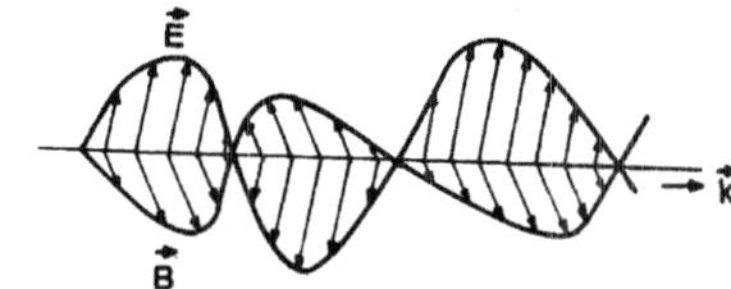

Figure 18.2. The electromagnetic wave at a given time. **E**, **B**, and **k** (the wave vector) are mutually perpendicular.

Feeding this into the wave equation we find

$$\omega^2 = k^2c^2$$

or

$$\omega = kc \tag{18.4.18}$$

The gauge condition tells us that

$$0 = \nabla \cdot \mathbf{A} = -(\mathbf{k} \cdot \mathbf{A}_0) \sin(\mathbf{k} \cdot \mathbf{r} - \omega t)$$

or

$$\mathbf{k} \cdot \mathbf{A}_0 = 0 \tag{18.4.19}$$

This means that **A** must lie in a plane perpendicular to the direction of propagation, i.e., that electromagnetic waves are *transverse*. The electric and magnetic fields corresponding to this solution are

$$\mathbf{E} = -\frac{1}{c}\frac{\partial \mathbf{A}}{\partial t} = -\left(\frac{\omega}{c}\right)\mathbf{A}_0 \sin(\mathbf{k} \cdot \mathbf{r} - \omega t) \tag{18.4.20}$$

$$\mathbf{B} = \nabla \times \mathbf{A} = -(\mathbf{k} \times \mathbf{A}_0) \sin(\mathbf{k} \cdot \mathbf{r} - \omega t) \tag{18.4.21}$$

Thus **E** and **B** are mutually perpendicular and perpendicular to **k** (i.e., they are also transverse)—see Fig. 18.2. They have the same magnitude:

$$|\mathbf{E}| = |\mathbf{B}| \tag{18.4.22}$$

The energy flow across unit area (placed normal to **k**) per second is (from any standard text)

$$|\mathbf{S}| = \frac{c}{4\pi} |(\mathbf{E} \times \mathbf{B})| = \frac{\omega^2}{4\pi c} |\mathbf{A}_0|^2 \sin^2(\mathbf{k} \cdot \mathbf{r} - \omega t) \tag{18.4.23a}$$

The time average over a cycle is

$$S_{\mathrm{av}} = \frac{\omega^2}{8\pi c}|\mathbf{A}_0|^2 \tag{18.4.23b}$$

The energy per unit volume is

$$u = (1/8\pi)\cdot[|\mathbf{E}|^2 + |\mathbf{B}|^2] \tag{18.4.24}$$

Notice that $|\mathbf{S}|$ equals the energy density times the velocity of wave propagation.

The Potentials in Quantum Theory

We now ask if quantum mechanics also is invariant under gauge transformations of the potentials. Let us seek the answer to this question in the path integral approach. Recall that

$$U(\mathbf{r}t, \mathbf{r}'t') = N \sum_{\text{paths}} \exp[iS/\hbar] \tag{18.4.25}$$

where N is a normalization factor and the action

$$S = \int_{t'}^{t} \mathscr{L}\, dt'' = \int_{t'}^{t} \left(\frac{1}{2}m|\dot{\mathbf{r}}|^2 + \frac{q}{c}\mathbf{v}\cdot\mathbf{A} - q\phi\right) dt'' \tag{18.4.26}$$

is to be evaluated along each path P that connects $(\mathbf{r}', t')$ and $(\mathbf{r}, t)$. Suppose we perform a gauge transformation of the potentials. Then

$$S \to S_\Lambda = S - \int_{t'}^{t} \frac{q}{c}\left(\mathbf{v}\cdot\nabla\Lambda + \frac{\partial\Lambda}{\partial t''}\right) dt'' \tag{18.4.27}$$

But

$$\mathbf{v}\cdot\nabla\Lambda + \frac{\partial\Lambda}{\partial t''} = \frac{d\Lambda}{dt''} \tag{18.4.28}$$

is the *total derivative* along the trajectory. Consequently

$$S_\Lambda = S + \frac{q}{c}[\Lambda(\mathbf{r}', t') - \Lambda(\mathbf{r}, t)] \tag{18.4.29}$$

It is clear that S and S_Λ imply the same classical dynamics: varying S and varying S_Λ (to find the path of least action) are equivalent, since S and S_Λ differ only by

$(q/c)\Lambda$ at the end points, and the latter are held fixed in the variation. Going on to the quantum case, we find from Eqs. (18.4.25) and (18.4.29) that

$$U \to U_\Lambda = U \cdot \exp\left\{\frac{iq}{\hbar c}[\Lambda(\mathbf{r}', t') - \Lambda(\mathbf{r}, t)]\right\} \tag{18.4.30}$$

Since

$$U(\mathbf{r}, t; \mathbf{r}', t') = \langle \mathbf{r}| U(t, t')|\mathbf{r}'\rangle \tag{18.4.31}$$

we see that effect of the gauge transformation is equivalent to a change in the coordinate basis:

$$|\mathbf{r}\rangle \to |\mathbf{r}_\Lambda\rangle = e^{(iq\Lambda/\hbar c)}|\mathbf{r}\rangle \tag{18.4.32}$$

which of course cannot change the physics. (Recall, however, the discussion in Section 7.4.) The change in the wave function under the gauge transformation is

$$\psi = \langle \mathbf{r}|\psi\rangle \to \psi_\Lambda = \langle \mathbf{r}_\Lambda|\psi\rangle = e^{-iq\Lambda(\mathbf{r},t)/\hbar c}\psi \tag{18.4.33}$$

This result may also be obtained within the Schrödinger approach (see the following exercise).

Exercise 18.4.4 (Proof of Gauge Invariance in the Schrödinger Approach). (1) Write H for a particle in the potentials $(\mathbf{A}, \phi)$.

(2) Write down H_Λ, the Hamiltonian obtained by gauge transforming the potentials.

(3) Show that if $\psi(\mathbf{r}, t)$ is a solution to Schrödinger's equation with the Hamiltonian H, then $\psi_\Lambda(\mathbf{r}, t)$ given in Eq. (18.4.33) is the corresponding solution with $H \to H_\Lambda$.

Although quantum mechanics is similar to classical mechanics in that it is insensitive to gauge transformations of the potentials, it is different in the status it assigns to the potentials. This is dramatically illustrated in the *Aharonov–Bohm effect*, depicted schematically in Fig. 18.3.‡ The experiment is just the double-slit experiment with one change: there is a small shaded region ($B \neq 0$) where magnetic fluxes comes out of the paper. (You may imagine a tiny solenoid coming out of the paper, inside which are confined the flux lines. These lines must of course return to the other end of the solenoid, but this is arranged not to happen in the experimental region.) The vector potential (in Coulomb gauge) is shown by closed loops surrounding the coil. At a classical level, this variation in the double-slit experiment is expected to make no change in the outcome, for there is no magnetic field along the classical paths P_1 and P_2. There is, of course, an $\mathbf{A}$ field along P_1 and P_2, but the potential has no direct significance in classical physics. Its curl, which is significant, vanishes there.

Consider now the quantum case. In the path integral approach, a particle emitted by the source has the following amplitude to end up at a point $\mathbf{r}$ on the screen, before $\mathbf{B}$ is turned on:

$$\psi(\mathbf{r}) \simeq \psi_{P_1}(\mathbf{r}) + \psi_{P_2}(\mathbf{r}) \tag{18.4.34}$$

‡ For the actual experiment see R. G. Chambers, *Phys. Rev. Lett.*, **5**, 3 (1960).

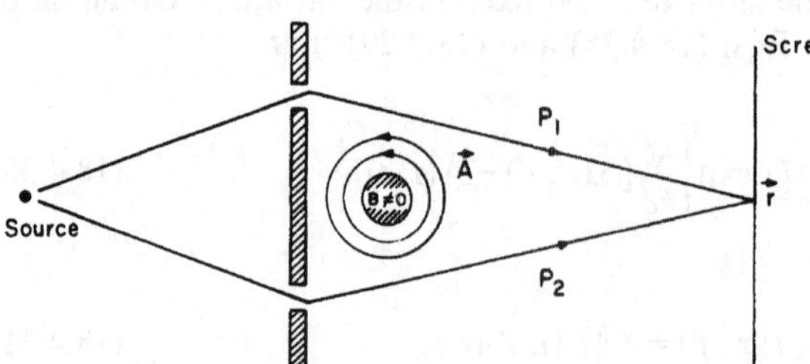

Figure 18.3. An experiment (schematic) that displays the Aharonov–Bohm effect. It is just the double-slit experiment but for the small coil coming out of the paper carrying magnetic flux (indicated by the shaded region marked $B \neq 0$).

where ψ_{P_i} $(i=1, 2)$ is the contribution from the classical path P_i and its immediate neighbors. The interference between these two contributions produces the usual interference pattern. Let us turn on **B**. Now each path gets an extra factor

$$\exp\left[\frac{iq}{\hbar c}\int_{t'}^{t}(\mathbf{v}\cdot\mathbf{A})\,dt''\right]=\exp\left(\frac{iq}{\hbar c}\int_{\text{source}}^{\mathbf{r}}\mathbf{A}\cdot d\mathbf{r}''\right) \tag{18.4.35}$$

Since $\nabla\times\mathbf{A}=0$ near P_1 and P_2, by Stoke's theorem the integral is the same for P_1 and its neighbors and P_2 and its neighbors. But the integral on P_1 is not the same as the integral on P_2, for these paths surround the coil and

$$\begin{aligned}\int_{P_2}\mathbf{A}\cdot d\mathbf{r}-\int_{P_1}\mathbf{A}\cdot d\mathbf{r}=\oint\mathbf{A}\cdot d\mathbf{r}&=\int_S(\nabla\times\mathbf{A})\cdot d\mathbf{s}\\&=\int\mathbf{B}\cdot d\mathbf{s}=\Phi\neq 0\end{aligned} \tag{18.4.36}$$

where s is any surface bounded by the closed loop P_1+P_2, and Φ is the flux crossing it, i.e., coming out of the paper in Fig. 18.3. Bearing this in mind, we get

$$\psi(\mathbf{r})=\exp\left(\frac{iq}{\hbar c}\int_{P_1}\mathbf{A}\cdot d\mathbf{r}''\right)\psi_{P_1}(\mathbf{r})+\exp\left(\frac{iq}{\hbar c}\int_{P_2}\mathbf{A}\cdot d\mathbf{r}''\right)\psi_{P_2}(\mathbf{r}) \tag{18.4.37}$$

Pulling out an overall phase factor, which does not affect the interference pattern, we get

$$\begin{aligned}\psi(\mathbf{r})&=\begin{pmatrix}\text{overall}\\\text{factor}\end{pmatrix}\left[\psi_{P_1}(\mathbf{r})+\exp\left(\frac{iq}{\hbar c}\oint\mathbf{A}\cdot d\mathbf{r}\right)\psi_{P_2}(\mathbf{r})\right]\\&=\begin{pmatrix}\text{overall}\\\text{factor}\end{pmatrix}[\psi_{P_1}(\mathbf{r})+\exp[iq\Phi/\hbar c]\psi_{P_2}(\mathbf{r})]\end{aligned} \tag{18.4.38}$$

By varying **B** (and hence Φ) we change the relative phase between the contributions from the two paths and move the interference pattern up and down. Whenever $(q\Phi/\hbar c)=2n\pi$, the pattern will return to its initial form, as if there were no field.

In other words, an integral multiple of the *flux quantum*

$$\Phi_0 = \frac{2\pi\hbar c}{q} \tag{18.4.39}$$

will not make any observable difference to the quantum mechanics of the particle. This idea is very frequently invoked; we shall do so in Chapter 21.

Let us understand how the particle discerns the magnetic field even though the dominant paths all lie in the $\mathbf{B}=0$ region. Suppose I show you Fig. 18.3 but cover the region where the coil is (the shaded region marked $B\neq 0$); will you know there is magnetic flux coming out of the paper? Yes, because the circulating $\mathbf{A}$ lines will tell you that $\oint \mathbf{A}\cdot d\mathbf{r} = \int \mathbf{B}\cdot d\mathbf{s} \neq 0$.‡ The classical particle, however, moves along P_1 *or* P_2, and can have no knowledge of $\oint \mathbf{A}\cdot d\mathbf{r}$. The best it can do is measure $\nabla\times\mathbf{A}$ locally, and that always equals zero. The quantum particle, on the other hand, "goes along P_1 *and* P_2" (in the path integral sense) and by piecing together what happens along P_1 and P_2 (i.e., by comparing the relative phase of the contributions from the two paths) it can deduce not only the existence of $\mathbf{B}$, but also the total flux. Notice that although the particle responds to $\mathbf{A}$ and not directly to $\mathbf{B}$, the response is gauge invariant.

18.5. Interaction of Atoms with Electromagnetic Radiation

We will make no attempt to do justice to this enormous field. We will consider just two illustrative examples. The first is the photoelectric effect in hydrogen (in which the incident radiation knocks the electron out of the atom). The second is the *spontaneous* decay of hydrogen from an excited state to the ground state (decay in the absence of external fields), which can be understood only if the electromagnetic field is treated as a quantum system.

Photoelectric Effect in Hydrogen

Consider a hydrogen atom in its ground state $|100\rangle$ centered at the origin, and on which is incident the wave

$$\mathbf{A}(\mathbf{r}, t) = \mathbf{A}_0 \cos(\mathbf{k}\cdot\mathbf{r} - \omega t) \tag{18.5.1}$$

For energies $\hbar\omega$ sufficiently large, the bound electron can be liberated and will come flying out. We would like to calculate the rate for this process using Fermi's

‡ This is like saying that you can infer the existence of a pole in the complex plane and its residue, without actually going near it, by evaluating $1/2\pi i \oint f(z)\, dz$ on a path that encloses it.

golden rule:

$$R_{i\to f} = \text{rate of transition } i\to f = \frac{2\pi}{\hbar}|\langle f^0|H^1|i^0\rangle|^2\delta(E_f^0 - E_i^0 - \hbar\omega) \quad (18.5.2)$$

Two points need to be explained before the application of this rule:

(1) For the final state, we must use a positive energy eigenstate of the Coulomb Hamiltonian $H^0 = P^2/2m - e^2/r$. Now we argue on intuitive grounds that if the ejected electron is very energetic, we must be able to ignore the pull of the proton on it and describe it by a plane wave $|\mathbf{p}_f\rangle$ in Eq. (18.5.2), with negligible error. While this happens to be the case here, there is a subtle point that is worth noting. If we view the Coulomb attraction of the proton as a perturbation relative to the free-particle Hamiltonian $P^2/2m$, we can write the eigenstate of H^0 as a perturbation series:

$$|f^0\rangle = |\mathbf{p}_f\rangle + \text{higher-order terms}$$

We are certainly right in guessing that $|\mathbf{p}_f\rangle$ dominates the expansion at high energies. But we are assuming more: we are assuming that when we evaluate the matrix element in Eq. (18.5.2) the leading term $|\mathbf{p}_f\rangle$ will continue to dominate the higher-order terms. Clearly, the validity of this assumption depends also on the initial state $|i^0\rangle$ and the operator H^1. Now it turns out that if the initial state is an s state (as in the present case) the higher-order terms are indeed negligible in computing the matrix element, but not otherwise. For instance if the initial state is a p state, the contribution of the first-order term to the matrix element would be comparable to the contribution from the leading term $|\mathbf{p}_f\rangle$. For more details, you must consult a book that is devoted to the subject.‡

(2) The rule applied for potentials of the form $H^1(t) = H^1 e^{-i\omega t}$, whereas here [recall Eq. (14.4.11)],§

$$\begin{aligned} H^1(t) &= \frac{-(-e)}{2mc}(\mathbf{A}\cdot\mathbf{P} + \mathbf{P}\cdot\mathbf{A}) \\ &= \frac{e}{mc}\mathbf{A}\cdot\mathbf{P} \quad (\text{because} \quad \nabla\cdot\mathbf{A} = 0) \\ &= \frac{e}{mc}\cos(\mathbf{k}\cdot\mathbf{r} - \omega t)\mathbf{A}_0\cdot\mathbf{P} \\ &= \frac{e}{2mc}[e^{i(\mathbf{k}\cdot\mathbf{r} - \omega t)} + e^{-i(\mathbf{k}\cdot\mathbf{r} - \omega t)}]\mathbf{A}_0\cdot\mathbf{P} \end{aligned} \quad (18.5.3)$$

‡ For example, Section 70 of H. Bethe and E. Salpeter, *Quantum Mechanics of One and Two Electron Atoms*, Plenum, New York (1977). This is also a good place to look for other data on this subject. For instance if you want to know what the expectation value of r^{-4} is in the state $|nlm\rangle$ of hydrogen, you will find it here.

§ We do not include in H^1 the term proportional to $|\mathbf{A}|^2$, which is of second order. The spin interaction $-\gamma\mathbf{S}\cdot\mathbf{B}$ *is* of the first order, but negligible in the kinematical region we will focus on. This will be demonstrated shortly.

Of the two pieces, only the first has the correct time dependence to induce the transition $i \to f$ with $E_f > E_i$; the second will be killed by the energy-conserving delta function. Hereafter we ignore the second term and let

$$H^1(t) = \frac{e}{2mc} e^{i\mathbf{k}\cdot\mathbf{r}} \mathbf{A}_0 \cdot \mathbf{P}\, e^{-i\omega t}$$

$$= H^1\, e^{-i\omega t} \qquad (18.5.4)$$

With these two points out of the way, we can proceed to evaluate the transition matrix element in the coordinate basis:

$$H^1_{fi} = \frac{e}{2mc} \frac{1}{(2\pi\hbar)^{3/2}} \left(\frac{1}{\pi a_0^3}\right)^{1/2} \int e^{-i\mathbf{p}_f\cdot\mathbf{r}/\hbar}\, e^{i\mathbf{k}\cdot\mathbf{r}} \mathbf{A}_0 \cdot (-i\hbar\nabla)\, e^{-r/a_0}\, d^3\mathbf{r} \qquad (18.5.5)$$

Consider the factor $e^{i\mathbf{k}\cdot\mathbf{r}}$. Recall from Chapter 5 that multiplication of a wave function by $e^{i\mathbf{p}_0\cdot\mathbf{r}/\hbar}$ adds to the state a momentum $\mathbf{p}_0$. Thus the factor $e^{i\mathbf{k}\cdot\mathbf{r}}$ represents the fact that a momentum $\hbar\mathbf{k}$ is imparted by the radiation to the atom.‡ For any transition *between atomic levels*, this momentum transferred is neglible compared to the typical momentum p of the electron. We see this as follows. The energy transferred is of the order of a Rydberg:

$$\hbar\omega \sim e^2/a_0 \qquad (18.5.6)$$

so that the photon momentum is

$$\hbar k = \frac{\hbar\omega}{c} \simeq \frac{e^2}{a_0 c} \qquad (18.5.7)$$

On the other hand, the typical momentum of the electron, estimated from the uncertainty principle, is

$$p \sim \frac{\hbar}{a_0} \qquad (18.5.8)$$

Thus

$$\frac{\hbar k}{p} \simeq \frac{e^2}{\hbar c} \simeq \frac{1}{137} \qquad (18.5.9)$$

In the present case $\hbar\omega$ is a lot higher because we have a liberated, high-energy electron. But there is still a wide range of ω over which $\hbar k/p \ll 1$. We will work in

‡ You may be worried that there is the $(-i\hbar\nabla)$ operator between $e^{i\mathbf{k}\cdot\mathbf{r}}$ and the atomic wave function. But since $\nabla\cdot\mathbf{A}=0$, we can also write $\mathbf{A}\cdot\mathbf{P}$ as $\mathbf{P}\cdot\mathbf{A}$, in which case the $e^{i\mathbf{k}\cdot\mathbf{r}}$ will be right next to the atomic wave function.

this domain. In this domain, the ratio of the spin interaction we neglected, to the orbital interaction we are considering, is roughly

$$\frac{\langle (e/2mc)\mathbf{S}\cdot\mathbf{B}\rangle}{\langle (e/mc)\mathbf{A}\cdot\mathbf{P}\rangle} \simeq \frac{\langle \hbar\boldsymbol{\sigma}\cdot\nabla\times\mathbf{A}\rangle}{\langle \mathbf{A}\cdot\mathbf{P}\rangle} \simeq \frac{\hbar k}{p} \ll 1 \tag{18.5.10}$$

which justifies our neglect.

The domain we are working in may also be described by

$$ka_0 \ll 1 \tag{18.5.11}$$

[Eq. (18.5.9)]. This means that the phase of the wave changes little over the size of the atom. Since the integral in Eq. (18.5.5) is rapidly cut off beyond $r \simeq a_0$ by the wave function e^{-r/a_0}, we may approximate $e^{i\mathbf{k}\cdot\mathbf{r}}$ in the integral as

$$e^{i\mathbf{k}\cdot\mathbf{r}} \simeq 1 \tag{18.5.12}$$

This is called the *electric dipole approximation.*‡ The reason is that in this approximation, the atom sees a *spatially* constant electric field,

$$\begin{aligned}\mathbf{E} &= \frac{-1}{c}\frac{\partial \mathbf{A}}{\partial t} \\ &= \frac{-1}{c}\frac{\partial}{\partial t}\left(\frac{\mathbf{A}_0}{2}e^{-i\omega t}\right)\S \\ &= \frac{i\omega}{2c}\mathbf{A}_0 e^{-i\omega t}\end{aligned} \tag{18.5.13}$$

and couples to it via its electric dipole moment $\boldsymbol{\mu} = -e\mathbf{R}$:

$$H^1(t) = -\boldsymbol{\mu}\cdot\mathbf{E} = \frac{i\omega e}{2c}\mathbf{A}_0\cdot\mathbf{R}e^{-i\omega t} \tag{18.5.14}$$

This must of course coincide with Eq. (18.5.3) in this approximation:

$$H^1(t) = \frac{e}{2mc}\mathbf{A}_0\cdot\mathbf{P}e^{-i\omega t} \tag{18.5.15}$$

‡ By keeping higher powers of $\mathbf{k}\cdot\mathbf{r}$ in the expansion, one gets terms known as electric quadrupole, magnetic dipole, electric octupole, magnetic quadrupole, etc. contributions.

§ We ignore the "wrong" frequency part of $\mathbf{A}$.

The equivalence of Eqs. (18.5.14) and (18.5.15) can be demonstrated in a general situation as follows. Since for any

$$H^0 = \frac{|\mathbf{P}|^2}{2m} + V(\mathbf{R}) \tag{18.5.16a}$$

it is true that

$$[\mathbf{R}, H^0] = \frac{i\hbar}{m}\mathbf{P} \tag{18.5.16b}$$

we find

$$\begin{aligned}\langle f^0|\mathbf{P}|i^0\rangle &= \frac{m}{i\hbar}\langle f^0|\mathbf{R}H^0 - H^0\mathbf{R}|i^0\rangle \\ &= \frac{m}{i\hbar}(E_i^0 - E_f^0)\langle f^0|\mathbf{R}|i^0\rangle \\ &= im\omega\langle f^0|\mathbf{R}|i^0\rangle \end{aligned} \tag{18.5.17}$$

so that

$$\begin{aligned}\langle f^0|\frac{e}{2mc}\mathbf{A}_0\cdot\mathbf{P}|i^0\rangle &= \frac{ie\omega}{2c}\mathbf{A}_0\cdot\langle f^0|\mathbf{R}|i^0\rangle \\ &= \langle f^0|(-\boldsymbol{\mu}\cdot\mathbf{E})|i^0\rangle \qquad \text{[by Eq. (18.5.14)]}\end{aligned} \tag{18.5.18}$$

Consider now the evaluation of the matrix element H^1_{fi} in the dipole approximation:

$$H^1_{fi} = N\int e^{-i\mathbf{p}_f\cdot\mathbf{r}/\hbar}\mathbf{A}_0\cdot(-i\hbar\boldsymbol{\nabla})\,e^{-r/a_0}\,d^3\mathbf{r} \tag{18.5.19}$$

where N is a constant given by

$$N = \left(\frac{e}{2mc}\right)\left(\frac{1}{2\pi\hbar}\right)^{3/2}\left(\frac{1}{\pi a_0^3}\right)^{1/2} \tag{18.5.20}$$

If we integrate the $\boldsymbol{\nabla}$ by parts, we get

$$H^1_{fi} = N\mathbf{A}_0\cdot\mathbf{p}_f\int e^{-i\mathbf{p}_f\cdot\mathbf{r}/\hbar}\,e^{-r/a_0}\,d^3\mathbf{r} \tag{18.5.21}$$

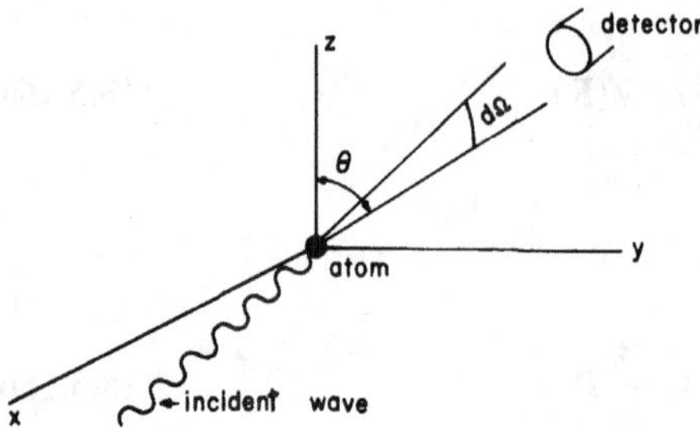

Figure 18.4. The photoelectric effect. In any realistic experiment, the resolution in energy and angle are finite. One asks how many electrons come into the cone of solid angle $d\Omega$ with magnitude of momentum between p and $p+dp$.

(It should now be clear why we prefer the $\mathbf{A}_0\cdot\mathbf{P}$ form of H^1 to the $\mathbf{A}_0\cdot\mathbf{R}$ form.) If we choose the z axis along $\mathbf{p}_f$, the $\mathbf{r}$ integral becomes

$$\int_0^\infty\int_{-1}^1\int_0^{2\pi} e^{-ip_fr\cos\theta/\hbar}\, e^{-r/a_0}r^2\, dr\, d(\cos\theta)\, d\phi$$

$$=2\pi\int_0^\infty\left(\frac{e^{-ip_fr/\hbar}-e^{ip_fr/\hbar}}{-ip_fr/\hbar}\right)e^{-r/a_0}r^2\, dr$$

$$=\frac{2\pi\hbar i}{p_f}\left[-\frac{\partial}{\partial(1/a_0)}\right]\int_0^\infty[e^{-(1/a_0+ip_f/\hbar)r}-e^{-(1/a_0-ip_f/\hbar)r}]\, dr$$

$$=\frac{8\pi/a_0}{[(1/a_0)^2+(p_f/\hbar)^2]^2} \tag{18.5.22}$$

Feeding this into Eq. (18.5.5), and the resulting expression into the golden rule, we get the transition rate

$$R_{i\to f}=\frac{2\pi}{\hbar}\left(\frac{e}{2mc}\right)^2\frac{1}{8\pi^3\hbar^3}\frac{1}{\pi a_0^3}\frac{|\mathbf{A}_0\cdot\mathbf{p}_f|^2 64\pi^2 a_0^6}{[1+(p_fa_0/\hbar)^2]^4}\times\delta(E_f^0-E_i^0-\hbar\omega) \tag{18.5.23}$$

Now the time has come to tackle the δ function. The δ function gives a singular probability distribution for finding a final electron in a state of *mathematically precise momentum* $\mathbf{p}_f$. This probability is of little interest in practice, where one sets up a detector with a finite opening angle $d\Omega$ and asks how many electrons come into it with magnitude of momentum between p_f and p_f+dp_f (see Fig. 18.4). The δ function tells us that electron momenta are concentrated at

$$\frac{p_f^2}{2m}=E_i^0+\hbar\omega$$

The contribution from this region is obtained by integrating the δ function over p_f. Using

$$\delta\left(\frac{p_f^2}{2m}-E_i^0-\hbar\omega\right)=\frac{m}{p_f}\,\delta\{p_f-[2m(E_i^0+\hbar\omega)]^{1/2}\} \tag{18.5.24}$$

we get the rate of transition into the detector to be

$$\begin{aligned} R_{i\to d\Omega}&=\frac{2\pi}{\hbar}\,|H_{fi}^1|^2 mp_f\,d\Omega \\ &=\frac{4a_0^3e^2p_f|\mathbf{A}_0\cdot\mathbf{p}_f|^2}{m\pi\hbar^4c^2[1+(p_fa_0/\hbar)^2]^4}\,d\Omega \end{aligned} \tag{18.5.25}$$

{In this and all following expressions, $p_f=[2m(E_i^0+\hbar\omega)]^{1/2}$.} Note that the rate depends only on the magnitude of the applied field $\mathbf{A}_0$, the angle between the polarization $\mathbf{A}_0$ and the outgoing momentum, and the magnitude of $\mathbf{p}_f$ or equivalently ω, the frequency of radiation. The formula above tells us that the electron likes to come parallel to $\mathbf{A}_0$, that is, to the electric field which rips it out of the atom. The direction of the incident radiation does not appear because we set $e^{i\mathbf{k}\cdot\mathbf{r}}=1$. If we keep the $e^{i\mathbf{k}\cdot\mathbf{r}}$ factor it will be seen that the electron momentum is also biased toward $\mathbf{k}$, reflecting the $\hbar\mathbf{k}$ momentum input.

*Exercise 18.5.1.** (1) By going through the derivation, argue that we can take the $e^{i\mathbf{k}\cdot\mathbf{r}}$ factor into account exactly, by replacing $\mathbf{p}_f$ by $\mathbf{p}_f-\hbar\mathbf{k}$ in Eq. (18.5.19).

(2) Verify the claim made above about the electron momentum distribution.

If we integrate $R_{i\to d\Omega}$ over all angles, we get the total rate for ionization. Choosing $\mathbf{A}_0$ along the z axis for convenience, we find

$$\begin{aligned} R_{i\to\text{all}}&=\frac{4a_0^3e^2p_f^3|\mathbf{A}_0|^2}{m\pi\hbar^4c^2[1+(p_fa_0/\hbar)^2]^4}\iint\cos^2\theta\,d(\cos\theta)\,d\phi \\ &=\frac{16a_0^3e^2p_f^3|\mathbf{A}_0|^2}{3m\hbar^4c^2[1+(p_fa_0/\hbar)^2]^4} \end{aligned} \tag{18.5.26}$$

Since this is the rate of ionization, and each ionization takes energy $\hbar\omega$ from the beam, the energy absorption rate is

$$\frac{dE_{\text{abs}}}{dt}=\hbar\omega\cdot R_{i\to\text{all}} \tag{18.5.27}$$

Now the beam brings in energy at the rate $\omega^2|\mathbf{A}_0|^2/8\pi c$ per unit area. Suppose we place, transverse to this beam, a perfectly absorbing disk of area σ. It will absorb energy at the rate

$$\frac{dE_{\text{abs}}}{dt}=\frac{\sigma|\mathbf{A}_0|^2\omega^2}{8\pi c} \tag{18.5.28}$$

By comparing Eqs. (18.5.27) and (18.5.28), we see that we can associate with the atom a *photoelectric cross section*

$$\sigma=\frac{8\pi c}{|\mathbf{A}_0|^2\omega^2}\cdot\hbar\omega\cdot R_{i\to\text{all}} \tag{18.5.29}$$

$$=\frac{128a_0^3\pi e^2p_f^3}{3m\hbar^3\omega c[1+p_f^2a_0^2/\hbar^2]^4} \tag{18.5.30}$$

in the sense that if an ensemble of N (N large) nonoverlapping (separation $\gg a_0$) hydrogen atoms is placed in the way of the beam, the ensemble will absorb energy like a perfectly absorbent disk of area $N\sigma$. We can also associate a *differential cross section* $d\sigma/d\Omega$, with the energy flowing into a solid angle $d\Omega$:

$$\frac{d\sigma}{d\Omega}=\frac{8\pi c}{|\mathbf{A}_0|^2\omega^2}\hbar\omega R_{i\to d\Omega}$$

$$=\frac{32a_0^3e^2p_f^3\cos^2\theta}{mc\omega\hbar^3[1+p_f^2a_0^2/\hbar^2]^4} \tag{18.5.31}$$

In the region where $p_fa_0/\hbar\gg 1$, the formula simplifies to

$$\frac{d\sigma}{d\Omega}=\frac{32e^2\hbar^5\cos^2\theta}{mc\omega p_f^5a_0^5} \tag{18.5.32}$$

*Exercise 18.5.2.** (1) Estimate the photoelectric cross section when the ejected electron has a kinetic energy of 10 Ry. Compare it to the atom's geometric cross section $\simeq\pi a_0^2$.

(2) Show that if we consider photoemission from the $1s$ state of a charge Z atom, $\sigma\propto Z^5$, in the limit $p_fa_0/Z\hbar\gg 1$.

Field Quantization‡

The general formalism, illustrated by the preceding example, may be applied to a host of other phenomena involving the interaction of atoms with radiation. The results are always in splendid agreement with experiment as long the electromagnetic field is of macroscopic strength. The breakdown of the above formalism for weak fields is most dramatically illustrated by the following example. Consider a hydrogen atom in free space (the extreme case of weak field) in the state $|2, l, m\rangle$. What is the rate of decay to the ground state? Our formalism gives an unambiguous answer of

‡ The treatment of this advanced topic will be somewhat concise. You are urged to work out the missing steps if you want to follow it in depth.

zero, for free space corresponds to $\mathbf{A}=0$ (in the Coulomb gauge), so that $H^1=0$ and the atom should be in the stationary state $|2, l, m\rangle$ forever. But it is found experimentally that the atom decays at a rate $R \simeq 10^9$ second^{-1}, or has a mean lifetime $\tau \simeq 10^{-9}$ second. In fact, all excited atoms are found to decay spontaneously in free space to their ground states. This phenomenon cannot be explained within our formalism.

So are we to conclude that our description of free space (which should be the simplest thing to describe) is inadequate? Yes! The description of free space by $\mathbf{A}=\dot{\mathbf{A}}=0$ is *classical*; it is like saying that the ground state of the oscillator is given by $x=p=0$. Now, we know that if the oscillator is treated quantum mechanically, only the *average* quantities $\langle 0|X|0\rangle$ and $\langle 0|P|0\rangle$ vanish in the ground state, and that there are nonzero fluctuations $(\Delta X)^2=\langle 0|X^2|0\rangle$ and $(\Delta P)^2=\langle 0|P^2|0\rangle$ about these mean values. In the same way, if the electromagnetic field is treated quantum mechanically, it will be found that free space (which is the ground state of the field) is described by $\langle \mathbf{A}\rangle=\langle\dot{\mathbf{A}}\rangle=0$ (where $\mathbf{A}$ and $\dot{\mathbf{A}}$ are *operators*)‡ with nonvanishing fluctuations $(\Delta\mathbf{A})^2$, $(\Delta\dot{\mathbf{A}})^2$. The free space is dormant only in the average sense; there are always quantum fluctuations of the fields about these mean values. It is these fluctuations that trigger spontaneous decay.

As long as we restrict ourselves to macroscopic fields, the quantum and classical descriptions of the field become indistinguishable. This is why in going from classical to quantum mechanics, i.e., in going from $\mathscr{H}^1=(e/mc)\mathbf{A}\cdot\mathbf{p}$ to $H^1=(e/mc)\mathbf{A}\cdot\mathbf{P}$, we merely promoted $\mathbf{p}$ to the operator $\mathbf{P}$, but let $\mathbf{A}$ continue to be the classical field. For this reason, this treatment is called the *semiclassical treatment*. We now turn to the full quantum mechanical treatment in which $\mathbf{A}$ will become an operator as well.

The basic idea behind quantizing the field is familiar: one finds a complete set of *canonical* coordinates and momenta to describe the classical field, and promotes them to operators obeying canonical commutation relations. One then takes $\mathscr{H}$, which is just the field energy written in terms of the canonical variables, and obtains H by the usual substitution rule. But there are many obstacles, as we shall see.

Let us start with the coordinates of the field. If we decide to describe it in terms of the potentials, we have, at each point in space $\mathbf{r}$, four real coordinates $(\phi(\mathbf{r}), \mathbf{A}(\mathbf{r}))$.§ Now, we already know that these coordinates are not entirely physical, in that they can be gauge transformed with no observable consequences. For them to be physical, we must constrain them to a point where there is no residual gauge freedom, say by imposing the Coulomb gauge conditions. Although we shall do so eventually, we treat them as genuine coordinates for the present.

What are the momenta conjugate to these coordinates? To find out, we turn to the Lagrangian:

$$\mathscr{L}=\frac{1}{8\pi}\int[|\mathbf{E}|^2-|\mathbf{B}|^2]d^3\mathbf{r}=\frac{1}{8\pi}\int\left[\left|-\frac{1}{c}\frac{\partial\mathbf{A}}{\partial t}-\nabla\phi\right|^2-|\nabla\times\mathbf{A}|^2\right]d^3\mathbf{r}\| \tag{18.5.33}$$

‡ We depart from our convention here and denote classical and quantum field variables by the *same symbols*, because this is what everyone does in this case.

§ Now $\mathbf{r}$ is just a label on the field coordinates and not a dynamical variable.

‖ If you are unfamiliar with this $\mathscr{L}$: Recall that the field energy is $\int(1/8\pi)[|\mathbf{E}|^2+|\mathbf{B}|^2]d^3\mathbf{r}$, Eq. (18.4.24). Write this in the gauge $\phi=0$ and change the sign of the term that corresponds to "potential energy." The above result is just the generalization to the gauge with $\phi\neq 0$.

which, when varied with respect to the potentials, gives Maxwell's equations.‡ The momentum conjugate to each "coordinate" is the derivative of $\mathscr{L}$ with respect to the corresponding "velocity." It follows that the momentum conjugate to $\phi(\mathbf{r})$ vanishes (at each point $\mathbf{r}$ in space), for $\dot{\phi}(\mathbf{r})$ does not appear in $\mathscr{L}$. The fact that we are dealing with a coordinate whose conjugate momentum vanishes *identically* tells us that we can not follow the canonical route. But fortunately for us, we have the freedom to work in a gauge where $\phi=0$. So hereafter we can forget all about ϕ and its vanishing conjugate momentum. In particular, we can set $\phi=0$ in Eq. (18.5.33).

Consider now the coordinates $\mathbf{A}(\mathbf{r})$. To find $\Pi_i(\mathbf{r}_0)$, the momentum conjugate to $A_i(\mathbf{r}_0)$, we use the relation

$$\Pi_i(\mathbf{r}_0)=\frac{\partial\mathscr{L}}{\partial\dot{A}_i(\mathbf{r}_0)}$$

In differentiating $\mathscr{L}$ with respect to $\dot{A}_i(\mathbf{r}_0)$, we treat the integral in Eq. (18.5.33) over $\mathbf{r}$ as a sum over the continuous index $\mathbf{r}$. The partial derivative picks out just the term in the sum carrying the index $\mathbf{r}=\mathbf{r}_0$ (because the velocities at different points are independent variables) and gives§

$$\Pi_i(\mathbf{r}_0)=\frac{1}{4\pi c^2}\dot{A}_i(\mathbf{r}_0)=\frac{-E_i(\mathbf{r}_0)}{4\pi c} \tag{18.5.34}$$

or in vector form (dropping the subscript 0 on $\mathbf{r}$)

$$\mathbf{\Pi}(\mathbf{r})=\frac{1}{4\pi c^2}\dot{\mathbf{A}}=-\frac{\mathbf{E}}{4\pi c} \tag{18.5.35}$$

Note that $\mathbf{\Pi}$ is essentially the electric field.

The natural thing to do at this point would be to promote $\mathbf{A}$ and $\mathbf{\Pi}$ to quantum operators obeying canonical commutation rules, and obtain the quantum Hamiltonian H by the substitution rule. But if we did this, we would not be dealing with

‡ See for example, H. Goldstein, *Classical Mechanics*, Addison-Wesley, Reading, Massachusetts (1965), page 366.

§ A more formal treatment is the following. If, say, $\mathscr{L}=\sum_i \dot{q}_i^2$, then we know

$$p_j=\frac{\partial\mathscr{L}}{\partial\dot{q}_j}=\sum_i 2\dot{q}_i\frac{\partial\dot{q}_i}{\partial\dot{q}_j}=\sum_i 2\dot{q}_i\delta_{ij}=2\dot{q}_j$$

Likewise if

$$\mathscr{L}=\int\sum_i \dot{A}_i^2(\mathbf{r})\,d^3\mathbf{r}$$

$$\frac{\partial\mathscr{L}}{\partial\dot{A}_j(r_0)}=\int\sum_i 2\dot{A}_i(\mathbf{r})\frac{\partial\dot{A}_i(\mathbf{r})}{\partial\dot{A}_j(\mathbf{r}_0)}\,d^3\mathbf{r}$$

$$=\int\sum_i 2\dot{A}_i(\mathbf{r})\delta_{ij}\delta^3(\mathbf{r}-\mathbf{r}_0)\,d^3\mathbf{r}=2\dot{A}_j(\mathbf{r}_0)$$

electrodynamics. The reason is classical and is as follows. Consider the Lagrangian in Eq. (18.5.33) with ϕ set equal to 0. By varying it with respect to the components of $\mathbf{A}$ we get the vector equation

$$\nabla^2\mathbf{A}-\frac{1}{c^2}\frac{\partial^2\mathbf{A}}{\partial t^2}-\nabla(\nabla\cdot\mathbf{A})=0 \tag{18.5.36}$$

which is just

$$\nabla\times\mathbf{B}-\frac{1}{c}\frac{\partial\mathbf{E}}{\partial t}=0$$

in the gauge with $\phi=0$. Two other equations

$$\nabla\cdot\mathbf{B}=0$$

$$\nabla\times\mathbf{E}+\frac{1}{c}\frac{\partial\mathbf{B}}{\partial t}=0$$

are identically satisfied if we write $\mathbf{E}$ and $\mathbf{B}$ in terms of $\mathbf{A}$. (Recall how the potentials were introduced in the first place.) As for the other Maxwell equation, Gauss's law,

$$\nabla\cdot\mathbf{E}=0$$

it does not follow from anything. (We would get this if we varied $\mathscr{L}$ with respect to ϕ, but we have eliminated ϕ from the picture.) It must therefore be appended as an *equation of constraint* on the momentum $\mathbf{\Pi}$, which is just $\mathbf{E}$ times a constant. (In contrast to an equation of motion, which has time derivatives in it, an equation of constraint is a relation among the variables at a given time. It signifies that the variables are not independent.) The constraint

$$\nabla\cdot\mathbf{\Pi}=0 \tag{18.5.37}$$

tells us that the components of momenta at nearby points are not independent. (Think of the derivatives in ∇ as differences.) We deduce an important feature of the constraint if we take the divergence of Eq. (18.5.36):

$$0=\nabla\cdot\nabla^2\mathbf{A}-\frac{1}{c^2}\frac{\partial^2}{\partial t^2}\nabla\cdot\mathbf{A}-\nabla^2\nabla\cdot\mathbf{A}\rightarrow\frac{\partial}{\partial t}(\nabla\cdot\mathbf{\Pi})=0 \tag{18.5.38}$$

In other words, the theory without the constraint has a conserved quantity $\nabla\cdot\mathbf{\Pi}$, and electrodynamics corresponds to the subset of trajectories in which this constant of motion is zero. Furthermore, if we limit ourselves to these trajectories, we see that $\nabla\cdot\mathbf{A}$ is also a constant of motion. [Write Eq. (18.5.37) as $\nabla\cdot\dot{\mathbf{A}}=0$.] We shall choose this constant to be zero, i.e., work in Coulomb gauge.

How are we to quantize this theory? One way is to ignore the constraints and to quantize the general theory and then try to pick out the subset of solutions (in the quantum theory) that correspond to electrodynamics. This can be done but is very hard. Let us therefore tackle the constraints at the classical level. The first problem they pose is that they render the variables **A** and **Π** noncanonical, and we do not have a recipe for quantizing noncanonical variables. Let us verify that the constraints indeed imply that **A** and **Π** are noncanonical. Had they been canonical, they would have obeyed the generalizations of

$$\{q_i, p_j\} = \delta_{ij}$$

(with all other PB zero), namely,

$$\{A_i(\mathbf{r}), \Pi_j(\mathbf{r}')\} = \delta_{ij}\delta^3(\mathbf{r}-\mathbf{r}') \tag{18.5.39}$$

(with all other PB zero). But if we use these PB's to evaluate the PB of $\nabla\cdot\mathbf{A}$ with $\mathbf{\Pi}$ or $\nabla\cdot\mathbf{\Pi}$ with **A** we do not get zero as required.

What we would like to do is the following. We would like to trade **A** and **Π** for a new set of variables that are fewer in number but have the constraints built into them. (This would be like trading the variables x, y, and z constrained by $x^2+y^2+z^2=a^2$ for the angels θ and ϕ on a sphere of radius a.) These variables and the corresponding momenta would be canonical and would automatically reproduce electrodynamics if we start with $\mathscr{H}$ written in terms of these. To quantize, we promote these variables to operators obeying canonical commutation rules. The Hamiltonian and other operators would then be obtained by the substitution rule.

Now the problem with the constraints

$$\nabla\cdot\mathbf{A}=0, \qquad \nabla\cdot\mathbf{\Pi}=0 \tag{18.5.40}$$

called *transversality constraints* (for a reason that will follow) is that they are not algebraic, but differential equations. To render them algebraic, we will trade **A** and **Π** for their Fourier transforms, since differential equations in coordinate space become algebraic when Fourier transformed. It is our hope that the algebraic constraints among the Fourier coefficient will be easier to implement. We will find that this is indeed the case. We will also find a bonus when we are done: the Fourier coefficients are normal coordinates; i.e., when we express the Hamiltonian

$$\mathscr{H}=\frac{1}{8\pi}\int[16\pi^2c^2|\mathbf{\Pi}|^2+|\nabla\times\mathbf{A}|^2]d^3\mathbf{r} \tag{18.5.41}$$

(which is obtained from $\mathscr{L}$ by changing the sign of the potential energy term and eliminating $\dot{\mathbf{A}}$ in favor of **Π**) in terms of these, it becomes a sum over oscillator Hamiltonians of decoupled oscillators. This result could have been anticipated for the following reason. If we use the relation $|\nabla\times\mathbf{A}|^2=-\mathbf{A}\cdot\nabla^2\mathbf{A}$, valid when $\nabla\cdot\mathbf{A}=0$, we get

$$\mathscr{H}=\frac{1}{8\pi}\int\sum_i\sum_j[16\pi^2c^2\Pi_i(\mathbf{r})\delta_{ij}\Pi_j(\mathbf{r})-A_i(\mathbf{r})\nabla^2\delta_{ij}A_j(\mathbf{r})]\,d^3\mathbf{r} \tag{18.5.42}$$

which is of the same form as Eq. (7.1.10). [Remember that when we sandwich the derivative operator or the identity operator between two elements of function space, there will be only one (explicit) sum over the continuous index **r**, the other one being eaten up by the delta functions in the matrix elements.] As the normal modes are the eigenvectors of ∇^2 (which we know are plane waves) the passage to the Fourier coefficients is the passage to normal coordinates.

With all these preliminaries out of the way, let us turn to the Fourier transform of the *unconstrained* A:

$$\mathbf{A}(\mathbf{r})=\int [\mathbf{a}(\mathbf{k})\, e^{i\mathbf{k}\cdot\mathbf{r}}+\mathbf{a}^*(\mathbf{k})\, e^{-i\mathbf{k}\cdot\mathbf{r}}]\, d^3\mathbf{k} \tag{18.5.43}$$

This expansion deserves a few comments.

(1) Since we are Fourier transforming a vector **A**, the Fourier coefficients are vectors **a**(**k**). [You may view Eq. (18.5.43) as giving three Fourier expansions, one for each component of **A**.]

(2) Since **A**(**r**) is a *real* function, the Fourier coefficient at **k** and −**k** must be complex conjugates. Our expansion makes this apparent. Stated differently, one *real* vector function **A** in coordinate space cannot specify one *complex* vector function **a**(**k**) in **k** space: if we multiply both sides with $e^{-i\mathbf{k}_0\cdot\mathbf{r}}$ and integrate over **r**, we find that this is indeed the case:

$$\int e^{-i\mathbf{k}_0\cdot\mathbf{r}}\mathbf{A}(\mathbf{r})\, d^3\mathbf{r}=(2\pi)^3[\mathbf{a}(\mathbf{k}_0)+\mathbf{a}^*(-\mathbf{k}_0)] \tag{18.5.44}$$

i.e., **A**(**r**) is seen to determine only the combination **a**(**k**) + **a***(−**k**). We shall exploit this point shortly.

(3) There is no time argument shown in Eq. (18.5.43) because we view it as linear relations between two sets of coordinates, such as the relations

$$x_1=\frac{x_{\rm I}+x_{\rm II}}{2^{1/2}}$$

$$x_2=\frac{x_{\rm I}-x_{\rm II}}{2^{1/2}}$$

which are understood to be true at all times. The discrete labels 1, 2, I, and II are replaced here by the continuous labels **r** and **k**.

We similarly expand **Π** (before the transversality constraint is imposed) as

$$\mathbf{\Pi}(\mathbf{r})=\frac{1}{4\pi ic}\int k[\mathbf{a}(\mathbf{k})\, e^{i\mathbf{k}\cdot\mathbf{r}}-\mathbf{a}^*(\mathbf{k})\, e^{-i\mathbf{k}\cdot\mathbf{r}}]\, d^3\mathbf{k} \tag{18.5.45}$$

The factor $(k/4\pi ic)$ is pulled out to simplify future manipulations. Note that the same function **a**(**k**) appears here. There is no conflict, since **Π**(**r**) determines a different combination:

$$\int \mathbf{\Pi}(\mathbf{r})\, e^{-i\mathbf{k}_0\cdot\mathbf{r}}\, d^3\mathbf{r} = \frac{(2\pi)^3}{4\pi i c}\, k_0[\mathbf{a}(\mathbf{k}_0) - \mathbf{a}^*(-\mathbf{k}_0)] \tag{18.5.46}$$

It is clear that Eqs. (18.5.44) and (18.5.46) may be solved for $\mathbf{a}(\mathbf{k})$ in terms of $\mathbf{A}$ and $\mathbf{\Pi}$: the two real vector functions $\mathbf{A}(\mathbf{r})$ and $\mathbf{\Pi}(\mathbf{r})$ determine one complex vector function $\mathbf{a}(\mathbf{k})$. Consider now the vector $\mathbf{a}(\mathbf{k})$ at a given $\mathbf{k}$. We can expand it in terms of any three orthonormal vectors. Rather than choose them to be the unit vectors along the x, y, and z directions, let us choose them (with an eye on the constraints) as a function of $\mathbf{k}$, in the following way:

$$\left.\begin{matrix}\boldsymbol{\varepsilon}(\mathbf{k}1)\\ \boldsymbol{\varepsilon}(\mathbf{k}2)\end{matrix}\right\}\ \text{orthonormal vectors in the plane perpendicular to } \mathbf{k}$$

$$\boldsymbol{\varepsilon}(\mathbf{k}3)\quad \text{a unit vector parallel to } \mathbf{k} \tag{18.5.47}$$

If we now expand $\mathbf{a}(\mathbf{k})$ (at each $\mathbf{k}$) as

$$\mathbf{a}(\mathbf{k}) = \sum_{\lambda=1}^{3} (c^2/4\pi^2\omega)^{1/2} a(\mathbf{k}\lambda)\boldsymbol{\varepsilon}(\mathbf{k}\lambda) \tag{18.5.48}$$

(where $\omega = kc$) and feed this into the expansions for $\mathbf{A}$ and $\mathbf{\Pi}$, we get

$$\mathbf{A}(\mathbf{r}) = \sum_\lambda \int \left(\frac{c^2}{4\pi^2\omega}\right)^{1/2} [a(\mathbf{k}\lambda)\boldsymbol{\varepsilon}(\mathbf{k}\lambda)\, e^{i\mathbf{k}\cdot\mathbf{r}} + a^*(\mathbf{k}\lambda)\boldsymbol{\varepsilon}(\mathbf{k}\lambda)\, e^{-i\mathbf{k}\cdot\mathbf{r}}]\, d^3\mathbf{k} \tag{18.5.49a}$$

$$\mathbf{\Pi}(\mathbf{r}) = \sum_\lambda \int \frac{1}{i}\left(\frac{\omega}{64\pi^4 c^2}\right)^{1/2} [a(\mathbf{k}\lambda)\boldsymbol{\varepsilon}(\mathbf{k}\lambda)\, e^{i\mathbf{k}\cdot\mathbf{r}} - a^*(\mathbf{k}\lambda)\boldsymbol{\varepsilon}(\mathbf{k}\lambda)\, e^{-i\mathbf{k}\cdot\mathbf{r}}]\, d^3\mathbf{k} \tag{18.5.49b}$$

These equations relate the old coordinates—three real components of $\mathbf{A}$ and three real components of $\mathbf{\Pi}$ at each point in $\mathbf{r}$ space—to three complex components of $\mathbf{a}$ at each point in $\mathbf{k}$ space. Since $\mathbf{A}$ and $\mathbf{\Pi}$ are canonical variables before we impose transversality, their PB are

$$\begin{aligned}\{A_i(\mathbf{r}), A_j(\mathbf{r}')\} &= 0\\ \{\Pi_i(\mathbf{r}), \Pi_j(\mathbf{r}')\} &= 0\\ \{A_i(\mathbf{r}), \Pi_j(\mathbf{r}')\} &= \delta_{ij}\delta^3(\mathbf{r}-\mathbf{r}')\end{aligned} \tag{18.5.50}$$

From these we may deduce (after some hard work) that

$$\begin{aligned}\{a(\mathbf{k}\lambda), a(\mathbf{k}'\lambda')\} &= 0 = \{a^*(\mathbf{k}\lambda), a^*(\mathbf{k}'\lambda')\}\\ \{a(\mathbf{k}\lambda), a^*(\mathbf{k}'\lambda')\} &= -i\delta_{\lambda\lambda'}\delta^3(\mathbf{k}-\mathbf{k}')\end{aligned} \tag{18.5.51}$$

We now address the problem of imposing the constraints, i.e., of regaining electrodynamics. The conditions $\nabla\cdot\mathbf{A}=0$ and $\nabla\cdot\boldsymbol{\Pi}=0$ tell us [when we apply them to Eqs. (18.5.43) and (18.5.45) and project both sides onto some given $\mathbf{k}$],

$$\mathbf{k}\cdot[\mathbf{a}(\mathbf{k})+\mathbf{a}^*(-\mathbf{k})]=0$$
$$\mathbf{k}\cdot[\mathbf{a}(\mathbf{k})-\mathbf{a}^*(-\mathbf{k})]=0$$

from which we deduce that

$$\mathbf{k}\cdot\mathbf{a}(\mathbf{k})=0 \tag{18.5.52}$$

The two differential equations of constraint have reduced, as anticipated, to (a complex) algebraic constraint. Imposing it on Eq. (18.5.48), we find [using $\mathbf{k}\cdot\boldsymbol{\varepsilon}$ ($\mathbf{k}$, 1 or 2) $=0$],

$$a(\mathbf{k}3)=0 \tag{18.5.53}$$

Thus the constraint tells us something very simple: every $a(\mathbf{k}3)$ is zero. (Since it forces $\mathbf{a}(\mathbf{k})$ to lie in a plane transverse to $\mathbf{k}$, we call it the transversality constraint). Implementation of the transversality constraint is very simple in momentum space: hereafter we let λ take on only the values 1 and 2. Also, setting $a(\mathbf{k}3)=0$ does not change the PB between the remaining a's. Equation (18.5.49) for $\mathbf{A}$ and $\boldsymbol{\Pi}$ continues to hold, with λ so restricted. However, these fields are now guaranteed to meet the transversality conditions.

Now for the other nice feature of these conditions. If we express $\mathscr{H}$ in terms of these, we get

$$\mathscr{H}=\sum_{\lambda=1}^{2}\int\omega[a^*(\mathbf{k}\lambda)a(\mathbf{k}\lambda)]\,d^3\mathbf{k} \tag{18.5.54}$$

Thus $a(\mathbf{k}\lambda)$ are normal coordinates in the sense that $\mathscr{H}$ contains no cross terms between a's carrying different labels. If we want to get the familiar oscillators, we define real variables

$$\begin{aligned} q(\mathbf{k}\lambda)&=\frac{1}{(2\omega)^{1/2}}[a(\mathbf{k}\lambda)+a^*(\mathbf{k}\lambda)]\\ p(\mathbf{k}\lambda)&=\frac{1}{i}\left(\frac{\omega}{2}\right)^{1/2}[a(\mathbf{k}\lambda)-a^*(\mathbf{k}\lambda)]\end{aligned} \tag{18.5.55}$$

which satisfy the canonical PB relations [as you may verify by combining Eqs. (18.5.51) and (18.5.55)]. In terms of these variables

$$\mathscr{H}=\sum_{\lambda}\int\left[\frac{1}{2}p^2(\mathbf{k}\lambda)+\frac{\omega^2}{2}q^2(\mathbf{k}\lambda)\right]d^3\mathbf{k} \tag{18.5.56}$$

Thus we find that the radiation field is equivalent to a collection of decoupled oscillators: there is an oscillator at each $\mathbf{k}$ and λ (=1 or 2) with frequency $\omega = kc$. The quantization of the radiation field then reduces to the quantization of the oscillator, which has already been accomplished in Chapter 7.

Since $q(\mathbf{k}\lambda)$ and $p(\mathbf{k}\lambda)$ are independent canonical coordinates describing the field, we can quantize the field by promoting these to operators Q and P obeying canonical commutation rules:

$$[Q(\mathbf{k}\lambda), P(\mathbf{k}'\lambda')] = i\hbar\{q, p\} = i\hbar\delta_{\lambda\lambda'}\delta^3(\mathbf{k}-\mathbf{k}')$$

with all other commutators vanishing. As in the case of a single oscillator, it proves useful to work with the combination

$$a(\mathbf{k}\lambda) = \left(\frac{\omega}{2\hbar}\right)^{1/2} Q + i\left(\frac{1}{2\omega\hbar}\right)^{1/2} P$$

and its adjoint

$$a^\dagger(\mathbf{k}\lambda) = \left(\frac{\omega}{2\hbar}\right)^{1/2} Q - i\left(\frac{1}{2\omega\hbar}\right)^{1/2} P \tag{18.5.57}$$‡

which obey

$$[a(\mathbf{k}\lambda), a^\dagger(\mathbf{k}'\lambda')] = \delta_{\lambda\lambda'}\delta^3(\mathbf{k}-\mathbf{k}') \tag{18.5.58}$$

and in terms of which $\mathbf{A}$ and $\mathbf{\Pi}$,§ which are now Hermitian operators, are given by

$$\mathbf{A} = \sum_\lambda \int \left(\frac{\hbar c^2}{4\pi^2\omega}\right)^{1/2} [a(\mathbf{k}\lambda)\boldsymbol{\varepsilon}(\mathbf{k}\lambda)\, e^{i\mathbf{k}\cdot\mathbf{r}} + a^\dagger(\mathbf{k}\lambda)\boldsymbol{\varepsilon}(\mathbf{k}\lambda)\, e^{-i\mathbf{k}\cdot\mathbf{r}}]\, d^3\mathbf{k} \tag{18.5.59a}$$

$$\mathbf{\Pi} = \sum_\lambda \int \frac{1}{i}\left(\frac{\hbar\omega}{64\pi^4 c^2}\right)^{1/2} [a(\mathbf{k}\lambda)\boldsymbol{\varepsilon}(\mathbf{k}\lambda)\, e^{i\mathbf{k}\cdot\mathbf{r}} - a^\dagger(\mathbf{k}\lambda)\boldsymbol{\varepsilon}(\mathbf{k}\lambda)\, e^{-i\mathbf{k}\cdot\mathbf{r}}]\, d^3\mathbf{k} \tag{18.5.59b}$$

To find H, we first symmetrize $\mathscr{H}$, i.e., $a^*a \to \frac{1}{2}(a^*a + aa^*)$, make the operator substitution, and use Eq. (18.5.58), to get

$$H = \sum_\lambda \int [a^\dagger(\mathbf{k}\lambda)a(\mathbf{k}\lambda) + \tfrac{1}{2}]\hbar\omega\, d^3\mathbf{k} \tag{18.5.60}$$

‡ A small point, in case you are following all the details: a and $a^\dagger$ above are the operators corresponding to the classical variables $a/\hbar^{1/2}$ and $a^*/\hbar^{1/2}$. To see this, invert Eq. (18.5.55). All we need hereafter are Eqs. (18.5.57)–(18.5.59).

§ We use the same symbols for the classical and quantum variables in order to follow a widely used convention in this case. It should be clear from the context which is which.

Let us now consider the eigenstates of H. In the field ground state $|0\rangle$, all the oscillators are in their respective ground states. Thus any lowering operator will annihilate $|0\rangle$:

$$a(\mathbf{k}\lambda)|0\rangle=0 \qquad \text{for all } \mathbf{k}, \lambda \tag{18.5.61}$$

The energy of this state, called the vacuum state or simply *vacuum*, is

$$E_0=\sum_\lambda \int \frac{\hbar\omega}{2}\, d^3\mathbf{k} \tag{18.5.62}$$

which is the sum over the zero point energies of the oscillators. This constant energy E_0 has no physical consequences.

We now verify the results claimed earlier. In this ground state

$$\begin{aligned}\langle 0|\mathbf{A}|0\rangle &\sim \langle 0|(a+a^\dagger)|0\rangle=0\\ \langle 0|\mathbf{\Pi}|0\rangle &\sim \langle 0|(a-a^\dagger)|0\rangle=0\end{aligned} \tag{18.5.63}$$

In the above equation we have omitted a lot of irrelevant factors; only the central idea—that $\mathbf{A}$ and $\mathbf{\Pi}$ are linear combinations of creation and destruction operators and hence have no diagonal matrix elements in $|0\rangle$—is emphasized. On the other hand,

$$\begin{aligned}\langle 0|\,|\mathbf{A}|^2|0\rangle &\neq 0\\ \langle 0|\,|\mathbf{\Pi}|^2|0\rangle &\neq 0\end{aligned} \tag{18.5.64}$$

for the same reason that $\langle X^2\rangle \neq 0$, $\langle P^2\rangle \neq 0$ for a single oscillator.

If we act on $|0\rangle$ with one of the raising operators, we get

$$a^\dagger(\mathbf{k}\lambda)|0\rangle=|\mathbf{k}\lambda\rangle \tag{18.5.65}$$

where the labels $\mathbf{k}$ and λ tell us that the oscillator bearing that label has gone to its first excited level. This state has energy $\hbar\omega=\hbar kc$ above E_0 as may be verified by letting H act on it and using Eqs. (18.5.58) and (18.5.61). What about the momentum content? Any standard textbook on electrodynamics will tell us that the momentum of the field is given, in classical physics, by

$$\mathscr{P}=\frac{1}{4\pi c}\int (\mathbf{E}\times\mathbf{B})\, d^3\mathbf{r} \tag{18.5.66}$$

If we calculate the corresponding quantum operator we will find that it is given by

$$\mathbf{P}=\sum_\lambda \int [a^\dagger(\mathbf{k}\lambda)a(\mathbf{k}\lambda)]\hbar\mathbf{k}\, d^3\mathbf{k} \tag{18.5.67}$$

It is clear on inspection or explicit operation that

$$\mathbf{P}|\mathbf{k}\lambda\rangle = \hbar\mathbf{k}|\mathbf{k}\lambda\rangle \tag{18.5.68}$$

Thus the state $|\mathbf{k}\lambda\rangle$ has momentum $\hbar\mathbf{k}$.

If we apply $a^\dagger(\mathbf{k}\lambda)$ on the vacuum n times, we will create a state with energy $n\hbar\omega$ and momentum $n\hbar\mathbf{k}$. This allows us to view the action of $a^\dagger(\mathbf{k}\lambda)$ as the creation of particles of momenta $\hbar\mathbf{k}$ and energy $\hbar\omega$. These particles, called photons, are massless since

$$m^2c^4 = E^2 - c^2p^2 = (\hbar\omega)^2 - (\hbar kc)^2 = 0 \tag{18.5.69}$$

In terms of photons, we have the correspondence

$$\begin{Bmatrix}\text{quantum state}\\ \text{of field}\end{Bmatrix} \leftrightarrow \begin{Bmatrix}\text{quantum state of}\\ \text{each oscillator}\end{Bmatrix} \leftrightarrow \begin{Bmatrix}\text{number of photons}\\ \text{at each } \mathbf{k} \text{ and } \lambda\end{Bmatrix}$$

For future use, let us obtain the wave function of the photon in the state $(\mathbf{k}, \lambda)$. We begin by deducing the normalization of the states. Combining Eqs. (18.5.65) and (18.5.58) we get

$$\begin{aligned}\langle \mathbf{k}'\lambda'|\mathbf{k}\lambda\rangle &= \langle 0|a(\mathbf{k}'\lambda')a^\dagger(\mathbf{k}\lambda)|0\rangle\\ &= \langle 0|a^\dagger a + \delta_{\lambda\lambda'}\delta^3(\mathbf{k}-\mathbf{k}')|0\rangle\\ &= \delta_{\lambda\lambda'}\delta^3(\mathbf{k}-\mathbf{k}')\end{aligned} \tag{18.5.70}$$

(assuming $\langle 0|0\rangle = 1$). The $\delta^3(\mathbf{k}-\mathbf{k}')$ factor and the fact that $\hbar\mathbf{k}$ is the momentum of the state tell us that the wave function corresponding to $|\mathbf{k}, \lambda\rangle$ is

$$\psi \sim \frac{1}{(2\pi)^{3/2}} e^{i\mathbf{k}\cdot\mathbf{r}} \tag{18.5.71}$$

We use the $\sim$ sign instead of the $\rightarrow$ sign because λ has not entered the wave function yet. From the $\delta_{\lambda\lambda'}$ factor and the way λ entered the picture in the first place, we conclude that λ represents the polarization vector:

$$|\mathbf{k}\lambda\rangle \rightarrow \frac{\boldsymbol{\varepsilon}(\mathbf{k}\lambda)\, e^{i\mathbf{k}\cdot\mathbf{r}}}{(2\pi)^{3/2}} \tag{18.5.72}$$

You may be unhappy over the fact that unlike the $e^{i\mathbf{k}\cdot\mathbf{r}}/(2\pi)^{3/2}$ factor, which followed from analyzing the momentum content of the state [i.e., from the analysis of Eq. (18.5.68)], the $\boldsymbol{\varepsilon}$ was pulled out of a hat. It too may be deduced, starting with angular momentum considerations. We do not do so here.

Since the wave function of the photon is not a scalar, it has spin. Furthermore, since $\boldsymbol{\varepsilon}$ is a three-component object, the spin is unity. However, the requirement that $\mathbf{k}\cdot\boldsymbol{\varepsilon} = 0$ imposes a constraint on the possible orientations of photon spin. Consider, for example, a photon moving along the z axis. The condition $\mathbf{k}\cdot\boldsymbol{\varepsilon} = 0$ tells us that

$\boldsymbol{\varepsilon}$ cannot have a component along the z axis. What does this mean? The component of $\boldsymbol{\varepsilon}$ parallel to the z axis is characterized by the fact that it remains invariant under rotations around the z axis, i.e., transforms like an $s_z=0$ state. So we conclude that the photon can have only $s_z=\pm\hbar$, but not $s_z=0$. More generally, the spin of the photon can only take values $\pm\hbar$ parallel to its momentum. The component of spin parallel to momentum is called *helicity*. The transversality condition restricts the helicity to be $\pm\hbar$—it precludes helicity zero.‡

We consider one last feature of photons before turning to the problem that started this inquiry, namely, spontaneous decay. Consider a state with one photon in $(\mathbf{k}\lambda)$ and another in $(\mathbf{k}'\lambda')$:

$$|\mathbf{k}\lambda, \mathbf{k}'\lambda'\rangle = a^\dagger(\mathbf{k}\lambda)a^\dagger(\mathbf{k}'\lambda')|0\rangle \tag{18.5.73}$$

If we exchange the photon states we get the state

$$|\mathbf{k}'\lambda', \mathbf{k}\lambda\rangle = a^\dagger(\mathbf{k}'\lambda')a^\dagger(\mathbf{k}\lambda)|0\rangle \tag{18.5.74}$$

But since $[a^\dagger, a^\dagger]=0$, the two state vectors coincide, as they should for identical bosons.

Spontaneous Decay

Consider the spontaneous decay of the hydrogen atom from $|2lm\rangle$ to $|100\rangle$. The perturbing Hamiltonian is still given by the substitution rule

$$\mathscr{H}^1 = \frac{e}{mc}\mathbf{A}\cdot\mathbf{p} \to H^1 = \frac{e}{mc}\mathbf{A}\cdot\mathbf{P} \tag{18.5.75}$$

but the $\mathbf{A}$ in H^1 is now the operator in Eq. (18.5.59a).

The initial state of the system (atom + field) is

$$|i^0\rangle = |2lm\rangle \otimes |0\rangle \tag{18.5.76}$$

The final state is

$$|f^0\rangle = |100\rangle \otimes |\mathbf{k}\lambda\rangle \tag{18.5.77}$$

The perturbation H^1 is time independent ($\mathbf{A}$ is the operator in the Schrödinger picture) and

$$E_f^0 - E_i^0 = E_{100} + \hbar\omega - E_{2lm} \tag{18.5.78}$$

‡ The graviton, which is massless and has spin 2, also has only two helicity states, $\pm 2\hbar$. This is a general feature of massless bosons with spin.

From Fermi's golden rule, we get‡

$$R_{i\to f}=\frac{2\pi}{\hbar}\left|\langle f^0|\frac{e}{mc}\mathbf{A}\cdot\mathbf{P}|i^0\rangle\right|^2\delta(E_{100}+\hbar\omega-E_{2lm}) \tag{18.5.79}$$

Consider

$$\langle f^0|\mathbf{A}\cdot\mathbf{P}|i^0\rangle=\langle 100|\langle\mathbf{k}\lambda|\mathbf{A}|0\rangle\cdot\mathbf{P}|2lm\rangle \tag{18.5.80}$$

Now, $\mathbf{A}$ is a sum over a's and $a^\dagger$'s with different labels. The only relevant one is $a^\dagger(\mathbf{k}\lambda)$, which raises $|0\rangle$ to $|\mathbf{k}\lambda\rangle$. Thus, including the factors that accompany $a^\dagger(\mathbf{k}\lambda)$,

$$\langle\mathbf{k}\lambda|\mathbf{A}|0\rangle=\left(\frac{\hbar c^2}{4\pi^2\omega}\right)^{1/2}\boldsymbol{\varepsilon}(\mathbf{k}\lambda)\,e^{i\mathbf{k}\cdot\mathbf{r}} \tag{18.5.81}$$

so that

$$\langle f^0|\mathbf{A}\cdot\mathbf{P}|i^0\rangle=\left(\frac{\hbar c^2}{4\pi^2\omega}\right)^{1/2}\int\psi^*_{100}\,e^{i\mathbf{k}\cdot\mathbf{r}}\boldsymbol{\varepsilon}\cdot(-i\hbar\nabla)\psi_{2lm}\,d^3\mathbf{r}$$

In the dipole approximation, this becomes, upon using Eq. (18.5.17),§

$$\langle f^0|\mathbf{A}\cdot\mathbf{P}|i^0\rangle=\left(\frac{\hbar c^2}{4\pi^2\omega}\right)^{1/2}(im\omega)\int\psi^*_{100}\boldsymbol{\varepsilon}\cdot\mathbf{r}\psi_{2lm}\,d^3\mathbf{r} \tag{18.5.82}$$

From parity considerations, it is clear that only $l=1$ is relevant. Writing $\boldsymbol{\varepsilon}\cdot\mathbf{r}$ in the spherical basis (recall Exercise 15.3.2),

$$\begin{aligned}\boldsymbol{\varepsilon}\cdot\mathbf{r}&=\sum_{-1}^{+1}(-1)^q\varepsilon_1^q r_1^{-q}\\&=-\varepsilon_1^1 r_1^{-1}+\varepsilon_1^0 r_1^0-\varepsilon_1^{-1}r_1^{+1}\end{aligned} \tag{18.5.83}$$

where

$$\varepsilon_1^{\pm1}=\mp\frac{\varepsilon_x\pm i\varepsilon_y}{2^{1/2}},\qquad \varepsilon_1^0=\varepsilon_z \tag{18.5.84}$$

‡ In the photoelectric effect, the field is treated as an external time-dependent perturbation that acts on the atom, and the $\hbar\omega$ in the delta function reflects this time dependence. In the present case, the field is part of the system and the $\hbar\omega$ stands for the change in its energy.

§ We are unfortunately forced to use the symbol m for the mass as well as the z component of angular momentum. It should be clear from the context what m stands for.

and from Eq. (12.5.42),

$$r_1^{\pm 1} = \left(\frac{4\pi}{3}\right)^{1/2} rY_1^{\pm 1}, \qquad r_1^0 = \left(\frac{4\pi}{3}\right)^{1/2} rY_1^0 \tag{18.5.85}$$

we get

$$\begin{aligned}\int \psi_{100}^* \boldsymbol{\varepsilon} \cdot \mathbf{r} \psi_{21m} d^3\mathbf{r} &= \left(\frac{4\pi}{3}\right)^{1/2} \int R_{10} r R_{21} r^2 dr \\ &\quad \times \left[\int Y_0^{0*} \left(-\varepsilon_1^1 Y_1^{-1} + \varepsilon_1^0 Y_1^0 - \varepsilon_1^{-1} Y_1^{+1}\right) Y_1^m d\Omega\right] \\ &= \left(\frac{3}{2}\right)^{1/2} \frac{2^8}{3^5} \frac{a_0}{3^{1/2}} \left(\varepsilon_1^1 \delta_{m,+1} + \varepsilon_1^0 \delta_{m,0} + \varepsilon_1^{-1} \delta_{m,-1}\right)\end{aligned} \tag{18.5.86}$$

The evaluation of the integrals (like so many other steps in this high-speed treatment) is left as an exercise. The modulus squared of the above quantity is

$$\frac{3}{2} \frac{2^{16}}{3^{10}} \frac{a_0^2}{3} \left[\left|\varepsilon_1^1\right|^2 \delta_{m,-1} + \left|\varepsilon_1^0\right|^2 \delta_{m,0} + \left|\varepsilon_1^{-1}\right|^2 \delta_{m,1}\right]$$

If we average over the three initial m's (i.e., over an ensemble of such atoms randomly distributed with respect to m), this reduces to

$$\frac{3}{2} \frac{2^{16}}{3^{10}} \frac{a_0^2}{3} \frac{1}{3} \left(\varepsilon_x^2 + \varepsilon_y^2 + \varepsilon_z^2\right) = \frac{2^{15} a_0^2}{3^{11}} \tag{18.5.87}$$

Notice that the result is independent of the direction of $\boldsymbol{\varepsilon}$. This is to be expected since the atom has no sense of direction after the angular (m) averaging. The transition rate is

$$R_{\bar{i} \to f} = \frac{2\pi}{\hbar} \left(\frac{e}{mc}\right)^2 \frac{\hbar c^2}{4\pi^2 \omega} m^2 \omega^2 \frac{2^{15} a_0^2}{3^{11}} \delta(E_{100} + \hbar\omega - E_{2lm}) \tag{18.5.88}$$

where $\bar{i}$ means the initial state is averaged over all orientations.

If we sum over all possible photon momenta and two possible polarizations at each momentum, we get, using

$$\int \delta(E_{100} + \hbar\omega - E_{2lm}) k^2 \, dk \, d\Omega = \frac{4\pi k^2}{\hbar c}$$

where

$$k=\frac{\omega}{c}=\frac{E_{2lm}-E_{100}}{\hbar c}=\frac{e^2}{2a_0\hbar c}\left(1-\frac{1}{4}\right)=\frac{3e^2}{8a_0\hbar c}$$

the total decay rate

$$R_{\bar{i}\to\text{all}}=\left(\frac{2}{3}\right)^8\alpha^5\frac{mc^2}{\hbar} \tag{18.5.89}$$

Recall that

$$\frac{mc^2}{\hbar}=\frac{mc^2c}{\hbar c}\simeq\frac{0.5\times10^6\text{ eV }c}{2000\text{ eV Å}}$$
$$\simeq 0.25\times10^3\text{ Å}^{-1}\,c$$

Now $c=3\times10^{10}$ cm/sec $=3\times10^{18}$ Å/sec. So

$$\frac{mc^2}{\hbar}\simeq10^{21}\text{ sec}^{-1}$$

and

$$R_{\bar{i}\to\text{all}}\simeq(0.67)^8\left(\frac{1}{137}\right)^5 10^{21}\text{ seconds}^{-1}$$
$$\simeq 0.6\times10^9\text{ seconds}^{-1}$$

The corresponding mean lifetime is

$$\tau=1/R\simeq1.6\times10^{-9}\text{ seconds} \tag{18.5.90}$$

in excellent agreement with experiment.

Even if the fields are macroscopic, we can use the full quantum theory, though the semiclassical treatment will give virtually identical results. The relation of the two approaches may be described as follows. Consider a process in which an atom goes from the state i_a to the state f_a and the field goes from the state with n photons in $(\mathbf{k}, \lambda)$ to $n+1$ photons in $(\mathbf{k}, \lambda)$.‡ The result we get in the quantum mechanical treatment of this process, which involves the *emission* of a photon, will agree with the semiclassical calculation if we use a classical field $\mathbf{A}$ whose energy density§ is the same as that of $(n+1)$ photons in $(\mathbf{k}, \lambda)$. The 1 in $n+1$ is all important at small n, and contains the key to spontaneous decay. If we consider a process where a photon is *absorbed*, so that $n\to n-1$, the semiclassical method gives the correct answer if we

‡ We do not concern ourselves with other modes, which are spectators.
§ The wavelength and polarization are of course the same as that of the photons.

use a classical field **A** such that the energy density is that of the n photons. The appearance of the $(n+1)$ and n factors is easy to understand in the oscillator language. When a photon is created, the amplitude goes as

$$\langle n+1|a^{\dagger}|n\rangle=(n+1)^{1/2}\langle n+1|n+1\rangle \tag{18.5.91}$$

which gives the factor $(n+1)$ in the probability, while if it is destroyed,

$$\langle n-1|a|n\rangle=n^{1/2}\langle n-1|n-1\rangle \tag{18.5.92}$$

which gives a factor n in the probability.

It is conventional to separate the emission probability proportional to $n+1$ into the *probability for induced emission*, proportional to n, and the *probability for spontaneous emission*, proportional to 1. The induced emission is induced by the preexisting photons, and the spontaneous emission is—well, spontaneous.

The $(n+1)$ factor in the emission probability is a feature of bosons in general: the probability of a system emitting a boson into a quantum state already occupied by n bosons (of the same kind), is $(n+1)$ times larger than the probability of emission into that state if it is initially unoccupied. This principle is exploited in a laser, which contains a cavity full of atoms in an excited state, ready to emit photons of a fixed frequency but arbitrary directions for $\mathbf{k}$ and λ. The geometry of the cavity is such that photons of a certain $\mathbf{k}$ and λ get trapped in it. Consequently, these trapped photons stay back to influence more and more atoms to emit into the mode $(\mathbf{k}\lambda)$. This is why we call it *l*ight *a*mplification by *s*timulated *e*mission of *r*adiation. (This general principle, in modified form, is exploited in television also: this is the whole idea behind canned laughter.)

19

Scattering Theory

19.1. Introduction

One of the best ways to understand the structure of particles and the forces between them is to scatter them off each other. This is particularly true at the quantum level where the systems cannot be seen in the literal sense and must be probed by indirect means. The scattering process gives us information about the projectile, the target, and the forces between them. A natural way to proceed (when possible) is to consider cases where two of these are known and learn about the third. Consider, for example, experiments at the Stanford Linear Accelerator Center in which high-energy photons were used to bombard static neutrons. The structure of the photon and its coupling to matter are well understood—the photon is a point particle to an excellent approximation and couples to electric charge in a way we have studied in some detail. It therefore serves as an excellent probe of the neutron. For instance, the very fact that the neutron, which is electrically neutral, interacts with the photon tells us that the neutron is built out of charged constituents (whose total charge add up to zero). These scattering experiments also revealed that the neutron's constituents have spin $\frac{1}{2}$, and fractional charges ($\frac{2}{3}e$, $-\frac{1}{3}e$), a picture that had been arrived at from another independent line of reasoning. Furthermore they also indicated that the interaction between these constituents (called quarks) gets very weak as they get close. This information has allowed us to choose, from innumerable possible models of the interquark force, one that is now considered most likely to succeed, and goes by the name of quantum chromodynamics (QCD), a subject that is being vigorously investigated by many particle physicists today.

Scattering theory is a very extensive subject and this chapter aims at giving you just the flavor of the basic ideas. For more information, you must consult books devoted to this subject.‡

A general scattering event is of the form

$$a(\alpha)+b(\beta)+\cdots \rightarrow f(\gamma)+g(\delta)+\cdots$$

where $\{a, b, \ldots\}$ are particle names and $\{\alpha, \beta, \gamma, \ldots\}$ are the kinematical variables

‡ See, for example, the excellent book by J. R. Taylor, *Scattering Theory*, Wiley, New York (1971). Any details omitted here due to lack of space may be found there.

specifying their states, such as momentum, spin, etc. We are concerned only with nonrelativistic, elastic scattering of structureless spinless particles.

In the next three sections, we deal with a formalism that describes a single particle scattering from a potential $V(\mathbf{r})$. As it stands, the formalism describes a particle colliding with an immobile target whose only role is to provide the potential. (This picture provides a good approximation to processes where a light particle collides with a very heavy one, say an α particle colliding with a heavy nucleus.) In Section 19.6 we see how, upon proper interpretation, the same formalism describes two-body collisions in the CM frame. In that section we will also see how the description of the scattering process in the CM frame can be translated to another frame, called the lab frame, where the target is initially at rest. It is important to know how to pass from one frame to the other, since theoretical calculations are most easily done in the CM frame, whereas most experiments are done in the lab frame.

19.2. Recapitulation of One-Dimensional Scattering and Overview

Although we are concerned here with scattering in three dimensions, we begin by recalling one-dimensional scattering, for it shares many common features with its three-dimensional counterpart. The practical question one asks is the following: If a beam of nearly monoenergetic particles with mean momenta $\langle P\rangle=\hbar k_0$ are incident from the far left ($x\to-\infty$) on a potential $V(x)$ which tends to zero as $|x|\to\infty$, what fraction T will get transmitted and what fraction R will get reflected?‡ It is not *a priori* obvious that the above question can be answered, since the mean momentum does not specify the quantum states of the incoming particles. But it turns out that if the individual momentum space wave functions are sharply peaked at $\hbar k_0$, the reflection and transmission probabilities depend only on k_0 and not on the detailed shapes of the wave functions. Thus it is possible to calculate $R(k_0)$ and $T(k_0)$ that apply to every particle in the beam. Let us recall some of the details.

(1) We start with some wave packet, say a Gaussian, with $\langle P\rangle=\hbar k_0$ and $\langle X\rangle\to-\infty$.

(2) We expand this packet in terms of the eigenfunctions ψ_k of $H=T+V$ with coefficients $a(k)$. The functions ψ_k have the following property:

$$\psi_k \xrightarrow[x\to-\infty]{} A\,e^{-ikx}+B\,e^{ikx}$$

$$\xrightarrow[x\to\infty]{} C\,e^{ikx} \tag{19.2.1}$$

In other words, the asymptotic form of ψ_k contains an incident wave $A\,e^{ikx}$ and a reflected wave $B\,e^{-ikx}$ as $x\to-\infty$, and just a transmitted wave $C\,e^{ikx}$ as $x\to\infty$. Although the most general solution also contains a $D\,e^{-ikx}$ piece as $x\to\infty$, we set

‡ In general, the particle can come in from the far right as well. Also $V(x)$ need not tend to zero at both ends, but to constants V_+ and V_- as $x\to\pm\infty$. We assume $V_+=V_-=0$ for simplicity. We also assume $|xV(x)|\to 0$ as $|x|\to\infty$, so that the particle is asymptotically free ($\psi\sim e^{\pm ikx}$).

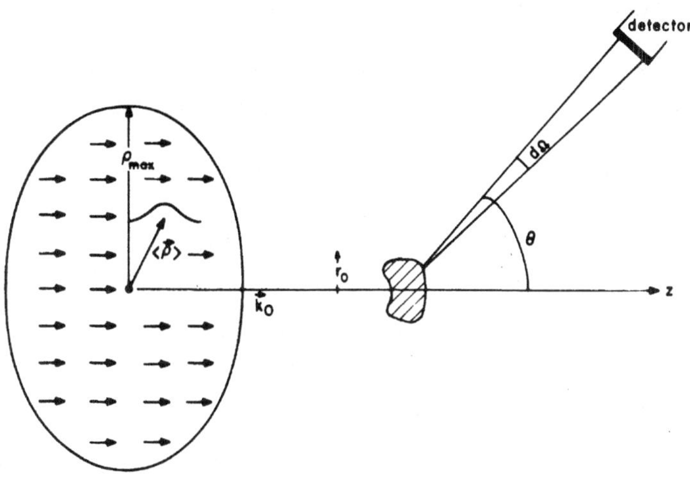

Figure 19.1. A schematic description of scattering. The incident particles, shown by arrows, are really described by wave packets (only one is shown) with mean momentum $\langle \mathbf{P} \rangle = \langle \hbar \mathbf{k}_0 \rangle$ and mean impact parameter $\langle \boldsymbol{\rho} \rangle$ uniformly distributed in the $\boldsymbol{\rho}$-plane out to $\boldsymbol{\rho}_{\max} \gg r_0$, the range of the potential. The shaded region near the origin stands for the domain where the potential is effective. The detector catches all particles that emerge in the cone of opening angle $d\Omega$. The beam is assumed to be coming in along the z axis.

$D=0$ on physical grounds: the incident wave $A\,e^{ikx}$ can only produce a right-going wave as $x\to\infty$.

(3) We propagate the wave packet in time by attaching to the expansion coefficients $a(k)$ the time dependence $e^{-iEt/\hbar}$, where $E=\hbar^2k^2/2\mu$. We examine the resulting solution as $t\to\infty$ and identify the reflected and transmitted packets. From the norms of these we get R and T respectively.

(4) We find at this stage that if the incident packet is sharply peaked in momentum space at $\hbar k_0$, R and T depend only on k_0 and not on the detailed shape of the wave function. Thus the answer to the question raised at the outset is that a fraction $R(k_0)$ of the incident particles will get reflected and a fraction $T(k_0)$ will get transmitted.

(5) Having done all this hard work, we find at the end that the same result could have been obtained by considering just one eigenfunction ψ_{k_0} and taking the ratios of the transmitted and reflected current densities to the incident current density.

The scattering problem in three dimensions has many similarities with its one-dimensional counterpart and also several differences that inevitably accompany the increase in dimensionality. First of all, the incident particles (coming out of the accelerator) are characterized, not by just the mean momentum $\langle \mathbf{P} \rangle = \hbar \mathbf{k}_0$, but also by the fact that they are uniformly distributed in the *impact parameter* $\boldsymbol{\rho}$, which is the coordinate in the plane perpendicular to $\mathbf{k}_0$ (Fig. 19.1). The distribution is of course not uniform out to $\rho\to\infty$, but only up to $\rho_{\max}\gg r_0$, where r_0, the *range of the potential*, is the distance scale beyond which the potential is negligible. [For instance, if $V(r)=e^{-r^2/a^2}$, the range $r_0\cong a$.] The problem is to calculate the rate at which particles get scattered into a far away detector that subtends a solid angle $d\Omega$ in the direction (θ, ϕ) measured relative to the beam direction (Fig. 19.1). To be precise, one wants the *differential cross section* $d\sigma/d\Omega$ defined as follows:

$$\frac{d\sigma(\theta,\phi)}{d\Omega}d\Omega = \frac{\text{number of particles scattered into } d\Omega/\text{ sec}}{\text{number incident/ sec /area in the } \boldsymbol{\rho} \text{ plane}} \tag{19.2.2}$$

The calculation of $d\sigma/d\Omega$ proceeds as follows.‡

(1) One takes some initial wave packet with mean momentum $\langle \mathbf{P}\rangle = \hbar\mathbf{k}_0$ and mean impact parameter $\langle \boldsymbol{\rho}\rangle$. The mean coordinate in the beam direction is not relevant, as long as it is far away from the origin.

(2) One expands the wave packet in terms of the eigenfunctions $\psi_{\mathbf{k}}$ of $H = T + V$ which are of the form

$$\psi_{\mathbf{k}} = \psi_{\text{inc}} + \psi_{\text{sc}} \tag{19.2.3}$$

where ψ_{inc} is the incident wave $e^{i\,\mathbf{k}\cdot\mathbf{r}}$ and ψ_{sc} is the scattered wave. One takes only those solutions in which ψ_{sc} is purely outgoing. We shall have more to say about ψ_{sc} in a moment.

(3) One propagates the wave packet by attaching the time-dependence factor $e^{-iEt/\hbar}(E = \hbar^2k^2/2\mu)$ to each coefficient $a(\mathbf{k})$ in the expansion.

(4) One identifies the scattered wave as $t \to \infty$, and calculates the probability current density associated with it. One integrates the total flow of probability into the cone $d\Omega$ at (θ, ϕ). This gives the probability that the incident particle goes into the detector at (θ, ϕ). One finds that if the momentum space wave function of the incident wave packet is sharply peaked at $\langle \mathbf{P}\rangle = \hbar\mathbf{k}_0$, the probability of going into $d\Omega$ depends only on $\hbar\mathbf{k}_0$ and $\langle\boldsymbol{\rho}\rangle$. Call this probability $P(\boldsymbol{\rho}, \mathbf{k}_0 \to d\Omega)$.

(5) One considers next a beam of particle with $\eta(\boldsymbol{\rho})$ particles per second per unit area in the $\boldsymbol{\rho}$ plane. The number scattering into $d\Omega$ per second is

$$\eta(d\Omega) = \int P(\boldsymbol{\rho}, \mathbf{k}_0 \to d\Omega)\eta(\boldsymbol{\rho})\, d^2\boldsymbol{\rho} \tag{19.2.4}$$

Since in the experiment $\eta(\boldsymbol{\rho}) = \eta$, a constant, we have from Eq. (19.2.2)

$$\frac{d\sigma}{d\Omega}d\Omega = \frac{\eta(d\Omega)}{\eta} = \int P(\boldsymbol{\rho}, \mathbf{k}_0 \to d\Omega)\, d^2\boldsymbol{\rho} \tag{19.2.5}$$

(6) After all this work is done one finds that $d\sigma/d\Omega$ could have been calculated from considering just the static solution $\psi_{\mathbf{k}_0}$ and computing in the limit $r \to \infty$, the ratio of the probability flow per second into $d\Omega$ associated with ψ_{sc}, to the incident probability current density associated with $e^{i\mathbf{k}_0\cdot\mathbf{r}}$. The reason the time-dependent picture reduces to the time-independent picture is the same as in one dimension: as we broaden the incident wave packet more and more in coordinate space, the incident and scattered waves begin to coexist in a steady-state configuration, $\psi_{\mathbf{k}_0}$. What about

‡ We do not consider the details here, for they are quite similar to the one-dimensional case. The few differences alone are discussed. See Taylor's book for the details.

the average over $\langle \boldsymbol{\rho} \rangle$? This is no longer necessary, since the incident packet is now a plane wave $e^{i\mathbf{k}_0\cdot\mathbf{r}}$ which is already uniform in $\boldsymbol{\rho}$.‡

Let us consider some details of extracting $d\sigma/d\Omega$ from $\psi_{\mathbf{k}_0}$. Choosing the z axis parallel to $\mathbf{k}_0$ and dropping the subscript 0, we obtain

$$\psi_{\mathbf{k}} = e^{ikz} + \psi_{sc}(r, \theta, \phi) \tag{19.2.6}$$

where θ and ϕ are defined in Fig. 19.1. Although the detailed form of ψ_{sc} depends on the potential, we know that far from the origin it satisfies the free-particle equation [assuming $rV(r)\to 0$ as $r\to\infty$].

$$(\nabla^2 + k^2)\psi_{sc} = 0 \qquad (r\to\infty) \tag{19.2.7}$$

and is purely outgoing.

Recalling the general solution to the free-particle equation (in a region that excludes the origin) we get

$$\psi_{sc} \underset{r\to\infty}{\longrightarrow} \sum_l \sum_m (A_l j_l(kr) + B_l n_l(kr)) Y_l^m(\theta, \phi) \tag{19.2.8}$$

Notice that we do not exclude the Neumann functions because they are perfectly well behaved as $r\to\infty$. Since

$$\begin{aligned} j_l(kr) &\underset{r\to\infty}{\longrightarrow} \sin(kr - l\pi/2)/(kr) \\ n_l(kr) &\underset{r\to\infty}{\longrightarrow} -\cos(kr - l\pi/2)/(kr) \end{aligned} \tag{19.2.9}$$

it must be that $A_l/B_l = -i$, so that we get a purely outgoing wave e^{ikr}/kr. With this condition, the asymptotic form of the scattered wave is

$$\psi_{sc} \underset{r\to\infty}{\longrightarrow} \frac{e^{ikr}}{kr} \sum_l \sum_m (-i)^l (-B_l) Y_l^m(\theta, \phi) \tag{19.2.10}$$

or

$$\psi_{sc} \underset{r\to\infty}{\longrightarrow} \frac{e^{ikr}}{r} f(\theta, \phi)\S \tag{19.2.11}$$

and

$$\psi_{\mathbf{k}} \underset{r\to\infty}{\longrightarrow} e^{ikz} + f(\theta, \phi)\frac{e^{ikr}}{r} \tag{19.2.12}$$

where f is called the *scattering amplitude*.

‡ Let us note, as we did in one dimension, that a wave packet does not simply become a plane wave as we broaden it, for the former has norm unity and the latter has norm $\delta^3(0)$. So it is assumed that as the packet is broadened, its norm is steadily increased in such a way that we end up with a plane wave. In any case, the overall norm has no significance.

§ Actually f also depends on k; this dependence is not shown explicitly.

To get the differential cross section, we need the ratio of the probability flowing into $d\Omega$ per second to the incident current density. So what are $\mathbf{j}_{sc}$ and $\mathbf{j}_{inc}$, the incident and scattered current densities? Though we have repeatedly spoken of these quantities, they are not well defined unless we invoke further physical ideas. This is because there is only one current density $\mathbf{j}$ associated with $\psi_{\mathbf{k}}$ and it is *quadratic* in $\psi_{\mathbf{k}}$. So $\mathbf{j}$ is not just a sum of two pieces, one due to e^{ikz} and one due to ψ_{sc}; there are cross terms.‡ We get around this problem as follows. We note that as $r\to\infty$, ψ_{sc} is negligible compared to e^{ikz} because of the $1/r$ factor. So we calculate the incident current due to e^{ikz} to be

$$|j_{inc}|=\left|\frac{\hbar}{2\mu i}(e^{-ikz}\nabla e^{ikz}-e^{ikz}\nabla e^{-ikz})\right|$$
$$=\frac{\hbar k}{\mu} \tag{19.2.13}$$

We cannot use this trick to calculate $\mathbf{j}_{sc}$ into $d\Omega$ because ψ_{sc} never dominates over e^{ikz}. So we use another trick. We say that e^{ikz} is really an abstraction for a wave that is limited in the transverse direction by some $\rho_{max}(\gg r_0)$. Thus in any realistic description, only ψ_{sc} will survive as $r\to\infty$ for $\theta\neq 0$.§ (For a given ρ_{max}, the incident wave is present only for $\delta\theta\lesssim\rho_{max}/r$. We can make $\delta\theta$ arbitrarily small by increasing the r at which the detector is located.) With this in mind we calculate (for $\theta\neq 0$)

$$\mathbf{j}_{sc}=\frac{\hbar}{2\mu i}(\psi_{sc}^*\nabla\psi_{sc}-\psi_{sc}\nabla\psi_{sc}^*) \tag{19.2.14}$$

Now

$$\nabla=\mathbf{e}_r\frac{\partial}{\partial r}+\mathbf{e}_\theta\frac{\partial}{r\,\partial\theta}+\mathbf{e}_\phi\frac{1}{r\sin\theta}\frac{\partial}{\partial\phi} \tag{19.2.15}$$

The last two pieces in ∇ are irrelevant as $r\to\infty$. When the first acts on the asymptotic ψ_{sc},

$$\frac{\partial}{\partial r}f(\theta,\phi)\frac{e^{ikr}}{r}=f(\theta,\phi)ik\frac{e^{ikr}}{r}+O\left(\frac{1}{r^2}\right)$$

so that

$$\mathbf{j}_{sc}=\frac{\mathbf{e}_r}{r^2}|f|^2\frac{\hbar k}{\mu} \tag{19.2.16}$$

‡ We did not have to worry about this in one dimension because j due to $A\,e^{ikx}+B\,e^{-ikx}$ is $(\hbar k/\mu)(|A|^2-|B|^2)=j_{inc}+j_{ref}$ with no cross terms.

§ In fact, only in this more realistic picture is it sensible to say that the particles entering the detectors at $\theta\neq 0$ are scattered (and not unscattered incident) particles. At $\theta=0$, there is no way (operationally) to separate the incident and scattered particles. To compare theory with experiment, one extracts $f(\theta=0)$ by extrapolating $f(\theta)$ from $\theta\neq 0$.

Probability flows into $d\Omega$ at the rate

$$R(d\Omega)=\mathbf{j}_{sc}\cdot\mathbf{e}_r r^2 d\Omega$$
$$=|f|^2\frac{\hbar k}{\mu}d\Omega \tag{19.2.17}$$

Since it arrives at the rate

$$j_{\text{inc}}=\hbar k/\mu \text{ sec}^{-1}\text{ area}^{-1}$$

$$\frac{d\sigma}{d\Omega}d\Omega=\frac{R(d\Omega)}{j_{\text{inc}}}=|f|^2 d\Omega$$

so that finally

$$\frac{d\sigma}{d\Omega}=|f(\theta,\phi)|^2 \tag{19.2.18}$$

Thus, in the time-independent picture, the calculation of $d\sigma/d\Omega$ reduces to the calculation of $f(\theta,\phi)$.

After this general discussion, we turn to specific calculations. In the next section the calculation of $d\sigma/d\Omega$ is carried out in the time-dependent picture *to first order*. In Section 4, we calculate $d\sigma/d\Omega$ to first order in the time-independent picture. (The two results agree, of course.) In Section 5, we go beyond perturbation theory and discuss some general features of f for spherically symmetric potentials. Two-particle scattering is discussed in Section 6.

19.3. The Born Approximation (Time-Dependent Description)

Consider an initial wave packet that is so broad that it can be approximated by a plane wave $|\mathbf{p}_i\rangle$. Its fate after scattering is determined by the propagator $U(t_f\to\infty, t_i\to-\infty)$, that is, by the operator

$$S=\lim_{\substack{t_f\to\infty\\ t_i\to-\infty}} U(t_f,t_i)$$

which is called the *S matrix*. The probability of the particle entering the detector in the direction (θ,ϕ) with opening angle $d\Omega$ is the probability that the final momentum $\mathbf{p}_f$ lies in a cone of opening angle $d\Omega$ in the direction (θ,ϕ):

$$P(\mathbf{p}_i\to d\Omega)=\sum_{\mathbf{p}_f\text{ in }d\Omega}|\langle\mathbf{p}_f|S|\mathbf{p}_i\rangle|^2$$

If we evaluate S or U to first order, treating V as a perturbation, the problem reduces to the use of Fermi's Golden Rule, which tells us that the transition rate is

$$R_{i\to d\Omega}=\frac{dP(\mathbf{p}_i\to d\Omega)}{dt}$$

$$=\frac{2\pi}{\hbar}\left[\int_0^\infty |\langle\mathbf{p}_f|V|\mathbf{p}_i\rangle|^2\delta\left(\frac{p_f^2}{2\mu}-\frac{p_i^2}{2\mu}\right)p_f^2dp_f\right]d\Omega \tag{19.3.1}$$

$$=\frac{2\pi}{\hbar}|\langle\mathbf{p}_f|V|\mathbf{p}_i\rangle|^2\mu\mathbf{p}_i d\Omega \tag{19.3.2}$$

(Hereafter $p_f=p_i=p=\hbar k$ is understood.) This transition rate is just the rate of the flow of probability into $d\Omega$. Since the probability comes in at a rate [recall $j=\rho v|\mathbf{p}_i\rangle\to(2\pi\hbar)^{-3/2}e^{i\mathbf{p}_i\cdot\mathbf{r}/\hbar}$]

$$j_{\text{inc}}=\frac{\hbar k}{\mu}\left(\frac{1}{2\pi\hbar}\right)^3 \tag{19.3.3}$$

in the direction $\mathbf{p}_i$, the differential cross section, which measures the rate at which probability is intercepted (and channeled off to $d\Omega$), is

$$\frac{d\sigma}{d\Omega}d\Omega=\frac{R_{i\to d\Omega}}{j_{\text{inc}}}=(2\pi)^4\mu^2\hbar^2|\langle\mathbf{p}_j|V|\mathbf{p}_i\rangle|^2d\Omega$$

$$\frac{d\sigma}{d\Omega}=\left|\frac{\mu}{2\pi\hbar^2}\int e^{-i\mathbf{q}\cdot\mathbf{r}'}V(\mathbf{r}')d^3\mathbf{r}'\right|^2 \tag{19.3.4}$$

where

$$\hbar\mathbf{q}=\mathbf{p}_f-\mathbf{p}_i \tag{19.3.5}$$

is the *momentum transferred* to the particle. For later reference note that

$$|\mathbf{q}|^2=|\mathbf{k}_f-\mathbf{k}_i|^2=2k^2(1-\cos\theta)=4k^2\sin^2(\theta/2) \tag{19.3.6}$$

Thus the dependence of $d\sigma/d\Omega$ on the incident energy and the scattering angle is through the combination $|\mathbf{q}|\equiv q=2k\ \sin(\theta/2)$.

By comparing Eqs. (19.3.4) and (19.2.18) we can get $f(\theta)$, up to a phase factor of unit modulus (relative to the incident wave). We shall see later that this factor is -1. So,

$$f(\theta,\phi)=\frac{-\mu}{2\pi\hbar^2}\int e^{-i\mathbf{q}\cdot\mathbf{r}'}V(\mathbf{r}')d^3\mathbf{r}' \tag{19.3.7}$$

Thus, in this *Born approximation, $f(\theta,\phi)=f(\mathbf{q})$ is just the Fourier transform of the potential with respect to momentum transfer* (*up to a constant factor*).

Hereafter we focus on potentials that are spherically symmetric: $V(\mathbf{r})=V(r)$. In this case, we can choose the z' direction parallel to $\mathbf{q}$ in the $d^3\mathbf{r}'$ integration, so that

$$f(\theta,\phi)=\frac{-\mu}{2\pi\hbar^2}\int e^{-iqr'\cos\theta'}V(r')\,d(\cos\theta')\,d\phi' r'^2\,dr'$$

$$=\frac{-2\mu}{\hbar^2}\int\frac{\sin qr'}{q}V(r')r'\,dr'$$

$$=f(\theta) \tag{19.3.8}$$

That f should be independent of ϕ in this case could have been anticipated. The incident wave e^{ikz} is insensitive to a change in ϕ, i.e., to rotations around the z axis. The potential, being spherically symmetric, also knows nothing about ϕ. It follows that f cannot pick up any dependence on ϕ. In the language of angular momentum, the incident wave has no l_z and this feature is preserved in the scattering. Consequently the scattered wave must also have no l_z, i.e., be independent of ϕ.

Let us calculate $f(\theta)$ for the *Yukawa potential*

$$V(r)=\frac{g\,e^{-\mu_0 r}}{r} \tag{19.3.9}$$

From Eq. (19.3.8),

$$f(\theta)=-\frac{2\mu g}{\hbar^2 q}\int_0^\infty\frac{e^{iqr'}-e^{-iqr'}}{2i}e^{-\mu_0 r'}\,dr'$$

$$=\frac{-2\mu g}{\hbar^2(\mu_0^2+q^2)} \tag{19.3.10}$$

$$\frac{d\sigma}{d\Omega}=\frac{4\mu^2 g^2}{\hbar^4[\mu_0^2+4k^2\sin^2(\theta/2)]^2} \tag{19.3.11}$$

If we now set $g=Ze^2$, $\mu_0=0$, we get the cross section for *Coulomb scattering* of a particle of charge e on a potential $\phi=Ze/r$ (or $V=Ze^2/r$):

$$\left.\frac{d\sigma}{d\Omega}\right|_{\text{Coulomb}}=\frac{\mu^2(Ze^2)^2}{4p^4\sin^4(\theta/2)}$$

$$=\frac{(Ze^2)^2}{16E^2\sin^4(\theta/2)} \tag{19.3.12}$$

where $E=p^2/2\mu$ is the kinetic energy of the incident particle. This answer happens to be exact quantum mechanically as well as classically. (It was calculated classically by Rutherford and is called the *Rutherford cross section.*) Although we managed to get the right $d\sigma/d\Omega$ by taking the $\mu_0\to 0$ limit of the Yukawa potential calculation, there are some fine points to note. First of all, the Coulomb potential cannot be handled by the formulation we have developed, since the potential does not vanish faster than r^{-1}. In other words, the asymptotic form

$$\psi \xrightarrow[r\to\infty]{} e^{ikz} + f(\theta, \phi)\frac{e^{ikr}}{r}$$

is not applicable here since the particle is never free from the influence of the potential. (This manifests itself in the fact that the total cross section is infinite: if we try to integrate $d\sigma/d\Omega$ over θ, the integral diverges as $\int d\theta/\theta^3$ as $\theta\to 0$.) It is, however, possible to define a scattering amplitude $f_c(\theta)$ in the following sense. One finds that as $r\to\infty$, there are positive energy eigensolutions to the Coulomb Hamiltonian of the form‡

$$\psi \xrightarrow[r\to\infty]{} \widetilde{e^{ikz}} + f_c(\theta)\left(\widetilde{\frac{e^{ikr}}{r}}\right) \tag{19.3.13}$$

where the tilde tells us that these are not actually plane or spherical waves, but rather these objects modified by the long-range Coulomb force. For example

$$\widetilde{\frac{e^{ikr}}{r}} = \frac{e^{i(kr-\gamma\ln kr)}}{r} \tag{19.3.14}$$

$$\gamma = \frac{Ze^2\mu}{\hbar^2 k} \tag{19.3.15}$$

is the distorted spherical wave, familiar to us from Section 12.6. By comparing the ratio of flux into $d\Omega$ to flux coming in (due to these distorted waves) one finds that

$$\frac{d\sigma}{d\Omega} = |f_c|^2$$

where

$$f_c(\theta) = -\frac{\gamma}{2k(\sin\theta/2)^2}\exp(-i\gamma\ln\sin^2\theta/2 + \text{const}) \tag{19.3.16}$$

and where the constant is *purely imaginary*. Comparing this to the Yukawa amplitude, Eq. (19.3.10), after setting $\mu_0 = 0$, $g = Ze^2$, we find agreement up to the exponential phase factor. This difference does not show up in $d\sigma/d\Omega$, but will show up when we consider identical-particle scattering later in this chapter.

‡ See A. Messiah, *Quantum Mechanics*, Wiley, New York (1966), page 422.

*Exercise 19.3.1.** Show that

$$\sigma_{\text{Yukawa}} = 16\pi r_0^2 \left(\frac{g\mu r_0}{\hbar^2}\right)^2 \frac{1}{1+4k^2 r_0^2}$$

where $r_0 = 1/\mu_0$ is the range. Compare σ to the geometrical cross section associated with this range.

*Exercise 19.3.2.** (1) Show that if $V(r) = -V_0\theta(r_0 - r)$,

$$\frac{d\sigma}{d\Omega} = 4r_0^2 \left(\frac{\mu V_0 r_0^2}{\hbar^2}\right)^2 \frac{(\sin\, qr_0 - qr_0\, \cos\, qr_0)^2}{(qr_0)^6}$$

(2) Show that as $kr_0 \to 0$, the scattering becomes isotropic and

$$\sigma \cong \frac{16\pi r_0^2}{9}\left(\frac{\mu\ V_0 r_0^2}{\hbar^2}\right)^2$$

*Exercise 19.3.3.** Show that for the *Gaussian potential*, $V(r) = V_0 e^{-r^2/r_0^2}$,

$$\frac{d\sigma}{d\Omega} = \frac{\pi r_0^2}{4}\left(\frac{\mu V_0 r_0^2}{\hbar^2}\right)^2 e^{-q^2 r_0^2/2}$$

$$\sigma = \frac{\pi^2}{2k^2}\left(\frac{\mu V_0 r_0^2}{\hbar^2}\right)^2 (1 - e^{-2k^2 r_0^2})$$

[Hint: Since $q^2 = 2k^2(1-\cos\theta), d(\cos\theta) = -d(q^2)/2k^2$.]

Let us end this section by examining some general properties of $f(\theta)$. We see from Eq. (19.3.7) that at *low energies* $(k \to 0), q = 2k\sin(\theta/2) \to 0$ and

$$\begin{aligned} f(\theta) &\sim -\frac{\mu}{2\pi\hbar^2}\int V(\mathbf{r}')d^3\mathbf{r}' \\ &\cong -\frac{\mu V_0 r_0^3}{\hbar^2} \end{aligned} \tag{19.3.17}$$

where V_0 is some effective height of V, and r_0 is some effective range. At *high energies*, the exponential factor $e^{-iqr'\cos\theta'}$ oscillates rapidly. This means that the scattered waves coming from different points $\mathbf{r}'$ add with essentially random phases, except in the small range where the phase is stationary:

$$\begin{aligned} qr'\cos\theta' &\lesssim \pi \\ 2k\sin(\theta/2)r_0 &\lesssim \pi \qquad (\text{since } r'\cos\theta' \cong r_0) \\ k\theta r_0 &\lesssim \pi \quad (\sin\theta/2 \simeq \theta/2) \end{aligned}$$

Thus the scattering amplitude is appreciable only in a small forward cone of angle (dropping constants of order unity)

$$\theta \lesssim \frac{1}{kr_0} \tag{19.3.18}$$

These arguments assume $V(r')$ is regular near $r'=0$. But in some singular cases $[V \propto (r')^{-3}$, say] the r' integral is dominated by small r' and $kr' \cos\theta'$ is not necessarily a large phase. Both the Yukawa and Gaussian potential (Exercise 19.3.3) are free of such pathologies and exhibit this forward peak at high energies.

Exercise 19.3.4. Verify the above claim for the Gaussian potential.

When can we trust the Born approximation? Since we treated the potential as a perturbation, our guess would be that it is reliable at high energies. We shall see in the next section that this is indeed correct, but that the Born approximation can also work at low energies provided a more stringent condition is satisfied.

19.4. Born Again (The Time-Independent Description)

In this approach, the central problem is to find solutions to the full Schrödinger equation

$$(\nabla^2 + k^2)\psi_{\mathbf{k}} = \frac{2\mu}{\hbar^2} V\psi_{\mathbf{k}} \tag{19.4.1}$$

of the form

$$\psi_{\mathbf{k}} = e^{i\mathbf{k}\cdot\mathbf{r}} + \psi_{sc} \tag{19.4.2a}$$

where

$$\psi_{sc} \xrightarrow[r\to\infty]{} f(\theta, \phi)\frac{e^{ikr}}{r} \tag{19.4.2b}$$

In the above, θ and ϕ are measured relative to $\mathbf{k}$, chosen along the z axis (Fig. 19.1). One approaches the problem as follows. One finds a *Green's function* $G^0(\mathbf{r}, \mathbf{r}')$ which satisfies

$$(\nabla^2 + k^2)G^0(\mathbf{r}, \mathbf{r}') = \delta^3(\mathbf{r} - \mathbf{r}') \tag{19.4.3}$$

in terms of which the *formal* general solution to Eq. (19.4.1) is

$$\psi_{\mathbf{k}}(\mathbf{r}) = \psi^0(\mathbf{r}) + \frac{2\mu}{\hbar^2}\int G^0(\mathbf{r}, \mathbf{r}')V(\mathbf{r}')\psi_{\mathbf{k}}(\mathbf{r}')\, d^3\mathbf{r}' \tag{19.4.4}$$

where $\psi^0(\mathbf{r})$ is an arbitrary free-particle solution of energy $\hbar^2k^2/2\mu$:

$$(\nabla^2 + k^2)\psi^0 = 0 \tag{19.4.5}$$

We will soon nail down ψ^0 using the boundary conditions.

Applying $\nabla^2 + k^2$ to both sides of Eq. (19.4.4) one may easily verify that $\psi_{\mathbf{k}}$ indeed is a solution to Eq. (19.4.1). The idea here is quite similar to what is employed in solving Poisson's equation for the electrostatic potential in terms of the charge density ρ:

$$\nabla^2\phi = -4\pi\rho$$

One first finds G, the response to a point charge at $\mathbf{r}'$:

$$\nabla^2 G = -4\pi\delta^3(\mathbf{r} - \mathbf{r}')$$

Exercise 12.6.4 tells us that

$$G(\mathbf{r}, \mathbf{r}') = G(\mathbf{r} - \mathbf{r}') = \frac{1}{|\mathbf{r} - \mathbf{r}'|}$$

One then views ρ as a superposition of point charges and, since Poisson's equation is linear, obtains ϕ as the sum of ϕ's produced by these charges:

$$\phi(\mathbf{r}) = \int G(\mathbf{r} - \mathbf{r}')\rho(\mathbf{r}')\, d^3\mathbf{r}' = \int \frac{\rho(\mathbf{r}')}{|\mathbf{r} - \mathbf{r}'|}\, d^3\mathbf{r}'$$

(By acting on both sides with ∇^2 and using $\nabla^2 G = -4\pi\delta^3$, you may verify that ϕ satisfies Poisson's equation.)

One can add to this $\phi(\mathbf{r})$ any ϕ^0 that satisfies $\nabla^2\phi^0 = 0$. Using the boundary condition $\phi = 0$ when $\rho = 0$, we get rid of ϕ^0.

In the scattering problem we pretend that the right-hand side of Eq. (19.4.1) is some given source and write Eq. (19.4.4) for $\psi_{\mathbf{k}}$ in terms of the Green's function. The only catch is that the source for $\psi_{\mathbf{k}}$ is $\psi_{\mathbf{k}}$ itself. Thus Eq. (19.4.4) is really not a solution, but an integral equation for $\psi_{\mathbf{k}}$. The motivation for converting the differential equation to an integral equation is similar to that in the case of $U_{\mathrm{I}}(t, t_0)$: to obtain a perturbative expansion for $\psi_{\mathbf{k}}$ in powers of V. *To zeroth order in* V, Eq. (19.4.2a) tells us that $\psi_{\mathbf{k}}$ is $e^{i\mathbf{k}\cdot\mathbf{r}}$, since there is no scattered wave if V is neglected; whereas Eq. (19.4.4) tells us that $\psi_{\mathbf{k}} = \psi^0$, since the integral over $\mathbf{r}'$ has an explicit power of V in it while ψ^0 has no dependence on V [since it is the solution to Eq. (19.4.5)]. We are thus able to nail down the arbitrary function ψ^0 in Eq. (19.4.4):

$$\psi^0 = e^{i\mathbf{k}\cdot\mathbf{r}} \tag{19.4.6}$$

and conclude that in the present scattering problem

$$\psi_{\mathbf{k}}=e^{i\mathbf{k}\cdot\mathbf{r}}+\frac{2\mu}{\hbar^2}\int G^0(\mathbf{r},\mathbf{r}')V(\mathbf{r}')\psi_{\mathbf{k}}(\mathbf{r}')\,d^3\mathbf{r}' \tag{19.4.7}$$

Upon comparing this to Eq. (19.4.2a) we see that we are associating the second piece with the scattered wave. For consistency of interpretation, it must contain purely outgoing waves at spatial infinity. Since $G^0(\mathbf{r},\mathbf{r}')$ is the scattered wave produced by a point source at $\mathbf{r}'$, it is necessary that $G^0(\mathbf{r},\mathbf{r}')$ be purely outgoing asymptotically. This is an additional physical constraint on G^0 over and above Eq. (19.4.3). As we shall see, this constraint, together with Eq. (19.4.3), will determine G^0 for us uniquely.

Imagine that we have found this G^0. We are now in a position to obtain a perturbative solution for $\psi_{\mathbf{k}}$ starting with Eq. (19.4.7). To zeroth order we have seen that $\psi_{\mathbf{k}}=e^{i\mathbf{k}\cdot\mathbf{r}}$. To go to first order, we feed the zeroth-order $\psi_{\mathbf{k}}$ into the right-hand side and obtain

$$\psi_{\mathbf{k}}=e^{i\mathbf{k}\cdot\mathbf{r}}+\frac{2\mu}{\hbar^2}\int G^0(\mathbf{r},\mathbf{r}')V(\mathbf{r}')\,e^{i\mathbf{k}\cdot\mathbf{r}'}\,d^3\mathbf{r}'+O(V^2) \tag{19.4.8}$$

If we feed this first-order result back into the right-hand side of Eq. (19.4.7), we get (in symbolic form) the result good to second order:

$$\psi_{\mathbf{k}}=\psi^0+\frac{2\mu}{\hbar^2}G^0V\psi^0+\left(\frac{2\mu}{\hbar^2}\right)^2G^0VG^0V\psi^0+O(V^3)$$

and so on.

Let us now turn to the determination of G^0, starting with Eq. (19.4.3):

$$(\nabla^2+k^2)G^0(\mathbf{r},\mathbf{r}')=\delta^3(\mathbf{r}-\mathbf{r}')$$

We note that this equation does not have a unique solution, since, given any solution, we can get another by adding to it a function η^0 that obeys the homogeneous equation

$$(\nabla^2+k^2)\eta^0=0$$

Conversely, any two columns G^0 and $G^{0\prime}$ can differ only by some η^0. So we will first find the simplest G^0 we can, and then add whatever η^0 it takes to make the sum purely outgoing.

Since (∇^2+k^2) and $\delta^3(\mathbf{r}-\mathbf{r}')$ are invariant under the overall translation of $\mathbf{r}$ *and* $\mathbf{r}'$, we know the equation admits translationally invariant solutions‡:

$$G^0(\mathbf{r},\mathbf{r}')=G^0(\mathbf{r}-\mathbf{r}')$$

‡ Note that if an equation has some symmetry, like rotational invariance, it means only that rotationally invariant solutions exist, and not that all solutions are rotationally invariant. For example, the hydrogen atom Hamiltonian is rotationally invariant, but the eigenfunctions are not in general. But there are some (with $l=m=0$) which are.

Replace $\mathbf{r}-\mathbf{r}'$ by $\mathbf{r}$ for convenience. [Once we find $G^0(\mathbf{r})$, we can replace $\mathbf{r}$ by $\mathbf{r}-\mathbf{r}'$.] So we want to solve

$$(\nabla^2+k^2)G^0(\mathbf{r})=\delta^3(\mathbf{r}) \tag{19.4.9}$$

For similar reasons as above, we look for a rotationally invariant solution

$$G^0(\mathbf{r})=G^0(r)$$

Writing

$$G^0(r)=\frac{U(r)}{r}$$

we find that *for* $r\neq 0$, $U(r)$ satisfies

$$\frac{d^2U}{dr^2}+k^2U=0$$

the general solution to which is

$$U(r)=A\,e^{ikr}+B\,e^{-ikr}$$

or

$$G^0(r)=\frac{A\,e^{ikr}}{r}+\frac{B\,e^{-ikr}}{r} \tag{19.4.10}$$

where A and B are arbitrary constants at this point. Since we want G^0 to be purely outgoing we set $B=0$:

$$G^0(r)=\frac{A\,e^{ikr}}{r} \tag{19.4.11}$$

We find A by calculating $(\nabla^2+k^2)G^0(r)$ as $r\to 0$‡

$$(\nabla^2+k^2)G^0(r)\xrightarrow[r\to 0]{}-4\pi A\delta^3(\mathbf{r}) \tag{19.4.12}$$

‡ We use $\nabla^2(\psi\chi)=\psi\nabla^2\chi+\chi\nabla^2\psi+2\nabla\psi\cdot\nabla\chi$ and $\nabla^2=r^{-2}(\partial/\partial r)r^2\,\partial/\partial r$ on a function of r alone.

which gives us

$$G^0(r) = -\frac{e^{ikr}}{4\pi r} \tag{19.4.13}$$

We cannot add any η^0 to this solution, without destroying its purely outgoing nature, since the general form of the free-particle solution, regular in all space, is

$$\eta^0(\mathbf{r}) = \sum_{l=0}^{\infty} \sum_{m=-l}^{l} C_{lm} j_l(kr) Y_l^m(\theta, \phi) \tag{19.4.14}$$

and since, as $r \to \infty$, the spherical Bessel functions are made up of incoming and outgoing waves of equal amplitude

$$j_l(kr) \underset{r\to\infty}{\longrightarrow} \frac{\sin(kr - l\pi/2)}{kr} = \frac{e^{i(kr-l\pi/2)} - e^{-i(kr-l\pi/2)}}{2ikr} \tag{19.4.15}$$

Let us now feed

$$G^0(\mathbf{r}, \mathbf{r}') = G^0(\mathbf{r} - \mathbf{r}') = -\frac{e^{ik|\mathbf{r}-\mathbf{r}'|}}{4\pi|\mathbf{r}-\mathbf{r}'|} \tag{19.4.16}$$

into Eq. (19.4.7) to obtain

$$\begin{aligned} \psi_{\mathbf{k}} &= e^{i\mathbf{k}\cdot\mathbf{r}} - \frac{2\mu}{4\pi\hbar^2} \int \frac{e^{ik|\mathbf{r}-\mathbf{r}'|}}{|\mathbf{r}-\mathbf{r}'|} V(\mathbf{r}')\psi_{\mathbf{k}}(\mathbf{r}')\, d^3\mathbf{r}' \\ &= e^{i\mathbf{k}\cdot\mathbf{r}} + \psi_{\mathrm{sc}} \end{aligned} \tag{19.4.17}$$

Let us now verify that as $r \to \infty$, ψ_{sc} has the desired form $f(\theta, \phi)\, e^{ikr}/r$. Our first instinct may be to approximate as follows:

$$\frac{e^{ik|\mathbf{r}-\mathbf{r}'|}}{|\mathbf{r}-\mathbf{r}'|} \cong \frac{e^{ikr}}{r}$$

in the $\mathbf{r}'$ integral since $\mathbf{r}'$ is confined to $|\mathbf{r}'| \lesssim r_0$ (the range), whereas $r \to \infty$. That this is wrong is clear from the fact that if we do so, the corresponding f has no θ or ϕ dependence. Let us be more careful. We first approximate

$$\begin{aligned} |\mathbf{r}-\mathbf{r}'| &= (r^2 + r'^2 - 2\mathbf{r}\cdot\mathbf{r}')^{1/2} \\ &= r\left[1 + \left(\frac{r'}{r}\right)^2 - 2\frac{\mathbf{r}\cdot\mathbf{r}'}{r^2}\right]^{1/2} = \end{aligned}$$

$$\simeq r\left(1-2\frac{\mathbf{r}\cdot\mathbf{r}'}{r^2}\right)^{1/2}$$

$$\cong r\left(1-\frac{\mathbf{r}\cdot\mathbf{r}'}{r^2}\right) \qquad (19.4.18)$$

We have thrown away the term quadratic in (r'/r) and used the approximation $(1+x)^n \cong 1+nx$ for small x. So

$$\frac{1}{|\mathbf{r}-\mathbf{r}'|}=\frac{1}{r[1-(\mathbf{r}\cdot\mathbf{r}')/r^2]}\cong\frac{1}{r}\left(1+\frac{\mathbf{r}\cdot\mathbf{r}'}{r^2}\right) \qquad (19.4.19)$$

Whereas replacing $|\mathbf{r}-\mathbf{r}'|^{-1}$ in the integral leads to errors which vanish as $r\to\infty$, this is not so for the factor $e^{ik|\mathbf{r}-\mathbf{r}'|}$. We have

$$\begin{aligned} k|\mathbf{r}-\mathbf{r}'| &= kr\left(1-\frac{\mathbf{r}\cdot\mathbf{r}'}{r^2}\right)\\ &= kr-k\hat{r}\cdot\mathbf{r}'\\ &= kr-\mathbf{k}_f\cdot\mathbf{r}' \end{aligned} \qquad (19.4.20)$$

where $\mathbf{k}_f$ is the wave vector of the detected particle: it has the same magnitude (k) as the incident particle and points in the direction $(\hat{r})$ of observation (Fig. 19.2). Consequently

$$\frac{e^{ik|\mathbf{r}-\mathbf{r}'|}}{|\mathbf{r}-\mathbf{r}'|}\cong\frac{e^{ikr}}{r}e^{-i\mathbf{k}_f\mathbf{r}'} \qquad (19.4.21)$$

and

$$\psi_{\mathbf{k}}\xrightarrow[r\to\infty]{} e^{i\mathbf{k}\cdot\mathbf{r}}-\frac{e^{ikr}}{r}\frac{2\mu}{4\pi\hbar^2}\int e^{-i\mathbf{k}_f\cdot\mathbf{r}'}V(\mathbf{r}')\psi_{\mathbf{k}}(\mathbf{r}')\,d^3\mathbf{r}' \qquad (19.4.22)$$

Thus the solution we have found has the desired form as $r\to\infty$. Equation (19.4.22) of course does not determine $f(\theta,\phi)$ since $\psi_{\mathbf{k}}$ is present in the $\mathbf{r}'$ integration.

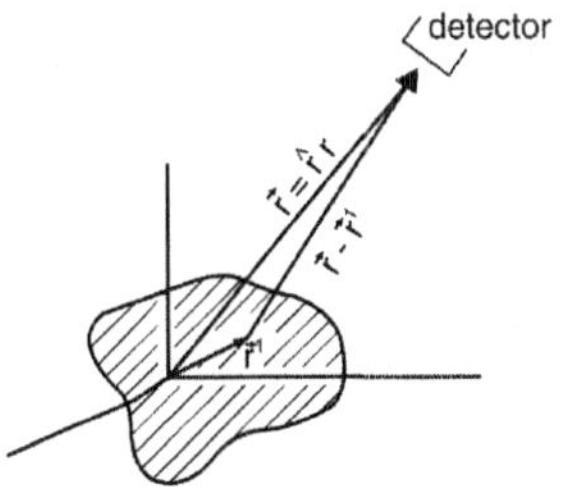

Figure 19.2. The particle is observed at the point $\mathbf{r}$. The $\mathbf{r}'$ integration is restricted to the shaded region which symbolizes the range of the potential.

However, to any desired order this $\psi_{\mathbf{k}}$ can be replaced by the *calculable* lower-order approximation. In particular, to first order,

$$f(\theta,\phi)=-\frac{2\mu}{4\pi\hbar^2}\int e^{-i\mathbf{k}_f\cdot\mathbf{r}'}V(\mathbf{r}')\,e^{i\mathbf{k}_i\cdot\mathbf{r}'}\,d^3\mathbf{r}' \tag{19.4.23}$$

where we have added a subscript i to $\mathbf{k}$ to remind us that it is the initial or incident wave vector. We recognize $f(\theta,\phi)$ to be just the Born approximation calculated in the last section [Eq. (19.3.7)]. The phase factor -1, relative to the incident wave was simply assumed there. The agreement between the time-dependent and time-independent calculations of f persists to all orders in the perturbation expansion.

There is another way (involving Cauchy's theorem) to solve

$$(\nabla^2+k^2)G^0(\mathbf{r})=\delta^3(\mathbf{r}) \tag{19.4.24}$$

Fourier transforming both sides, we get

$$\left(\frac{1}{2\pi}\right)^{3/2}\int e^{-i\mathbf{q}\cdot\mathbf{r}}(\nabla^2+k^2)G^0(\mathbf{r})\,d^3\mathbf{r}=\left(\frac{1}{2\pi}\right)^{3/2} \tag{19.4.25}$$

If we let ∇^2 act to the left (remember it is Hermitian) we get

$$(k^2-q^2)\left(\frac{1}{2\pi}\right)^{3/2}\int e^{-i\mathbf{q}\cdot\mathbf{r}}G^0(\mathbf{r})\,d^3\mathbf{r}=\left(\frac{1}{2\pi}\right)^{3/2} \tag{19.4.26}$$

$$(k^2-q^2)G^0(\mathbf{q})=\left(\frac{1}{2\pi}\right)^{3/2} \tag{19.4.27}$$

As always, going to momentum space has reduced the differential equation to an algebraic equation. The solution is

$$G^0(\mathbf{q})=\frac{1}{(2\pi)^{3/2}(k^2-q^2)} \tag{19.4.28}$$

except at the point $q=k$ where $G^0(\mathbf{q})$ diverges. The reason for this divergence is the following. Equation (19.4.24) is the coordinate space version of the abstract equation

$$(D^2+k^2)G^0=I \tag{19.4.29}$$

where

$$D^2=D_x^2+D_y^2+D_z^2 \tag{19.4.30}$$

(D_x is just the x derivative operator D introduced in Section 1.10, and D_y and D_z are y and z derivative operators.) Thus G^0 is the inverse of (D^2+k^2):

$$G^0=(D^2+k^2)^{-1} \tag{19.4.31}$$

Now, we know that we cannot invert an operator that has a vanishing determinant or equivalently (for a Hermitian operator, since it can be diagonalized) a zero eigenvalue. The operator (D^2+k^2) has a zero eigenvalue since

$$(\nabla^2+k^2)\psi=0 \tag{19.4.32}$$

has nontrivial (plane wave) solutions. We therefore consider a slightly different operator, $D^2+k^2+i\varepsilon$, where ε is positive and infinitesimal. This too has a zero eigenvalue, but the corresponding eigenfunctions are plane waves of *complex wave number*. Such functions are not part of the space we are restricting ourselves to, namely, the space of functions normalized to unity or the Dirac delta function.‡ Thus $D^2+k^2+i\varepsilon$ may be inverted *within* the *physical Hilbert space*. Let us call the corresponding Green's function G_ε^0. At the end of the calculation we will send ε to zero.§

Clearly

$$G_\varepsilon^0(\mathbf{q})=\frac{1}{(2\pi)^{3/2}}\frac{1}{k^2+i\varepsilon-q^2} \tag{19.4.33}$$

The coordinate space function is given by the inverse transform:

$$G_\varepsilon^0(\mathbf{r})=\frac{1}{(2\pi)^3}\int\frac{e^{i\mathbf{q}\cdot\mathbf{r}}}{k^2+i\varepsilon-q^2}\,d^3\mathbf{q} \tag{19.4.34}$$

We choose the q_z axis parallel to $\mathbf{r}$. If θ and ϕ are the angles in $\mathbf{q}$ space,

$$G_\varepsilon^0(\mathbf{r})=\frac{1}{8\pi^3}\int\frac{e^{iqr\cos\theta}}{k^2+i\varepsilon-q^2}\,d(\cos\theta)\,d\phi q^2\,dq$$

$$=\frac{1}{4\pi^2}\int_0^\infty\frac{e^{iqr}-e^{-iqr}}{iqr}\,\frac{q^2\,dq}{k^2+i\varepsilon-q^2} \tag{19.4.35a}$$

$$=\frac{1}{4\pi^2}\int_{-\infty}^\infty\frac{e^{iqr}}{iqr}\,\frac{q^2\,dq}{k^2+i\varepsilon-q^2} \tag{19.4.35b}$$

$$=\frac{-i}{4\pi^2 r}\int_{-\infty}^\infty\frac{e^{iqr}q\,dq}{k^2+i\varepsilon-q^2} \tag{19.4.35c}$$

‡ Recall from Section 1.10 that if k is complex, the norm diverges exponentially.

§ This is called the "$i\varepsilon$ prescription." Throughout the analysis ε will be considered only to first order.

[In going from (19.4.35a) to (19.4.35b) above, we changed q to $-q$ in the e^{-iqr} piece.]

We proceed to evaluate the above integral by means of Cauchy's residue theorem, which states that for any analytic function $f(z)$ of a complex variable z,

$$\oint f(z)\,dz = 2\pi i \sum_j R(z_j) \tag{19.4.36}$$

where $\oint$ denotes integration around a closed contour in the complex z plane and $R(z_j)$ is the residue of the pole at the point z_j lying inside the contour.‡

Let us view q as a complex variable which happens to be taking only real values ($-\infty$ to $+\infty$) in Eq. (19.4.35).

We are trying to evaluate the integral of the function

$$w(q) = \frac{-i}{4\pi^2 r}\frac{e^{iqr}q}{k^2+i\varepsilon-q^2} \tag{19.4.37}$$

along the real axis from $-\infty$ to $+\infty$.

This function has poles where

$$k^2+i\varepsilon-q^2=0$$

or (to first order in ε),

$$(k+q+i\eta)(k-q+i\eta)=0 \qquad (\eta\cong\varepsilon/2k) \tag{19.4.38}$$

These poles are shown in Fig. 19.3.

We are not yet ready to use Cauchy's theorem because we do not have a closed contour. Let us now close the contour via a large semicircle C_ρ whose radius $\rho\to\infty$. Now we can use Cauchy's theorem, but haven't we changed the quantity we wanted to calculate? No, because C_ρ does not contribute to the integral as $\rho\to\infty$. To see this, let us write $q=\rho\, e^{i\theta}$ on C_ρ. Then

$$w(q) \xrightarrow[\rho\to\infty]{} (\text{const})\frac{e^{iqr}}{q} \tag{19.4.39}$$

and

‡ Recall that if

$$f(z) \xrightarrow[z\to z_j]{} \frac{R(z_j)}{z-z_j}$$

then

$$R(z_j)=\lim_{z\to z_j} f(z)(z-z_j)$$

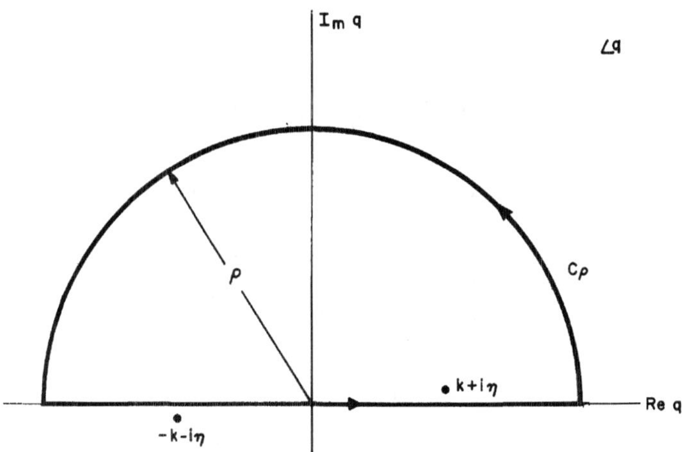

Figure 19.3. The poles of the function $w(q)$ in the complex q plane. We want the integral along the real axis from $-\infty$ to $+\infty$. We add to it the contribution along C_ρ (which vanishes as ρ tends to ∞) in order to close the contour of integration and to use Cauchy's theorem.

$$\int_{C_\rho} w(q)\, dq \sim \int_{C_\rho} e^{iqr} \frac{dq}{q} = \int_0^\pi e^{i\rho r(\cos\theta + i\sin\theta)} i\, d\theta \tag{19.4.40}$$

Except for the tiny region near $\theta = 0$ (which contributes only an infinitesimal amount) the integral vanishes since $e^{-\rho r\sin\theta} \to 0$ as $\rho \to \infty$. We now use Cauchy's theorem. The only pole enclosed is at $q = k + i\eta$. The residue here is

$$R(k+i\eta) = \lim_{q\to k+i\eta} (q-k-i\eta)w(q) = \frac{i}{8\pi^2 r} e^{i(k+i\eta)r} \tag{19.4.41}$$

and

$$G^0(\mathbf{r}) = \lim_{\eta\to 0} 2\pi i R = -\frac{e^{ikr}}{4\pi r} \tag{19.4.42}$$

Notice that although the $i\varepsilon$ ($\varepsilon > 0$) prescription happens to give the right answer here, there are other ways to evaluate the integral, which may be appropriate in other contexts. For example if we choose $\varepsilon < 0$, we get a purely incoming wave, since η changes sign and the pole near $q \cong -k$ gets into the contour.

Validity of the Born Approximation

Since in the Born approximation one replaces $\psi_{\mathbf{k}} = e^{i\mathbf{k}\cdot\mathbf{r}'} + \psi_{sc}$ by just $e^{i\mathbf{k}\cdot\mathbf{r}'}$ in the right-hand side of the integral Eq. (19.4.17), it is a good approximation only if $|\psi_{sc}| \ll |e^{i\mathbf{k}\cdot\mathbf{r}'}|$ in the region $|\mathbf{r}'| \lesssim r_0$. Since we expect ψ_{sc} to be largest near the origin, let us perform a comparison there, using Eq. (19.4.17) itself to evaluate $\psi_{sc}(0)$:

$$\frac{|\psi_{sc}(0)|}{|e^{ikz}(0)|}=|\psi_{sc}(0)|=\left|\frac{2\mu}{4\pi\hbar^2}\int\frac{e^{ikr'}}{r'}V(\mathbf{r}')\,e^{-i\mathbf{k}_i\cdot\mathbf{r}'}\,d^3\mathbf{r}'\right| \tag{19.4.43}$$

Let us assume $V(\mathbf{r})=V(r)$. In this case a rough criterion for the validity of the Born approximation is

$$\frac{2\mu}{\hbar^2 k}\left|\int e^{ikr'}\sin kr'\,V(r')\,dr'\right|\ll 1 \tag{19.4.44}$$

Exercise 19.4.1. Derive the inequality (19.4.44).

At low energies, $kr'\to 0$, $e^{ikr'}\to 1$, $\sin kr'\to kr'$ and we get the condition

$$\frac{2\mu}{\hbar^2}\left|\int r'V(r')\,dr'\right|\ll 1 \tag{19.4.45}$$

If $V(r)$ has an effective depth (or height) V_0 and range r_0, the condition becomes (dropping constants of order unity)

$$\frac{\mu V_0 r_0^2}{\hbar^2}\ll 1 \tag{19.4.46}$$

The low energy condition may be written as

$$V_0\ll\frac{\hbar^2}{\mu r_0^2}$$

Now a particle confined to a well of dimension r_0 must have a momentum of order $\hbar/r_0$ and a kinetic energy of order $\hbar^2/\mu r_0^2$. The above inequality says that if the Born approximation is to work at low energies, the potential must be too shallow to bind a particle confined to a region of size r_0.

At high energies when $kr_0\gg 1$ let us write inside the integral in Eq. (19.4.44)

$$e^{ikr'}\sin kr'=\frac{e^{2ikr'}-1}{2i}$$

and drop the exponential which will be oscillating too rapidly within the range of the potential and keep just the -1 part to get the following condition:

$$\frac{\mu}{\hbar^2 k}\left|\int V(r')\,dr'\right|\ll 1 \tag{19.4.47}$$

which could be rewritten as

$$\frac{\mu V_0 r_0^2}{\hbar^2} \ll k r_0 \tag{19.4.48}$$

We found that the Born approximation can be good even at low energies if the inequality (19.4.46) is satisfied. In fact, if it is, the Born approximation is good at all energies, i.e., Eq. (19.4.48) is automatically satisfied.

19.5. The Partial Wave Expansion

We have noted that if $V(\mathbf{r}) = V(r)$, $f(\theta, \phi) = f(\theta)$. Actually f is also a function of the energy $E = \hbar^2 k^2 / 2\mu$, though this dependence was never displayed explicitly. Since any function of θ can be expanded in terms of the Legendre polynomials

$$P_l(\cos\theta) = \left(\frac{4\pi}{2l+1}\right)^{1/2} Y_l^0 \tag{19.5.1}$$

we can expand $f(\theta, k)$ in terms of $P_l(\cos\theta)$ with k-dependent coefficients:

$$f(\theta, k) = \sum_{l=0}^{\infty} (2l+1) a_l(k) P_l(\cos\theta) \tag{19.5.2}$$

One calls $a_l(k)$ the *lth partial wave amplitude*. It has the following significance. The incident plane wave e^{ikz} is composed of states of all angular momenta [from Eq. (12.6.41)]:

$$e^{ikz} = e^{ikr\cos\theta} = \sum_{l=0}^{\infty} i^l (2l+1) j_l(kr) P_l(\cos\theta) \tag{19.5.3}$$

Since the potential conserves angular momentum, each angular momentum component scatters independently. The amplitude a_l is a measure of the scattering in the angular momentum l sector.

As it stands, the expansion in Eq. (19.5.2) has not done anything for us: we have traded one function of the two variables (θ and k) for an infinite number of functions $a_l(k)$ of one variable k. What makes the expansion useful is that at low energies, only the first few $a_l(k)$ are appreciably different from zero. In this case, one manages to describe the scattering in terms of just a few functions $a_0, a_1, \ldots$ of one variable. The following heuristic argument (corroborated by explicit calculations) is usually given to explain why the scattering is restricted to a few low l values at low k. Coming out of the accelerator is a uniform beam of particles moving along the z axis. All particles in a cylinder of radius ρ and thickness $d\rho$ (ρ is the impact parameter) have angular momentum

$$\hbar l \cong \hbar k \rho \tag{19.5.4}$$

If the potential has range r_0, particles with $\rho > r_0$ will "miss" the target. Thus there will be scattering only up to

$$l_{max} = k\rho_{max} \cong kr_0 \tag{19.5.5}$$

[Conversely, by measuring l_{max} (from the angular dependence of f) we can deduce the range of the potential.]

*Exercise 19.5.1.** Show that for a 100-MeV (kinetic energy) neutron incident on a fixed nucleus, $l_{max} \cong 2$. (Hint: The range of the nuclear force is roughly a Fermi $= 10^{-5}$ Å. Also $\hbar c \cong 200$ MeV F is a more useful mnemonic for nuclear physics.)

Given a potential $V(r)$, how does one calculate $a_l(k)$ in terms of it? In other words, how is a_l related to the solution to the Schrödinger equation for angular momentum l? We begin by considering a free particle. Using

$$j_l(kr) \underset{r\to\infty}{\longrightarrow} \frac{\sin(kr - l\pi/2)}{kr} \tag{19.5.6}$$

We get, from Eq. (19.5.3),

$$e^{ikz} \underset{r\to\infty}{\longrightarrow} \frac{1}{2ik} \sum_{l=0}^{\infty} i^l (2l+1) \left(\frac{e^{i(kr-l\pi/2)}}{r} - \frac{e^{-i(kr-l\pi/2)}}{r} \right) P_l(\cos\theta) \tag{19.5.7}$$

$$= \frac{1}{2ik} \sum_{l=0}^{\infty} (2l+1) \left(\frac{e^{ikr}}{r} - \frac{e^{-i(kr-l\pi)}}{r} \right) P_l(\cos\theta) \tag{19.5.8}$$

upon using $i = e^{i\pi/2}$. Thus at each angular momentum we have incoming and outgoing waves of the same amplitude. (Their phases differ by $l\pi$ because the repulsive centrifugal barrier potential is present at $l \neq 0$ even for a free particle.) The probability currents associated with the two waves are equal and opposite.‡ This equality is expected since in this steady state there should be no net probability flux flowing into the origin or coming out of it. (This balance should occur separately for each l, since scattering in each l is independent due to angular momentum conservation.)

What happens if we turn on a potential? As $r \to \infty$, the radial wave functions must reduce to the free-particle wave function, although there can be a *phase shift* $\delta_l(k)$ due to the potential:

$$R_l(r) = \frac{U_l(r)}{r} \underset{r\to\infty}{\longrightarrow} \frac{A_l \sin[kr - l\pi/2 + \delta_l(k)]}{r} \tag{19.5.9}$$

‡ Once again, can we speak of the current associated with a given l and also with the incoming and outgoing waves at a given l? Yes. If we calculate the total **j** (which will have only a radial part as $r \to \infty$) and integrate over all angles, the orthogonality of P_l's will eliminate all interference terms between different l's. There will also be no interference between the incoming and outgoing waves. [See footnote related to Eq. (19.2.13).]

where A_l is some constant. So

$$\psi_{\mathbf{k}}(\mathbf{r}) \underset{r\to\infty}{\longrightarrow} \sum_{l=0}^{\infty} A_l \frac{(e^{i(kr-l\pi/2+\delta_l)} - e^{-i(kr-l\pi/2+\delta_l)})P_l(\cos\theta)}{r} \tag{19.5.10}$$

To find A_l, we note that since $V(r)$ produces only an outgoing wave, the *incoming* waves must be the same for $\psi_{\mathbf{k}}$ and the plane wave $e^{i\mathbf{k}\cdot\mathbf{r}} = e^{ikz}$. Comparing the coefficients of e^{-ikr}/r in Eqs. (19.5.8) and (19.5.10), we get

$$A_l = \frac{2l+1}{2ik} e^{i(l\pi/2+\delta_l)} \tag{19.5.11}$$

Feeding this into Eq. (19.5.10) we get

$$\psi_{\mathbf{k}}(\mathbf{r}) \underset{r\to\infty}{\longrightarrow} \frac{1}{2ikr} \sum_{l=0}^{\infty} (2l+1)[e^{ikr}e^{2i\delta_l} - e^{-i(kr-l\pi)}]P_l(\cos\theta) \tag{19.5.12}$$

$$= e^{ikz} + \left[\sum_{l=0}^{\infty} (2l+1)\left(\frac{e^{2i\delta_l}-1}{2ik}\right)P_l(\cos\theta)\right]\frac{e^{ikr}}{r} \tag{19.5.13}$$

Comparing this to Eq. (19.5.2) we get

$$a_l(k) = \frac{e^{2i\delta_l}-1}{2ik} \tag{19.5.14}$$

Thus, to calculate $a_l(k)$, one must calculate the phase shift δ_l in the asymptotic wave function.

A comparison of Eqs. (19.5.12) and (19.5.8) tells us that the effect of the potential is to attach a phase factor $e^{2i\delta_l}$ to the outgoing wave. This factor does not change the probability current associated with it and the balance between the total incoming and outgoing currents is preserved. This does not mean there is no scattering, since the angular distribution is altered by this phase shift.

One calls

$$S_l(k) = e^{2i\delta_l(k)} \tag{19.5.15}$$

the *partial wave S matrix element* or the S matrix for angular momentum l. Recall that the S matrix is just the $t\to\infty$ limit of $U(t, -t)$. It is therefore a function of the Hamiltonian. Since in this problem $\mathbf{L}$ is conserved, S (like H) will be diagonal in the common eigenbasis of energy ($E = \hbar^2k^2/2\mu$), angular momentum (l), and z component of angular momentum ($m=0$). Since S is unitary (for U is), its eigenvalues $S_l(k)$ must be of the form $e^{i\theta}$ and here $\theta = 2\delta_l$. If we go to some other basis, say the $|\mathbf{p}\rangle$ basis, $\langle\mathbf{p}'|S|\mathbf{p}\rangle$ will still be elements of a unitary matrix, but no longer diagonal, for $\mathbf{p}$ is not conserved in the scattering process.

If we rewrite $a_l(k)$ as

$$a_l(k)=\frac{e^{2i\delta_l}-1}{2ik}=\frac{e^{i\delta_l}\sin\delta_l}{k} \tag{19.5.16}$$

we get

$$f(\theta)=\frac{1}{k}\sum_{l=0}^{\infty}(2l+1)\,e^{i\delta_l}\sin\delta_l P_l(\cos\theta) \tag{19.5.17}$$

The total cross section

$$\sigma=\int|f|^2\,d\Omega$$

is given by

$$\sigma=\frac{4\pi}{k^2}\sum_{l=0}^{\infty}(2l+1)\sin^2\delta_l \tag{19.5.18}$$

upon using the orthogonality relations for the Legendre polynomials

$$\int P_l(\cos\theta)P_{l'}(\cos\theta)\,d(\cos\theta)=\frac{2}{2l+1}\,\delta_{ll'}$$

Note that σ is a sum of partial cross sections at each l:

$$\sigma=\sum_{l=0}^{\infty}\sigma_l,\qquad \sigma_l=\frac{4\pi}{k^2}(2l+1)\sin^2\delta_l \tag{19.5.19}$$

Each σ_l has an upper bound $\sigma_l^{\max}$, called the *unitarity bound*

$$\sigma_l<\sigma_l^{\max}=\frac{4\pi}{k^2}(2l+1) \tag{19.5.20}$$

The bound is saturated when $\delta_l=n\pi/2$, n odd.

Comparing Eqs. (19.5.17) and (19.5.18) and using $P_l(\cos\theta)=1$ at $\theta=0$, we get

$$\sigma=\frac{4\pi}{k}\,\mathrm{Im}\,f(0) \tag{19.5.21}$$

This is called the *optical theorem*. It is not too surprising that there exists a relation between the total cross section and the forward amplitude, for the following reason. The incident plane wave brings in some current density in the z direction. Some of it gets scattered into the various directions. This must reflect itself in the form of a decrease in current density behind the target, i.e., in the $\theta=0$ direction. The decrease can only occur because the incident plane wave and the scattered wave in the forward direction interfere destructively. It is of course not obvious why just the imaginary part of $f(0)$ is relevant or where the factor $4\pi/k$ comes from. To find out, you must do Exercise 19.5.6.

A Model Calculation of δ_l: The Hard Sphere

Consider a hard sphere, which is represented by

$$\begin{aligned} V(r) &= \infty, \qquad r<r_0 \\ &= 0, \qquad r>r_0 \end{aligned} \tag{19.5.22}$$

We now proceed to solve the radial Schrödinger equation, look at the solution as $r\to\infty$, and identify the phase shift. Clearly the (unnormalized) radial function $R_l(r)$ vanishes inside $r\le r_0$. Outside, it is given by the free-particle function:

$$R_l(r) = A_l j_l(kr) + B_l n_l(kr) \tag{19.5.23}$$

(We keep the n_l function since it is regular for $r>0$.) The coefficients A_l and B_l must be chosen such that

$$R_l(r_0)=0 \tag{19.5.24}$$

to ensure the continuity of the wave function at $r=r_0$. Thus

$$\frac{B_l}{A_l} = -\frac{j_l(kr_0)}{n_l(kr_0)} \tag{19.5.25}$$

From Eq. (12.6.32), which gives the asymptotic form of j_l and n_l,

$$\begin{aligned} R_l(r) &\xrightarrow[r\to\infty]{} \frac{1}{kr}\left[A_l \sin(kr-l\pi/2) - B_l \cos(kr-l\pi/2)\right] \\ &= \frac{(A_l^2+B_l^2)^{1/2}}{kr}\left[\sin\left(kr-\frac{l\pi}{2}+\delta_l\right)\right] \end{aligned} \tag{19.5.26}$$

where

$$\delta_l = \tan^{-1}\left(\frac{-B_l}{A_l}\right) = \tan^{-1}\left[\frac{j_l(kr_0)}{n_l(kr_0)}\right] \tag{19.5.27}$$

For instance [from Eq. (12.6.31)]

$$\delta_0 = \tan^{-1}\left[\frac{\sin(kr_0)/kr_0}{-\cos(kr_0)/kr_0}\right]$$
$$= -\tan^{-1}\tan(kr_0)$$
$$= -kr_0 \tag{19.5.28}$$

It is easy to understand the result: the hard sphere has pushed out the wave function, forcing it to start its sinusoidal oscillations at $r=r_0$ instead of $r=0$. In general, repulsive potentials give negative phase shifts (since they slow down the particle and reduce the phase shift per unit length) while attractive potentials give positive phase shifts (for the opposite reason). This correspondence is of course true only if δ is small, since δ is defined only modulo π. For instance, if the phase shift $kr_0=\pi$, a_0 vanishes and s-wave scattering does not expose the hard sphere centered at the origin.

Consider the hard sphere phase shift as $k\to 0$. Using

$$j_l(x) \underset{x\to 0}{\longrightarrow} x^l/(2l+1)!!$$
$$n_l(x) \underset{x\to 0}{\longrightarrow} -x^{-(l+1)}(2l-1)!!$$

we get

$$\tan\delta_l \underset{k\to 0}{\cong} \delta_l \propto (kr_0)^{2l+1} \tag{19.5.29}$$

This agrees with the intuitive expectation that at low energies there should be negligible scattering in the high angular momentum states. The above $(kr_0)^{2l+1}$ dependence of δ_l at low energies is true for any reasonable potential, with r_0 being some length scale characterizing the range. [Since there is no hard and fast definition of range, we can *define* the range of any potential to be the r_0 that appears in Eq. (19.5.29).] Notice that although $\delta_0 \propto k^1$, the partial cross section does not vanish because $\sigma_0 \propto k^{-2}\sin^2\delta_l \sim k^{-2}\delta_l^2 \nrightarrow 0$, as $k\to 0$.

Resonances

The partial cross section σ_l is generally very small at low energies since $\delta_l \propto (k)^{2l+1}$ as $k\to 0$. But it sometimes happens that δ_l rises very rapidly from 0 to π [or more generally, from $n\pi$ to $(n+1)\pi$] in a very small range of k or E. In this region, near $k=k_0$ or $E=E_0$, we may describe δ_l by

$$\delta_l = \delta_b + \tan^{-1}\left(\frac{\Gamma/2}{E_0 - E}\right) \tag{19.5.30}$$

where δ_b is some *background phase* ($\cong n\pi$) that varies very little. The corresponding cross section, neglecting δ_b, is

$$\sigma_l = \frac{4\pi}{k^2}(2l+1)\sin^2\delta_l$$

$$\underset{E\cong E_0}{=} \frac{4\pi}{k^2}(2l+1)\frac{(\Gamma/2)^2}{(E_0-E)^2+(\Gamma/2)^2} \tag{19.5.31}$$

σ_l is described by a bell-shaped curve, called the *Breit–Wigner* form, with a maximum height $\sigma_l^{\max}$ (the unitarity bound) and a half-width $\Gamma/2$. This phenomenon is called a *resonance*.

In Eq. (19.5.31) for σ_l, valid only near E_0, we have treated Γ as a constant. Its k dependence may be deduced by noting that as $k\to 0$, we have [from Eq. (19.5.29)],

$$\sigma_l \sim \frac{1}{k^2}\sin^2\delta_l \cong \frac{1}{k^2}\delta_l^2 \cong \frac{(kr_0)^{4l+2}}{k^2}$$

which implies

$$\Gamma/2 = (kr_0)^{2l+1}\gamma \tag{19.5.32}$$

where γ is some constant with dimensions of energy. Thus the expression for σ_l that is valid over a wider range is

$$\sigma_l = \frac{4\pi}{k^2}(2l+1)\frac{[\gamma(kr_0)^{2l+1}]^2}{(E-E_0)^2+[\gamma(kr_0)^{2l+1}]^2} \tag{19.5.33}$$

For any $l\neq 0$, σ_l is damped in the entire low-energy region by the net k^{4l} factor, except near E_0, where a similar factor in the denominator neutralizes it. Clearly, as l goes up, the resonances get sharper. The situation at $l=0$ (where σ_0 starts out nonzero at $k=0$) depends on the potential. More on this later.

We would now like to gain some insight into the dynamics of resonances. We ask what exactly is going on at a resonance, in terms of the underlying Schrödinger equation. We choose to analyze the problem through the S matrix. Near a resonance we have

$$S_l(k) = e^{2i\delta_l} = \frac{e^{i\delta_l}}{e^{-i\delta_l}} = \frac{1+i\tan\delta_l}{1-i\tan\delta_l} = \frac{E-E_0-i\Gamma/2}{E-E_0+i\Gamma/2} \tag{19.5.34}$$

Although k and E are real in any experiment (and in our analysis so far), let us think of $S_l(k)$ as a function of complex E or k. Then we find that the resonance corresponds to a pole in S_l at a complex point,

$$E = E_0 - i\Gamma/2 \tag{19.5.35}$$

or

$$k = k_0 - i\eta/2 \tag{19.5.36}$$

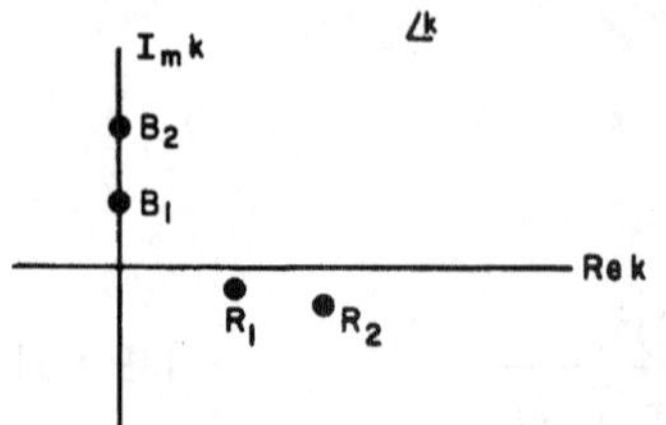

Figure 19.4. Some of the singularities of $S_l(k)$ in the complex k plane. The dots on the positive imaginary axis stand for bound state poles and the dots below the real axis stand for resonance poles. The physical or experimentally accessible region is along the real axis, where S_l has the form $e^{2i\delta_l}$.

where $E_0=\hbar^2k_0^2/2\mu$ and $\Gamma=\eta\hbar^2k_0/\mu$ (for small η and Γ). Since Γ and η are small, the pole is very close to the real axis, which is why we trust the form of S_l that is valid near the point $E=E_0$ on the real axis.

What is the implication of the statement that the resonance corresponds to a (nearby) pole in $S_l(k)$? To find out, we take a new look at bound states in terms of the S matrix. Recall that for k real and positive, if

$$R_{kl}(r)\underset{r\to\infty}{\longrightarrow}\frac{A\,e^{ikr}}{r}+\frac{B\,e^{-ikr}}{r} \tag{19.5.37}$$

then [from Eqs. (19.5.9) and (19.5.10) or Eq. (19.5.12)],

$$e^{2i\delta_l}=S_l(k)=\frac{A}{B}=\frac{\text{outgoing wave amplitude}}{\text{incoming wave amplitude}} \tag{19.5.38}$$

(up to a constant factor i^{2l}). We now define $S_l(k)$ for complex k as follows: solve the radial equation with k set equal to a complex number, find $R(r\to\infty)$, and take the ratio A/B. Consider now the case $k=i\kappa\,(\kappa>0)$, which corresponds to E real and negative. Here we will find

$$R_{kl}(r)\underset{r\to\infty}{\longrightarrow}\frac{A\,e^{-\kappa r}}{r}+\frac{B\,e^{\kappa r}}{r} \tag{19.5.39}$$

Whereas $S_l(k=i\kappa)$ is well defined, the corresponding R_{kl} does not interest us, since it is not normalizable. But recall that for some special values of k, R_{kl} is exponentially damped and describes the wave function of a bound state. These bound states correspond to k such that $B=0$, or $S_l(k)=\infty$. Thus poles of $S_l(k)$ at $k=i\kappa$ correspond to bound states.

So a resonance, which is a pole at $k=k_0-i\eta$ must also be some kind of bound state. (See Fig. 19.4 for poles of the S matrix.) We next argue heuristically as follows.‡ Since the bound state at $E=E_B$ (a negative number) has the time dependence

$$e^{-iE_Bt/\hbar}$$

‡ This result may be established rigorously.

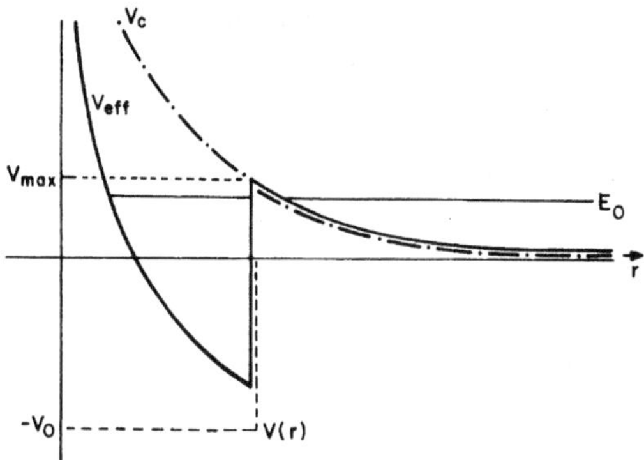

Figure 19.5. A typical potential that can sustain resonances. The centrifugal repulsion V_c (dot–dash line) plus the actual attractive potential (dotted line) gives the effective potential V_{eff} (solid line). The figure shows an example where there would have been a bound state at E_0 but for tunneling. But because of tunneling the particle can leak out, and by the same token, a particle can come from outside with positive energy E_0, form a metastable bound state (with a lifetime inversely proportional to the tunneling probability), and then escape. This is called resonance.

the resonance must have a time dependence

$$e^{-i(E_0 - i\Gamma/2)t/\hbar} = e^{-iE_0 t/\hbar}\, e^{-\Gamma t/2\hbar}$$

This describes a state of positive energy E_0, but whose norm falls exponentially with a half-life $t \sim \hbar/\Gamma$. Thus, a resonance, corresponding to a pole at $E = E_0 - i\Gamma/2$, describes a *metastable* bound state of energy E_0 and lifetime $t = \hbar/\Gamma$.‡

So we must next understand how a positive-energy particle manages to form a metastable bound state. Consider the case where $V(r)$ is attractive, say a square well of depth V_0 and range r_0. The potential appearing in the radial equation is $V_{\text{eff}} = V + V_c$, where V_c is the centrifugal repulsion (Fig. 19.5). The main point is that V_{eff} is attractive at short distances and repulsive at long distances. Consider now a particle with energy $E_0 < V_{\max}$, such that if tunneling is ignored, the particle can form a bound state inside the attractive region, i.e., we can fit in an integral number of half-wavelengths. But tunneling is of course present and the particle can escape to infinity as a free particle of energy E_0. Conversely, a free particle of energy E_0 shot into the potential can penetrate the barrier and form a metastable bound state and leak out again. This is when we say resonance is formed. This picture also explains why the resonances get narrower as l increases: as l increases, V_c grows, tunneling is suppressed more, and the lifetime of the metastable state grows. We can also see why $l = 0$ is different: there is no repulsive barrier due to V_c. If $V = V_{\text{eff}}$ is purely attractive, only genuine (negative energy) bound states are possible. The closest thing to a resonance is the formation of a bound state near zero energy (Exercise 19.5.4). If, however, V itself has the form of V_{eff} in Fig. 19.5, resonances are possible.

*Exercise 19.5.2.** Derive Eq. (19.5.18) and provide the missing steps leading to the optical theorem, Eq. (19.5.21).

‡ The energy is not strictly E_0 because the uncertainty principle does not allow us to define a precise energy for a state of finite lifetime. E_0 is the mean energy.

Exercise 19.5.3. (1) Show that $\sigma_0 \to 4\pi r_0^2$ for a hard sphere as $k \to 0$.

(2) Consider the other extreme of kr_0 very large. From Eq. (19.5.27) and the asymptotic forms of j_l and n_l show that

$$\sin^2 \delta_l \xrightarrow[kr_0 \to \infty]{} \sin^2(kr_0 - l\pi/2)$$

so that

$$\sigma = \sum_{l=0}^{l_{\max}=kr_0} \sigma_l \cong \frac{4\pi}{k^2} \int_0^{kr_0} (2l) \sin^2 \delta_l \, dl$$

$$\cong 2\pi r_0^2$$

if we approximate the sum over l by an integral, $2l+1$ by $2l$, and the oscillating function $\sin^2 \delta$ by its mean value of $1/2$.

*Exercise 19.5.4.** Show that the s-wave phase shift for a square well of depth V_0 and range r_0 is

$$\delta_0 = -kr_0 + \tan^{-1}\left(\frac{k}{k'} \tan k' r_0\right)$$

where k' and k are the wave numbers inside and outside the well. For k small, kr_0 is some small number and we ignore it. Let us see what happens to δ_0 as we vary the depth of the well, i.e., change k'. Show that whenever $k' \simeq k'_n = (2n+1)\pi/2r_0$, δ_0 takes on the resonant form Eq. (19.5.30) with $\Gamma/2 = \hbar^2 k_n / \mu r_0$, where k_n is the value of k when $k' = k'_n$. Starting with a well that is too shallow to have any bound state, show k'_1 corresponds to the well developing its first bound state, at zero energy. (See Exercise 12.6.9.) (Note: A zero-energy bound state corresponds to $k=0$.) As the well is deepened further, this level moves down, and soon, at k'_2, another zero-energy bound state is formed, and so on.

Exercise 19.5.5. Show that even if a potential absorbs particles, we can describe it by

$$S_l(k) = \eta_l(k)\, e^{2i\delta_l}$$

where $\eta(<1)$, is called the *inelasticity factor*.

(1) By considering probability currents, show that

$$\sigma_{\text{inel}} = \frac{\pi}{k^2} \sum_{l=0}^{\infty} (2l+1)[1-\eta_l^2]$$

$$\sigma_{\text{el}} = \frac{\pi}{k^2} \sum_{l=0}^{\infty} (2l+1)(1+\eta_l^2 - 2\eta_l \cos 2\delta_l)$$

and that once again

$$\sigma_{\text{tot}} = \frac{4\pi}{k} \operatorname{Im} f(0)$$

(2) Consider a "black disk" which absorbs everything for $r \leq r_0$ and is ineffective beyond. Idealize it by $\eta = 0$ for $l \leq kr_0$; $\eta = 1$, $\delta = 0$ for $l > kr_0$. Show that $\sigma_{el} = \sigma_{inel} \simeq \pi r_0^2$. Replace the sum by an integral and assume $kr_0 \gg 1$. (See Exercise 19.5.3.) Why is σ_{inel} always accompanied by σ_{el}?

Exercise 19.5.6. (The Optical Theorem). (1) Show that the radial component of the current density due to interference between the incident and scattered waves is

$$j_r^{\text{int}} \underset{r\to\infty}{\sim} \left(\frac{\hbar k}{\mu}\right) \frac{1}{r} \operatorname{Im}[i\, e^{ikr(\cos\theta - 1)} f^*(\theta) \cos\theta + i\, e^{ikr(1-\cos\theta)} f(\theta)]$$

(2) Argue that as long as $\theta \neq 0$, the average of j_r^{int} over any small solid angle is zero because $r \to \infty$. [Assume $f(\theta)$ is a smooth function.]

(3) Integrate j_r^{int} over a tiny cone in the forward direction and show that (see hint)

$$\int_{\text{forward cone}} j_r^{\text{int}} r^2 \, d\Omega = -\left(\frac{\hbar k}{\mu}\right) \frac{4\pi}{k} \operatorname{Im} f(0)$$

Thus, if we integrate the total current in the region behind the target, we find that the interference term (important only in the near-forward direction, behind the target) produces a depletion of particles, casting a "shadow." The total number of particles (per second) missing in the shadow region is given by the above expression for the integrated flux. Equating this loss to the product of the incident flux $\hbar k/\mu$ and the cross section σ, we regain the optical theorem. (Hint: Since θ is small, set $\sin\theta \simeq \theta$, $\cos\theta = 1$ or $1-\theta^2/2$ using the judgment. In evaluating the upper limit in the θ integration, use the idea introduced in Chapter 1, namely, that the limit of a function that oscillates as its argument approaches infinity is equal to its average value.)

19.6. Two-Particle Scattering

In this section we will see how the differential cross section for two-body scattering may be extracted from the solution of the Schrödinger equation for the relative coordinate with a potential $V(\mathbf{r} = \mathbf{r}_1 - \mathbf{r}_2)$. Let us begin by considering the total and differential cross sections for two-body scattering. Let σ be the total cross section for the scattering of the two particles. Imagine a beam of projectiles with density ρ_1 and *magnitude* of velocity v_1 colliding *head on* with the beam of targets with parameters ρ_2 and v_2. How many collisions will there be per second? We know that if there is only one target and it is at rest,

$$\begin{aligned} \text{No. of collisions/sec} &= \sigma \times \text{incident projectiles/sec/area} \\ &= \sigma \rho_1 v_1 \end{aligned} \tag{19.6.1}$$

Here we modify this result to take into account that (1) there are ρ_2 targets per *unit volume* (ρ_2 is assumed so small that the targets scatter independently of each other), and (2) the targets are moving toward the projectiles at a relative velocity $v_{\text{rel}} = v_1 + v_2$. Consequently we have

$$\begin{aligned}\text{No. of collisions/sec/volume of interaction} &= \sigma\rho_1(v_1+v_2)\rho_2 \\ &= \sigma\rho_1\rho_2 v_{\text{rel}} \end{aligned} \tag{19.6.2}$$

Note that σ is the same for all observers moving along the beam–target axis.

What about the differential cross section? It *will* depend on the frame. In the lab frame, where the target is initially at rest, we define, in analogy with Eq. (19.6.2),

$$\begin{aligned}&\text{No. of projectiles scattered into } d(\cos\theta_L)\, d\phi_L/\text{sec/vol} \\ &\quad = \frac{d\sigma}{d\Omega_L}\, d\Omega_L \rho_1 \rho_2 v_{\text{rel}}\end{aligned} \tag{19.6.3}$$

Here v_{rel} is just the projectile velocity and θ_L and ϕ_L are angles in the lab frame measured relative to the projectile direction. (We can also define a $d\sigma/d\Omega_L$ in terms of how many *target* particles are scattered into $d\Omega_L$, but it would not be an independent quantity since momentum conservation will fix the fate of the target, given the fate of the projectile.) The only other frame we consider is the CM frame, where $(d\sigma/d\Omega)\, d\Omega$ is defined as in Eq. (19.6.3).‡ We relate $d\sigma/d\Omega$ to $d\sigma/d\Omega_L$ by the following argument. Imagine a detector in the lab frame at (θ_L, ϕ_L) which subtends an angle $d\Omega_L$. The number of counts it registers is an absolute, frame-independent quantity, although its orientation and acceptance angle $d\Omega$ may vary from frame to frame. (For example, a particle coming at right angles to the beam axis in the lab frame will be tilted forward in a frame moving backward.) So we deduce the following equality from Eq. (19.6.2) after noting the frame invariance of $\rho_1\rho_2 v_{\text{rel}}$:

$$\frac{d\sigma}{d\Omega_L}\, d\Omega_L = \frac{d\sigma}{d\Omega}\, d\Omega \tag{19.6.4}$$

or

$$\frac{d\sigma}{d\Omega_L} = \frac{d\sigma}{d\Omega}\,\frac{d\Omega}{d\Omega_L} \tag{19.6.5}$$

We will consider first the calculation of $d\sigma/d\Omega$, and then $d\Omega/d\Omega_L$.

Let us represent the state of the two colliding particles, long before they begin to interact, by the product wave function (in some general frame):

$$\psi_{\text{inc}} = e^{i\mathbf{k}_1\cdot\mathbf{r}_1}\, e^{i\mathbf{k}_2\cdot\mathbf{r}_2} \tag{19.6.6}$$

‡ The CM variables will carry no subscripts.

We should remember that these plane waves are idealized forms of broad wave packets. Assuming both are moving along the z axis,

$$\begin{aligned}\psi_{\text{inc}} &= e^{ik_1z_1}\, e^{ik_2z_2} \\ &= \exp\left[i(k_1+k_2)\left(\frac{z_1+z_2}{2}\right)\right]\exp\left[i\left(\frac{k_1-k_2}{2}\right)(z_1-z_2)\right] \\ &= \psi_{\text{inc}}^{\text{CM}}(z_{\text{CM}})\,\psi_{\text{inc}}^{\text{rel}}(z) \end{aligned} \tag{19.6.7}$$

Since the potential affects only the relative coordinate, the plane wave describes the CM completely; there is no scattering for the CM as a whole. On the other hand, $\psi_{\text{inc}}^{\text{rel}}(z)$ will develop a scattered wave and become

$$\begin{aligned}\psi(z) &= e^{ikz} + \psi_{\text{sc}}(\mathbf{r}) \\ &\xrightarrow[r\to\infty]{} e^{ikz} + f(\theta,\phi)\, e^{ikr}/r \end{aligned} \tag{19.6.8}$$

where we have dropped the superscript "rel," since the argument z makes it obvious, and set $(k_1-k_2)/2$ equal to k. Thus the static solution for the entire system is

$$\begin{aligned}\psi_{\text{system}}(\mathbf{r}_1,\mathbf{r}_2) &= \psi^{\text{CM}}(z_{\text{CM}})[e^{ikz} + \psi_{\text{sc}}(\mathbf{r})] \\ &\xrightarrow[r\to\infty]{} \psi^{\text{CM}}(z_{\text{CM}})[e^{ikz} + f(\theta,\phi)\, e^{ikr}/r] \end{aligned} \tag{19.6.9}$$

If we go to the CM frame, $\psi^{\text{CM}}(z_{\text{CM}}) = e^{i(k_1+k_2)\cdot z_{\text{CM}}} = 1$, since $k_1+k_2=0$ defines this frame. So we can forget all about the CM coordinate. The scattering in the CM frame is depicted in Fig. 19.6. The classical trajectories are not to be taken literally;

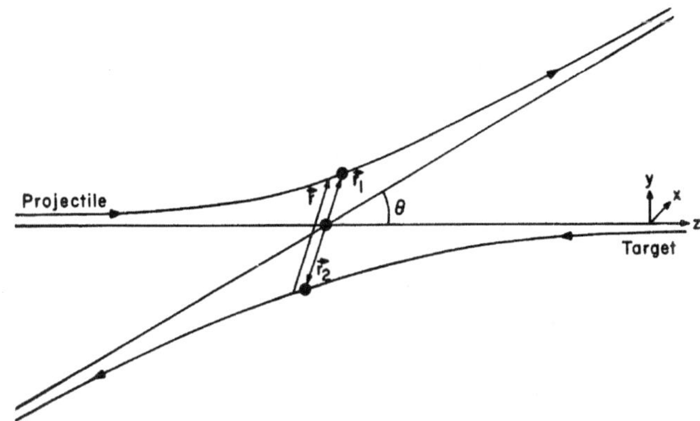

Figure 19.6. Two-body scattering in the CM frame. The projectile and target coordinates are $\mathbf{r}_1$ and $\mathbf{r}_2$, respectively. The relative coordinate $\mathbf{r}=\mathbf{r}_1-\mathbf{r}_2$ is slightly displaced in the figure for clarity. Since $\mathbf{r}_1$ and $\mathbf{r}$ are always parallel, the probability that the projectile scatters into $d\Omega$ is the same as the probability that the fictitious particle described by $\mathbf{r}$ scatters into $d\Omega$. To find the latter, we must solve the Schrödinger equation for $\mathbf{r}$.

they merely define the relative coordinate $\mathbf{r}$ and the individual coordinates $\mathbf{r}_1$(projectile) and $\mathbf{r}_2$(target).

What we want is the rate at which the projectile scatters into $d\Omega$. But since $\mathbf{r}_1$ is parallel to $\mathbf{r}$, this equals the rate at which the fictitious particle described by $\mathbf{r}$ scatters into solid angle $d\Omega$. We find this rate by solving the Schrödinger equation for the relative coordinate. Having done so, and having found

$$\psi(\mathbf{r}) \underset{r\to\infty}{\longrightarrow} e^{ikz} + f(\theta, \phi)\frac{e^{ikr}}{r} \tag{19.6.10}$$

we recall from Eq. (19.2.17) the rate of scattering into $d\Omega$:

$$R_{i\to d\Omega} = |f(\theta, \phi)|^2 \frac{\hbar k}{\mu} d\Omega \tag{19.6.11}$$

Note that is the rate *per unit volume* of target–beam interaction, since the probability density for the CM is unity. To extract $d\sigma/d\Omega$ from $R_{i\to d\Omega}$ above we turn to Eq. (19.6.3) which defines $d\sigma/d\Omega$ (upon dropping the subscript L). Since the definition makes sense only for a flux of wave packets and since we are dealing with plane waves here, we replace the number scattered into $d\Omega$ per second by the probability flowing into $d\Omega$ per second, and the particle densities ρ_1 and ρ_2 by probability densities of the colliding beams. Since the colliding beams ($e^{ikz} = e^{ik(z_1 - z_2)} = e^{ikz_1} \cdot e^{-ikz_2}$) are plane waves of unit modulus, $\rho_1 = \rho_2 = 1$. How about v_{rel}? Remember that in the CM frame

$$m_1 v_1 = m_2 v_2$$

so

$$v_{\text{rel}} = v_1 + v_2 = v_1\left(1 + \frac{m_1}{m_2}\right) = v_1\left(\frac{m_2 + m_1}{m_2}\right) = m_1 v_1\left(\frac{m_2 + m_1}{m_1 m_2}\right)$$

$$= \hbar k\left(\frac{m_2 + m_1}{m_1 m_2}\right) = \frac{\hbar k}{\mu} \tag{19.6.12}$$

So

$$\frac{d\sigma}{d\Omega} d\Omega = \frac{R_{i\to d\Omega}}{\rho_1 \rho_2 (v_1 + v_2)} = \frac{|f|^2 (\hbar k/\mu)}{\hbar k/\mu} d\Omega$$

or

$$\frac{d\sigma}{d\Omega} = |f|^2 \tag{19.6.13}$$

Thus the $d\sigma/d\Omega$ we calculated in the previous sections for a single particle scattering off a potential $V(\mathbf{r})$ can also be interpreted as the CM cross section for two bodies interacting via a potential $V(\mathbf{r}=\mathbf{r}_1-\mathbf{r}_2)$.

Passage to the Lab Frame

We now consider the passage to the lab frame, i.e., the calculation of $d\Omega/d\Omega_L$. We discuss the equal mass case, leaving the unequal mass case as an exercise. Figure 19.7a shows the particles coming in with momenta p and $-p$ along the z axis in the CM frame. If $\mathbf{p}'$ is the final momentum of the projectile,

$$\tan\theta=\frac{(p_x'^2+p_y'^2)^{1/2}}{p_z'}\equiv\frac{p_\perp'}{p_z'} \tag{19.6.14a}$$

$$\tan\phi=p_y'/p_x' \tag{19.6.14b}$$

To go to the lab frame we must move leftward at a speed p/m. In this frame, all momenta get an increment in the z *direction* (only) equal to p. (Thus T, the target, will be at rest before collision.) The scattering angles in the lab frame are given by

$$\tan\theta_L=p_\perp'/(p_z'+p) \tag{19.6.15a}$$

$$\tan\phi_L=p_y'/p_x' \tag{19.6.15b}$$

Comparing Eqs. (19.6.14) and (19.6.15) we get

$$\phi_L=\phi \tag{19.6.16}$$

$$\begin{aligned}\tan\theta_L&=\frac{p_\perp'}{p_z'+p}=\frac{p_\perp'/p}{p_z'/p+1}=\frac{\sin\theta}{\cos\theta+1}\\&=\tan(\theta/2)\quad(\text{using }|\mathbf{p}'|=p)\end{aligned}$$

So

$$\theta_L=\theta/2 \tag{19.6.17}$$

One consequence of the result is that $\theta_L\le\pi/2$. Given Eqs. (19.6.16) and (19.6.17) it is a simple matter to relate $d\sigma/d\Omega$ to $d\sigma/d\Omega_L$.

*Exercise 19.6.1.** (1) Starting with Eqs. (19.6.16) and (19.6.17), show that the relation between $d\sigma/d\Omega$ and $d\sigma/d\Omega_L$ is

$$\left.\frac{d\sigma}{d\Omega_L}\right|_{\theta_0}=\left.\frac{d\sigma}{d\Omega}\right|_{2\theta_0}4\cos\theta_0$$

(2) Show that $\theta_L\le\pi/2$ by using just energy and momentum conservation.

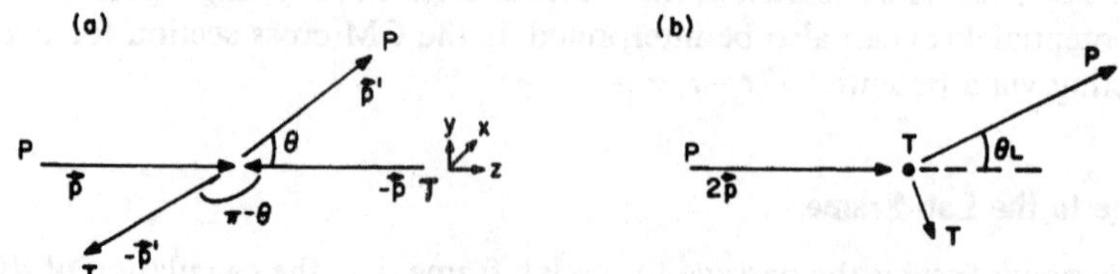

Figure 19.7. (a) Collision of two equal masses in the CM frame. The labels P and T refer to projectile and target. The angle ϕ equals $\frac{1}{2}\pi$ in the figure. (b) The same collision in the lab frame (where T is initially at rest).

(3) For unequal mass scattering, show that

$$\tan \theta_L = \frac{\sin \theta}{\cos \theta + (m_1/m_2)}$$

where m_2 is the target mass.

Scattering of Identical Particles

Consider the scattering of two identical spin-zero bosons in their CM frame. We must describe them by a symmetrized wave function. Under the exchange $\mathbf{r}_1 \leftrightarrow \mathbf{r}_2$; $\mathbf{r}_{CM} = (\mathbf{r}_1 + \mathbf{r}_2)/2$ is invariant while $\mathbf{r} = \mathbf{r}_1 - \mathbf{r}_2$ changes sign. So $\psi^{CM}(\mathbf{r}_{CM})$ is automatically symmetric. We must symmetrize $\psi(\mathbf{r})$ by hand:

$$\psi_{sym}(\mathbf{r}) \underset{r\to\infty}{\longrightarrow} (e^{ikz} + e^{-ikz}) + [f(\theta, \phi) + f(\pi - \theta, \phi + \pi)]\, e^{ikr}/r \qquad (19.6.18)$$

We have used the fact that under $\mathbf{r} \to -\mathbf{r}$, $\theta \to \pi - \theta$ and $\theta \to \phi + \pi$. The scattering amplitude is thus

$$f_{sym}(\theta, \phi) = f(\theta, \phi) + f(\pi - \theta, \phi + \pi) \qquad (19.6.19)$$

Note that f_{sym} is consistent with the fact that since the particles are identical, one cannot say which one scattered into (θ, ϕ) and which one into $(\pi - \theta, \phi + \pi)$ (Fig. 19.7). The differential cross section is

$$\begin{aligned}\frac{d\sigma}{d\Omega} &= |f(\theta, \phi) + f(\pi - \theta, \phi + \pi)|^2 \\ &= |f(\theta, \phi)|^2 + |f(\pi - \theta, \phi + \pi)|^2 + 2\,\mathrm{Re}[f(\theta, \phi) f^*(\pi - \theta, \phi + \pi)] \qquad (19.6.20)\end{aligned}$$

The first two terms are what we would get if we had two distinguishable particles and asked for the rate at which one or the other comes into $d\Omega$. The third term gives the usual quantum mechanical interference that accompanies identical particles. There are two features worth noting about Eq. (19.6.20):

(1) To find σ, we must integrate over only 2π radians and not 4π radians (if not, we will count each *distinguishable* event twice).

(2) Recall that when we obtained the Rutherford cross section by taking the $\mu_0 \to 0$ limit of the Yukawa cross section, we got the right answer although $f(\theta)$ was not right: it did not contain the exponential phase factor that comes from a careful treatment of the Coulomb potential [see Eq. (19.3.16) and the sentences following it.] When we consider the Coulomb scattering of identical bosons (of charge e, say) the interference terms expose the inadequacy of the $\mu_0 \to 0$ approach. The correct cross section is‡

$$\frac{d\sigma}{d\Omega}=\left(\frac{e^2}{4E}\right)^2\left[\frac{1}{\sin^4\theta/2}+\frac{1}{\cos^4\theta/2}+\frac{2\cos(\gamma\ln\tan^2\theta/2)}{\sin^2\theta/2\cos^2\theta/2}\right] \tag{19.6.21}$$

whereas the $\mu_0 \to 0$ trick would not have given the $\cos(\gamma \ln \tan^2 \theta/2)$ factor. (The classical Rutherford treatment would not give the third term at all. Notice, however, that as $\hbar \to 0$, it oscillates wildly and averages to zero over any realistic detector.)

Consider now the scattering of two identical spin-1/2 fermions, say electrons. Let us assume that the spin variables are spectators, except for their role in the statistics: in the triplet state the spatial function is antisymmetric, while in the singlet it is symmetric. If the electrons are assumed to come in with random values of s_z, the triplet is three times as likely as the singlet and the average cross section will be

$$\frac{d\sigma}{d\Omega}=\frac{3}{4}|f(\theta,\phi)-f(\pi-\theta,\phi+\pi)|^2 + \frac{1}{4}|f(\theta,\phi)+f(\pi-\theta,\phi+\pi)|^2 \tag{19.6.22}$$

For Coulomb scattering of electrons this becomes

$$\frac{d\sigma}{d\Omega}=\left(\frac{e^2}{4E}\right)^2\left[\frac{1}{\sin^4\theta/2}+\frac{1}{\cos^4\theta/2}-\frac{\cos(\gamma\ln\tan^2\theta/2)}{\sin^2\theta/2\cos^2\theta/2}\right] \tag{19.6.23}$$

Exercise 19.6.2. Derive Eq. (19.6.21) using Eq. (19.3.16) for $f_c(\theta)$.

Exercise 19.6.3. Assuming $f=f(\theta)$ show that $(d\sigma/d\Omega)_{\pi/2}=0$ for fermions in the triplet state.

‡ $\gamma=e^2\mu/\hbar^2 k$ here.

20

The Dirac Equation

Nonrelativistic quantum mechanics, which was developed in the previous chapters, is very successful when applied to problems like the hydrogen atom, where the typical velocity (speaking semiclassically) is small compared to c. (Recall $v/c=\beta=\alpha\cong 1/137$ in the ground state.) But even in this case, there are measurable (fine-structure) corrections of the order of $(v/c)^4$ which have to be put in by hand. If these corrections are to emerge naturally and if relativistic systems (high-Z atoms, for example) are to be described well, it is clear that we need an equation for the electron that has relativity built into it from the start. Such an equation was discovered by Dirac. We study it here with the *main goal* of seeing the coherent emergence of several concepts that were introduced disjointly at various stages—the spin of the electron, its magnetic moment ($g=2$), the spin–orbit, and other fine-structure corrections.

In the last section we address some general questions that accompany Dirac's formulation and indicate the need for quantum field theory.

20.1. The Free-Particle Dirac Equation

Let us consider the simplest case, of a free particle. We start by stating the relation between classical mechanics and the free-particle Schrödinger equation in a way that facilitates generalization. If we start with the nonrelativistic relation

$$\mathscr{H}=\frac{|\mathbf{p}|^2}{2m}=\frac{p^2}{2m} \tag{20.1.1}$$

and make the substitution

$$\begin{aligned}\mathbf{p}&\to\mathbf{P}\\ \mathscr{H}&\to i\hbar\frac{\partial}{\partial t}\end{aligned} \tag{20.1.2}$$

and let both sides act on a state vector $|\psi\rangle$, we get Schrödinger's equation

$$i\hbar\frac{\partial|\psi\rangle}{\partial t}=\frac{P^2}{2m}|\psi\rangle \tag{20.1.3}$$

A natural starting point for the relativistic equation is the corresponding relation due to Einstein

$$\mathscr{H}=(c^2p^2+m^2c^4)^{1/2} \tag{20.1.4}$$

If we make the substitution mentioned above, we get

$$i\hbar\frac{\partial|\psi\rangle}{\partial t}=(c^2P^2+m^2c^4)^{1/2}|\psi\rangle \tag{20.1.5}$$

This equation is undesirable because it treats space and time asymmetrically. To see this, we first go to the momentum basis, where $\mathbf{P}$ is just $\mathbf{p}$ and the square root may be expanded in a series:

$$i\hbar\frac{\partial\psi(\mathbf{p},t)}{\partial t}=mc^2\left(1+\frac{p^2}{2m^2c^2}-\frac{p^4}{8m^4c^4}+\cdots\right)\psi(\mathbf{p},t) \tag{20.1.6}$$

If we now transform to the coordinate basis, each p^2 becomes $(-\hbar^2\nabla^2)$ and the asymmetry between space and time is manifest. What we want is an equation that is of the same order in both space and time.

There are two ways out. One is to replace Eq. (20.1.4) by

$$\mathscr{H}^2=c^2p^2+m^2c^4 \tag{20.1.7}$$

and obtain, upon making the operator substitution,

$$\frac{\partial^2|\psi\rangle}{\partial t^2}=\left(-\frac{c^2P^2}{\hbar^2}-\frac{m^2c^4}{\hbar^2}\right)|\psi\rangle \tag{20.1.8a}$$

In the coordinate basis this becomes

$$\left[\frac{1}{c^2}\frac{\partial^2}{\partial t^2}-\nabla^2+\left(\frac{mc}{\hbar}\right)^2\right]\psi=0 \tag{20.1.8b}$$

This is called the *Klein-Gordon equation* and has the desired symmetry between space and time. But we move along, since ψ here is a scalar and cannot describe the electron. It is, however, a good candidate for pions, kaons, etc., which are spinless.

The second alternative, due to Dirac, is the following. Let us suppose that the quantity in the square root in Eq. (20.1.5) can be written as a perfect square of a quantity that is linear in $\mathbf{P}$. We can then take the square root (which will give us our Hamiltonian) and obtain an equation that is of the first order in time and space. So let us write

$$\begin{aligned} c^2P^2+m^2c^4 &= (c\alpha_x P_x + c\alpha_y P_y + c\alpha_z P_z + \beta mc^2)^2 \\ &= (c\boldsymbol{\alpha}\cdot\mathbf{P}+\beta mc^2)^2 \end{aligned} \tag{20.1.9}$$

where $\boldsymbol{\alpha}$ and β are to be determined by matching both sides of

$$\begin{aligned} &c^2(P_x^2+P_y^2+P_z^2)+m^2c^4 \\ &\quad = [c^2(\alpha_x^2P_x^2+\alpha_y^2P_y^2+\alpha_z^2P_z^2)+\beta^2m^2c^4] \\ &\qquad + [c^2P_xP_y(\alpha_x\alpha_y+\alpha_y\alpha_x)+\text{ and cyclic permutations}] \\ &\qquad + [mc^3P_x(\alpha_x\beta+\beta\alpha_x)+x\to y+x\to z] \end{aligned} \tag{20.1.10}$$

(We have assumed that $\boldsymbol{\alpha}$ and β are space independent, which is a reasonable assumption for a free particle.) These equations tell us that

$$\begin{aligned} &\alpha_i^2=\beta^2=1 \qquad (i=x,y,z) \\ &\alpha_i\alpha_j+\alpha_j\alpha_i=[\alpha_i,\alpha_j]_+=0 \qquad (i\neq j) \\ &\alpha_i\beta+\beta\alpha_i=[\alpha_i,\beta]_+=0 \end{aligned} \tag{20.1.11}$$

It is evident that $\boldsymbol{\alpha}$ and β are not c numbers. They are matrices and furthermore Hermitian (so that the Hamiltonian $H=c\boldsymbol{\alpha}\cdot\mathbf{P}+\beta mc^2$ is Hermitian), traceless, and have eigenvalues ±1. (Recall the results of Exercise 1.8.8). They must also be even dimensional if the last two properties are to be compatible. They cannot be 2×2 matrices, since, as we saw in Exercise 14.3.8, the set of three Pauli matrices with these properties cannot be enlarged to include a fourth. So they must be 4×4 matrices. They are not unique (since $\alpha\to S^\dagger\alpha S$, $\beta\to S^\dagger\beta S$ preserves the desired properties if S is unitary.) The following four are frequently used and will be used by us:

$$\boldsymbol{\alpha}=\begin{bmatrix}0 & \boldsymbol{\sigma}\\ \boldsymbol{\sigma} & 0\end{bmatrix}, \qquad \beta=\begin{bmatrix}I & 0\\ 0 & -I\end{bmatrix} \tag{20.1.12}$$

In the above, $\boldsymbol{\sigma}$ and I are 2×2 matrices.‡ We now have the *Dirac equation*:

$$i\hbar\frac{\partial|\psi\rangle}{\partial t}=(c\boldsymbol{\alpha}\cdot\mathbf{P}+\beta mc^2)|\psi\rangle \tag{20.1.13}$$

with $\boldsymbol{\alpha}$ and β known. Hereafter we work exclusively in the coordinate basis. However, we depart from our convention and use the symbol $\mathbf{P}$, reserved for the momentum operator in the abstract, to represent it in the coordinate basis (instead of using $-i\hbar\boldsymbol{\nabla}$). This is done to simplify the notation in what follows.

‡ For example, β is a 4×4 diagonal matrix with the first two entries $+1$ and the next two entries -1.

The fact that $\boldsymbol{\alpha}$ and β in

$$i\hbar\frac{\partial\psi}{\partial t}=(c\boldsymbol{\alpha}\cdot\mathbf{P}+\beta mc^2)\psi \tag{20.1.14}$$

are 4×4 matrices implies that ψ is a four-component object. It is called a *Lorentz spinor*. Our reaction is mixed. We are happy that relativity, plus the requirement that the equation be first order in time and space, have led naturally to a multicomponent wave function. But we are distressed that ψ has four components instead of two. In the next two sections we will see how, despite this apparent problem, the Dirac equation describes electrons.

For later use, let us note that since the Hamiltonian is Hermitian, the norm of the state is conserved. In the coordinate basis this means

$$\int \psi^\dagger\psi\, d^3\mathbf{r}=\text{const} \tag{20.1.15}$$

Just as in the nonrelativistic case, this global conservation law has a local version also. (See exercise below.)

*Exercise 20.1.1.** Derive the continuity equation

$$\frac{\partial P}{\partial t}+\nabla\cdot\mathbf{j}=0$$

where $P=\psi^\dagger\psi$ and $\mathbf{j}=c\psi^\dagger\boldsymbol{\alpha}\psi$.

20.2. Electromagnetic Interaction of the Dirac Particle

In this central section, we see how several properties of the electron emerge naturally from the Dirac equation. As a first step, we couple the particle to the potential $(\mathbf{A}, \phi)$. We then consider the equation to order $(v/c)^2$ and show that the particle can be described by a two-component wave function and that it has $g=2$. Finally we consider the equation to order $(v/c)^4$ and see the fine-structure emerge.

The coupling of the electromagnetic potentials is suggested by the classical Hamiltonian for a particle of charge q:

$$\mathscr{H}=[(\mathbf{p}-q\mathbf{A}/c)^2c^2+m^2c^4]^{1/2}+q\phi \tag{20.2.1}$$

which leads us to

$$i\hbar\frac{\partial\psi}{\partial t}=[c\boldsymbol{\alpha}\cdot(\mathbf{P}-q\mathbf{A}/c)+\beta mc^2+q\phi]\psi \tag{20.2.2}$$

To see just these two features emerge, we can set $\phi=0$ and work to order $(v/c)^2$. If we look for energy eigenstates

$$\psi(t)=\psi\, e^{-iEt/\hbar}$$

of Eq. (20.2.2), we get

$$E\psi=(c\boldsymbol{\alpha}\cdot\boldsymbol{\pi}+\beta mc^2)\psi \tag{20.2.3}$$

where

$$\boldsymbol{\pi}=\mathbf{P}-q\mathbf{A}/c \tag{20.2.4}$$

is the kinetic ($m\mathbf{v}$) momentum operator. We now write ψ as

$$\psi=\begin{bmatrix}\chi\\ \Phi\end{bmatrix} \tag{20.2.5}$$

where χ *and* Φ *are two-component spinors*. Equation (20.2.3), with $\boldsymbol{\alpha}$ and β explicitly written, becomes

$$\begin{bmatrix}E-mc^2 & -c\boldsymbol{\sigma}\cdot\boldsymbol{\pi}\\ -c\boldsymbol{\sigma}\cdot\boldsymbol{\pi} & E+mc^2\end{bmatrix}\begin{bmatrix}\chi\\ \Phi\end{bmatrix}=\begin{bmatrix}0\\ 0\end{bmatrix} \tag{20.2.6}$$

which means

$$(E-mc^2)\chi-c\boldsymbol{\sigma}\cdot\boldsymbol{\pi}\Phi=0 \tag{20.2.7}$$

and

$$(E+mc^2)\Phi-c\boldsymbol{\sigma}\cdot\boldsymbol{\pi}\chi=0 \tag{20.2.8}$$

The second equation tells us that

$$\Phi=\left(\frac{c\boldsymbol{\sigma}\cdot\boldsymbol{\pi}}{E+mc^2}\right)\chi \tag{20.2.9}$$

Let us examine the term in brackets at low velocities. The denominator is

$$E+mc^2=E_S+2mc^2 \tag{20.2.10}$$

where $E_S = E - mc^2$ is the energy that appears in Schrödinger's equation. At low velocities, since $E_S \ll mc^2$‡

$$E + mc^2 \cong 2mc^2 \tag{20.2.11}$$

The numerator is of the order mvc, where mv is the typical momentum of the state. So

$$\left|\frac{\Phi}{\chi}\right| \cong \frac{1}{2}\left(\frac{v}{c}\right) \ll 1 \tag{20.2.12}$$

For this reason χ and Φ are called the *large and small components*, respectively. The terminology is of course appropriate only in the nonrelativistic domain. In this domain

$$\Phi \cong \frac{\boldsymbol{\sigma}\cdot\boldsymbol{\pi}}{2mc}\chi \tag{20.2.13}$$

and Eq. (20.2.7) becomes

$$E_S\chi = c\boldsymbol{\sigma}\cdot\boldsymbol{\pi}\Phi = \frac{(\boldsymbol{\sigma}\cdot\boldsymbol{\pi})(\boldsymbol{\sigma}\cdot\boldsymbol{\pi})}{2m}\chi \tag{20.2.14}$$

This is called the *Pauli equation*.§ If we use the identity

$$\boldsymbol{\sigma}\cdot\mathbf{A}\boldsymbol{\sigma}\cdot\mathbf{B} = \mathbf{A}\cdot\mathbf{B} + i\boldsymbol{\sigma}\cdot\mathbf{A}\times\mathbf{B} \tag{20.2.15}$$

and

$$\boldsymbol{\pi}\times\boldsymbol{\pi} = \frac{iq\hbar}{c}\mathbf{B} \tag{20.2.16}$$

we get

$$\left[\frac{(\mathbf{P} - q\mathbf{A}/c)^2}{2m} - \frac{q\hbar}{2mc}\boldsymbol{\sigma}\cdot\mathbf{B}\right]\chi = E_S\chi \tag{20.2.17}$$

It is evident that this equation describes a spin-$\frac{1}{2}$ particle *with* $g=2$. It is therefore appropriate to electrons. {Although $g=2$ emerges so naturally from Dirac theory,

‡ $E_S = T + V = \frac{\pi^2}{2m} + V \underset{\text{(virial theorem)}}{\cong} O\left(\frac{\pi^2}{m}\right) = mv^2, \quad \frac{E_S}{mc^2} \cong \left(\frac{v}{c}\right)^2 \ll 1$

§ Actually the Pauli equation is the time-dependent version, with $i\hbar\dot{\chi}$ on the left-hand side.

it is incorrect to say that we need relativity to get this result. If we write the free-particle Schrödinger equation as

$$\frac{(\boldsymbol{\sigma}\cdot\mathbf{P})^2}{2m}\chi = E_S\chi$$

[since $(\boldsymbol{\sigma}\cdot\mathbf{P})^2 = P^2$] and then couple the vector potential $\mathbf{A}$ as prescribed by *nonrelativistic* mechanics ($\mathbf{P}\to\mathbf{P}-q\mathbf{A}/c$), we get $g=2$. Of course spin is introduced artificially here, but $g=2$ is not.}

*Exercise 20.2.1.** Derive Eq. (20.2.16).

*Exercise 20.2.2.** Solve for the exact levels of the Dirac particle in a uniform magnetic field $\mathbf{B}=B_0\mathbf{k}$. Assume $\mathbf{A}=(B_0/2)(-y\mathbf{i}+x\mathbf{j})$. Consult Exercise 12.3.8. (Write the equation for χ.)

Hydrogen Fine Structure

We now apply the Dirac equation to the case

$$V = e\phi = -e^2/r \tag{20.2.18}$$

that is to say, the electron in the hydrogen atom. (The proton is assumed to be fixed, i.e., infinitely massive.) The small and big components obey the following coupled equations:

$$(E-V-mc^2)\chi - c\boldsymbol{\sigma}\cdot\mathbf{P}\Phi = 0 \tag{20.2.19}$$

$$(E-V+mc^2)\Phi - c\boldsymbol{\sigma}\cdot\mathbf{P}\chi = 0 \tag{20.2.20}$$

The second one tells us that

$$\Phi = (E-V+mc^2)^{-1}c\boldsymbol{\sigma}\cdot\mathbf{P}\chi \tag{20.2.21}$$

(Since $\mathbf{P}$ can differentiate V, the order of the factors is important.) If we feed this into the first, we get

$$(E-V-mc^2)\chi = c\boldsymbol{\sigma}\cdot\mathbf{P}\left[\frac{1}{E-V+mc^2}\right]c\boldsymbol{\sigma}\cdot\mathbf{P}\chi \tag{20.2.22}$$

If we approximate $E-V+mc^2$ on the right-hand side as $2mc^2$, we get

$$\begin{aligned} E_S\chi &= \left[\frac{(\boldsymbol{\sigma}\cdot\mathbf{P})^2}{2m}+V\right]\chi \\ &= \left[\frac{P^2}{2m}+V\right]\chi \end{aligned} \tag{20.2.23}$$

This is just the nonrelativistic Schrödinger equation we solved in Chapter 13. Notice that the Hamiltonian is order $(v/c)^2$ since it is quadratic in the momentum. To see the fine structure, we must go to order $(v/c)^4$. We do this by expanding $(E-V+mc^2)^{-1}$ on the right-hand side to one more order in v^2/c^2:

$$\frac{1}{E-V+mc^2}=\frac{1}{2mc^2+E_S-V}=\frac{1}{2mc^2}\left(1+\frac{E_S-V}{2mc^2}\right)^{-1}$$

$$\cong\frac{1}{2mc^2}\left(1-\frac{E_S-V}{2mc^2}\right)=\frac{1}{2mc^2}-\frac{E_S-V}{4m^2c^4} \tag{20.2.24}$$

Equation (20.2.22) now becomes

$$E_S\chi=\left[\frac{P^2}{2m}+V-\frac{\boldsymbol{\sigma}\cdot\mathbf{P}(E_S-V)\boldsymbol{\sigma}\cdot\mathbf{P}}{4m^2c^2}\right]\chi \tag{20.2.25}$$

We cannot view this as the time-independent Schrödinger equation (i.e., as $E_S\chi=H\chi$) since E_S appears on *both sides*. By now even our spinal column knows how to respond to such a crisis. The right-hand side is a power series in v^2/c^2. The first two terms are of the order v^2/c^2, and the third is expected to be of order v^4/c^4. Now the two $\boldsymbol{\sigma}\cdot\mathbf{P}$ factors in the third form use up a factor v^2/c^2. So we need E_S-V only to order v^2/c^2. This we get from the same equation truncated to this order:

$$(E_S-V)\chi=\frac{P^2}{2m}\chi \tag{20.2.26}$$

We cannot use this result directly in Eq. (20.2.25) since E_S-V there does not act on χ directly; there is $\boldsymbol{\sigma}\cdot\mathbf{P}$ in the way. So we do the following:

$$(E_S-V)\boldsymbol{\sigma}\cdot\mathbf{P}\chi=\boldsymbol{\sigma}\cdot\mathbf{P}(E_S-V)\chi+\boldsymbol{\sigma}\cdot[E_S-V,\mathbf{P}]\chi$$

$$=(\boldsymbol{\sigma}\cdot\mathbf{P})\frac{P^2}{2m}\chi+\boldsymbol{\sigma}\cdot[\mathbf{P},V]\chi \tag{20.2.27}$$

Feeding this into Eq. (20.2.25) we get

$$E_S\chi=\left\{\frac{P^2}{2m}+V-\frac{P^4}{8m^3c^2}-\frac{(\boldsymbol{\sigma}\cdot\mathbf{P})(\boldsymbol{\sigma}\cdot[\mathbf{P},V])}{4m^2c^2}\right\}\chi$$

$$=\left\{\frac{P^2}{2m}+V-\frac{P^4}{8m^3c^2}-\frac{i\boldsymbol{\sigma}\cdot\mathbf{P}\times[\mathbf{P},V]}{4m^2c^2}-\frac{\mathbf{P}\cdot[\mathbf{P},V]}{4m^2c^2}\right\}\chi$$

$$=H\chi \tag{20.2.28}$$

using once again the identity (20.2.15). We recognize the third term to be just the relativistic correction to the kinetic energy. It is just [recall Eq. (17.3.6)]

$$H_T = -\frac{P^4}{8m^3c^2} \tag{20.2 29}$$

The fourth term is the spin–orbit interaction, $H_{\text{s.o.}}$ of Eq. (17.3.16):

$$\begin{aligned}
&\frac{-i\boldsymbol{\sigma}\cdot\mathbf{P}\times[\mathbf{P}, V]}{4m^2c^2} \\
&= \frac{-i\boldsymbol{\sigma}\cdot\mathbf{P}\times[-i\hbar\nabla(-e^2/r)]}{4m^2c^2} \qquad \left\{\text{using} \quad [P, f(x)] = -i\hbar\frac{df}{dx}\right\} \\
&= \frac{-\hbar e^2\boldsymbol{\sigma}\cdot\mathbf{P}\times\mathbf{r}}{4m^2c^2r^3} = \frac{\hbar e^2}{4m^2c^2r^3}\boldsymbol{\sigma}\cdot\mathbf{r}\times\mathbf{P}\ddagger \\
&= \frac{e^2}{2m^2c^2r^3}\mathbf{S}\cdot\mathbf{L} = H_{\text{s.o.}}
\end{aligned} \tag{20.2.30}$$

Notice that the Thomas factor is built in.

Consider now the fifth and last term. It upsets the whole interpretation because *it is not Hermitian* (check this). So if the quantity in brackets in Eq. (20.2.28) is used as a Hamiltonian we will find

$$\int |\chi|^2 \, d^3\mathbf{r} \neq \text{const in time}$$

But this is not surprising, since the conservation law that comes from the Dirac equation is

$$\int \psi^\dagger \psi \, d^3\mathbf{r} = \int [|\chi|^2 + |\Phi|^2] \, d^3\mathbf{r} = \text{const} \tag{20.2.31}$$

It follows that χ is not a good candidate for the Schrödinger wave function to this order in v/c. [It was all right when we worked to order $(v/c)^2$.] We find the right one as follows. Note that

$$\Phi = \frac{c\boldsymbol{\sigma}\cdot\mathbf{P}}{E-V+mc^2}\chi = \frac{c\boldsymbol{\sigma}\cdot\mathbf{P}}{2mc^2+E_S-V}\chi \cong \frac{\boldsymbol{\sigma}\cdot\mathbf{P}}{2mc}\chi \tag{20.2.32}$$

‡ Although $\mathbf{P}$ is a differential operator, $\mathbf{P}\times\mathbf{r} = -\mathbf{r}\times\mathbf{P}$, just as if $\mathbf{P}$ and $\mathbf{r}$ were c numbers, because the cross product never involves the product of a given coordinate and its conjugate momentum. This point was made earlier in the book when it was stated that there was no ordering ambiguity in passing from $\mathbf{l} = \mathbf{r}\times\mathbf{p}$ to $\mathbf{L}$.

(The neglected terms make corrections of order v^6/c^6 in the end.) Consequently

$$|\Phi|^2 = \frac{\chi^\dagger(\boldsymbol{\sigma}\cdot\mathbf{P})(\boldsymbol{\sigma}\cdot\mathbf{P})\chi}{(2mc)^2} = \chi^\dagger \frac{P^2}{4m^2c^2}\chi$$

and so, from Eq. (20.2.31),

$$\begin{aligned}\int \chi^\dagger\left(1+\frac{P^2}{4m^2c^2}\right)\chi d^3\mathbf{r} &= \int \left[\left(1+\frac{P^2}{8m^2c^2}\right)\chi\right]^\dagger \cdot \left(1+\frac{P^2}{8m^2c^2}\right)\chi d^3\mathbf{r} \\ &= \text{const}\end{aligned} \qquad (20.2.33)$$

using $(1+x)=(1+x/2)(1+x/2)+O(x^2)$ and the Hermiticity of P^2. Consequently, the candidate for the Schrödinger wave function is

$$\chi_s = \left(1+\frac{P^2}{8m^2c^2}\right)\chi \qquad (20.2.34)$$

for it will have a time-independent norm. (To the present accuracy, that is. If we go to higher and higher orders in v^2/c^2, $\boldsymbol{\Phi}$ will creep in more and more.)

The equation for χ_s is obtained by eliminating χ in Eq. (20.2.28):

$$E_s\left(1+\frac{P^2}{8m^2c^2}\right)^{-1}\chi_s = H\left(1+\frac{P^2}{8m^2c^2}\right)^{-1}\chi_s$$

$$\begin{aligned}E_s\chi_s &= \left(1+\frac{P^2}{8m^2c^2}\right)H\left(1-\frac{P^2}{8m^2c^2}\right)\chi_s \\ &= \left(H+\left[\frac{P^2}{8m^2c^2}, H\right]\right)\chi_s \qquad \text{(to this order in } v/c\text{)} \\ &= H_s\chi_s.\end{aligned} \qquad (20.2.35)$$

In evaluating the commutator, we need consider just the v^2/c^2 part of H, since $P^2/8m^2c^2$ is $O(v^2/c^2)$ and we are working to order v^4/c^4. So

$$H_s = H + \left[\frac{\mathbf{P}\cdot\mathbf{P}}{8m^2c^2}, V\right]$$

is the desired Schrödinger Hamiltonian. The extra piece the above analysis yields combines with the non-Hermitian piece in Eq. (20.2.28) to form the *Darwin* term H_D:

$$\begin{aligned} H_D &= \frac{1}{8m^2c^2}(-2\mathbf{P}\cdot[\mathbf{P}, V] + [\mathbf{P}\cdot\mathbf{P}, V]) \\ &= \frac{-1}{8m^2c^2}[\mathbf{P}, \cdot [\mathbf{P}, V]] \quad \text{(using the chain rule for commutators of products)} \\ &= \frac{\hbar^2}{8m^2c^2}\nabla^2 V \quad \{\text{using } [P, f(x)] = -i\hbar\, df/dx \text{ twice}\} \\ &= \frac{e^2\hbar^2\pi}{2m^2c^2}\delta^3(\mathbf{r}) \end{aligned} \tag{20.2.37}$$

Thus the Darwin term affects only the *s states*.‡ In the ground state, for example,

$$\langle 100|H_D|100\rangle = \frac{e^2\hbar^2\pi}{2m^2c^2}\frac{1}{\pi a_0^3} = \frac{1}{2}mc^2\alpha^4$$

and in general

$$\langle n00|H_D|n00\rangle = \frac{1}{2}\frac{mc^2\alpha^4}{n^3} \tag{20.2.38}$$

Recall that in our previous treatment of fine structure we obtained a spin–orbit shift valid only for $l \neq 0$ and then applied it to $l = 0$ as well, without any real justification. The result we got for $l = 0$ is just what H_D generated above, which was our reason for doing what we did then. Thus $H_{\text{s.o.}}$ (relevant for $l \neq 0$) and H_D (relevant only for $l = 0$) *together* conspire to produce a fine-structure shift that is smooth in l. The physics behind the Darwin term has nothing to do with spin–orbit coupling (for there is no such thing for $l = 0$). Rather, it reflects the fact that in a relativistic theory, the particle cannot be localized to better than its Compton wavelength $\hbar/mc$. Thus the potential that is relevant is not $V(\mathbf{r})$ but some smeared average around the point $\mathbf{r}$:

$$\begin{aligned} \overline{V(\mathbf{r})} &= V(\mathbf{r}) + \sum_i \overline{\frac{\partial V}{\partial r_i}\delta r_i} + \frac{1}{2!}\sum_i\sum_j \overline{\frac{\partial^2 V}{\partial r_i \partial r_j}\delta r_i \delta r_j} + O(\delta r^3) \\ &= V(\mathbf{r}) + \frac{1}{6}(\delta r)^2\nabla^2 V + O(\delta r^3)^3 \end{aligned} \tag{20.2.39}$$

where, in the averaging, we have assumed that fluctuations in the various directions are uncorrelated and spherically symmetric. If we now feed in $\delta r \cong \hbar/mc$, we get the right sign and almost the right magnitude for the Darwin term [see Eq. (20.2.37)].

‡ Recall that only in these states is ψ nonzero at the origin.

Although we chose to work to order v^4/c^4, the Dirac equation can be solved exactly in the Coulomb case, $V=-e^2/r$. The resulting energy spectrum is

$$E_{nj}=mc^2\left[1+\left(\frac{\alpha}{n-(j+\frac{1}{2})+[(j+\frac{1}{2})^2-\alpha^2]^{1/2}}\right)^2\right]^{-1/2} \tag{20.2.40}$$

If we expand this in powers of α, we get the rest energy, the Schrödinger energy, the fine-structure energy, and so on. Notice that the states of a given n and j are degenerate to all orders in α.

Whereas the above formula is in fantastic agreement with experiment,‡ it is not the last word. For example, very precise measurements show that the $2S_{1/2}$ level is above the $2P_{1/2}$ level. This phenomenon, called the *Lamb shift*, can be understood only if the electromagnetic field is treated quantum mechanically.

20.3. More on Relativistic Quantum Mechanics

With the principal goal of this chapter achieved in the last section, we direct our attention to certain phenomena that come out of Dirac theory but were not apparent in the last few pages. Let us first note that the union of relativity and quantum mechanics produces the following problem: relativity allows particle production given enough energy, and quantum mechanics allows arbitrarily large energy violations over short times. Consequently the degrees of freedom of a relativistic system are neither fixed nor finite; a system that initially has one particle can evolve into a state with 15 of them. Why doesn't this problem appear in the Dirac theory, which seems like a single-particle theory? The answer is that it does appear, but in the guise of *negative-energy solutions*. Let us see what these are and how they lead to proliferation of the degrees of freedom.

Consider the free-particle Dirac equation (with $\hbar=c=1$)

$$i\frac{\partial\psi}{\partial t}=(\boldsymbol{\alpha}\cdot\mathbf{P}+\beta m)\psi \tag{20.3.1}$$

Let us look for plane wave solutions

$$\psi=w(\mathbf{p})\,e^{i(\mathbf{p}\cdot\mathbf{r}-Et)} \tag{20.3.2}$$

where $w(\mathbf{p})$ is a spinor that has no space-time dependence. It satisfies

$$Ew=(\boldsymbol{\alpha}\cdot\mathbf{p}+\beta m)w \tag{20.3.3}$$

‡ After hyperfine interactions are taken into account.

Figure 20.1. In the Dirac theory there are two continuous bands of energy available to the free particle; one goes from $+m$ up to ∞ and the other goes from $-m$ down to $-\infty$.

or in terms of χ and Φ,

$$\begin{bmatrix} E-m & -\boldsymbol{\sigma}\cdot\mathbf{p} \\ -\boldsymbol{\sigma}\cdot\mathbf{p} & E+m \end{bmatrix}\begin{bmatrix} \chi \\ \Phi \end{bmatrix}=\begin{bmatrix} 0 \\ 0 \end{bmatrix} \tag{20.3.4}$$

If $\mathbf{p}=0$, χ and Φ decouple. The equation for χ is

$$(E-m)\chi=0 \rightarrow E=m \tag{20.3.5}$$

which is fine. It says a particle at rest has energy $E=m$ and is described by an arbitrary two-component spinor which we identify as the spin degree of freedom.

The equation for Φ is

$$(E+m)\Phi=0 \rightarrow E=-m \tag{20.3.6}$$

Now even a layperson will tell you that E is supposed to be mc^2 not $-mc^2$. The significance of the four components of ψ are evident in the rest frame: there are two possible spin orientations and two signs of the energy. The problem persists for $\mathbf{p}\neq 0$ as well. Here we find

$$\chi=\frac{\boldsymbol{\sigma}\cdot\mathbf{p}}{E-m}\Phi \tag{20.3.37}$$

$$\Phi=\frac{\boldsymbol{\sigma}\cdot\mathbf{p}}{E+m}\chi \tag{20.3.8}$$

These are consistent only if

$$\frac{p^2}{E^2-m^2}=1$$

or

$$E^2=p^2+m^2$$

or

$$E=\pm(p^2+m^2)^{1/2} \tag{20.3.9}$$

The energy levels corresponding to these two options are shown in Fig. 20.1. What do we do with the negative-energy solutions? If there are no interactions, positive-energy electrons will stay where they are and we can postulate that there are no negative-energy electrons. But there are always some perturbations acting on all electrons and these can induce all positive-energy electrons to cascade down to the negative-energy states. How do we understand the stability of positive-energy electrons?

There are two ways out, one due to Dirac and one due to Feynman.‡ Dirac postulated that the negative-energy states are all occupied—that what we call the vacuum is really the occupied (but *unobservable*) sea of negative-energy electrons. If we accept this, the stability problem is solved by the exclusion principle, which prevents the positive-energy electrons from decaying to the occupied negative-energy states. This picture has some profound consequences. Suppose we give a negative-energy electron enough energy (at least $2m$) for it to come to a positive-energy state. Now we have a positive-energy, charge $-e$ object. But we also have created a hole in the "Dirac sea." Since the filled Dirac sea was postulated to be unobservable, the hole is observable; it represents an increase in charge by $+e$ (the disappearance of $-e=$ appearance of $+e$), and an increase in energy by $|E|$, if $-|E|$ was the energy of the electron ejected from the sea.§ Thus the hole, which has charge $+e$ and *positive* energy, is created along with the electron. It is called a *positron*. Its mass can easily be shown to be m. Positrons were observed a few years after Dirac's theory of holes was published.

When an electron meets a positron, i.e., a hole in the sea, it jumps in and we lose both particles, though some energy (at least equal to $2m$) will be liberated in the form of photons. (Hereafter we will occasionally refer to these particles as e^-, e^+, and γ, respectively.)

The trouble with Dirac's solution is that it doesn't apply to spinless particles, which don't obey the Pauli principle but which do have the same problem of negative-energy solutions, as one can see by plugging a plane wave solution into the Klein-Gordon equation. (In fact this was the reason the Klein-Gordon equation was rejected in the early days and Dirac sought a first-order equation.) So let us turn to Feynman's resolution, which applies to bosons *and* fermions.

Feynman's idea is the following: *negative-energy particles can only travel backward in time*. Let us see first how this resolves the problem and then how the statement is actually implemented in quantum theory. Consider a negative-energy particle that is created at the space-time point c and travels backward to d, where it is destroyed (Fig. 20.2a). To us, who move forward in time and see space-time in equal-time slices, this is what will seem to be happening:

(1) $t<t_d$ Nothing anywhere.

(2) $t=t_d$ Negative energy $-|E|$ and charge $-e$ are destroyed, i.e., world energy goes up by $|E|$ and charge goes up by e relative to the past. A positron is born.

(3) $t=t_c$ Negative energy is created, charge $-e$ is created. This wipes out the positron.

(4) $t>t_c$ Nothing anywhere.

‡ In its basic form, the idea exploited by Feynman was pointed out by Stueckelberg.

§ Recall the story of the fellow who got so used to the midnight express going past his house that one day when it failed to show up, he woke up screaming "What's that noise?"

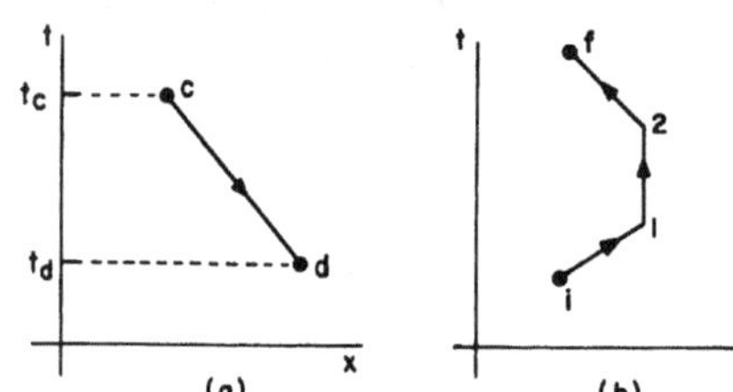

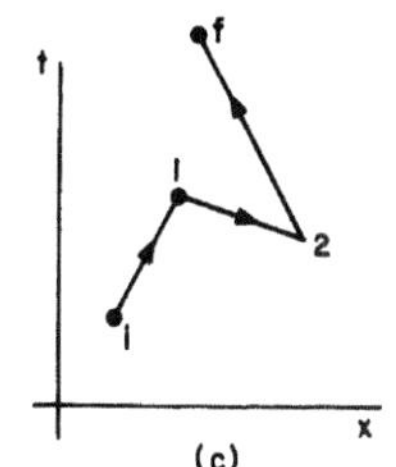

Figure 20.2. (a) A negative-energy particle is created at c, travels back in time to d, where it is destroyed. To us, who move forward in time, it will seem as though an antiparticle of positive energy is created at d and destroyed at c. (b) A normal second order scattering process. (c) A second-order process that involves back-scattering in time. Between times 2 and 1 we will see a particle–antiparticle pair in addition to the original particle.

Thus the process makes perfect sense and represents a positron *created at d* and *destroyed at c*.

How does Feynman ensure that negative-energy states propagate backward? Here is a sketchy description. Recall that the Schrödinger propagator we have used so far is (in the coordinate basis)

$$U_S(\mathbf{r}, t; \mathbf{r}', t') = \sum_n \psi_n(\mathbf{r})\psi_n^*(\mathbf{r}')\, e^{-iE_n(t-t')} \tag{20.3.10}$$

where ψ_n is an energy eigenfunction labeled by a generic quantum number n.

Since every term in the sum satisfies the Schrödinger equation, it is clear that

$$\left(i\frac{\partial}{\partial t} - H\right)U_S = 0 \tag{20.3.11}$$

given this U_S and $\psi(t')$ at some initial time, we can get $\psi(t)$ at a later time $(t > t')$:

$$\psi(t) = U_S\,\psi(t') \quad \text{(schematic)} \tag{20.3.12}$$

Now note that although we use U_S to propagate ψ forward in time, it can also propagate it backward, since $U_S \neq 0$ for $t < t'$. To avoid this possibility explicitly, let us work with

$$G_S(\mathbf{r}t, \mathbf{r}'t') = \theta(t-t')\,U_S(\mathbf{r}t, \mathbf{r}'t') \tag{20.3.13}$$

which simply cannot propagate ψ backward. The equation satisfied by G_S is

$$\begin{aligned}\left(i\frac{\partial}{\partial t} - H\right)G_S &= \left[i\frac{\partial}{\partial t}\,\theta(t-t')\right]\sum_n \psi_n(\mathbf{r})\,\psi_n^*(\mathbf{r}')\, e^{-iE_n(t-t')}\\ &= i\delta(t-t')\delta^3(\mathbf{r}-\mathbf{r}')\\ &= i\delta^4(x-x') \qquad [x = (t, \mathbf{r})]\end{aligned} \tag{20.3.14}$$

[We have used the completeness of the eigenfunctions, and $\dot{\theta}(t-t')=\delta(t-t')$.]

The propagator in Dirac theory, G_D, obeys a similar equation. Consider the *free-particle case*. Here

$$\left(i\frac{\partial}{\partial t}-H^0\right)G_D^0=i\delta^4(x-x') \tag{20.3.15}$$

with H^0 the free-particle Dirac Hamiltonian. The solution is

$$G_D^0(x,x')=\theta(t-t')\left(\sum_{n+}+\sum_{n-}\right) \tag{20.3.16}$$

where $\sum_{n\pm}$ denote sums over *positive- and negative*-energy eigenfunctions, respectively. [If we throw away $\sum_{n-}$ we lose completeness and won't get $i\delta^4$ on the right-hand side of Eq. (20.3.15).] Although G_D^0 satisfies the requisite equation, it has the negative-energy solutions propagating forward in time. Now here is the trick. G_D^0 is not a unique solution to Eq. (20.3.15); we can add or subtract any solution to the free-Dirac equation, provided we subtract it for all times. (If we subtract it only for $t>0$, say, we are subtracting a θ function times the solution, which doesn't obey the homogeneous equation.) Let us subtract all negative-energy solutions for all times. This gives us *Feynman's propagator*

$$G_F^0(x,x')=\theta(t-t')\sum_{n+}-\theta(t'-t)\sum_{n-} \tag{20.3.17}$$

Consider now some initial state $\psi_i(t')$ which is composed of just positive-energy solutions. G_F^0 will propagate it forward in time, since $\psi_i(t')$ is orthogonal to every term in $\sum_{n-}$. Thus $G_F^0\psi_i(t')=\psi_f(t)$ contains only positive-energy components and keeps moving forward. On the other hand if $\psi_i(t')$ is built out of negative-energy components only, it is orthogonal to every term in $\sum_{n+}$ and gets propagated backwards from t' to t. We will see it as a positron propagating from t to t'.

Consider now the electron in some external potential V. The exact propagation of the electron can be described by a perturbation series based on G_F^0; and in schematic form,

$$\psi_f(t)=G_F^0(t,t')\psi_i(t')+\sum_{t''}G_F^0(t,t'')V(t'')G_F^0(t'',t')\psi_i(t')+\cdots$$

We can represent these multiple scattering events by diagrams very much like the ones in Section 18.3. There is just one difference. Consider a second-order process. There is of course the usual double scattering in which the electron just gets scattered forward in time (Fig. 20.2b). But now there is also the possibility that the potential scatters it backward in time at 1 and then forward at 2 (Fig. 20.2c). As we move forward in time, we first see the electron, then an e^+e^- pair created at 2, then the annihilation of the e^+ with the original e^- at 1 and finally the arrival of the created e^- at f. Since the electron can wiggle and jiggle any number of times (as we go to higher orders in the expansion) the intermediate stages can contain any number of

e^+e^- pairs. This is how the degrees of freedom proliferate in a relativistic theory. Even though we started with a one-particle equation, particle production creeps in through the negative-energy solutions—either because the latter imply an infinite sea of sleeping particles which can be awakened or because they allow a single electron to go back and forth in time, thereby becoming many particles at a given time. Although particle production (at least pair production) can be handled in the present formulation, it is time to learn quantum field theory, which provides a natural framework for handling the creation and destruction of particles. We have already seen one example, namely, the quantized electromagnetic field, whose quanta, the photons, can be created and destroyed by operators $a^\dagger$ and a. We need a theory in which particles like electrons and positrons can also be created and destroyed. You are ready for that subject.‡

‡ See for example J. D. Bjorken and S. D. Drell, *Relativistic Quantum Mechanics* and *Relativistic Quantum Fields*, McGraw-Hill, New York (1964), or C. Itzykson and J. B. Zuber, *Quantum Field Theory*, McGraw-Hill, New York (1980).

21

Path Integrals: Part II

In this chapter we return to path integrals for a more detailed and advanced treatment. The tools described here are so widely used in so many branches of physics, that it makes sense to include them in a book such as this. This chapter will be different from the earlier ones in that it will try to introduce you to a variety of new topics without giving all the derivations in the same detail as before. It also has a list of references to help you pursue any topic that attracts you. The list is not exhaustive and consists mostly of pedagogical reviews or books. From the references these references contain, you can pursue any given topic in greater depth. All this will facilitate the transition from course work to research.

In Chapter 8 the path integral formula for the propagator was simply postulated and shown to lead to the same results as the operator methods either by direct evaluation of the propagator (in the free particle case) or by showing once and for all that the Schrödinger equation followed from the path integral prescription for computing the time evolution.

We begin this chapter by doing the reverse: we start with the operator Hamiltonian $H=P^2/2m+V$ and derive the propagator for it as a path integral. We shall see that there are many types of path integrals one can derive. We will discuss

- The configuration space path integral, discussed in Chapter 8.
- The phase space path integral.
- The coherent state path integral.

You will see that the existence of many path integrals is tied to the existence of many resolutions of the identity, i.e., to the existence of many bases.

Following this we will discuss two applications: to the Quantum Hall Effect (QHE) and a recent development called the Berry Phase.

We then turn to imaginary time quantum mechanics and its relation to statistical mechanics (classical and quantum) as well the calculation of tunneling amplitudes by a semiclassical approximation. You will learn about instantons, the transfer matrix formulation, and so on.

Finally, we discuss path integrals for two problems with no classical limit: a spin Hamiltonian and a fermionic oscillator.

21.1. Derivation of the Path Integral

Let us assume that the Hamiltonian is time-independent and has the form

$$H=\frac{P^2}{2m}+V(X) \tag{21.1.1}$$

The propagator is defined by

$$U(xt;x'0)\equiv U(x,x',t)=\langle x|\exp\left(-\frac{i}{\hbar}Ht\right)|x'\rangle \tag{21.1.2}$$

It was stated in Chapter 8 that U may be written as a sum over paths going from $(x'0)$ to (xt). We will now see how this comes about.

First, it is evident that we may write

$$\exp\left(-\frac{i}{\hbar}Ht\right)=\left[\exp\left(-\frac{i}{\hbar}H\frac{t}{N}\right)\right]^N \tag{21.1.3}$$

for any N. This merely states that $U(t)$, the propagator for a time t, is the product of N propagators $U(t/N)$. Let us define

$$\varepsilon=\frac{t}{N} \tag{21.1.4}$$

and consider the limit $N\to\infty$. Now we can write

$$\exp\left(-\frac{i\varepsilon}{\hbar}(P^2/2m+V(X))\right)\simeq\exp\left(-\frac{i\varepsilon}{2m\hbar}P^2\right)\cdot\exp\left(-\frac{i\varepsilon}{\hbar}V(X)\right) \tag{21.1.5}$$

because of the fact that

$$e^Ae^B=e^{A+B+1/2[A,B]+\cdots} \tag{21.1.6}$$

which allows us to drop the commutator shown (and other higher-order nested commutators not shown) on the grounds that they are proportional to higher powers of ε which is going to 0. While all this is fine if A and B are finite dimensional matrices with finite matrix elements, it is clearly more delicate for operators in Hilbert space which could have large or even singular matrix elements. We will simply assume that in the limit $\varepsilon\to 0$ the $\simeq$ sign in Eq. (21.1.5) will become the equality sign for the purpose of computing any reasonable physical quantity.

So we have to compute

$$\underbrace{\langle x|\exp\left(-\frac{i\varepsilon}{2m\hbar}P^2\right)\cdot\exp\left(-\frac{i\varepsilon}{\hbar}V(X)\right)\cdot\exp\left(-\frac{i\varepsilon}{2m\hbar}P^2\right)\cdot\exp\left(-\frac{i\varepsilon}{\hbar}V(X)\right)\ldots|x'\rangle}_{N\text{ times}} \tag{21.1.7}$$

The next step is to introduce the resolution of the identity:

$$I=\int_{-\infty}^{\infty}dx|x\rangle\langle x| \tag{21.1.8}$$

between every two adjacent factors of $U(t/N)$. Let us illustrate the outcome by considering $N=3$. We find (upon renaming x, x' as x_3, x_0 for reasons that will be clear soon)

$$\begin{aligned}U(x_3, x_0, t)=\int\prod_{n=1}^{2}dx_n\langle x_3|\exp\left(-\frac{i\varepsilon}{2m\hbar}P^2\right)\exp\left(-\frac{i\varepsilon}{\hbar}V(X)\right)|x_2\rangle\\ \times\langle x_2|\exp\left(-\frac{i\varepsilon}{2m\hbar}P^2\right)\exp\left(-\frac{i\varepsilon}{\hbar}V(X)\right)|x_1\rangle\\ \times\langle x_1|\exp\left(-\frac{i\varepsilon}{2m\hbar}P^2\right)\exp\left(-\frac{i\varepsilon}{\hbar}V(X)\right)|x_0\rangle\end{aligned} \tag{21.1.9}$$

Consider now the evaluation of the matrix element

$$\langle x_n|\exp\left(-\frac{i\varepsilon}{2m\hbar}P^2\right)\cdot\exp\left(-\frac{i\varepsilon}{\hbar}V(X)\right)|x_{n-1}\rangle \tag{21.1.10}$$

When the rightmost exponential operates on the ket to its right, the operator X gets replaced by the eigenvalue x_{n-1}. Thus,

$$\begin{aligned}&\langle x_n|\exp\left(-\frac{i\varepsilon}{2m\hbar}P^2\right)\cdot\exp\left(-\frac{i\varepsilon}{\hbar}V(X)\right)|x_{n-1}\rangle\\ &\quad=\langle x_n|\exp\left(-\frac{i\varepsilon}{2m\hbar}P^2\right)|x_{n-1}\rangle\exp\left(-\frac{i\varepsilon}{\hbar}V(x_{n-1})\right)\end{aligned} \tag{21.1.11}$$

Consider now the remaining matrix element. It is simply the free particle propagator from x_{n-1} to x_n in time ε. We know what it is [say from Eq. (5.1.10)] or the following exercise

$$\langle x_n|\exp\left(\frac{-i\varepsilon}{2m\hbar}P^2\right)|x_{n-1}\rangle=\left[\frac{m}{2\pi i\hbar\varepsilon}\right]^{1/2}\exp\left[\frac{im(x_n-x_{n-1})^2}{2\hbar\varepsilon}\right] \tag{21.1.12}$$

Exercise 21.1.1. Derive the above result independently of Eq. (5.1.10) by introducing a resolution of the identity in terms of momentum states between the exponential operator and the position eigenket in the left-hand side of Eq. (21.1.12). That is, use

$$I=\int_{-\infty}^{\infty}\frac{dp}{2\pi\hbar}|p\rangle\langle p| \tag{21.1.13}$$

where the plane wave states have a wave function given by

$$\langle x|p\rangle=e^{ipx/\hbar} \tag{21.1.14}$$

which explains the measure for the p integration.

Resuming our derivation, we now have

$$\begin{aligned}\langle x_n|\exp\left(-\frac{i\varepsilon}{2m\hbar}P^2\right)\cdot\exp\left(-\frac{i\varepsilon}{\hbar}V(X)\right)|x_{n-1}\rangle\\ =\left[\frac{m}{2\pi i\hbar\varepsilon}\right]^{1/2}\exp\left[\frac{im(x_n-x_{n-1})^2}{2\hbar\varepsilon}\right]\exp\left(-\frac{i\varepsilon}{\hbar}V(x_{n-1})\right)\end{aligned} \tag{21.1.15}$$

Collecting all such factors (there are just three in this case with $N=3$), we can readily see that for general N

$$\begin{aligned}U(x_N,x_0,t)=\left(\frac{m}{2\pi i\hbar\varepsilon}\right)^{1/2}\left[\int\prod_{n=1}^{N-1}\left(\frac{m}{2\pi i\hbar\varepsilon}\right)^{1/2}dx_n\right]\\ \times\exp\left[\sum_{n=1}^{N}\frac{im(x_n-x_{n-1})^2}{2\hbar\varepsilon}-\frac{i\varepsilon}{\hbar}V(x_{n-1})\right]\end{aligned} \tag{21.1.16}$$

If we drop the V terms we see that this is in exact agreement with the free particle path integral of Chapter 8. For example, the measure for integration has exactly N factors of B^{-1} as per Eq. (8.4.8), of which $N-1$ accompany the x-integrals. With the V term, the integrand is just the discretized version of $\exp(iS/\hbar)$:

$$\begin{aligned}\exp\left[\sum_{n=1}^{N}\frac{im(x_n-x_{n-1})^2}{2\hbar\varepsilon}-\frac{i\varepsilon}{\hbar}V(x_{n-1})\right]\\ =\exp\frac{i}{\hbar}\varepsilon\sum_{n=1}^{N}\left[\frac{m(x_n-x_{n-1})^2}{2\varepsilon^2}-V(x_{n-1})\right]\end{aligned} \tag{21.1.17}$$

We can go back to the continuum notation and write all this as follows:

$$U(x, x', t) = \int [\mathscr{D}x] \exp\left[\frac{i}{\hbar}\int_0^t \mathscr{L}(x, \dot{x})\, dt\right] \tag{21.1.18}$$

where

$$\int [\mathscr{D}x] = \lim_{N\to\infty} \left(\frac{m}{2\pi i\hbar\varepsilon}\right)^{1/2} \int \left[\prod_{n=1}^{N-1} \left(\frac{m}{2\pi i\hbar\varepsilon}\right)^{1/2} dx_n\right] \tag{21.1.19}$$

The continuum notation is really a schematic for the discretized version that preceded it, and we need the latter to define what one means by the path integral. It is easy to make many mistakes if one forgets this. In particular, there is no reason to believe that replacing differences by derivatives is always legitimate. For example, in this problem, in a time ε, the variable being integrated over typically changes by $\mathcal{O}(\varepsilon^{1/2})$ and not $\mathcal{O}(\varepsilon)$, as explained in the discussion before Eq. (8.5.6). The works in the Bibliography at the end of this chapter discuss some of the subtleties. The continuum version is, however, very useful to bear in mind since it exposes some aspects of the theory that would not be so transparent otherwise. It is also very useful for getting the picture at the semiclassical level and for finding whatever connection there is between the macroscopic world of smooth paths and the quantum world. We will take up some examples later.

The path integral derived above is called the *Configuration Space* path integral or simply the path integral. We now consider another one. Let us go back to

$$\langle x_N|\underbrace{\exp\left(-\frac{i\varepsilon}{2m\hbar}P^2\right)\cdot\exp\left(-\frac{i\varepsilon}{\hbar}V(X)\right)\cdot\exp\left(-\frac{i\varepsilon}{2m\hbar}P^2\right)\cdot\exp\left(-\frac{i\varepsilon}{\hbar}V(X)\right)\ldots}_{N \text{ times}}|x_0\rangle \tag{21.1.20}$$

Let us now introduce resolutions of the identity between *every exponential* and the next. We need two versions

$$I = \int_{-\infty}^{\infty} dx|x\rangle\langle x| \tag{21.1.21}$$

$$I = \int_{-\infty}^{\infty} \frac{dp}{2\pi\hbar}|p\rangle\langle p| \tag{21.1.22}$$

where the plane wave states have a wave function given by

$$\langle x|p\rangle = e^{ipx/\hbar} \tag{21.1.23}$$

Let us first set $N=3$ and insert three resolutions of the identity in terms of p-states and two in terms of x-states with x and p resolutions alternating. This gives us

$$U(x_3, x_0, t) = \int [\mathscr{D}p\mathscr{D}x]\langle x_3|\exp\left(-\frac{i\varepsilon}{2m\hbar}P^2\right)|p_3\rangle$$

$$\times\langle p_3|\exp\left(-\frac{i\varepsilon}{\hbar}V(X)\right)|x_2\rangle\langle x_2|\exp\left(-\frac{i\varepsilon}{2m\hbar}P^2\right)|p_2\rangle$$

$$\times\langle p_2|\exp\left(-\frac{i\varepsilon}{\hbar}V(X)\right)|x_1\rangle\langle x_1|\exp\left(-\frac{i\varepsilon}{2m\hbar}P^2\right)|p_1\rangle$$

$$\times\langle p_1|\exp\left(-\frac{i\varepsilon}{\hbar}V(X)\right)|x_0\rangle \tag{21.1.24}$$

where

$$\int[\mathscr{D}p\mathscr{D}x] = \underbrace{\int_{-\infty}^{\infty}\int_{-\infty}^{\infty}\int_{-\infty}^{\infty}\int_{-\infty}^{\infty}\cdots\int_{-\infty}^{\infty}}_{2N-1\text{ times}}\prod_{n=1}^{N}\frac{dp_n}{2\pi\hbar}\prod_{n=1}^{N-1}dx_n \tag{21.1.25}$$

Evaluating all the matrix elements of the exponential operators is trivial since each operator can act on the eigenstate to its right and get replaced by the eigenvalue. Collecting all the factors (a strongly recommended exercise for you) we obtain

$$U(x, x', t) = \int[\mathscr{D}p\mathscr{D}x]\exp\left[\sum_{i=1}^{N}\left(\frac{-i\varepsilon}{2m\hbar}p_n^2 + \frac{i}{\hbar}p_n(x_n - x_{n-1}) - \frac{i\varepsilon}{\hbar}V(x_{n-1})\right)\right] \tag{21.1.26}$$

This formula derived for $N=3$ is obviously true for any N. In the limit $N\to\infty$, i.e., $\varepsilon\to 0$, we write schematically in continuous time (upon multiplying and dividing the middle term by ε), the following continuum version:

$$U(x, x', t) = \int[\mathscr{D}p\mathscr{D}x]\exp\left[\frac{i}{\hbar}\int_0^t[p\dot{x} - \mathscr{H}(x, p)]\,dt\right] \tag{21.1.27}$$

where $\mathscr{H} = p^2/2m + V(x)$ and $(x(t), p(t))$ are now written as functions of a continuous variable t. This is the *Phase Space Path Integral* for the propagator. The continuum version is very pretty [with the Lagrangian in the exponent, but expressed in terms of (x, p)] but is only a schematic for the discretized version preceding it.

In our problem, since p enters the Hamiltonian quadratically, it is possible to integrate out all the N variables p_n. Going back to the discretized form, we isolate the part that depends on just p's and do the integrals:

$$\prod_{1}^{N}\int_{-\infty}^{\infty}\frac{dp_n}{2\pi\hbar}\exp\left[\left(\frac{-i\varepsilon}{2m\hbar}p_n^2+\frac{i}{\hbar}p_n(x_n-x_{n-1})\right)\right]$$

$$=\prod_{1}^{N}\left(\frac{m}{2\pi i\hbar\varepsilon}\right)^{1/2}\exp\left[\frac{im(x_n-x_{n-1})^2}{2\hbar\varepsilon}\right] \tag{21.1.28}$$

If we now bring in the x-integrals we find that this gives us exactly the configuration space path integral, as it should.

Note that if p does not enter the Hamiltonian in a separable quadratic way, it will not be possible to integrate it out and get a path integral over just x, in that we do not know how to do non-Gaussian integrals. In that case we can only write down the phase space path integral.

We now turn to two applications that deal with the path integrals just discussed.

The Landau Levels

Let us discuss a problem that is of great theoretical interest in the study of QHE (see Girvin and Prange). We now explore some aspects of it, not all having to do with functional integrals. Consider a particle of mass μ and charge q in the $x-y$ plane with a uniform magnetic field B along the z-axis. This is a problem we discussed in Exercise (12.3.8). Using a vector potential

$$\mathbf{A}=\frac{B}{2}(-y\mathbf{i}+x\mathbf{j}) \tag{21.1.29}$$

we obtained a Hamiltonian

$$H=\frac{[P_x+qYB/2c]^2}{2\mu}+\frac{[P_y-qXB/2c]^2}{2\mu} \tag{21.1.30}$$

You were asked to verify that

$$Q=\frac{(cP_x+qYB/2)}{qB}\qquad P=(P_y-qBX/2c) \tag{21.1.31}$$

were canonical variables with $[Q,P]=i\hbar$. It followed that H was given by the formula

$$H=\frac{P^2}{2\mu}+\frac{1}{2}\mu\omega_0^2Q^2 \tag{21.1.32}$$

and had a harmonic oscillator spectrum with spacing $\hbar\omega_0$, where

$$\omega_0=qB/\mu c \tag{21.1.33}$$

is the *cyclotron frequency*. In terms of

$$a=\left(\frac{\mu\omega_0}{2\hbar}\right)^{1/2} Q+i\left(\frac{1}{2\mu\omega_0\hbar}\right)^{1/2} P \tag{21.1.34}$$

and its adjoint, we can write

$$H=[a^\dagger a+\tfrac{1}{2}]\hbar\omega_0 \tag{21.1.35}$$

We seem to have gone from a problem in two dimensions to a one-dimensional oscillator problem. How can that be? The point is that there is another canonical pair

$$P'=\frac{(cP_x-qYB/2)}{qB} \qquad Q'=(P_y+qBX/2c) \tag{21.1.36}$$

which commutes with Q, P and does not enter H.

Exercise 21.1.2. If you do not recall the details of Exercise (12.3.8), provide all the missing steps in the derivation starting at Eq. (21.1.29) and ending with Eq. (21.1.35). Check the advertised commutation rules for (Q', P').

The cyclic character of (Q', P') is reflected in the fact that the levels of the oscillator, called *Landau Levels*, are infinitely degenerate. To see this degeneracy consider the *Lowest Landau Level*, abbreviated LLL. The states in this level obey the equation

$$a|0\rangle=0 \tag{21.1.37}$$

which becomes in the coordinate representation

$$\left[\frac{\partial}{\partial z^*}+\frac{qB}{4\hbar c}z\right]\psi_0(z, z^*)=0 \tag{21.1.38}$$

wherein we have switched to complex coordinates

$$z=x+iy \qquad z^*=x-iy \tag{21.1.39}$$

If we make the ansatz

$$\psi_0(z, z^*)=\exp\left[-\frac{qB}{4\hbar c}zz^*\right]u(z, z^*) \tag{21.1.40}$$

we find the beautiful result

$$\frac{\partial}{\partial z^*}u(z, z^*)=0 \tag{21.1.41}$$

as the defining rule for the LLL. Thus u is any *analytic function*, i.e., function of the combination $x+iy$. The family of such functions is clearly infinitely large, with the monomials $[z^m|m=0, 1, 2, \ldots]$ serving as a linearly independent basis. Thus the ground state function ψ_0 is not a unique function as in the case of the truly one dimensional oscillator but a superposition of functions of the form

$$\psi_{0,m}=z^m \exp\left[-\frac{qB}{4\hbar c}zz^*\right] \tag{21.1.42}$$

I now make the following assertions:

- For large m the probability density for the particle is concentrated at some radius $r_m=\sqrt{2m}r_0$ where $r_0=\sqrt{c\hbar/qB}$ is called the *magnetic length.*
- If the system is not infinite in size, but is a disc of radius R, the biggest value of m that can fit in, and hence N, the number of LLL states that fit into the disc, is given by

$$N=\frac{\pi R^2 B}{\Phi_0} \tag{21.1.43}$$

where the numerator is the flux through the sample and the denominator is the flux quantum of Eq. (18.4.39):

$$\Phi_0=\frac{2\pi\hbar c}{q} \tag{21.1.44}$$

*Exercise 21.1.3.** (Mandatory if you wish to follow the discussion of the QHE.) Derive the equation for the LLL in the coordinate representation by providing the missing steps in the derivation. Prove the above assertions. Note that N, the number of states in the LLL, is given by the flux through the sample in units of the flux quantum.

In the following discussion we will hold N, i.e., the field and sample dimensions, fixed.

In the study of the QHE one is interested in the problems of an electron gas designed to live in two dimensions. (Since the electron charge is $q=-e$, our formulas will hold with $q=e$ if the sign of the vector potential and field are reversed at the outset. Henceforth imagine this has been done and that q stands for the magnitude of the electron charge.) The electron spin is frozen along the applied field and has no interesting dynamics. In particular it is the burden of the orbital wave function to ensure antisymmetry. In a real-life problem one is also required to consider the interaction between the electrons as well as interaction between the electrons and any external scalar potential $V(x, y)$ due to the background medium. It is assumed that both these interactions have a scale much smaller than the gap $\hbar qB/\mu c$ between Landau levels. Thus at low temperatures, one would like a simplified problem with the Hilbert space restricted to the LLL. What does this problem look like?

The path integral can tell us that. We will work out the answer for the case where electron–electron interaction is zero. (Only then do the electrons propagate independently and we can write out a functional integral for just one electron.) The action is

$$S=\int\left[\frac{\mu}{2}(\dot{x}^2+\dot{y}^2)+\frac{qB}{2c}(-y\dot{x}+x\dot{y})-V(x,y)\right]dt \tag{21.1.45}$$

where the terms linear in velocity represent the $(q/c)\mathbf{v}\cdot\mathbf{A}$ in the Lagrangian in the gauge we are using. To get the low-energy physics we must banish the higher Landau levels. *Since the gap to the higher levels is $\hbar qB/\mu c$ this is readily done by taking the limit $\mu\to 0$.* (In this limit the zero point energy of the oscillator, which gives the energy of the LLL, diverges. It is assumed this constant is subtracted out of the Hamiltonian.) This gives us the low-energy action

$$S_{LLL}=\int\left[\frac{qB}{c}x\dot{y}-V(x,y)\right]dt \tag{21.1.46}$$

where we have done an integration by parts to combine the two terms linear in velocity. (The surface term will not affect the equations of motion.)

Notice the interesting result that the action is that of a *phase space path integral with y and $\partial\mathscr{L}/\partial\dot{y}=(qB/c)x\equiv\bar{x}$ as canonically conjugate variables.* $V(y,\bar{x})$ now plays the role of the Hamiltonian for this problem. Since we have just one coordinate and one momentum, the problem of the LLL is essentially one-dimensional.

In the semiclassical picture, the orbits will obey Hamilton's equations:

$$\dot{y}=\frac{\partial V}{\partial\bar{x}} \qquad \dot{\bar{x}}=-\frac{\partial V}{\partial y} \tag{21.1.47}$$

and one can try to do Bohr-Sommerfeld quantization. At the quantum level, V can become a complicated differential operator since $\bar{x}$ will turn into the y-derivative. I leave the details and applications of the semiclassical picture to the references.

Now you might object that if we did not have the operator solution telling us that the levels of the problem go as μ^{-1} it might not occur to us to consider the limit $\mu\to 0$ in order to isolate the low energy physics. This is not so. We will simply argue that in the limit of low energies, i.e., low frequencies, terms in the action with higher time derivatives can be neglected compared to those with fewer ones. This would allow us to throw out the same kinetic energy term. (Now you can do this even in a problem without the magnetic field, but this would leave you with very little interesting dynamics. Here we have some linear derivatives left over, i.e., here the low-energy physics is the physics of the entire infinitely degenerate LLL.) In problems where such nontrivial dynamics is left, one usually finds that variables that used to commute become canonically conjugate.

Exercise 21.1.4. Study the semiclassical orbits and show that the motion is on contours of constant V. (Hint: Consider the gradient of V.)

How can X and Y suddenly become noncommuting when by postulate they are commuting? The answer is simply that if two matrices commute in a given space (the full Hilbert space), their truncations to a subspace (here the states of the LLL) need not. What is nice is that the commutator of X and Y, instead of being something ugly is a constant, making the pair canonically conjugate (upon trivial rescaling).

Exercise 21.1.5. Consider the commuting 3×3 matrices Ω and Λ from Exercise (1.8.10). If you truncate them by dropping the third row and column, show that the 2×2 truncations do not commute.

Consider a finite system with N electrons, i.e., a system with a fully filled LLL, *with one electron per state in the LLL.* Ignore all interactions between the electrons or with the medium. What is its ground state? Since their spins are polarized along the field, the spatial wave function must be antisymmetric in the electron spatial coordinates and be analytic. An unnormalized product wave function for the N particles is

$$\Psi_P = z_1^0 z_2^1 z_3^2 \cdots z_N^{N-1} \exp\left(-\frac{qB}{4\hbar c}\sum_i z_i^* z_i\right) \equiv u_P \exp\left(-\frac{qB}{4\hbar c}\sum_i z_i^* z_i\right) \quad (21.1.48)$$

When antisymmetrized, this leads to

$$u_A = \prod_{i=1}^{N}\prod_{j=1}^{i-1}(z_i - z_j) \quad (21.1.49)$$

Exercise 21.1.6. Verify the above equation for the three particle case. Show this also by writing out the (3×3) determinant as in Eq. (10.3.36). (In all these manipulations, the exponential factor in the wave function, which is totally symmetric in the coordinates, plays no part.)

This wave function is unique since there is just one way to place N (spin polarized) electrons in N states. So we know the unique ground state for the fully filled LLL in the noninteracting limit. But even if we consider the interactions between electrons, this is the only antisymmetric wave function we can write for this problem where the number of states equals the number of electrons, *if we do not want to go above the LLL.*

Now, the really interesting problem is one where in the same field and sample, we have a smaller number νN of electrons where $1/\nu$ is an odd integer. (This is one of the cases where the experiments show surprising results.) We say the system has a *filling factor* ν meaning it has ν times the maximum allowed number of particles in the LLL. The fully filled LLL is thus given by $\nu = 1$. Whereas previously we put an electron in each LLL state (and there was just one way to do it and hence one antisymmetric wave function), now there is more than one way and hence many possible superpositions of LLL wave functions that can be candidates for the ground state of a system of electrons interacting via the Coulomb potential. Laughlin proposed the following wave function:

$$u_\nu = \prod_{i=1}^{\nu N} \prod_{j=1}^{i-1} (z_i - z_j)^{1/\nu} \tag{21.1.50}$$

Let us verify that this wave function fits the description. Pick any one particle coordinate, say z_1 (since the particles are identical) and observe that the highest power that occurs is (for large N)

$$z_1^{1/\nu \cdot \nu N} = z_1^N$$

Thus the size of the biggest wave function precisely matches the sample size. Next note that the function is antisymmetric under exchange of any two coordinates since $1/\nu$ is odd. Lastly note that the electrons nicely avoid each other (due to the high-order zero when any two coordinates approach each other) thereby minimizing their repulsive interaction. Not surprisingly, this happens to be an excellent ground state wave function at these filling factors, (for small $1/\nu$).

The Berry Phase

The problem in question has to do with the adiabatic approximation. Recall the example of the particle in a box of size L. Let us say it is in the ground state. Suppose the box slowly expands with time as some function $L(t)$. The adiabatic principle states that if the expansion is slow enough, the particle will be in the ground state of the box of size $L(T)$ at time T. Likewise the particle that starts out in the state $|n(L(0))\rangle$ will find itself in the instantaneous eigenstate $|n(L(t))\rangle$ at time t.

More generally, if the particle Hamiltonian is given by $H(R(t))$ where R is some external coordinate which changes slowly and appears parametrically in H, the adiabatic principle tells us that the particle will sit in the nth instantaneous eigenket of $H(R(t))$ at a time t if it started out in the nth eigenstate of $H(R(0))$.

What is the solution to the Schrödinger equation in this approximation? Here is a reasonable guess:

$$|\psi(t)\rangle = \exp\left(-\frac{i}{\hbar}\int_0^t E_n(t')\, dt'\right)|n(t)\rangle \tag{21.1.51}$$

where

$$H(t)|n(t)\rangle = E_n(t)|n(t)\rangle \tag{21.1.52}$$

First note that if H does not vary with time, the above answer is clearly correct, with the phase factor appropriate to energy E_n. The above formula recognizes that the instantaneous energy varies with time and gives the accumulated phase shift, just as the WKB wave function gives the phase as the spatial integral of a position-dependent momentum for a particle moving in a nonconstant $V(x)$.

Over the years, many people, notably Herzberg and Longuet-Higgins and Mead and Truhlar, recognized various problems with this formula and found ways to fix them. The whole problem was brought into sharp focus and synthesized by Berry. You are urged to read his very lucid writings and the collection of related papers (with helpful commentary) edited by Shapere and Wilczek, referred to in the Bibliography at the end of the chapter.

To see what is missing in the above ansatz, let us modify it as follows:

$$|\psi(t)\rangle = c(t)\exp\left(-\frac{i}{\hbar}\int_0^t E_n(t')\,dt'\right)|n(t)\rangle \tag{21.1.53}$$

where the extra factor $c(t)$ must be equal to unity if the old ansatz is right. Let us apply the Schrödinger equation to this state:

$$\left(i\hbar\frac{\partial}{\partial t} - H(t)\right)|\psi(t)\rangle = 0 \tag{21.1.54}$$

When the time derivative acts, it generates three terms: one from the derivative of the accumulated phase factor (which neutralizes the action of H on the eigenket), one from the derivative of $c(t)$ and *one from the derivative of the instantaneous eigenket*. The last two terms lead to the following equation (on dotting both sides with the instantaneous bra):

$$\dot{c}(t) = -c(t)\langle n(t)|\frac{d}{dt}|n(t)\rangle \tag{21.1.55}$$

with a solution

$$c(t) = c(0)\exp\left(-\int_0^t \langle n(t')|\frac{d}{dt'}|n(t')\rangle\,dt'\right) = c(0)\,e^{i\gamma} \tag{21.1.56}$$

$$\gamma = i\int_0^t \langle n(t')|\frac{d}{dt'}|n(t')\rangle\,dt' \tag{21.1.57}$$

The impressive thing is not to find this extra phase, called the *Berry phase* or the *geometric phase*, but to recognize that it can have measurable consequences. After all, we have been learning all along that the phase of a ket makes no difference to any measurable quantity. Since the instantaneous kets themselves are defined only up to a phase factor, we can choose a new set and modify the extra phase. If we choose

$$|n'(t)\rangle = e^{i\chi(t)}|n(t)\rangle \tag{21.1.58}$$

then we find

$$i\langle n'(t)|\frac{d}{dt}|n'(t)\rangle = i\langle n(t)|\frac{d}{dt}|n(t)\rangle - \frac{d\chi(t)}{dt} \tag{21.1.59}$$

suggesting that perhaps we could choose $\chi(t)$ so as to completely neutralize the extra phase. It had been generally assumed that such a choice could always be made and the extra phase forgotten.

Suppose now that the parameter that changes with time and causes the Hamiltonian to change returns to its starting value after time T so that:

$$H(T) = H(0) \tag{21.1.60}$$

Now it is no longer obvious that we can get rid of the extra phase. We find

$$i\oint \langle n'(t)|\frac{d}{dt}|n'(t)\rangle\, dt = i\oint \langle n(t)|\frac{d}{dt}|n(t)\rangle\, dt - (\chi(T) - \chi(0)) \tag{21.1.61}$$

Now the choice of phase factors is quite arbitrary, but it must meet the requirement that the assignment is single-valued, at least within the region containing the closed loop in question. (A single-valued choice in the entire parameter space will generally be impossible. This is a subtle topic, reserved for Exercise (21.1.15).) So let us start with such a basis $|n(t)\rangle$ and make a switch to another one $|n'(t)\rangle = e^{i\chi(t)}|n(t)\rangle$. Since the new basis is by assumption single-valued, so must be the additional phase factor. In other words, $(\chi(T) - \chi(0)) = 2m\pi$, where m is an integer. This in turn means that the prefactor $e^{i\gamma} \equiv \exp[i\oint \langle n(t)|i\,(d/dt)|n(t)\rangle\, dt]$ arising in a closed circuit cannot be altered by a choice of basis. Note also that since dt cancels out in any of the integrals, we cannot shake this phase by slowing down the rate of change of the parameter. The phase factor depends only on the path in parameter space, which explains the name "geometric phase." Note that we have not shown that $e^{i\gamma} \neq 1$, but only that its value is not affected by redefinition of phases for the state vectors.

So let us suppose we have a nonzero γ. What exactly does it do? To see this, let us consider a problem where the box is not really a box, but the potential of some heavy object. For example, let R be the coordinate of some nucleus and r that of an electron that is orbiting around it. In this discussion we will deviate from our usual notation: capital letters will stand for nuclear coordinates and momenta (classical or quantum) and lowercase letters will represent the electron. We will also temporarily ignore the vector nature of these variables. The box here is the Coulomb well created by the nucleus. As the nucleus moves, the box moves, rather than change size, but the issues are the same. As the nucleus crawls from place to place, the nimble electron stays in the instantaneous eigenstate. Even though we have paid no attention to the dynamics of the nucleus, we shall see one is generated by the Berry phase. Let us rewrite the phase factor as follows:

$$\exp\left(-\int_0^t \langle n(t')|\frac{d}{dt'}|n(t')\rangle\, dt'\right)$$

$$=\exp\left(\frac{i}{\hbar} i\hbar \int_0^t \langle n(t')|\frac{d}{dt'}|n(t')\rangle\, dt'\right) \tag{21.1.62}$$

$$=\exp\left(\frac{i}{\hbar}\int_0^t i\hbar\langle n(R(t'))|\frac{d}{dR}|n(R(t'))\rangle \frac{dR}{dt'}\, dt'\right) \tag{21.1.63}$$

$$=\exp\left(\frac{i}{\hbar}\int_0^t A^n(R)\frac{dR}{dt'}\, dt'\right) \quad \text{where} \tag{21.1.64}$$

$$A^n(R)=i\hbar\langle n(R)|\frac{d}{dR}|n(R)\rangle \tag{21.1.65}$$

Thus we see that the slow nuclear degree of freedom has a velocity coupling to a vector potential $A^n(R)$, called the *Berry potential*. The potential depends on which quantum state $|n\rangle$ the electronic degree of freedom is in. When the state vectors are redefined by phase transformations, this vector potential undergoes a gauge transformation:

$$|n(R)\rangle \rightarrow e^{i\chi(R)}|n(R)\rangle \tag{21.1.66}$$

$$A^n(R)\rightarrow A^n(R)-\hbar\frac{d\chi}{dR} \tag{21.1.67}$$

However, its line integral around a closed loop is gauge invariant and could be nonzero. To ignore this would be to get the wrong dynamics for the nucleus.

Now some of you may feel a little unhappy and say: "I know how the vector potential is supposed to enter the Lagrangian or action, but you pulled it out of a phase factor in the wave function of the fast coordinates." This is a fair objection and in answering it in some detail we will learn that there is also a scalar potential besides the vector potential.

We begin by constructing a path integral for the nuclear degrees of freedom. What resolution of the identity should we use? The one appropriate to our problem is this:

$$I=\int dR\sum_n |R, n(R)\rangle\langle n(R), R| \tag{21.1.68}$$

where $|R, n(R)\rangle \equiv |R\rangle \otimes |n(R)\rangle$. In other words, at each R, we pick a basis for the electrons that diagonalizes the instantaneous electronic Hamiltonian $H_e(R, r, p)$

$$H_e(R, r, p)|R, n(R)\rangle = E_n(R)|R, n(R)\rangle \tag{21.1.69}$$

Of course, you can pick a basis for the electrons that has no correlation to the nuclear coordinates. While this is mathematically correct, it is not wise for the adiabatic approximation. For the latter, we now make the approximation that if the electron starts out at some value of n, it stays there and all other values can be ignored. Thus we write:

$$I \simeq \int dR |R, n(R)\rangle\langle n(R), R| \tag{21.1.70}$$

where the sum on n has been dropped. The derivation of the configuration space path integral in R proceeds as usual. A typical factor in the path-integrand will be

$$\langle n(R(t+\varepsilon)), R(t+\varepsilon)| \exp\left[-\frac{i\varepsilon}{\hbar} H(R, P)\right] \exp\left[-\frac{i\varepsilon}{\hbar} H_e(R, r, p)\right] |n(R(t)), R(t)\rangle \tag{21.1.71}$$

The nuclear part, sandwiched between nuclear coordinate eigenstates, will give the usual factor

$$\langle R(t+\varepsilon)| \exp\left[-\frac{i\varepsilon}{\hbar} H(R, P)\right] |R(t)\rangle$$

$$= \sqrt{\frac{m}{2\pi\hbar i\varepsilon}} \exp\left[\frac{i\varepsilon}{\hbar}\left[\frac{m}{2\varepsilon^2}(R(t+\varepsilon) - R(t))^2 - V(R)\right]\right] \tag{21.1.72}$$

while the electronic exponential will act on its eigenket to the right and give a factor $\exp[-(i\varepsilon/\hbar)E_n(R)]$ which will change the nuclear potential by $E_n(R)$. This is how Born and Oppenheimer analyzed molecules, where there is a clear separation of fast (electronic) and slow (nuclear) degrees of freedom: fix the slow ones, solve for the fast ones at this value, and use the fast eigenenergies as an additional potential for the slow problem which is then solved.

But this is not the full story. After the electronic exponential has acted on its eigenket to the right, yielding the exponential phase factor $\exp[-(i\varepsilon/\hbar)E_n(R)]$, we are still left with the following dot product which multiplies everything:

$$\langle n(R(t+\varepsilon))|n(R(t))\rangle \equiv \langle n(R')|n(R)\rangle \tag{21.1.73}$$

All the results will follow from an analysis of this factor. First, it is true that when $R = R'$ this factor equals unity. We are going to perform a Taylor expansion of this product in the difference $R - R' = \eta$. How far should we go? The answer is clear if we recall Chapter 8 where we derived the Schrödinger equation from the path integral by considering the propagator for infinitesimal times, i.e., one time slice of width ε. I reproduce the relevant formula Eq. (8.5.7) with two changes. I drop all interactions and keep just the free particle propagator but I append the dot product $\langle n(R')|n(R)\rangle$. This yields for the nucleus

$$\psi(R',\varepsilon)=\left(\frac{m}{2\pi\hbar i\varepsilon}\right)^{1/2}\int_{-\infty}^{\infty} e^{im\eta^2/2\hbar\varepsilon}\langle n(R')|n(R'+\eta)\rangle\psi(R'+\eta,0)\,d\eta \tag{21.1.74}$$

The exponential allows η to fluctuate by (recall Eq. (8.5.6))

$$|\eta|\simeq\left(\frac{2\pi\hbar\varepsilon}{m}\right)^{1/2} \tag{21.1.75}$$

This means we must go to order η^2 since we want to go to order ε to derive the Schrödinger equation. So we expand ψ and $\langle n(R')|n(R'+\eta)\rangle$ to this order:

$$\psi(R'+\eta,0)=\psi(R',0)+\eta\frac{\partial\psi}{\partial\eta}+\frac{\eta^2}{2}\frac{\partial^2\psi}{\partial\eta^2}+\cdots \tag{21.1.76}$$

$$\langle n(R')|n(R'+\eta)\rangle=1+\eta\langle n|\partial n\rangle+\frac{\eta^2}{2}\langle n|\partial^2 n\rangle+\cdots \tag{21.1.77}$$

where all derivatives are taken at the point R' and $|\partial n\rangle$ is the derivative of $|n\rangle$ with respect to R' and so on. If we now inject these expansions into Eq. (21.1.74), and keep just the even powers of η as we did in Chapter 8, we find upon doing the Gaussian integrals and dropping the prime on R'

$$i\hbar(\psi(R,\varepsilon)-\psi(R,0))=\varepsilon\left[-\frac{\hbar^2}{2m}\frac{\partial^2\psi}{\partial R^2}-\frac{\hbar^2}{m}\langle n|\partial n\rangle\frac{\partial\psi}{\partial R}-\frac{\hbar^2}{2m}\langle n|\partial^2 n\rangle\psi\right] \tag{21.1.78}$$

*Exercise 21.1.7.** Provide the missing steps leading to the above equation.

The Hamiltonian can be read off the above:

$$H=\frac{1}{2m}(P-A^n)^2+\Phi^n \tag{21.1.79}$$

$$A^n=i\hbar\langle n|\partial n\rangle \tag{21.1.80}$$

$$\Phi^n=\frac{\hbar^2}{2m}[\langle\partial n|\partial n\rangle-\langle\partial n|n\rangle\langle n|\partial n\rangle] \tag{21.1.81}$$

*Exercise 21.1.8.** Providing the missing steps. Use $\langle n|\partial n\rangle=-\langle\partial n|n\rangle$ which follows from $\partial\langle n|n\rangle=0$. The potential Φ^n arises from adding and subtracting the $(A^n)^2$ term which isn't there to begin with.

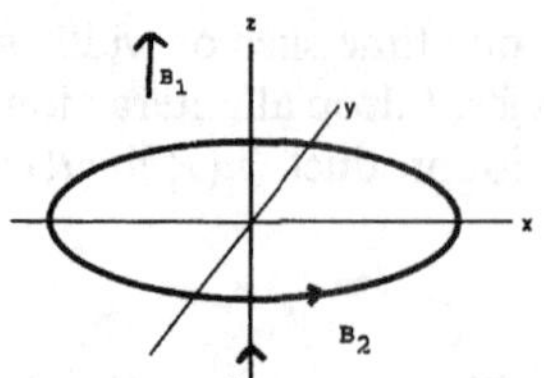

Figure 21.1. The field B_2 and electron motion are along the circle. The particle spin is up or down the local magnetic field which is the sum of B_1 and B_2. The current I produces B_2.

The (discretized) action function which will give exactly these results will have the $v\cdot A''$ term (with A'' evaluated at the midpoint) and the extra scalar potential Φ''. We will not write that down since we have the Hamiltonian. The following exercise considers this point more carefully.

Exercise 21.1.9. Suppose we do not derive the Hamiltonian as above (by invoking the wave function) but want to determine the correct discretized action function starting with Eq. (21.1.71) and expanding $\langle n(R')|n(R)\rangle$ to quadratic order in $R'-R$ as per Eq. (21.1.77) and exponentiating the result. Do all of the above and show that the argument of the vector potential that arises is not at the midpoint to begin with, as it should to represent the effect correctly [Exercise (8.6.4)]. Fix this with a Taylor series, combine the term quadratic in $R'-R$ that arises, with the one you had to begin with, to obtain (for one time slice)

$$S=\frac{im(R'-R)^2}{2\hbar\varepsilon}+\frac{i}{\hbar}(R'-R)A''\left(\frac{R+R'}{2}\right)$$
$$-\frac{(R'-R)^2}{2}\langle\partial n|(I-|n\rangle\langle n|)|\partial n\rangle \tag{21.1.82}$$

Let us now ask what continuum form this describes. Multiplying and dividing by ε converts the first term into the kinetic energy and the middle term to the vector potential coupling. The last term needs to be multiplied and divided by ε^2 to become the square of the velocity. But this would leave it with an extra ε in the continuum action. Despite this, the term is important since the square of the velocity is very singular. The effect of the term is best revealed by noting that the factor $(R'-R)^2$ is going to be replaced by $i\varepsilon\hbar/m$ when the functional integral is done, (because of the kinetic energy term in the action that controls the variance of $R-R'$), make this replacement now, and convert this term to the scalar potential Φ'', which we know describes the right Hamiltonian. The role of such terms, naively vanishing in the continuum limit has been discussed by Klauder. Klauder and Skagerstam (1985).

It should be clear that the preceding results generalize with R and A replaced by vectors $\mathbf{R}$, $\mathbf{A}$, more fast and slow degrees of freedom, etc.

We turn to a simple problem where the Berry potential makes a difference.‡ Consider the situation in Fig. 21.1.

A spinless, electrically neutral particle of mass M is restricted to move on a circle of radius a. This motion is going to be the slow degree of freedom in our problem. The orbit is penetrated by a flux due to a field $B_1\mathbf{k}$ along the z-axis. In addition, a wire carrying some current along the z-axis is introduced at the center.

‡ I thank Ady Stern for suggesting a variant of this example. He is not responsible for any errors in my presentation.

It produces an azimuthal field of strength B_2. The total field makes an angle

$$\theta = \arctan B_2/B_1$$

with respect to the z-axis and has a magnitude $B = \sqrt{B_1^2 + B_2^2}$. When the particle coordinate is ϕ, the field B_2 is tangent to the circle, i.e., has an azimuthal angle $\phi + \pi/2$ in **B**-space. The Hamiltonian of the particle (not yet coupled to **B**) is:

$$H = \frac{L^2}{2I} \tag{21.1.83}$$

where $I = Ma^2$ is the moment of inertia, set equal to 1/2 from now on and $L = -i\hbar\,\partial/\partial\phi$ is the angular momentum operator. The energy eigenvalues are

$$E_m = \hbar^2 m^2 \qquad m = 0, \pm 1, \pm 2 \ldots \tag{21.1.84}$$

We now bring in the fast degree of freedom. Imagine that the particle has spin 1/2. As the particle goes around the circle, the spin will see a varying magnetic field, **B**, which is the vector sum of the fixed field B_1 along the z-axis and the azimuthal field B_2. We modify H as follows:

$$H = L^2 - C\boldsymbol{\sigma}\cdot\mathbf{B}(\phi) \tag{21.1.85}$$

where C and hence the splitting between the two spin states is assumed to be so large (as is the frequency associated with the splitting) that the spin is truly a fast degree of freedom which will not jump between its states as the particle crawls around the loop.

What will the allowed energies be? The naive answer is

$$E_m = \hbar^2 m^2 \mp CB \tag{21.1.86}$$

where $B = \sqrt{B_1^2 + B_2^2}$ and the two signs correspond to the spin pointing up/down the local magnetic field as the particle goes round and round. This is however wrong and one must take into account the Berry potentials $A(\phi)$ and Φ. Let us focus on the lower-energy solution in which the spin points up the local field. We choose the spinor to be

$$|\theta\phi\rangle = \begin{bmatrix} \cos\dfrac{\theta}{2} \\ i\sin\dfrac{\theta}{2}\,e^{i\phi} \end{bmatrix} \tag{21.1.87}$$

(The additional i in the lower component is due to the fact that orbital angle ϕ differs from the azimuthal angle of the field by $\pi/2$ as is clear from Fig. 21.1.) It is readily found that

$$A^{+}(\phi)=i\hbar\langle\theta\phi|\frac{\partial}{\partial\phi}|\theta\phi\rangle=-\hbar\sin^{2}\frac{\theta}{2} \tag{21.1.88}$$

which is independent of ϕ, and that the scalar Berry potential is

$$\Phi=\frac{\hbar^{2}\sin^{2}\theta}{4} \tag{21.1.89}$$

which is independent of whether the spin is pointing up or down the local field. Since θ is fixed in this problem, Φ can be eliminated by a choice of reference energy, and we no longer consider it.

Exercise 21.1.10. Prove the above equations for the vector and scalar potentials.

Since the effect of the vector potential is $L\to L-A^{+}$, it follows that if we solve

$$\left[-i\hbar\frac{\partial}{\partial\phi}-A^{+}\right]\psi=\lambda\psi \tag{21.1.90}$$

the energy is given by

$$E^{+}=\lambda^{2}-BC \tag{21.1.91}$$

The orbital eigenfunctions are once again

$$\psi=e^{im\phi}\qquad m=0,\pm1,\pm2,\ldots \tag{21.1.92}$$

so that

$$\lambda=m\hbar-A^{+}=\left(m+\sin^{2}\frac{\theta}{2}\right)\hbar \tag{21.1.93}$$

and the energy of the spin up state is

$$E^{+}=\left(m+\sin^{2}\frac{\theta}{2}\right)^{2}\hbar^{2}-BC \tag{21.1.94}$$

It is evident that without the vector potential we would get the wrong answer. For example, without it, there would be a twofold degeneracy under $m\to-m$.

Exercise 21.1.11. Find the potential for the other (spin down) state and the energy eigenvalues.

Let us rederive the scalar and vector potentials of Eqs. (21.1.79–21.1.81) without path integrals, by extracting the effective Hamiltonian that acts on the slow degrees of freedom R. Now the latter need not be in an eigenstate of position, it could be in a superposition $\psi(R)$:

$$|\psi\rangle = \int \psi(R)|R, n(R)\rangle\, dR \tag{21.1.95}$$

Note that $|\psi\rangle$ is a ket in the direct product space of the slow and fast degrees of freedom. Usually the coefficients in such a superposition would depend on both labels. But in our problem the fast degree of freedom is slaved to the slow one, so that the amplitude for the slow one to be in $|R\rangle$ is the same as the amplitude for the entire system to be in $|R, n(R)\rangle$. We are going to find the Hamiltonian in the coordinate representation by calculating

$$(H\psi)(R') \equiv \langle R', n(R')|H|\psi\rangle \tag{21.1.96}$$

$$= \int \langle R', n(R')|H|R, n(R)\rangle\langle R, n(R)|\psi\rangle\, dR \tag{21.1.97}$$

$$= \int \langle R', n(R')|H|R, n(R)\rangle\psi(R)\, dR \tag{21.1.98}$$

Let

$$H = P^2/2M + V(R) + H_f(r, p, R) \tag{21.1.99}$$

It is evident that the fast Hamiltonian H_f, acting to the right on its eigenket, will give $E_n(R)$ and that this will join with $V(R)$ to provide a potential energy term. We focus therefore on just the $P^2/2M$ since here is where the action is. Let us recall that

$$\langle R'|\frac{P^2}{2M}|R\rangle = -\frac{\hbar^2}{2M}\delta''(R'-R) \tag{21.1.100}$$

and insert it into Eq. (21.1.98) to obtain

$$\begin{aligned}(H\psi)(R') &= -\frac{\hbar^2}{2M}\int \langle n(R')|n(R)\rangle\delta''(R'-R)\psi(R)\, dR \\ &= -\frac{\hbar^2}{2M}\langle n(R')|[|\partial^2 n(R)\rangle\psi(R) + 2|\partial n(R)\rangle\partial\psi(R) + |n(R)\rangle\partial^2\psi(R)]_{R=R'} \\ &= -\frac{\hbar^2}{2M}[\langle n|\partial^2 n\rangle\psi(R') + 2\langle n|\partial n\rangle\, \partial\psi(R') + \partial^2\psi(R')]\end{aligned} \tag{21.1.101}$$

where ∂ denotes derivatives respect to R'. It is now straightforward to show that the operator on the right-hand side is indeed the one in Eq. (21.1.79). The details are left to the following exercise.

Exercise 21.1.12. Provide the missing details. Suggestion: Start with Eq. (21.1.79) and expand out the $(P-A)^2$. Note that when P comes to the left of A, it differentiates both A and the wave function ψ that is imagined to be sitting to the right of the Hamiltonian. Now go to Eq. (21.1.101), add and subtract the A^2 term and regroup the terms using relations like $\partial\langle n|\partial n\rangle = \langle\partial n|\partial n\rangle + \langle n|\partial^2 n\rangle$.

Now that we accept the reality of the Berry vector potential, let us understand it a little better. Normally when we have a vector potential, we take its curl and the corresponding magnetic field has as its origin some current. Had there been magnetic monopoles, the source could have been a monopole. What is producing the Berry potential? Let us first appreciate that the source of the potential does not lie in the configuration space of the fast degree of freedom, but in the space of parameters that are slowly varying in the fast Hamiltonian H_f. Of course, this slow parameter could itself be a real live degree of freedom (as in our ring example) but this is not our focus. We simply treat the slow variables as external parameters that define H_f. Thus if we consider a spin-1/2 object with

$$H = -\boldsymbol{\sigma}\cdot\mathbf{B} \tag{21.1.102}$$

then the Berry potential lives in $\mathbf{B}$ space. (Since we focus on just the fast variables, we drop the subscript on H_f.) To ease our thinking we are going to rename $\mathbf{B}$ space as $\mathbf{R}$ space, but you should not forget this fact. So we write

$$H = -\boldsymbol{\sigma}\cdot\mathbf{R} \tag{21.1.103}$$

Every point in $\mathbf{R}$ space defines a possible spin Hamiltonian. We have managed to define in this space a vector potential. It is derived from the nth quantum state of the above Hamiltonian and is given by

$$\mathbf{A}^n = i\hbar\langle n(\mathbf{R})|\nabla|n(\mathbf{R})\rangle \tag{21.1.104}$$

What is its curl? To figure this out, we need a little groundwork. Using

$$0 = \nabla\langle n|H|m\rangle \qquad m \neq n \tag{21.1.105}$$

we find on differentiating all three factors and shifting a derivative from bra to ket at the cost of sign change (thanks to $\nabla\langle n|m\rangle = 0$),

$$\langle n|\nabla|m\rangle = \frac{\langle n|(\nabla H)|m\rangle}{E_m - E_n} \tag{21.1.106}$$

It is now easy to find a formula for the field tensor F_{ij} associated with the Berry potential:

$$\begin{aligned} F_{ij}^n &= \partial_i A_j^n - \partial_j A_i^n \\ &= i\hbar[\partial_i \langle n|\partial_j n\rangle - \partial_j \langle n|\partial_i n\rangle] \\ &= i\hbar \sum_{m\neq n} \frac{\langle n|(\partial_i H)|m\rangle\langle m|(\partial_j H)|n\rangle - \langle n|(\partial_j H)|m\rangle\langle m|(\partial_i H)|n\rangle}{(E_m - E_n)^2} \\ &\left(\partial_j = \frac{\partial}{\partial R_j}\right) \end{aligned} \tag{21.1.107}$$

where m labels a complete set of states we introduce along the way. (The $m=n$ terms drop out due to a cancellation.) This formula is valid in general (for any H) and we now apply it to our problem.

In our problem there are many simplifying features:

- $\partial H/\partial R_j = -\sigma_j$
- There are only two states and hence only one term in the sum over m. The energy denominator squared is $4R^2$ since $2R$ is the difference between up and down spin states. (Remember R is now the magnitude of the magnetic field!)
- So we pull out this denominator, which is independent of m, add a term with $m=n$ (which vanishes by antisymmetry in i and j), use completeness to eliminate the intermediate states, use the commutation relations for the Pauli matrices, and finally the fact that $\langle n|\boldsymbol{\sigma}|n\rangle = \pm\hat{\mathbf{R}}$ (for the states up/down the field).

Rather than state the field in terms of the tensor F_{ij}^n, we write in terms of the more familiar magnetic field defined by $\mathscr{B}_k^n = F_{ij}^n$ (where the indices i, j, k run cyclically):

$$\mathscr{B}^n = \mp\hbar\frac{\hat{\mathbf{R}}}{2R^2} \tag{21.1.108}$$

This is the field of a monopole of strength $-\hbar/2$ sitting at the origin, which is the point of degeneracy of the Hamiltonian.

Exercise 21.1.13. Furnish the missing steps in the above derivation.

Note that there are two different magnetic fields in the problem. The first is a real one **B** which couples to the electron spin and resides in real space. It is produced by currents in real space. (There are no known monopole sources for such fields.) The second field is the curl of the Berry vector potential that resides in parameter space. Its components are denoted by $\mathscr{B}_k^n$ which happens, in our problem, to describe a monopole in parameter space. We will now see that the Berry monopole will arise in any problem where the Hamiltonian (not necessarily containing magnetic fields) becomes doubly degenerate.

Assuming the parameter space is three-dimensional, let us focus on just the two nearly degenerate levels. Now, any 2×2 Hermitian operator can be written as

$$H=\sum_{\mu=0}^{3}\sigma_\mu f_\mu \tag{21.1.109}$$

where $\sigma_0=I$ is the fourth partner to the Pauli matrices, and f_μ are four functions of the three independent coordinates of parameter space. The eigenvalues of H are clearly

$$E=f_0\pm\sqrt{f_x^2+f_y^2+f_z^2}. \tag{21.1.110}$$

The degeneracy occurs at $f_x=f_y=f_z=0$ which we choose to be the origin of coordinates. We also shift the overall zero of energy so that the degenerate eigenvalue $f_0(0)$ vanishes. Let us now use the three f's themselves as the new coordinates in which case f_0 will be some function of these coordinate and vanish at the origin. Thus

$$H=f_0(\mathbf{f})I+\boldsymbol{\sigma}\cdot\mathbf{f} \tag{21.1.111}$$

in obvious notation. Note that f_0 vanishes at the origin but not necessarily elsewhere. Let us repeat the same analysis we used in the spin problem, starting with

$$\partial_i H=\partial_i f_0 I+\sigma_i \tag{21.1.112}$$

If we next evaluate the field tensor as per Eq. (21.1.107), we see that the part proportional to the identity does not matter (since $\langle m|n\rangle=0$ for $m\neq n$), the problem becomes isomorphic to the one in Eq. (21.1.103) and we get just the monopole at the origin.

Exercise 21.1.14. Take another look at the problem we studied, of a particle moving around in a loop with fields in the azimuthal and z-directions. As the particle goes once around the circle, the line integral of the vector potential A^+ is

$$\oint A^+\,d\phi=-2\pi\hbar\sin^2\frac{\theta}{2}$$

Let us now look at the same closed orbit in **B**-space where it is a loop of fixed radius B_2 at a fixed height B_1 above the B_x-B_y plane. Thus it defines the co-latitude (at angle θ measured from the north pole) of a sphere of radius $\sqrt{B_1^2+B_2^2}$. In this space we have a monopole of strength $-\hbar/2$ at the origin according to Eq. (21.1.108). The flux through this loop is then the monopole flux penetrating the area of the cap bounded by this latitude. Using Stoke's theorem show that this flux equals $-2\pi\hbar\sin^2\theta/2$ as it should. (Note that the Berry vector potential is different in real space and parameter space. Its line integral over a closed loop, which measures the accumulated phase change per revolution, is of course the same. Consider in general a map from manifold X with points labeled x, to Y with points labeled y, such that each x goes into a unique y. If $A(y)$ is a vector potential in Y, we can import it to X by defining a vector potential $A(x)$ such that (suppressing indices)

$$A(x)\,dx=A(y)\,dy \tag{21.1.113}$$

By construction, closed loops in X go to closed loops in Y. The line integral of $A(x)$ around a closed loop in X will then equal the line integral of $A(y)$ around the image loop in Y.)

Execise 21.1.15. Let us discuss the question of assigning phases to state vectors in parameter space through an example. Let $\mathbf{R} \equiv (R, \theta, \phi)$ be the coordinate in parameter space. Consider the Hamiltonian $H = -\boldsymbol{\sigma} \cdot \mathbf{R}$. Let us write down the ground state for this problem for all points. It is the one where the spin points radially outward everywhere. A choice for the spinor is

$$|+, \theta, \phi\rangle = \begin{bmatrix} \cos\frac{\theta}{2} \\ \sin\frac{\theta}{2} e^{i\phi} \end{bmatrix}$$

This is just the ket we used in the problem of the electron going around in a loop (except for the factor i in the lower component which arose due the $\pi/2$ difference between the azimuthal angles in real and parameter space). Since the spinor has no R dependence let us look at it on a unit sphere $R = 1$. Observe that the lower component does not approach a unique value as we approach the south pole from different directions. (This problem does not exist at the north pole since $\sin \theta/2 = 0$ there.) Thus we really have not defined the spinor globally. If we multiply the whole spinor by the single-valued phase factor $e^{-i\phi}$, we now have a spinor well defined near the south pole, but singular at the north pole. It follows that we can only define the spinor in patches of parameter space. In our problem two patches will do, one excluding the north pole and one excluding the south.

Since we found the Berry potential by taking derivatives of the ket, it follows that the former is also defined only in the patches and not globally. In other words, Eq. (21.1.88) for A^+ is to be used away from $\theta = \pi$. To describe the south pole, we can use, for example, the potential coming from the spinor with good behavior at the south pole, but bad behavior at the north pole.

I will now argue that attempts to find a global vector potential in the presence of a monopole are doomed. Say we had a global nonsingular vector potential. Consider its line integral along the direction of increasing ϕ on a latitude near the north pole on a unit sphere surrounding the monopole. By Stokes's theorem this equals the flux through the cap above this latitude. If we enlarge the loop and go past the equator, the line integral will monotonically increase. Finally, let us shrink the loop to an infinitesimal one around the south pole. As this loop shrinks, the line integral does not vanish; it equals the full monopole flux. It follows there must be a singularity at the south pole since the integral of a nonsingular potential around an infinitesimal loop must be infinitesimal and vanish with loop size. (It is also possible that the singularity is elsewhere on the sphere, but it has to exist by similar reasoning.)

Starting with the gradient in spherical coordinates, show that the vector potential associated with $|+, \theta, \phi\rangle$ is given by

$$\mathbf{A} = -\frac{\hbar}{2} \mathbf{e}_\phi \frac{(1 - \cos\theta)}{R \sin\theta}$$

Observe the singularity at the south pole. This is called the *Dirac string*. Show that its line integral around a tiny loop surrounding the south pole is the full monopole flux. What is happening is this. This vector potential describes not a monopole at the origin, but one where a tiny tube (the Dirac string) comes up the negative z-axis, smuggling in the entire flux to the origin, from which point it emanates radially. The string flux is the reason the tiny loop around the south pole gives a nonzero answer equal to the total flux.

Now there is nothing special about the south pole when we look at the monopole, since it is spherically symmetric. This is reflected in the fact that the Dirac string can be moved around by a gauge transformation. Calculate the vector potential $\mathbf{A}'$ with the spinor obtained by multiplying both components of $|+, \theta, \phi\rangle$ by $e^{-i\phi}$. Show that it has troubles at the north pole and that the two vector potentials are related by the gauge transformation associated with the redefinition $|+, \theta, \phi\rangle \to e^{-i\phi}|+, \theta, \phi\rangle$.

If we are allowed to use words instead of equations, we can describe the effect of the monopole without any strings: when the charged particle goes around in a loop, it picks up a phase proportional to the solid angle the loop subtends at the origin (where the monopole is). The vector potential is the analytical way to generate the solid angle via Stokes's theorem, but it cannot do it globally.

Now Dirac ran into this problem trying to ask how we would describe a real (not Berry) monopole of charge g in real space. It has a radial field that falls off as g/R^2. No problem there. *But quantum mechanics forces us to work with vector potentials.* Now any vector potential we can come up with has a string. As usual, Dirac turned a potential disaster into a dazzling prediction by arguing that *if there is a monopole and we have no choice but to describe it with a vector potential, it must be that the string is unobservable.* The line integral of the vector potential around the string at the south pole is $4\pi g$, the total flux of the monopole. For a particle of charge q, this will enter the dynamics via the factor

$$e^{4\pi i q g/\hbar c}$$

as per Eq. (18.4.38). (Think of an Aharaonov–Bohm experiment in which a particle goes on either side of the string.) If this factor is to be unobservable we require that

$$q = \frac{\hbar n c}{2g}$$

where n is any integer. This remarkable argument tells us that *even if there is a single monopole in the universe, it forces all electric charges to be multiples of $\hbar c/2g$.* This explains, for example, why the proton and electron have exactly the same charge. However no monopole has yet been seen. But, the argument is so attractive I for one am sure at least one monopole exists. If not, nature would have missed a wonderful opportunity, to paraphrase Einstein.

In modern treatments, one uses two patches, say one without the south pole and one without the north pole, with a different vector potential in each. By demanding that where the patches overlap, say the equator, the two potentials differ by a single-valued gauge transformation, one recovers Dirac's quantization condition. (You may provide the proof yourself if you remember that (1) the difference of the line integrals of the two patch potentials around the equator is the integral over the whole sphere of the outgoing flux; (2) when the wave function of a particle of charge q is changed by a phase factor $\psi \to e^{i\chi}\psi$, vector potential changes as per $A \to A + \hbar c/q\, \partial\chi$; (3) the change in χ around a closed loop must be an integral multiple of 2π.)

In the Berry phase problem we looked at, the vector potential had q/c, the factor multiplying $\mathbf{A}$ in the Hamiltonian, equal to unity, $g = \hbar/2$, and hence $n = 1$.

As another application of the Berry phase, let us return to the Hall effect. Laughlin proposed that the excited state (above the ground state), called the *quasihole* state, be given by

$$u_{qh} = \prod_{i=1,}^{\nu N} (z_i - z_0) u_\nu \tag{21.1.114}$$

Clearly this describes a situation where the wave function is modified in the vicinity of z_0. We say it describes a quasihole centered at z_0. Note that electrons avoid the point z_0 due to the extra zeros of the form $z - z_0$. This means the charge density near this point is below normal. If one integrates the charge deficit due to this modification in the wave function (which is the charge of the quasihole) one finds it is νq, where q is the elementary charge e. Thus a theory with elementary charges that are integers (electrons) has excitations which have fractional charge! The fractional charge can also be demonstrated as follows. First note that the location z_0 of the quasihole is arbitrary. Assume there is some substrate potential underneath the electron gas whose minimum selects out some preferred location. Suppose we slowly vary the potential and drag the coordinate z_0 in u_{qh} around some closed loop and calculate the accumulated Berry phase for this closed orbit. (Since we know the wave function explicitly for any z_0, this is easily done.) This must equal the flux (due to the external magnetic field B that produces the Landau levels) enclosed times $\bar{q}/\hbar c$ where $\bar{q}$ is the quasihole charge. The calculation gives a charge ν times the elementary charge. Similarly, one may show that the quasiholes are neither bosons nor fermions, but *anyons* (a term coined by Wilczek; see Bibliography): they acquire a phase factor $e^{i\nu\pi}$ under exchange, by taking a state with two quasiholes (located at z_0 and z_0') and adiabatically exchanging them (i.e., their centers) and computing the Berry phase change in the wave function. The adiabatic analysis is valid since the quasihole states are separated by a gap from other states. For details, see Shapere and Wilczek (1990).

We conclude with some history.

Why did Born and Oppenheimer miss the Berry phase? The reason was quite subtle. They were working with a real Hamiltonian whose wave functions could be chosen real. They assumed such a choice had been made and that the choice was nonsingular. While this is correct for any open curve in parameter space, there exists the possibility that in closed curves, one could be forced to return to minus the starting wave function. Berry considered complex Hamiltonians (isomorphic to the spin example) which allowed a continuum of possible values for the phase (instead of just ± 1) and made the phenomenon more transparent.

Finally, although we have discussed the Berry phase in connection with quantum mechanics, it was discovered in optics many decades earlier by Pancharatnam (1958) who considered a polarized beam of light rather than a quantum state going on a closed path in parameter space (see Bibliography). For a fascinating review of even earlier precursors, see Berry's article in *Physics Today* (see Bibliography).

Coherent State Path Integral

Now we discuss yet another resolution of the identity and the associated path integral. These are based on *coherent states* defined to be eigenstates of the destruction operator in the harmonic oscillator problem.

Each coherent state carries a complex label z and is given by

$$|z\rangle = \exp[za^\dagger]|0\rangle \tag{21.1.115}$$

where $|0\rangle$ is the ground state of the oscillator. If we recall that

$$|n\rangle=\frac{(a^\dagger)^n}{\sqrt{n!}}|0\rangle \tag{21.1.116}$$

we see that

$$|z\rangle=\sum_0^\infty \frac{z^n}{\sqrt{n!}}|n\rangle \tag{21.1.117}$$

States labeled by different values of z are not orthonormal. We should have expected nonorthogonality since the basis $|n\rangle$ labeled by the positive integers n forms a complete basis and here we have one state for every complex number z! So they couldn't all be orthogonal. It is also possible that despite their large number, they are not a complete set. We shall, however, see that they are an *overcomplete basis*, i.e., a basis with enough vectors to expand any vector but with more than the smallest number one could have gotten away with.

Now we will establish the key property

$$a|z\rangle=z|z\rangle \tag{21.1.118}$$

as follows:

$$a|z\rangle=a\sum_0^\infty \frac{z^n}{\sqrt{n!}}|n\rangle \tag{21.1.119}$$

$$=\sum_1^\infty \frac{z^n\sqrt{n}}{\sqrt{n!}}|n-1\rangle \tag{21.1.120}$$

$$=z|z\rangle \tag{21.1.121}$$

where, in going to the last line, we have redefined a dummy label $n'=n-1$ which runs from 0 to ∞.

Likewise, by taking the adjoint of Eq. (21.1.118), the coherent state bra

$$\langle z|=\langle 0|\exp[z^*a] \tag{21.1.122}$$

is seen to obey

$$\langle z|a^\dagger=\langle z|z^* \tag{21.1.123}$$

Let us now consider the inner product

$$\langle z_2|z_1\rangle=\langle 0|\exp[z_2^*a]\exp[z_1a^\dagger]|0\rangle \tag{21.1.124}$$

If we use the identity

$$e^A e^B = e^B e^A e^{[A,B]} \qquad (21.1.125)$$

which is valid if $[A, B]$ commutes with A and B, we see

$$\langle z_2 \mid z_1 \rangle = e^{z_2^* z_1} \qquad (21.1.126)$$

upon noting that when the exponentials are exchanged and expanded out, only the first term with no a's acting to the right or $a^\dagger$'s acting to the left survives.

Completeness is shown by proving the following resolution of the identity

$$I = \int \frac{dx\, dy}{\pi} |z\rangle\langle z| e^{-z^* z} \equiv \int \frac{dz\, dz^*}{2\pi i} |z\rangle\langle z| e^{-z^* z} \qquad (21.1.127)$$

where $z = x + iy$ and $z^* = x - iy$. Note that the integral is over the entire $x - y$ plane, and after replacing every z and z^* in the integrand by $x \pm iy$, may be carried out using any other coordinates. For example, in Exercise (21.1.16) polar coordinates are recommended in verifying the above completeness relation. One can also formally go from (x, y) to (z, z^*) (after inserting a Jacobian $1/2i$), but integration over (z, z^*) is a subtle question we will not get into. We indicate that measure in terms of (z, z^*) anyway (now and later) so you will know what it means if you ever run into it again.

To show Eq. (21.1.127), one uses

$$|z\rangle = \sum_0^\infty \frac{z^n}{\sqrt{n!}} |n\rangle \qquad (21.1.128)$$

and its adjoint, does the $dx\, dy$ integral in polar coordinates, and recovers the usual sum over $|n\rangle\langle n|$.

Exercise 21.1.16. Verify the above resolution of the identity. Consult Appendix A2 for the Gamma function integral.

Since the coherent states are right eigenstates of a and left eigenstates of $a^\dagger$,

$$\langle z_2| : H(a^\dagger, a) : |z_1\rangle = \langle z_2| H(z_2^*, z_1) |z_1\rangle \qquad (21.1.129)$$

where: H: is any *normal ordered expression* i.e., an expression with all the destruction operators to the right and creation operators to the left. Thus, $a^\dagger a^2$ is a normal ordered expression while $a^2 a^\dagger$ is not. Given any expression we can always normal order it by pushing the a's to the right, keeping track of commutators.

Exercise 21.1.17. Show that $a^2 a^\dagger =: a^2 a^\dagger : + 2a$. (Push one power of a at a time to the right, or use $[AB, C] = A[B, C] + [A, C]B$.)

We now prove the following remarkable result: if H is the oscillator Hamiltonian,

$$H = \hbar\omega a^\dagger a \tag{21.1.130}$$

(we drop the constant zero-point energy for this discussion), then

$$U(t)|z\rangle = U(t)\exp[a^\dagger z]U^\dagger(t)U(t)|0\rangle = \exp[a^\dagger e^{-i\omega t}z]|0\rangle = |ze^{-i\omega t}\rangle \tag{21.1.131}$$

where we have used the Heisenberg equations of motion for $a^\dagger$. (In the Heisenberg picture $U^\dagger(t)\Omega U(t) = \Omega(t)$. Here $U^\dagger(t) = U(-t)$ appears in place of $U(t)$. We use the result $a^\dagger(t) = a^\dagger(0)e^{i\omega t}$ and reverse the sign of t.)

It is remarkable that *under time evolution the coherent state remains a coherent state, but with a new label.* This was one of the reasons one got interested in them in the first place. They have far too many interesting properties for us to discuss them all here. Instead you are directed to the reference on this subject.

Exercise 21.1.18. Show that the wave function of the coherent state is

$$\psi_z(x) = \langle x|z\rangle = \left(\frac{m\omega}{\pi\hbar}\right)^{1/4} e^{-z^2/2}e^{-(m\omega/2\hbar)x^2}e^{\sqrt{(2m\omega/\hbar)}zx} \tag{21.1.132}$$

Start by using $a|z\rangle = z|z\rangle$ in the coordinate representation. Fix the normalization by demanding that $\langle z'|z\rangle = e^{z'^*z}$. Read off its mean momentum and position. Show that these evolve with time like classical coordinates given that $|z\rangle \to |ze^{-i\omega t}\rangle$. Suggestion: Look at Eq. (9.3.7) and parametrize z as $z = \sqrt{(m\omega/2\hbar)}x_0 + i\sqrt{(1/2m\omega\hbar)}p_0$.

It is very easy to find the propagator for the oscillator in this basis:

$$U(z_N, z_0, t) = \langle z_N|U(t)|z_0\rangle = \langle z_N|z_0e^{-i\omega t}\rangle = \exp\left[z_N^* z_0 e^{-i\omega t}\right] \tag{21.1.133}$$

where the subscripts on the end point anticipates the following discussion.

Consider the path integral representation for the propagator. Let us first imagine that there are just three intermediate time slices (so that $\varepsilon = t/4$) and three resolutions of the identity operator are used, giving us

$$\begin{aligned}
&\langle z_4|U^4(t/4)|z_0\rangle \\
&= \int [\mathscr{D}z\mathscr{D}z^*] < z_4|\left(I - \frac{i\varepsilon}{\hbar}H(a^\dagger a)\right)|z_3\rangle e^{-z_3^*z_3}\langle z_3|\left(I - \frac{i\varepsilon}{\hbar}H(a^\dagger a)\right)|z_2\rangle e^{-z_2^*z_2}\langle z_2| \\
&\qquad \times \left(I - \frac{i\varepsilon}{\hbar}H(a^\dagger a)\right)|z_1\rangle e^{-z_1^*z_1}\langle z_1|\left(I - \frac{i\varepsilon}{\hbar}H(a^\dagger a)\right)|z_0\rangle
\end{aligned}$$

where

$$[\mathscr{D}z\mathscr{D}z^*] = \prod_1^{N-1}\frac{dz_i dz_i^*}{2\pi i} = \prod_1^{N-1}\frac{dx_i dy_i}{\pi} \tag{21.1.134}$$

A typical factor we run into is as follows:

$$\langle z_{n+1}|\left(I-\frac{i\varepsilon}{\hbar}H(a^\dagger a)\right)|z_n\rangle=\exp\left(-\frac{i\varepsilon}{\hbar}H(z_{n+1}^*,z_n)\right)\langle z_{n+1}|z_n\rangle \qquad (21.1.135)$$

$$=\exp\left(-\frac{i\varepsilon}{\hbar}H(z_{n+1}^*,z_n)\right)\exp(z_{n+1}^*z_n) \qquad (21.1.136)$$

where we have treated ε as infinitesimal since eventually it will be, as we let $N\to\infty$. If we assemble all the exponential factors together, there will be a piece related to the Hamiltonian which clearly gives a factor

$$\exp\left(-\frac{i}{\hbar}\int_0^t \hbar\omega z^*(t)z(t)\,dt\right) \qquad (21.1.137)$$

in the continuum notation, where z_n has become $z(t=n\varepsilon)$. (We also made the approximation $H(z^*(t+\varepsilon), z(t))\simeq H(z^*(t), z(t))$.)

The other factor in the exponent is

$$z_4^*z_3-z_3^*z_3+z_3^*z_2-z_2^*z_2+z_2^*z_1-z_1^*z_1+z_1^*z_0 \qquad (21.1.138)$$

$$=(z_4^*-z_3^*)z_3+(z_3^*-z_2^*)z_2+(z_2^*-z_1^*)z_1+z_1^*z_0 \qquad (21.1.139)$$

which we write in continuum notation as

$$\frac{i}{\hbar}\left[\int_0^t(-i\hbar)\frac{dz^*}{dt'}z\,dt'\right]+z^*(0)z(0) \qquad (21.1.140)$$

where $z(0)=z_0$ and $z^*(0)=\lim_{\varepsilon\to 0}z^*(\varepsilon)$. In other words, in the discretized version z_0 was defined but not z_0^*. Only in the continuum picture, where we focus on smooth trajectories, is this object defined as the above limit.

The sum in Eq. (21.1.139) can also be rearranged to give

$$\left[\frac{i}{\hbar}\int_0^t(i\hbar)\left(z^*\frac{dz}{dt}\right)dt\right]+z^*(t)z(t) \qquad (21.1.141)$$

where $z(t)$ is again extraneously introduced as a limit $z(t)=\lim_{\varepsilon\to 0}z(t-\varepsilon)$.

One usually sees the two schemes averaged to give the following final form of the continuum result:

$$\langle z_f|U(t)|z_i\rangle=\exp\left[\frac{z_f^*z_f+z_i^*z_i}{2}+\frac{i}{\hbar}\int_0^t\left[\frac{i\hbar}{2}\left(z^*\frac{dz}{dt}-\frac{dz^*}{dt}z\right)-H(z^*,z)\right]dt\right] \qquad (21.1.142)$$

We will use the asymmetric form obtained by doing an integration by parts:

$$\langle z_f|U(t)|z_i\rangle=\exp\left[z_f^*z_f+\frac{i}{\hbar}\left[\int_0^t\left[i\hbar z^*\frac{dz}{dt}-H(z^*,z)\right]dt\right]\right] \quad (21.1.143)$$

The warning that this is just a schematic for the previous discretized expression is all the more true here since there is very little in the action to guarantee smooth paths. However, in the limit $\hbar\to 0$, the integral is asymptotically approximated by smooth paths. Let us evaluate this integral in such a limit by finding the stationary point of the action, i.e., the classical solution. It is clear from the action, which has the phase space form $(p\dot{x}-\mathscr{H})$ that z and $i\hbar z^*$ are canonically conjugate variables. Given this action, if one were asked to quantize, one would promote them to operators obeying commutation relations

$$[Z, i\hbar Z^\dagger]=i\hbar \quad (21.1.144)$$

which we see are just the commutation rules for a and $a^\dagger$. Of course, we are not trying to construct the quantum theory from the classical one, but the reverse. The Hamiltonian equation is

$$\dot{z}=\frac{\partial(\hbar\omega z^*z)}{\partial(i\hbar z^*)}=-i\omega z \quad (21.1.145)$$

which is solved to give

$$z(t)=z(0)\,e^{-i\omega t} \quad (21.1.146)$$

Similarly, we find

$$z^*(t)=z^*(0)\,e^{i\omega t} \quad (21.1.147)$$

To evaluate

$$\langle z_f|U(T)|z_i\rangle \quad (21.1.148)$$

in the semiclassical approximation, we need to find a solution that obeys

$$z(0)=z_i \quad (21.1.149)$$

$$z^*(T)=z_f^* \quad (21.1.150)$$

Now we see a problem that we did not have in the configuration space version: since the equations here are first order in time, z_i determines $z(t)$ for all times. How can we get $z^*(T)$ to equal an independently given z_f? The answer is that we must regard

z and z^* as independent and restrict $z(t)$ at $t=0$ and $z^*(t)$ at $t=T$. The solutions then are

$$z(t)=z_i\, e^{-i\omega t} \tag{21.1.151}$$

$$z^*(t)=z_f^*\, e^{i\omega(t-T)} \tag{21.1.152}$$

Note that $z^*(T)$ is not the complex conjugate of $z(T)$. This means that x and y invoked in the definition $z=x+iy$ are not real on this trajectory. However, a Gaussian integral is given by its saddle point even if the point is off the original axis of integration. This point is explained in Faddeev's lectures (see Bibliography).

If we feed this solution into the action we find that the t-integral gives zero due to a cancellation between the two terms in the integrand and the only piece that survives is

$$z^*(T)z(T)=z_f^*\, z_i\, e^{-i\omega T}$$

giving us

$$\langle z_f|\, U(T)|z_i\rangle = \exp(z_f^* z_i\, e^{-i\omega T}) \tag{21.1.153}$$

which is the exact answer!

Exercise 21.1.19. Evaluate the action for the above path and check the answer given.

Exercise 21.1.20. Consider the Gaussian integrals in Eqs. (A.2.4–A.2.5.) Show that if we want just the exponential dependence of the answer, it is given by finding the exponential where the exponent is stationary. This is a general feature of Gaussian integrals.

Exercise 21.1.21. A good take-home problem. Rederive the oscillator propagator $\langle x_2|\, U(T)|x_1\rangle$ given $\langle z_f|\, U(T)|z_i\rangle=\exp[z_f^*\, z_i\, e^{-i\omega T}]$. Introduce two resolutions of the identity on either side of $U(T)$ in $\langle x_2|\, U(T)|x_1\rangle$. Use the suitably normalized wave functions $\langle x|z\rangle$ from Exercise (21.1.18). You will have to do a Gaussian integral over the two pairs of intermediate coherent state variables. Do the integral by saddle point, i.e., find the stationary point of the action and evaluate the integrand there. Focus on just the exponential factor and show that you get the answer to Exercise (8.6.2).

21.2. Imaginary Time Formalism

Consider the *imaginary time propagator*

$$U(\tau)=\exp\left(-\frac{1}{\hbar}H\tau\right) \tag{21.2.1}$$

This is obtained by setting

$$t=-i\tau \tag{21.2.2}$$

in the usual propagator. In other words, if the Schrödinger equation had been

$$-\hbar\frac{d}{d\tau}|\psi(\tau)\rangle = H|\psi(\tau)\rangle \tag{21.2.3}$$

this would have been the propagator.

The reasons for looking at this operator will be clear as we go along. But first let us note that we can write down the formula for it at once:

$$U(\tau) = \sum |n\rangle\langle n| \exp\left(-\frac{1}{\hbar}E_n\tau\right) \tag{21.2.4}$$

where

$$H|n\rangle = E_n|n\rangle \tag{21.2.5}$$

The main point to note is that even *though the time is now imaginary, the eigenvalues and eigenfunctions that enter into the formula for* $U(\tau)$ *are the usual ones*. Conversely, if we knew $U(\tau)$, we could extract the former.

Path Integral for the Imaginary Time Propagator

Consider the matrix element

$$U(x, x', \tau) = \langle x|U(\tau)|x'\rangle \tag{21.2.6}$$

We can write down a path integral for it following exactly the same steps as before. The final answer in continuum notation is

$$\langle x|U(\tau)|x'\rangle = U(x, x', \tau) = \int[\mathscr{D}x]\exp\left[-\frac{1}{\hbar}\int_0^\tau \mathscr{L}_E(x, \dot{x})d\tau\right] \tag{21.2.7}$$

$$\int[\mathscr{D}x] = \lim_{N\to\infty}\left(\frac{m}{2\pi\hbar\varepsilon}\right)^{1/2}\prod_0^{N-1}\left(\frac{m}{2\pi\hbar\varepsilon}\right)^{1/2}dx_i \tag{21.2.8}$$

$$\mathscr{L}_E = \frac{m}{2}\left(\frac{dx}{d\tau}\right)^2 + V(x) \tag{21.2.9}$$

where $\varepsilon = \tau/N$ and $\mathscr{L}_E$ is called the *euclidean Lagrangian*. The adjective "euclidean" means that space and time now behave alike—the minus signs of Minkowski space in the formula for invariants are gone. For example, the invariant $x^2 - c^2t^2$ now becomes $x^2 + c^2\tau^2$. Notice that $\mathscr{L}_E$ is the *sum* of the euclidean kinetic energy and real-time potential energy. *Thus the particle obeying the euclidean equations of motion will see the potential turned upside down*. This will be exploited later.

We have emphasized that the continuum form of the path integral is a shorthand for the discrete version. It is true here also, but of all the path integrals, this is the best behaved. Rapidly varying paths are suppressed by the falling (rather than rapidly oscillating) exponential factor.

Suppose we want to calculate the euclidean path integral for a free particle. We can proceed as we did in Chapter 8 and obtain

$$\langle x|U(\tau)|x'\rangle = \left(\frac{m}{2\pi\hbar\tau}\right)^{1/2} \exp\left[-\frac{m(x-x')^2}{2\hbar\tau}\right] \tag{21.2.10}$$

If someone gave us this propagator, we could get the Minkowski space answer by setting

$$\tau = it \tag{21.2.11}$$

This is called *analytic continuation.*

A very important feature of euclidean quantum mechanics is that the operator $U(\tau)$ is not unitary but Hermitian. Thus the norm of the state is not preserved in time. In fact what happens is that after a long time every state evolves into the ground state $|0\rangle$:

$$\lim_{\tau\to\infty} \langle x|U(\tau)|x'\rangle = \lim_{\tau\to\infty} \sum \langle x|n\rangle\langle n|x'\rangle \exp\left(-\frac{1}{\hbar}E_n\tau\right) \tag{21.2.12}$$

$$\simeq \langle x|0\rangle\langle 0|x'\rangle \exp\left(-\frac{1}{\hbar}E_0\tau\right) \tag{21.2.13}$$

$$= \psi_0(x)\psi_0^*(x') \exp\left(-\frac{1}{\hbar}E_0\tau\right) \tag{21.2.14}$$

Thus all states lead to the ground state as long as the starting point has some overlap with it. This is one way to find the ground state in any problem: take any initial state and let it evolve for a long time. You should hit the ground state unless you had chosen an initial state orthogonal to the ground state. (Sometimes you may do this on purpose to find the first excited state. For instance if the problem has parity invariance and you choose an initial state odd under parity, you will hit an excited state.)

For example, the propagator for the oscillator is

$$U(x, x', \tau) = A(\tau) \exp\left(-\frac{m\omega}{2\hbar \sinh \omega\tau}[(x^2+x'^2)\cosh \omega\tau - 2xx']\right) \tag{21.2.15}$$

obtained, say by analytic continuation from real times of the answer in Exercise (8.6.2). Note that as $\tau\to\infty$ this becomes proportional to the product of ground state wave functions. The prefactor is left to the following exercise.

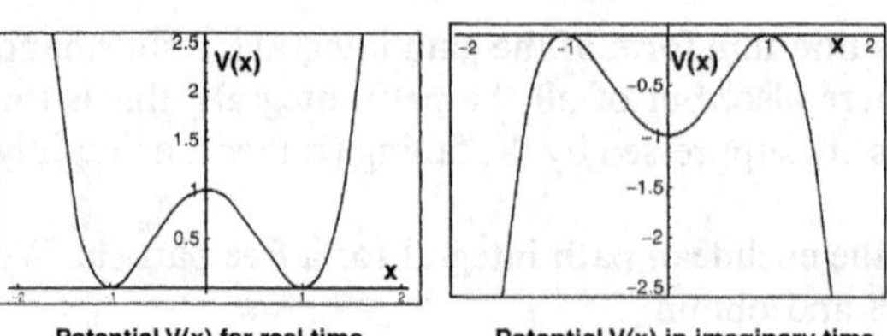

Figure 21.2. The double-well potential in real and imaginary time.

Exercise 21.2.1. Obtain $A(t)$ from Exercise (8.6.3) and continue to imaginary time, and verify that in the large τ limit, it yields the right prefactor.

Tunneling by Path Integrals: Well, well!

We now consider one application of the euclidean formalism. We have seen how one can derive the WKB wave function for nonbound states by using path integrals. This procedure does not work for tunneling amplitudes across barriers since we cannot find a classical path that goes over the barrier. On the other hand, in the euclidean dynamics the potential is turned upside down and what is forbidden in Minkowski space is suddenly allowed in the euclidean region!

Here is a problem that illustrates this point and many more. Consider a particle in a double-well potential

$$V(x) = A^2(x^2 - a^2)^2 \tag{A.4.16}$$

The classical minima are at

$$x_{L/R} = \pm a \tag{A.4.17}$$

Figure 21.2 shows a graph in Minkowski and euclidean space for the case $a = A = 1$.

Notice that in the euclidean problem the double-well has been inverted into the double-hill.

What is the ground state of the system? The classical ground state is doubly-degenerate: the particle can be sitting at either of the two minima. In the semiclassical approximation, we can broaden these out to Gaussians that are ground states $|\pm a\rangle$ in the harmonic oscillatorlike potential around each minimum at $x = \pm a$. This will shift each degenerate ground state by $\frac{1}{2}\hbar\omega$ where ω measures the curvature of the potential near the minimum. We can go to higher-order approximations that recognize that the bottom of the well is not exactly quadratic and shift the ground state energies by higher powers of $\hbar$. However, none of this will split the degeneracy of the ground states since whatever we find at the left minimum we will find at the right by symmetry under reflection. Lifting of the degeneracy will happen only if we take into account tunneling between the two wells. So we study this problem in the following stripped-down version. First we drop all but the degenerate ground states $|\pm a\rangle$. (The Gaussians centered around the two minima are not quite orthogonal.

Assume they have been orthogonalized by a Gram–Schmidt procedure.) The approximate Hamiltonian looks like this in this subspace:

$$H=\begin{bmatrix} E_0 & 0 \\ 0 & E_0 \end{bmatrix} \tag{21.2.18}$$

Let us shift our reference energy so that $E_0=0$.

Note that there are no off-diagonal matrix elements. If this were an exact result, it should mean that if a particle starts out in one well it will never be found at the other. But we know from the wave function approach that if it starts at one side, it can tunnel to the other. This means that there is effectively a nonzero matrix off-diagonal matrix element $H_{+-}=H_{-+}=\langle a|H|-a\rangle$ in this basis. The challenge is to find that element in the semiclassical approximation. Once we find it, it is evident that the energy levels will be split into

$$E=\pm H_{+-} \tag{21.2.19}$$

and the eigenstates will be $|S/A\rangle$, the sum and difference of $|\pm a\rangle$.

Consider

$$\langle a|\, U(\tau)|-a\rangle=\langle a|\, \exp\left(-\frac{1}{\hbar} H\tau\right)|-a\rangle \tag{21.2.20}$$

In this discussion of tunneling, $U(\tau)$ is the propagator from $-\tau/2$ to $\tau/2$ and not from 0 *to* τ. Note that the term linear in τ gives us the off-diagonal matrix element:

$$\langle a|\, \exp\left(-\frac{1}{\hbar} H\tau\right)|-a\rangle\simeq 0-\frac{1}{\hbar}\tau\langle a|H|-a\rangle+\mathcal{O}\tau^2 \tag{21.2.21}$$

We shall calculate $\langle a|\, e^{-(1/\hbar)H\tau}|-a\rangle$ by the semiclassical approximation to the euclidean path integral and extract the approximate matrix element H_{+-}. *Once again, as in the real-time semiclassical approximation, we focus on just the exponential factor and ignore all prefactors.* In the semiclassical approximation,

$$\langle a|\, \exp\left(-\frac{1}{\hbar} H\tau\right)|-a\rangle\simeq\exp\left(-\frac{1}{\hbar} S_{\mathrm{cl}}\right) \tag{21.2.22}$$

where S_{cl} is the euclidean action for the classical path connecting the left hill to the right. *The key point, of course, is that in the double-hill potential of euclidean mechanics the classical ground states are not separated by a barrier, so that there will be no problem finding a classical path going from one hill to the other.*

The euclidean equations of motion are the same as the real times ones, *except for the reversal of the potential.* Thus there will be a conserved energy E_e given by

$$E_e = \frac{m}{2}\left(\frac{dx}{d\tau}\right)^2 - V(x) \tag{21.2.23}$$

Using this we can solve for the trajectory by quadrature:

$$\int_{x_1}^{x_2} \frac{\sqrt{m}\,dx}{\sqrt{2(E_e + V(x))}} = \int_{t_1}^{t_2} d\tau \tag{21.2.24}$$

Now we want the tunneling from the state $|-a\rangle$ to the state $|a\rangle$. These are not eigenstates of position, but Gaussians centered at $x = \mp a$. We shall however calculate the amplitude to tunnel from the *position eigenstate* $x = -a$ to the position eigenstate $x = a$. Except for the overlaps $\langle x = a | a\rangle$ and $\langle -a | x = -a\rangle$ this is the same as $\langle a | U | -a\rangle$. These overlaps know nothing about the tunneling barrier. They will constitute undetermined prefactors in front of the exponential dependence on the barrier which alone we are after. To extract the extreme low energy physics we must let $\tau \to \infty$. To this end, let us consider the trajectory that has $E_e = 0$. It is given by doing the above integral with $E_e = 0$:

$$x(\tau) = a \tanh\left[\sqrt{\frac{2}{m}}\, aA\tau\right] \tag{21.2.25}$$

Notice that in this trajectory the particle starts out at the left *maximum* (Fig. 21.2) at $\tau \to -\infty$ and rolls down the hill and only reaching of the right maximum as $\tau \to \infty$. If the starting point and ending point are exactly $x = \mp a$, tunneling takes infinite time since only in this limit does the tanh take its limiting value of $\pm a$. Physically, it takes forever since the particle must start from rest at the left end to have zero euclidean energy. On the other hand, for points which are slightly below the maximum at each end, the time of travel will be finite since the particle can start with nonzero velocity. Since these points will also have roughly the same overlap with the states $|\pm a\rangle$ we can start with them instead of $x = \pm a$ in which case the tunneling will take place in finite time. This will be understood in what follows.

The action for the above solution is (using $T = V$ for the zero energy solution),

$$S_{\text{cl}} = \int (T + V)\,d\tau = \int 2T\,d\tau = \int_{-a}^{a} p(x)\,dx = \int_{-a}^{a} \sqrt{2mV(x)}\,dx \tag{21.2.26}$$

and the tunneling amplitude is (ignoring prefactors)

$$\langle a | U | -a\rangle \simeq \exp\left(-\frac{1}{\hbar}\int_{-a}^{a} \sqrt{2mV(x)}\,dx\right) \tag{21.2.27}$$

in agreement with tunneling result in the Schrödinger approach, Eq. (16.2.24) with $E = 0$.

Now, onward to extract the matrix element by looking for the term linear in τ in the answer. But we see no such explicit τ dependence in the answer! The resolution can be stated in two ways.

- The first is tied to the fact that in the limit of large τ, the problem becomes translationally invariant in time. In other words, if we stare at the classical solution above, we see that the tanh is close to $\pm a$ most of the time and jumps rapidly from $-a$ to a in a short time centered around $\tau=0$. Pictorially, the particle takes a long time to roll off the top, but once it gets going, it rolls down very quickly to a point close to the other end point. (For this reason this solution is called an *instanton*, a term coined by 't Hooft: except for the brief "instant" when tunneling takes place, the system is essentially in one of its classical ground states.) If we draw a new trajectory in which the same tunneling takes place in the same time interval, but is centered around a time $\tau=\tau_0\neq 0$, this too will be close to being a minimum of the action. (It will have exactly the same action as $\tau\to\infty$.) In other words, the solution we found has many companions, all of nearly the same action, but different tunneling instants τ_0. We must sum over all these paths, i.e., integrate over the instant of tunneling τ_0. Since they all have nearly the same action, the effect is to multiply the answer by τ since τ_0 is forced to lie within the period $-\tau/2<\tau_0<\tau/2$.
- The second way to argue is that once we find one classical path, we must integrate the functional over all fluctuations $\delta x(\tau)=x(\tau)-x_{\text{cl}}(\tau)$. (See Section 8.6.) If we expand the action near x_{cl}, there will be no linear term since the action is stationary here and we will start with a quadratic expression in $\delta x(\tau)$. By diagonalizing this quadratic form we can get the answer as a product of Gaussian integrals. Consider the one-dimensional example of some function approximated by a Gaussian centered at $x=0$:

$$I(a)=\int_{x_1}^{x_2} e^{-\alpha x^2}\,dx \tag{21.2.28}$$

If $\alpha>0$ we can assume the limtis can be pushed to infinity and the answer approximated by

$$I(a)\simeq\sqrt{\pi/\alpha} \tag{21.2.29}$$

What happens when $\alpha\to 0$? The approximate answer diverges but we know the real answer is

$$I=\lim_{\alpha\to 0}\int_{x_1}^{x_2} e^{-\alpha x^2}\,dx=x_2-x_1 \tag{21.2.30}$$

This is essentially what happens in the functional integral. Say $x(\tau)$ is a classical solution. Then $x(\tau-\tau_0)$ is also a solution, and

$$\delta x(\tau)=x(\tau-\tau_0)-x(\tau) \tag{21.2.31}$$

is a fluctuation that costs no extra action, i.e., the Gaussian that is supposed to damp out this fluctuation has $\alpha \to 0$. The Gaussian integral is then replaced by the range of integration corresponding to this degree of freedom, which is just $\int d\tau_0 \simeq \tau$.

So we have argued for a prefactor of τ which came from considering a fluctuation about the classical solution. We were forced to consider it since it reflected an exact symmetry (under time-translation) as result of which it had no α in the Gaussian to cut it off. We do, however, ignore the Gaussian integrals over the rest of the fluctuations since they cut off by nonzero αs.

With the prefactor τ in front of

$$\langle -a|\, U(\tau)|a\rangle = \tau \exp\left(-\frac{1}{\hbar} S_{\text{cl}}\right) \tag{21.2.32}$$

we are ready to compare to

$$\langle a|\exp\left(-\frac{1}{\hbar} H\tau\right)|-a\rangle \simeq 0 - \frac{1}{\hbar}\tau\langle a|H|-a\rangle + \mathcal{O}\tau^2 \tag{21.2.33}$$

and read off

$$H_{-+} \simeq -\exp\left(-\frac{1}{\hbar} S_{\text{cl}}\right) \tag{21.2.34}$$

where once again we have dropped all prefactors except for the sign which is important. (All euclidean transition amplitudes are positive since the functional is positive. The minus sign comes from $e^{-(1/\hbar)H\tau}$.)

It is now clear that with H_{-+} negative, the new eigenstates and energies are as follows:

$$|S\rangle = \sqrt{\tfrac{1}{2}}[|+a\rangle + |-a\rangle] \qquad E_S = -\exp\left(-\frac{1}{\hbar} S_{\text{cl}}\right) \tag{21.2.35}$$

$$|A\rangle = \sqrt{\tfrac{1}{2}}\,[|+a\rangle - |-a\rangle] \qquad E_S = \exp\left(-\frac{1}{\hbar} S_{\text{cl}}\right) \tag{21.2.36}$$

Spontaneous Symmetry Breaking

Why are we interested in a term that vanishes exponentially fast as $\hbar \to 0$ when we ignored all the perturbative corrections to the states $|\pm a\rangle$ which vanished as finite powers of $\hbar$? The reason is that the exponentially small term is the leading term in the *splitting* of the two classically degenerate ground states.

But there is another very significant implication of the tunneling calculation. This has to do with the phenomenon of *spontaneous symmetry breaking* which will now be described.

Consider a Hamiltonian which has a symmetry, say under parity. *If the lowest energy state of the problem is itself not invariant under the symmetry, we say symmetry is spontaneously broken.*

Spontaneous symmetry breaking occurs quite readily in classical mechanics. Consider the single-well oscillator. The Hamiltonian is invariant under parity. The ground state is a particle sitting at the bottom of the well. This state respects the symmetry: the effect of parity on this state gives back the state. Now consider the double-well with minima at $x=\pm a$. There are two lowest energy configurations available to the particle: sitting still at the bottom of either well. No matter which choice it makes, it breaks the symmetry. The breakdown is spontaneous in that there was nothing in the Hamiltonian that tilted the scales. Once the particle has made a choice (based on accidents of initial conditions) the other option does not enter its dynamics. Let us note the twin signatures of symmetry breaking: there is more than one ground state, and these states are not invariant under the symmetry (some observable, not invariant under the symmetry has a nonzero value), but instead get mapped into each other by the symmetry operation.

Now consider the quantum case of the double well, but with an infinite barrier between the wells. (I mean a barrier across which tunneling is impossible either in the path integrals or wave function approach. So a delta function spike is not such a barrier.) Once again the particle has two choices, these being Gaussian-like functions centered at the two troughs: $|\pm a\rangle$. They show the twin features of symmetry breaking: they are degenerate and noninvariant under parity ($\langle X\rangle\neq 0$). But here is a twist. In quantum theory a particle can be in two places at the same time. In particular, we can form the combinations of these degenerate eigenvectors

$$|S/A\rangle=\frac{[|+a\rangle\pm|-a\rangle]}{\sqrt{2}} \tag{21.2.37}$$

$$\Pi|S/A\rangle=\pm|S/A\rangle \tag{21.2.38}$$

which are eigenstates of parity. Indeed, in quantum theory the relation

$$[\Pi, H]=0 \tag{21.2.39}$$

guarantees that such parity eigenstates *can* be formed. But *should* they be formed? The answer is negative in this problem due to the infinite barrier. The reason is this. Suppose the particle in question is sighted in one side during a measurement. Then there is no way for its wave function to develop any support in the other side. (One says the motion is not ergodic.) Even in quantum theory, where energy can be nonconserved over small times, barrier penetration is forbidden if the barrier is infinite. This means in particular that the symmetric and antisymmetric functions will never be realized by any particle that has ever been seen on either side. The correct thing to do then is to build a Hilbert space of functions with support on just one side. That every state so built has a degenerate partner in the inaccessible well across the barrier, is academic. The particle will not even know a parallel universe just like its own exists. Real life will not be symmetric in such a problem and the symmetric and antisymmetric wave functions (with zero $\langle X\rangle$) represent unrealizable situations. Symmetry *is* spontaneously broken.

Now for the more typical problem with a finite barrier. In this case, a particle once seen in the left side can later be seen in the right side and vice versa. Symmetric and antisymmetric wave functions are physically sensible and we can choose energy eigenstates which are also parity eigenstates. These states will no longer be degenerate. In normal problems, the symmetric state, or more generally the state with eigenvalue unity for the symmetry operation, the one invariant under the symmetry operation, will be the unique ground state. Recall that in the oscillator problem the ground state not only had definite parity, it was invariant under parity. Likewise, in the hydrogen atom, the ground state not only had definite angular momentum, the angular momentum was zero and was invariant under rotations. However, in both these problems there was no multiplicity of classical ground states and no real chance of symmetry breakdown. (The oscillator had just one classical ground state at the bottom of the well, and the hydrogen atom had one infinitely deep within the Coulomb well.) What the instanton calculation tells us is that the double well, despite having two classical ground states that break symmetry, has, in the quantum theory, a unique, symmetric, ground state.

Thus, even though the tunneling calculation was very crude and approximate, it led to a very profound conclusion: the symmetry of the Hamiltonian is the symmetry of the ground state, symmetry breaking does not take place in the double-well problem.

This concept of symmetry restoration by tunneling (which in turn is tied to the existence of classical euclidean solutions with finite action going from one putative degenerate ground state to another) is very deep and plays a big role in many problems. There have been problems (quantum chromodynamics) where one did not even realize that the minimum one had assumed was unique for years was one of an infinite family of degenerate minima, till an instanton (of finite action) connecting the two classical minima was found and interpreted. We discuss a simpler example to illustrate the generality of the notion: a particle in a periodic potential $V(x)=1-\cos 2\pi x$. The minima are at $x=n$, where n is any integer. The symmetry of the problem is the discrete translation $x\to x+1$. The approximate states, $|n\rangle$, which are Gaussians centered around the classical minima, break the symmetry and are converted to each other by T, the operator that translates $x\to x+1$

$$T|n\rangle=|n+1\rangle \tag{21.2.40}$$

However, adjacent classical minima are connected by a nonzero tunneling amplitude of the type we just calculated and H has off-diagonal amplitudes between $|n\rangle$ and $|n\pm 1\rangle$. (There are also solutions describing tunneling to next-nearest-neighbor minima, but these have roughly double the action as the nearest-neighbor tunneling process and lead to an off-diagonal matrix element that is roughly the square of the one due to nearest-neighbor tunneling.) Suppose the one-dimensional world were finite and formed a closed ring of size N, so that there were N degenerate classical minima. These would evolve into N nondegenerate levels (the analogs of $|S/A\rangle$) due to the mixing due to tunneling. The ground state would be a symmetric combination:

$$|S\rangle=\frac{1}{\sqrt{N}}\sum_1^N|n\rangle \tag{21.2.41}$$

The details are left to the following exercise.

Exercise 21.2.2. (Very important)
Assume that

$$H=\sum_1^N E_0|n\rangle\langle n|-t(|n\rangle\langle n+1|+|n+1\rangle\langle n|) \tag{21.2.42}$$

describes the low-energy Hamiltonian of a particle in a periodic potential with minima at integers n. The integers n go from 1 to N since it is assumed the world is a ring of length N so that the $N+1$th point is the first. Thus the problem has symmetry under translation by one site despite the finite length of the world. The first term in H represents the energy of the Gaussian state centered at $x=n$. The second represents the tunneling to adjacent minima with tunneling amplitude t. Consider the state

$$|\theta\rangle=\frac{1}{\sqrt{N}}\sum_1^N e^{in\theta}|n\rangle \tag{21.2.43}$$

Show that it is an eigenstate of T. Find the eigenvalue. Use the condition $T^N=I$ to restrict the allowed values of θ and make sure that we still have just N states. Show that $|\theta\rangle$ is an eigenstate of H with eigenvalue $E(\theta)=E_0-2t\cos\theta$. Consider $N=2$ and regain the double-well result. (You might have some trouble with a factor of 2 in front of the $\cos\theta$ term. Remember that in a ring with just two sites, each site is both ahead and behind the other and H couples them twice.)

Will the ground state always be invariant under the symmetry operation that commutes with H? The answer is yes, as long as the barrier height is finite, or more precisely, as long as there is a finite action solution to the euclidean equations of motion linking classical minima. This is usually the case for quantum mechanics of finite number of degrees of freedom with finite parameters in the Hamiltonian. On the other hand, if $V_0\to\infty$ in the periodic potential, there really will be N degenerate minima with particles living in any one minimum trapped there forever. In quantum field theory, where there are infinitely many degrees of freedom, even if the parameters are finite, the barrier is often infinitely high if all degrees of freedom try to jump over a barrier. In other words, symmetry breaking can take place.

For a more complete discussion of the tunneling question, you must consult the Bibliography, especially the works by Coleman and Rajaraman. These references will also answer other questions you might have such as: What about solutions where the particle rattles back and forth between the two hilltops in the inverted double-well potential? (These give contributions where the prefactors go as higher powers of τ.) Is there a way to read off the splitting between $|S/A\rangle$ directly from $\langle a|U(\tau)|-a\rangle$ without picking off the term linear in τ? (Yes, by summing over an infinite amount of rattling back and forth.) You will find many interesting points to ponder, but our result will prove to be correct to leading order in the exponentially small quantity $e^{-(1/\hbar)S_{cl}}$.

Imaginary Time Path Integrals and Quantum Statistical Mechanics

We now discuss two other reasons for studying imaginary time path integrals. The first concerns quantum statistical mechanics and the second classical statistical mechanics.

Consider the partition function for a quantum system:

$$Z=\sum_n e^{-\beta E_n} \tag{21.2.44}$$

where the temperature T and Boltzmann's constant k appear in the combination $\beta=1/kT$ and where E_n is the energy of the nth eigenstate of the Hamiltonian H. We can rewrite this as

$$Z=\mathrm{Tr}\, e^{-\beta H} \tag{21.2.45}$$

where the trace is taken in the eigenbasis of H. Now we exploit the fact that the trace is invariant under a unitary change of basis and switch to the x-basis to obtain

$$Z=\int_{-\infty}^{\infty} \langle x|\, e^{-\beta H}\, |x\rangle\, dx \tag{21.2.46}$$

The integrand is of course familiar to us now:

$$\langle x|\exp(-\beta H)|x\rangle=\langle x|\exp\left(-\frac{1}{\hbar}\beta\hbar H\right)|x\rangle=U(x,x,\beta\hbar) \tag{21.2.47}$$

In other words, Z is the sum over amplitudes to go from the point x back to the point x in imaginary time $\tau=\beta\hbar$, in other words, over closed paths.

Exercise 21.2.3. Starting with $U(x, x, \tau)$ for the oscillator (see Eq. (21.2.15) and Exercise (21.2.1)) do the integral over x to obtain Z. Compare this to the sum

$$Z=\sum_0^\infty e^{-\beta\hbar\omega(n+1/2)} \tag{21.2.48}$$

This connection between quantum statistical mechanics and imaginary time quantum mechanics is the starting point for a whole industry. Some applications are discussed in the book by Feynman and Hibbs. It would take us too far astray to get into any of these in depth. I will merely show how we take the classical limit of this formula. Consider a single particle of mass m in a potential $V(x)$. Then

$$Z(\beta)=\int dx \int_x^x [\mathscr{D}x]\exp\left[-\frac{1}{\hbar}\int_0^{\beta\hbar}\left[\frac{m}{2}\left(\frac{dx}{d\tau}\right)^2+V(x(\tau))\right]d\tau\right] \tag{21.2.49}$$

where the limits on the functional integral remind us to consider paths starting and ending at the same point x, which is then integrated over, via the ordinary integral. Consider the limit $\beta\hbar \to 0$ either due to high temperatures or vanishing $\hbar$ (the classical limit). Look at any one value of x. We need to sum over paths that start at x, go somewhere and come back to x in a very short time $\beta\hbar$. If the particle wanders off a distance Δx, the typical kinetic energy is $m(\Delta x/\beta\hbar)^2$ and the suppression factor is

$$\simeq \exp\left(-\frac{1}{\hbar} m(\Delta x/\beta\hbar)^2 \beta\hbar\right) \tag{21.2.50}$$

from which it follows that

$$\Delta x \simeq \sqrt{\frac{\beta}{m}}\,\hbar \tag{21.2.51}$$

If the potential does not vary over such a length scale [called the *thermal wavelength*, see Exercise (21.2.4)] we can approximate it by a constant equal to its value at the starting point x and write

$$\begin{aligned} Z(\beta) &\simeq \int dx\, e^{-\beta V(x)} \int_x^x [\mathscr{D}x] \exp\left[-\frac{1}{\hbar}\int_0^{\beta\hbar}\left[\frac{m}{2}\left(\frac{dx}{d\tau}\right)^2\right] d\tau\right] \\ &= \int dx\, e^{-\beta V(x)} \sqrt{\frac{m}{2\pi\hbar\beta\hbar}} \end{aligned} \tag{21.2.52}$$

where in the last step we have used the fact that with $V(x)$ pulled out, the functional integral is just the amplitude for a free particle to go from x to x in time $\beta\hbar$. How does this compare with classical statistical mechanics? There the sum over states is replaced by an integral over phase space:

$$Z = A\int dx \int dp \exp\left[-\beta\left(\frac{p^2}{2m} + V(x)\right)\right] \tag{21.2.53}$$

where the arbitrary prefactor A reflects one's freedom to multiply Z by a constant without changing anything physical since Z is a sum over relative probabilities and any prefactor will drop out in any averaging process. Equivalently it corresponds to the fact that the number of classical states in a region $dx\,dp$ of phase space is not uniquely defined. If we do the p integral and compare to the classical limit of the path integral we see that quantum theory fixes

$$A = \frac{1}{2\pi\hbar} \tag{21.2.54}$$

in accordance with the uncertainty principle which associates an area of order $\Delta X \Delta P \simeq \hbar$ in phase space with each quantum state.

Exercise 21.2.4. Consider a particle at temperature T, with mean energy of order kT. Assuming all the energy is kinetic, estimate its momentum and convert to the de Broglie wavelength. Show that this gives us a number of the order of the thermal wavelength. This is the minimum size over which the particle can be localized.

Relation to Classical Statistical Mechanics

So far we have discussed the relation of the imaginary time path integral to quantum statistical mechanics. Now we consider its relation to classical statistical mechanics. Consider a classical system with $N+1$ sites and a degree of freedom x_n at each site. The variables at the end of the chain, called x_0 and x_N, are fixed. Then

$$Z=\int_{-\infty}^{\infty}\prod_{1}^{N-1} dx_i \exp\left(-\frac{1}{kT}E(x_0,\ldots,x_N)\right) \tag{21.2.55}$$

where E is the energy function and we have written β in terms of the more familiar temperature variable as $\beta=(1/kT)$. Let E have the form

$$E=\sum_{1}^{N-1}[K_1(x_n-x_{n-1})^2+K_2x_n^2] \tag{21.2.56}$$

where the first term represents the springlike coupling between nearest neighbors that forces them to maintain a fixed separation and the second one provides a quadratic potential that discourages each x from wandering off its neutral position $x=0$. If we compare this to the discretized imaginary time Feynman path integral for the quantum oscillator

$$U(x_0,x_N,\tau)=\int_{-\infty}^{\infty}\prod_{1}^{N-1} dx_i \exp\left[-\frac{1}{\hbar}\sum_{1}^{N-1}\varepsilon\left(\frac{m}{2}\frac{(x_n-x_{n-1})^2}{\varepsilon^2}+\frac{m\omega^2}{2}x_n^2\right)\right] \tag{21.2.57}$$

we see the following correspondence:

- The Feynman path integral from x_0 to x_N is identical in form to a classical partition function of a system of $N+1$ coordinates x_n with the boundary condition that the first and last be fixed at x_0 and x_N. The variables x_n are interpreted as intermediate state labels of the quantum problem (in the repeated resolution of the identity) and as the classical variables summed over in the partition function.
- The role of the action in the Feynman integral is played by the energy in the partition function.
- The role of $\hbar$ is played by T. In particular, as either variable goes to zero, the sum over configurations is dominated by the minimum of action or energy and fluctuations are suppressed.
- The parameters in the classical and quantum problems can be mapped into each other. For example, $\beta K_1=m/2\hbar\varepsilon$ and $\beta K_2=m\omega^2\varepsilon/2\hbar$.
- Since $\varepsilon\to 0$ in the quantum problem, the parameters of the classical problem must take some limiting values ($K_1\to\infty$ and $K_2\to 0$ in a special way) to really be in correspondence with the quantum problem with $H=P^2/2m+m\omega^2x^2/2$.

- The single quantum degree of freedom is traded for a one dimensional array of classical degrees of freedom. This is a general feature: the dimensionality goes up by 1 as we go from the quantum to the classical problem. For example, a one-dimensional array of *quantum* oscillators would map on to the partition function of a two-dimensional array of classical variables. The latter array would be labeled by the time slice n as well as the quantum oscillator whose intermediate state label it stands for.

Our emphasis has been on the notion that the quantum oscillator problem can be written as a path integral which we now see is also a classical partition function. It is just as interesting to take a classical problem and translate it back to the operator version. In the classical problem we are interested in the free energy and thermal averages over the Boltzmann distribution, i.e., correlation functions like

$$\langle x_{12}x_{78}\rangle = \frac{\int_{-\infty}^{\infty}\prod_1^{N-1} dx_i\, x_{12}x_{78}\, e^{-\beta E(x_0,\ldots,x_N)}}{\int_{-\infty}^{\infty}\prod_1^{N-1} dx_i\, e^{-\beta E(x_0,\ldots,x_N)}} \tag{21.2.58}$$

where we use wedgy brackets to represent thermal averages as we did quantum averages, hoping you will be able to keep track of what is meant from the context. In the quantum theory we are interested in eigenstates of H, especially the ground state, Heisenberg operators, etc. We now develop the dictionary between the two approaches. Rather than use the oscillator, we turn to a problem with a simpler Hilbert space: that of a spin-1/2 problem.

For this purpose consider the *Ising model* in one dimension. The lattice now is an array of $N+1$ dots numbered 0 to N. At each point lies an *Ising spin* which can take only two values, $s=\pm 1$. The partition function is

$$Z=\sum_{s_i=\pm 1}\exp\left[\sum_{i=0}^{N-1} K(s_i s_{i+1}-1)\right] \tag{21.2.59}$$

where K contains the factor $-\beta$. For the case we are interested in, $K>0$, the Boltzmann weight is large when $s_i=s_{i+1}$ and small when $s_i=-s_{i+1}$. Thus the nearest-neighbor coupling represents the ferromagnetic tendency of the spins to be aligned with their neighbors. The additional, spin independent energy of minus $-K$ per site is a shift in energy made for convenience. Given this formula for Z, we can answer all thermodynamic questions. This is our classical problem. We will first solve for the free energy and correlation function viewing the problem classically. Then we will map this into a quantum problem and rederive the same results and our dictionary.

Let us first keep s_0 fixed at one value and define a relative variable:

$$t_i = s_i s_{i+1} \tag{21.2.60}$$

It is clear that given s_0 and t_i, we can reconstruct the state of the system. Thus, we can write

$$Z=\sum_{t_i} \exp\left[\sum_{i=0}^{N-1} K(t_i-1)\right]=\sum_{t_i}\prod_i e^{K(t_i-1)} \tag{21.2.61}$$

Since the exponential factorizes into a product over i, we can do the sums over each t_i and obtain (after appending a factor of 2 for the two possible choices of s_0)

$$Z=2(1+e^{-2K})^N \tag{21.2.62}$$

One is generally interested in the free energy per site in the *thermodynamic limit* $N\to\infty$:

$$f(K)=\lim_{N\to\infty}\frac{1}{N}\ln Z \tag{21.2.63}$$

(This definition of f differs by a factor $-\beta$ from the more traditional one. I use the present one to reduce the clutter.) We see that

$$f(K)=\ln(1+e^{-2K}) \tag{21.2.64}$$

where we have dropped $\ln 2/N$ in the thermodynamic limit. Had we chosen to fix s_0 at one of the two values, the factor 2 would have been missing in Eq. (21.2.62) but there would have been no difference in Eq. (21.2.64) for the free energy per site. Boundary conditions are unimportant in the thermodynamic limit in this sense.

Consider next the *correlation function* (which measure the likelihood that spins s_i and s_j are parallel):

$$\langle s_j s_i\rangle=\frac{\sum_{s_k} s_j s_i \exp\left[\sum_k K(s_k s_{k+1}-1)\right]}{Z} \tag{21.2.65}$$

for $j>i$. Using the fact that $s_i^2\equiv 1$, we can write

$$s_j s_i = s_i s_{i+1} s_{i+1} s_{i+2}\cdots s_{j-1} s_j = t_i t_{i+1}\cdots t_{j-1} \tag{21.2.66}$$

Thus

$$\langle s_i s_j\rangle=\langle t_i\rangle\langle t_{i+1}\rangle\cdots\langle t_{j-1}\rangle \tag{21.2.67}$$

where the answer factorizes over i since the Boltzmann weight factorizes over i when written in terms of t_i. The average for any one t is easy

$$\langle t\rangle=\frac{1\,e^{0\cdot K}-1\,e^{-2K}}{e^{0\cdot K}+e^{-2K}}=\tanh K \tag{21.2.68}$$

so that finally

$$\langle s_j s_i \rangle = (\tanh K)^{j-i} = \exp[(j-i) \ln \tanh K] \tag{21.2.69}$$

Note that the result depends on just the difference in coordinates. *This is not a generic result but a peculiarity of this model.* The reason is that the problem of $N+1$ points (for any finite N) is not translationally invariant. Correlations between two spins could, and generally do, depend on where the two points are in relation to the ends. On the other hand, in all models we expect that as $N \to \infty$, we will see translational invariance far from the ends and deep in the interior. To have translational invariance in a finite system, we must use periodic boundary conditions: now the world has the shape of a ring and every point is equivalent to every other. Correlation functions will now depend only on the difference between the two coordinates but they will not decay monotonically with separation! This is because as one point starts moving away from the other, it eventually starts approaching the first point from the other side! Thus the correlation function will be a sum of two terms, one of which grows as $j-i$ increases to values of order N. However, if we promise never to consider separations comparable to N, this complication can be ignored [see Exercise (21.2.9)]. (Our calculation of correlations in terms of t_i must be amended in the face of periodic boundary conditions to ensure that the sum over t_i is restricted to configurations for which the product of t_i's over the ring equals unity.)

The *correlation length* ξ is defined by the formula

$$\lim_{j-i \to \infty} \langle s_i s_j \rangle \to e^{-(j-i)/\xi} \tag{21.2.70}$$

Thus in our problem

$$\xi^{-1} = -\ln \tanh K \tag{21.2.71}$$

(We have assumed $j > i$ in our analysis. In general $j-i$ is to be replaced by $|j-i|$ in these definitions. Also the model in question shows the exponential behavior for all separations and not just in the limit $|j-i| \to \infty$. This too is peculiar to our model and stems from the fact that the model is in one spatial dimension and the Ising spin can take only two values.)

We will now rederive these results in the quantum version. If Z stands for a path integral, the Ising variables must be the intermediate state labels that occur in the resolution of the identity for a quantum problem. Clearly the quantum problem is that of a spin-1/2.

To proceed, let us take another look at

$$Z = \sum_{s_i} \prod_i e^{K(s_i s_{i+1} - 1)} \tag{21.2.72}$$

Each exponential factor is labeled by two discrete indices which can take two values each. Furthermore, the second label for any factor is the first label for the next.

Finally, these labels are being summed over. It is clear that we are seeing here a matrix product. (We are simply undoing the resolution of the identity.) So we write

$$Z=\sum_{s_i} T_{s_N s_{N-1}} \cdots T_{s_2 s_1} T_{s_1 s_0} \tag{21.2.73}$$

where we have introduced a 2×2 matrix T whose rows and columns are labeled by a pair of spins and whose element $T_{ss'}$ equals the Boltzmann weight associated with a pair of neighboring spins in the state s, s'. Thus

$$T_{++}=T_{--}=1,\ T_{+-}=T_{-+}=\exp(-2K)$$

Thus this matrix, called the *Transfer Matrix*, is given by

$$T=I+e^{-2K}\sigma_1 \tag{21.2.74}$$

and

$$Z=\langle s_N|T^N|s_0\rangle \tag{21.2.75}$$

for the case of fixed boundary conditions (which we will focus on) where the first spin is fixed at s_0 and the last at s_N. If we sum over the end spins (free boundary conditions)

$$Z=\sum_{s_0 s_N} \langle s_N|T^N|s_0\rangle \tag{21.2.76}$$

If we consider periodic boundary conditions where $s_0=s_N$ and one sums over these,

$$Z=\mathrm{Tr}\ T^N \tag{21.2.77}$$

We will now show the insensitivity of the free energy per site to boundary conditions in the thermodynamic limit. Suppose we used fixed boundary conditions. Then if we write

$$T=\lambda_0|0\rangle\langle 0|+\lambda_1|1\rangle\langle 1| \tag{21.2.78}$$

where $|i\rangle$, λ_i $[i=0, 1]$ are the eigenvectors (assumed orthonormal) and eigenvalues of T, then

$$T^N=\lambda_0^N|0\rangle\langle 0|+\lambda_1^N|1\rangle\langle 1| \tag{21.2.79}$$

Assuming λ_0 is the bigger of the two eigenvalues,

$$T^N \lim_{N\to\infty} \simeq \lambda_0^N|0\rangle\langle 0|\left(1+\mathcal{O}\left(\frac{\lambda_1}{\lambda_0}\right)^N\right) \tag{21.2.80}$$

and

$$Z \simeq \langle s_N|0\rangle\langle 0|s_0\rangle\lambda_0^N\left(1+\mathcal{O}\left(\frac{\lambda_1}{\lambda_0}\right)^N\right) \tag{21.2.81}$$

and the free energy per site in the infinite volume limit,

$$f=\lambda_0+\frac{1}{N}\ln(\langle s_N|0\rangle\langle 0|s_0\rangle)+\cdots \tag{21.2.82}$$

is clearly independent of the boundary spins as long as $\langle 0|s_0\rangle$ and $\langle s_N|0\rangle$ do not vanish.

Exercise 21.2.5. Check this claim for periodic boundary conditions starting with Eq. (21.2.75).

Let us rewrite T as follows. Consider the identity

$$e^{K^*\sigma_1}=\cosh K^*+\sinh K^*\sigma_1 \tag{21.2.83}$$

$$=\cosh K^*(I+\tanh K^*\sigma_1) \tag{21.2.84}$$

where K^* is presently unrelated to K; in particular, it is not the conjugate! If we choose

$$\tanh K^*=e^{-2K} \tag{21.2.85}$$

we see from Eq. (21.2.74) that up to a prefactor $\cosh K^*$,

$$T=e^{K^*\sigma_1} \tag{21.2.86}$$

We will temporarily drop this prefactor but remember to subtract $\ln\cosh K^*$ from the free energy per site. It does not, however, affect the correlation function which will be seen to depend only on the ratios of eigenvalues of T. Note that K^*, called the *dual of K*, is large when K is small and vice versa.

For later reference, let us note that in the present case, the eigenvalues of T are $e^{\pm K^*}$ and the corresponding eigenvectors are

$$|0\rangle, |1\rangle=\frac{1}{\sqrt{2}}\begin{bmatrix}1\\ \pm 1\end{bmatrix} \tag{21.2.87}$$

Suppose we write

$$T=e^{-H} \tag{21.2.88}$$

Then T can be interpreted as the time evolution operator for one time step in the imaginary time direction. The spatial site index i of the classical problem has become the discrete imaginary time index for the quantum problem. The free energy is simply related to E_0, the ground state energy of H:

$$H=-K^*\sigma_1 \tag{21.2.89}$$

$$f=-E_0=K^* \tag{21.2.90}$$

Exercise 21.2.6. Show that f above agrees with Eq. (21.2.64) upon remembering to subtract ln cosh K^* and using the definition of K^*.

Consider next the correlation function $\langle s_j s_i\rangle$ for $j>i$. I claim that if the boundary spins are fixed at s_0 and s_N,

$$\langle s_j s_i\rangle=\frac{\langle s_N|T^{N-j}\sigma_3 T^{j-i}\sigma_3 T^i|s_0\rangle}{\langle s_N|T^N|s_0\rangle} \tag{21.2.91}$$

To see the correctness of this, look at the numerator. Retrace our derivation by introducing a complete set of σ_3 eigenstates between every factor of T. Reading from right to left, we get just the Boltzmann weights till we get to site i. There the σ_3 acting on its eigenstate, gives s_i, the value of the spin there. Then we proceed as usual to j, repeat this and go to the Nth site. (The dependence of $\langle s_j s_i\rangle$ on the boundary conditions will be seen to disappear in the thermodynamic limit.) Let us rewrite Eq. (21.2.89) another way. Define *Heisenberg operators*

$$\sigma_3(n)=T^{-n}\sigma_3 T^n \tag{21.2.92}$$

In terms of these

$$\langle s_j s_i\rangle=\frac{\langle s_N|T^N\sigma_3(j)\sigma_3(i)|s_0\rangle}{\langle s_N|T^N|s_0\rangle} \tag{21.2.93}$$

Consider now the limit as $N\to\infty$, *i and j fixed at values far from the end points labeled* 0 *and N* so that $N-j$ and i are large, and we may approximate

$$T^\alpha\simeq|0\rangle\langle 0|\lambda_0^\alpha \qquad \alpha=N, N-j, i \tag{21.2.94}$$

In this limit, we have from Eq. (21.2.91)

$$\langle s_j s_i\rangle=\frac{\langle s_N|0\rangle\langle 0|\lambda_0^{N-j}\sigma_3 T^{j-i}\sigma_3\lambda_0^i|0\rangle\langle 0|s_0\rangle}{\langle s_N|0\rangle\lambda_0^N\langle 0|s_0\rangle}=\langle 0|\sigma_3(j)\sigma_3(i)|0\rangle \tag{21.2.95}$$

and the dependence on the boundary has dropped out. For the case $i>j$, we will get the operators in the other order. In general then,

$$\langle s_j s_i\rangle=\langle 0|\mathcal{T}(\sigma_3(j)\sigma_3(i))|0\rangle \tag{21.2.96}$$

where the *time-ordering symbol* $\mathscr{T}$ will order the operators with time increasing from the right to left:

$$\mathscr{T}(\sigma_3(j)\sigma_3(i))=\theta(j-i)(\sigma_3(j)\sigma_3(i))+\theta(i-j)(\sigma_3(i)\sigma_3(j)) \qquad (21.2.97)$$

We will pursue the evaluation of this correlation function using the eigenvectors of T. But first let us replace $\sigma_3(j)$ by the unit operator in the above derivation to obtain the mean magnetization as

$$\langle s_i\rangle=\langle 0|\sigma_3(0)|0\rangle \qquad (21.2.98)$$

In our example, $|0\rangle$ is the eigenket of σ_1 so that there is no mean magnetization. The only exception is at zero temperature or zero K^*: now the eigenvalues are equal and we can form linear combinations corresponding to either of the fully ordered (up or down) σ_3 eigenstates.

Let us compare symmetry breaking and its restoration in the Ising problem to what happened in the double well.

- In the limit $\hbar\to 0$, the particle in the double-well seeks the minimum of the euclidean action:

$$S_E=\int\left(\frac{m}{2}(dx/d\tau)^2+V(x(\tau))\right)d\tau \qquad (21.2.99)$$

which is given by $(dx/d\tau)=0$, $x=\pm a$, the minima of the double-well potential. There is degeneracy and symmetry breaking in the ground state. A particle that starts out in one well will not ever go to the other in the course of time. Even though Π commutes with H, we do not form parity eigenstates, instead we form eigenstates of position (or more accurately, well index, left or right). In the Ising problem, in the limit of zero temperature, the partition function is dominated by the state of minimum energy, with all spins up or all spins down on all sites (which can be viewed as discrete points in imaginary time of the spin-1/2 problem). In the operator language, T and H commute with σ_1 in general and eigenstates of H are chosen to be eigenstates of σ_1 as well. But at zero K^*, the two eigenstates become degenerate and we form combinations which are chosen to be eigenstates of σ_3. This is because a state starting out up/down with respect to σ_3 will stay that way forever. (In classical statistical mechanics terms, if the spin at one of the chain is up/down, all will be up/down at zero temperature.)
- For nonzero $\hbar$, there is tunneling between the wells, degeneracy is lifted and symmetry is restored in the ground state. This is thanks to an instanton configuration that has finite action and connects the two classical ground states. In the Ising problem, for nonzero K^*, i.e., nonzero temperature, there exist instantonlike configurations in which the spin starts out up at one end of the chain (i.e., the distant past in the imaginary time interpretation) and at some point flips down and vice versa. This has finite energy (only one pair of nearest-neighbor spins is antiparallel and the additional energy cost is $2K$). The eigenstates of the transfer matrix (or the spin Hamiltonian) are now the symmetric and antisymmetric

combinations of the up and down states, i.e., eigenstates of σ_1. The ground state is unique and symmetric.

Exercise 21.2.7. Consider the Hamiltonian of the spin-1/2 problem that arises in the transfer matrix treatment of the Ising chain

$$H=-K^*\sigma_1 \tag{21.2.100}$$

The off-diagonal matrix element (after pulling out the sign), i.e., K^*, must represent the tunneling amplitude (for going from up to down ground state) per unit time in the low-temperature limit (which you recall is like the $\hbar\to 0$ limit). The preceding discussion tells us it is just e^{-2K} where $2K$ is the energy cost of the interface of the up and down ground states. Verify that these two results agree for low temperatures by going back to the definition of K^*.

Let us return to Eq. (21.2.96). Even though it appears that everything depends on just the ground state, a knowledge of all states is required even in the infinite volume limit to evaluate the correlation. Going to Eq. (21.2.96) for the case $j>i$, let us insert the complete set of (two) eigenvectors of T between the Pauli matrices. When we insert $|0\rangle\langle 0|$ we get $\langle s\rangle^2$, the square of the magnetization which happens to vanish here. Moving it to the left-hand side, we get the *connected correlation function*

$$\langle s_j s_i\rangle_c \equiv \langle s_j s_i\rangle - \langle s\rangle^2 = \langle 0|T^{-j}\sigma_3(0)T^{j-i}|1\rangle\langle 1|\sigma_3(0)T^i|0\rangle \tag{21.2.101}$$

$$=\left(\frac{\lambda_1}{\lambda_0}\right)^{j-i}|\langle 0|\sigma_3|1\rangle|^2 \tag{21.2.102}$$

$$=e^{-2K^*(j-i)}|\langle 0|\sigma_3|1\rangle|^2 \tag{21.2.103}$$

Let us note that

- The correlation depends only on ratios of the eigenvalues of T and falls exponentially with distance with a coefficient $2K^*$. Now $2K^*$ is just the gap to the first excited state of the Hamiltonian H defined by $T=e^{-H}$ which in our example is $-K^*\sigma_1$. The result

$$\xi^{-1}=E_1-E_0\equiv m \tag{21.2.104}$$

  is also very general. The reason one uses the symbol m for the gap (called the mass gap) is that in a field theory the lowest energy state above the vacuum is a single particle at rest and this has energy m (in units where $c=1$).
- The connected correlation function is determined by the matrix element of the operator in question (σ_3) between the ground state and the next excited state. This is also a general feature. If this matrix element vanishes, we must go up in the levels till we find a state that is connected to the ground state by the action of the operator. (In this problem we know $|\langle 0|\sigma_3|1\rangle|^2=1$ since σ_3 is the spin-flip operator for the eigenstates of σ_1.)

This simple example has revealed most of the general features of the problem. The only difference is that for a bigger transfer matrix, the sum over states will have more than two terms. Thus the correlation function will be a sum of decaying exponentials and a unique correlation length will emerge only asymptotically when the smallest mass gap dominates. Also in the more complex problems (in higher dimensions) there may not be any finite action instantons connecting the multiple classical minima and there can be many ground states of H with broken symmetry. Assuming this happens, as it does in the two-dimensional Ising model (below some temperature T_c), you can ask: how does the ground state choose between spin up and spin down since there is no bias in the Boltzmann weight to make the choice? The answer is that indeed, if we do not set any bias, the system will always pick a mean magnetization of zero. How then do we know that the system is ready to magnetize? We use a principle called *clustering*. It states that as i and j separate, $\langle s_j s_i\rangle \to \langle s_j\rangle\langle s_i\rangle$. The idea is that if i lies in our galaxy and j lies in another they become statistically independent. Consider now the two-dimensional Ising model below T_c. In zero field we will find that $\langle s_j s_i\rangle$ does not approach $\langle s_i\rangle\langle s_j\rangle$ (which is zero since we gave the system no reason to choose one value of magnetization over its opposite) but that instead $\langle s_i s_j\rangle$ approaches the square of the magnetization the system will have if you would only give it the slightest reason for choosing one sign over the other. At this point, having seen the breakdown of clustering for the spin variable, you are to modify the partition function to restore clustering in two equivalent ways. One is to limit the sum over states to those with a net positive (or negative) magnetization. Then $\langle s\rangle \neq 0$ any more and you will find that $\langle s_i\rangle\langle s_j\rangle \to \langle s\rangle^2$. The other option is to apply a small field, calculate the magnetization, and let the field go to zero. (This too essentially kills half the states in the sum. Both recipes reflect the fact that a magnet below its T_c will not be able to dynamically evolve from pointing up to pointing down. Recall the particle trapped on one side of the infinite barrier between the two wells. Thus summing over things the system cannot do is a mistake.) Now, the magnetization is the derivative of the free energy with respect to the applied field h. It is easy to show that it is an even function of h. [See Exercise 21.2.8).] If the system does not want to magnetize, you will find that $f \sim h^2$, so that $df/dh \to 0$ as $h \to 0$. On the other hand if it wants to magnetize you will find $f \sim |h|$ and $df/dh \sim \text{sign}\, h$.

Exercise 21.2.8. Consider the Ising model in a magnetic field by adding a term $h\sum s_i$ to the exponent in Eq. (21.2.59). Show that $Z(h) = Z(-h)$. Show that the transfer matrix $T = e^{K^*\sigma_1}\, e^{h\sigma_3} \equiv T_K T_h$ reproduces the Boltzmann weight. Note that T is not Hermitian. By splitting the coupling to h into two factors, show that $T_h^{1/2} T_K T_h^{1/2}$ is just as good and also Hermitian. Find its eigenvalues and eigenvectors and show that there is degeneracy only for $h = K^* = 0$. Find the magnetization as a function of h by evaluating $\langle s\rangle = \langle 0|\sigma_3|0\rangle$. Starting with the partition function, show that

$$\langle s\rangle = \frac{1}{N}\frac{\partial \ln Z}{\partial h} = \frac{\partial f}{\partial h}$$

Evaluate f from the largest eigenvalue of T and regain the answer for $\langle s\rangle$ found from $\langle s\rangle = \langle 0|\sigma_3|0\rangle$.

Exercise 21.2.9. Consider the correlation function for the problem with periodic boundary conditions and write it as a ratio of two traces. Saturate the denominator with the largest eigenket, but keep both eigenvectors in the numerator and show that the answer is invariant under $j-i \leftrightarrow N-(j-i)$. Using the fact that σ_3 exchanges $|0\rangle$ and $|1\rangle$ should speed things up. Provide the interpretation. Argue that as long as $j-i$ is much smaller than N, only one term is needed.

Exercise 21.2.10. Recall the remarkable fact that the correlation function $\langle s_j s_i \rangle$ in the Ising model was translationally invariant in the finite open chain with one end fixed at s_0. Derive this result using the transfer matrix formalism as follows.

Explicitly evaluate $\sigma_3(j)$ by evaluating $T^{-j}\sigma_3 T^j$ in terms of σ_3 and σ_1. Show that $\sigma_3(j)\sigma_3(i)$ is a function only of $j-i$ by using some identities for hyperbolic functions. Keep going till you explicitly have the correlation function. It might help to use $\sum_{s_N} |s_N\rangle = (I+\sigma_1)|s_0\rangle$.

21.3. Spin and Fermion Path Integrals

Now we turn to path integrals for two systems with no classical limit: a spin S system and a fermionic oscillator, to be described later. The fermion problem will be somewhat abstract at this stage, but it is in here because you are likely to see it in many different branches of physics.

Spin Coherent States and Path Integral

Consider a spin S degree of freedom. The Hilbert space is $2S+1$ dimensional. Choosing S_z eigenstates as our basis we can write the propagator $U(t)$ as a sum over configurations by using the resolution

$$I = \sum_{-S}^{S} |S_z\rangle\langle S_z| \tag{21.3.1}$$

The intermediate states will have discrete labels (as in the Ising model).

We consider here an alternate scheme in which an overcomplete basis is used. Consider the *spin coherent state*

$$|\Omega\rangle \equiv |\theta, \phi\rangle = U(R(\Omega))|SS\rangle \tag{21.3.2}$$

where $|\Omega\rangle$ denotes the state obtained by rotating the normalized, fully polarized state, $|SS\rangle$ by an angle θ around the x-axis and then by ϕ around the z-axis using the unitary rotation operator $U(R(\Omega))$.

Given that

$$\langle SS|\mathbf{S}|SS\rangle = \mathbf{k}S \tag{21.3.3}$$

it is clear (say by considering $U^\dagger \mathbf{S} U$) that

$$\langle \Omega | \mathbf{S} | \Omega \rangle = S(\mathbf{i} \sin\theta \cos\phi + \mathbf{j} \sin\theta \sin\phi + \mathbf{k} \cos\theta) \tag{21.3.4}$$

Note that our spin operators are not defined with an $\hbar$. Thus for spin-1, the eigenvalues of S_z are 0, ± 1.

Exercise 21.3.1. Show the above result by invoking Eq. (12.4.13).

The *coherent state* is one in which the spin operator has a nice expectation value: equal to a classical spin of length S pointing along the direction of Ω. *It is not an eigenvector of the spin operator* (not expected anyway since the three components of spin do not commute) and higher powers of the spin operators do not have expectation values equal to the corresponding powers of the classical spin. For example, $\langle \Omega | S_x^2 | \Omega \rangle \neq S^2 \sin^2\theta \cos^2\phi$. However, the difference between this wrong answer and the right one is of order S. Generally the nth power of the spin operator will have an expectation value equal to the nth power of the expectation value of that operator plus corrections that are of order S^{n-1}. If S is large, they may be ignored. This is so when one usually uses the present formalism.

Let us now examine the equation

$$\langle \Omega_2 | \Omega_1 \rangle = \left(\cos\frac{\theta_2}{2} \cos\frac{\theta_1}{2} + e^{i(\phi_1 - \phi_2)} \sin\frac{\theta_2}{2} \sin\frac{\theta_1}{2} \right)^{2S} \tag{21.3.5}$$

The result is obviously true for $S = 1/2$, given that the up spinor along the direction $\theta\phi$ is

$$|\Omega\rangle \equiv |\theta\phi\rangle = \cos\frac{\theta}{2} |1/2, 1/2\rangle + e^{i\phi} \sin\frac{\theta}{2} |1/2, -1/2\rangle \tag{21.3.6}$$

As for higher spin, imagine $2S$ spin-1/2 particles joining to form a spin S state. There is only one direct product state with $S_z = S$: where all the spin-1/2's are pointing up. Thus the normalized fully polarized state is

$$|SS\rangle = |1/2, 1/2\rangle \otimes |1/2, 1/2\rangle \otimes \cdots |1/2, 1/2\rangle \tag{21.3.7}$$

If we now rotate this state, it becomes a tensor product of rotated states and when we form the inner product in the left-hand side of Eq. (21.3.5), we obtain the right-hand side.

The resolution of the identity in terms of these states is

$$I = \frac{2S+1}{4\pi} \int d\Omega |\Omega\rangle\langle\Omega| \tag{21.3.8}$$

where $d\Omega = d\cos\theta\, d\phi$. The proof can be found in the references. You are urged to do the following exercise that deals with $S = 1/2$.

Exercise 21.3.2. Prove the completeness relation for $S=1/2$ by carrying out the integral over Ω using Eq. (21.3.6).

When we work out the path integral we will get a product of factors like the following:

$$\cdots\langle\Omega(t+\varepsilon)|I-\frac{i\varepsilon}{\hbar}H(\mathbf{S})|\Omega(t)\rangle\cdots \tag{21.3.9}$$

We work to order ε. Since H already has a factor of ε in front of it, we set

$$\langle\Omega(t+\varepsilon)|-\frac{i\varepsilon}{\hbar}H(\mathbf{S})|\Omega(t)\rangle\simeq-\frac{i\varepsilon}{\hbar}\langle\Omega(t)|H(\mathbf{S})|\Omega(t)\rangle\equiv-i\varepsilon\mathscr{H}(\Omega) \tag{21.3.10}$$

If the Hamiltonian is linear in S, we simply replace the quantum spin operator by the classical vector pointing along θ, ϕ and if not, we can replace the operator by the suitable expectation value in the state $|\Omega(t)\rangle$. This is what we called $\hbar\mathscr{H}(\Omega)$ in the preceding equation.

Next we turn to the product

$$\langle\Omega(t+\varepsilon)|\Omega(t)\rangle\simeq 1-i\varepsilon S(1-\cos\theta)\dot{\phi}\simeq e^{iS(\cos\theta-1)\dot{\phi}\varepsilon} \tag{21.3.11}$$

where we have expanded Eq. (21.3.5) to first order in $\Delta\theta$ and $\Delta\phi$. This gives us the following representation of the propagator in the continuum limit:

$$\langle\Omega_f|U(t)|\Omega_i\rangle=\int\mathscr{D}\Omega\exp\left[i\int_{t_1}^{t_2}[S\cos\theta\dot{\phi}-\mathscr{H}(\Omega)]\,dt\right] \tag{21.3.12}$$

where a total derivative in ϕ has been dropped and $\int\mathscr{D}\Omega$ is the measure with all factors of π in it.

Even by the standards of continuum functional integrals we have hit a new low, when we replaced differences by derivatives as if the paths are smooth. In the configuration path integral, we saw that between one time and the next the fluctuation in x was of the order $\varepsilon^{1/2}$ which is why we had to expand $\langle n(R')|n(R)\rangle$ to order $(R'-R)^2$ in the Berry calculation of the effective interaction. The factor that provided any kind of damping on the variation in the coordinate was the kinetic energy term $\exp[im(x'-x)^2/2\hbar\varepsilon]$. *In the present problem there is no such term.* There is no reason why the difference in Ω from one time to another should be treated as a small quantity. Thus although the discretized functional integral is never wrong (since all we use is the resolution of the identity) any further assumptions about the smallness of the change in Ω from one time to the next are suspect. There is one exception. Suppose $S\to\infty$. Then we see from Eq. (21.3.5) that the overlap is unity if the two states are equal and falls rapidly if they are different. (It is easier to consider the case $\phi_2=\phi_1$.) This is usually the limit ($S\to\infty$) in which one uses this formalism.

We now consider two simple applications. First let

$$H=\hbar S_z \tag{21.3.13}$$

We know the allowed eigenvalues are $\hbar(-S, -S+1, \ldots, S)$. Let us derive this from the continuum path integral.

Given $\langle\Omega|H|\Omega\rangle = \hbar S\cos\theta$, it follows that $\mathscr{H} = S\cos\theta$, and that the functional integral is

$$\left[\int \mathscr{D}\cos\theta\,\mathscr{D}\phi\right]\exp\left[iS\int(\cos\theta\,\dot{\phi} - \cos\theta)\,dt\right] \tag{21.3.14}$$

We note that

- This is a phase space path integral with $\cos\theta$ *as the momentum conjugate to* ϕ!
- Phase space is compact here (the unit sphere), as compared to the problem of a particle moving on a sphere for which configuration space is compact but all momenta are allowed and phase space is infinite in extent.
- The spin S plays the role of $1/\hbar$.
- The Hamiltonian for the dynamics is $\cos\theta$ since we pulled out the S to the front. In particular, this means that $\cos\theta$ is a constant of motion, i.e., the orbits will be along fixed latitude.

Recall the WKB quantization rule

$$\oint p\,dq = 2\pi n\hbar \tag{21.3.15}$$

for a problem with no turning points. In our problem, $p = \cos\theta$ is just the conserved energy E. Of all the classical orbits along constant latitude lines, the ones chosen by WKB obey

$$\oint E\,d\phi = 2\pi n S^{-1} \tag{21.3.16}$$

since S^{-1} *plays the role of* $\hbar$. The allowed energies are

$$E_n = \frac{n}{S} \qquad [-S \le n \le S] \tag{21.3.17}$$

Note that there is exactly enough room in this compact phase space for $2S+1$ orbits and that the allowed values of E translate into the allowed values of H when we reinstate the factor of $\hbar S$ that was pulled out along the way.

So we got lucky with this problem. In general, if H is more complicated we cannot hope for much luck unless S is large. Now you may ask why we bother with this formalism given that spins of real systems are very small. Here is at least one reason, based on a problem I am familiar with. In nuclear physics one introduces a *pseudospin* formalism in which a proton is called spin up and the neutron is called spin down. A big nucleus can have a large pseudospin, say 25. The Hamiltonian for the problem can be written in terms of the pseudospin operators and they can be

50×50 matrices. Finding the energy levels analytically is hopeless. But we can turn the large S in our favor by doing a WKB quantization using the appropriate H.

Coherent states are also very useful in the study of interacting quantum spins. For example, in the one-dimensional Heisenberg model, the Hamiltonian is a sum of dot products of nearest neighbor spin operators on a line of points. Since each spin operator appears linearly, the Hamiltonian in the action is just the quantum one with **S** replaced by a classical vector of length S. Even though the spin is never very large in these problems, one studies the large S limit to get a feeling for the subject and to make controlled approximations in $1/S$.

Fermion Oscillator and Coherent States

Let us recall that in the case of the harmonic oscillator the fact that the energy levels were uniformly spaced

$$E = n\hbar\omega \tag{21.3.18}$$

(dropping zero point motion) allowed one to introduce the notion of quanta. Rather than saying the oscillator was in the nth state we could say there was one quantum level of energy $\hbar\omega$ and there were n quanta in it. This is how phonons, photons, etc., are viewed, and it is a very seminal idea.

That the level could be occupied by any number of quanta meant they were bosons. Indeed our perception of a classical electric or magnetic field is thanks to this feature.

Consider now a variant of the problem wherein the quanta are fermions. Thus the level can contain one or no quanta. There can be no macroscopic field associated this state, which is why the fermion problem is unfamiliar to us at first. We now develop the theory of a fermionic oscillator.

We start by writing down the Hamiltonian:

$$H_0 = \Psi^\dagger \Psi \Omega_0 \tag{21.3.19}$$

What distinguishes this problem from the bosonic one are the *anticommutation relations*:

$$\{\Psi^\dagger, \Psi\} = \Psi^\dagger \Psi + \Psi \Psi^\dagger = 1 \tag{21.3.20}$$

$$\{\Psi, \Psi\} = \{\Psi^\dagger, \Psi^\dagger\} = 0 \tag{21.3.21}$$

Note that the last equation tells us

$$\Psi^{\dagger 2} = \Psi^2 = 0 \tag{21.3.22}$$

This equation will be used all the time without explicit warning. We shall see that it represents the Pauli principle forbidding double occupancy. The *number operator*

$$N = \Psi^\dagger \Psi \tag{21.3.23}$$

obeys

$$N^2 = \Psi^\dagger\Psi\Psi^\dagger\Psi = \Psi^\dagger(1-\Psi^\dagger\Psi)\Psi = N \tag{21.3.24}$$

Thus the eigenvalues of N can only be 0 or 1. The corresponding normalized eigenstates obey

$$N|0\rangle = 0|0\rangle \tag{21.3.25}$$

$$N|1\rangle = 1|1\rangle \tag{21.3.26}$$

We will now prove that

$$\Psi^\dagger|0\rangle = |1\rangle \tag{21.3.27}$$

$$\Psi|1\rangle = |0\rangle \tag{21.3.28}$$

As for the first,

$$N\Psi^\dagger|0\rangle = \Psi^\dagger\Psi\Psi^\dagger|0\rangle = \Psi^\dagger(1-\Psi^\dagger\Psi)|0\rangle = \Psi^\dagger|0\rangle \tag{21.3.29}$$

which shows that $\Psi^\dagger|0\rangle$ has $N=1$. Its norm is unity:

$$|\Psi^\dagger|0\rangle|^2 = \langle 0|\Psi\Psi^\dagger|0\rangle = \langle 0|(1-\Psi^\dagger\Psi)|0\rangle = \langle 0|0\rangle = 1 \tag{21.3.30}$$

It can be similarly shown that $\Psi|1\rangle = |0\rangle$ after first verifying that $\Psi|1\rangle$ is not a null vector, that it has unit norm.

There are no other vectors in the Hilbert space: any attempts to produce more states are thwarted by $\Psi^2 = \Psi^{\dagger 2} = 0$. In other words, the Pauli principle rules out more vectors: the state is either empty or singly occupied.

Thus the Fermi oscillator Hamiltonian

$$H_0 = \Omega_0 \Psi^\dagger\Psi \tag{21.3.31}$$

has eigenvalues 0 and Ω_0.

We will work not with H_0 but with

$$H = H_0 - \mu N \tag{21.3.32}$$

where μ is called the *chemical potential*. For the oscillator, since

$$H = (\Omega_0 - \mu)\Psi^\dagger\Psi \tag{21.3.33}$$

this merely amounts to measuring all energies relative to the chemical potential. The role of the chemical potential will be apparent soon.

Let us now turn to thermodynamics. The central object here is the *grand partition function*, defined to be

$$Z = \text{Tr}\, e^{-\beta(H_0 - \mu N)} = e^{A(\mu,\beta)} \tag{21.3.34}$$

where the trace is over any complete set of eigenstates, β is the inverse temperature $1/kT$, and A is the free energy (different from the traditional one by a factor $-\beta$). The term *grand partition function* signifies that we are summing over states with a different number of particles or quanta. For this reason the free energy is denoted by A and not f. Just as β controls the amounts of energy the system takes from the reservoir, μ controls the number of particles. (This description is also possible for the bosonic oscillator. Instead of saying that we have just one oscillator which can be in any state labeled by n, and viewing the sum over states as the partition function of one oscillator, we can focus on the quanta and say that we are summing over states with variable number of quanta and interpret the usual sum over states as a grand partition function.)

If we use the N basis, this sum is trivial:

$$Z = 1 + e^{-\beta(\Omega_0 - \mu)} \tag{21.3.35}$$

All thermodynamic quantities can be deduced from this function. For example, it is clear from Eq. (21.3.34) that the mean occupation number is

$$\langle N \rangle = \frac{1}{\beta Z}\frac{\partial Z}{\partial \mu} = \frac{1}{\beta}\frac{\partial \ln Z}{\partial \mu} = \frac{1}{e^{\beta(\Omega_0 - \mu)} + 1} \tag{21.3.36}$$

Exercise 21.3.3. Prove the formula for $\langle N \rangle$ in general, starting with Eq. (21.3.34). (Write out the trace in a basis common to H and N, as a sum over energy levels at any one N, followed by a sum over N.)

At zero temperature we find from Eq. (21.3.36)

$$\langle N \rangle = \theta(\mu - \Omega_0) \tag{21.3.37}$$

i.e., the fermion is present if its energy is below chemical potential and absent if it is not. At finite temperatures the mean number varies more smoothly with μ.

We will now develop a path integral formula for the partition function.

We proceed in analogy with the bosonic oscillator by trying to find a *fermion coherent state* $|\psi\rangle$ which is an eigenstate of the destruction operator

$$\Psi|\psi\rangle = \psi|\psi\rangle \tag{21.3.38}$$

The eigenvalue ψ is a peculiar beast because if we act once more with Ψ we find

$$\psi^2 = 0 \tag{21.3.39}$$

since $\Psi^2 = 0$. Any ordinary variable whose square is zero is itself zero. But this ψ is no ordinary variable, it is a *Grassmann variable*. These variables *anticommute* with each other and with all fermionic creation and destruction operators. (They will therefore commute with a string containing an even number of such operators.) That is how they are defined. The variable ψ is rather abstract and defined by its anticommuting nature. There are no big or small Grassmann variables. You will get used to them and even learn to love them just as you did the complex numbers. (Surely when you first heard it, you did not readily embrace the notion that $4i$ was an honest solution to the question "What number times itself gives -16?" You probably felt that it may be the right answer, but it sure wasn't a number.)

We now write down the coherent state. It is

$$|\psi\rangle=|0\rangle-\psi|1\rangle \tag{21.3.40}$$

where ψ is a Grassmann number. This state obeys:

$$\Psi|\psi\rangle=\Psi|0\rangle-\Psi\psi|1\rangle \tag{21.3.41}$$

$$=0+\psi\Psi|1\rangle \tag{21.3.42}$$

$$=\psi|0\rangle \tag{21.3.43}$$

$$=\psi(|0\rangle-\psi|1\rangle) \tag{21.3.44}$$

$$=\psi|\psi\rangle \tag{21.3.45}$$

where we have appealed to the fact that ψ anticommutes with Ψ and that $\psi^2=0$. If we act on both sides of Eq. (21.3.45) with Ψ, the left vanishes due to $\Psi^2=0$ and the right due to $\psi^2=0$.

It may be similarly verified that

$$\langle\bar{\psi}|\Psi^\dagger=\langle\bar{\psi}|\bar{\psi} \tag{21.3.46}$$

where

$$\langle\bar{\psi}|=\langle 0|-\langle 1|\bar{\psi}=\langle 0|+\bar{\psi}\langle 1| \tag{21.3.47}$$

Please note two points. First, the coherent state vectors are not the usual vectors from a complex vector space since they are linear combinations with Grassmann coefficients. Second, $\bar{\psi}$ is not in any sense the complex conjugate of ψ and $\langle\bar{\psi}|$ is not the adjoint of $|\psi\rangle$. You should therefore be prepared to see a change of Grassmann variables in which ψ and $\bar{\psi}$ undergo totally unrelated transformations.

The inner product of two coherent states is

$$\langle\bar{\psi}|\psi\rangle=(\langle 0|-\langle 1|\bar{\psi})(|0\rangle-\psi|1\rangle) \tag{21.3.48}$$

$$=\langle 0|0\rangle+\langle 1|\bar{\psi}\psi|1\rangle \tag{21.3.49}$$

$$=1+\bar{\psi}\psi \tag{21.3.50}$$

$$=e^{\bar{\psi}\psi} \tag{21.3.51}$$

Any function of a Grassmann variable can be expanded as follows:

$$F(\psi)=F_0+F_1\psi \tag{21.3.52}$$

there being no higher powers possible.

We will now define integrals over Grassmann numbers. (Don't throw up your hands: it will be over in no time.) These have no geometric significance (as areas or volumes) and are formally defined. We just have to know how to integrate 1 and ψ since that takes care of all possible functions. Here is the list of integrals:

$$\int \psi \, d\psi = 1 \tag{21.3.53}$$

$$\int 1 \, d\psi = 0 \tag{21.3.54}$$

That's it! As you can see, a table of Grassmann integrals is not going to be a best-seller. (For those of you have trouble remembering all these integrals, here is a useful mnemonic: the integral of any function is the same as the derivative! Verify this.) There are no limits on these integrals. Integration is assumed to be a linear operation. The differential $d\psi$ is also a Grassmann number. Thus $\int d\psi\psi = -1$. The integrals for $\bar{\psi}$ or any other Grassmann variable are identical. These integrals are simply assigned these values. They are very important since we see for the first-time ordinary numbers on the right-hand side. Anything numerical we calculate in this theory goes back to these integrals.

A result we will use often is this:

$$\int \bar{\psi}\psi \, d\psi \, d\bar{\psi} = 1 \tag{21.3.55}$$

Note that if the differentials or variables come in any other order there can be a change of sign. For example, we will also invoke the result

$$\int \bar{\psi}\psi \, d\bar{\psi} \, d\psi = -1 \tag{21.3.56}$$

Let us now consider some Gaussian integrals. You are urged to show the following:

$$\int e^{-a\bar{\psi}\psi} \, d\bar{\psi} \, d\psi = a \tag{21.3.57}$$

$$\int e^{-\bar{\psi}M\psi} \, [d\bar{\psi} \, d\psi] = \det M \tag{21.3.58}$$

where in the second formula M is a 2-by-2 matrix, ψ is a column vector with entries ψ_1 and ψ_2, $\bar{\psi}$ a column vector with entries $\bar{\psi}_1$ and $\bar{\psi}_2$ and $[d\bar{\psi}\, d\psi]= d\bar{\psi}_1\, d\psi_1\, d\bar{\psi}_2\, d\psi_2$. This result is true for matrices of any size. To prove these simply expand the exponential and do the integrals.

Exercise 21.3.4. Prove the above two equations.

Consider next the "averages" over the Gaussian measure:

$$\langle\bar{\psi}\psi\rangle=\frac{\int \bar{\psi}\psi\, e^{a\bar{\psi}\psi}\, d\bar{\psi}\, d\psi}{\int e^{a\bar{\psi}\psi}\, d\bar{\psi}\, d\psi}=\frac{1}{a}=-\langle\psi\bar{\psi}\rangle \tag{21.3.59}$$

The proof is straightforward and left as an exercise.

Exercise 21.3.5. Provide the missing details in the evaluation of the above integral.

Exercise 21.3.6. Jacobians for Grassmann change of variables are the inverses of what you expect. Start with $\int a\phi\, d\phi=a$. Define $\chi=a\phi$, write $d\phi=J(\phi/\chi)\, d\chi$ and show that $J(\phi/\chi)=a$ and not $1/a$. (Treat the Jacobian as a constant that can be pulled out of the integral.) Evaluate Eq. (21.3.57) by introducing $\chi=a\psi$. Remember there is no need to change $\bar{\psi}$.

Consider now two sets of Grassmann variables (labeled 1 and 2). It is readily shown that

$$\langle\bar{\psi}_i\psi_j\rangle=\frac{\int \bar{\psi}_i\psi_j\, e^{a_1\bar{\psi}_1\psi_1+a_2\bar{\psi}_2\psi_2}\, d\bar{\psi}_1\, d\psi_1\, d\bar{\psi}_2\, d\psi_2}{\int e^{a_1\bar{\psi}_1\psi_1+a_2\bar{\psi}_2\psi_2}\, d\bar{\psi}_1\, d\psi_1\, d\bar{\psi}_2\, d\psi_2} \tag{21.3.60}$$

$$=\frac{\delta_{ij}}{a_i}\equiv\langle\bar{i}j\rangle \tag{21.3.61}$$

Exercise 21.3.7. Prove the above result.

Exercise 21.3.8. Show that

$$\langle\bar{\psi}_i\bar{\psi}_j\psi_k\psi_l\rangle=\frac{\int \bar{\psi}_i\bar{\psi}_j\psi_k\psi_l\, e^{a_1\bar{\psi}_1\psi_1+a_2\bar{\psi}_2\psi_2}\, d\bar{\psi}_1\, d\psi_1\, d\bar{\psi}_2\, d\psi_2}{\int e^{a_1\bar{\psi}_1\psi_1+a_2\bar{\psi}_2\psi_2}\, d\bar{\psi}_1\, d\psi_1\, d\bar{\psi}_2\, d\psi_2} \tag{21.3.62}$$

$$=\frac{\delta_{il}}{a_i}\frac{\delta_{jk}}{a_j}-\frac{\delta_{ik}}{a_i}\frac{\delta_{jl}}{a_j} \tag{21.3.63}$$

$$\equiv\langle\bar{i}l\rangle\langle\bar{j}k\rangle-\langle\bar{i}k\rangle\langle\bar{j}l\rangle \tag{21.3.64}$$

This is called *Wick's theorem* and is very useful in field theory and many-body theory.

We need two more results before we can write down the path integral. The first is the resolution of the identity:

$$I = \int |\psi\rangle\langle\bar{\psi}|\, e^{-\bar{\psi}\psi}\, d\bar{\psi}\, d\psi \tag{21.3.65}$$

In the following proof of this result we will use all the previously described properties and drop terms that are not going to survive integration. (Recall that only $\bar{\psi}\psi = -\psi\bar{\psi}$ has a nonzero integral.)

$$\begin{aligned}
\int |\psi\rangle\langle\bar{\psi}|\, e^{-\bar{\psi}\psi}\, d\bar{\psi}\, d\psi &= \int |\psi\rangle\langle\bar{\psi}|(1-\bar{\psi}\psi)\, d\bar{\psi}\, d\psi \\
&= \int (|0\rangle - \psi|1\rangle)(\langle 0| - \langle 1|\bar{\psi})(1-\bar{\psi}\psi)\, d\bar{\psi}\, d\psi \\
&= \int (|0\rangle\langle 0| + \psi|1\rangle\langle 1|\bar{\psi})(1-\bar{\psi}\psi)\, d\bar{\psi}\, d\psi \\
&= |0\rangle\langle 0| \int (-\bar{\psi}\psi)\, d\bar{\psi}\, d\psi + |1\rangle\langle 1| \int \psi\bar{\psi}\, d\bar{\psi}\, d\psi \\
&= I
\end{aligned} \tag{21.3.66}$$

The final result we need is that for any bosonic operator (an operator made of an even number of Fermi operators)

$$\operatorname{Tr} \Omega = \int \langle -\bar{\psi}|\Omega|\psi\rangle\, e^{-\bar{\psi}\psi}\, d\bar{\psi}\, d\psi \tag{21.3.67}$$

The proof is very much like the one just given and is left as an exercise.

Exercise 21.3.9. Prove the above formula for the trace.

The Fermionic Path Integral

Consider the partition function for a single oscillator:

$$Z = \operatorname{Tr} e^{-\beta(\Omega_0-\mu)\Psi^\dagger\Psi} \tag{21.3.68}$$

$$= \int \langle -\bar{\psi}| e^{-\beta(\Omega_0-\mu)\Psi^\dagger\Psi} |\psi\rangle\, e^{-\bar{\psi}\psi}\, d\bar{\psi}\, d\psi \tag{21.3.69}$$

You cannot simply replace $\Psi^\dagger$ and Ψ by $-\bar{\psi}$ and ψ, respectively, in the exponential. This is because when we expand out the exponential not all the Ψs will be acting to the right on their eigenstates and neither will all $\Psi^\dagger$s be acting to the left on their eigenstates. (Remember that we are now dealing with operators, not Grassmann numbers. The exponential will have an infinite number of terms in its expansion.) We need to convert the exponential to its *normal ordered form* in which all the creation operators stand to the left and all the destruction operators to the right. Luckily we can write down the answer by inspection:

$$e^{-\beta(\Omega_0-\mu)\Psi^\dagger\Psi}=1+(e^{-\beta(\Omega_0-\mu)}-1)\Psi^\dagger\Psi \tag{21.3.70}$$

whose correctness we can verify by considering the two possible values of $\Psi^\dagger\Psi$. (Alternatively, you can expand the exponential and use the fact that $N^k=N$ for any nonzero k.) Now we may write

$$Z=\int\langle-\bar\psi|1+(e^{-\beta(\Omega_0-\mu)}-1)\Psi^\dagger\Psi|\psi\rangle\, e^{-\bar\psi\psi}\,d\bar\psi\, d\psi \tag{21.3.71}$$

$$=\int\langle-\bar\psi|\psi\rangle(1+(e^{-\beta(\Omega_0-\mu)}-1)(-\bar\psi\psi))\, e^{-\bar\psi\psi}\,d\bar\psi\, d\psi \tag{21.3.72}$$

$$=\int(1-(e^{-\beta(\Omega_0-\mu)}-1)\bar\psi\psi)\, e^{-2\bar\psi\psi}\,d\bar\psi\, d\psi \tag{21.3.73}$$

$$=1+e^{-\beta(\Omega_0-\mu)} \tag{21.3.74}$$

as expected. While this is the right answer, this is not the path integral approach. It does, however, confirm the correctness of all the Grassmannian integration and minus signs. As for the path integral approach the procedure is the usual one. Consider

$$Z=\text{Tr}\, e^{-\beta H} \tag{21.3.75}$$

where H is a normal ordered operator $H(\Psi^\dagger, \Psi)$. We write the exponential as follows:

$$e^{-\beta H}=\lim_{N\to\infty}\left(\exp\left(-\frac{\beta}{N}H\right)\right)^N \tag{21.3.76}$$

$$=\underbrace{(1-\varepsilon H)\dots(1-\varepsilon H)}_{N\text{ times}}\qquad \varepsilon=\beta/N \tag{21.3.77}$$

take the trace as per Eq. (21.3.67) by integrating over $\bar\psi_0\psi_0$, and introduce the resolution of the identity $N-1$ times:

$$Z=\int\langle-\bar\psi_0|(1-\varepsilon H)|\psi_{N-1}\rangle\, e^{-\bar\psi_{N-1}\psi_{N-1}}\langle\bar\psi_{N-1}|(1-\varepsilon H)|\psi_{N-2}\rangle\, e^{-\bar\psi_{N-2}\psi_{N-2}}$$

$$\times\langle\bar\psi_{N-2}|\cdots|\psi_1\rangle\, e^{-\bar\psi_1\psi_1}\langle\bar\psi_1|(1-\varepsilon H)|\psi_0\rangle\, e^{-\bar\psi_0\psi_0}\prod_{i=0}^{N-1}d\bar\psi_i\, d\psi_i \tag{21.3.78}$$

Now we may legitimately make the replacement

$$\langle\bar{\psi}_{i+1}|1-\varepsilon H(\Psi^{\dagger},\Psi)|\psi_i\rangle=\langle\bar{\psi}_{i+1}|1-\varepsilon H(\bar{\psi}_{i+1},\psi_i)|\psi_i\rangle$$
$$=e^{\bar{\psi}_{i+1}\psi_i}e^{-\varepsilon H(\bar{\psi}_{i+1},\psi_i)} \tag{21.3.79}$$

where in the last step we are anticipating the limit of infinitesimal ε. Let us now *define* an additional pair of variables (not to be integrated over)

$$\bar{\psi}_N=-\bar{\psi}_0 \tag{21.3.80}$$

$$\psi_N=-\psi_0 \tag{21.3.81}$$

The first of these equations allows us to replace the leftmost bra in Eq. (21.3.78), $\langle-\bar{\psi}_0|$, by $\langle\bar{\psi}_N|$. The reason for introducing ψ_N will follow soon.

Putting together all the factors (including the overlap of coherent states) we end up with

$$Z=\int\prod_{i=0}^{N-1}e^{\bar{\psi}_{i+1}\psi_i}e^{-\varepsilon H(\bar{\psi}_{i+1},\psi_i)}e^{-\bar{\psi}_i\psi_i}\,d\bar{\psi}_i\,d\psi_i \tag{21.3.82}$$

$$=\int\prod_{i=0}^{N-1}\exp\left[\left[\left(\frac{(\bar{\psi}_{i+1}-\bar{\psi}_i)}{\varepsilon}\psi_i-H(\bar{\psi}_{i+1},\psi_i)\right)\right]\varepsilon\right]d\bar{\psi}_i\,d\psi_i \tag{21.3.83}$$

$$\simeq\int\exp\left(\int_0^{\beta}\bar{\psi}(\tau)\left(-\frac{\partial}{\partial\tau}-\Omega_0+\mu\right)\psi(\tau)\,d\tau\right)[\mathscr{D}\bar{\psi}\,\mathscr{D}\psi] \tag{21.3.84}$$

where the last step needs some explanation. With all the factors of ε in place we do seem to get the continuum expression in the last formula. However, the notion of replacing differences by derivatives is purely symbolic for Grassmann variables. There is no sense in which $\bar{\psi}_{i+1}-\bar{\psi}_i$ is small, in fact the objects have no numerical values. What this really means here is the following. In a while we will trade $\psi(\tau)$ for $\psi(\omega)$ related by Fourier transformation. At that stage we will replace $-\partial/\partial\tau$ by $i\omega$ while the exact answer is $e^{i\omega}-1$. If we do not make this replacement, the Grassmann integral, when evaluated in terms of ordinary numbers, will give exact results for anything one wants to calculate, say the free energy. With this approximation, only quantities insensitive to high frequencies will be given correctly. The free energy will come out wrong but the correlation functions will be correctly reproduced. (This is because the latter are given by derivatives of the free energy and these derivatives make the integrals sufficiently insensitive to high frequencies.) Notice also that we are replacing $H(\bar{\psi}_{i+1},\psi_i)=H(\bar{\psi}(\tau+\varepsilon),\psi(\tau))$ by $H(\bar{\psi}(\tau),\psi(\tau))$ in the same spirit.

Now turn to the Fourier expansions alluded to above. Let us write

$$\bar{\psi}(\tau)=\sum_n\frac{e^{i\omega_n\tau}}{\beta}\bar{\psi}(\omega) \tag{21.3.85}$$

$$\psi(\tau)=\sum_n\frac{e^{-i\omega_n\tau}}{\beta}\psi(\omega) \tag{21.3.86}$$

where the allowed frequencies, called *Matsubara frequencies*, are chosen to satisfy the antisymmetric boundary conditions in Eqs. (21.3.80–21.3.81). Thus

$$\omega_n = \frac{(2n+1)\pi}{\beta} \tag{21.3.87}$$

where n is an integer. Note that we have chosen the Fourier expansions as if ψ and $\bar{\psi}$ were complex conjugates, which they are not. This choice, however, makes the calculations easy.

For future reference note that if $\beta \to \infty$, it follows from Eq. (21.3.87) that when n increases by unity, ω_n changes by $d\omega = 2\pi/\beta$. Thus

$$\frac{1}{\beta}\sum_n \to \int \frac{d\omega}{2\pi} \tag{21.3.88}$$

The inverse transformations are

$$\psi(\omega) = \int_0^\beta \psi(\tau)\, e^{i\omega_n \tau}\, d\tau \tag{21.3.89}$$

$$\bar{\psi}(\omega) = \int_0^\beta \bar{\psi}(\tau)\, e^{-i\omega_n \tau}\, d\tau \tag{21.3.90}$$

where we use the orthogonality property

$$\int_0^\beta e^{i\omega_n \tau}\, e^{-i\omega_m \tau}\, d\tau = \frac{e^{i(\omega_n - \omega_m)\beta} - 1}{i(\omega_n - \omega_m)} = \beta \delta_{mn} \tag{21.3.91}$$

Performing the Fourier transforms in the action and changing the functional integration variables to $\psi(\omega)$ and $\bar{\psi}(\omega)$ (the Jacobian is unity) and going to the limit $\beta \to \infty$, which converts sums over discrete frequencies to integrals over a continuous ω, as per Eq. (21.3.88), we end up with

$$Z = \int \exp\left[\int_{-\infty}^{\infty} \frac{d\omega}{2\pi}\, \bar{\psi}(\omega)(i\omega - \Omega_0 + \mu)\psi(\omega)\right][\mathscr{D}\bar{\psi}(\omega)\, \mathscr{D}\psi(\omega)] \tag{21.3.92}$$

Although β has disappeared from the picture it will appear as $2\pi\delta(0)$, which we know stands for the total imaginary time β. (Recall Fermi's golden rule calculations.) An example will follow shortly.

Let us first note that the frequency space correlation function is related to the integral over just a single pair of variables [Eq. (21.3.59)] and is given by:

$$\langle\bar{\psi}(\omega_1)\psi(\omega_2)\rangle$$

$$=\frac{\int\bar{\psi}(\omega_1)\psi(\omega_2)\exp\left[\int_{-\infty}^{\infty}\frac{d\omega}{2\pi}\bar{\psi}(\omega)(i\omega-\Omega_0+\mu)\psi(\omega)\right][\mathscr{D}\bar{\psi}(\omega)\mathscr{D}\psi(\omega)]}{\int\exp\left[\int_{-\infty}^{\infty}\frac{d\omega}{2\pi}\bar{\psi}(\omega)(i\omega-\Omega_0+\mu)\psi(\omega)\right][\mathscr{D}\bar{\psi}(\omega)\mathscr{D}\psi(\omega)]}$$

$$=\frac{2\pi\delta(\omega_1-\omega_2)}{i\omega_1-\Omega_0+\mu}\tag{21.3.93}$$

In particular,

$$\langle\bar{\psi}(\omega)\psi(\omega)\rangle=\frac{2\pi\delta(0)}{i\omega-\Omega_0+\mu}=\frac{\beta}{i\omega-\Omega_0+\mu}\tag{21.3.94}$$

Exercise 21.3.10. Try to demonstrate the above two equations. Note first of all that unless $\omega_1=\omega_2$, we get zero since only a $\bar{\psi}\psi$ pair has a chance of having a nonzero integral. This explains the δ-function. As for the 2π, go back to the stage where we had a sum over frequencies and not an integral, i.e., go against the arrow in Eq. (21.3.88) and use it in the exponent of Eq. (21.3.93).

Let us now calculate the mean occupation number $\langle N\rangle$:

$$\langle N\rangle=\frac{1}{\beta Z}\frac{\partial Z}{\partial\mu}\tag{21.3.95}$$

$$=\frac{1}{\beta}\int_{-\infty}^{\infty}\frac{d\omega}{2\pi}\langle\bar{\psi}(\omega)\psi(\omega)\rangle\tag{21.3.96}$$

$$=\int_{-\infty}^{\infty}\frac{d\omega}{2\pi}\frac{e^{i\omega o^+}}{i\omega-\Omega_0+\mu}\tag{21.3.97}$$

$$=\theta(\mu-\Omega_0)\tag{21.3.98}$$

as in the operator approach.

Notice that we had to introduce the factor $e^{i\omega o^+}$ into the ω integral. We understand this as follows. If we had done the calculation using time τ instead of frequency ω, we would have calculated the average of $\Psi^\dagger\Psi$. This would automatically have turned into $\bar{\psi}(\tau+\varepsilon)\psi(\tau)$ when introduced into the path integral since the coherent state bra to the left of the operator would have come from the next time slice compared to the ket at the right. [Remember how $H(\Psi^\dagger,\Psi)$ turned into $H(\bar{\psi}(i+1),\psi(i))$.] Notice that the integral over ω was not convergent, varying as $d\omega/\omega$. It was therefore sensitive to the high frequencies and we had to intervene with the factor $e^{i\omega o^+}$. This factor allows us to close the contour in the upper half-plane. If $\mu > \Omega_0$, the pole of the integrand lies in that half-plane and makes a contribution. If not we get zero. In correlation functions that involve integrals that have two or more powers of ω in the denominator and are hence convergent, we will not introduce this factor.

Exercise 21.3.11. Advanced
In field theory and many-body physics one is interested in the Green's function:

$$G(\tau) = \langle \mathcal{T}(\Psi(\tau)\Psi^\dagger(0))\rangle \tag{21.3.99}$$

where $\langle\,\rangle$ denotes the average with respect to Z,

$$\Omega(\tau) = e^{H\tau}\,\Omega\, e^{-H\tau}$$

is the Heisenberg operator, and $\mathcal{T}$ the time-ordering symbol for fermionic operators:

$$\mathcal{T}(\Psi(\tau)\Psi^\dagger(0)) = \theta(\tau)\Psi(\tau)\Psi^\dagger(0) - \theta(-\tau)\Psi^\dagger(0)\Psi(\tau) \tag{21.3.100}$$

Note the minus sign when the order of operators is reversed. Show that $\Psi(\tau) = \Psi\, e^{-(\Omega_0-\mu)\tau}$ for our problem of the single oscillator.

Show, using the operator formalism that in our problem

$$G(\tau) = \frac{\theta(\tau)\, e^{-(\Omega_0-\mu)\tau} - \theta(-\tau)\, e^{-(\Omega_0-\mu)(\tau+\beta)}}{1+e^{-\beta(\Omega_0-\mu)}} \tag{21.3.101}$$

and that in the zero-temperature limit this reduces to

$$G(\tau) = \theta(\tau)\, e^{-(\Omega_0-\mu)\tau} \qquad \mu < \Omega_0 \tag{21.3.102}$$

$$= -\theta(-\tau)\, e^{-(\Omega_0-\mu)\tau} \qquad \mu > \Omega_0 \tag{21.3.103}$$

Let us define the pair of transforms:

$$G(\omega) = \int_{-\infty}^{\infty} G(\tau)\, e^{i\omega\tau}\, d\tau \tag{21.3.104}$$

$$G(\tau) = \int_{-\infty}^{\infty} G(\omega)\, e^{-i\omega\tau}\, \frac{d\omega}{2\pi} \tag{21.3.105}$$

Show that

$$G(\omega) = \frac{1}{\Omega_0 - \mu - i\omega} \tag{21.3.106}$$

independent of which of Ω_0 or μ is greater.

We saw in the study of the Ising model that the two-point correlation function in the functional integral translates into ground state average of the time-ordered product (for infinitely long system in the imaginary time direction) and vice versa. (If the parenthetical condition is not met, there will not be enough time for the system to relax into the ground state before we stick in the operators being averaged.)

It is likewise true here that

$$\langle \mathcal{T}(\Psi(\tau)\Psi^\dagger(0))\rangle = \langle \psi(\tau)\bar{\psi}(0)\rangle \tag{21.3.107}$$

where the average on the right-hand side is done by the Grassmann functional integral. Working at zero temperature, verify this for the frequency transform of both sides. (In the right-hand side write $\psi(\tau)$ in terms of $\psi(\omega)$, etc., using the zero temperature version of Eqs. (21.3.85–21.3.86) and Eq. (21.3.93).)

This brings us to the end of the discussion of fermionic path integrals. Clearly this is just the beginning and our discussion has been just an introduction.

21.4. Summary

Let us survey what has been done in this chapter. We started by learning how to use different resolutions of the identity to derive different path integrals. We looked at the configuration space, phase space, and coherent state path integrals. We realized that, while the introduction of the resolution of the identity is not an approximation, any assumption that changes in the coordinates being integrated over were small between time slices was to be carefully examined. In configuration space integrals the kinetic energy term provided a damping of fluctuations to something of order $\varepsilon^{1/2}$. In other integrals there was no such assurance. In particular, the continuum forms of the action were purely formal objects and only the discrete version defined the path integral, assuming the limit of infinite number of integrals existed. Despite this, the path integrals were very useful for seeing the theory as a whole before us, as a constructive solution to the quantum evolution problem. In particular, in the classical limit the smallness of $\hbar$ allowed us to think in terms of smooth paths. The study of the LLL (in connection with the QHE) and the Berry phase analysis illustrated some correct uses of the path integral.

We then turned to imaginary time quantum mechanics. We showed that from it one could extract the real-time energies and wave functions. In addition, imaginary time path integrals directly defined quantum statistical mechanics and were formally similar to classical statistical mechanics. The transfer matrix played the role of the discrete imaginary time evolution operator. Symmetry breaking was analyzed from many angles.

Finally, we studied two systems with no classical limit: the quantum spins and fermion oscillators. Although we studied just one fermionic oscillator, the generalization to many is direct and you should have no trouble following that topic when you get to it. Grassmann integrals are undoubtedly the most abstract notion in this book. But there is no doubt that as you use them (comparing them to the operator solution as a check) you will soon learn to think directly in terms of them. But remember this: there is no real notion of a semiclassical analysis here since the action is not a number-valued object and cannot be said to be stationary at any point. Note also that every Grassmann integral you write is eventually equal to an ordinary number though the integrand and integration measure are not. These numbers correspond to physical entities like the ground energy or correlation function of a fermion system.

The only functional integral we evaluated was the Gaussian integral. This is essentially all we know how to do. What if the action is not quadratic but has quartic terms? Then we do perturbation theory. We bring down the quartic term from the exponential (in the form of an infinite series) and evaluate term by term since we know how to integrate x^n times a Gaussian. Recall Appendix A.2 as well as the Wick's theorem for fermions in Exercise (21.3.64). But that's another story.

Bibliography

M. Berry, *Proc. R. Soc. Lond., Ser.* A392 **45**, 1984. A fascinating account of the history of the subject may be found in M. Berry, *Physics Today*, **43**, 12 1990.

S. Coleman, *Aspects of Symmetry* (Cambridge University Press, New York, 1985). Included here for the article on instantons.

L. D. Faddeev, in *Methods in Field Theory*, Les Houches Lectures, 1975 (R. Balian and J. Zinn-Justin, eds.) (North-Holland/World Scientific, Singapore, 1981). Look in particular at the discussion of holomorphic form of the functional integral, i.e., the coherent state integral.

M. Gutzwiller, *Chaos in Classical and Quantum Mechanics* (Springer Verlag, New York, 1990). For a solid introduction to many aspects of path integrals, especially the semiclassical limit.

J. R. Klauder and B. Skagerstam, eds., *Coherent States* (World Scientific, Singapore, 1985). Everything you ever wanted to know about coherent states.

S. Pancharatnam, *Ind. Acad. Sci.*, **44**(5), Sec. A, 1958. Reprinted in Shapere and Wilczek (1989) (see below). Discusses the geometric phase in the context of optics.

R. E. Prange and S. M. Girvin, eds., *The Quantum Hall Effect* (Springer, New York, 1987). Has many contributed papers by leaders in the field as well as helpful commentaries.

R. Rajaraman, *Instantons and Solitons* (North Holland, New York, 1982). Extremely clear discussion of the subject, usually starting with a warmup toy example from elementary quantum mechanics. Very few details are "left to the reader."

L. S. Schulman, *Techniques and Applications of Path Integrals* (Wiley Interscience, New York, 1981). A very readable and clear discussion of functional integrals, and pitfalls and fine points (such as the midpoint prescription for vector potential coupling).

A. Shapere and F. Wilczek, eds., *Geometric Phases in Physics* (World Scientific, Singapore, 1989). Collection of all key papers and some very good introductions to each subtopic. Saves countless trips to the library.

M. Stone, ed., *The Quantum Hall Effect* (World Scientific, Singapore, 1992). A nice collection of reprints with commentary.

't Hooft, *Phys. Rev. Lett.*, **37**, 8 (1976).

F. Wilczek, ed., *Fractional Statistics and Anyon Superconductivity* (World Scientific, Singapore, 1990). Referenced here for its applications to the Quantum Hall Effect. However, other topics like fractional statistics discussed there should be within your reach after this book.

Appendix

A.1. Matrix Inversion

This brief section is included only to help you understand Eq. (1.8.5) in the main text and is by no means comprehensive.

Consider the inversion of a 3×3 matrix

$$M=\begin{bmatrix} a_1 & a_2 & a_3 \\ b_1 & b_2 & b_3 \\ c_1 & c_2 & c_3 \end{bmatrix} \tag{A.1.1}$$

The elements of M have been named in this way rather than as M_{ij}, for in the following discussion we will treat the rows as components of the vectors **A**, **B**, and **C**, i.e., in the notation of vector analysis (which we will follow in this section),

$$\mathbf{A}=a_1\mathbf{i}+a_2\mathbf{j}+a_3\mathbf{k} \text{ and so on}$$

Consider next a triplet of vectors

$$\begin{aligned} \mathbf{A}_R &= \mathbf{B}\times\mathbf{C} \\ \mathbf{B}_R &= \mathbf{C}\times\mathbf{A} \\ \mathbf{C}_R &= \mathbf{A}\times\mathbf{B} \end{aligned} \tag{A.1.2}$$

which are said to be *reciprocal* to **A**, **B**, and **C**. In general,

$$\mathbf{A}\cdot\mathbf{A}_R\neq 0,\qquad \mathbf{A}\cdot\mathbf{B}_R=\mathbf{A}\cdot\mathbf{C}_R=0\qquad\text{and cyclic permutations}\tag{A.1.3}$$

If we construct now a matrix $\bar{\mathrm{M}}$ (called the *cofactor* transpose of M) whose *columns* are the reciprocal vectors,

$$\bar{\mathrm{M}}=\begin{bmatrix}(a_R)_1 & (b_R)_1 & (c_R)_1\\ (a_R)_2 & (b_R)_2 & (c_R)_2\\ (a_R)_3 & (b_R)_3 & (c_R)_3\end{bmatrix}$$

then

$$M\cdot\bar{\mathrm{M}}=\begin{bmatrix}\mathbf{A}\cdot\mathbf{A}_R & \mathbf{A}\cdot\mathbf{B}_R & \mathbf{A}\cdot\mathbf{C}_R\\ \mathbf{B}\cdot\mathbf{A}_R & \mathbf{B}\cdot\mathbf{B}_R & \mathbf{B}\cdot\mathbf{C}_R\\ \mathbf{C}\cdot\mathbf{A}_R & \mathbf{C}\cdot\mathbf{B}_R & \mathbf{C}\cdot\mathbf{C}_R\end{bmatrix}=\begin{bmatrix}\mathbf{A}\cdot\mathbf{A}_R & 0 & 0\\ 0 & \mathbf{B}\cdot\mathbf{B}_R & 0\\ 0 & 0 & \mathbf{C}\cdot\mathbf{C}_R\end{bmatrix}\tag{A.1.4}$$

Now all three diagonal elements are equal:

$$\begin{aligned}\mathbf{A}\cdot\mathbf{A}_R&=\mathbf{A}\cdot(\mathbf{B}\times\mathbf{C})=\mathbf{B}\cdot(\mathbf{C}\times\mathbf{A})=\mathbf{B}\cdot\mathbf{B}_R=\mathbf{C}\cdot(\mathbf{A}\times\mathbf{B})=\mathbf{C}\cdot\mathbf{C}_R\\ &=\det M\end{aligned}\tag{A.1.5}$$

where the last equality follows from the fact that the cross product may be written as a determinant:

$$\mathbf{B}\times\mathbf{C}=\begin{vmatrix}\mathbf{i} & \mathbf{j} & \mathbf{k}\\ b_1 & b_2 & b_3\\ c_1 & c_2 & c_3\end{vmatrix}\tag{A.1.6}$$

(We shall follow the convention of using two vertical lines to denote a determinant.)

Hence the inverse of the matrix M is given by

$$M^{-1}=\frac{\bar{\mathrm{M}}}{\det M}\tag{A.1.7}$$

When does $\det M$ vanish? If one of the vectors, say **C**, is a linear combination of the other two; for if

$$\mathbf{C}=\alpha\mathbf{A}+\beta\mathbf{B}$$

then

$$\mathbf{A}\cdot(\mathbf{B}\times\mathbf{C})=\mathbf{A}\cdot(\mathbf{B}\times\alpha\mathbf{A})+\mathbf{A}\cdot(\mathbf{B}\times\beta\mathbf{B})=\mathbf{B}\cdot(\alpha\mathbf{A}\times\mathbf{A})=0$$

Thus the determinant vanishes if the rows of the matrix are not linearly independent (LI) and vice versa. If the matrix is used to represent three simultaneous equations, it means not all three equations are independent. The method can be generalized for inverting $n\times n$ matrices, with real or complex elements. One defines a cross product of $n-1$ vectors as

$$\mathbf{A}_1\times\mathbf{A}_2\times\cdots\mathbf{A}_{n-1}=\begin{vmatrix}\mathbf{i} & \mathbf{j} & \mathbf{k} & \cdots & \\ (a_1)_1 & (a_1)_2 & & \cdots & \\ & & & \vdots & \\ (a_{n-1})_1 & (a_{n-1})_2 & & \cdots & (a_{n-1})_n\end{vmatrix}\tag{A.1.8}$$

The resulting vector is orthogonal to the ones in the product, changes sign when we interchange any two of the adjacent ones, and so on, just like its three-dimensional counterpart. If we have a matrix M, whose n rows may be identified with n vectors, $\mathbf{A}_1, \mathbf{A}_2, \ldots, \mathbf{A}_n$, then the cofactor transpose has as its columns the reciprocal vectors $\mathbf{A}_{1R}, \ldots, \mathbf{A}_{nR}$, where

$$\mathbf{A}_{jR}=\mathbf{A}_{j+1}\times\mathbf{A}_{j+2}\times\cdots\mathbf{A}_n\times\mathbf{A}_1\times\cdots\mathbf{A}_{j-1}\tag{A.1.9}$$

One tricky point: the cross product is defined to be orthogonal to the vectors in the product with respect to an inner product

$$\mathbf{A}\cdot\mathbf{B}=\sum A_iB_i$$

and *not*

$$\mathbf{A}\cdot\mathbf{B}=\sum A_i^*B_i$$

even when the components of $\mathbf{A}$ are complex. There is no contradiction here, for the vectors $\mathbf{A}_1, \ldots, \mathbf{A}_n$ are fictitious objects that enter a mnemonic and not the elements of the space $\mathbb{V}^n(C)$ on which the operator acts.

Exercise A.1.1. Using the method described above, show that

$$\begin{bmatrix}2 & 1 & 3\\ 0 & 1 & 2\\ -1 & 1 & 1\end{bmatrix}^{-1}=\begin{bmatrix}1 & -2 & 1\\ 2 & -5 & 4\\ -1 & 3 & -2\end{bmatrix}$$

and

$$\begin{bmatrix}2 & 1 & 3\\ 4 & 1 & 2\\ 0 & -1 & 2\end{bmatrix}^{-1}=\frac{1}{12}\begin{bmatrix}-4 & 5 & 1\\ 8 & -4 & -8\\ 4 & -2 & 2\end{bmatrix}$$

Theorem A.1.1. If $\Omega|V\rangle=|0\rangle$ implies $|V\rangle=|0\rangle$ then Ω^{-1} exists.

Proof. Let $|V_1\rangle, \ldots, |V_n\rangle$ be a LI basis in $\mathbb{V}^n$. Then another LI basis is generated by the action of Ω, i.e., $\Omega|V_1\rangle, \ldots, \Omega|V_n\rangle$ is also a LI basis. To see this, let us assume the contrary, that there exists a relation of the form

$$\sum_i \alpha_i \Omega|V_i\rangle=0$$

with not all $\alpha_i=0$. Upon pulling out Ω, because it is linear, we get

$$\Omega\left(\sum_i \alpha_i|V_i\rangle\right)=0$$

which, when combined with the assumed property of Ω, implies that

$$\sum_i \alpha_i|V_i\rangle=|0\rangle$$

with not all $\alpha_i=0$, which is not true. So we can conclude that every vector $|V'\rangle$ in $\mathbb{V}^n$ may be written as a *unique* linear combination in the new basis generated by Ω as

$$|V'\rangle=\sum_i \alpha_i \Omega|V_i\rangle$$

In terms of $|V\rangle=\sum \alpha_i|V_i\rangle$, we see that *every* $|V'\rangle$ in $\mathbb{V}^n$ may be written as

$$|V'\rangle=\Omega|V\rangle$$

where $|V\rangle$ is *unique*. In other words, we can think of *every* $|V'\rangle$ in $\mathbb{V}^n$ as arising from a *unique* source $|V\rangle$ in $\mathbb{V}^n$ under the action of Ω. Define an operator Λ whose action on any vector $|V'\rangle$ in $\mathbb{V}^n$ is to take it back to its unique source $|V\rangle$. (If the source of $|V'\rangle$ were not unique—say, because there are two vectors $|V_1\rangle$ and $|V_2\rangle$ that are mapped into $|V'\rangle$ by Ω—then we could not define Λ, for acting on $|V'\rangle$, it would not know whether to give $|V_1\rangle$ or $|V_2\rangle$.) The action of Λ is then

$$\Lambda|V'\rangle=|V\rangle, \quad \text{where } |V'\rangle=\Omega|V\rangle$$

We may identify Λ as the inverse of Ω,

$$\Lambda=\Omega^{-1} \quad \text{or} \quad \Lambda\Omega=I$$

since for any $|V'\rangle$ in $\mathbb{V}^n$

$$\Lambda|V'\rangle=\Lambda\Omega|V\rangle=|V\rangle \quad \text{Q.E.D.}$$

A.2. Gaussian Integrals

We discuss here all the Gaussian integrals that we will need. Consider

$$I_0(\alpha)=\int_{-\infty}^{\infty} e^{-\alpha x^2}\,dx, \qquad \alpha>0 \tag{A.2.1}$$

This integral cannot be evaluated by conventional methods. The trick is to consider

$$I_0^2(\alpha)=\int_{-\infty}^{\infty} e^{-\alpha x^2}\,dx\int_{-\infty}^{\infty} e^{-\alpha y^2}\,dy=\int_{-\infty}^{\infty}\int_{-\infty}^{\infty} e^{-\alpha(x^2+y^2)}\,dx\,dy$$

Switching to polar coordinates in the x–y plane,

$$\begin{aligned}I_0^2(\alpha)&=\int_0^{\infty}\int_0^{2\pi} e^{-\alpha\rho^2}\rho\,d\rho\,d\phi\\&=\pi/\alpha\end{aligned}$$

Therefore

$$I_0(\alpha)=(\pi/\alpha)^{1/2} \tag{A.2.2}$$

By differentiating with respect to α we can get all the integrals of the form

$$I_{2n}(\alpha)=\int_{-\infty}^{\infty} x^{2n}\,e^{-\alpha x^2}\,dx$$

For example,

$$\begin{aligned}I_2(\alpha)&=\int_{-\infty}^{\infty} x^2\,e^{-\alpha x^2}\,dx=-\frac{\partial}{\partial\alpha}\int_{-\infty}^{\infty} e^{-\alpha x^2}\,dx\\&=-\frac{\partial}{\partial\alpha}I_0(\alpha)=\frac{1}{2\alpha}\left(\frac{\pi}{\alpha}\right)^{1/2}\end{aligned} \tag{A.2.3}$$

The integrals $I_{2n+1}(\alpha)$ vanish because these are integrals of odd functions over an even interval $-\infty$ to $+\infty$. Equations (A.2.2) and (A.2.3) are valid even if α is purely imaginary.

Consider next

$$I_0(\alpha,\beta)=\int_{-\infty}^{\infty} e^{-\alpha x^2+\beta x}\,dx \tag{A.2.4}$$

By completing the square on the exponent, we get

$$I_0(\alpha, \beta) = e^{\beta^2/4\alpha} \int_{-\infty}^{\infty} e^{-\alpha(x-\beta/2\alpha)^2} dx = e^{\beta^2/4\alpha} \left(\frac{\pi}{\alpha}\right)^{1/2} \tag{A.2.5}$$

These results are valid even if α and β are complex, provided Re $\alpha > 0$. Finally, by applying to both sides of the equation

$$\int_0^{\infty} e^{-\alpha r} dr = \frac{1}{\alpha}$$

the operator $(-d/d\alpha)^n$, we obtain

$$\int_0^{\infty} r^n e^{-\alpha r} dr = \frac{n!}{\alpha^{n+1}}$$

Consider this integral with $\alpha = 1$ and n replaced by $z - 1$, where z is an arbitrary complex number. This defines the *gamma function* $\Gamma(z)$

$$\Gamma(z) = \int_0^{\infty} r^{z-1} e^{-r} dr$$

For real, positive and integral z,

$$\Gamma(z) = (z-1)!$$

A.3. Complex Numbers

A complex variable z can be written in terms of two real variables x and y, and $i = (-1)^{1/2}$, as

$$z = x + iy \tag{A.3.1}$$

Its *complex conjugate* z^* is defined to be

$$z^* = x - iy \tag{A.3.2}$$

One may invert these two equations to express the *real and imaginary parts*, x and y, as

$$x = \tfrac{1}{2}(z + z^*), \qquad y = (z - z^*)/2i \tag{A.3.3}$$

The *modulus squared* of z, defined to be zz^*, equals

$$zz^* \equiv |z|^2 = (x+iy)(x-iy) = x^2+y^2 \tag{A.3.4}$$

You may verify that $z=z'$ implies that $x=x'$ and $y=y'$ by considering the modulus of $z-z'$.

From the power-series expansions

$$\sin x = x - x^3/3! + x^5/5! - \cdots$$
$$\cos x = 1 - x^2/2! + x^4/4! - \cdots$$

one can deduce that

$$e^{ix} = \cos x + i \sin x \tag{A.3.5}$$

It is clear that e^{ix} has unit modulus (x is real).

The expression $z=x+iy$ gives z in *Cartesian form*. The *polar form* is

$$\begin{aligned} z = x+iy &= (x^2+y^2)^{1/2}\left[\frac{x}{(x^2+y^2)^{1/2}} + i\frac{y}{(x^2+y^2)^{1/2}}\right] \\ &= \rho(\cos\theta + i\sin\theta) \\ &= \rho\, e^{i\theta} \end{aligned}$$

where

$$\rho = (x^2+y^2)^{1/2} \quad \text{and} \quad \theta = \tan^{-1}(y/x) \tag{A.3.6}$$

Clearly

$$|z| = \rho \tag{A.3.7}$$

Each complex number $z=x+iy$ may be visualized as a point (x, y) in the x–y plane. This plane is also called the *complex z plane*.

A.4. The $i\varepsilon$ Prescription

We will now derive and interpret the formula

$$\frac{1}{x \mp i\varepsilon} = \mathscr{P}\frac{1}{x} \pm i\pi\delta(x) \tag{A.4.1}$$

where $\varepsilon \to 0$ is a positive infinitesimally small quantity. Consider an integral of the form

$$I = \lim_{\varepsilon \to 0} \int_{-\infty}^{\infty} \frac{f(x)\,dx}{x - i\varepsilon}. \tag{A.4.2}$$

Viewing this as the integral on the real axis of the complex $z = x + iy$ plane, we see that the integrand has an explicit pole at $z = i\varepsilon$ in addition to any singularities f might have. We assume f has no singularities on or infinitesimally close to the real axis. A long as ε is fixed, there is no problem with the integral. For example, if f has some poles in the upper half-plane and vanishes fast enough to permit our closing the contour in the upper half-plane, the integral equals $2\pi i$ times the sum of the residues of the poles of f and the pole at $z = i\varepsilon$. Likewise, if we change the sign of the ε term, we simply drop the contribution from the explicit pole, which is now in the lower half-plane.

What if $\varepsilon \to 0$? Now the pole is going to ram (from above) into our contour which runs along the x-axis. So we prepare for this as follows. Since the only singularity near the real axis is the explicit pole at $z = i\varepsilon$, we make the following deformation of the contour without changing the value of I: the contour runs along the real axis from $-\infty$ to $-\varepsilon'$, (ε' is another positive infinitesimal) goes around counterclockwise, below the origin in a semicircle of radius ε', and resumes along the real axis from $x = \varepsilon'$ to ∞. The nice thing is that we can now set $\varepsilon = 0$, which brings the pole to the origin. The three parts of the integration contour contribute as follows:

$$\begin{aligned} I &= \lim_{\varepsilon' \to 0} \left[\int_{-\infty}^{-\varepsilon'} \frac{f(x)\,dx}{x} + \int_{\varepsilon'}^{\infty} \frac{f(x)\,dx}{x} + i\pi f(0) \right] \\ &\equiv \mathscr{P} \int_{-\infty}^{\infty} \frac{f(x)\,dx}{x} + i\pi f(0). \end{aligned} \tag{A.4.3}$$

The sum of the two integrals in the limit $\varepsilon' \to 0$ is defined as the *principal value integral* denoted by the symbol $\mathscr{P}$. In the last term, which is restricted to the infinitesimal neighbourhood of the origin, we have set the argument of the smooth function f to zero and done the integral of dz/z counterclockwise around the *semicircle* to get $i\pi$.

Eq. (A.4.1) is a compact way to say all this. It is understood that Eq. (A.4.1) is to be used inside an integral only and that inside an integral the factor $1/(x - i\varepsilon)$ leads to two terms: the first, $\mathscr{P}(1/x)$, leads to the principal value integral, and the second, $i\pi\delta(x)$, leads to $i\pi f(0)$.

It is clear that if we reverse the sign of the ε term, we change the sign of the delta function since the semicircle now goes around the pole in the clockwise direction. The principal part is not sensitive to this change of direction and is unaffected.

It is clear that if we replace x by $x - a$ the pole moves from the origin to $x = a$ and $f(0)$ gets replaced by $f(a)$ so that we may write

$$\frac{1}{(x-a) \mp i\varepsilon} = \mathscr{P} \frac{1}{(x-a)} \pm i\pi\delta(x-a) \tag{A.4.4}$$

It is clear that the limits on x need not be $\pm\infty$ for the formula to work.

Finally, note that according to Eq. (A.4.4) the difference between the integrals with two signs of ε is just $2\pi i f(a)$. This too agrees with the present analysis in terms of the integral I in Eq. (A.4.2) since in the difference of the two integrals the contribution along the real axis cancels due to opposite directions of travel except for the part near the pole where the difference of the two semicircles (one going above and going below the pole) is a circle around the pole.

Answers to Selected Exercises

Chapter 1

1.8.1. (1) $|\omega=1\rangle\to\begin{bmatrix}1\\0\\0\end{bmatrix}$, $|\omega=2\rangle\to\frac{1}{(30)^{1/2}}\begin{bmatrix}-5\\-2\\1\end{bmatrix}$, $|\omega=4\rangle\to\frac{1}{(10)^{1/2}}\begin{bmatrix}1\\0\\3\end{bmatrix}$

(2) No, no.

1.8.2. (1) Yes

(2) $|\omega=0\rangle\to\begin{bmatrix}0\\1\\0\end{bmatrix}$, $|\omega=1\rangle\to\frac{1}{2^{1/2}}\begin{bmatrix}1\\0\\1\end{bmatrix}$, $|\omega=-1\rangle\to\frac{1}{2^{1/2}}\begin{bmatrix}-1\\0\\1\end{bmatrix}$

1.8.10. $\omega=0, 0, 2;\ \lambda=2, 3, -1.$

Chapter 4

4.2.1. (1) 1, 0, −1

(2) $\langle L_x\rangle=0$, $\langle L_x^2\rangle=1/2$, $\Delta L_x=1/2^{1/2}$

(3) $|L_x=1\rangle\to\begin{bmatrix}1/2\\1/2^{1/2}\\1/2\end{bmatrix}$, $|L_x=0\rangle\to\begin{bmatrix}-1/2^{1/2}\\0\\1/2^{1/2}\end{bmatrix}$,

$|L_x=-1\rangle\to\begin{bmatrix}1/2\\-1/2^{1/2}\\1/2\end{bmatrix}$

(4) $P(L_x=1)=1/4, \quad P(L_x=0)=1/2, \quad P(L_x=-1)=1/4$

(5) $|\psi\rangle \to \frac{1}{(1/4+1/2)^{1/2}}\begin{bmatrix}1/2\\0\\1/2^{1/2}\end{bmatrix}$ = projection of $|\psi\rangle$ on the $L_z^2=1$ eigenspace. $P(L_z^2=1)=3/4$. If L_z is measured $P(L_z=1)=1/3$, $P(L_z=-1)=2/3$. Yes, the state changes.

(6) No. To see this right away note that if $\delta_1=\delta_2=\delta_3=0$, $|\psi\rangle=1|L_x=1\rangle$ and if $\delta_1=\delta_3=0$ and $\delta_2=\pi$, $|\psi\rangle=|L_x=-1\rangle$. [See answer to part (3).] The vectors $|\psi\rangle$ and $e^{i\theta}|\psi\rangle$ are physically equivalent only in the sense that they generate the same probability distribution for any observable. This does not mean that when the vector $|\psi\rangle$ appears as a part of a linear combination it can be multiplied by an arbitrary phase factor. In our example one can only say, for instance, that

$$|\psi\rangle'=e^{-i\delta_1}|\psi\rangle$$

$$=\frac{1}{2}|L_z=1\rangle+\frac{e^{i(\delta_2-\delta_1)}}{2^{1/2}}|L_z=0\rangle+\frac{e^{i(\delta_3-\delta_1)}}{2}|L_z=-i\rangle$$

is physically equivalent to $|\psi\rangle$. Although $|\psi\rangle'$ has different coefficients from $|\psi\rangle$ in the linear expansion, it has the same "direction" as $|\psi\rangle$. In summary, then, the relative phases $\delta_2-\delta_1$ and $\delta_3-\delta_1$ are physically relevant but the overall phase is not, as you will have seen in the calculation of $P(L_x=0)$.

Chapter 5

5.4.2. (a) $R=(maV_0)^2/(\hbar^4k^2+m^2a^2V_0^2)$; $T=1-R$

(b) $T=(\cosh^2 2\kappa a+\alpha^2\sinh^2 2\kappa a)^{-1}$ where $i\kappa$ is the complex wave number for $|x|\le a$ and $\alpha=(V_0-2E)/[4E(V_0-E)]^{1/2}$.

Chapter 7

7.4.2. 0, 0, $(n+1/2)\hbar/m\omega$, $(n+1/2)m\omega\hbar$, $(n+1/2)\hbar$. Note that the recipe $m\omega\to(m\omega)^{-1}$ is at work here.

7.4.5. (1) $(1/2^{1/2})(|0\rangle e^{-i\omega t/2}+|1\rangle e^{-3i\omega t/2})$

(2) $\langle X(t)\rangle=(\hbar/2m\omega)^{1/2}\cos\omega t$, $\langle P(t)\rangle=-(m\omega\hbar/2)^{1/2}\sin\omega t$

(3) $\langle\dot{X}(t)\rangle=(i\hbar)^{-1}\langle[X,H]\rangle=\langle P(t)\rangle/m$, $\langle\dot{P}(t)\rangle=-m\omega^2\langle X(t)\rangle$. By eliminating $\langle\dot{P}\rangle$ we can get an equation for $\langle X(t)\rangle$ and vice versa and solve it using the initial values $\langle X(0)\rangle$ and $\langle P(0)\rangle$, e.g., $\langle X(t)\rangle=\langle X(0)\rangle\cos\omega t+[\langle P(0)\rangle/m\omega]\sin\omega t$.

Chapter 10

10.3.2. $3^{-1/2}[|334\rangle + |343\rangle + |433\rangle]$

Chapter 12

12.6.1. $E = -\hbar^2/2\mu a_0^2, \quad V = -\hbar^2/\mu a_0 r$

Chapter 13

13.3.1. Roughly 200 MeV.

13.3.2. Roughly 1 Å.

Chapter 14

14.3.5. $M = \left(\frac{\alpha+\delta}{2}\right) I + \left(\frac{\beta+\gamma}{2}\right)\sigma_x + i\left(\frac{\beta-\gamma}{2}\right)\sigma_y + \left(\frac{\alpha-\delta}{2}\right)\sigma_z$

14.3.7. (1) $2^{1/4}(\cos\pi/8 + i(\sin\pi/8)\sigma_x)$.

(2) $2/3I - 1/3\sigma_x$.

(3) σ_x

14.4.4. Roughly 2×10^{-9} second.

14.4.6. $(e\hbar/2mc)\tanh(e\hbar B/2mckT)\mathbf{k}$

14.5.2. (1) Roughly one part in a million.

(2) 10^{10} G.

14.5.3. 1/2, 1/4, 0.

14.5.4. $\left(\frac{1+\cos\theta}{2}\right)^2$

Chapter 15

15.2.2. (1) $\langle 1\ 1, 1/2(-1/2)|3/2\ 1/2\rangle = (1/3)^{1/2}$
$\langle 1\ 0, 1/2\ 1/2|3/2\ 1/2\rangle = (2/3)^{1/2}$
$\langle 1\ 1, 1/2(-1/2)|1/2\ 1/2\rangle = (2/3)^{1/2}$
$\langle 1\ 0, 1/2\ 1/2|1/2\ 1/2\rangle = -(1/3)^{1/2}$

(2) $|jm\rangle = |2, 1\rangle = 2^{-1/2}|m_1 = 1, m_2 = 0\rangle + 2^{-1/2}|m_1 = 0, m_2 = 1\rangle$
$|2, 0\rangle = 6^{-1/2}|1, -1\rangle + (\tfrac{2}{3})^{1/2}|0, 0\rangle + (\tfrac{1}{6})^{1/2}|-1, 1\rangle$
$|1, 1\rangle = 2^{-1/2}|1, 0\rangle - 2^{-1/2}|0, 1\rangle$
$|1, 0\rangle = 2^{-1/2}|1, -1\rangle - 2^{-1/2}|-1, 1\rangle$
$|0, 0\rangle = 3^{-1/2}|1, -1\rangle - 3^{-1/2}|0, 0\rangle + 3^{-1/2}|-1, 1\rangle$

The others are either zero, obvious, or follow from Eq. (15.2.11).

15.2.6. $\mathbb{P}_+ = \dfrac{(2\mathbf{L}\cdot\mathbf{S})/\hbar^2 + l + 1}{2l+1}, \qquad \mathbb{P}_- = \dfrac{l - (2\mathbf{L}\cdot\mathbf{S})/\hbar^2}{2l+1}$

Chapter 16

16.1.2. $E(a_0) = 10E_0/\pi^2$

16.1.3. $-ma_0^2 V_0^2/\pi\hbar^2$

16.1.4. $E(a_0) = \frac{1}{2}\hbar\omega(\frac{12}{11})^{1/2}$

16.2.4. Roughly 1.5×10^{17} seconds or 10^{10} years.

Table of Constants

$$\hbar c = 1973.3 \text{ eV Å}$$

$$\alpha = e^2/\hbar c = 1/137.04$$

$$mc^2 = 0.511 \text{ MeV} \quad (m \text{ is the electron mass})$$

$$Mc^2 = 938.28 \text{ MeV} \quad (M \text{ is the proton mass})$$

$$a_0 = \hbar^2/me^2 = 0.511 \text{ Å}$$

$$e\hbar/2mc = 0.58 \times 10^{-8} \text{ eV/G} \quad \text{(Bohr magneton)}$$

$$k = 8.62 \times 10^{-5} \text{ eV/K}$$

$$kT \simeq 1/40 \text{ eV at } T = 300 \text{ K} \quad \text{(room temperature)}$$

$$1 \text{ eV} = 1.6 \times 10^{-12} \text{ erg}$$

Mnemonics for Hydrogen

In the ground state,

$$v/c \equiv \beta = \alpha$$

$$E_1 = -T = -\tfrac{1}{2}mv^2 = -\tfrac{1}{2}mc^2\alpha^2$$

$$mva_0 = \hbar$$

In higher states, $E_n = E_1/n^2$.

Index

图书在版编目（CIP）数据

Shankar 量子力学原理 : 第 2 版 = Principles of Quantum Mechanics, Second Edition : 英文 / (美) 拉马穆蒂 · 香卡 (R. Shankar) 著 . — 北京 : 世界图书出版有限公司北京分公司 , 2023.1
ISBN 978-7-5192-9684-1

Ⅰ . ① S··· Ⅱ . ①拉··· Ⅲ . ①量子力学—英文 Ⅳ . O413.1

中国版本图书馆 CIP 数据核字（2022）第 131077 号

中文书名　Shankar 量子力学原理（第 2 版）
英文书名　Principles of Quantum Mechanics *2nd edition*
著　　者　［美］拉马穆蒂 · 香卡（R. Shankar）
策划编辑　陈　亮
责任编辑　陈　亮

出版发行　世界图书出版有限公司北京分公司
地　　址　北京市东城区朝内大街 137 号
邮　　编　100010
电　　话　010-64038355（发行）　64033507（总编室）
网　　址　http://www.wpcbj.com.cn
邮　　箱　wpcbjst@vip.163.com
销　　售　新华书店
印　　刷　北京建宏印刷有限公司
开　　本　710 mm × 1000 mm　1/16
印　　张　43.5
字　　数　676 千字
版　　次　2023 年 1 月第 1 版
印　　次　2023 年 1 月第 1 次印刷
版权登记　01-2022-3387
国际书号　ISBN 978-7-5192-9684-1
定　　价　179.00 元